A Short Social and Cultural Anthropology of the
NORTHERN LUO
of South Sudan

"BUILDING COMMUNITY OF SHARED CULTURES, BELIEFS AND DESTINY"

"Anthropology is the science of man and of his culture at various levels of development. It includes the study of the human frame, of racial distinctions, of civilization, of social structure, and of man's mental reactions to his environment." (C. G. Seligman 1932)

Saturnino Onyala

FIRST EDITION JUNE 2019

A Short Social and Cultural Anthropology
of the Northern Luo of South Sudan
Saturnino Onyala

First published in June 2019

June, 2019
Melbourne, Australia

Copyright © Saturnino Onyala 2019

ISBN: 978-0-6484367-1-3 (pbk)
ISBN: 978-0-6484367-2-0 (hbk)

Printed and bound by Lightning Source Australia.

About myself

Saturnino is a South Sudanese Australian who came to Australia as a refugee in 2003. By obtaining a Master degree in Social Science, specializing in International Development from RMIT University Melbourne, the author was exposed to global cultural trends. This in turn, inspired him to create conversations around hidden and undocumented aspects of African cultures and historical milestones. His first book 'The History and Expressive Cultures of the Acholi of South Sudan' was a reflection of valuable knowledge gained through a number of research and conversations with elders in the community. His second book 'A Short Social and Cultural Anthropology of the Northern Luo of South Sudan' explores the cultural frameworks of South Sudanese Lou people and their untold stories.

Acknowledgments

Many individuals and institutions have contributed to the publication of this book. The book is the product of a Luo meeting held on 20th June 2015, in Footscray, Melbourne, Australia. The meeting explored *who are the Northern Luo and what strategies could be used to reunify the Luo groups of South Sudan, both in Diaspora and on the ground in South Sudan*. This was followed by a series of meetings that led to the formation of the Greater Luo Community Association in Victoria (GLCAIV).

I am grateful to individuals and institutions that have generously supported me in my long journey in documenting this important Luo oral history and culture. This book would not have been possible without the generous support of individuals and institutions that have preserved and protected historical imagery, objects and information. I am grateful to the following individuals who I thought it was appropriate to mention by name: Banydhuro Oyay, Paul Jok Dwong, Tarizah Chol Otor, Stephen Deng Lual, Fashak Deng, John Kuldit, Simon Okyij, Simon Kur, Okwom Nyikang and Adhwok Yoleker from Shilluk; and from Anywak Friday Awow, William Dimo, Gora Hassen. From Pari (Lokoro): David Akwai; and from Jo Luo of Bhar El Ghazal, William Dimo, Alberto Rimo Akot Dimo, William Gol, Deng Apiny Alany, Lau Uling Bouk, Andrea Lual and William Apai. These people have helped me with answering questionnaires and, in addition, I have used them in focus groups.

My special thanks goes to Mrs Michelle Mestry for reading through the MS and editing the book. I also thank Randika Wijekoon, from Migrant Information Centre, Eastern Melbourne, for his assistance in photographing (diagramming). I also want to thank Rev Joel Mestry for his support in producing the first drafts of the front and back covers of the book, which made the redesigning work of Luke Harrison easier. Last but not least, I would like to give my special thanks to Luke Harris from Working Type Design, Eltham North, Melbourne Victoria, for his assistance in the layout and design of the book, and in helping me get the book into Amazon. This book would have not come about without such technical support.

The publication of this book was only financially possible by leaving almost insufficient funds for the members of my family. I have done this because of the importance of disseminating information about the Luo cultures to people who might not have access to the information.

Phonetic Spelling of Native Words and Names

Abel-tek	sleeping animal skin
Adull	temporary huts/slavery huts
Ajiga	father or owner of jwok, priest
Agem	village revolution
Akaya	stomach
Angudho	razor blade
Athurci	
Atoja	oath at the wedding
Awal	calabash bowl
Awope or Awobe	youngsters
Bach tuko	palm leaves
Bang-Nyireth	servant of Reth
Bang Reth	the possessed class, owned by the King (First group are Wives and widows of the dead Kings and second groups are descendants of slaves captured
Bang Reth Kwadhker	descendants of retainers
Bang Reth Nyakwac	Reth's grand-daughter?
Bany	principal singer and president of war
Bareth	servants
Bareth Nyikang	servants of Nyikang
Bul	drum
Cen	bad spirit
Challa	day a widow or a widower removed black-mourning-cloths from the body
Cidong	elders
Cien	ghostly vengeance
Ci-jwok	evil eye
Chor-paabo	a narrow road leading to a throne (kwom) at Nyikang's shrine
Collo	the class that includes the majority of Shulluk people

Cong	final initiation dance
Daawi	magical charms
Dang	ceremonial bow
Dem spears	spears used by Anyuak for marriage
Diel	owner lineage or original lineage
Dura	sorghum
Dumoi	spear in Anyuak land used for marriage
Dwar	hunting
Ipuura	marching with song
Jak	the important men of a village
Juga	brother-in-law
Jwok	God, Spirit or divinity
Ger	Northern Shullukland
Gikwer	first dowry
Goats	Diek
Gony-challa	removing black mourning cloth from the body
Gol Dhiang	ordinary Chollo
Gol Nyikang	Chollo from the line of Nyikang
Gyiel	Arm ring???
Gyiek	water buckets
Kaalo	cattle enclosure made of wood
Kali	round, spotted hard beans
Kengo (Nyikang)	the royal grave shrine of Nyikang
Kii	abomination
Kona	feathers
Konggi Waato	beer sacrifice
Koor	collective invocations
Kudu	(Kudu horn?)
Kwa	descendants
Kwai-ngam	the owner of the land
Kwanyikang	grandfather of Nyikang
Kwa-Dikwor	grandfather of Dikwor
Kwa-Nyicu	grandfather of Nyicu
Kwanyiker	grandfather of Nyiker
Kwano	to confess or to count
Kwa-wang	grandfather of Wang
Kwa-Owango	grandfather of Owango
Kwanyikwom	army of Nyikang

Kwanyimongo	grandfather of Nyimongo
Kwa-Ocwako	grandfather of Ocwako
Kwar-Nyireth	grandsons of Nyireth
Kwar-Reth	royal clan formed by descendants of Nyikang
Kwon-cak	food eaten with milk
Kwom	stool
Kwom-jago	stool or chair of a chief
Kuiny-bur	digging a hole (grave)
Lapidi	nurse-child
Lakwena	messenger
Lam	invocation, prayer
Lamo	to bless (v.)
Lau	cloth
Libangga	milk porridge and unripe sorghum used for sacrifice
Likweri	peacemakers
Lipul	God, hill or jwok
Luak/Lwak	Southern Shullukland
Ma	my mother
Magala	earrings
Mal	ewe
Mojomiji	the ruling age groups
Morro-Mwomo	a Shulluk place near Kosti
Mur or Amur	duiker, dik-dik
Mwor abutha	stick for old people
Ngol	cutting ceremony
Nya-kwer	girl for taxes
Nyalam	New Year celebration
Nyareth	king's daughter
Ngat-gam	traditional birth attendant
Nyi-dhok	messenger
Nyinyireth	Nyikang of ocoro?
Nyikwer	great grand-children of Reth
Nyireth	sons and daughters of the king(s)
Nyiye	noble clan
Ober Owango	red flamingo's feathers
Okwen	spears used for marriage among the Anyuak
Opoko	gourd

Ororo	a branch of the Royal line that has lost its place in the line of succession
Pala	ochre
Paare	a village where the elected king was born
Pa Nyireth	residential area of the king
Paneyo	relative of the wife
Pawego	relatives of the husband
Reth	king
Rom	ram
Rony	installation or coronation
Ruda	waterbuck
Rwath or Rwot	chief
Rwodhi kot	rain chief
Shauweir	largish bead *with red head used by the Anyuak for marriage*
Tiar-bul	dancing ground
Tii	necklace beads
Teng	dried sorghum straw (stick)
Tet bead	Anyuak dark greenish colour used for marriage
Thou	Balamites aegiptica or ???
Thworo	dancing ground
Tol	robe
Tuk	smallest ant hills
Tuong-duong	political lineage
Wang yomo	the place where Nyikang disappeared
Wath-Owono	bull from a rich man given to his son for initiation
Wee	stomach contents
Wo-jwok	father or owner of jwok, a priest
Wora	my father
Wudo	ostrich
Ubel	special spear for chief of Wiatuo
Ucuok	necklace
Ulawi /Alau	spear inherited from forefather Dim
Ywero	Husband's brother or wife's sister

Saturnino Onyala, the Man

"Autobiographical sketch"

I was born on the 01/01/1951 in Licari-Obbo. Licari is situated on a plateau. People settled here because the soil is very fertile. Secondly, there is a good number of games which provide meat for the inhabitants – food staff in Laciri is adequate. There are no schools in Licari. All children study in a 'Bush school' in Oyere, which is about four to five miles from Licari. My father was a great hunter and farmer. He also worked in coffee farm at Obbo between 1960 and 1962. Later, he went to work as a game ranger in Kasese-Uganda.

I started Primary One in a bush school in 1960 when I was already ten years old. Joining school was a problem to me because teachers said I was above the age of starting school. Fortunately, the God of my father had "lain stomach up". A police officer, Lazario the brother of Adik-Diko-moi, of Licari, who headed the police who were keeping peace at the time of enrolling school pupil told the teachers to register me in Primary One. With the authority of Lazario, I was enrolled in Primary One. Although my father was working in coffee farm at Obbo, we did not have a bicycle. So I always walked to school. Sometimes I met elephants and buffaloes on my way to school. My 'Bush school' teachers were Opeko of Obbo, and Teodoro Lo-um, (he was often called by his nickname, Otyien) of Pajok. As the Acholi said, it is the children from poor families are the ones who are brighter in class. Truly, I was a bright child and all my teachers liked me.

In 1963, I passed the examinations which were for the pupils of Palwar, Pajok, Obbo, Magwi, Omeo and Panyikwara. At that time, Acholi children from different families completed their primary two from village schools, after which they did the examinations to join Palotaka Elementary School. From every bush school, teachers chose forty five Primary Two pupils who scored high marks to join Palotaka Elementary School. I studied in Palotaka Elementary School for six months. Later, government transferred one teacher from Juba to Palotaka. This teacher had two boys who were in Primary Three. That year, Mr. Labuk was the headmaster of Palotaka Elementary School. Shortly, the new teacher asked the headmaster to remove two pupils from Primary Three so that there was space for his boys in the class. I do not know how the headmaster's eyes functioned! His sight fell on me and one of my friends, Michael Oyuru of Pajok Paitenge. The headmaster called both of us in his office and informed us that each of us should go back to study in the village schools. Michael Oyuru went back to study in Pajok village school while I went to Oyere village school.

The Mwony Anya-nya rebel had started in 1962. In 1964, my mother, Laura Onek, fled with my two sisters and I to Palabek Akeli-kongo in Uganda. Refugee life was not easy. A short while later, the Palabek people started calling us *Agono ki Lokung* (I came via Lukung). In 1964, I resumed Primary Three in Palabek Kal Primary School. All the same, life was hard because the wife of my Uncle, Dr Mario Tokwaro, with whom I lived, used to deny me breakfast every morning. She did not want me to wash my face with the water in her pot. At the same time, I was also a mass boy at Palabek Mission. When it was time for me to go to school or church, I washed my face from a bore hole, or with dew using cassava leaves. In 1965 I went to study in Palabek Padwat where I lived with my mother's sister called Mrs. Imerigiana Lalaa. When I realised that I may not have a good place to study from, I joined *Pakele Minor Seminary* in 1966 when Father Leopoldo Anywar of Amika Magwi was the Rector of Pakele Minor Seminary. When Father Father Leopoldo identified me as a bright boy, he appointed me as a storekeeper of student's food store. Later, Father Hillary Loswat Oboma of Lokoro, realised that the Parish Priest of Pakele Mission wanted to remove me out of the seminary because I gave food to students whom he wanted to punish by starvation. It is true that when the Parish Priest ordered that there was no food for students the following day, I gave much food staff to be prepared for their meals because I thought that if they eat much today, they would not feel very hungry the following day. In 1967 Father Hillary Loswat Oboma, transferred me to *Nadiket Minor Seminary* in Moroto, where I completed Primary seven. In 1969, I went to study in *Lacor Senior Seminary* Gulu-Uganda. It was from there that I began to think of writing a book. I wanted to write a book titled *All Acholi Traditional Names Have Meanings*. In this book, I wanted to explain the meaning of Acoli names like Okeny, Onen, Opiyo, Acen, Olum, Oryiem, Acayo, Abalo, Akwero, Onek and many others that I cannot list all here. I found that I had a lot of academic work, so I suspended the writing project because I wanted to score high marks at the seminary.

In 1973, when the Government of Sudan and the leaders of Anya-nya One rebel group signed a peace deal at Adiss Ababa in Ethiopia, Bishop of Sudan said that all Sudanese seminarians who were in Uganda should go back to Sudan. In 1973 I started studying Philosophy in *Juba St Paul National Major Seminary*. Father Joseph Nyekindi, who later became a Bishop of Yambio, was a Rector. I completed Philosophy in 1975, and in 1976 I joined *St Paul National Seminary* in *Bussere National Major Seminary*, where I studied Theology. Eventually, I discovered that academic victory was not in the classroom because we were only three theologians in the class. For this reason, we wrote to the Bishops to allow us to study at *St. Thomas Aquinas National Major Seminary* in Kenya. Unfortunately, the lecturers at Bussere National Major Seminary were Italian and American fathers. They looked at our request as disrepute to them. One evening the fathers talked this issue over their supper. They said I was the one who enticed the other two students to think of going to study at St. Thomas Aquinas National Major Seminary. As a result, they resolved and sent me on a one year probation. Bishop Gabriel Ziber, who headed Wau Diocese, came to the Seminary to investigate the matter. He found out that I was innocent. He then asked the lecturers if they could allow me to resume studies at Bussere. Unfortunately, none of the lecturers agreed with the Bishop. In 1977 they sent me on the one year probation in Juba so that I could stay out of the seminary. When I was on probation, I asked Bishop Ereneo Dut of Juba Diocese if I could look for a job so that I could take care of my welfare before going back to the seminary. He did not object to my request. I was employed in the Ministry of Finance as an Office Clerk.

I worked in the Ministry of Finance for six months before the government appointed me Community Development Officer in the Regional Ministry of Cooperative and Community Development. The government transferred me to Torit District where I lived with Father Julius Igaa in *Torit Mission*. The one year probation elapsed but the Rector of Bussere National Seminary did not call me back to the Seminary. Subsequent to this, I left Torit Mission and rented a house in Malakia. In 1978, I married my senior wife, Mrs. Margarete Ayaa of Oruku (Madi). In December 1978, God blessed us with a child whom we named Emmanuel Onen (which means, behold God is with us). I served in Torit from 1978- to 1982, and then the government sent me back to Juba as Assistant Director of Community Development. In that year, *Agency for Cooperation and Research Development* (ACORD) began to work in partnership with Department of Community Development. The government saw my efficiency at work and promoted me to Coordinator of Community Development Support Unit where I supervised Community Officers of Equatoria Region, Upper Nile Region and Bahr El Ghazal Region.

In 1986 I started writing the book, *The History and Expressive Cultures of the Acholi of South Sudan*. The Dr. John Garang de Mabior, insurgency did not give me time to visit Acholi informants in different places. In 1990, when I was piecing up my work, my Uncle (my father's brother), Justice John Onge Kassiba observed that the book I was writing "would be useful to future generations." unfortunately, I did not know where I could get the money to publish the book.

In 1991 when the war in Juba intensified, I took refuge in Khartoum together with my family. Because of my good work record and efficiency CONCERN SUDAN (Irish International Organisation) employed me as a Refugee Program Manager at Kosti where I worked from 1991 to 1994. Unfortunately, before I finished writing the book, my wife, Mrs. Margarete Ayaa passed away in 1993 from Kosti. I had five children with her one daughter and four sons. But one of the sons called John Onge, later died from Juba. My second wife Santa Alal was in Juba, South Sudfan, when we were in Kosti. In 1996 I went to Khartoum, and worked as Logistic Assistant Officer with Action Contra La Faim (ACF) at Jaborona Regugee Camps in Omdurman. In 1996 I married the third wife, Mrs. Faima Doka. I had the first child with her and called her Anna Achiro Olaa. In 1999 I went to study Diploma Primary Health Care at Halliah University Omdurman.

In 2001, the security situation in Khartoum grew tense. The soldiers of President Omer El Bashir were secretly killing them. Many people lost their lives. I then fled to Cairo-Egypt. Before I reached Cairo, I thought Cairo would be a resting place, but when I reached there I found life in Cairo as hard as sitting on fire. Sudanese traveled fearfully. Refugees slept in fear. The Egyptians stated calling us *Chokoleta*. Egyptians also increased rent for Sudanese refugees up to LS 500 for a three room house. Yet an Egyptian rented it at only Ls 20 to 30. In addition to that, most UNHCR workers were Egyptians. As a result, passing interviews in the office of UNHCR became very hard for Sudanese refugees. To get out of Egypt, refugees must for the decision of UNHCR for two or seven years. If they allowed you to go, they conditioned you to go to America, Canada, Finland, and Australia. Some Sudanese have never got their way out of Egypt. Some people went back to Sudan, but those who realised that there was no peace in Sudan are still in Egypt.

In 2003 I came to Australia, and I thought it would be easy for me to get a job here because of my vast experience. Unfortunately, Australian government did not recognise degrees of Sudanese universities. This made it a little difficult for me to get a job here. In 2004 I began to study Community Development in Dandenong TAFE, and graduated with Diploma in Community Development in 2005. From 2006 to 2008, I obtained B.A. in Human Services, Victoria University. When I was studying at Victoria University, I worked with Migrant Information Centre, Eastern Melbourne, on the terms of three days a week. On completion of the course, the Manager of Migrant Information Centre gave me Case Management job which was on the basis of five days per week. Although most of the employees of Migrant Information Centre were women, there was high level of solidarity among them. This made my work at Migrant Information Centre easy. When I was search- ing for other jobs from the Internet, I discovered that most employers wanted people with Master Degree. For this reason, I began to pursue Master in International Development at RMIT University-Melbourne campus. I completed this course in 2012 since I was a part time student and, at the same time, working to support my children.

In 2010, when I was sending the manuscript of this book, *The History and Expressive Cultures of the Acholi of South Sudan* to Mwaka Emmanuel Lutukumoi, I thought of writing another book in English. It will be titled *The Short Social Anthropology of the Northern Luo of South Sudan*.

Introduction

'Social structure' refers to the pattern of social relationships in a given society. Such structures, regardless of the interventions among members of the society, provide guidelines within the cultural norms for understanding cultural values and social relationships. Thus, social relations among the Northern Luo of South Sudan are governed by rules of kinship, beliefs, gender and age.

The geopolitics, social interactions, intermarriage, boundaries, climates, destination and resources have made it difficult for scholars to study the anthropology of the Northern Luo, who live in South Sudan. Some scholars have attempted to study Northern Luo ethnic groups, but have focused on limited geographical areas, emphasising a particular Northern Group such as Shulluk, Acholi, Pari, Anyuak or Jur-Chol, instead of the Northern Luo groups as a whole. For this reason, this book will attempt to put all the information about various Luo groups of South Sudan in one place, which will make it easy for a person to compare the information about different Northern Luo groups. The information of the Northern Luo of South Sudan, in this book, will cover their migration, settlements, political organisations, marriages, religion, expressive cultures and much more.

The distribution of Luo groups in different African countries has left many with unanswered questions. As Murrdock Peter George (1959) puts it, the history of the migration of ethnic Luo groups and their consequent ethnographic distribution over six African countries (Kenya, Tanzania, Uganda, Congo, South Sudan and Ethiopia) have left most social scientists, including Luo people themselves, wondering how the Luo disintegrated into their current groupings. Many Luo lack basic facts about their African country of origin or ancestry. Indeed, the limitations of written resources about Northern Luo has made it somewhat difficult for me to write this book.

Due to mass migration among the Luo, the Northern Luo ethnic groups have reconstructed themselves as independent groups and clans: some (like Mabaan and Jumjum) are tempted to assimilate with other ethnic groups like the Dinka.

As such, the process of tracing Northern Luo ethnic ancestry remains a major task and challenge.

The Northern Luo of South Sudan did not appear out of nowhere. This book will therefore examine each article discussed from an anthropological perspective. In order to be more objective, I will focus on the description of the Northern Luo rather than consider all the South Sudanese, recounting their origin, migration and settlements; their ancestors; their norms, beliefs and customs; their language; their political structure; and their methods of conflict resolution. In addition, the book will examine the ancient Kingdom of Kush and the Funj Kingdom, as a platform to link and understand the relationship between the Luo people and the Nubians.

South Sudan is a vast country of 619,745 square kilometres. The country is inhabited by black Africans with different cultural backgrounds. However, we shall closely examine only the demographic and geographic locations of each of the Northern Luo groups within the Republic of South Sudan. The Northern Luo, even though they are scattered throughout South Sudan, still have similar customs, traditions and beliefs, and share a sense of belonging.

Accounts of contemporary South Sudan fall short of explaining the origin, cultures and social relationships of the Northern Luo. When South Sudan became an independent State on 9th July 2011, the Luo groups had hoped that the newly independent country would be built on principles of ethnic equality, democracy, rule of law and federalism, and that cultural diversity would be promoted in the country. People were - and are still - optimistic about the future of South Sudan because they had high hopes that South Sudan would become a paradise of equality and cultural diversity. This hope was based on the belief that the people of South Sudan had, for over fifty-five years (from 1956-1973 and from 1983-2015), endured bitter struggles for equality and for the recognition of cultural diversity in the old Sudan.

A good deal has been written at one time or another about various tribes inhabiting South Sudan, but less information has been accurately recorded about the Northern Luo. Moreover, some of the little information recorded about the Northern Luo is somewhat fragmentary and not reliable. However, some writers, like Professor Seligman (1932), M.E.C. Pumphrey (1941), P.P. Howell (19412), E.E. Evans Pritchard (1941 & 1948), David Graeber (1996), Deidrick Wassermann (1912) and J.P. Crazzolara (1951 & 1954) have now given us a general overview of the Northern Luo, their traditions, their culture, and their ways of thought, which are of great importance for future researchers.

The new nation lacks socio-economic development. As a result, with the exception of the road between Juba/Nimule and Gulu in Uganda, the means of communication are very poor. To help us understand the nature of current socio-political developments and the structure of the South Sudanese government, we need to revisit the Luo history and cultures in the country. At this juncture, I would like to remind the readers that Luo history tends to be overlooked – if not forgotten - by the government of South Sudan. This is one of the reasons that prompted me to pick up my pen and to start to write this valuable book: to raise awareness about Northern Luo both for people who are within and for those who are outside the country.

Luo initiation rites are fundamental to human growth and development. Luo ancestors originally established these rites in order to link the individual Luo family to specific Luo communities, and to the broader community in general. These initiation rites are a natural and necessary part of a community, just as arms and legs are natural and necessary extensions to the human body.

Initiation is fundamentally connected with transformation. It has been a central component of Luo traditional cultures since time immemorial. Although history tells us that the Luo groups of South Sudan are from one grandfather named Kola, the father of Omara, nevertheless, details of rites are basic components of Luo groups. They help guide, maintain and promote a person from one stage of life to the next – that is, from birth to death and beyond.

Interestingly, current political views tend to confuse Luo groups, which at times makes them forget that they are one people and from one male ancestor. The driving force behind this is that (1) people want positions in the Government and are ready to give up community relationships for the positions; (2) some people think that history and cultures are old-fashioned, and therefore it is unnecessary to maintain them; and (3) the separation of Luo groups happened approximately five or six hundred years ago, that makes it difficult for the Luo to see an overall picture of how the communities link to each other.

Consequently, the ruling parties in South Sudan have taken hold of this to marginalize the Northern Luo; the parties even think that the Luo as such do not exist - that there is no tribe called Luo. This is reinforced by the simple fact that the Northern Luo do not have documentation to substantiate their existence and their relationships to their ancestors. It is my hope that this book will demonstrate to the readers and to the Government of South Sudan that combined, the Luo tribe is the largest tribe in South Sudan - as opposed to what is now being claimed: that the Dinka is the largest tribe in the New Nation. The reality is that the Luo is the largest tribe in South Sudan, followed by the Dinka and the Nuer tribes.

The marginalization of the Luo people goes back to 1956 - following independence of Sudan from Anglo-Egyptians – or even further. Therefore, it is important for the Luo groups to understand and appreciate that they are one people and the largest tribe in South Sudan. This book offers an opportunity to understand and appreciate the importance of the reunification of the Luo groups regardless of where they are located in South Sudan. Moreover, it also highlights the impacts disunity has had on socio-political development in Luo areas.

The Northern Luo are proud people: aloof, tenacious of their old beliefs and ideas, intensely religious, and by far the most introvert of the people of South Sudan. Desiring nothing from the white men of previous or current governments, they want to be left alone, and when this is not granted they are ready to oppose the relevant authorities.

The Luo have an array of myths and legends that speak of their historical unity at the land of *Kero (Karro)* before the separation of Nyikang and his brother Dim. There are historians who disagree about the common origin of the Luo, but sons and daughters of Luo should not listen to them because they do not know what they are talking about – the facts speak for themselves. Like other South Sudanese, the Luo have been part of the South Sudanese body since as early as the 1820s, when the Nation State was beginning to take shape - or indeed since the Ottoman invasion from Egypt in the 14th century. More importantly, the Luo have remained part of the South Sudanese body since the independence of South Sudan on 9th July 2011.

South Sudan is inhabited by 64 tribes, who have become voiles following dictatorship of the government. Since the Addis Ababa Agreement in 1972, the Dinka have gradually come to dominate the region in terms of government, and by taking over land from the local people. The taking of land by the Government has been experienced by the Shulluk in the Upper Nile region, and by the Bari, Acholi and Madi in Central and Eastern Equatoria.

Luo people are rarely given political positions in the Government of South Sudan. This can be traced back to the formation and running of the SPLA/M and SPML-IO. The bottom line is that the majority of Northern Luo, who participated in the Movements of *Anya-nya-one* and SPLS/M and SPLA/M–IO, have not enjoyed the fruits of what they fought for. The Addis Ababa Agreement and the Comprehensive Peace Agreement (CPA) signed in 2005 in Nairobi, Kenya, have mainly benefited the dominant tribe. The question for the Luo to answer is what they should now do to be recognized and to participate effectively in the government structure. Those of us who know that, in South Sudan, people are ruled on the basis of majority, would put the answer simply: *"The Northern Luo must reunite and accept their common ancestral links, and use this as a mirror to reflect the current socio-political development in the country. Only by doing this will the Luo enjoy equal opportunity and have a card in their hands to play for the future president of The Republic of South Sudan."*

Among much that is of interest and value in this book, I would draw the reader's attention to Chapters 3, 4, 5 and 19. The author here discusses in depth the origin, separation, and family trees of various Luo groups that are found in South Sudan. In subsequent articles about the Northern Luo, all of them receive comprehensive treatment, and in every case I have provided references from contemporary scholars, quoting their materials to substantiate my arguments and to support my research.

Tribes of South Sudan in Alphabetic Order

1. Acholi
2. Aja
3. Anyuak
4. Atuot (Reel)
5. Avukaya
6. Azande
7. Bai
8. Baka
9. Balanda-Boor
10. Balanda-Viri
11. Banda
12. Bari
13. Binga
14. Bongo
15. Boya (Larim is their village)
16. Didinga
17. Dinka
18. Dongotona
19. Falata (Arab Nomads)
20. Feroghe
21. Gollo
22. Horyok
23. Indri
24. Jiye (Jie)
25. Jur (Beli & Modo)
26. Jurchol
27. Kakwa
28. Kara
29. Keliko
30. Kresh
31. Kuku
32. Lango
33. Lokoya
34. Lopit
35. Lugbwara
36. Lulu'bo
37. Maban
38. Madi
39. Makaraka
40. Mananger
41. Mangayat
42. Muru
43. Moro Kodo
44. Mundari
45. Mundu
46. Murle
47. Ndogo
48. Ngulngule
49. Nuer
50. Nyangaton
51. Nyangwara
52. Otuho
53. Pari
54. Pojulu
55. Sere
56. Shaaya
57. Shatt
58. Shilluk (Chollo)
59. Suri (Kachipo)
60. Tid
61. Toposa
62. Uduk
63. Woro
64. Yulu

STT

Table of Contents

About myself iii

Phonetic Spelling of Native Words and Names vii

Saturnino Onyala, the Man xi

Introduction xv

Chapter 1: **The Ancient Kingdom of Kush (Nubia/Luo)** 1

    Cursed be Canaan: The Luo people 4

Chapter 2: **The Funj Kingdom 1504-1821** 9

    Funj are the first inhabitants of the present Shulluk homeland 9

    Trashed to Ancestor Beni Omaya downwards 12

    List of Funj Kings 13

Chapter 3: **The Origin of Luo** 17

    Reasons why the Luo groups left their homeland 19

    The Split and Movement of the Luo Groups 20

Chapter 4: **Tiko Separated Dumo and Nyikang** 21

    Disputes over the Kingdom 21

    Nyikang in Wau, and marrying the daughter of Chief Dimo 28

    Nyikang settled temporarily in El Duem 31

    The Blood of Ubogo Enabled Nyikang to cross the River Blue Nile 32

    Nyikang Arrived in Malakal 34

    Nyikang gave one of his daughters to Dak for a wife 36

    Descendants of Nyikang and the Royal Line... 38

Chapter 5: **Luo Family Tree** 41

    The Northern Luo are descendants of Adam from Lwor through Okwa 42

    Luo Custom of Moral Expectation 44

Chapter 6: **The Northern Luo: Demography and Geographic Locations** 45

    Nyikang's descendants migrated from Karoo/Kara in Lake Albert, Uganda............ 45

Chapter 7: **The Northern Luo Migration Challenges** **49**

Migration Challenges 49

Where did the people get their food and medicine? 49

How did the Luo defend themselves during the march? 50

How the Luo people treated prisoners of war 50

Did the Luo people move in a block or in smaller groups? 51

How did the prisoners of war assimilate into Luo communities? 51

Chapter 8: **The origin of civilization** **53**

The Luo People are the Founders of World Civilisation 53

The Luo Cradle-land at Tekidi (Napata): The Grand Court of Kush 58

Rwot (King) Menya: the first King of Egypt 62

How the Luo people influenced Ancient Greek Civilisation 64

The Luo shared gods with the Ancient Greeks and Egyptians 66

The belief of the Luo about the origins of death 67

The Northern Luo today practise some Ancient Egyptian Beliefs 69

Chapter 9: **Gender and Peace Building: The Roles of Northern Luo Women
in peace building** **73**

Chapter 10: **Luo Groups** **77**

Acholi 77

Anyuak 81

Balanda Boori and Balanda Bviri 99

Jo-Luo of Bhar El Ghazal (Jur Chol) 104

Jumjum 115

Mabaan 117

Pari (Lokoro) 120

Shulluk 166

Shatt or Thuri 247

Chapter 11: **Saturnino Obwoya Opio: The Hero and Initiator of Liberation
of South Sudan Independent** 251

Birth of Saturnino Opio 252

Root Cause of Torit Mutiny 252

Torit Mutiny 18th August 1955 260

The death of Saturnino: Who killed him? 264

Chapter 12: **Religion Beliefs and Cultures** 267

Christian history in South Sudan 269

Acholi Religion and Expressive Culture 274

Anyuak Religion and Expressive cultures...289

Balanda Religion and Expressive Cultures..291

Jo-Luo of Bhar El Ghazal Religion and Expressive Cultures................................ 292

Jumjum Religion and Expressive Cultures...301

Mabaan Religion and Expressive Culture 303

Pari Religion and Expressive Culture..303

Shulluk Region and Expressive Culture...304

Shatt (Thuri) Religion and Expressive Cultures 325

Chapter 13: **Language** **329**

Locations of Luo Language in South Sudan 330

The Acholi Language 331

The Anyuak Language...332

The Balanda Language...334

The Jo-Luo of Bhar El Ghazal Language..337

The Mabaan Language 340

The Pari (Lokoro) Language 341

The Colo Language (Dhocollo) 345

The Thuri/Shatt Language 349

Chapter 14: **Reconciliation: Means to Healing the Inner Wounds**...................**353**

What is forgiveness and why forgive others? 355

What is reconciliation and how it works? 356

The seven signs of genuine repentance 359

What does it mean to heal through reconciliation?...359

The Ten Principles of Truth and Reconciliation.. 361

Traditional Justice and Reconciliation..362

The Rite of Reconciliation Called Mato Oput 365

Anyuak Traditional Justice and Reconciliations 384

Collo Traditional Justice and Reconciliations: Forms of Compensations 385

Jo-Luo Traditional Justice and Reconciliation...401

Pari Traditional Justice and Reconciliations 408

Riek Machar attempt for Reconciliation 410

Reconciliation and Healing in South Sudan 412

South Sudan's Kings and Chiefs in South Africa, Botswana and Ghana............... 426

Bewitch (hex) and South Sudan Politics 435

Chapter 15: **Kinship and Family Life**... 437

    Acholi Kinships and Family Life 440

    Anyuak Kinship and Family 446

    Jo-Luo of Bhar El Ghazal Kinship and Family Life 458

    Pari Kinship and family Life 469

    Shulluk Kinship and family Life...480

Chapter 16: **Arts, Crafts and Music**.. 497

    Arts and Crafts 497

    Music instruments 498

    Blacksmiths 499

    Acholi 500

    Anyuak Arts, Handicrafts and Music 506

    Arts and Handcarts 506

    Anyuak traditional Musical instruments 508

    Balanda Arts, Music and Handicrafts.......................................510

    Jo-Luo Height and Dresses... 513

    Shatt/Thuri Arts and Handicrafts..514

    Shulluk Arts and Music 515

Chapter 17: **Economy and Livelihoods** 529

    The Background to Northern Luo Economy and Livelihoods.............. 529

    Agriculture and Sources of Food... 531

    General Sources of Income 533

    Acholi Economy and Livelihoods...534

    Anyuak Economy and Livelihoods...538

    Balanda Economy and Livelihoods 549

    Jo-Luo of Bhar El Ghazal Economy and Livelihoods 551

    Mabaan Economy and Livelihoods 558

    Pari Economy and Livelihoods...562

    Shatt Economy and Livelihoods 563

    Shulluk Economy and Livelihoods 564

Chapter 18: **Marriages**... 587

    What is marriage? 587

    The top ten secrets one need to know about marriage 592

    Purposes of Marriage 598

    What constitute true marriage?.. 598

Polygamy 599

What should a wife stop doing if she wants to improve her marriages?.............600

What should a husbands stop doing if he wants to improve his marriage? 603

What are the things a man should not do after getting married?......................... 605

What should I do when my wife doesn't want to have sex with me?.................... 607

Reasons why a wife doesn't want to have sex 608

Narcissistic Partner 609

Who are narcissists: Can they be found in Luo marriages?............................... 610

Am I married to a Narcissistic partner? 611

What should I keep on watching for from my partner? 612

Case Study: Narcissistic Partner 614

Acholi Traditional Marriage and Family..616

Anyuak Traditional Marriage and Family.. 626

Balanda Traditional Marriage and Family 634

Jo-Luo of Bhar El Ghazal Traditional Marriage and Family 638

Mabaan Traditional Marriage and Family...657

Pari Traditional Marriage and Family... 658

Shatt/Thuri Traditional Marriage and Family 668

Shulluk Traditional Marriage and Family 670

Chapter 19: **Reproduction**..691

Introduction to Reproduction Process 691

Acholi Reproduction Process 693

Child Naming and Ceremony 696

Anyuak Reproduction Process 702

Child Naming and Ceremony 704

Jo Luo of Bhar El Ghazal Production Process 706

Jo Luo of Bhar El Ghazal Child Naming..711

Mabaan Child Naming 717

Shatt Reproduction Process... 717

Child Naming and Ceremony 718

Shulluk Reproduction Process..718

Child Naming and Ceremony 721

Pari Production process 723

Child Naming and Ceremony 725

Chapter 20: **Female Genital Cutting (F.G.C.)** 729

Definition 729

The Short and Long Term Health issues Associated with FGC 731

Types of Female Genital Cutting 732

Practices of Female Genital Cutting among the Northern Luo 735

The Law and the FGC 737

Chapter 21: **Inheriting the Wife of the Dead Brother** 739

Inheritance of Widow of Deceased Brother 739

Significant of Death 740

Mourning in the Northern Luo Culture 742

Widow Inheritance and Levirate 743

Challenges of Inheritance of a Widow 746

Widowhood Empowerment programme 747

Chapter 22: **Deaths and Burial Ceremonies among the Luo of South Sudan** 749

Background 749

The Luo's Concept of Death 750

Death in Acholi 753

Acholi Rituals for the Dead 753

Anyuak Rituals for the Dead 763

Balanda Rituals for the Dead 766

Jo-Luo of Bhar El Ghazal Rituals for the Dead 771

Mabaan Ritual for the Dead 778

Pari Rituals for the Dead 781

Shulluk Rituals for the Dead 784

Conclusion 793

Bibliography 799

Appendix 805

Chapter 1:
The Ancient Kingdom of Kush (Nubia/Luo)

The Luo were called Phoenicians before they entered Egypt
The Phoenicians were Semitic people who moved from the east (from Canaan - see map, figure 1). Some of them went and settled in Lebanon, while others went and settled in western Syria, and still others settled in Kush (in Lower Egypt, present-day North Sudan). These latter are the Luo people - the Northern Luo, the Central Luo and the Southern Luo. Those who settled in Lebanon and Syria lost their relationship with the group of people who settled in Lower Egypt.

The name 'Phoenician' was given to them by the Greeks. It was derived from the Greek word for the purple dye for which Phoenicia was famous. The society was identified with its signature dye, such that the name 'Phoenician' is believed to have come from the Greek word 'phoenix,' which means "purple-red". In Phoenician times, purple garments were a marker of elite status. Before this group became known as Phoenician, they were called (in the Bible) 'Sidonians,' as we will see later. From here, we can see how people changed their names from place to place and from generation to generation.

Phoenicians were known for their trade throughout the Mediterranean world. According to Lereon (19??), the Phoenicians developed into an alliance of coastal cities around 1550 B.C. and established colonies as far as Iberia, but they never coalesced into a

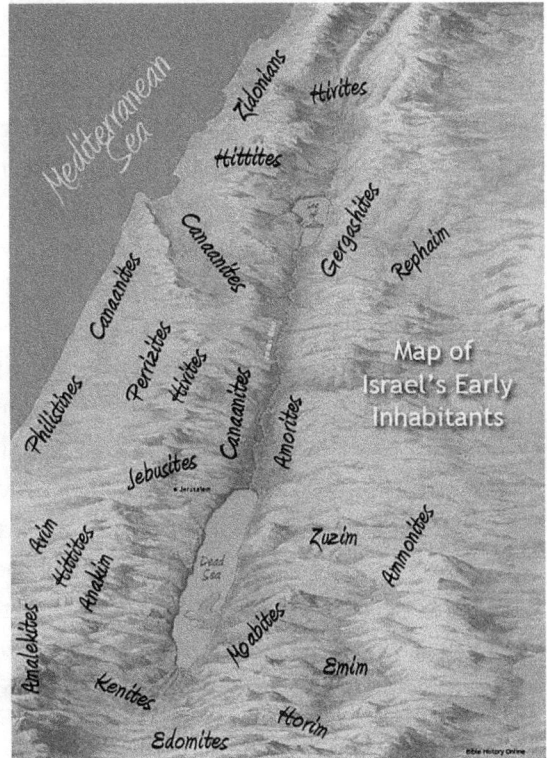

Figure 1: Map of Israel's Early Inhabitants

1

nation. Their trade as merchant sailors led the Phoenicians to develop various innovations and modernizations which are still in use today.

The maritime nations and tribes along the coast included the Sidonians (Luo/Africans), who were given the name 'Phoenician' by the Greeks. These people dwelt in the north; the Canaanites dwelt near Mount Carmel and down the Jordan valley, and the Philistines dwelt along the south-western border of the land.

During the earliest times, many people and tribes migrated into the land of ancient Israel, called in the Bible "The Land of Canaan." There were races of people who conquered other people and became powerful nations. However, it is unfortunate that the information about the earliest inhabitants of ancient Canaan before the conquest of Israel is somewhat limited in exact detail although there is evidence of several migrations.

The Bible names six people groups, each with their own identity and territory, who migrated into ancient Canaan. Most of these groups were later called "giants."

In those former times, it was widely accepted that the Phoenicians (Luo) exerted considerable influence on the Greeks. According to Herodotus (1972), the most important accomplishment of the Phoenicians was the introduction of the alphabet. The modern western alphabet originated from a set of letters that the Phoenicians devised, and that the Greeks and Romans later adopted and modified. The Phoenicians were seeking a simpler system to record commercial transactions. More important was the introduction of writing and art, which were presumably unknown to the Greeks. Many writers like John C. Trever (1971) argue that there exist a relationship between early Greek writing and Hebrew writing, because the Greeks derived alphabets from those of the Phoenicians (Luo), which later became the basis of modern scripts.

However, not all scholars accept this view on the alphabet. Marsh (1885), for example, asserted that the Phoenicians (Luo) did not bring any fruitful idea into the world. He moreover said that the art of Phoenicians does not deserve to be called art, and that the Phoenicians were mostly traders and therefore had nothing to do with the world of thoughts. Their architecture, sculpture and paintings were of the most unimaginative sort. Their religion, as far as we know, was entirely an appeal to the senses.

Even after the Phoenicians introduced to the Greeks divinities and the art of writing, Greek writers like Diop (1974) tried extensively to avoid accepting or acknowledging that the Phoenicians were the ones who introduced the art of writing to the Greeks. However, this is a clear confirmation of what I stated before that Western scholars never believe any good thing or invention can come from the black Africans (Luo people). People like Holm (1886) accepted that some Phoenicians migrated to Greece, but he rejected their influence on the Greeks.

According to Genesis 9:18-19, Noah went into the ark with his three sons: Shem, Ham and Japheth. After the flood, Noah came out with his three sons. The Bible tells us that the people now in the world are descendants of the three sons, and that from these people the whole world was populated (verse 19). The Bible in Genesis 10:6 tells us that Ham became the father of four sons, namely Cush, Egypt (Mizraim), Put and Canaan. When Canaan grew to manhood, he became the father of Sidon his firstborn and Heth the second born (Genesis 10:15). Scripture also says that the area south of present-day Egypt was inhabited by the descendants of Cush or Kush (Genesis 10:6). This is probably where "Kush Kingdom" derived its name, and the descendants of Put became the inhabitants of Libya.

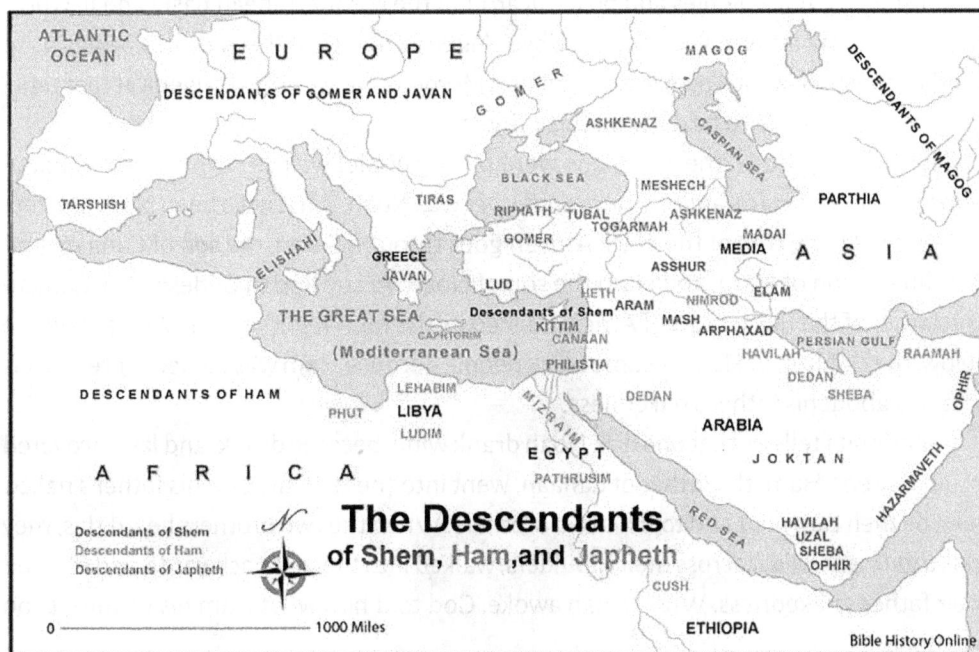

Figure 2: The Nations of Genesis 10

Afterwards, the clans of the Canaanites were scattered from Canaan; it is believed that the Egyptian clan and some of the Sidonian clan migrated to Cush (south of Egypt, which is the present North Sudan). On arrival in Lower Egypt, the Greeks called the Sidon Phoenicians the 'Kush clan' or *Kushites*. Herodotus (1972) points out that, at that time, Lower Egypt was inhabited by Kushites, well before they began mixing with Greeks, Romans and French. The Hamitic black races were all considered Kushitic in ancient times. We could thus conclude that the original land of the Luo people was in Canaan (early Israel's land). According to Herodotus (1972), the Luo people conquered and ruled Israel for a short period of time, although not much detail could be traced. Diop (1974) has pointed out that the Phoenicians were actually Canaanites and Africans. Houston (1985) has taken the matter further by arguing that, in the days of Jesus Christ, the Phoenicians called themselves Ethiopians (p.22). We discuss this in Chapter 8, as we argue that the word

"Ethiopian" is a corruption of the word "Itiyopian" which in the Luo Nilotic word (*pariance*) means 'people who have pledged collective loyalty to the god Anu.'

In as far as the language spoken at that time is concerned, the Bible tells us that the people living in Canaan spoke the Semitic language. It is believed that Semitic, Kushitic, Arabic and Egyptian are all part of the same language family. We could therefore conclude that the ancient Phoenicians were black people or dark skinned people. While still in Canaan, the black people (the Luo and Nubians) lived near the Mediterranean Sea. They established trading colonies and posts all around the Mediterranean basin and in North Africa, including in Morocco, Tunisia, Libya, Algeria and Egypt (Persia). This brings us to another question: Which people were living in Morocco, Tunisia, Libya, Algeria at that time? More research needs to be done on this subject.

If, as argued above, the Sidonians were black people, we can surmise that Canaan, whose father was Ham, and whose grandfather was Noah, must have been black. In other words, the family tree of the black African goes through Sidon, the son of Canaan, and Canaan the son of Ham, and Ham the son of Noah. According to Genesis 9:18, Canaan (the father of the black races of Africa) is listed as the fourth son of Ham. As we shall see below, the children of Ham became black people because Ham was cursed for telling his brothers about his father's nakedness.

Genesis 9:21 tells us that one day, Noah drank wine, became drunk, and lay uncovered inside his tent. Ham, the father of Canaan, went into the tent and saw his father's nakedness; he then came out and told his two brothers. When the two brothers heard this, they took a garment, laid it across their shoulders, walked into the tent backwards, and covered their father's nakedness. When Noah awoke, God told him what Ham his youngest son had done.

Cursed be Canaan: The Luo people

Holy Bible maintains that disrespect of Ham (the father of the black people) brought cursed on his descendants. Genesis 9: 20-27 tells us that one day Noah drank some wine and eventually he became drunk and lied uncovered in his tent. His son called Ham entered the tent at the time and saw his father's nakedness; he failed to cover him but just came out of the tent and told his two brothers that their father was lying naked in the tent. According to the Hebrew Bible, Canaan (Luo) was a son of Ham and the grandson of Noah –the wife of Ham was called Naeltamauk. And Canaan (Luo) became the father of Canaanites (Luo people) who were all cursed because of the sin of their grandfather (Ham). Genesis 9:25 precisely states, *"A servant of servants shall be to his brother"*

God then cursed all the descendants of Canaan, which means the Africans are cursed: they will remain slaves to the North (the western world). This might even explain why many western people look at Africans as irrational animals, uncivilized and undeveloped, despite their introduction of the alphabet to the world. Many contemporary scholars do not want to accept that the Sidonians/Phoenicians were black people, because they do not want to acknowledge that ancient civilisation came from the black people.

The empire of Kush, also known as Nubia, emerged in the second millennium B.C. to the 4th century A.D. in what is now called Northern Sudan. It was located south of Egypt, at the base of the mountains at the start of River Nile. Its three major cities were Kerma, Napata and Meroe (see figure 3). The land has good soil with plenty of rainfall all year long to keep things fresh, growing and productive, with only a few areas that rely on irrigation systems.

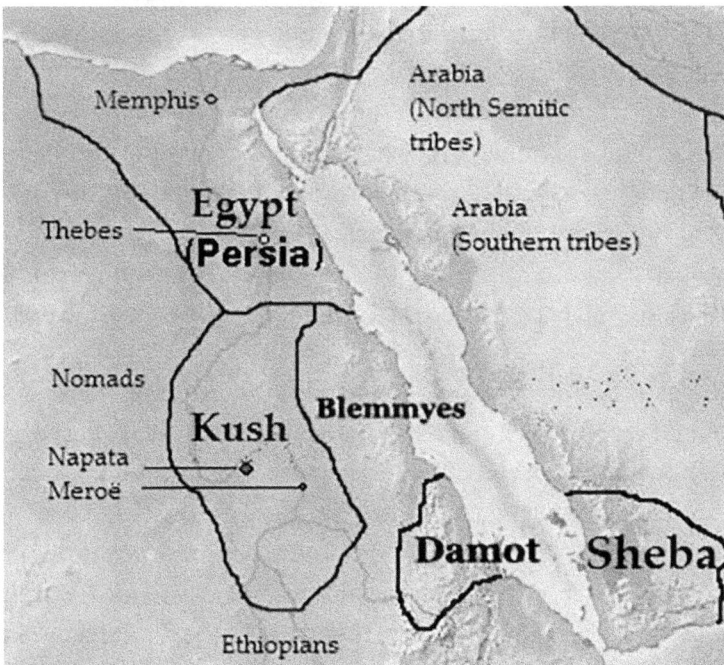

Figure 3: The Kingdom of Kush

Shortly after black people settled in Kush, King Kashta, a Kushite, invaded Egypt. In the 8th century B.C., he conquered the Egyptians (who were known as Persians) and ruled the country for a century, as Pharaohs of the 25th dynasty, until expelled by Psamtik I in 656B.C.

The population of Kush were farmers and were proud of their villages. Each village had a leader, but the leaders were not kings or queens or chiefs - they were more facilitators of decisions. This implies that the traditional village leaders of Kush did not rule; they facilitated decision-making, a process normally done through villagers. All village leaders were under the administration of one king, called King Kashta. The Kush Empire developed a system which stated that all members of the community must meet in one and the same place. It was clearly stipulated in the traditional policy that for a person to attend the meeting, he or she must be invited; thus if a person had not received a formal invitation yet still came to the meeting place, he or she would not be permitted to enter the meeting place.

The Kush Empire had tremendous natural wealth such as gold mines, ivory, incense and iron ore. It is worth mentioning that this region of Africa produced more gold than anywhere in the world at that time. As a result, some of the kingdoms neighbouring Kush, like Egypt, were very jealous of the wealth and often tried to conquer Kush so that they could take the wealth for themselves. However, the Kush did not allow this to happen.

The Kush people lived in similar houses to those of the Egyptians; they ate the same food as the Egyptians and they worshiped the same gods as the ancient Egyptians. The outspoken gods that the Kush worshiped were Amun and Osiris (see figures 21 and 23), but there were other gods beside these two. When living in Lower Egypt, the Kush identified themselves as Egyptians, but the Egyptians did not recognise the Kushites as Egyptians. The Kushites buried their monarchs along with their courtiers in mass graves. They would dig a pit and put stones around them in a circle.

Unlike the Egyptians, after the death of King Kashta, the Kush Empire was ruled by queens, rather than kings or pharaohs. They mummified their dead and built tombs with flat roofs.

The Kushites used iron ore to make iron weapons and iron tools. To produce iron from ore, they required wood. Many scholars believe that Moroe was chosen as the capital because the region had enough woodlands to provide fuel for the iron works. Before long, however, the woods were depleted. Kush was the centre of iron trade in ancient Africa, but when the Kushites ran out of woodland, they turned their attention to other trade goods to survive. Soon, they learned of wonderful gold mines in West Africa. The place was very far from their homeland, and to reach there they had to travel through the Sahara Desert.

Thus, around 750 B.C., Kush tried using camels and camel trains to cross the sea of sand. These were dangerous trips, but the Kushites had no alternative. Before long, the Kush traders discovered that travelling through the Sahara Desert could be done with very little danger. Consequently, camels and camel trains opened the trade between West and East African. As people traded across Africa, towns sprang up where there were oases. Kushite traders passed ivory, ebony, incense and other exotic goods from the south to the Egyptians, who in turn traded with other Mediterranean people. The city of Kerma, with its strategic location on the River Nile, controlled the trade between lands further south of Kush, and Egypt in the north. We can say without doubt that this trade influenced the relationship between Kush and Egypt.

Although many historians consider Egypt to be the most famous ancient civilisation on the African continent, this does not mean that it was the only ancient civilisation. Indeed the Kush civilisation was the first of the early ancient civilisations. However, having said this, it should be noted that Kush battled both the Egyptians and the Romans. There is no fight reported between Kush and the Assyrians; they seem to have lived in peace.

The Kush Empire finally collapsed in the 4th century, following persistent attacks of nomads from the south and the lands around. As mentioned before, the Kushite imperial capital was at Meroe. By the first century AD, the Beja Dynasty tried to revive the Empire, but it failed to recapture Meroe. Before long, the Kushite capital (Meroe) was eventually captured and burnt down by the Kingdom of Axum of Ethiopia, thus making the end of Kush.

This is a brief history of Kush, outlining where some scholars assert Luo groups have come from. However, as we have seen, the Luo groups came from a place beyond Kush – that is, Canaan in early Israel.

Chapter 2:
The Funj Kingdom 1504-1821

Funj are the first inhabitants of the present Shulluk homeland

According to Diedrich Westermann (1912), the Funj country, which is also known as Dar Fung, stretches on both sides of the Blue Nile. Its boundary to the north extends to Jebel (mount) Gereiwa and Rera; and to the east, it extends to Jebel (mount) Agadi and the Fazogli district. In the south, it extends to the Abyssinian frontier, and includes the district of Keili and the northern part of Burun Country; while westwards, the land extends toward the Dinkas of the White Nile.

After been expelled from the Upper Nile by Shulluk in 1504, the Funj were driven to Sennar, north of the present Malakal. They defeated Wed Ageed, who was the King of Sennar at that time, and forced a treaty upon him. Subsequently, the Sennar Kingdom became subject to the Funj and eventually, they took possession of the whole of Gezira. At that time, Amara Dunkas declared himself King of all the Funj tribes, having found small groups of black people called Nuba living in the Sennar Country. Some of the Nubians, after conquest, remained in the country, but others migrated to the mountains of Fazogli and Kordofan. Those who remained embraced Islam, intermarried with their conquerors, and lost their language and nationality.

In contrast, the Shulluk tradition remembers the Funj as the previous inhabitants of the present Shulluk homeland. Moreover, many tourists who visited Sennar in the eighteenth and nineteenth centuries were told that the Funj had come from the White Nile (the Upper Nile Region). According to Shulluk tradition, the early Funj lived along the White Nile approximately between Rank and Malakal. Archaeological evidence - pottery found on the White Nile mounds - has revealed that the Funj were a southern Nuba people. This then suggests that the rise of the Funj Kingdom at Sennar did not start in 1504, but rather began in 1493, when Amara Dunkas, the Sheikh of a sub-section of the Funj, declared himself king of all the Funj shrines; all these districts were already inhabited by the Nuba tribes.

The origin of the Funj remains unclear. Holt (1963) has argued that the origin of the Funj is one of the most controversial issues in Sudanese history. There are three hypotheses which have crystallised out the controversy:-

The first hypothesis, asserted by Crawford and Al-Shatir Busayli, was that the Funj people migrated from Abyssinia into the Upper Nile regions.

The second hypothesis, originally propounded in the eighteen century by James Bruce, holds that the Funj people are the original inhabitants of the present Shulluk land, and that they were a subsection of the Shulluk people. This theory suggests that the Funj were dislodged from their motherland by war, and that then the land was recaptured by the Collo. The Funj were forced to cross the White Nile, where they conquered Sennar. However, when the Funj were still in the land of the Shulluk, they were called *"Shillook"*. More significantly, they were pagans, but when they conquered Sennar and founded the Sennar Kingdom, they accepted Mohammedanism and took the name "Funj," which literally means "free citizen" –this could be interpreted as *they were free from the rule of the Shulluk Kingdom*. This hypothesis was defended by Arkell (1932) in his article "Funj Origins."

However, in a subsequent article entitled "More about Funj Origins" (1946), Arkell developed the third hypothesis about the origin of the Funj. This hypothesis suggests that the Funj came from Bornu and settled among the Shulluk (Holt, 1963, pp.39-55).

This third hypothesis - that the Funj came from Bornu - lacks any supporting documentation or evidence in the sources originating in the Nilotic Sudan. Therefore, in this book we are more inclined to support the second hypothesis stating that the Funj were warriors of Shulluk, and that they came from a tribal homeland of the White Nile – that is, that they are the aboriginal people of the present Shulluk land.

According to Seligman (1932), the population of Funj Hills has been exposed to foreign influence far more consistently and for a far longer period. Foreigners travelled east-west through the land. Thus the region between the two Niles has been the seat of the Kingdom of the Funj for several hundred years. The Kingdom or Empire stretched from beyond the Abyssinian border to the neighbourhood of the third Cataract, and for a short period included much of Kordofan; and it extended from the confluence of the White and Blue Niles to the swamps area north of the Sobat. Now the Funj Province is limited to the southern portion of the old Empire, lying between the two rivers, between 13 30' and 9 30' north (p.413-4).

The throwing stick (*trombash*), which resembles a large non-returning Australian boomerang, is characteristic of the Funj Hills, that spread westward to the South Nuba in 1910. It is only in the south, among the Burun, that the bow and poisoned arrows are found (see Tribal map of Southern Sudan, Figure 26, Page 78).

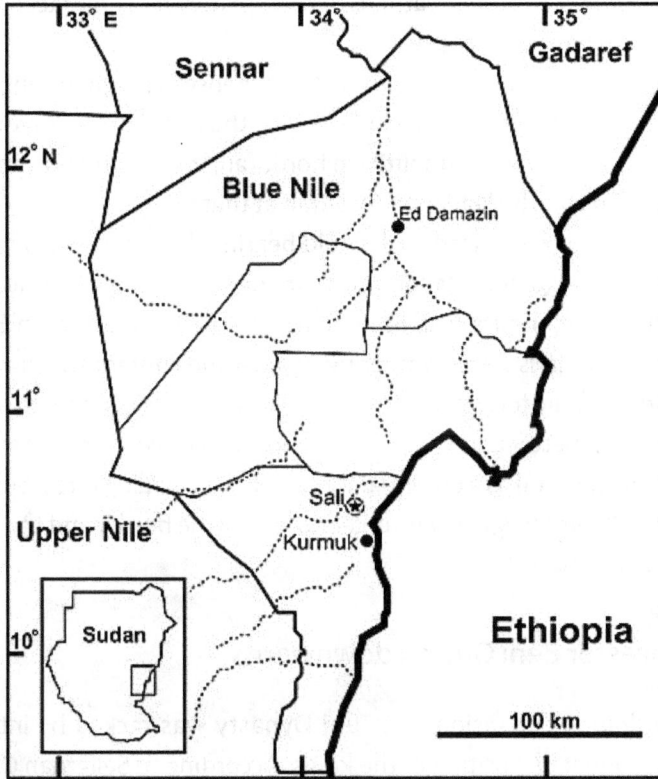

Figure 4

Shulluk and Funj traditions acknowledge the small groups of Nuba in Sennar to have come from the eastern region, possibly from the Nuba Mountains. In the 15th century, the part of Nubia formerly controlled by Makuria was home to many small states that were often attacked by the desert nomads. The area was reunited under Abdallah Jumma, who received people coming from the eastern regions. Abdallah was a hero, as shown by traditions depicting him marrying the daughter of Alshikh Hamd Abou Dunana. These small groups who were under the leadership of Abdallah Jumaa grew powerful, but they were disempowered when Amara Dunkas and his groups fought and defeated them. Amara Dunkas and his people then set up their own kingdom in 1504-1505, with its capital at the city of Sennar, on the left bank of the Blue Nile. These districts were already inhabited by Nuba tribes, some of whom after the conquest remained in the mountains of Fazogli and Kordofan. Those who remained in the region became Muslim and they were encouraged to intermarry with their conquerors. Thus they eventually lost their language and nationality, becoming known collectively under the name of Funj. Although the remaining Nuba in Sennar later embraced Islam, the records in the 15th century reveal that the Nuba people were not true Muslim, but rather heathens; although they had embraced Islam formally, in their hearts, they were still pagans.

From Sennar, the Funj expanded rapidly at the expense of neighbouring states. In 1554, they extended westward across the south Gezira, the Butana, the Bayuda, the hills of Sakadi and into south Kordofan. In southern Kordofan, the Funj met resistance from the armies of Dai 1V, who was the leader of Kordofan at that time.

Southward, the Funj progressed to the gold bearing districts of Fazoghli. As the kingdom grew stronger and extended its control in various parts of the region, it caused tension with its immediate neighbours. The Ethiopians and the New Ottoman Egypt felt very threatened. The Ethiopians were competing against the Funj for the lowlands, but the Funj conquests eventually forced the Ethiopians to move their capital to nearby Gondar, from where they were better able to secure their influence over disputed areas. As for the Egyptians, they saw the Funj as a potential threat and an invading force, but they failed to conquer the area; Ottoman Egypt was forced to fortify the border and thus consolidated their hold on north Nuba.

Trashed to Ancestor Beni Omaya downwards

Despite the continuing expansion, the Funj Dynasty was racked by internal conflict, marked by the frequent disposition of the kings. According to Seligman C.G and Brenda Z. Seligman (1932), some of King Beni Omayya's descendants who survived the massacre of their family by the Abbasids fled to Abyssinia in Ethiopia. They went and settled in the hills towards the present eastern border and intermarried with the black inhabitants, becoming known as a Christian Kingdom. Shortly thereafter, Amara Dunkas overthrew the Christian Kingdom of Alwa and conquered its southern portion, and in 1504 made his capital at Sennar. Sennar became part of the trade route from Cairo to Abyssinia and from the Red Sea to Central Africa. Among its exports were gum, slaves, ivory, gold, ostrich feathers, ebony and hippopotamus hides (p.414). We could assume that King Beni Omayya was not included in the below list of Funj kings because he was a Christian.

According to Diedrich Westernmann (1912), the average reign of each king in the Funj Kingdom was a little more than thirteen years. King Baadi Abu Dign 11, who reigned from approximately 1644-1681 (some argue he reigned from 1635-1671), attacked the Shulluk people, who at this time inhabited the country on both sides of the White Nile south of Kawa. The King took a large number of the Shulluk people as slaves (slaces). After attacking the Shulluk people, King Baadi Abu Dign 11 also attacked the Nuba people on the mountains of Tagale, taking a large number of them as slaves. On his returned from Shulluk land and Kordofan to Sennar, King Baadi 11 built a number of villages for his slaves and prisoners. It is said that the prisoners named some of these villages with names similar to those of villages they left behind; for example, they called one village Jebel Nuba, and another they called Tagale.

Having said this, it is important to recall that Amara Dukan's Empire was short-lived. Although researchers in social anthropology of the Northern Luo have not indicated who took over from Amara Dunkas, I would like to suggest that it was Sultan Baadi 1. Many writers talk about Baadi 11, Baadi 111, Baadi 1V and Baadi V. This implies that there was a Baadi 1, who would fit in with my suggestion. History tells us that Sennar became under sultan Baadi 11, who later defeated the kingdom of Tagali in the west.

List of Funj Kings

(1) Amara Dunqas 1503-1533/4
(2) Nayil 1533-1550
(3) Abd al-Qadir I 1550-1557
(4) Abu Sakikin 1557-1568
(5) Dakin 1568-1585
(6) Dawra 1585-1587
(7) Tayyib 1587-1591
(8) Unsa I 1591-1603
(9) Abd al-Qadir II 1603-1606
(10) Adlan I 1606-1611
(11) Badi I 1611-1616
(12) Rabat I 1616-1644
(13) Baadi II 1644-1681
(14) Unsa II 1681-1692
(15) Baadi III 1692-1716
(16) Unsa III 1719-1720
(17) Nul 1720-1724
(18) Baadi IV 1724-1762
(19) Nasir 1762-1769
(20) Isma'il 1768-1769
(21) Adlan II 1776-1789
(22) Awkal 1787-1788
(23) Tayyib II 1788-1790
(24) Baadi V 1790
(25) Nawwar 1790-1791
(26) Baadi VI 1791-1798
(27) Ranfi 1798-1804
(28) Aghan 1804-1805
(29) Baadi VII 1805-1821

Source: bluenlstate.blkogspot.com.au 11/08/2015

At that time, the armies of Sennar relied mostly on slaves who had been captured in wars with neighbours. These armies included horsemen drawn from nobility. The riders were armoured with chain mail, while the horses were covered in thick quilts and copper headgear. Most of the troops were armoured infantry carrying swords, and they were highly effective against their less organised rivals.

The Funj Kingdom stretches on both sides of the Blue Nile. The present border stands on the north in Mount Gereiwa and Bera; on the east it stands on Mount Agadi and the Fazogli district; on the south it extends to the Abyssinian frontier, including the district of Keili and the northern Burun region; and to the west it extends as far as the land of the Dinka of the Upper Nile, who settled in the White Nile.

In the days when the Funj were a great power in the region, their areas included part of Ethiopia and large districts of the White Nile. This explains why we find Funj in Ethiopia today.

At that time, the Funj State practised loses confederation of Sultanates and dependent tribal chiefs. The wealth and power of the sultans had long rested on the control of the economy. All caravans were controlled by the monarchy, as well as the gold supply that functioned as the state's main currency. The sultans also depended on the role played by the Funj in the slave trade. Farming and herding moreover thrived in Gezira and the southern rainforests. The Sultan similarly derived income from crown lands set aside for his use in each village. Within a short time, however, foreign currencies became widely used by the merchants, breaking the power of the monarchy to closely control the economy. The thriving trade created a wealthy class of educated and literate merchants who later became concerned about the lack of orthodoxy in the kingdom. The monarchy of Sennar had long been regarded a semi-divine, in keeping with ancient traditions where festivals and rituals involved massive consumption of food and alcohol. This was somewhat distressing to the Muslims.

The Sultan appointed chiefs in each village and received tribute. He levied tax and, in return, asked chiefs to supply troops in the time of war. It is worth mentioning that the population of Sennar, at that time, relied on the Sultan to resolve internal disputes.

The presence of the Funj Kingdom in Sennar helped to stabilise the region by creating a big military block between the Arabs and Abyssinians in the north and east respectively, and the non-Muslim blacks in the south.

However, we cannot deny that Sennar was heavily divided along geographic and racial or ethnic lines. The society was divided into six racial groups as follows: The blue, the green, the yellow, the red, the green mixed with yellow and the slaves. There was a sharp division between the heirs of the ancient kingdom of Alodia and the rest of Sennar. The Alodians adopted the mantle of the defeated Abdallah Jumma and were thus generally known as the Abdallah. Nevertheless, Sennar was prosperous through trade; as a result, numbers of representatives from the Middle East and Africa were stationed in the city of Sennar.

Nevertheless, the internal divisions weakened the Funj Kingdom to such an extent that it was possible for Muhammad Ali Ottoman Khedive of Egypt to lead his armies into Sennar in 1821. On arrival in Sennar, they did not face resistance from the people because of the sharp divisions within the community of Funj. Subsequently, Muhammad Ali defeated the Funj and absorbed a good number of its population into Ottoman Egypt. In 1956, following the independence of the Sudan, the region was absorbed into Anglo-Egyptian Sudan.

The Funj had originally practised a religious mix of African traditional religion and

Christianity. They were only converted to Islam in 1523. Although the Funj accepted Islam, many of them continued to practise animism and Christianity. It has been argued that even though the Funj claimed to be Muslim, fifteenth century records show that they were heathens – that is, they are people who claimed to practise Islam but their actions showed that their hearts were pagan. Although Islam became an official religion in the region, it was not widely practised.

Unlike the Shulluk, the Funj did not want their kings to die a natural death. When the king was sick, the executioner of the royal court went to the king's palace and put him to death. In contrast, Shulluk Divine Kingship narrates how, when the Shulluk king Reth was sick, the chief's wife entered the palace and strangled him to death. Traditional laws did not allow anybody to stay around the king's palace when the king was being put to death.

Sennar was at the peak of its power in the mid-seventeenth century; it repulsed the northward advance of the Shulluk up the White Nile and compelled many of them to submit to Funj authority as slaves. In this year the Shulluk inhabited the lands on both sides of the White Nile south of Kawa. After this victory, the sultan Baadi 11 Abu Dign, who reigned from 1642-1681, sought to centralize the government of the confederacy at Sennar. Baadi introduced a standing army of slave soldiers that freed Sennar from needing to depend on vassal sultans supplying the military. This provided the sultan with the means to enforce his will through the tribes. This is one of the reasons for the decline of Sennar in the mid-seventeenth century. The second reason was that the chiefs of a non-Funj tributary tribe were by this time managing their court affairs independently.

The Funj even invaded the mountain of Tagale, destroyed Kordofan, and took a number of them as slaves. On return from the battlefields, Sultan Baadi Abu Ding built a number of villages in the district to accommodate prisoners of war, including Jebel Nuba and Tagale. These similar names were given in memory of the villages prisoners of war left behind. But in the seventeenth century, Sennar began to decline as the power of the monarchy was eroded, when merchants funded the Ulema groups to undermine justice in the region.

Until the 1810s, the kingdom was home to the largest army in East Africa, stationed at garrisons and ready to fight the surrounding enemies. Clashes with the Shulluk continued until such a time as the Funj and Shulluk realised that their common enemies were the Dinka. They then united to fight against the Dinka, the so called largest tribe in South Sudan.

In 1762 when under the rule of Sultan Badi 1V, the Funj were invaded from the northeast of Sennar by Abu Likayik of the red Hamaj. Sultan Badi 1V was defeated and Abu Likayik installed another person from the Funj royal family as his puppet sultan. It is said by some Funj people that, although power was given to another person from the royal family, the Funj Dynasty continued normally but with no real authority. The new king could not exercise his power as he was not installed according to Funj traditions and customs. However, this began a long conflict between the Funj sultans and the Hamaj, as the sultans attempted to reassert their independence and authority, and the Hamaj sought to remain in control of the true power of the state.

These internal divisions greatly weakened the state and, in the late eighteenth century, Sultan Adlan 11, son of Sultan Taifara, took power. At this time, the Turkish ruler Al-Tahir Agha married Khadeeja a daughter of sultan Adlan 11, thus establishing a presence in the Funj kingdom. This paved the way for the assimilation of the Funj into the Ottoman Empire. Baadi V11 was the last ruler of the kingdom: he could not resist the Ottoman, as the kingdom was already too weak to fight foreign invaders.

Thus in 1821, when General Ismail bin Muhammad Ali, son of Ottoman Khedive of Egypt, led an army into Sennar, there was no resistance from the indigenous army. As a result, Sennar was considered defeated and subsequently absorbed into the Anglo-Egyptian Sudan, when Sudan got independence in 1956. This end was the end of the Funj Sultanate.

Chapter 3:
The Origin of Luo

. . . .

"History is a mirror of society which should reflect its past and must acts as a guide for the future. We as Luo are poor learners form the history. The past may be past, but facts will always remain sacred. History is not a playthings; it cannot be tailor-made to suite one group or the other. Nor can it be tempered with to promote a specific viewpoint in order to annihilate another group" (Olara, 2013).

There are conflicting reports about the origin of the Luo people: there is no regional agreement on where the Luo people have exactly originated from on the face of the earth. However, it is important to understand that the Luo people did not fall from the sky. They have originated from somewhere on the face of the earth. Thus to know the Luo as they are and where they have come from, it is of paramount importance to trace their origin back through their parents, grandparents and their exodus from the great land or cradle-land. According to Crazzolara (1950), the cradle-land of the Luo is in the south-east of Wau and south of the Bhar El Ghazal, somewhere in the region where 70 North Latitude meets 30 East longitude. The very importance glue to the cradle land of Luo is found in the Shulluk tradition; they call the country Podhi Luo, meaning 'country of the Luo' (p31).

From my research, I believe that all the Northern Luo have originated from Okwa, the father of Nyikang, Dumo or Duwaat, Anyuak, Gilo, and Achol. In the earliest times, the Northern Luo of South Sudan was one of the tribes who settled in ancient Israel. The Bible tells us that the people who migrated into ancient Canaan were named in the bible as "six peoples," each having its own identity and territory. We could then also assert that the great grandfather of Okwa was born in ancient Canaan, and might have died at the eve of Luo migration from Canaan to Lower Egypt. When the Luo were still in Canaan, they were known as Phoenicians or Semitic people, who lived in ancient Israel. They spoke their own language known as "*Dho-Luo*" or "Luo Language." Interestingly, when the Phoenicians (Luo) settled in Lower Egypt, they abandoned their true God and started worshipping gods of the Egyptians. These Egyptian gods include Apolo, Dianysas and Osiris. The Phoenicians/Luo shared the worshipping of gods with the Egyptians, so that the Egyptians would feel that they were part of them, but in actual fact, the Egyptians did not recognise the Luo as Egyptians.

Oral history recounts that before long, the Phoenicians/Luo fought the Egyptians and conquered them. When the Luo conquered the Egyptians, a man called Menya became the first King of the Egyptian Dynasty 1. Since the Europeans spoke different languages, they found it difficult to pronounce Luo words with the consonant "ny." As a result, the Europeans completely omitted the letter "y" and the word "Menya" came to be read as "Mena" or "Menes". Houston Dunjee (1985) confirms that the first King of Egypt was not called "Mena" or "Menes" as is written in many European texts, but "Menya" (p.69).

Many attempts have been made by different scholars, anthropologists and researchers to find the origin of the Luo people. Some writers like Omolo (2013) and Odhiambo (2014) assert that the Luo people originated from Siala in Egypt around 749 B.C. They suggest that in Egypt, the Luo people were known by the name "Lado Nation" or black Egyptians. They moved from Egypt with their cattle and settled in the Nuba Mountains (Te Kidi), in northern Sudan. Soon after settling in the Nuba Hills, the Luo formed two Kingdoms, the kingdoms of Nuba and Makuria. It is from the kingdom of Makuria that the Luo groups started migrating away from the Nuba Hills between the year 640 A.D and 1270 A.D. They came to the River Nile and walked along the Nile, searching for good farming, grazing and metal deposit land. They moved and settled in Wau, near the confluence of the Meride and Sue Rivers (Bhar El Ghazal), in southern Sudan. From Sue Rivers, the Luo groups moved to different parts of countries bordering Sudan. The Northern Luo, under discussion in this book, moved and settled in the areas they occupy today, where there are vast fertile and rich mineral areas. The Central Luo migrated and settled in northern Uganda, in the Gulu and Kitgum Districts, around 1500 A.D and 1800 A.D. The Southern Luo migrated into western Kenya via today's eastern Uganda (Moroto Districts, Karamajong Land) and settled there, but other groups continued to Tanzania and settled in Nyanza. Traditional history indicates that the Luo of Kenya moved into the host country in five waves at different times. The first and largest group was the people of Joka-Jok, who migrated from Achloi lands (Pajok in Eastern Equatoria, South Sudan). The second group was the people of Alur, who migrated from the Alur in Uganda. The third group was the Owing people, who migrated from Padhola in Uganda. The fourth group was the people of Jok Omol, who migrated from Pawir in Uganda and the fifth group was the people of Abasuba, who migrated from Bhar El Ghazal to Kenya via Uganda. The present day Kenyan Luo traditionally consisted of twenty three sub-tribes, each of which comprised various clans and sub-clans.

However, other writers like Onyala (2014) and Onyango-ku-Odong J.B. Webster (1976) argue that the Luo people did not originate from Egypt, but from the Great Lakes (Dog Nam) in the north eastern region of Sudan. This place is known today as Kassala. The Luo people migrated from Kassala around 300 B.C. and settled in Wau, near the confluence of the Meride and Sue Rivers, in Bhar El Ghazal. The Luo people lived in Bhar El Ghazal for approximately eight hundred years. Between 1200 A.D. and 1450 A.D., the Luo people moved to nearly all the countries neighbouring Sudan, resulting in separate groups with variations in language and sometimes traditions, as each group moved further away from their kin.

Onyala (2014) says that, after moving to many neighbouring countries, some Luo groups such as the Central and Southern Luo resettled in Bhar El Ghaza, but the Northern Luo settled around Lake Albert, in Uganda, in a place commonly known as Karoo (Kara) or 'homeland'. Other episodes record the Northern Luo cradle-land as being in Rumbek (Bhar El Ghazal). According to Crazzolara (1950), Shulluk tradition calls this place *podhi Luo*, which means 'Land of the Luo'; or *Dowaat/Dhimo*, which means 'Country of Dhimo' (p.31). The descendants of Omara lived here for many years. They eventually left Karoo (Kara) because of a dispute between Dimo and Nyikang.

Reasons why the Luo groups left their homeland

Due to insufficient records about Luo groups, we do not know the exact reason why the Luo groups left their homeland. We can therefore only guess as to what first caused the Luo to leave their original home. Writers like Saturnino Onyala (2014) argue that the separation of Dimo and Nyikang was due to bead and spear, as we will see later in this book. Other writers like Crazzolara (1950) argue that Nyikang decided to migrate from their homeland because Dhimo, his half-brother, was elected by the community to the position of the *Jago* (chief). Traditionally, this position had been occupied by Nyikang, in accordance with the will of their late father Okwaa. Thus when Nyikang realised he had fewer supporters and that his weak position had become precarious after open discord, he fled by night with his groups and went over to Rwot Dimo, leaving Dhimo behind in the home land. Shortly after, the homeland became known as *Pa-Dhimo*, which means 'place of Dhimo.' Another reason for the migration of the Luo groups might have been because their numbers were growing and they were facing pressure from the Dinka and Nuer, who were expanding pasturelands for their herds (p.31).

At this point, it is important for us to remember that Luo groups were living independently from each other when they were in the homeland. Crozzalara (1950) says, *"In their country of origin we must assume that the Luo consisted, politically, of a number of independent groups. A clear proof of this we find in the system of separation. Throughout the whole course of their migration we do not come across a single fact that would indicate the existence of a central unit which would try to keep the various tribal groups together; each group on the country appears to have enjoyed full independence"* (P.6).

The Split and Movement of the Luo Groups

Thus we could conclude that the Luo never migrated as a block but rather in small groups, and that each group went its own way and at its own risk. Luo tradition tells us that there were three main divisions that occurred among the Luo groups.

The first division that broke away from the Luo groups were the Boor group or Balanda, who separated and went to Bhar El Ghazal in the present South Sudan. According to Jur Chol tradition, Boor was a brother of Odak and son of Nyikang. Although there is evidence that there was a serious quarrel between Boor and his brother Odak, this does not necessarily explain why Boor decided to leave the homeland. The reality seems to be that Boor separated on his own accord, leaving his brother Odak behind. Shilluk tradition also says that when Boor was separating from the main Luo groups in the homeland, he took a *dekagi* (a long stick used by the Shilluk for showing durra) and threw it at Nyikang with the following words of imprecation: "Take this *dekagi* to bury your people with." At this, Nyikang allegedly replied, "Never mind; I shall bury my people, but my people shall not perish for that" (Crozzalara, 1950, p.36).

The second group that left the homeland following the example of Boor were the Jo-Luo (Wau), now known as Jur Chol, who the Shulluk called the *Wuate Dhimo*. The Jo-Luo groups migrated together with other small Luo groups called Thuri and Bwodho. When they reached Bhar El Ghazal, the Thuri and Bwodho groups broke away from the Jur Luo and settled in the land bordering Raja; their traditional history tells us that ever since then they have been living peacefully, but independently (Crozzalara, 1950).

The third group who left the homeland were the Nyikang Groups. This is also known as the "The Main group's First Movement." There are two reasons given to justify the movement of the Nyikang from the homeland. The first reason, as mentioned above, was that Dhimo was elected to the position of Chief, and Nyikang was expected to submit to him. The second reason was the case of bead and spear, which we shall see later in this book.

Nothing has been written about the movement of Dhimo from the homeland to southern Sudan, although we can surmise that Dhimo might have followed his brother Nyikang. However, where he ended up or where he was buried remains a mystery and will ever remain a mystery for the anthropologists to come as they will never find any related information about the movement of Dhimo from the homeland.

Chapter 4:
Tiko Separated Dumo and Nyikang

Disputes over the Kingdom

After the death of Okwa, Nyikang, the Divine King, took over the kingdom of his late father. Nyikang lived peacefully with his three brothers and one sister (see Nyikang Geanology, p.214). The children of Okwa all spoke one language. They considered themselves to be one tribe.

Luo oral history differs from some of the writings about the separation of Nyikang and his brother Dumo. There are three narratives about the separation of the two brothers.

The first group of Luo elders, and writers like Crazzolara (1952), say it was only the issue of spear (*tong*) and bead (*tiko*) that caused division among the children of Okwa.

However, the second group of Luo elders, and writers like Diedrich Westernmann (1912), say the two brothers separated not because of 'spear and bead' but rather because the calves of Nyikang ate the "dura" (sorghum) of Dumo, (who is also known as Duwadh). When Dumo learned that the calves of Nyikang had eaten his dura (sorghum), Dumo went and beat the calves. Nyikang became angry with Duwadh and left him in the country (*Kerau*) (p.167).

The third narrative comes from Diedrich (1912), who has argued, "Nyikang and Duwadh (Dumo) were twins, they lived far away to the south (possibly at Lake Albert). Okwa died and his village was deserted, so the people asked, "Who shall we elect as king?" Some people said, "We will elect Nyikang," but others said, "We will elect Duwadh/Dumo." In the end, a majority of people elected Dumo to be their new king, but Nyikang refused to swear allegiance. Subsequently there was war and the people divided. Nyikang then turned away from the land of Kerau/Karoo and went to Potherthura (in Wau, Bhar El Ghazal), where he married the daughter of chief Dimo, and a son, Dak, was born (p.157).

The fourth narrative is from my informants who have said that, after the death of Okwa, Nyikang went to war with his brother Duwadh, who was considered the legitimate successor to the throne (kwom). Nyikang was defeated and, as a result, he left his homeland and migrated north-east with his followers, who included: (1) Ju Wad Okwa, his step-brother; (2) Otono Wad Okwa – of Kwa-okollo of Willinyang; (3) Anongo, his son; (4) Ogyelo wad Nyikang of Mwomo; (5) Okell of Kwa-kell; (6) Okumo, who became known as Kwa-gweno of Didigo-Panyidway; (7) Ogawi-Lawi of Pathworo-Panyidway and his descendants "Kwa-jango"; (8) Obugo and his family members; (9) Bur, his son; and (10) Shall or Cal, his son. They travelled until they reached Wau in Bahr-el-Ghazal (meaning *River of gazelles* in Arabic). Here, Nyikang married the daughter of Dimo, a divine man, and begot a son called Dak.

Examining these four narratives, we can clearly see that the two brothers separated, but the root cause of their separation is uncertain. However, from my extensive research, I conclude that the two brothers separated because of 'bead and spear.' Many Luo oral histories speak of this.

It is said that one evening, Dumo sat in front of his house to put beads on a string (*rubo tiko*). That same evening, the wife of Nyikang went to chat with Dumo, the brother of her husband. She sat by Dumo's side and left her small child to play on the ground.

Before long, the child took a bead, put it into his mouth and swallowed the bead. When Dumo saw that the child of his brother had swallowed his bead, he went to his brother Nyikang and told him, "Ho! Brother, your child has swallowed my bead. I do not want much talk; I only want my bead".

Figure 5: Beads

Nyikang responded to his brother with a soft voice, saying, "Brother, as the child has already swallowed your bead, I kindly ask if we could wait until the morning. If the child defecates, we will scatter his stools, take your bead and give it back to you."

Dumo answered his brother with a rough and dishonest voice, saying, "I cannot wait for tomorrow; I want by bead now, now, now!"

Figure 6: Dumo putting beads on a string

Nyikang was not happy with his brother's response. He entered his house, took a spear, and brought the child to his brother Dumo. Nyikang then opened the stomach of the child with the spear. He found the bead in his child's stomach and gave it to Dumo, saying, "My brother, here is your bead. Take it; I have killed my child because of your bead."

Figure 6: Nyikang opens the stomach of his child

Dumo was very happy; he took the bead with a big smile and returned to his house. Unfortunately, Dumo had forgotten that he was a good hunter. He was a hunter but he only had one spear. Before long, Dumo took his spear and went hunting. When he reached deep into the bush, Dumo found many elephants standing under a big tree. He hid himself, crept close to the elephants, and speared a big mother elephant. The mother elephant did not die but ran away with Dumo's spear in her body. Dumo did not know what to do, as he was left without any spears.

Figure 7: Dumo speared an elephant

Dumo returned home and one day thought again of going to hunt animals. Since he had no spear, he went to his brother Nyikang and asked, "Brother, an animal has taken the only spear I have; could you lend me one of your spears, so that I can go hunting? When I return from hunting, I will return it to you".

Nyikang did not refuse his brother's request: he entered his house and took the very spear with which he had opened the stomach of his child, and gave to Dumo.

Dumo took the spear and went hunting, as usual, in the bush. Dumo had not gone far before he found a herd of elephants. He was very happy and, coming close to the elephants, speared one of them. The elephant cried aloud, but did not fall down or die, but ran away with the spear in her body.

The Luo people believe that in ancient times, elephants could talk to human beings and to each other, just as people today talk to and understand one another. The oral history of the Luo people tells us that when the speared elephant reached home, her brother asked her, "Who speared you like this?"

The speared elephant answered, "There was a small, short man who speared me."

Dumo went back home to his brother Nyikang and begged for forgiveness, saying, "Brother, excuse me! I speared an elephant with the spear which you lent me, but the elephant did not die: it left with the spear in its body."

Nyikang immediately answered his brother, "Brother, I also do not want much talk; I have already killed my child because of your bead, and now I want my spear."

Dumo could not do something different, so early in the morning, he took some food (peke) and went into the bush. He walked very far and eventually came to a herd of elephants. He did not know how to approach the elephants. Dumo looked around and saw a big tree; he went and sat under the big tree. He was just wondering how he could reach the elephants.

Figure 8: The elephant asked Dumo, "What do you want here"?

When the elephants saw Dumo sitting under a big tree, they sent one elephant to ask him, "Man, what you want here?"

Dumo answered, "I want my spear with which I have speared one of your sisters."

The elephant returned to the group and told them, "That man wants his spear with which he has speared one of our sisters."

Many modern Luo and scholars wonder how an animal could talk to a human being. The reality is that in former times, animals did talk to people in human voices. Proof of this can be found in the Gospel Transformation Bible, in Numbers chapter 22:27-30, which reads, "When the donkey saw the angel of the Lord, she lay down under Balaam. And Balaam's anger was kindled, and he struck the donkey with his staff. Then the Lord opened the mouth of the donkey, and she said to Balaam, *"What have I done to you that you have struck me these three times?"*...and the donkey said to Balaam, *"Am I not your donkey, on which you have ridden all your life long to this day? Is it my habit to treat you this way?"* Similarly, it is written in 2 Peter:2:16,*"but was rebuked for his own transgression; a speechless donkey spoke with a human voice and restrained the prophet's madness."* Today, human beings have developed a way of communicating with animals in different ways.

There were a number of spears that the elephants had kept, so one elephant took the very spear that Dumo was looking for and gave it to the small elephant, to give to Dumo. He was very happy when he saw that it really was his brother's spear. Dumo received the spear with a smile and returned home. As he neared his home, he called his brother and gave him the spear, saying, "Good brother, I have brought your spear that was taken by the elephant. Here is your spear, take it".

Figure: 9, The lost spear

Nyikang received the spear from his brother and immediately told him, "Good brother, to avoid future confrontation and wrong-doing between you and me, I want to inform you that, as from today, I will leave you in the land of Karoo, near Lake Albert in Uganda. I will look for a new place where I can settle with my people (followers).'

That very evening, Nyikang gathered his people and the people of his two brothers Anyuak and Gilo and told them what had happened between him and Dumo. When he had finished telling the stories, he then told the groups, "As from today, I am going to leave Karoo to my brother Dumo and his people. If any of you would like to come with me, he/she must prepare food (peke) for the journey."

The groups of Anyuak and Gilo agreed to follow their brother. Nyikang then told them, "Tomorrow morning, we shall go north towards Sudan until we find fertile land for cultivation and for our animals. There we will settle."

Early in the morning, Nyikang and his groups started moving towards the north. Before they had gone far, Dumo followed them and called to Nyikang, "My brother, look behind you." When Nyikang heard the voice, he turned his head and looked over his shoulder to see who was calling him.

Dumo then threw a sharp stick (*tir*) after Nyikang, cursing him and the groups, saying, "I have thrown this sharp stick (*tir*) to indicate that, as you have decided to leave us in Karoo, where there is no death, you and your people must meet death in the new land, where you are going to settle. You and your children and grandchildren must meet death. This "tir" must be used to dig tombs where you will be buried."

Figure 10: Tir (sharp stick)

In his first account, Diedrick (1912) says, "*In ancient time, people were living in a place called Kerau/Karoo and this is the place Nyikang separated from his brother Dumo (Duwat). Nyikang told his brother, "You better stay here, I am going away". Dumo asked, "Nyikang, where are you going?" He replied, I am going to that place there". Again Dumo asked, "Nyikang look behind," And Nyikang turned his round, and looked back, and he saw a stick for planting "dura", which Dumo had thrown to him. Nyikang then came back to take it and he asked, "What is that?" Dumo replied, "Go, that is a thing with which to dig the ground of your village". Nyikang picked up the stick and went until he came to a country called Turo (Potherthura). This is the country of his son Dak.*" (p.159). In contrast, in his second version, Diedrick (1912) asserts, "*Nyikang said to his brother, 'Dumo stay here! I'll go.' And Nyikang went away walking. Dumo ran after Nyikang, saying: 'Nyikang, stop!' But Nyikang refused. Again, he called, 'Look!' And Nyikang looked behind, and Dumo threw a digging stick towards him, saying, 'Take this stick to bury your people with!' ,*" (p.167).

Nyikang watched all the dirty games Dumo was playing. When Dumo finished cursing, Nyikang returned and picked up the "*tir*" and said to Dumo, "Brother, I am now going, but not all of us will die; some people will die but others will survive. Those who survive will multiply and fill the new land in which we are going to settle."

One man saw the bitter curses between Dumo and his brother. He went to Nyikang and begged him to return. When Dumo saw this man speaking to Nyikang, he shouted to the man from a distance, "You must choose one thing: either stay with me or go with Nyikang!" Nyikang turned slowly to the man and asked him to go back with Dumo. It remains unclear whether this man returned with Dumo or proceeded with Nyikang.

Nyikang and his groups left Lake Albert around 1550 A.D., travelling with his two sons Bur and Shall. Cal (Shall) was accompanied by his wife Ungwedo/Ongwedo, his two uncles Moiny Nywado and Juok, and by three of his workers. The workers' names were Ubogo/Obogo, Ujul and Mielo. Oral tradition says that Nyikang did not only travel with these people, but with a multitude of people.

Nyikang in Wau, and marrying the daughter of Chief Dimo

From Karoo/Kerau near Lake Albert, Nyikang went to Pothethura (Turo), a place near the town of Wau in Bhar El Ghazal. This is where Nyikang found Rwot Dimo, the chief of Potherthura country. Luo tradition tells us that Dimo, the chief of Potherthura, had some divine power.

Collo tradition says that, while in Potherthura, Nyikang married Rwot Dimo's daughter. According to Luo culture, if they were related, Nyikang would have not married this woman – therefore, we can conclude that there was no blood relationship between Nyikang and Chief Dimo. After some time, the daughter of Dimo gave birth to a baby boy who they named Dak.

Dak grew to be a man of hot temper; he liked fighting, and showed no respect to the elders in the communities. However, he was very respectful to his grandfather Rwot Dimo.

One day, Rwot Dimo challenged and insulted Nyikang, saying to him, "You are not smart - your son Dak is cleverer than you!"

The following morning, Rwot Dimo used his divine power to hide all fire from the families in Potherthura country, so that no one could cook. Luo oral history indicates that Dak shared the divine power of his grandfather. Thus, when Dak realised there were no fires in the entire area of Potherthura, he knew it was his grandfather who had done this. Dak also used his divine power: he blinded the eyes of everybody who was under the leadership of Rwot Dimo. The only people who were not blinded were Rwot Dimo and his wives, and the people of Nyikang. Rwot Dimo understood that it was Dak who had made the eyes of his people blind. This matter created confusion between Nyikang and his son Dak, because some people thought it was Nyikang who had done this, but Rwot Dimo was certain that Dak was responsible.

The blind people then requested Rwot Dimo to rescue them. Dimo asked his wives, "Why has Dak made people blind?" The wives did not answer.

Rwot Dimo then called Dak and said to him, "I know very well that you are the one who made my people blind; if this is the case, I am hereby returning fires to the people, so that they can cook". When Dak saw that there was fire in every house, he also cured the blind people.

However, Rwot Dimo did not learn from Dak's actions. Three days later, Rwot Dimo stopped rains from falling on the fields of Dak and his father Nyikang. When Dak realised there were no rains in his fields or in those of his father, he made all the cows of Rwot Dimo blind. The cows could not see the grass when they were taken for grazing and, as a result, all the cows lost weight and there was not enough milk for Rwot Dimo and his wives.

When this happened, Rwot Dimo again asked his wives, "Why did Dak and Nyikang make the eyes of my cows blind?"

The wives answered, "And why did you stop the rains from falling on the fields of Dak and Nyikang?"

Rwot Dimo had no choice: he returned rains to the fields of Dak and his father. When Dak saw there was rain in his and his fathers' fields, he in turn opened the eyes of Dimo's cows. These matters brought big challenges and competition between Rwot Dimo and his grandson, Dak.

As a result, Rwot Dimo summoned all the elders of *"gang kal"* (from the clan). The elders came and met under big tamarin tree (*yat cwaa*) that was right in the middle of Dimo's compound. Dimo opened the meeting with a word of welcome and told them, "I am happy that you have turned up to attend this important meeting. I want to inform you that Nyikang is not smart but Dak is smart and, because of this, I want to kill Dak. I called you here so that you can show me the best way that I can kill him."

After a short deliberation, it was resolved that Dak should be killed in the evening after he had blown his 'tum" (trumpet) or *"opuk,"* as he was accustomed to doing. However, the elders wanted the hands of Rwot Dimo to be clean from the blood of Dak, (that is, they did not want Dimo to be accountable for the death of Dak), so they decided to kill him by themselves.

The traditional history of the Northern Luo of South Sudan tells us that when the elders were discussing ways to kill Dak, there was a foreigner (*ja-jur*) among them. This foreigner was called Obogi. Deidrick (1912) recounts, "When the council of elders were speaking about the killing of Dak, a man called Obogi was sitting there silently" (p.157). This foreigner pretended to be deaf, but in fact was listening to all the discussions and resolutions discussed. The council of elders asked him, "Do you understand what we are talking about?" Obogi replied, "Ah!" with surprise, like a deaf and dumb person. They concluded, "This man cannot hear."

However, the elders knew very well that this foreigner was a very close friend of Dak. The elders thought that it would be better for a foreigner to kill Dak, so they asked him to carry out the murder. Making signs to him, the elders asked the supposedly deaf man to go and kill Dak in his house. The man did not refuse the order, but, before evening had fallen and the time for killing Dak had been reached, the man went to his friend Dak and told him everything the elders had been discussing.

One version the episode says that when Dak learned of this arrangement, he prepared statues that looked exactly like him, dressing the statues in his usual white evening clothes and putting them in the middle of his house. When evening fell, Dak left his home and started blowing his *"twom"* or *"opuk,"* while simultaneously watching his house from a distance.

The other version of the story, written by Deidrick (1912), says that the elders, who were also Dak's uncles, were very jealous of Dak. In the meeting, they asked themselves, "Is this country to be ruled by Dak?" They all said *no; Dak must be killed.* However, Obogi went and reported the matter to Dak, saying, "You are going to be killed by your uncles." Obogi then went to Nyikang and told him that Dak's uncles wanted to kill Dak. Nyikang did not want his son to die, so he went to fetch an "ambach." He hewed it and made a wooden statue that look like a man, and placed it in the middle of Dak's house. Dak used to sit on the ashes of the village to play the "tom" (opuk). When evening came, Dak went and sat in his place as usual, and began playing his instrument. When he had finished playing, he took off his bracelet and put it on the wooden statue in his house. When the uncles heard no more music, they thought Dak might have already entered his house. They came and speared the wooden statue, thinking that it was Dak. Dak, however, remained unhurt.

When the uncles thought they had killed Dak, they said, "He is dead; good," and they went away. Nyikang then came and said to the people, "My son has been killed by his uncles." When the uncles of Dak heard this, they were afraid and said to themselves, "Let everybody stay at home for four days; after four days we may mourn him." After the fourth day had passed, the people gathered to mourn Dak. They told the people at the mourning site, "We have killed a dog." When they had finished, suddenly Dak came out from his enclosure and went to dance the "mado dance" (funeral dance) with the people. Nyikang said to the people, "We will leave you; we are going to look for a fertile land where we will grow our sorghum." When his uncles saw this, they ran away from the dancing ground, and the mourning came to an end (p.157-159).

When the elders heard Dak was blowing his *"opuk/twom"*, they immediately thought he was already in his house —as usual. Shortly, elders selected four men to go and spear Dak. The four men took spears and went to his house, as they reached closer to the house; they saw a man sitting in the middle of the house wearing the white clothes. They thought that was Dak, so the four men speared at the status at the same time and it felt down. When they saw status falling down, they thought they have killed the man (Dak). Without verifying that this was really Dak, the four men went back to Rwot Dimo and his other people, and told them the man has been killed.

When the people learned that Dak has been killed, they organised a big traditional dance to celebrate his death.

Rwot Dimo was extremely happy about the death of his grandson Dak, because he considered Dak to be more of an enemy than a grandson. While the people were dancing, Dimo said to them, "We are mourning Dak."

The oral history of the Northern Luo says that everybody at the dancing ground was happy that Dak had died. The people then loudly sang the following song:

Lakwara-latin pa nyara!

Dak Latin pa Nyikang otyieko too! ·

which means:-

My grandson!

Dak, the son of Nyikang is dead!

As the people were dancing, suddenly Dak appeared; when the dancers saw Dak, their blood ran cold. There were two things that brought fear into people's hearts: firstly, they thought this was God's Spirit (*tipu Juok*); secondly, they thought that if this was really Dak, they would be unable to escape from his wrath. The people of Nyikang then decided to leave Wau County for a new settlement. In about 1550 A.D., Nyikang and his groups migrated to Malakal.

Nyikang settled temporarily in El Duem

According to some Collo elders, after leaving Wau, Nyikang and his groups travelled to El Duem, 280km south of Khartoum. They stayed here for a few months and then proceeded to the Upper Nile Region, where they found the Funj.

In contrast, Diedrick (1912) asserts that, from Potherthura "Tura" (Wau-Bhar El Ghazal), Nyikang and his people travelled and came to a place called Tongs "Acieta gwok or Achyiete-guok." According to this theory, they did not reach El Duem, instead travelling directly to the Upper Nile Region. Here at Tonga, Nyikang found the white men (Funj people with light colour) occupying the land. He thought it was not possible for his people to settle among the white people at Achyiete-guok, so he decided to go back to where they had come from. They then went back and settled on the other side of the River Sobat in a place called Pijo, which is just at the head of the River Sobat. Nyikang built his homes at Nyilwal, Pepwojo, Adwalo and Tedigo, and some of his people went and built at Wau-Shilluk, Oshoro, Panyikang Otego, Akuruwar, Moro and at Oryiang. All these villages were under the Kingdom of Nyikang. Dak, however, passed straight through and settled with his people in a place called Falol or Fanyidwai Pac or Wij-Palo (p.160).

The Blood of Ubogo Enabled Nyikang to cross the River Blue Nile

When Ubogo/Obogo learned about the attempt to kill Dak, and, more importantly, what had happened in the dancing ground, he went to Nyikang and said to him, "The Potherthura (Tura) is not a good land for your people to live in. I am greatly concerned about what Dimo has done to his grandson. If possible, we should proceed ahead. Your two sons, Bur and Shall (Cal) are the cause of this problem. Be mindful that tomorrow, if there is a fight, it will be you and Dak who will face the battle. It is likely that Dak could die in this battle."

Nyikang reflected on this advice and appreciated the suggestions offered by Ubogo. The following morning, Nyikang took his livestock (including cows, goats and sheep), children, wives and other people and left the land of Tura/Turo. Having travelled for thirty days, they came to the River Blue Nile, at the meeting point of the River Sobat and the River Blue Nile. The water was covered with so many floating plants and grass that it was impossible to pass through. Deidrick (1912) confirms this in his book entitled "The Shulluk People: Their Language and Folklore," when he says that, from this place (Patherthura/Turo/Tura), Nyikang decided to move and look for a new settlement. He walked along the river and came to the river source. They wanted to cross the river with boats and their animals, but the river was full of sudd (plants and grass). Nyikang stood still and said, "Where have this sudd come from, and what shall we do to remove it, so that we can cross the river?" The way was blocked and the people could go no further (p.160).

Ubogo/Obogo then told Nyikang, "One of the men could be slaughtered to appease the River Blue Nile, so that ways could be opened for the groups to cross the river." Nyikang asked Dak if he could give one of his servants to be slaughtered, but Dak was not happy about his father's request, responding, "Are my servants chicken to be slaughtered on the route to settlements?" When Ubogo/Obogo heard Dak make such rude remarks, he told Nyikang, "To avoid a dispute between you and your son, I offer myself to be sacrificed, to appease the river so that your people can cross. But there is one important thing I want to tell you: if am killed, you must look after my children and wife that I will leave behind."

Nyikang accepted Ubogo's suggestion and asked some men to slaughter him. Ubogo/Obogo went and lay by the riverbank where he was slaughtered, his blood pouring into the water. The plants and grass were removed and the people and their animals crossed the River Blue Nile. According to Diedrick (1912), Obogo had gone to Reth Nyikang and said, "Nyikang, I have finished eating. Spear me under the sudd, and if you come to any place where the sudd is, just go in after it." Thus Ubogo/Obogo was stabbed under the sudd, the sudd broke away leaving the way open, and they crossed the river (p.160).

Figure 11: Picture of Ubogo being slaughtered

Other writers report that Obogo did not sacrifice himself; he instead took Nyikang's sickle and entered the blocked river, cutting the grass with the sickle as he waded through. However, in the process of cutting the grass, the sickle injured him just below his armpit and blood was seen flowing on the surface of the water. Obogo managed to make an opening through the sudd and the whole group crossed to the other side. In this version, Obogo did not die at the water's edge - he went with the group until they reached the present Shulluk-land. Obogo died later in the village of Wuobo and a shrine was erected in his memory.

Yet another version, reported by Diedrich Westernmann (1912), suggests that when Nyikang arrived with his followers at the River Atulfi, he found the river was shut up by the sudd and he could not find a way through. Obogo (who is believed to have been an albino) asked Nyikang, "Why do you stop? Is it because you do not find a way through?"

Nyikang replied, "Yes, I do not see a way where we can pass."

Obogo said, "When I have finished eating, I shall come; I will be killed with a spear, my blood will flow into the river, and the sudd will break away." When Obogo finished eating, he came and Nyikang speared him, his blood flowed into the river, and the sudd broke away. Thus, Nyikang then found a way through (p.81).

Nyikang Arrived in Malakal

In Malakal, Nyikang found the Funj people, who are the traditional (aboriginal) owners of the land. Nyikang fought the aboriginal people and conquered them. Seth Nyikang wanted them to come under his Kingdom, but the Funj people declined and moved northward, settling in their present day home in Northern Sudan. Luo oral history says that the Funj people are Luo groups and that they were the first Luo groups who settled in Malakal. It is not clear when and where the Funj people separated from the main Luo groups, nor in which year they settled in Malakal. Nevertheless, the oral history of the Northern Luo group tells us that Nyikang arrived in Malakal with only a few people and that, although the numbers of Funj people were rather greater than Nyikang's groups, because they were good fighters, Nyikang and his people fought and defeated the Funj. (An alternative version reports that when Nyikang arrived in Malakal with his group, he did not find anybody on the land).

Malakal is a vast land, so Nyikang soon started to hunt for people to be brought into his kingdom. Nyikang fought people around Malakal and brought many men, women and children under his Kingdom. Apparently, the king often sent his warriors upstream to Dinka land, where they fought, captured and brought back Dinka as slaves or prisoners. Thus the king of Shulluk increased his possessions and his influence on the Shulluk country through his armed men.

There are many narratives which show how Nyikang tried to hunt for people and to bring them under his kingdom. We will here outline four of the most striking narratives, as they appear in the Shulluk culture. Rev. D.S. Oyler (2011) records how one day, Nyikang was sitting under a big tree near the river when he saw big fish swimming in the river. He thought that they were not fish, but people who might have changed into fish. Nyikang went close to the bank of the river and tried to catch the fish, but he could not catch even one. When Nyikang realised he could not catch even one fish, he called Dak and told him to catch the fish.

Dak entered into the water and caught a fish. As Dak was emerging from the riverbank with this fish in his hand, the fish started changing: first it changed into a snake and then into a man. Dak took this man and gave him to his father. Nyikang took the man and placed him under his kingdom.

The second narrative is from the Shulluk elders, who say that one day, Nyikang was walking around an ant hill near his house. As he walked past the anthill, ants bit his toes. When he bent down to pick up the ants, he could not see them, and subsequently could not pick up even one ant. Nyikang called Dak and said to him, "Son, ants bit my toes when I was walking around the ant hill. I want you to look for these ants and, if you find any, bring them to me."

That evening, Dak went to the anthill and waited for the ants to come out from their holes. Soon, many ants came out for conversation. Dak rushed and caught many ants, put them into an *"opoko"* (gourd) and took them home to Nyikang. When Dak arrived home, he poured the ants out of the gourd and all of them changed into men.

Another narrative, also from the elders, says that one day Nyikang took his spears and went hunting. When he reached some burnt grass (*lyiek*) far away from home, he came across some people who were also hunting. These people went hunting with dogs and without spears. Nyikang bravely went to them but, as he was approaching, all the hunters and their dogs disappeared into the ground. After some minutes, these people with their dogs again appeared in front of Nyikang. He tried again to approach them but they once again entered into the ground and reappeared at a distance from Nyikang. This event was repeated three or four times and in the end, Nyikang realised he could not get hold of these people and their dogs.

Nyikang returned home and asked Dak to help bring these people home. Early in the morning, Dak took three dogs and started marching to the place where people kept on entering into the ground. He did not pursue the people; he let his dogs find them. However, when the dogs from the other side saw Dak's dogs, they fought and killed them all. Dak was left without any dogs.

Dak was angry but, not knowing what to do, he went back home. The following morning, he took a boat that travels both on land as well as on water. He went to pursue those people, and shortly came across them. Dak approached the people but, as he came closer to them, they recognised him and entered into the ground with their dogs as usual. Dak hurried to the place and sat down. After a few seconds, the people emerged from the ground with their dogs. They did not know that Dak was sitting and waiting for them. There was nothing the people could do, so they surrendered. Dak put all the men with their dogs into the boat. As the dogs entered the boats, they all changed to women. Dak then discovered that the dogs were actually the men's wives.

The fourth narrative is also from the Shulluk elders, who say that one day, Nyikang again went fishing, but as he put his hook into the water, he caught a man. The man asked Nyikang, "Why do you catch men instead of catching fish?" Nyikang answered, "If I do not catch men, how would I fill my vast, new land?" At that very moment, the man changed into six different forms: a snake, then a crocodile, a lion, a leopard, a hyena and finally back to a man. Nyikang watched these changes taking place in the man, but he was not afraid. He then used these different forms to populate his new land in Malakal.

In this new settlement, Bur and his brother Shall often quarrelled. When Nyikang realised this, he decided to separate the two sons. To facilitate the separation, Nyikang took his wife, Dak's mother, and gave her to Bur for a wife. Furious, Bur left Kofal/Kodok and looked for a new place to settle.

Before Bur had travelled far from Kofal, Dak decided to follow him. He was secretly thinking that if he found Bur, he would kill him. However, Dak's mother knew what was running through the mind of her son. Thus, when they were walking, Dak's mother tied Bur on to her back like a child. As Dak drew nearer to his mother and saw that Bur was tied on to her back, he realised that if he speared Bur from behind, his mother might also be hurt. Dak decided not to spear Bur, but returned back to Nyikang at Kodok. When Dak arrived home, he went straight to Nyikang and asked him, "Why did you give my mother to Bur for a wife?"

Nyikang gave one of his daughters to Dak for a wife

Nyikang realised that Dak was very angry about what had happened to his mother. He then tried to appease Dak in two different ways. Firstly, Nyikang offered some of his wives to Dak, but Dak refused. Secondly, Nyikang gave Dak numerous cattle, but Dak refused that offer too. Nyikang eventually asked Dak, "What do you really want?" Dak answered, "I want one of your daughters for a wife."

Luo traditions and customs do not allow children of the same father to marry, even if they have different mothers. Nyikang was compelled to test the traditional law by giving one of his daughters to Dak as a wife. It is commonly said that Nyikang did this with the hope that Dak would respect the Luo traditional laws. Nyikang also believed that, based on tradition, Dak would not have sexual intercourse with his sister. Nyikang deceived himself further by believing that even if Dak did have intercourse with his sister, they would not be able to produce a child – according to traditional beliefs.

However, the girl became pregnant and gave birth to a baby boy. Nyikang was surprised that Dak and his sister had managed to produce a child despite the Luo traditional beliefs. One day, Nyikang called Dak and reprimanded him, "Son, I thought you would respect your sister and not have a sexual relationship with her." Dak replied, "What do you think your daughter is - is she a female or a male?"

The marriage of Dak to his sister brought huge problems and arguments among the Luo elders. One evening, the elders met to reinstate the traditional laws. They passed a resolution that children of the same chief should not marry among themselves, but could only marry children of the other chiefs. According to Crazzolara (1950), quarrels between Nyikang and Dak (or Odaak) happened at Wipac and not in Malakal. Dak had married Akec, a daughter of Thurro, who was the son of Dimo.

Shulluk tradition reports that the leader Garo, the son of Cang, was a rich man with many wives. He lived in the country through which Nyikang and Dak had to pass. Garo wore a silver ring on one of his fingers, and this ring was visible from a distance. When Dak and his father were passing through the region, Dak saw the silver ring and determined to get it by any means. Nyikang kept aloof.

One day, Dak went with his people towards Garo's country. The people of Garo lived along the river, at the side of a small lake. Dam and his men came to a pool where he found some women bathing. He pretended to be thirsty and asked them if they could give him some water to drink. The women were annoyed at the stranger's request and asked, "Who are you?"

He answered, "I want some water."

They told him they were of the people of Garo, and that their people would soon arrive to give him a lesson he would never forget. The women sent a message to the village saying there was a stranger disturbing them in the river. Garo soon appeared with his people and started fighting Dak. As their men watched, both men threw spears at each other but missed. As they had no more spears at hand, they started wrestling and hitting each other. Dak threw Garo to the ground and cut off the finger with the bright ring. Garo ran home with his companions to alert the villagers, while Dak and his followers drove off Garo's cattle.

Garo and his people went and reorganised themselves for battle; soon the men of Garo, together with his father, Cang, arrived. Nyikang took an adze (*koco*) and struck it against the Garo's legs. After a short fight, Garo and his father withdrew, but they drove Shulluk's cattle before them. The silver ring "Atego" remained in Dak's possession, and was taken to the Phodi Collo. This famous silver ring, together with other silver, was used to make a bracelet which was given to the King, "Reth," at his enthronement by the functioning members of the *Nyikwom* clan of *Akurwa* district (they are the seat of Nyikang). Only a king may wear such a wristlet. A silver ring round the neck of a spear denotes the bearer to the king "Reth" of Shulluk (Crazzolara, 1950, p.39).

The Collo oral history says that between the reign of King Ocollo Wad Dak (1615-1635) and 1861, the Shulluk tried to expand their northern border militarily. By 1630 the Dinka people were becoming a threat to the Shulluk Kingdom and Sultanate of Sennar. Before long, the Shulluk and Funj united against the Dinka and defeated them militarily. This era marked the beginning of Shulluk economic ties to groups such as the Funj, the Arabs, European merchants and the Mahdists.

During the 18th century, the Sultanate of Sennar declined in power. The Shulluk Kings took the disappearance of Sennar from the political scene as an opportunity to strengthen their position on the northern frontier. The caravans were under the influence of the Shulluk kings and were enriched by the shuttle service made available by the Shulluk to merchants wishing to cross the White Nile to Asalaya when travelling between Sennar and El-Obeid.

Descendants of Nyikang and the Royal Line

According to Collo traditions, illness is usually caused by the attachés of royal spirits, and only Dak and those who are posed by the spirit of Nyikang are the common carriers. To cure the sick person, an elder must call the spirit of Dak or Nyikang; once the spirit has been called, then the sick person will be cured. However, although the sick people are cured, yet in the end they will still die.

In Collo tradition and customs, when a sick person dies he/she is buried. There were, however, four kings un Shulluk who, according to traditional history, did not experience physical death. Instead, each vanished and their bodies were never recovered; their bodies were replaced by an effigy. It is said that Nyikang is the first of the four Kings who did not experience death. He simply disappeared into the river, and was replaced by his timid elder son Cal. This son also disappeared in unknown circumstances, and was replaced by his brother Dak. Surprisingly, Dak also vanished in a fit of frustration over popular grumbling about his endless wars of conquest. Dak was succeeded by his son Nyidoro. This king marks a point of transition in that he was the last king who disappeared without experiencing physical death. (There are, however, accounts of Nyidoro being murdered by his younger brother Odak, and later on his body magically disappearing.) When Nyidoro disappeared, his brother Odak became the king of the Shulluk kingdom. Traditional history tells us that when Odak became king, he fought and brought many outsiders under his Kingdom and the outsiders became known as Ororo. Besides this, Odak was blessed with many sons, most of whose names the elders of Shulluk cannot remember. The only son the elders of Shulluk could remember they called him Duwat, who has become direct ancestor of the current Royal family line in Shulluk line. .

Odak continued warring with his neighbours until eventually he was defeated in a battle with the Dinka and the Funj. Although Odak was blessed with many sons, unfortunately all his sons died (possibly in the battle fields) and he was left only with one son called Duwat, whom he constantly belittled. Angry, Duwat vowed to degrade all his brothers' children from royalty to commoners.

One day, Odak took the sacred spears of Nyikang and threw them in the river in a gesture of despair, crying "now all my sons are dead!" His father's crying hurt Duwat's feelings, and this was the last straw for him. When Duwat saw all the sacred spears of Nyikang lying in the river, he snatched up one of the spears and single-handledly drove the enemy (Dinka/Fung) away from the Shulluk land. Thus Duwat became the king of Shulluk. As king, one of his first acts was to degrade the descendants of his brothers to a lower status than the royal clan. Subsequently, the descendants of his brothers became the Ororo, excluded from succession, but who nonetheless play a key role in royal ritual (David Graeber, 1996, pp.25-26).

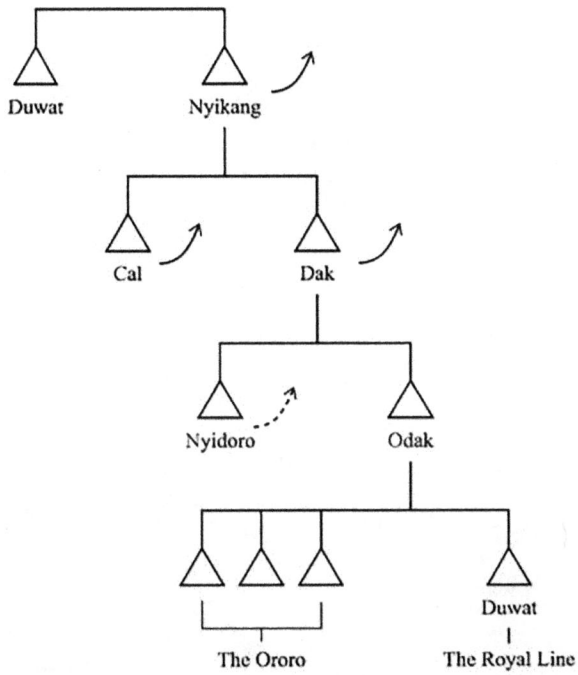

Figure 12: The origin of the Ororo and the royal line

Chapter 5: Luo Family Tree

"Family is like the branches of a tree: we grow in different directions, but our roots remain as one. Oral Tradition remains relevant particularly if backed by Written Sources"
Source: Safira Anek – Alero/Uganda

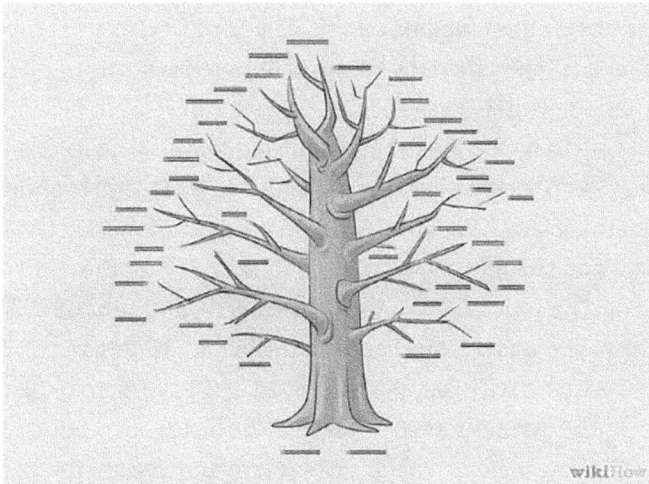
Figure 13: Family Tree

In the beginning, God created the heavens and the earth...On the sixth day, God said, "Let us make man in our image, after our likeness, so that they may rule over the fish of the sea and over the birds of the sky...and over all creatures that move on the ground." Thus, God created mankind in his own image, in the image of God he created male and female. God blessed them and said to them, "Be fruitful and multiply; fill the earth and subdue it" (Genesis 1:1,26-28). As there are no documented records after God ordered Adam and Eve to multiply and fill the earth, we cannot trace the Luo family tree from Lwor or Omara back to Adam. Nevertheless, the whole world, from east to west and from north to south, knows that Adam and his wife made love, she conceived and bore their first son Cain, then their sons Abel and Seth, and then many other sons and daughters which are not chronologically recorded in the Bible (Gen. 4:1-2, 5:3-4).

The Northern Luo are descendants of Adam from Lwor through Okwa

Although we cannot trace the Northern Luo family tree from Omara back to Adam, the bottom line is that all human beings are descendants of Adam. The Northern Luo traditional history indicates that Dyiang Kulu or Dhyang Adugo (which literally means 'cow of the river') bore a gourd. When it was torn, a man named Omara arose. Omaro fathered Kolo; Kolo begot Wat Mol and Wat Mol begot Okwa. We can therefore conclude that from Omaro through to Okwa were descendants of Adam and Eve. Although there is no academic document to support hypotheses that Dyiang Kulu and Omera are the children of Adam and Eve, but yet the Bible tells us that all human beings are descendants of Adam and Eve. Genesis 1:26 tells us that God said, "Let us make man in our image, according to our likeness." Genesis 2:7 goes further, saying "The Lord God formed the man out of the dust from the ground and breathed the breath of life into his nostrils, and the man became a living being...then the Lord God said, "It is not good for man to be alone. I will make a helper corresponding to him" (Genesis 2:18). Subsequently, in Genesis 2:21-22, "The Lord God caused a deep sleep to come over the man, and he slept. God took one of his ribs and closed the flesh at that place." Then the Lord God used the rib he had taken from the man to make a woman, and brought the woman to the man. After this, God blessed them in Genesis 1:28, saying, "Be fruitful, multiply, fill the earth, and subdue it. The man named his wife Eve because she was the mother of all the living" (Genesis 3:20). Before long, the man was intimate with his wife Eve; she conceived and gave birth to Cain. She also gave birth to his brother Abel (Genesis 4:1-2). After this, the history of productions went on and on. Therefore, with these views in mind, since Adam and Eve are the father and mother of all human kinds, we can logically say that Dyiang Kulu and Okwa are descendants of Adam and Eve.

As discussed in chapter one and in accordance with Genesis 9:18 and 21, *Ham is the father of the black people*. With this in mind, we are led to believe that Lwor through Dyiang Kulu to Kolo all died in Canaan; and it was Wad Mol who led the Northern Luo people out of Canaan. Thus, the genealogy of Okwa, the son of Dyiang Kulu, is as follows:

Dyiang Kulu begot Omara;
Omara begot Kolo;
Kolo was the father of Wat Mol; and
Wat Mol was the father of Okwa.

Okwa is said to have visited the riverbank of an unknown river. There, he saw two beautiful women, Nyakayo (or Nikaayo) and Ongwak, coming out of the water. Okwa took them for wives and paid a high bride price to their father, Dunyel. Oral Shulluk history says that Okwa married other wives besides Nyakayo and Ongwak. Before long, Okwa's first wife bore his firstborn son, Dumo, and his second wife bore his second son, Nyikang. More sons were born: his third wife bore Anyuak; the fourth bore Luo; and the fifth bore Gilo. Okwa's sixth and youngest wife then bore the only daughter in the family, Achol.

Deidrick's account (1912) differs somewhat from this traditional understanding. According to Deidrick, Okwa was the son of Omara from heaven. Okwa had six sons: twin sons Nyikang and Dowad (Dumo) from his wife Nyikaya; and Ju, Okil, Otin and Moi, probably all from his wife Ungwet (p.157). Deidrick appears mistaken in not mentioning the other sons (Anyuak, Luo, Gilo) or their sister Achol. Deidrick is also remiss in saying that Okwa was the son of Omara; Okwa was in fact the grandson of Omara since, as noted above, Omara was the father of Kolo, Kolo was the father of Mol and Mol then become the father of Okwa. As many Northern Luo recall, this is the Okwa's correct genealogy.

Therefore, we can say without hesitation that Lwor is the father of all Luo people. When we look back five generations, we are able to trace today's Northern Luo family tree as follows:

I.   Dumo is the ancestor of the Balanda Bor, Balanda Viri, and Jur Chol in Bhar El Ghazal, South Sudan. Shilluk tradition called them *Wuate Dhimo*.

I.   Nyikang is the ancestor of the Shulluk in the Upper Nile, South Sudan.

I.   Nyikang also is the ancestor of Thuri/Shatt in western Bhar El Ghazal, south Sudan. They are also known as the Shilluk of Bhar El Ghazal.

I.   Anyuak is the ancestor of the Anyuak people in the present Jongolei State in the Upper Nile, and in South Sudan and in Ethiopia (Gambella).

I.   Gilo is the ancestor of the Pari/Lokoro in Eastern Equatoria, South Sudan.

I.   …became the ancestor of Jumjum/Berin, in present Bentiu State in the North Upper Nile, South Sudan.

I.   Achol and Olum became the ancestor of the Acholi in Eastern Equatoria, and also of other Acholi in northern Uganda.

I.   Rwot Ramogi is the ancestor of the Southern Luo in Kenya and Tanzania
      Source: Onyala, (2014): Tekwaro Pa Acholi Me South Sudan

Luo mythology and oral history cite "Gilo" as the youngest brother of Dumo, Nyikango, Anyuak, Luo and Achol. This account of oral history within Luo groups in South Sudan, Uganda, Congo, Chad, Ethiopia, Kenya and Tanzania refers to Dumo as the older brother of the five siblings, and the most humble and intelligent sibling in the family. It is commonly understood that because of his humility, he played the case of 'spear and bead' in a cool and intelligent way. Based on Luo tradition and customs, a simple conflict resolution took place in the family, which consequently led to the break-up of the children of Okwa.

As detailed in the previous chapter, Dumo and Nyikang separated from each other because of a dispute about a bead. This dispute subsequently led to the social disintegration of the entire family. The dispute is mostly cited as one between Dumo and Nyikang, with little (if any) reference to Anyuak, Luo, Gilo and Achol.

Luo Custom of Moral Expectation

In Luo tradition and culture, children are expected to give due respect to their parents and to any elderly persons in the community. Children are also expected to respect each other according to their birth order. Madut (2012) states, "According to Luo culture, there is a custom of moral expectation and line of respect that governs family members according to the birth order, which indicates that older siblings are not be questioned or confronted, even if they are wrong" (p.2). This is one of the reasons why the other siblings are not mentioned in the dispute between Dumo and Nyikang.

This is the reason why Dumo was the one to determine how the dispute was to be resolved. This also explains why Gilo and Anyuak left with Nyikang and settled in the Upper Nile Region in South Sudan. In their adulthood, Gilo and Anyuak later broke away from Nyikang and formed their own kingdoms. Anyuak formed his kingdom in the present Jongolei State, with its capital at Pachala; and Gilo formed his kingdom around Lafon Mountain, in the present Eastern Equatoria State, with its capital at Lafon town.

Ancestral linkages are not considered important among most of the Luo groups in South Sudan, largely because of an ignorance of their usefulness. Only the Luo Bhar El Ghazal have taken ancestral linkages seriously, and this is reflected in their culture and customs. They have an understanding of ancestry through oral history, and this makes tracing their ancestral line to other Luo groups much easier. Like other Northern Luo, they have managed to keep alive their language and their ethnic Luo names, such as Dimo, Kany, Ukeilo (Okello) and Gilo. These names continue to be given to children in memory of the great ancestors. The Acholi in Eastern Equatoria give the name 'Okello,' and the Anyuak in Jongolei State use the name 'Didumo.'

Chapter 6:
The Northern Luo:
Demography and Geographic Locations

Nyikang's descendants migrated from Karoo/Kara in Lake Albert, Uganda

The descendants of Nyikang (Shulluk) migrated from Karoo (Lake Albert in Uganda) and settled in Malakal in the Upper Nile Region, with their capital at Fashoda. The descendants of Gilo (Anyuak) migrated from Karoo (Lake Albert) and settled in the current Jongolei State, in the Upper Nile Region, with their capital at Pachala. Other groups of Anyuak were later separated (for political reasons) from the main groups and attached to Gambella, Ethiopia. The other descendants of Gilo (the Pari/Lokoro) migrated and settled around Lafon Mountain, in Eastern Equatoria State, with their capital at Lafon town. The descendants of Achol and Olum (the Acholi) migrated and settled in Eastern Equatoria State, with their capital at Magwi. These are known as the Acholi of South Sudan. The other groups of the Acholi settled in the Gulu and Kitgum Districts in Northern Uganda, with their capitals at Gulu and Kitgum. Mabaan people migrated from Wi-pac and settled along the bank of White River Nile, in Unity State in the Upper Nile Region, with their capital at Bunj; while other groups of Mabaan people settled in Ethiopia, in a place called Assosa, east of Blue Nile. It is said that when the Mabaan people reached Ethiopia they called themselves Oromo and they live next to Amhara people. The descendants of Jumjum (Jumjum or Berin people) migrated and settled along the River Jumjum, in North Upper Nile Province, with their capital at Mayak. The descendants of Dimo (Balanda Bor and Balanda Vri) migrated from Karoo (Lake Albert) and settled in Bhar El Ghazal, with their capital at Bussere Town. Other descendants of Dimo (the Jur Chol) migrated and settled in Bhar El Ghazan, with their capital at Mbili. The descendants of Nyikango, the Thuri people, separated from the rest of the Shulluk under Nyikango and settled in western Bhar El Ghazal, with their capital in Wau.

Early writers like Crazzolara (1950) have claimed that the Luo people migrated from

Rumbek, Bhar El Ghazal, in South Sudan. Alternatively, Wasonga (2012) suggests in his article entitled, *"The Luo Route from the North,"* that the Luo groups originated from Persia in Iran and later on moved to Egypt on their way to the good farming, grazing and metal areas of today. However, there are no supporting documents to justify the theory that Luo groups originated from Persia, as claimed by Wasonga.

One thing, however, is clear: in accordance with my hypothesis, there is no regional agreement about the origin of the Luo people. I maintain that the Luo groups originated in Canaan, moved to Kush in Lower Egypt, and from there travelled to the Great Lakes in Uganda.

The Northern Luo of South Sudan are the Nilotic people. They consist of nine groups: the Acholi, Anyuak, Balanda, Jur Chol, Jumjum, Funj, Mabaan, Pari, Shulluk and Thuri (Shatt). Nevertheless, some writers like Evans Pritchard (1934) refer to Dinka and Nuer as Nilotic people. The Luo people, however, do not consider the Dinka or Nuer to be their kin. The characters of the Northern Luo and their social behaviours are distinct from those of the Dinka and the Nuer. The Northern Luo are gentler than the Dinka and Nuer in dealing with social issues. Some of my Dinka and Nuer friends asked me, "Why are the Dinka and Nuer so socially aggressive?" I answered them, "The Dinka and Nuer don't look at social issues through the same lens that Northern Luo look through. The Northern Luo people are naturally friendly and non-aggressive, as opposed to the other two tribes."

Presumably, various anthropologists and historians have referred to Dinka and Nuer as Nilotic people because of some words that the two tribes share with the Luo, such as *"cak"* (milk) and *"mac"* (fire). This, however, does not necessarily mean that the Dinka and Nuer belong to the Luo groups. Perhaps, during the Stone Age, there were waves of moments among various tribes in Sudan. During these, the Luo groups might have lived together with the Dinka and Nuer in a temporary settlement. This would have offered an opportunity for the two tribes to learn some words from the Luo. It is even possible that these new words were previously non-existent in the Dinka and Nuer languages.

It is important to understand that the Luo-speaking people find their dialects from the language that the southern Luo used as their national name, which unites not only the Southern Luo, but the entire ancient kin in South Sudan, Uganda, the Congo, Chad, Ethiopia, Kenya and Tanzania. The origin, migration, language, settlement and clans of the Northern Luo will be discussed further in Chapter 10.

Thus, in this book, when we speak of the Luo, we refer only to the nine groups of the Northern Luo mentioned above. The case of Dinka and Nuer being additional Luo groups deserves further study.

Although there has not been an official, effective census conducted in South Sudan since the signing of the Comprehensive Peace Agreement (CPA) in 2005, it is widely understood that today, the combined Northern Luo population is approximately fifty percent of the total population of South Sudan.

The journey of Dumo and his people from Kerau/Karoo in Lake Albert to South Sudan remains a mystery up to the present time. Both former and contemporary scholars have dwelt more on Nyikang travelling from Lake Albert to Wau in Bahr El Gazzal and from there to Malakal in South Sudan. Little if anything has been mentioned about the travels of Dum (Duwat) from Lake Albert in Uganda to his place of permanent settlement in South Sudan. Diedrich (1912) took the argument further when he says, "The Shulluk call the Jur Chol as Jur Dimo, which means descendants of Dimo. Now Dumo is the brother of Nyikang, whom the latter left at Kerau/Karoo near Lake Albert in Uganda. According to Crozzalara (1950) Dhimo is not a blood brother of Nyikang, he rather a half-brother of Nyikang (p.33). All the Shulluk traditions are unique in the assertion that Nyikang did not go northwards together with his brother Dumo. So this implies that Jur Chol never wandered into the White Nile country (Malakal), but went their ways directly westward into their present seats" (P.LI).

Chapter 7:
The Northern Luo Migration Challenges

Migration Challenges

When I began thinking about writing *A Short Social Anthropology of the Northern Luo of South Sudan*, it occurred to me that the people might have met many challenges during their migration from their homeland to their present day locations. Therefore, I thought it would be good if documented some of the migration challenges the people might have faced during their march.

Prompted by reading a number of articles and books about the Luo of South Sudan, I then asked myself, "How can we learn about the challenges faced by the Luo people in their march?" This prompted me to ask five main questions: (1) During the march, where did the people get their food and medication? (2) The people travelled long distances and, in the course of their movement, they must certainly have met a number of enemies. How did they defend themselves? (3) How did the Luo people treat their prisoners of war? (4) Did the Luo people move in one large block, or in a number of smaller groups? (5) What happened to the prisoners of war? Let us now answer these questions one by one.

Where did the people get their food and medicine?

Some writers like Crazzolara (1950) have told us that the Luo people marched in small independent groups of between 3,000 to 20,000 people. We understand a 'community group' to consist of men, women and children. Each household could be expected to have a large number of family members, including children and grandchildren. Some people also continued to add new wives to the household. All these people needed to be fed each day, and sick people needed to be treated with medication. Each clan aspired to grow in number through births or adoptions.

If the Luo defeated a group of enemies, they would plunder the food in the granaries, and animals such as cows, sheep and goats. At the conclusion of the fight, the people wanted to preserve everything they got hold of and every person captured in the battlefields. They made sure that prisoners of war would not swell the ranks of the Luo community. This solved the problems of food provision and also of manpower.

The Luo might also have obtained some of their food from hunting and fishing. This was particularly so when they were in areas of scarce population, where they could not rely on plundering the granaries of defeated people.

How did the Luo defend themselves during the march?

When people groups moved in those days, they faced the constant question of life and death – of how to save and protect people's lives. I am sure the problems of support and safety were equally important and were always in their minds. Both adult males and young boys involved in the move provided a strong defence against enemies. Just the look on these warriors' faces showed their enemies that the Luo groups were irresistible; in many cases, their appearance alone kept enemies away. The Luo never risked being defeated. Despite reading through many history books, we have never come across an account of the Luo people being defeated by another group of people.

The Luo groups maintained a high level of skill in fighting, to enable them to move freely. The groups did not want to be ruled by any other leader apart from their own leaders in the groups. The Luo groups were fearless, broad-minded and tolerant, but they were always ready to eliminate any intruders, or any armed resistance. Studies conducted by some anthropologists have revealed that on arrival in Kodok (Malakal), the Shulluk fought with the Nuer. They fought fearlessly, conquered the Nuer, and subsequently assimilated some of them into their community.

In general, the Luo people did not know what danger they would face the following day; they took their lives into their own hands all the time. The prisoners of war they captured were also fearless and brave, fighting alongside their masters. Because of this, the Luo groups developed a very strong trust and confidence in their prisoners of war. No prisoners of war ever turned against the Luo in the battlefield. In fact, while in their service, the prisoners enjoyed the same rights as their masters.

Luo people never despised a man because he is an alien. This momentum offered an opportunity for the Luo groups to plan, implement and evaluate their migratory march. This then became the basis of their growth when they finally settled in South Sudan.

How the Luo people treated prisoners of war

The Luo of the past and the present Luo have identical characteristics. It is a general practice of the Luo that any person captured in a fight, whether man, woman, or child, should not be killed; they should be taken as prisoners or slaves. The old and disabled persons, however, might be killed, as they cannot contribute as manpower. At first, the prisoners of war were certainly well guarded and any suspected movement on the part of the male prisoners was responded to with direct spears - that is, they were killed. Male prisoners were advised to forget about their wives and children and to fight wholeheartedly with their masters. These acts helped male prisoners feel better, both emotionally and psychologically. This is in conformity with the Luo principle that prisoners of war must feel at home.

Although boys who were captured during the march were not killed, they were suspected of being dangerous. As such, the Luo did not trust the young survivors, so when they were first brought from the battlefields, they were given useful information about the village. We could call this as an orientation course.

After walking a long distance from their homeland, the prisoners found it difficult and dangerous to escape, especially if their wives and children were walking beside them. They thought that, even if they were to escape, the chances of reaching their homeland were slim.

The female prisoners were engaged, along with the Luo women, in carrying the food and other household items. This is a very important job during the Luo migrations. We could, therefore, conjecture that male prisoners were also used for similar jobs, at least in the beginning: that is, for carrying weak or sick people, or for looking after the cattle. Later, as mentioned above, the men would join the fighters.

At this point, it is important to understand that the social position of slaves varied from family to family and from group to group. At that time, it would have been difficult for foreigners to have seen the difference between Luo and slaves, but the Luo would have known the difference. As Crozzalara (1950) has surmised, "It is hardly human to deny the possibility of an occasional discrimination between a slave and one's own child" (p.48).

Did the Luo people move in a block or in smaller groups?

As noted above, the Luo moved in small groups of between 3,000 and 20,000 people. During the migrations, the Luo had strict rules to protect people's lives when they were marching or hunting. The tribe groups, villagers and family groups were asked to keep close to one another. This traditional rule enabled them to assist each other. It also allowed every individual and all the smaller groups to feel assured that if there was hardship or danger, they was not alone; they had their brothers solidly on their side, ready to assist them and to defend them. The Luo considered this to be the most important - an indispensable - rule, as it offered an opportunity to protect their rights and blood.

How did the prisoners of war assimilate into Luo communities?

These newly and rather forcibly formed groups of Luo and slaves were soon united with common goals and interests. None of the prisoners could imagine going back to their homelands. They were totally assimilated into Luo culture. Nevertheless, the real Luo groups always remained the distinct masters - the ruling clans - while the assimilated war prisoners always remained descendants of ancient slaves. Although the assimilated aliens remained descendant slaves, they hardly felt inferior in any of the social gatherings in the communities. If, during social gatherings, somebody tried to interfere with their lives, their masters always defended them as their own men. Sometimes, however, when a Luo man picked a quarrel with the assimilated aliens, he insulted them by calling them *"guci"* or *"opi,"* meaning 'slave'. The Luo rules and customs also permitted the slaves to marry anybody they chose, whether Luo or foreigner.

Little by little, the assimilated groups became large and self-conscious, with strong feelings of independence.

We should also note that the life of adventure led by the Luo had a very strong attraction to many people. As such, when these small groups felt dissatisfied with their neighbours, they willingly joined the Luo groups.

Chapter 8:
The origin of civilization

This historical analysis in this book is a contribution towards the reclamation and linking of Luo and African history to that of ancient Itiyopianu and Egypt...Attempts by some western historians, Egyptologists and missionary scholars to conceal the Luo cradle-land, distort the identities of some Luo groups of peoples and their migration patterns were part of a strategy calculated to rob not only the Luos, but Africans of their historical heritage.
Source: Dr. Terence Okello Paito (2011, p.1)

The Luo People are the Founders of World Civilisation

To get to the root of the origins of the Luo civilisation, we need to ask ourselves some fundamental questions, such as "Where and when did the Luo civilisation begin? What were the driving forces of this civilisation? Where was the cradle-land of the Luo people? Was the Luo civilisation brought from the Western World? And if the Luo civilization started from within, how did it spread to other countries? By answering these questions, we shall get the facts about Luo origins.

Before we penetrate deep into our discussions, I want to make it clear to readers that the history curriculum at that time in most African education institutions was based on the experiences of Europeans and on their worldview. As we will see later, to the Europeans, nothing good could come out of the Africans. Nevertheless, this chapter will prove that the first world civilisation originated from the African kingdom known as Kush. Studies show that the reclamation of African heritage from ancient Egyptian civilisation has been a daunting endeavour. This is due to a great misconception about the identity of the ancient Egyptians, and also to the efforts of some euro-centric scholars to deny Africans their historical heritage, portraying them as a people without history.

In addition, the false perception of colonialism as a 'civilising' mission has misled Africans to embrace a euro-centric curriculum. According to Ochieng (2013), in the early 19th century, Euro-imperialism used 'civilisation or barbarism' as an ideological tool to explain the justification for colonialism. Throughout the colonial period in Africa and South

Sudan in particular, western anthropologists' sustained denigration of the African perso-nality and culture deterred Africans from considering the "African origin of civilisation." Some scholars believe that civilisation began in Egypt, but writers like Henri Frankfort (1948) clearly stated in *Kingship and the Gods* that "There are alive today in Africa groups of people who are the true survivors of that great East African substratum out of which the Egyptian culture arose...."(p.6).

As a result, the history of the Luo came under intense scrutiny, and was subjected to distortion and denigration. Writers like Richard (1969) puts it plainly when he states that none of the Nilotic people had achieved a centralised system of governance by the time Speke visited the Acholi tribe in 1862. He argues that the Luo were organised on a clan and lineage basis; the Nilotic, he asserts, were incapable of developing any social institution including that of the state. However, we can say that Richard's argument was drawn from the "Lineage Theory," which states that there was a dichotomy between societies with states and those without states. Thus, the Luo were considered to be a society with a state but without history, and could only be studied from prospective Kingship analysis.

According to Hegel (1956), western people at that time understood the word "African" to mean unhistorical, with an undeveloped spirit, involved in the conditions of mere nature, and only at the threshold of the world's history. He did not believe that Africans could contribute towards the world's development and, more importantly, he regarded any development that existed in Africa must have come from Asia or Europe.

Figure 14: Picture of Hegel

Other writers like James Henry Breasted (1937) argued further that there was no link between ancient Egypt and Africa (the Luo). Nevertheless, many historians believe that the aboriginals of Egypt were pure African Negro, although their bloodlines were later tinctured with Asian or European blood. Ochieng (2013) cites Budge (1911) as observing the similarities between the myths and rites of ancient Egypt and those of Sub-Sahara African, and concluding that the ancient Egyptians "are origin from indigenous Nilotic or Sudanic" (p. V11).

Ku Odongo (1976) argued that at that time, the Luo had developed a prosperous

Kingdom with the capital at grand court at the foot of a mountain (Te-Kidi). He stressed that the Luo Kingdom was making steady progress in many fields of human endeavour. Here, Ku Odongo was referring to the Kush Kingdom at Napata. According to Ku Odongo, the first brown men who met the Luo destroyed the Kush Kingdom; and the last Rwot or King of the Kush Kingdom was called Rwot Owing wod Pule Rac Koma (p.80). The people of Rwot Owing wod Pule Rac Koma lived peacefully in Kush Linmgdom for some years. Later on, a strange people who were thought to be jok (juok) or super-natural beings interrupted their peace. These strangers invaded the Luo settlement from the north. Some of the old Acholi men said some of these strangers were white and some were believed and known as red, brown or yellow. The invaders had black silky hair and untidy beards.

Figure 15: Picture of Onyango-Ku-Odongo

As we have seen, prejudices contributed significantly to the concealment of the truth about the history of the black Africans. For this reason, many attempts were made to separate Egypt from Africa and to deny the latter its past from widespread. Smith (1776) took the arguments further by de-linking ancient Egypt from the rest of the African continent. He wrote, "All the inland parts of Africa, seems in all ages of the world to have been in the same barbarous and uncivilised state in which we find them at present"(p.125). Hegel (1956) also shows enormous contempt for Africa when he says, "It...[is] surpris[ing] to find among them, in the vicinity of Africa, stupidity reflective intelligence, a tharal national organisation characterising all institutions and the most astonishing works of art" (p.199).

Thus in this current generation, African scholars should not continue these wrong perceptions that mislead the world; enough is enough. It is time for us to prove beyond doubt to the world that civilisation was not a product of the Western World, but rather originated in Africa and by the Africans. In the 19th century, there were African scholars who were prepared to examine the facts of ancient Egyptian history, but they were blocked by the strong western perception and ideology. The 21st century opens a window of opportunity to examine the facts about ancient Egyptian history and the origin of civilisation.

Diop (1981) has suggested that for us to trace out the Luo origin of civilisation, it would be better to do it through a linguistic approach, by linking the history of black Africa to that of ancient Egypt. Interestingly, Luo people were the first builders of the ancient Itiyo-pi-anu civilisation at the Kingdom known by them as the Kush Kingdom. It is said that, after some time, this civilisation slowly expanded into Upper Egypt, South Arabia, Mesopotamia, Spain and Phoenicia. Adam Smith (1776) noted: "The nations that appear to have been first civilised were those that dwelt around the cost of the Mediterranean Sea....of all the countries on the coast of the Mediterranean Sea, Southern Egypt (Kush) seems to have been the first in which either agriculture or manufacturing were cultivated and improved to any considerable degree" (p.124).

Figure 16: Picture of Diop

Who then were the first Egyptians who settled around the coast of the Mediterranean Sea? Were they the black Africans or the brown, yellow, red or white? Interestingly, to create doubts among the intellectual Africans and other non-European scholars, the Europeans introduced what is known as "Hermetic Theory," which stipulates that ".....the divine kingdom of ancient Egypt [was] derived from a pre-historic Caucasian Negro culture called Hamitic...this culture has given rise to many East African societies, including the kingdom of Shilluk and Anyuak in the south Sudan, and the kingdom of Bunyoro and Baganda further south in Uganda" (Seligman, C.G. 1966, p.100).

Having examined these arguments, this book will contend that the survivors of the ancient Egyptian culture - that is, the precursors of 'world civilisation' - are the Nilotic people, now commonly known as the Luo people in south Sudan, Uganda, Kenya, Tanzania and the Republic of Congo.

In early times, the Luo people and the Nubians worshipped one of the gods the Egyptians worshipped. This god is called "Anu" and the Luo have been referred to by historians and anthropologists as the true people of Anu (Itiyo-pi-anu). This is where we could discover the Luo origin of civilisation. But who was Anu and what does this mean for the Luo?

We have seen before that Diop (1981) suggests that to trace out Luo origins or history, it is better to do so through a linguistic approach. In the Luo language, the word "i" means you or thou; the word "tiyo" is a verb that means *to work* or *to dedicate services to*; and the word "pi" means *for*. Therefore, when we put all the words together as "Itiyo-pi" it means "you (people) who dedicate your services to". In the ancient history of Egypt, Anu was the primordial watery mass, the god of gods. According to Ochieng (2013), Anu means the "Father of gods". Thus, "Itiyo-pi-anu" literally means the people who dedicated their services to Anu, the father of gods.

Diop (1981) demonstrates that it was the Anu people who authored the various books and scripts associated with Egyptian religion and philosophies. The Anu people were also referred to as the "Agu". Houston (1985) writes, "The aboriginal race of Abyssinia was symbolised by the great Sphinx and the marvellous face of Cheops (see figure 15) and the "Agu" of the monuments represented this aboriginal race. They were ancestors of the Nubians (Luo), and were the ruling race of Egypt. This old race of the Upper Nile, The Agu or Anu of the ancient traditions, spread their arts from Egypt to the Aegean, from Sicily to Italy and to Spain" (pp.35-49).

Figure 17: Face of Cheops

According to Ochieng (2013), the Acholi people of South Sudan and Uganda maintained the worship of Agu until the 1960s. The Acholi brought with them the Agu as they migrated from Kush and placed it on the mountain in Lokung-Uganda; this mountain became known as Mountain Agu (Got Agu). The Acholi continued to offer their sacrifices to Agu on Mt Agu or *Got Agu* until the 1960s. This has been confirmed by people like Keny, son of Ocak, who went from Pari to Mt Agu (in Lokung Uganda) to offer sacrifices to Agu. After the offering, Keny did not return to Pari; he died on Mt Agu soon after the offering. The second person whom Luo oral history speaks of is Obaak, the son of Keny. Obaak also went to Mt Agu to offer sacrifices to Agu and died there after his offering. Likewise the third person: Atanga, son of Obaak, went to Mt Agu to offer sacrifices to Agu and died there soon after the offering (p.11). These are clears indications that god Agu was not only worshiped while the Luo were still in Kush, but the Acholi continued to worship it on Mt Agu in Lokung-Uganda.

Many anthropologists have written about the Anu people. Like Amelineau, Diop (1974) argues, "The Anu....were an agricultural people, raising cattle on a large scale along the Nile, shutting themselves up in walled cities for defensive purposes. Therefore, we can assert, without fear of error, that the most ancient Egyptian Books entitled "The Book of the Dead" and the texts of the pyramids and, consequently, all the myths or religious teachings, were written about them (the Anu people)" (p.77).

The Luo Cradle-land at Tekidi (Napata): The Grand Court of Kush

At this juncture, we cannot proceed without pointing out that most aspects of the period of the Luo evolution in the cradle-land have been well documented in both the oral traditions and in written sources. Although it has been asserted that the ancient Egyptians invented letters and writing, we could still say, without hesitation, that the false of Meroe (Napata) culminated in the migration of the Luo and it contributed to the loss of their agricultural tools, as well as to the ancient art of writing. The inability of Africans to read ancient scripts remains a big obstacle in their quest to re-discover their past. The analysis in this book also attempts to contribute towards the reclamation and linking of Luo and African history to that of ancient Itiyo-pi-anu and Egypt. We have also discussed before that the introduction of Hamitic Theory was undertaken to distort such an effort.

Interestingly, archaeologists have found the ancient Kushite Kingdom at Napata, where they found the same name 'Taharqa' written in many places (see figure 18 above). As G. Mokhatar (1990) writes, "Taharqa's name is found in numerous monuments through the whole length of the Nile Valley. He built his sanctuaries at the foot of the holy mountain of Dabal-Barkal, a kind of stone table formation, which dominates the large fertile basin of Napata...." (p.163); this place was considered the land of the gods. This is the evidence that the Luo Cradle-land was at Napata in North Sudan and not in the east at the Great Lakes or Rumbek in Bhar El Ghazal, as previously asserted by Wassermann and Fr. Crazzolara.

With these views in mind, we surmise that the reason for misrepresenting the location of the Luo Cradle-land was aimed at de-linking the Luo from and denying them any claims to the ancient Kingdom of Kush.

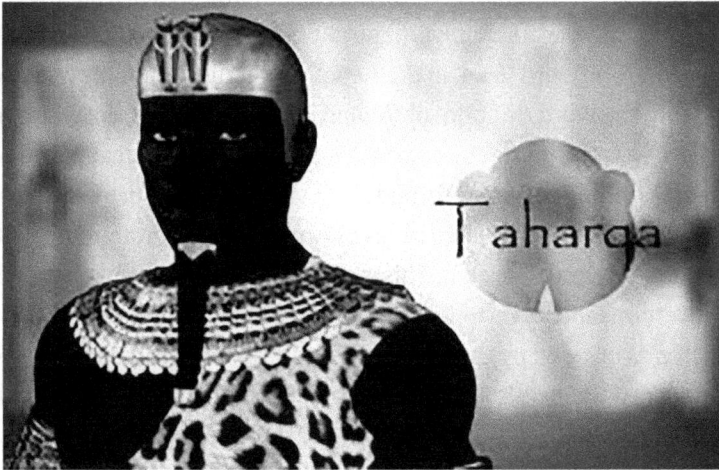

Figure 18: King Taharqa

However, an incursion for the control over Tekidi is still recorded in Acholi songs by musicians and royal bwola dancers. The song reads as follows:-

Lakila Oywayo mwony me Lanek Oyuro do!
Oyuro do!
I Tekidi Tong pa Oyuro odong, I Tekidi!
Iyee Lakila Oywayo mwong me lanek Oyuro Iyee!
I Tekidi oyuro ye!
Nok rac ocera Iweny!
Onek Oyuru ye!
I Tekidi Nok rac ocera Iweny!
-Which means:-
Lakila mobilised an army to annihilate Oyuro!
Oyuro ye!
At Tekidi Oyuro's spear was abandoned at Tekidi Iyee!
Lakila mobilised an army to annihilate Oyuro ye!
At Tekidi Oyuro ye!
Poor state of conscription made the war difficult!
Oyuro got killed at Tekidi!
Poor state of conscription made the war difficult!

Those of the Agoro State of the Acholi do indeed recall Lakila as being one of their great kings. More importantly, the Acholi use the word *"Yuro"* or *"yuru"* to describe the densely silky haired, which fits the descriptions of the untidy bearded invaders who were conquered at Tekidi. Once deconstructed, the proper name *"Bwomoono"* that exists among the Acholi does reveal the wars that were fought between the Luo people and the white/brown/red invaders. The word *"Bwo"* in the Acholi language means *to overcome* or *to defeat*. On the other hand, the word "Moono" or *(Munu)* in Acholi means "the white man" (Okello Paito, 2011). Therefore, the name *"Bwomoono"* is calling for the defeat of the white men.

Recent research findings have confirmed the invasion of Napata by a mercenary supported Egyptian military force led by Harmachis or Amasis. G. Mokhtar (1990) writes, "Aspelta was a contemporary of Psammetik 11. This is one of the few secure synchronisms, almost the only one in a thousand years of history. In -591 or the second year of the King's reign, the land of Kush was invaded by the Egyptian expedition, reinforced by Greek and Carina mercenaries, under two generals Amasis and Potasimto, and Napata was captured" (p.164).

According to Okello Paito (2011), Amasis, one of the leaders of the invaders, succeeded and became the last indigenous Egyptian Pharaoh. He became a hated figure among the Luo and he was humiliated at the hand of Cyrus, which has been captured in Acholi song and folklore as follows:

Napata was indeed the Grand capital of the ancient Kush Kingdom. The constant threats and incursions from Lower Egypt led to the relocation of the grand court of Kush from Napata to Meroe. G. Mokhtar (1990) writes, "It is undoubtedly to the Egyptian raid, whose importance has long been underestimated, that we must attribute the transfer from Napata to Meroe (p.174). As we can see, enough evidence exists to prove that the Merotic Kingdom was of the Luo origin.

Although Diop (1974) wrote much to convince the world that the Meroitic State was of Luo origin, nevertheless, in the end, he failed in his judgement when he suggested, "The name Meroe does not seem to derive from an African root. It is probably what foreigners used after Cambyses to designate the capital of Ethiopia (in Sudan)" (pp.143,288).

However, Diop (1974) in his article entitled "Origin in ancient Egyptians" noted that in ancient Egypt, there is the word "mer" which means "love" (p.29). This coincides with the same Luo word "Mer" which means 'harmony, or to be friendly with," while the word "Mar" in the Luo language means "Love." Despite these similarities, Diop declined to believe that the name Meroe arose from a Luo word "Mero" (or Meru), meaning "to cultivate a harmonious relationship." It would appear that inguistic evidence exists to demonstrate the affinity between the Luo language and the Meroitic Script.

A.A. Hakem (1990) has tried to explain this perception of Luo words through the usage of the Meroitic script, when he writes, "In Acholi, Luo dialect, "Ker" (Qer) means royalty or kingship." He cites as support Reuben Anywar's work (1954) *"Acholi ki ker megi,"* translated as "Kingship of the Acholi," and also that of Lacito Okech (1953), *Tekwaro ki ker Lobo West Acholi*," which means"History and Kingship in Western Acholi."

In the same Acholi dialect, the word "Ot Ker" denotes the royal family or household. This implies that the word "Meroitic", which means "Ker", is a Luo word. Moreover, but when we look deeper into the history, we find that in the ancient Egyptian, the sovereigns of Meroe were titled "Reth" or "Rwoth» (Rwot). A.C.A. Write (1940) concurs: "Reth is a frequent title in Demotic and is translated as "Inspector" or "Agent" (p.367). In Luo, the word "rt" means "Reth" and in Egyptian (as foumd in Coptic books), the word "rt" is written as "rwd," meaning "Agent." Today, the word "rt" (Reth) is used as a title for the Shulluk King, while "rwd" (Rwot) is the title of an Acholi King.

Among the Luo, gender equality was recognised, so women participated fully in governance. The title given to women rulers was derived from the ancient Egyptian word "Ktke" or "Kdke", which means "Queen mother." Thus, queens were addressed as "Meroitic Kdke." Oral Luo history tells us that there were only four queens known to have used this title: Amanirenas, Amanishekhete, Nawidemak and Maleqereabar (Hakem, 1990, p.174). Futhermore, In Acholi, the queen mother occupies a special place, and in ancient Egyptian, the word was written "is," which referred to either "Min Rwot" or "Dak Ker." Thus, Kdke in the Meroitic script simply refers to "Dak Ker."

Just as threats at Napata forced the relocation of the Kush Kingdom to Meroe, eventually the destruction of Meroe led to the decline and fall of the greatest civilisation of the Nile Valley, forcing the Luo to move to the great Lakes region and to other parts of Africa. In the Great Lakes region, the Luo soon began to build themselves up and founded the Kitara Empire.

The Nilotic rulers who founded that Empire still remembered Meroe. For example, when Major John Hammington Speke visited King Rumanika (who ruled Karagwe as a satellite of the Kitara) in the 19th century, he pointed out to him that the origin of the Kitarans were from the ancient Kingdom of Meroe or Meru in the Sudan. Rumanika also told Speke that Kitara had fragmented and was no longer the powerful empire it had formerly been. More importantly, he informed Speke that Kitara was the successor of the once powerful state known as Meru or Meroe, and that this state was also described as Ethiopia. This connection between Meroe and Itiyopi-anu (Ethiopia) proves significant for us.

Figure 19: Major John Hannington Speke

Rwot (King) Menya: the first King of Egypt

Kush was the original Itiyopian Kingdom that first emerged in the Nile Valley in the modern Republic of Sudan. The sovereigns of Kush extended Itiyopianan rule into the Lower Nile region, which later became known as Egypt. (The Upper and Lower Nile Valley was formerly nameless, though some people called it "Kemit." We shall later discuss how this land came to be named Egypt.) More specifically, it was Theban priests/kings, from the Upper Nile Valley of Kush, that established theocratic rule in ancient Egypt (Okello Paito, 2011, p.7 and Ochieng David, 2013, p.13).

Luo traditional history tells us that there was no theocratic rule in ancient Egypt. Religious leaders commonly known as Theban, who were from the Upper Nile valley of the Kush kingdom, introduced this rule later to the community. The first king of Dynasty 1 was called Rwot "Menya;" he was of Luo origins, and was a rich and powerful person. The Europeans found it difficult to pronounce African words with the consonant "ny" as in *Kenya* or *Nyerere*. In the case of "Menya," the Europeans completely omitted the letter "y" so that Menya is spelt and read as "Mena" or "Menes." Houston Dunjee (1985) pointed out to the Europeans that "The first king of Egypt was not called Mena or Menes, as has been depicted in many texts, but he was called Menya; he was said to have been originally a Thebban or one of the Itiyopian priests" (p.69).

At this juncture, it is good to remind readers that "Menya" is a Luo name and this name is very common among the Luo, especially among the Acholi today. In the Luo language, the word "Menya" means "shine on me" or "torch me". The Luo generally, and the Acholi in particular, still remember Menya as a rich and powerful person. In fact, an Acholi proverb is used to help new Luo generations know and remember the former riches and power of Rwot Menya: "Tong gweno oloyo Menya," which is translated by Okot P'Bitek (1985) as, "Menya failed to get the egg" (p.86). Menya was such a powerful figure, nothing was beyond his reach. However, on one fateful day, he was unable to get an egg: it was beyond his reach.

Traditional Luo oral history also tells us that Rwot Menya built the first temples in ancient Egypt. One of the many temples he built was called "Ptah," named in honour of the sun god. The sun god illuminates the entire world by the fire of its eyes. Therefore, the luminary shone on Rwot Menya. The word "Tah' (tar) here is also a Luo word which means "bright light" or "simple and luminary." The word "Ptah" in ancient Egyptian script means "of the light." The prefix "p" means "of," as in the case of Okot P'Bitek, P'Acholi and P'Collo.

Now let us examine the origin of the term "Egypt." There is a word in ancient Egyptian which reads: "A-gy-ptah" or "Ae-gy-ptah," which means, "I am the light." Some contemporary Luo scholars like Okello Patio (2011) and Ochieng David (2013) have confirmed that the name *Egypt* is not Kemit, as alleged by some people, but has come from the ancient term "Aegyptah." This word "Aegyptah" was later reconstructed by Rwot Menya (the first king of the land) as the word "Egypt." After the creation of the word "Egypt," Rwot Menya then instituted the first dynasty and theocratic monarchy in Egypt. Eventually, Rwot Menya lifted himself up to the position of Sun god, which has descended to the men in the Kush Kingdom/Egyptian region.

When Rwot Menya died in 1870? BC, his son "Aha" succeeded him. Here again, "Aha" (A-aa) is another Luo word which means, "I have risen." When Rwot Aha died in 1805BC, he was succeeded by Rwot Djer (Frankfort H., 1948, p.24). The word "Djer" is also a Luo word, which means "set back" (e.g. Dier cen). According to Collo, the last Rwot or King of the first Dynasty was called Dimo or Dhimmo. Dimo was the father-in-law of the Shilluk. As a matter of fact, Fr. Crazzolara (1951) wrote, "Nyikang fled to Dhimmo, who is said to have been Reth of the people whose country was distinct from that of Dowaat.....whose sister Akec, Nyikango had married. Apparently before his final clash with Dowaat" (p.123).

Around 750BC, the word Kabak, which in ancient Egyptian script was written as "Pi-ankh," reasserted Itiyopian rule over Egypt. It cannot be overemphasised that these Luo titles are significant among the Luo people. Scholar Okot P'Bitek provided the definition of the word "Baka" when he wrote, "I swear in the name of baka the Jok of Patiko chiefdom, that I shall speak the truth, without hiding anything from you, or tell a lie, but shall tell all the truth as I know it" (Okot P'Bitek, 1988, p.69). The point we are driving at here is that through Luo migration and relocation from the Napata region, the use of the royal title "Baka" for the Kush ruler continued in the post Meroitic era in the Great Lakes region (Okello Paito, 2011).

In the Nile valley today, the magnificent seats of government, such as "Thebes," and temples at Karnak and Luxor in Egypt, still carry Luo names. For example, the word *Thebbe* or *Tebbe* is an Acholi word for "seat of government." I can recall in 1976, after the Addis Ababa Agreement, I met Rwot Mathia Aburi Lolori, of the Obbo local government, and he described to me his headquarters as his "Tebbe." Rwot Oliya of the Atyiak local government in Acholi Uganda also described his headquarters to Lacito Okech (1953) as his "Tebbe" (p.87).

Although the people in the Nile Valley continue to use Luo words, we nevertheless find that they still distort these Luo words. For example, there is a town in Egypt called *"Ka-naka,"* which in the Luo language means "the place of the everlasting." *Luxor*, on the other hand, comes from the Luo words "Lu-kwor" or "Lu-kwo," meaning "the living."

As Lower Egypt increasingly came under foreign domination, the inhabitants of Lu-Kwor (Luxor) retreated to Kush at Napata, the grand capital of the Kush kingdom. Onyango-ku-Odongo (1976) notes, "Here the people of Lukwor prospered and made progress in many fields" (p.80). The last indigenous Egyptian Pharaoh was called *Amacic*, but other people called him *Harmachis* or *Amasis*. Cambysis put this last indigenous king to death in 525BC. King Amacic is still remembered among the Northern Luo today, even though he was considered a hated figure who participated in the destruction of Napata. He was regarded as a puppet king, and also as a traitor - for colluding with foreigners, and, more significantly, for his inability to stand firm in the face of flagrant aggression by Cyrus and his son Cambyses. Moreover, Cyrus wanted an Egyptian girl to serve him.

It is reported that the expert selected was resentful and suggested to Cambyses to demand Amacic's daughter for a wife. The humiliation experienced by Amacic at that time has been captured in an Acholi folk tale, and in a song that has been passed down for generations:

> *Got Amacic yee!* Mountain Amacic yee!
> *Got Amacic in immi dako, Got Amacic!* Mountain Amacic give a bride, Mountain Amacic!
> *Man Rwot ma ocwala got Amacic ni immi dako got Amacic!* I am being sent by the chief, Mt Amacic, he wants you, Mt Amacic, to give a bride!
> *Oh Got Amacic mi dako!* Oh Mt Amacic please provide a bride!
> *Got Amacic mi dako!* Mt Amacic provide a bride
> *Got Amacic, Rwot ocwala wek ater dako!* Mt Amacic, I am the King's messenger sent to collect the bride!

Which is translated in short by Okello Paito (2011) as *"Oh Amacic the mountain, Please provide a bride; Amacic the mountain, I am the king's messenger sent to collect the bride"* (p.8).

How the Luo people influenced Ancient Greek Civilisation

The identification of people groups who influenced Ancient Greek civilisation has been the subject of intense debate among scholars of anthropology. Few, however, have identified the Acholi people as one such influence.

Nevertheless, many scholars have traditionally considered Egypt as the original source of Greek civilisation. According to Okello Paito (2015), from the late 18th to early 19th centuries, Christians challenged this notion (p.1), arguing instead that it was rather the Ancient Greeks who contributed towards the Egyptian and Luo civilisations. This arose because, as noted previously, the common European view was that nothing good could come out of Africa.

Nevertheless, substantial evidence exists to suggest otherwise. Djait (1981), a Tunisian historian who participated in the drafting of *A General History of Africa*, points out that history leaves marks in a variety of forms that includes, but is not limited to, written documents, archaeology, linguistics and oral tradition. McNeal (1972) likewise suggests that the ancient history of the Aegean be studied through archaeological artefacts, language, skeleton material and myth and legend (p.20).

Archaeological artefacts, written documents and oral traditions reveal that the Luo (Acholi) people - who at that time were called black Egyptians - actually invented the stringed instruments of Ancient Greece, such as the lyre, the harp and the lute (Maa and Snyder, 1989). The people of Mesopotamia and Egypt (black Egyptians) introduced these instruments, and they were later adopted by the Ancient Greeks. An instrument which produces sound by way of vibrating string or of strings stretched between two points is known as a "chordophone." In practice, the strings could be set into motion through plucking (in the case of a harp), strumming (such as in the case of a guitar), or by rubbing with a bow (like a violin).

Some Acholi historians, like Okello Paito (2015), argue that the lyre belongs to the kithara (guitar) family. The kithara has strings of equal length attached to a cross bar supported by two arms. The two arms enter a sound box. The strings, which may number from five to eight, are fixed to a tall piece of wood. The kithara or guitar was the product of the Luo people. In Ancient Greece, the evidence for the kithara is recorded on painted vases, where the kithara is mostly depicted next to the god Apollo (Maa and Snyder, 1989).

Interestingly, many scholars around the world have wanted to explore the origins of Ancient Egyptian civilisation. This has resulted in collaboration between African/Luo scholars and western scholars. Writing from a western viewpoint, Seligman (1903), a Euro-American scholar, suggested that pastoral Europeans known as "Hamites" were the very builders of Egyptian civilisation. Subsequently, Seligman reminded Africanists of the time that he was opposed to the idea of the Nilotic-Luo being the founders of the Itiyopian and Egyptian civilisations.

This suggests that Seligman knew that, if Western scholars acknowledged the Nilotic Luo to be the founders of Itiyopian civilisation, they would thereby confirm that the Luo are the founders of ancient Egyptian civilisation - a civilisation that later spread to Greece and other parts of the modern world. In other words, Seligman was opposed to the idea that civilisation was introduced to the ancient and modern world by the Luo people. The Hamitic models were developed and promoted at the expense of the Nilotic-Luo, to complicate the identification of people who truly did influence Greek and Egyptian civilisation. Nevertheless, the reality is that, whether we believe it or not, the Nilotic Luo have influenced both Greek and Egyptian civilisations. As we have seen before, the Luo people are also the founders of the Itiyopian Kingdom, which expanded its rule first into Lower Egypt, then into ancient Greece, and eventually to the entire world. Thus, the colonisation and expansion of ancient Egypt took Luo influence - such as widespread linguistic marks - and civilisation into Greece (Okello Paito, 2015).

The Luo shared gods with the Ancient Greeks and Egyptians

As I mentioned early in this chapter, we are going to demonstrate that the Luo people and Nubians worshipped the same gods the Egyptians used to worship, and the names of these gods were drawn from Luo words. More importantly, we are going to demonstrate that in ancient times, the Luo were known by various names, such as *Egyptian* or *Phoenician*. The Greeks only came to worship the Luo/Egyptian gods between 250-225 B.C. This was the period in which the Luo lost control of Egypt to the Greeks. The Greeks ruled over Egypt and they adopted the existing Luo/Egyptian civilisation. Herodotus (1972) noted that it was the Egyptians and the Phoenicians (Luo) who introduced civilisation to the Greeks, and not the other way around.

The Luo, Egyptians and Greeks played musical instruments like the guitar (kithara), harp and lute, in honour and service to the gods. The ancient Luo, Greeks and Egyptians had many gods. Some of the Egyptian gods they worshipped were: Amum-Ra "the hidden one"; Mut "the mother goddess"; Osiris "the king of the living"; Anubis "the divine embalmer"; Ra "the god of sun and radiance"; Horus "the god of vengeance"; Thoth "the god of knowledge and wisdom"; and Hathor "the goddess of motherhood." Greek gods they worshipped include Hermes "a messenger among the Greek gods," who was the son of Zeus and Maia; Hephaestus; Ares; Cronos; Apollo "god of music"; Dionysus; Prometheus; and Poseidon.

For example, Apollo was a Greek god who was equivalent to the god Heru of Egypt. However, the Greek god "Apollo" bears a Luo name - *Apollo* is derived from the Luo word "pollo." In the Luo language, the prefix *a-* simply means 'of.' The term *pollo* has four meanings, so the word "Apollo" literally means *of the sky, of the heavens, of outer space* or *of lightening*.

To the Egyptian, the god Amun was considered the king of the gods, a supreme creator-god. He was the ancient Egyptian god of fertility and life. Most Egyptian gods were 'born' at various points in the history of the Egyptian nation, and many floated in and out of popularity. Amun appears to have been present in the mythology and culture of the Egyptian people almost from the very inception of this complex and mysterious nation (www.ask.com).

Figure 20:
the god Apollo

Figure 21: Head of Amun
Statue
Source: Mary Narrsch

The ancient Luo recognise the god Nyasaye as a supreme being. This god was commonly described as Nyakalaga, which means *the one who dwells everywhere*. He is the creator of the universe and he continues to support his creation in its totality. Nyasaye is considered a *bding withoy* physical body (he is spirit). He is powerful and intervenes directly in the daily activities of man. The Luo people strongly believe that Nyasaye can create and destroy man. He can send various diseases, disasters or punishments on his people if he is angered by them. Above all, he is the source of man's blessing "gueth or mukica" (Okoth, 2002).

The belief of the Luo about the origins of death

► As the creator of all things physical and metaphysical, Juok/Nyasaye is also the creator of death. The Luo therefore believe that death reflects Juok's/Nyasaye's wisdom, since it is inconceivable that life could go on without death. The earth would soon be totally filled. There would be insufficient pasture for the cattle, inadequate land on which to build houses, and a lack of fields to cultivate. The Luo have a common saying: *When a child is being trained or educated, it may think that it is being maltreated, but an adult understands the purpose. In the same way, one must see the relationship between Juok/Nyasaye and his creation, including man.* These beliefs and philosophies are mythological and are explained in Luo oral narratives (e.g. the chameleon, who condemned man to death.)

The ancient Luo, Greeks and Egyptians commemmorated the histories and identities of their gods through paintings. Herodotus (1972) observed, "In many paintings, Apollo appears in the company of gods and goddesses such as Zeus, Athena and Dionysus" (p.187). Zeus was considered the supreme god of the Greek people, in whose honour the Olympics was originally celebrated. The Egyptian and Luo method of celebrating the festival of Dionysus is much the same as that of the Greeks, where they sacrifice a hog (pig) before the door of their houses. According to Greek traditional worship, this is followed by choric dance.

Figure 22: the god Dionysus

In contrast, the Luo and Egyptians do not have choric dance during religious celebrations. Thus, the Greeks did not impart their culture on the Luo people. Herodotus (1972) argues that the Greeks did not export their customs of worship to their neighbours. On the one hand, Egyptian gods were worshipped under Greek names, but on the other hand, the Luo have exported words such as "Kithara" or "Bito" to the Greeks. This then brings us to the simple fact that the "Itiyopiani-sation" of the Greeks was the mechanism by which the gods of ancient Egypt or of the Luo people were worshipped under Greek names. Herodotus (1972) further noted that the Luo (Egyptians, from the name Itiyopian) worshipped Zeus and Dionysus.

The Northern Luo today practise some Ancient Egyptian Beliefs

Interestingly, among the Northern Luo people - and in particular among the Shulluk - the god Dionysus and Osiris were later referred to as "Nyikang." Frankfort (1948) noted that "Nyikang" was identified as Osiris when he wrote, "Among the Shulluk of the Upper Nile, who retain many traits recalling Egyptian usages and beliefs, the king became charged with the supernatural power of royalty by being enthroned on the sacred stool (See Figure 22). This normally supports the fetish of Nyikang, who, like the Osiris, is both a god and the ancestor of the new Monarch. In Egypt too, the central ceremony of the accession took place when the ruler (Pharaoh) was enthroned, when he received the diadems and sceptres" (pp.43-44).

Figure 23: the god Osiris

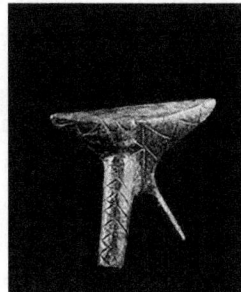

The Shulluk are a Nilotic group that, together with the Acholi, Anyuak, Jumjum, Jur Chol, Balanda, Thuri (Shatt) and Pari, comprise the Northern Luo, who retain vast knowledge of ancient Egyptian beliefs. Their ceremony of environment of *Rwot*, Reth or King was and is similar to that of ancient Egypt. The ceremony involves the introduction of the King-elect to Nyikang, in the case of the Shilluk, and of the King-elect to Giilo, in the case of the Anyuak.

As we shall see later, the Shulluk are obliged to present numerous gifts to newly-elected Reth, whose effigy is brought out by priests (royal elders) and transported to all the districts, accompanied by the new Reth's warriors/armies. On arrival at the headquarters, the elected king is placed on a royal stool.

Figure 24: royal stool

It is generally believed that, when the Reth is placed on the royal stool, this is also the time when the spirit of Nyikang enters the newly-elected Reth/King. The King-elect will sit on the royal stool for a while, to ensure that he is completely infused with the spirit of Nyikang. This arises from the belief among the Shulluk that Nyikang did not die; he 'returned to the river, from where his grandfather had come,' as they say. Prichard (1948) observed that to the Shulluk, regardless of who is the king-elect, Nyikang is always the King of Shulluk. When an elected king dies, his spirit is believed to have departed from his body to take up its abode in a new effigy especially made to accommodate his soul (spirit) at the shrine. Conversely, when the new king is elected, Nyikang's spirit enters him, he becomes the Divine King, and Nyikang thus continues ruling the Shulluk (p.28).

In contrast, the Acholi, Balanda, Jur Chol, Jumjum, Thuri and Pari, do not believe that when their kings die, they return to the river, because oral traditions of these Luo groups tell us that their kings did not originally come from the river; they are, rather, born of women. This leaves us with an unanswered question: "If all Luo groups are descendants of Okwa, grandson of Omara, grandson of Dyiang Kulu, why do these Luo groups not share the belief that their grandfather came from the river?" This area needs more research. Another (as yet) unsolved mystery is where the soul/spirit of the elected king goes as Nyikang's spirit enters his body.

The eternity of Nyikang is narrated as part of the power struggle among the princes in the Shilluk land. Thus, when an enthroned prince is killed by a contester, Nyikang's spirit will still rise from the dead and enter into the body of the incoming King/Reth. Fr. Crazzolara (1951) recounts Nyikang's death/disappearance as follows: "One such contender went one day and threw a spear at Nyikang hitting him in the chest. Nyikang said to his people, take me into a hut". He was taken into a hut. Shortly after, the roof of the hut opened by itself and Nyikang ascended to heaven, in the shape of smoke – *Keta mala irro*. The news of the ascension of Nyikang spread among the important chiefs and before long, they gathered around a hut where Nyikang had descended. The chiefs entered the hut to find out if Nyikang was there or not. They looked around and did not find him, because he had ascended into the sky as smoke. The family members of Nyikang were also standing around the hut. They then asked the chiefs, "What are you looking for? Is he not up there moving towards the heavens?" Therefore, when Nyikang reached a certain height from the ground, in the air, he looked down and saw the chiefs and members of his family gazing into the hut. He (Nyikang) said to them reprovingly, "My children, why are you looking for me?" He again spoke especially to his successors (children) saying, "You shall be buried in the earth; you shall not come to heaven after me." He then disappeared from their sight and went "mal" in the sky" (pp.126-127).

So often in the ancient world, the death of a king triggered furious struggles among princes as to who should succeed to the throne. Among the Shulluk people, this is demonstrated clearly in the case of Nyikang and his brother Dawaat (Dumo); and among the Luo who were left behind in Egypt, it was demonstrated in the case of Osiris and his brother Set.

Nyikang was to be enthroned as King of Shulluk following the death of his father, Okwaa, but instead, Dawaat (Dumo) was enthroned. Shulluk oral tradition says Dawaat did not kill his brother Nyikang, and that Nyikang never died, but disappeared in the shape of smoke (or 'returned to the river'). This narrative is supported by an African writer Okello Paito (2015), who has observed that many Shulluk do not believe that Nyikang ever died, believing instead that he simply returned to the water from which his grandfather had come. Among the Shulluk, Nyikang is a national hero and is revered as semi-divine (p.8).

In the case of the Luo group who was left in Egypt, a similar power struggle was evidenced between Osiris and his jealous brother Set. Oral tradition says that Osiris was to be enthroned as king of Egypt. Unfortunately, Set killed Osiris, whose spirit later rose from among the dead. This creates another uncertainty in our minds, as we ask ourselves, "Did the spirit of Osiris enter the body of Set or not?" Reading through various articles and books, I have been unable to find an answer to this question.

In conclusion, when captain E.N.T. Grove (1919) visited Acholi-land, he observed, "The Acholi have a strong idea, like that of the Greeks, that if you are too prosperous, the envy of the gods will fall on you" (p.179). Many scholars of anthropology did not take the assertion of Captain Grove seriously, but if we want to better understand Grove's assertion, it is important for us to revisit the Luo origin of the enterprising Phoenicians.

Chapter 9:
Gender and Peace Building: The Roles of Northern Luo Women in peace building

Women are amongst the most vulnerable victims in war and conflict situations, yet they are also often the ones that trigger peace mechanisms. What is exactly women's role in conflict resolution and peace building? How can we involve women more in peace building processes? What can be done to enhance men's role in contributing to peace processes?
By Wikigender University students, February 2011

During the civil wars in the then Sudan in 1956-1972, and in South Sudan in 1983-2005, the Northern Luo women and girls were among the most affected. From 1964 to 1972, many were abducted and forced to become wives of the *Anya-nya*. In the later conflict from 1983 to 2005, the Sudan People's Liberation Army (SPLA) forced their enemies to kill their own people. The killing of innocent people forced people to migrate into neighbouring countries. In the refugee camps in Uganda, Kenya, Ethiopia and the Congo, South Sudanese refugees suffered greatly, living in abject poverty with little support from the United Nations (UN). The camps were characterised by outbreaks of disease, malnutrition, racial discrimination and socio-cultural breakdown.

Figure 25: Women in Peace Building

In addition, many Luo women and girls were traumatised by gender-based sexual violence like rape, defilement, forced marriages and consequent unwanted pregnancies, and the spread of sexually transmitted diseases. The report of the International Rescue Committee's (IRC's) 2017 summary reads, "Up to 33% of women in South Sudan reported experiencing non-partner sexual violence during their lifetime."

In the face of these horrific events, women have not been empowered to take up leadership roles. They have thus been unable to attract the attention of international or even national communities, to inform them of their plight. Even when efforts were made to increase female participation in government structures to 25 percent, no Luo women were given positions in the South Sudan Government. This contrasts with the 10-25 percent representation experienced by women from other (mainly Dinka and Nuer) tribes.

Employment of these women in government structures is not based on merit or education, but on family background. 99.9 percent of women employed by the current South Sudan Government have family members who have previously held government positions - most commonly as President, a Minister or a Commissioner. No Luo women have ever been employed by the government, so the Northern Luo Women (NLW) are not represented in government, and are thereby deprived from contributing to nation-building and peace-building in South Sudan. The glaring question is, "*What is the South Sudanese Government doing to the Northern Luo Women?*"

It should be noted here that, with the signing of the peace agreement in Ethiopia on 12th September 2018, we hope that Luo women will be given opportunities to occupy positions in the South Sudan Government. If this is done, it would enable them to participate in the peace process and in the promotion of harmony and healing.

Nevertheless, as I am writing this book, Northern Luo women and girls in particular are still suffering from the effects of the civil wars in South Sudan. It is our hope and wish that they will be encouraged to take part in peace building processes initiated by local communities, the government, the Worker Trade Union (WTU), and international and local NGOs operating in the Upper Nile, Bhar El Ghazel and Eastern Equatorial regions.

It is important for the South Sudan Government to understand that women play an important role in holding societies together, by using traditional peace building mechanisms in their day-to-day lives. Unfortunately, their contributions to peace building processes have not been fully recognized by all stakeholders. My fear is that, even if Luo women organise themselves in peace building processes, they will still lack the financial resources needed to involve wider communities in their peace efforts of healing and reconciliation. Not only this, they may also face resistance from men who hold the traditional belief that "*women are to listen but are not to be heard.*" Nevertheless, we should promote gender equality and female empowerment across South Sudan.

Since South Sudan has just received its independence so recently, on 9th July 2011, and the political situation on the ground is still fragile, I would advocate that peace building, healing and reconciliation should be cultivated in the heart of the government, so that Luo women, girls, men and boys can start to recover from the unforgettable conflict. Partnership between South Sudanese political leaders and other peace-building actors is critical, to ensure that women and girls are architects and builders of peace for future generations.

I would also advocate in this book for all the stakeholders in South Sudan to take responsibility in addressing gender equality, and to condemn all acts of violence against women and girls, especially in Luo regions or counties.

Chapter 10:
Luo Groups

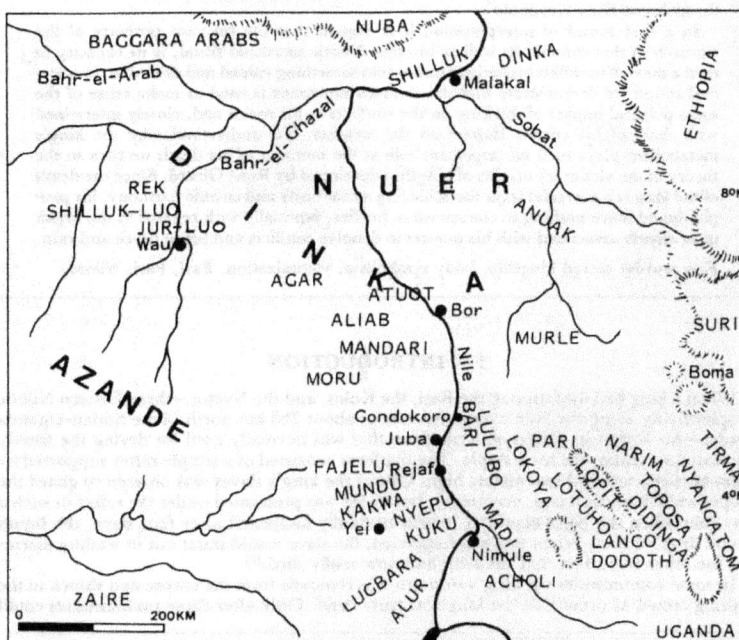

Fig. 1. Ethinic groups of southern Sudan.

Figure: 26, Source: Simonse (1994): The Burst of the Stomach, University of Leiden

Acholi

Acholi Geographic and Demographic
The Acholi are part of East Africa who lives predominantly in South Sudan and Northern Uganda. Acholi is an ethnic group from the district of Gulu, Kitgum, Agago, Amuru, Nwoya, Lamwo and Pader, in northern Uganda, and Magwi County in South Sudan. The Acholi in Magwi County composed of eight villages or clans, which include Agoro, Omeo, Magwi, Ofirika, Obbo, Panyikwara, Pajok and Palwar. The 1983 Sudan census counted 24,240

Acholi living in Sudan. But the estimated population of Acholi in 2010, following the election in South Sudan, they counted 600.000-700,000. And the 1991 Uganda census counted 746,796 Acholi living in Uganda and 45,000 Acholi were displaced (total Acholi Uganda is 791,796). Thus, when we put the total number of Acholi South Sudan and Northern Uganda they counted 1,362,296.

The Acholi-land, which is commonly known as Acholi Sub-Region, is an inexact term that refers to the region traditionally inhabited by the Acholi. To other scholars like (......) the region is composed of the present-day Ugandan districts of Gulu, Kitgum, Agago, Amuru, Nwoya, Lamwo and Pader. They neglected all together, the Acholi who live in South Sudan; and if you see from the below map, Acholi South Sudan are excluded from the political meaning of the term "Acholi land". However, if we do not examine this assertion carefully, it could easily mislead readers, they would easily conclude that Acholi are not in South Sudan. But the reality is that Acholi are in South Sudan.

Figure: 27, Picture of Okot P'Tek

Since the colonial period, the Acholi people developed distinct identity that characterizes them as Northern Luo of Sudan; this is something that set them apart from Southern Luo of Kenya, and Tanzania. Thus, the Acholi of South Sudan and Uganda were separated by the British in 1930, when the British were demarcating borders of their colonies. And the Acholi of Uganda called themselves as "Central Luo".

In about 1650 the Luo and the Acholi in particular, were living together as one people. In this year, the Acholi were not calling themselves as 'Acholi" but rather they called themselves collectively as "Sudanic" or "An-loco-li". As Doom (1999) noted, "Before colonialism, the people known today as Acholi, referred to themselves as "An-loco-li", which literarily means "I am a human being" (P.10). However, the word or label "An-loco-li" did not point at ethnic group or geographical boundary. Although the Acholi people collectively identify themselves as people of one family, of one culture and customs that governed their existence for thousands and thousands of years. "As a result of the Acholi people's self-understanding as human beings, they embraced peaceful coexistence among themselves and their neighbors" (P.11).

However, between the years 1698-1700, the Southern Luo started immigration to their present places in Kenya, Tanzania and Ethiopia. But the Acholi remain as a block, as Okot P'Bitek (1953????) wrote, *"Inyim bino pa jo British, jo Acholi me Uganda ki Sudan onongo gin tye acel. Ento teke jo British odonyo I lobo Acholi, gin opoko dano I dul pyero-adek ma bene tye I tee loc pa Rwodi pyero-adek. Dul acel-acel loyo kome kene. Ento lacen dul magi pyero-adek gin obino i tee loc acel pa District Commissioner (D.C) ma gang kal mere madit obedo i Gulu Uganda. Pi man Gulu odoko "Acholi" District.*

Which could be translated as:-

"Before the coming of the British, Acholi Uganda and Sudan were living together as one people. But when British entered Acholi-land, they divided the people into thirty groups or villages; each group or village was exercising self-autonomy, headed by a chief. Later on, the British brought these groups under one District Commissioner (D.C.) with its headquarters at Gulu Uganda. For this reason, Gulu became known as "Acholi" District.

Source: Saturnino Onyala, 2014, P. 48

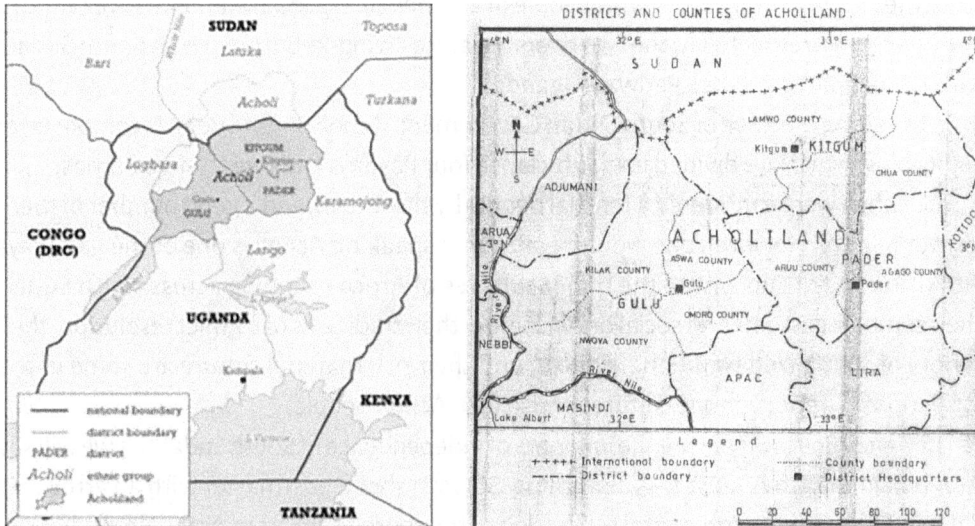

Figure: 28, Source: A New Macmillan Social Studies Atlas, 1998.

William J. House and Kwin D. Phillip Howard (1982?) argue that if the land of Acholi of South Sudan and Acholi of Northern Uganda are put together, it could encompasses about 28,112 sq. km, with total population of ...

The Acholi land extended as far south as Lake Victoria and to the southeast is extends to Lake Albert. To the east, it extends up to Lolubai, Agoro and Got Oromo; to the north it extends as far as "Kulu-wi-Jubi" or River Kulu-wi-jubi at Opari; it also extends to Koli River to Got Kapai (Mt. Kapai) near Imurok.

Meanwhile, the land of Acholi south Sudan alone encompasses 2,700 sq. km. with estimated population of between 600,000 – 700,000 individuals. The border of Acholi

South Sudan, with their neighbours, stands as follows: on the south, the border is at Limur River, the River separate Pajok (in South Sudan) and Lokung (in northern Uganda). From the southeast, the border is at River Lirringa, this river separates Palwar (in south Sudan) and Lokung (in northern Uganda). From the North East the border stands at mountain Kidikidi (Got Kidikidi), this mountain separates Palwar and Lotuko. From the north, the border stands at Lowai and Got Kapai, Lowai separate Omeo and Lokoya. Further more in the north, the border extends to Got Kapai; the Mt Kapai separates the Obbo people from Lotuko people. From the northeast, the border of Acholi with Lotuko is at River Koli, this River separate Magwi from Imurok. From the northwest, the border with Bari people is at Joseph Lagu's farm (that is at the corner of Ame to Magwi road and Ame to Nimule road). From the west, Acholi border with Bari is at Otur River, this River separates Panyikwara people from Bari people. From the south-west, the border of Acholi with Madi is at Kulu-wi-Jubi or River Wi-Jubi (at Opari) this River separates Panyikwara people from Madi people. From the south, Acholi border with people of Palabek (in Uganda) stands at River Limuru and River Aswa respectively, the two rivers separate owing-ki-bul people (in South Sudan) with the people at Palabek Padwat (Uganda).

In the administrative of South Sudan Government, Acholi South Sudan is composed of eight clans (which are divided into sub-clans), four Payams and twenty-five Bomas.

The Acholi were considered a martial people by the British, and a good number of them joined military. However, one would realize that speaking Acholi is one of the gateways into the world of Luo, one of the major cultures of Africa spanning across south Sudan. Their complex customs and social organization, their traditions of conflict resolution, their variety of specialized traditional dances, and their rich material culture are some of the attractions to reflect on the cultures and arts of Acholi people.

The Acholi South Sudan are the initiators of independence of South Sudan. As we will see later on, on the 6th August 1955, Saturnino Obwoya shot a postmaster with an arrow. He did this, because he wanted total separation of the southerners from Arabs North. It is sad that many of south Sudan politicians do not realize this fact and never publicly spoken of his intention of shorting a postmaster. Today, politicians speak of John Garang De Mabior as the initiator and founder of independence in South Sudan. They forget that Garang was rather a unionist than separatist. However, people like Riak Machar Atony, Lam Akol and Salva Kiir Mayardit were real separatists. How can a unionist bring independence of South Sudan?

From the Acholi of Sudan, apart from Saturnino Obwoya, there are other important people who deserved to be mentioned of their national standing. These include but not limited to Nicola Obwoya from Pajok, Nicola Obwoya from Panyikwara, Urbano Oyet from Agoro, Nataniel Oyet Polo from Pajok, Ogweta Opoka from Obbo, Gen. Juma Okot from Panyikwara, Charles Ogeno from Palwar, Pangrasio Ocieng from Omeo, and chief Mathia Aburi Lolori, from Obbo.

The Acholi land in South Sudan has been used as battlefield since 1964-1972 (by Anya-nya 1) and from 1985-2005 (by SPLA).

Anyuak

Origin and Anyuak Kingdom

As we have mentioned in chapter 3 above, there are conflicts reports about the origin of the Luo groups in general, and the origin of the Anyuak in particular. We will see later that some of the writers, like Lewis (1976 P. 158) asserted that Ukiro (Ocwudho) is the founder of the Anyuak nobility; he simply appeared out of the water and was captured and made ruler of Anyuak villages. But other writers like Crazzolara (1951 P. 122), Wamari (2006 P.5) argue that the noble clan of the Anyuak and the royal clan of the Shilluk can be traced to their semi-divine ancestry which have intimate ties with the river. By this semi-divine, they seem to referred to Dyiang Kulu (who is the father of Okwa) or to Nyikang. However, it is most likely that they referred to Dyiang Kulu because Anyuak are not descendants of Nyikang; rather, they are descendants of Dhimo. The bottom line is that no one tribe can originate from two distinct ancestries. Logic tells us that a tribe can originate from ancestry or that tribe never originated all.

Therefore based on the Luo oral history and some findings from anthropologists we can conclude that the origin of the Anyuak can be traced only to their grandfather Omaaro and Okwaa. As such, we cannot deny the truth that Anyuak people have common origin with the Acholi, Shilluk (Collo), Jumjum, Mabaan, Jur Chol, Pari, Balanda and Thuri (Shatt).

Anyuak traditions suggest that they have migrated to Upper Nile Region and Gambella (Ethiopia) from the northwest, some five to six hundred years ago. Many anthropologists suggested that the ecological variation that exists within the territory where the Anyuak settled has had a significant impact on their lives and settlements.

Lewis (1976) says, "The cultural affinities between the Anyuak and the Shilluk people of the Upper Nile have been known for some time. The close relationship between their languages has been established beyond doubt, and the historical traditions of both people tell of their common descent from two brothers who quarrelled and separated. There are many other social and cultural similarities as well. Frazer Lecture and Lienhardt in their articles of 1955:32 have suggested a common historical and structural relationship between the political systems of these people (P. 151)".

Thus, one of the purposes of this book is to examine further and in-depth the relation-ships of the Luo groups in south Sudan; and to prove how the Northern Luo have common origin.

Some anthropologists like Crazzolara (1954) confirmed that Anyuak people are descendants of Gilo (or Wuate Giilo), who is the brother of Nyikang; who are the children of Okwaa. Despite all these facts, there is still confusions about the origin of the Anyuak people. According to Crazzolara (1950 P.31-33) said Anyuak came from their cradle-land at Wau or *Wi Pac*. Somewhere in the region where 70 north Latitude meets 30 East longitude. He went on to say that by the fourteen century oral traditions firmly place the Luo in the vicinity of Rumbek in Bhar El Ghazal. In the fifteenth century, the Luo began to move again, more rapidly than the glacial speed of past centuries. Small cluster of Luo clans wondered north from Rumbek. This group in turn experienced further defections during the northward march. The Boor (Balanda) and Jur Chol (Luo Bhar El Ghazal) made their way west to the ironstone plateau south of Wau town. Another group led by Gilo also disengaged themselves from

Figure, 29, Picture of Fr. Crazzolara

the main body, migrated north and west to Pibor and Sobat River; and to the Adura and Mokwau rivers in the north. They were forced out of these areas by Nuer. However, some of the Anyuak remained in the Sobat River near Nassir and at Ajungnur near the mouth of the Pibor River. The Anyuak people who were forced out of Pibor and Sobat River continued upstream and settled at the base of the Ethiopia escarpment, in the valleys of the Baro Pibor. Today, they are known as Anyuak Gambella.

Some eight or ten generations ago, in the seventeenth century, a splinter group moved south from Anyuak land to Lafon Hills, where they became known as Pari; and the second clan the Pajok penetrated further south into Acholi territory at the border of south Sudan and Uganda.

The second narrative like Collins (1956 P.1) said the history of Anyuak seems to have began at the Gezira –near Khartoum in Sudan, a place thought to be their original homeland, and from here they moved to their present locations: Upper Nile and Gambella Ethiopia respectively.

The third narrative of Anyuak origin and descent do not believe Gilo is the brother of Nyikang –they consented that all the Anyuak account of their origin differs from that of other Luo groups including the Shilluk. The informant said one day, some women went to the fetch water from the Okira River and they discovered a mysterious person standing by the side of the river, with his fishing spear. When the women draw nearer to him, the man disappeared into the river. These same events happened about two to three times and in the last day, women managed to capture the mysterious man and took him in the village.

However, Crazzolara (1951, P. 147) narrated the story slightly different. According to him, Ocuudho (who was known as Kiro) was living with his wives and cattle in the rivers. One day, children were sitting in a circle round a pool and catching fish with their hands. Two of the children caught a fish at the same time and started quarrelling. One child was holding the fish on the head and the other child was holding fish on the tail. Ocuudho was sitting on a log of wood in the river watching them. As they continue to quarrel, Ocuudho intervened and decided that the fish belonged solely to the child that held its head.

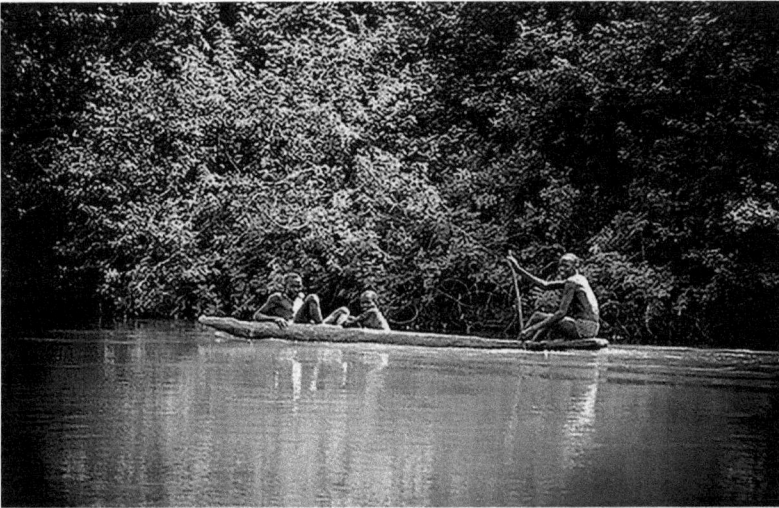

Figure: 30, Anyuak River People

The boy, who was holding the fish by the head took it and the children, went home. When the children reached home, they narrated to the villagers what happened when the two boys caught a fish at the same and how Ocuudho intervened. When the people of the village heard of this sentence, they told the children to bring Ocuudho home. Next day, two children and some young men went back to the river but young men remained in hiding near the river. They instructed the two boys to go by the riverside, and when they reach there, they should pretend to quarrel about the fish. The two children went as instructed and when they came by the riverside, they pretended to be quarrelling over the fish. Shortly, Ocuudho reappeared to give judgement; soon a group of young men came out from their hiding-place and caught him. He was brought before Cuai or Ocua – who was the headmen, of one, of the three main clan groups of the country. According to Crazzolara the villagers tried hard to persuade the man to stay with them but the man refused and went back to the river. Other narratives in contras stated that the villagers succeeded to persuade the man and he accepted to stay with them.

Anyuak tradition said the river man had no name and later they gave him the name Ocwudho (also known as Okiro) —which means the river man. When Ocuudho was given food and drinks he refused. When the villagers heard that the man refused to eat and drink, they were afraid that he might die of hunger and thirst. Cuai (the headman) then sent his daughter Koori to take care of Cuudho. When Koori gave food and drink to Cuudho, he accepted and shortly Koori became his wife. Before long, Koori became pregnant. Thus, when Ocuudho realised Koori was pregnant, he withdrew again into the river.

After nine months, the girl (Koori) gave birth to a baby boy, who was named Giilo or Gilo who later started the royal dynasty of Anyuak. Crozzalara (1951, P.147) reports that when Ocuudho learned that the wife has given birth to a baby boy, he came out of the river, wore the famous "*Ocuok*" (beads) round his neck and went to the village. He took with him "*Ocala*" (spear) and the *oduedi* (harpoon) with its "*odeege* (string) of "*dimui*" (beads).

Giilo was killed by his brother Oteeno, who was presumably expected to succeed him. However, Oteeno did not succeed Giilo —rather his son called Eno succeeded him. This son was saved by Akango, a sister of Oteeno. The Anyuak people were not happy with the act of Oteeno, subsequently he was sent out of the village. Where he (Oteeno) went after this has never been spoken of.

Thus, when Giilo died he was said to have returned to the river, the place of his father. As I mention before, the Anyuak origin has intimate ties to the river. The bottom line is that all the Anyuak believe they are descendants of Giilo —we cannot independently verify whether they believe in the Giilo the son of Cuudho or Giilo the brother of Nyikang. But for the sake of this book, we say without doubt that Anyauk are descents of Giilo the brother of Nyikang.

The list of Anyuak Kings, given by Professor E.E. Evans Pritchard, could help us understand the successors of each king but the list is fairly reliable. The Anyuak Kings are:

Names		Names	
1.	Okiro or Cuudho	2.	Giilo (wa-Koori)
3.	Opio	4.	Obok
5.	Openo	6.	Giilo (war-Akanga)
7.	Eno	8.	Adhiedhi

Names	Names
9. Odol (wi-Nyitok)	10. Nyanyuak (lady)
11. Googho (war-Adhiedhi)	12. Cam (war-Aweti)
13. Odola	14. Wango
15. Giilo (wi-Nyijaango)	16. Ajak (wi-Nyidekuno)
17. Goora	18. Olueth
19. Giilo	20. Obang (wi-Nyidoor
21. Akwei-wa-Cam	22. Sham Kwei
23. Agwa Akwon	24. Adongo Agada Akwai Cham
25. Akwai Agada Akwai Cham	

Figure: 32, Source: Crazzolara, 1951, P. 148

According to informants, Giilo No. 15 died in Anyuak homeland and his son Ajak led a wave of Anyuak to Upper Nile Region. In addition, king Goora, Cam, Googho (war-Adhiedhi) were all buried at Othuon on the Akobo.

Figure: 31, crowning of King Agada Akwai

King Akwai Agada Akwai Cham was crowned at Royal compound in Otalo on 25th April 2012, and has replaced his late brother King Adongo Agada Akwai Cham who died on 30th November 2011 in Nairobi. King Adongo Agada was born on 1st January 1959; he was a teacher in Sudan before taking refuge in Ottawa (Canada) where he was reduced to a factory-worker. He was called back to Sudan in 2001 by Anyuak community, to take the throne of his father who was king for 60 years (1931-2001). The people needed him to settle outstanding issues, such as their strained relation with neighbouring tribes and the difficulties faced by their fellow Anyual across the border in Ethiopia. He is the 23rd King of the Anyuak Nyiudola Royal Dynasty and is considered by some of the Anyuak to be a "demi god". According to Anyuak traditional law, the crowning of King Akwai Agada Akwai would have, taken place in the same month (November 2011) the former king (Adongo Agada Akwai Cham) passed away.

The most important political unit among the Anyuak is the village; it is viewed as the fundamental unit of defence and that no mechanisms exist for obtaining competition for deaths occurring in fighting between villages. The Anyuak are grouped into a number of non-exogamous clans, thus, clanship holds little significance in Anyuak life. Villages are usually associated with one lineage of a clan and there is little contact among lineages of the same clan in other villages. In each village there is a lineage which is said to be "the owner of the land" (*kwai ngam*), the original inhabitants and the founders of the village. In smaller villages there may be only one resident lineage; but larger villages may have more than one, the others being regarded as guests of the *kwai ngam*. Initially this lineage should be the dominant political lineage (*tuong duong*) from which the village headman is chosen. Only male members of this traditional headman's lineage whose biological fathers have served as headman are eligible for the election into the position.

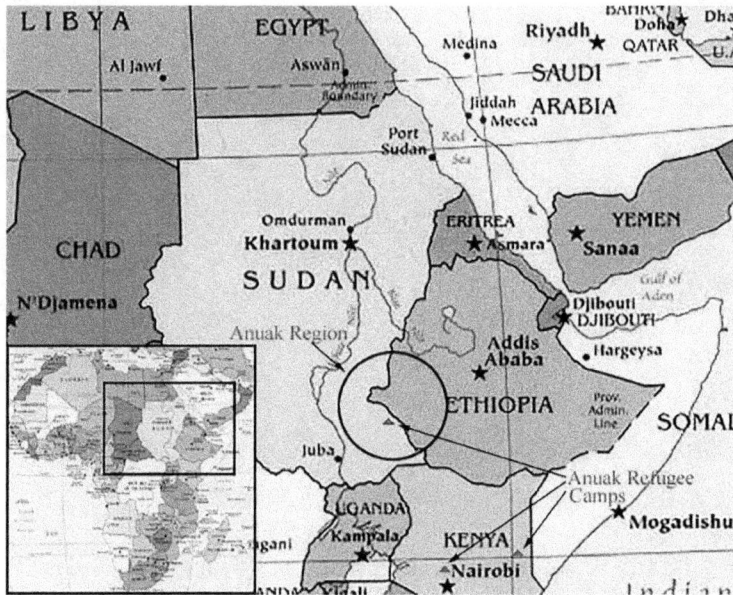

Figure: 32, Anyuak Region
Source: BBC

Anyuak Chieftainships

Lewis (1976) observes that since the village represents the limit of everyday Anyuak experience, it is not surprising that the village headman should be symbol of the village. He is the focal point for the village loyalty and expresses –the uniqueness, unity, and exclusiveness of each village vis-a-vis its neighbours. He controls the village drums and the beads of office. Great former respect is paid to him. He is never allowed to sit directly upon the ground. Special vocabulary exists for many of his possessions and the buildings of his homestead, which are decorated with carved ornamental poles and grass screens. Those who approach him must do so, on their knees and address him respectfully. The headman also has a number of deputies and court official attached to him, and the youths of the village often group themselves around him as temporary followers and retainer "*luak*" (P.152).

However, having said this, we should not misinterpret that headmen are autocratic rulers –they are not, they remain in the office only in as long as their rules are beneficial to the communities. The headman must be generous in the distribution of food and gifts to his people. The headman remains strong in as long as the majority of the population are with him –follow him. He guides his village largely by consensus and force of personality rather than inherent power. This implies that the headman must develop an efficient system within the village, failure to do this, ultimately leads to his downfall by exhausting his personal resources. Once he becomes poor the community will immediately replace him through "village revolution" what they called in Anyuak language *agem*.

However, among the Anyuak of the south-eastern, they have different political system. Although all villages have a headman, in the south-eastern of Anyuak County headmen must compete for power with a noble clan, known in Anyuak language as *"nyiye"*. The nobles are members of a single patrilineal clan spread throughout Anyuak-land who have replaced traditional headmen in many villages. According to Anyuak myth, this noble clan was founded by a mysterious man named Ukiro who appeared in the river one day and was captured by the Anyuak, who took him to their village and made him their ruler at the expense of an unpopular headman. Ukiro brought with him five bead necklaces, four spears, two stools, a spear-rest, a drum and a few others objects; the most important of these objects is the necklace (*ucuok*). To become eligible to take over the headman-ship of a village a noble must first be invested with these ritual emblems. Once invested, a member of the *Nyiye* may be invited to replace a headman in the aftermath of a village revolution (P.153).

All nobles are treated with respect by the Anyuak, but if a noble succeeds in establishing him as the new head of a village this respect will increase. He will be surrounded by large number of court officials and retainers that composed mainly of youths. The noble who have taken position of headman has to ensure that he supply people with enough food and gifts. When doing something in the village, he must move with the people, he cannot act on his own caprice. This is expensive as Evans-Pritchard (1947:78) puts it that one noble was forced to sacrifice eight oxen from his tiny herd within the first few months of his tenure in office in a bid to keep public support. When villages are tire of a noble they cannot dispose him by force; they can only persuade him to leave the position of headman on his own accord or to invite another noble to come and rule in his place. This noble then has the sanction to attack his rival and drive him out, or kill him.

SETTLEMENT

As mentioned earlier, the Anyuak and Collo traditions suggest that they have migrated into their present place from northwest. We have also seen that Giilo and Nyikang migrated from their homeland at Lake Albert in Uganda (the place commonly known as Kerau or Karoo). Anyuak people came to Bhar El Ghazal and from here they proceeded to Kodok, in Upper Nile Region. From Kodok, Giilo and his groups decided to leave Nyikang and they moved and settled in big villages along the Akobo and Baro as well as Gilo Rivers and in

Pibor and along Sobat River. Each village has sub-chief (kway-luak) who is in control of the social and administrative matters of the village. More impotantly, all villages are under the control of the King (Nyie).

The separation of Giilo's groups from Kodok could be traced along the following two lines:

(1) The Luo marched along the Sobat River and reached Nyanding River. Some of the groups went up this River and reached to Pibor, and other groups settled on the bank of Sobat River. Majority of the Luo continued and when they reached the junction of the two rivers Baro (or Peeno) and the Pibor, first group separated and followed the Peeno eastwards into Ethiopia, and along the Aluro. The main group decided to fol-low River Pibor and set out in a southerly course. Again some small groups broke off and followed the Giilo, Akobo and Sobat Rivers. A good number remained behind and settled there. The majority continued their movements southwards. They went and reached at Wi-Pari –which means the end of the land. Here, the groups resorted into deep discussions. Some of them have "wander-lust" –they still want to see more of the world, but some people were already tired of roaming. Those who were tired settled at Wi-Pari. They settled here for some 30 years and after that a group moved to Lafon and settled there and other groups returned to their brothers at Anyuak-land. Thus, Wi-Pari was completely deserted. And those (majority groups) who were still ambitious to explore more of the world continued to move southwards and they came to Lepfool Hill, the place today known as Lafon Hills –west of the Lopit moun-tain range (Crazzolara, 1950).

(2) Formerly, the Anyuak had big land, but they lost part of their land to the Nuer in the 18th to 20th century, following the migration of the Nuer into the region. Col-lins (1956) explains that in the mid-eighteen century, there was a massive eastwards flight of the Bul Nuer. They descended upon the Jikany Nuer that participated a domino effect driving the Jikany and some of the Anyuak eastward into Ethiopia. However, majority of the Anyuak people remained in the Sobat River and they be-came known as "Sobat Anyuak" and those who were driven into Ethiopia, became known as Gambella Anyuak.

(3) According to Ojulu Odola (2013), the Anyuak were one people up to 1902, following the signing of Anglo-Ethiopia Treaty by Emperor Menelik 11 of Ethiopia that estab-lished his western frontier with the Anglo-Egyptian Sudan. This treaty has divided the Anyuak land into two, portion in the Sudan and portion in Ethiopia. However, this was done without the consultation of the leaders. The south-western remain under the British control, as part of Sudan and eastern part became part of Ethiopia under the name Gambella. Gambella was named after an Anyuak chief, who went to Ethio-pia and lived there for one hundred years, and was visited by first Sudanese customer

inspector in 1905. Article 1v of the Anglo-Ethiopian Agreement defined the leased territory as 2,000 meters along the north bank of the Openo and no more than 4,000 acres beyond the river. The lease was to last in as long as Sudan is under Anglo-Egyptian control, and the enclave could not be used for any military or political purposes. In 1903, Major Charles W. Gwynn conducted a superficial boundary survey of the borderland between British and Ethiopian. He ignored the treaty signed in 1902. Major Gwynn found it more convenience to set boundaries by simply following the course of the various rivers. Despite the defence of his demarcation, the true of the matter is that the rivers do not make good international boundaries especially when people of the same ethnicity live on either bank as the Anyuak –under the jurisdiction of different governments. Thus if we have to agree with Major Gwynn by delimiting the boundary along the rivers openo, Pibor and Akobo, this would implies that we are giving out the territory surrounded by them and inhabited by the Anyuak to Ethiopia.

Thus, for us to get into the depth of the history of the two groups of Anyuak, we are going to discuss first about Sobat Anyuak and later discuss about Gambella Anyuak.

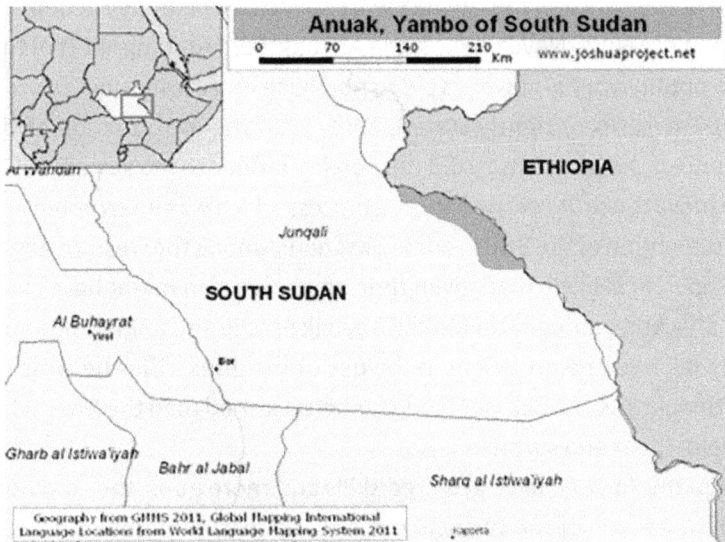

Source: Joshua Project
Figure: 33, Anyuak location in South Sudan

SOBAT ANYUAK

We can say without doubt that the Anyuak who live in Upper Nile Region are happy as they live in a grassy region that provides grazing for the cattle. In the north-western regions, the land is low and flat and is subject to seasonal flooding from the Pibor and Sobat rivers.

Lewis (1956) pointed out that during the raining season, this area floods, so that much of it becomes swampland with various channels of deep water running through them.

Although the Sobat Anyuak people were happy of the new settlement, the Nuer did not give chance for them to look after their cattle peacefully; there were series of fighting. And after many years, following the Nuer raids in 1880s, great crisis occurred in the valleys of the Pibor and Akobo where many Anyuak villages were destroyed by the Nuer, and these include the populous village of Ukadi. The Nuer then took over the land at Abwong, Adong and Akobo, and this lead to assimilation of some of the Anyuak into Nuer culture and population.

Anyuak traditions said the Nuer went fighting as far as Ubaa and the sacred rock-pools of Abula in the southeast. The Nuer would have settled here, but because the tsetse fly in the area that made their cattle suffer heavily; they were forced to move from Anyuak land to the treeless plains. The Lau (or Lao) Nuer moved to the west, and the Jikany moved to the north. Since that time, the Nuer never attempted to penetrate deeply into Anyuak country, but they continued to raid the western Anyuak with immunity.

By the end of the century, the Anyuak were saved because they acquired guns from Ethiopia to fight the Nuer. Nevertheless, the Anyuak were running short of fund but they succeeded to obtain ivory and were able to purchase more ammunition and then rifles. Collins (1956), Lewis (1976) rightly stated, "Only with the introduction of the firearms towards the end of the last century did the nobility influence the systems of Anyuak policies, begin to move towards centralisation; a process which was interrupted and then stopped by the intervention of the British administration. Among the western Anyuak, no such system developed. It is likely that, given time, the noble clan might have extended itself through that area and gradually united the Anyuak people into a single body politic" (P.161)

As the Anyuak became proficient in the use of the guns, rifles became increasingly available in Ethiopia, the Anyuak were soon far better armed than the Nuer, who continued to rely on shield, spear and surprise.

Even though the Anyuak have acquired skills and more guns, they did not defeat the Nuer in the first decade of the twenty century. The reason might have been that they were not united under one strong leader. However, in 1910, the Anyuak anointed Akwei-wa-Cam in the Royal Emblems. He united the Anyuak and in 1911, he launched concerted attacks against the Lau Nuer and the Jikany and dismantled them and this devastated the Nuer. They pursued the Nuer all the way to the Bhar-El-Zaraf and returned to Akobo with hundreds Nuer captives and thousands of cattle.

As we shall see in migration problems, the Luo generally do not kill female prisoners, children, disabled or men. This implies that the Nuer women, who were captured in this fight, were absorbed and became emerged in the Anyuak community. "The small girls were treated as children of the family and never as wives. After the girls grew to adults, they were settled in the community, they were given away in Marriage just like other Anyuak daughters. The Nuer men become prisoner slaves, added to clan-group, and treated as blood-relatives. These men were later given wives or cattle to marry, as members of family" (Crazzolara, 1950, P. 47).

The Sudan Government were not happy with over 10,000 guns possessed by the Anyuak, so in 1912 a large force was sent up to Akobo, under the command of major C.H. Leveson, they fought with Anyuak and after heavy losses on the side of the government troops, eventually the Anyuak were driven out from Akobo. Moreover, all the surrounding villages were burned down but the Anyuak were not subdued.

As the Sudan government could not penetrate into the inner Anyuak villages, they established a military garrisons at Akobo and Pibor in 1911; they also established small police stations between Anyuak and the Nuer residents. This was done to contain them but other people asserted that the Anyuak established it mainly to protect the Nuer from being slaughter. However, it was not possible to stop the raids, the Anyuak of Ciro, including the Adongo people under King Akwai –wa-Cam in 19---, raided the Nuer for cattle, women and children.

Nevertheless, this containment policy was partially successful because the British officials were only able to control the people around their posts, but people beyond the posts were free from control (Collins, 1956)

When king Akwei-wa-Cam died in 1920, his son, Sham Akwei, who was just twelve years old, took over the power (possessed emblems). This then opened the way for the British to penetrate deep into Anyuak villages. Lt Cornel C.R.K bacon was the first British official to enter into the heartland of the Anyuak, he went up to the remote area called Adonga region. When Bacon first reached Adonga region, he was deeply impressed with the respect accorded to this young boy (Sham Akwei) because he possessed the Emblems. For this reason, Bacon thought it would be good to make Anyuak people under a single chief and to consolidate this; he initiated a consistent policy of supporting Sham Akwei against any aspirants to power, this was then done. For the six years, relationship between the Anyuak population and Sham Akwei were cordial. And in 1921, Bacon made reconnaissance through Adonga and fourteen years later (in 1935), the Sudan government was able to send the first British District Commissioner to Adonga.

The British demarcated the boundary between Luo-Nuer and Anyuak of Akobo at Dwa-Achan and Wanga-Ading; and they demarcated boundary between Murle and Anyuak at Biem. These borders are known as the 1956 borders recognized by the then Sudan government when the British left the country.

By the year 1925, the Sudan government wanted military Commissioners in Soba and Pibor be replaced by civilian Commissioners. Surprisingly enough, Bacon without any consultation with the natives, said that Sham Akwei be given the position of Civilian Commissioner. He wanted this to be done, because he thought that Sham Akwei has a considerable influence over the whole of Anyuak and would be a great assistance in the formation and organisation of native courts in the areas. In addition, the native courts have especial vocabulary to use during the court hearing and this is known as 'royal language conventions" (Lewis, 1976).

Before long, in 1927 the District Commissioner at Akobo, Major G.W. Tunnicliffe, tried to change the system by giving the office of District Commissioner to direct descents of Oshoda, who claimed to equal right to possess the Emblems –this would then mean that the throne will be held for one year, and after this a new person will be elected by peers from that descendant. This created more problems than it could solve. Because of this, two holders of the Emblems refused to hand over their positions.

Thus, by the process of recognizing direct descendants, according to the customs, this brought a sharp division among the Anyuak people in 1932; and the initiative of Bacon to consolidate Nuer tribes and native administration under a single chief were destroyed.

However, Bacon did not give up, he went on with the election and the people elected Ahwa Akwon. He was given more permanent custody with salary payment of $1 (only one dollar) per month. Thus, Chief Agwa Akwon was expected to give strong support to the District Commissioner at Akobo.

GAMBELLA ANYUAK

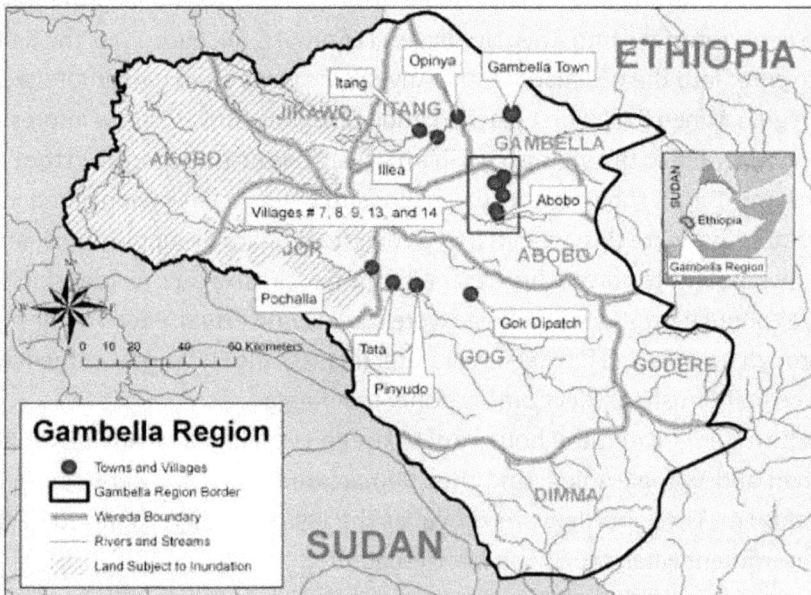

Source: Books.openedition.org
Figure: 34, Location of Gambella

94

Following the fight of Nuer against Anyuak at Sobat River in the 18th to 20th century, the Anyuak were divided into two one group was pushed to Ethiopia and they became known as Gambella Anyuak. They came and settled in Gambella, a place situated in the south-western part of Ethiopia bordering with Benishangul Gumuz to the North, bordering Oromo region to the west, bordering the south Ethiopia People Regional State to the south and bordering the Sudan to the west.

The region has the characteristic of marsh land, hot and tropical savannah forest, with rich fertile, well-watered soil coming from rivers, with annual rainfall about 800 mm. This has tremendous influence on the economy and life style of the Anyuak people. However, it is important to understand that the Anyuak people who live in the lowlands of Gambella, at the beginning of the settlement suffered from slavery. The slavery was abolished when the Gambella became under British rule in the early 20th century, and was restored when the region was ceded to Ethiopia after the Second World War. Nevertheless, after the slavery was officially abolished, many Anyuak people were illegally abducted and enslaved in Abyssinian households. Today, Anyuak people are still looked down upon as black slaves and they continue to complain of racial discrimination and marginalisation by other ethnic groups in Ethiopia. Subsequently, this has affected the Anyuak children access to education, health Care and other basic services in addition to limitation to socio-economic development of the area.

Moreover, this does not only affect the children, it also affects adults because they also have no access to employment in the Ethiopian Government or to the Private Sectors.

Figure: 35, Gambella: no employment people sitting under shades

At this point, we should know that Gambella became strategic place of trade between Sudan Government and Ethiopia Empire. This was followed by signing of Anglo-Ethiopia Agreement between Sudan Government and Emperor Menelik 11 of Ethiopia, in 1902; which allowed the Sudan to establish trading posts at territory 2,000 meters along the north bank of the Baro (Openo) and about 4,000 acres beyond the river. The least was to last as long as the Sudan was under Anglo-Egyptian control, and the enclave could not be used for any military or political purposes. Emperor Menelik was enthusiastic about the commercial posts, because it provided outlet for the expansion of the rich coffee grow-ing areas of Western Ethiopia that would generate revenue for the State and offered an exchange for Sudan salt. As we have seen above in 1903, Major Charles W. Gwynn conducted a superficial boundary survey to determine the border ignored at the treaty signed in 1902. He decided to demarcate boundary along the rivers Openo, Pibor and Akobo and this would give the territory inhabited by Anyuak to Ethiopia. Therefore, by the year 1920, Gambella was possessing a shade and a house furnished with a table and a bed for the customs inspector.

British regarded Gambella as a miserable place to live in because the place could not be reached from the surrounding in certain times of the year. The place could only be reached by steamer between June and November, when there is enough water in the rivers. Moreover, in dry season, which is between December and May, Gambella is cut off through these dry seasons.

The customs inspectors were not only collecting duty on coffee, but they were also used as judges. The customs inspector summons offenders both the Nuer and Anyuak to court. The local traders called on the government of Sudan and Ethiopia to reinforce security between Gore and Gambella. Merchants did not feel save when they travel between Gore and Gambella. The merchants also wanted much freedom from Ethiopian tax collectors. The customs inspector tried to work on these requests, but because they have very little judiciary power, consequently the people could not listen to them. Eventually, the security situation in the areas deteriorated.

When Marsh saw the security was going out of hand, he took over the roles of Police, storekeeper, chief clerk, judge, jury and financial controller. Although Marsh became a jack-of-all-trades and master of none, the battle over the tax at Gambella did not stopped, local Ethiopian officials continue to generate revenue for their own interest at the expense of local merchants. Despite the fact that British consul was stationed at Gore, and District Commissioner was appointed at Gambella to preserve the principles of free trade, the Ethiopians continued with the collection of taxes.

More importantly, the final blow to Gambella trade ended in January 1947, when all merchants operating in Gambella, were asked to obtain passports in person from Addis Ababa, to enable them continue with trade in Gambella. Thus, without passport, no trader was permitted to do business in the areas. This hardship combined with the currency regulations, effectively ended Sudanese trade with Ethiopians throughout the Enclave and destroyed any hope of revenue (Collins, 1956).

ANYUAK CLANS

Professor Evans Pritchard stated that the Anyuak has two main clans these are the Nyiye and the Jowatong. The Nyiye are the royal clan who descended from Giilo while the Jowatong are a branch of the royal clan, they have lost the rank of nobles and are became known commonly as "commoner" (*baangi*); in turn commoner are divided into thirteen clans bringing the total of Anyuak clans to fifteen. These fifteen clans form fifteen villages and each village consists of between 40,000 and 50,000 people. And each village is divided into sub-clans like Ciro Anyuak that is divided into eight sub-clans namely Nyikwaar, Dekole, Ogilo, Nyium, Bet, Alil, Ojwaa and Omill. The thirteen clans of the Jowatong are listed here below according to their size, starting from the largest clan down to the smallest clan. The old Anyuak tradition put the fifteen Anyuak clans as follows:

1. Nyiye
2. Jowatong
3. Jomat-Cuaa –with twenty eight villages
4. Jowat-Maaro –with thirteen villages
5. Jowat-Naadhi –with eleven villages
6. Jowat-Naamo –with teen villages
7. Jowat- Muoongo –with seven villages
8. Jowat-Yada (Jo-po-Onguu) –with seven villages
9. Jowat-Luaalo –with four villages
10. Jowat-Kaanyo –with two villages
11. Jowat-Bitho (Jo-wathitho or Jo-watYuua –with five villages
12. Jowat-Jaango –with two villages
13. Jowat-Caamo –with 2 villages
14. Jowat-Maalo –with three villages
15. Jowat-Biedo –with three villages

Figure37, Source: Crazzolara, 1954, P. 393)

But the contemporary Anyuak historians and elders state that Anyuak people are divided into eleven modern clans. They are as follows:

1. Ciro Anyuak is sub-divided into eight sub-clans, which include Nyikwaar, Dekole, Ogilo, Nyium, Bet, Alil, Ojwaa and Omill. They are the inhabitants of Akobo land,

the place known today as Akobo Gedim; they drink from Dikony, Agwei and Kikon rivers. Akobo was the first districts established in Anyuak land in 1911. Due to much flood of the area, the town was moved in 1912 to a place called Agii, which is found on the other side of Dikony River. It is assumed that most of the sub-clans in Ciro Anyuak clan might have migrated from some of the below clans, their names could suggest that.

2. Nyikwaar –they inhabited Dikon, Borawiil, Arini, Agulbool and Borajanga in the northeast of Akobo town.

3. Dekole –they inhabited the old Akobo, Nyikaani on the eastern side of Akobo town; and Okalla, Tungdol and Agii on the northern sides of Akobo town. Some of the Dekole clans are the inhabitants of Chan and Abworo in the south west of Akobo town.

4. Ogilo -they are the inhabitants of Dimma and Othil on the southeast of Akobo town and have become known as sub-clans.

5. Oboo –They are inhabitants of Ciban, Owit, Buodo, Alawi and the whole of Agwei and Burmath areas, south and southeast of Akobo town. These are also been known as sub-clans.

6. Nyium –They are inhabitants of Pakang located northeast of Akobo town. Still some of the Nyium clans live down the Sobat River to Jekaw and areas between Nasir and Akobo. Anyuak traditions acknowledged that some of the Nyium clans are living in Nasir. It is most likely that these could be the Anyuak people who were taken slaves by Nuer Jikany. The fact of the matter remains to be traced.

7. Bet –the clan are inhabitants of Ogalo and part of Pakang forming two sub-clans.

8. Alil –This clan inhabited Alali, Wibura, Baba, Ojoki, Burtuor, Abuk, Nyikuola and Odee.

9. Ojwaa –the clan are inhabitants of Ogak, Omeda, Anyang, Dier and Debango in the extreme southeast of Akobo; and some people settled along the Agwei River and into the part of Obooth.

10. Obeer –the people from Obeer clan are inhabitants of Aluali, Kudobwor, Barabala and Bur Ojou areas south of Akobo town.

11. Omilla –The people of Omilla clan are found at Omilla, Odiek, Awal and Ojaalo sub-clan in the east of Akobo town.

Local Government Administration

In the administrative of South Sudan Government, Anyuak South Sudan is composed of eight clans (which are divided into sub-clans), four Payams and twenty-five Bomas as follows:

Balanda Boori and Balanda Bviri

Demography and Geographic

Originally, Balanda people called themselves Bori Luo; they are composed of two main groups namely Boor and Bviri. The Balanda Boor inhabited west Bhar El Ghazal, nevertheless, some of them are found in Western Equatoria. The Boor are divided into two main groups known as "The river people" or "Jo-Kunam" —who are closely knit; and "the hill people" or "Jo-Ugot" —who are composed of three loose and independent tribal groups called Fugaya (Fi-Juga), Afaranga and the Mbene. Some people believe that their ancestor is Boor or Bwor, who was the elder son of Nyikango but others say that their ancestor, Boor, was the elder brother of Nyikang. Originally, the Mabaan people say they separated from Shulluk in the Sobat River area, but other people believe that they have separated from Shulluk in Wau (Bhar el Ghazal).

According to the Shulluk, Bwor did not on get well with his cousin brother, Dak, and was granted his father's permission to remain behind with his uncles when Nyikang and his people decided to migrate to Malakal. Bwor established the Boor lineage by marrying a Bviri woman. The Boor used to live closely alongside their Bviri neighbours until the Azande invaded them in 1860s. In Bhar El Ghazal the Boor live in Mbili, Raffili, Tirga, Baziz and Taba. Moreover, Balanda Bviri inhabited mainly the areas in western Equatoria. In other words, we can say the Boor and Bviri inhabited the stretch of territory lying south of Wau town and northeast of Tambura town in western Equatoria. They live in Nagero, Koma, Bangazegino and Tambura. According to Tim McKulka (2012) the Jo Kunam migrated into the region first and the Jo Ugot followed them later, they settled in the same region. They believe their ancestor, Boor/Bwor, was the eldest son of Nyikang —the founder of Shilluk Kingdom.

Balanda are parts of the Luo groups and linked more closely to the Jur-Chol, Shilluk, Thuri, Acholi, Anyuak, Pari (Lokoro) and Jumjum. According to the 1983 census, the Boor and Bviri population numbering between 8,000 and 10,000 (1983: SIL); but according to the 2009 census, their population was numbering between 40,000 and 50,000.

In the south, the Boor people are bordered by Zande, in the east they are bordered by Bongo, and in the west and north they are bordered by Ndogo.

Interestingly enough, although both the Boor and Bviri live in the same region, yet they speak two different languages, nevertheless, they do understand each other. The Bviri speak Pivri language, while the Boor kept their Luo language and the origin of their traditional names. Interestingly, some south Sudanese people believed that Boor people have lost their Luo language and now different Mabaan groups speak different Nilotic dialects related to Shulluk, Anyuak, Dinka and Nuer. The balanda Bviri, are considered ethnic Bantu group found in western Equatoria.

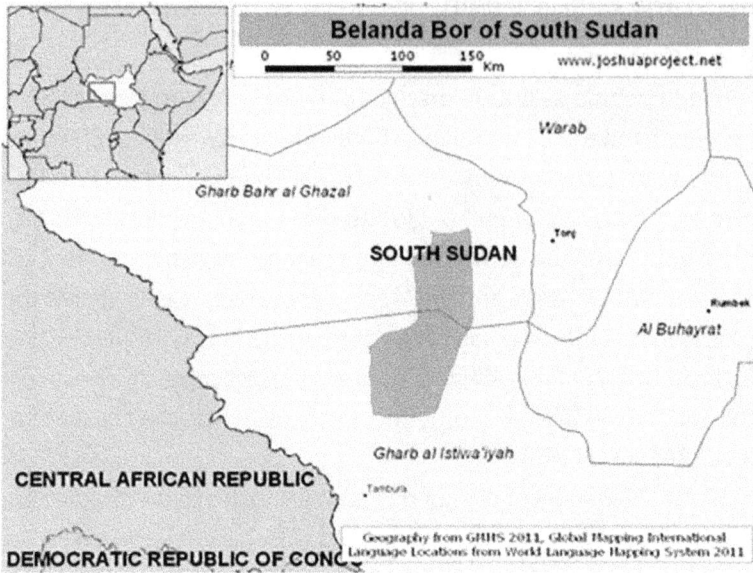

Source: Joshua Project
Figure: 36, Location of Balanda in South Sudan

Migration and Settlement

"The Shulluk element of the Balanda (Bor) have migrated southwards in successive waves. The last two waves: Mberidi and Fujiga must have broken away from the Luo during the last century, but the earlier waves may have moved southwards very much earlier, possibly even before Nyikang migrated from Bahr-el-Ghazal to the present Shulluk homeland (Malakal) in the sixteen century"

According to Crazzolara (1950), when the Luo people were in their place of origin, they were living in a number of independence groups. Each group could decide whether to remain in the motherland or to immigrate to other parts of the countries. Thus, the first group that separated from the main Luo groups were the Boor group; they were followed by the Jo-Luo Wau. They moved on their own way and risk. Boor, who is believed to be the elder son of Nyikang, headed them. They moved west first and then southwards and eventually settled on the Sue/Jur River in the region of the Raffili Rapids (Bhar El Ghazal). This small group of people called themselves "Ka-Boor" -that means descendants of Boor; which comprises an alien tribal group, Bviri.

The Jo-Luo Wau, who followed the footstep of the Boor people, came and settled further of Jur River and they named the Ka-Boor, we have seen above, "Belaane" or "Belanda". This tells us how the name of Luo Bori revolved.

This small group (Boor) continued to live independently alongside the Jo-Luo Wau, but sooner than later, the Government of Sudan ordered them to join the Jo-Luo Wau, under one chief or Ruoth for administrative purposes. The Boor people were not happy with the arrangement of the Sudan Government, so they preferred to move a long distance from the Jo-Luo Wau, to the south in order to re-unite their own group, the Boor.

According to Shilluk traditions, when Nyikang came to Bhar El Ghazal, he married a Bviri woman , and this woman gave birth to a son called Bwor or Bur. Shilluk tradition has it that Bwor did not get on well with his cousin brother, Daak. More importantly, the Shilluk tradition did not referred to Nyikang as leaving the land of Thuri with Bwor. Thus, he (Bwor) was granted his father's permission to remain behind with his uncles when Nyikang and his group decided to migrate. Nyikang and his groups migrated northward and from there to Malakal, in Upper Nile region. Bwor remain behind and established the Boor lineage by marrying a Bviri woman.

When their leader, Boor/Bwor died, his son Uthoo, became the chief (Rwot) of Ka-Boor. Let us not be confused when we read that Boor was the brother of Daak or Odaak. Boor was born to Nyikang from his first wife, Nyakaya, and Daak was born to Nyikang from his other wife a Bviri woman, married when Nyikang arrived the land of Rwot Dimo. This fit perfectly well according to Luo traditional laws, which says the two sons and daughters born of the same father but different mothers called themselves brother and sisters. Boor was born in the motherland at Lake Albert; while Daak was born in Bhar El Ghazal or "Wipac".

According to Jo-Luo Wau traditional history, a heavy quarrel broke out between Daak (OdaaK) and Uthoo; this quarrel was started by the women of the two brothers, Daak and Uthoo. Daak was said to have been supporting his wife. When the matter was brought before the clan elders, under Dimo and Nyikang, the elders found that Daak was wrong; he was supporting his wife without knowing the root cause of the quarrel. After thuaral examination of the case, the elders found the wife of Daak was mistaken. When Daak heard the pronouncement, he became disappointed with the decisions and decided to separate from the rest, he then moved southward. However, such a separation is against Collo/Shilluk traditional laws and cannot be taken into consideration.

Balanda Boor tradition tells us that before long, they separated with the Jo-Luo Wau. The separation followed a quarrel about the beautiful bead belonging presumably to the leader of the Jo Luo Wau. A baby from a family of Boor took the bead in its hand, then placed it in its mouth and swallowed. The owner of the bead, when he realised the boy has swallowed his bead, he requested his bead be returned at all cost; and the baby had to be killed in order to recover the bead.

The Jo-Luo Wau came and settled in Bhar El Ghazal. Here they split into two small groups: the Thuri and Bwodho broke away from the Jo-Luo but they continue to live peaceful with each other. The Thuri and Bwodho live independently but border on each other in north-westerly direction. It was said part that of the Thuri in turn broke away and went with Nyikang group into Shilluk County. They became known as Wuate Dimo-Thuri and still have a deep feeling of being a united and distinct from the group of Shilluk people (Crazzolara, 1950, p. 7).

However, according to Shilluk tradition, the separation between the Boor and Jo-Luo Wau did not happened because of beautiful bead, but rather the Boor decided to leave motherland when the situation had become precarious.

Balanda Boor are linked to their great ancestor called "Uthoo" who is a descendant of Dimo as a great grandfather. Crazzolara (1951) noted that the balanda boor know very little about their original migration that led them to western Bhar El Ghazal.

Today, when you ask any Boor elder, "Where have you originally immigrated from?" They will simply answer, "We have lived on the banks of the river Sue for many centuries". Hence, in their opinions the whole separation with the Jo Luo happened when the Jo-Luo Wau and Bongo crossed the river Jur to the eastern side —living balanda Boor in the western side of the Sue River. They also believe that before long, the Balanda boor crossed the River Sue and proceeded south some 150 miles and settled in the area called Kojali mountain range, which is located south of Wau Town.

Boor oral tradition tells us that when Balanda Boor settled at Kojali mountain range, they found today Balanda Bviri were already living in the southwest of the region; at that time the Balanda Bviri were known originally as "Gamba" —they acquired the name Balanda Bviri later on.

The two groups began to live side-by-side in most intimate relations, intermarrying freely. Consequently, both the Boor and Bviri became bilingual. Which means that that Boor speaks Luo language as well as Piviri language and vise versal. Anne Storch (2003) noted, "Unlike Luo-peaking groups, where bilingualism could have been sporadic for a long time and involved genetically related languages, the Balanda have developed into a strictly bilingual society where it was obligatory to speak at least one additional, generically unrelated language" (P.87).

After sometimes, the Bviri groups moved further west to Dem Zubeir and the north to Bussere-Mboro Road, subsequently separating from the Balanda Boor and forgetting the Luo language.

Interestingly, the Balanda Boor knows that their ancestors came from the 'toc" swamp in the north. They said, they arrived together with the Ko-Luo wau in the region of western Bhar El Ghazal. However, other informants say the Boor separated from the Luo before they reached the land of Bongo. Here, the Boor encountered strong opposition by the Bongo, who did not want them to settle in their territory.

The Bongo people were many in number and were very determined to keep the Boor out from the territory; but, as I made further research, I did not find details of the Boor invasion. Therefore, it is assumed that the Balanda Boor, being small in number, avoided the fight and went southward, looking for save haven. Northern Luo traditional history says, sooner than later the Boor went south, the Jo-Luo of Wau arrived in the territory, and fought the Bongo courageously, defeated them and snatched the land from Bongo and settled there.

Zande invasion of Boor

According to Northern Luo oral tradition, today Balanda Boor was originally called Beer-Laana –which means people of the mountains. The possibly symbolic relationship of Boor and Bviri lasted for at least two centuries. Thereafter, came enormous pressure from Zande, when they invaded Boor. At this time, there was wild spread of slave raids in the region; this led to the amalgamation of the groups. Thus, in 18 century the Zande, under chief Tambura, invaded Balanda Boor and Bviri and bought them under their rule. There is no much information about the invasion, due to lack of written documents. However, after the defeat, chief Tambura allowed the Boor to settle in various parts of Zande land. Moreover, the two groups (Boor and Bviri) continue to speak their languages, although they are under the rule of Zandi chief. Nevertheless, eventually they learned Zande language.

Crazzolara (1951) observed that after some years, the Sudan Government moved the Bviri north of the River Boo and far beyond. They went as far as river Sue/Jur, bordering the southern groups of the Abat Jo-Luo. Therefore, the three groups (boor, Bviri and Jo-Luo) speak their own language, yet they understand each other and intermarrying among themselves.

Northern Luo oral traditions reveal that some three to four centuries, Madi, Bangba and Barambo tribe had series of invasion of Balanda County at Ulelle. They drove some people before them and creating ethnic confusion on the west bank of the Nile. Following the Azande invasion, Moro were cut off from their relatives the Madi and Lugwara (Lugbwara).

As we have seen behind, the Balanda is composed of five main tribal groups, which include Kuunam, Fi-Juga, Mbene, Afarang-ga and Fi-Kuma. Each of these tribal groups, in turn are divided into clans as follows:

(1) Kauunam tribal group is divided into six clans, which include Baka, Fi-Nyi-Dumo, Fi-Nying-gen, Fi-yugo, Kobedi and Taban (Fi-Gem).

(2) Fi-Juga tribal group is divided into six clans, which include Ka-kongo, Fa-Gaayi, Fu-Fonyji, Fi-Reng-go, Fu-Ubyio (Fa-Bio), and Mbijo.

(3) Mbene tribal group is divided into eleven clans, which include Biling-go, Fi-Camayo, Fi-bilmogo, Fi-dao, Fi-Madi, Min-Imbara, Fi-Nyaro, Fi-Yang-go, Fi-Yugi, Kpono and Fi-Ngor.

(4) Afarang-ga tribal group is divided into four clans, which include Fi-Bong-gu, Fi-Mbirya, Fa-Yugi and Fa-Yio.

(5) Fi-Kuma tribal group is divided into six clans, which include Fi-Gwei, Fi-Bango, Fi-Ndogo, Fi-Juga, Fi-Bel and Fi-Jabo

Source: Crazzolara, 1951, p.115

Jo-Luo of Bhar El Ghazal (Jur Chol)

"The Luo Wau live around Wau, Tonj and Aweil in western Bahr el Ghazal, Warrap and Northern Bahr-el-Ghazal States, they are known as part of the larger Luo family which are made up of Acholi, Anyuak, Shilluk, Jumjum, Thuri, Mabaan and Pari (Lokoro) in South Sudan. They are also part of Luo family in Kenya, Uganda, Tanzania and the Republic of Congo. They speak dho or Jo-luo which is very close to Shilluk, Pari and Anyuak"
Source: Tim Mckulka, 2012

Demography and Geography

According to 1983 census the Jo-Luo Wau or Jur Chol population was numbering between 80,000 and 100,000 in the region. However, other informants like James Deng Dimo put the population between 100,000 and 900,000 (James Demg Dimo, 20....). The Dinka Bahr El Ghazal nick named the Jo-Luo Wau as Jur Chol —which means 'black stranger" -this is because the Jo-Luo were strangers to the Dinka. Today, many Jo-Luo of Wau do not want to be called with the name Jur Chol —because it sounds offensive to them. The Jo-Luo of South Sudan are the aboriginal of Wau town, Gaite, Kiango, Udici, Hana, Barmayen, Fongo, Aler up to Aweil. While the Jo-Luo in southeast of Wau Town are found in Waadlyiela, Barwoul, Pambili (Umbili), Aya, Kwajiena, Mapel, Barurood, Roc-Roc dong, Atida, Kangi up to Tonj Town (Thony) Their main towns include Wau, Mapel, Udici, Alel, Tonj, Barmayen, Kwajieno, Roc-Roc dong, Kangi, Kiango (Kayongo), Atido and Umbili.

Map of South Sudan Ju Luo Regions

Figure: 38, Location of Luo of Wau in South Sudan

According to Tim Mckulka (2012) the Luo traditionally believe they are the descendants of a legendary ancestor, Dimo, whose bothers Nyikang and Gilo left him behind in Bahr El Ghazal for Upper Nile region after a family feud and ended up founding the Shilluk Kingdom.

Jur (Beli and Modo)

Besides the main group of Jo-Luo of Wau, there is a distinct group of Jur known as Jur Beli and Jur Modo; this small group numbering approximately 100,000. The Jur Beli people are found in Bahr Gel, north of Cueibet, Wulu and Billing. Moverover, the Jur Modo people live in Mvolo, Bogri, Woko and Bahr Girindi near to Yirol. In addition, in between the Beli and the Modo there is a minor group called Lori and they are found in Woko and Gira. What remain mysteries up to the present day, "Are the Beli and Modo part of the Luo group? If so how did they split from the main Jur Chol?" This need further study to be conducted.

However, the Ku Modo are divided into two sub-sections (1) the Wira (which include the Yeri, Lesi and Dari) and (2) Nyamosa group which include part of Yeri, Bahr Gindi, Boyi and Kpelikiri. These sub-sections either mistakenly or deliberately counted as Moru clans. The Jur Beli and Jur Modo both speak languages that belong to the Baka, Bongo and Moru groups. In other words, the Beli and Modo speak Baka, Bongo and Moru and we do not know exactly which of the three, is their language. Alternatively, if Beli and Modo do not speak Luo, how can we call then Luo? Could we say, they have lost Luo language in the process of their migration and settlement? These are some of the information; we need to find out in my Second Edition of this book.

The Beli and Modo societies are organized in kinships, clans and families. The clans often come together when there are dances, marriage ceremony, funerals and funeral rituals as well as in hunting. The Beli and Mod removed their lower teeth at the age of 12, and if this is done to a young person, he or she is said to have joined the human race.

Soon after the removal of the lower teeth, girls and boys are taught their respective roles in the community. According to Jur Beli and Modo culture, like in Luo culture, one is expected to respect his/her in-laws; one should not use rude words or insulting in the presence of the in-laws. The young child must respect the older brothers and children by not looking straight into their eyes when the elder brothers or children are talking to him/her. If the young boy or girl looks straight into the eyes of elder boy, this is considered rude and sometimes subject to punishment.

There are no arranged marriages among the Beli and Modo; once the girl and the boy have reached an agreement (they are in love) the groom will ask his friends to accompany him to the bride's house. The tradition puts it plainly that they must arrive at bride's house before sunset. If they arrive, the bride's house after sunset, they might be fined, or not allowed to enter the compound. Since the girl is already in deep love with the groom, when the boys arrive, she well come the party and give them a seat.

The groom and his friends will spend the night there but he is not allowed to share a room with the bride. Thus, in the early hours of the morning, the groom and his friends will be seated in the warm morning sunlight for introduction.

Once the introduction is finished and the proposal is accepted by the bride's side (which they always do) the party is decorated with beads and flowers; for which they must carry home to show to the rest of the relatives that the engagement has been successful.

After one to two months following the introduction, customarily dowry must follows where some goats, sheep, hoes, money, bows and arrows are paid to the bride's father. How much the groom pays, will never be considered as the dowry is completed. After paying the dowry to the bride's father, the groom is left behind to work for a complete season in the in-laws' gardens and during this period he must build a new house for them (father and mother in-laws). At the end of the season, when all these have been accomplished, the relatives of the bride will then considered the dowry is completed and the boy can be released to his house. Now, it is worth mentioning that while the groom is working in the gardens of the bride's father, they are allowed to share a room with the bride. In cause of this, the groom often leaves the bride under pregnancy. When the boy reaches his home, he organises a big celebration and feasting. Until the ritual for "mouth opening" ceremony (Kpakopi) has been performed the groom not the bride cannot be seen eating in either in-laws' homes.

Migration and Settlement
According to Apai Gabriel (2005), the Luo was once of the biggest single tribes in Africa. Due to revolutionary movement, they broke up into sub-tribes. In the first movement all the Luo moved southwards following the River Nile until they reached Kenya. The Kenya Luo remained in Kenya and the Tanzania Luo proceeded to Tanzania, while the Northern Luo of South Sudan, returned to the then Sudan (PP. 2-3). The Northern Luo of South Sudan came from Kenya and settled first around Lake Albert, before proceeding to Thuri in Bhar el Ghazal (Wau).

The oral history of Jo-Luo of Bhar El Ghazal says they have come from directions between north and east, as we have discussed in the migration of the Balanda above. The Balanda also say their forefathers have come from the north, together with the Jo-Luo of Bahr el Ghazal. Their ancestor is Dimo, the brother of Nyikang, although Shilluk tradition does not speak of Dimo and Nyikang as brothers (Crazzolara, 1951, p. 116).

According to Jo-Luo of Bahr el Ghazal, Nyikang had two sons: Boor and Odak (or Daak). They say both Boor and his brother Odak had wives, but one day their wives started a quarrel. Shortly, Odak started supporting his wives without understand the root cause of the problem. When Boor saw Odak was supporting his wives, he even jumped into the quarrels and starting supporting his wives as well. Sub-sequent, the two brothers were in bitter discord too, and they came to blows.

When Nyikang and Dimo heard and saw the bitter fighting of the two brothers, they called both men and their wives and tried to resolve the case. As Nyikang and Dimo, plus other elders, examined the case, they found that the wives Odak were on the wrong side, and told informed him. The elders decided at the favour of Boor and his wives. This action angered and upset Odak that led him to make a decision to separate from the group and moved south (this is contrary to the Collo traditions and facts). However, Boor remained behind with his father Nyikang, and Boor's son called Uthoo or Othoo. Dimo, Nyikang and Boor (Uthoo) built a beautiful and spacious village around a big tree and they called Ongac. Their groups expanded quickly as did their cattle, goats and sheep. This event is most likely has happened in Thuri land, that was under the leadership of King Dimo, who became the mother-in-laws of Nyikang; because Nyikang married his daughter who has become the mother of Daak or Odak.

As we have seen above in the Balanda's traditional history, the Jo-Luo say they have separated from the Luo Bori or Balanda due to the pressure that was encountered from the Dinka, when they were still at Ongac or Wipac. The Dingka in the south and the Nuer in the north forced them, to leave Ongac (Wipac) and they moved westwards. Crazzolara (1951) observed that it was because of the grazing land that grave differences arose between the newly arrived and the Dinka and Nuer, who were already there. He went on to say that, it was of this movement that the Boor and the Jo-Luo separated. As heads of the groups at that time, Dimo with his children "who were Jo-Luo clans", Nyikang with his children "who were the Shilluk people", Boor and Othoo with their children "who were the Balanda" (P. 116).

The Jo-Luo of Bhar El Ghazal, like the Balanda, they know very little about the original migration that led them to their present country or region. According to Jo-Luo traditions, there were three groups of people on move: (1) the people of Dimo, (2) the people of Nyikang and (3) the people of Boor and his son Othoo. As they move, when they come to stop for the nights, the wives of Dimo and Nyikang used to prepared food for the entire groups. They prepared food with great care; the wives first pound the corn in mortars and neatly sifting the flour. When the wives finished cooking food, the food was appetizing but the problem was that, it always takes them a long time to prepare meals for the three groups. On the other, hand the wives of Othoo also busy-preparing meals. Othoo's wives used simply to grand the corns on stones and the flour is instantly ready for food. The meals of Othoo's wives usually come out first and the meals of Nyikang and Dimo's wives come out last. Every day, people have to wait for the food of Nyikang and Dimo's wives to come out before they eat, because they want to eat the meals together.

However, Othoo and his groups were not happy with the slowness of Nyikang and Dimo's wives. They were tired of late preparation of meals; one day, after the groups have eaten meals, Othoo's group sworn not to continue moving with the other groups –they then took their belongings and moved on ahead. This caused the Balanda to separate from Jo-Luo and Nyikang's people for good. As Crazzolara (1951) noted, it happened that the Boor separated from the Jo-Luo for good (P. 117). Othoo's group went and reached the Sue River (Jur River) during the dry season and crossed it. They then built their villages on the left bank of the Jur River –this is the place they occupy up to today; they even built alongside Wau Tobura Road stretching some 300 km.

When the Balanda had left the group, the Jo-Luo (Jur Chol) soon followed them. The Jo Luo left Nyikang and his brother Dimo at Wipac. The Jo-Luo moved westwards, they came to Bongo territory; this is where they found the Balanda. The northern Luo traditional history tells us that, as soon as the Jo-Luo arrived Bongo Territory, the Balanda moved immediately southwards. The Jo-Luo of Bhar El Ghazal then fought the Bongo courageously, defeated them, snatched the land from them, and settled there. They called their new place of settlement as the "Ka-Boor".

Dimo and Nyikang, who had remained behind, very soon separated also, because of a beautiful and precious bead of Nyikang, which one of the Dimo's babies had swallowed. Nyikang requested its return even at the cost of ripping open the belly of the baby. (For more information read Chapter 4: *Tiko separate Dimo and Nyikang*). When the bead was taken from the belly of the baby and handed to Nyikang, Dimo and his group parted from Nyikang and followed Othoo (Balanda). At this time, the Balanda had already settled on the left bank of Jo-Luo River (Jur River). Dimo and his group went and arrived at Abilek, a village about 50 miles northeast of the present Tonj Town. Here, they settled and stayed for many years. Dimo set up Kingdom similar to that of Nyikang and Anyuak (Apai Gabriel, 2005, P.5). This place became popular as a burial place for all descendants of Dimo, the father of the Jo-Luo. Due to lack of written documents about the Jo-Luo of Bhar el Ghazal, we do not know the names of the kings that followed Dimo; but what we know, is that the last King (*Jah*) of Abilek (Tonj) was Jah (King) Gir Ajangjok. According to writers like Apai Gabriel (2005) King Gir Ajangjok was killed by Dinka when they raided his Kingdom at Abilek. Dinka took his copper wrist ring that symbolised his Majesty. After the death of King Gir Ajangjok, no other Kings were crowned because of battles with Dinka who continue to attack the position of the Jo-Luo from time to time. Eventually, Jo-Luo scattered within Bhar El Ghazal Region and live in small villages according to the sections or clans. Each settlement had chief (*ruoth*) to govern the people.

The people of dead King Gir Ajangjok the left Abilek and moved southward, they came and settled at Ting-Aguar, Nyiel and Agota areas. When the survived Jo-Luo settled at Aguar, Nyiel and Agota, they organised themselves, captured some of the Dinka's girls and took them as captives. When the Dinka learned that their girls have been taken hostages, because of the copper ring of late King Gir Ajangjok, they gave back the copper ring to the Jo-Luo and their girls were released from prison (Apai Gabriel, 2005).

The Jo-Luo oral history tells us that Dimo came with a large herd of cattle, soon, Dinka started to raid tem. Luo tradition say there were constant attacks by the Dinka on the

Dimo group, which eventually forced them to move on, after staying many years in the region. If it were not the invasion of the Dinka, presumably Dimo and his group would have settle here permanently.

Nevertheless, oral history of Jo-Luo of Bhar El Ghazal reveals that they have tried to settle at various times and places, but were unable to drive away the actual occupants, like the Dinka in Tonj, and had to give way in each case and move further to look for a better opportunity. Therefore, when the Dimo group could not resist the attacks from the Dinka, they moved South West and reached a place called Guo, which is just ten miles away from Tonj, and there, they settled.

Here (at Guo), it was the strong tribe of the Bongo and the Azande that became hostile to the group of Dimo. The Jo-Luo could not resist them, as they were weak, so the Bongo and Azande forced them to move again. They went westwards and reached Tonj; they could not settle in Tonj but proceeded and reached river Jur. This time, it was a raining season, Jur River was full and the Jo-Luo could not cross it. Subsequently, they settled southwards along the river, mainly in the place called Abul that is east of river Bussere. This is the only place that the Jo-Luo had a test of peace of mind; they remained undisturbed for many years.

This has offered an opportunity for the Jo-Luo to grow and multiply. When the number of the Jo-Luo became bigger and bigger, they started to quarrel among them and this has resulted into split of the group into smaller groups of hunters; and these small groups shortly developed misunderstanding among the Dimno's groups.

Thus, it became evident that they should separate; subsequently, some of the groups separated and went eastwards, while other groups went westwards. The groups that went westwards, came to Sue River, crossed it, and continue moving northward directions.

The clans of Pi-Nyileo, Pi-Barjwok, Pi-Gwero, Pi-Kabo, Pi-Nyidwai, Ya-Gundwong and Ya-Bado become the inhabitants of Nyiduk, Kwakango, Acano (which are close to Mbili Catholic Church) and Diem Akwil —on the Wau-Meshra road in approximately 1860. The Dingka, of Abwok/Meshra with their hostilities made frequent raids against Abat-Dedom group, in search of hoes (kwer/kweri). This people were blacksmiths they would mold hoes from iron they mined. Following these frequent attackas, the small group called Akotdit from Kwa-Nyileo and Pi-Barjwol escaped to Kwac (Kayang-go); other small group called Akotdit from Kwadwongo escaped to Alur; while the group of Remtong took refugee among the Dinka beyond Tonj town; the small group called Pi-coo took refuge also among the Dingka (Kwac). Thus, when an Arab man, from Dongola came to Abat-Dedom region, he found few people left behind. The news of the coming of a Dongolawi spread very quickly, and people thought there was peace in the region, so some of them returned to the sides of their ancestors, but others did not.

Although there were divisions among the Jo-Luo, nevertheless, their cultural feeling and identity remained strong. As a result, the Jo-Luo could not continue to separate indefinitely; soon they united. They built in the west of Jur River, which is northwest of present Wau town, and some built in the east of present Wau town. The group, who built in the western bank of Jur River, became known as "Ya-Kwac"; and the group who built in the east of Wau town and Jur River as far as Tonj, became known as "Ya Abat". Apai Gabriel (2005) pointed out that some of the Jo-Luo (Bodho) settled around Jur River (Abat), while others settled west of Wau along the River Kuango up to Achana and beyond (P.3).

The Jo-Luo groups who settled in the western Wau along the River Kuango and Achana were raided by the Dinka Apuk and Dinka Kuac and many of the Jo-Luo were assimilated into Dinka culture. The Dinka then called the Jo-Luo of Bhar El Ghazal *"Jur-Chol"*, which literally means black strangers or black foreigners. They have given them this name to distinguish them (Jo-Luo) from other ethnic tribes who has light skin colour such as Bantu tribes and the Arabs (whom they called *"Jur-mathiang"* (which means stranger of the horse). However, the Jo-Luo do not like the name "Jur-Chol", they argue rightly that no one is a stranger in his/her own home. According to Apai Gabriel (2005) today, the Jo-Luo of Bgar El Ghazal are divided into two main groups namely Abannen and Bodhen (Chat). The Abannen are the people who live along the Jo-Luo River (Jur River) from the west to Alur in the Far East. The Bodhen (of Bodho) are people who live along the Kuango River to Bar Mayen, Abul, Areyo, and Acaana down to Chelkou in the west (P.11).

As we shall see below, the Ya-Kwac and Ya-Abat are divided geographically because of Sudan politics. In addition, each group in turn sub-divided into smaller groups under their local chiefs or Rwoth/Rwot.

It is generally believe that the Thuri (Shatt) and the Bwodho (Dembo) were part of the Lo-Luo. They separated from the main Dimo's group, when people divided into small groups of hunters. The Thuri and Bwodho then formed independent groups. However, these needs to be explore more.

Having discussed the migration of the Jo-Luo of Bhar-el-Ghazal, let us now examine how the Ya-Abat (Logo) and Ya-Kwac (Amac) reconstructed after their political and geographical divisions. Let us first examine Ya-Abat and later on, we can examine Ya-Kwac.

YA-ABAT (Logo)

Historically, Ya-Abat is made up of six political groups, which include (1) Abat-Dedom, (2) Akuer, (3) Athiro or Uthiro, (4) Alur, (5) Yau, and (6) Pinigiem.

For us to understand the political divisions and development of these six groups, it would be better to open them one by one.

(a) *Abat-Dedom clan* –the word Abat Dedom, originated from the arm of antelope (bat lay) used to been offered to Obindak (the King). Jo-Luo oral traditions say the Abat-Dedom groups are the descendants of Obindak. What remained unclear to many of us is that "Who is the father and grandfather of Obindak?" Oral history of Jo-Luo say Obindak had two wives; the first wife gave birth to five sons called Remtong, Kwabor, Burjwok, Nyileo and Coo respectively. The second wife also gave birth to four sons called Larpiny, Kabo, Nyidwai and Gundwong respectively. Jo-Luo oral history tells us that there was a man called Bado, from Nuer, this man came and joined the children of Obindak, eventually creating the ten members of Abat Demo and each became head of a clan group. When Obindak became old, and before he died, he appointed Kwabor, a son from the first wife, to be his successor (King or Jaa/Jago); and he appointed Kabo, a son from the second wife, to be a sub-chief (Rwoth/Rwot). At that time, sub-chief was to be the leader of war. Thus, Kabo was considered the "spears of the fathers" or *"Tong-weno"* armed with which he led his groups in the wars against the Bongo. This *tong-weno* has ever since been kept by the Barjwok clan (Crazzolara, 1951, P. 118).

(b) *Akuer clan* received their name from an event that happened during the separation at Abul –on the Jur River. Jur Chol oral traditional history says the ancestors of the *Akuer* refused (okwero/kwero) to help Abat. And this why they have been given this name Akuer because they refused to help the grandfather. The word "kuero" in Luo means, "refuse". They then settled between the rivers Bussere and Jur River (Jo-Luo River) –and they called the place "Logo" or "Loka" which means beyond the water. As time passed-by, they formed two clans called Ya-Gam and Ya-Pinyikuawa. The group of Ya-Gam are divided into fourteen sub-clans that include (1) Oling (2) Onyongo (3) Ocan (4) Balogo (5) Ocundit (6) Lao (7) Penyi (8) Ogudo- Ocu (9) Nyiboo (10) Aywek (11) Yamo (12) Odero (13) Goi and (14) Akaa. While the group of Ya-Pinyikuawa are divided into nine sub-clans, which include (1) Angwec (2) Demin (3) Okel (4) Kwol (5) Wol-Okel (6) Pi-Nyidimo (7) Pi-Omenyo (8) Pi-Omenyo and (9) Pi-Nyiwaro.

(c) *Athiro or Uthiro Clan* This clan draw their name from the grass used for plaiting baskets and this grass in Luo language called *"thiro"*. This group also found their name after separation from Abul. The Athiro group, led by a man called Demau, went with his group and built near a pond where many "Thiro grasses" were grown. The women used these grasses to plait their baskets. When other Jo-Luo saw this group used many *thiro* grasses, they called them "Ya-Athiro". The people of Ya-Athiro live along the Tonj-Wau Rroad about 26 miles from the today Wau Town. Athiro traditional history tells us that when they settled in their new land, two women called Alla and Angomo joined them. Not only these two women, there are other persons which include Ogwo, Akot, Royo (with his two sons called Bono and Otor), Ywo, Othero and Oriemo with their people joined the Athiro clan. The Golo, the Pi-Ojieno and Pi-Nadho people also came and joined Athiro group. The amalgamation of these people, into Athiro, added greatly to the numbers of the people (Athiro). The Jo-Luo oral history reveals that all these names later considered as heads of clan groups. Interestingly, Athiro consists of four main clans called (1) Golo-Pi-Kir, (2) Golo-Pi-Julo, (3) Pi-Ojieno and (4) Pi-Nadho. Thus, Golo-Pi-Kir clan consists of three sub-clans namely Robo, Bwolo and Ojwana. The Golo-Pi-Julo clan consists of three sub-clans called Julo, Nyigilo and Kwot. The Pi-Ojieno clan consists of seven sub-clans called (1) Ojieno (2) Ogwak (3) Awet-Anywon (4) Otho (5) Ojango (6) Demo and (7) Dor. The Pi-Nadho (descendants of the Nuer man) consists of five sub-clans that include (1) Onadho (2) Ocan (3) Opur (4) Anemo and (5) Tingo.

(d) *Alur* When the Jo-Luo separated at Abul, a man called Ngor took his group, they went eastwards into the direction they have come before. The people who were left in Abul, when they saw this group going eastwards, they started saying, *"Kej-ye ge luur piny kedo* –which means they are going back with big dust at their back- *luur/luri* in Luo language is dust. Hence, they were then given the name Alur. This group called Alur later on divided into eight sub-clans. These include(1) Ngor (2) Twol (3) Demau (4) Malek (5) Ocero (6) Nyikwawa (7) Oyaa {who are from the Nuer Pi-Yaa} and (8) Ojango {who are from the Nuer Pi-Ojano}

(e) *Yau* -when people separated at Abul, a man called Nyiywo and his brother Othero, took his group to Nyikijo, and changed the name Nyikijo to *Yau*. The place now is called Yau –which is about fourteen miles from Wau Town. There is no specific meaning to the name *Yau*. When the group of Nyiywo settled in the eastern side of Wau Town, it is said that some two groups came and joined them. These groups include the people of Nyikir (or Pi-Nyikir) and Demau (with his two sons Nyiywo and Ongwec). Before long, another group of Bodo came from Kayang-go (Kwac) and joined Yau clan. Yau, like the Athiro, with the amalgamation of these people, into Yau, added greatly to the numbers of the people of Yau.

(f) *Pi*-Nyiem - In my research I did not find any Luo traditional history that tells me who was the leader of this group called Pi-Nyiem. We are forced to assume that they have also separated from the main Jo-Luo group at Abul. What we know is that the Pi-Nyiem clan is divided into eight sub-clans as follows; (1) Kwello (2) Pi-Laa (3) Pi-Rwoth (4) Pi-Nyithindi (5) Pi-Nyikuro (6) Demau (7) Pi-Nyidol and (8) Pi-Obal

YA-KWAC (Amac)

According to Fr. S. Santandrea the Ya-Kwac political group is divided into east and west. He called the Jo-Luo on the western part of the region as Ya-Kwac; and he called the Jo-Luo on the eastern part of the region as Ya-Gony –however, these are also known as the two main clans of Ya-Kwac.

The two groups of the Jo-Luo are each headed by Jaa or Jago (King). Due to lack of written information, I am unable to provide the names of the *Jaa* or *Jago*; but I hope to include their names in my Second Edition of this book. However, the western Jo-Luo (Ya-Kwac) is divided into two clans, which are Kwac or Amac and Logo clans.

According to Crazzolara (1951) the two groups of the Jo Luo (the Ya-kwac and Ya-Gony) have been divided into sub-clans. Thus, Ya-Kwac is divided into two sub-clans namely Amac (or Kwac) and Logo clans. Crazzolara (1951) observed that the Amac (Kwac) sub-clan, in turn, has been divided into fourteen lineages as follows: (1) Aciek (2) Acyiek (3) Buno (4) Nyiwara (5) Pabur (6) Pejuadi (7) Pinyiburo (8) Pukuno (9) Pukwai (10) Pumol (11) Pungor (12) Purama (13) Puyany and (14) Tubo. Crazzolara (1951) went on to say that the Logo clan has been divided into twelve sub-clans as follows: (1) Kwelo (2) Nipadi (3) Nyimei (4) Nyipwoc (6) Pajol (7) Perpyiou (8) Pinyikwai (9) Pinyiwil (10) Pujulu (11) Punyano and (12) Ya-Thuri (Crazzolara, 1951, pp. 120-121).

According to Apai Gabriel (2005) the Jo-Luo tribal sections or clans are divided into two major sections/clans namely Amac clan and Logo clan and these clans are in turn divided into many small subjections with specific names and areas they occupy as shown below (Amac clan consuists of 24 subclans and Logo Clans consists of 20 subclans). First let us see the composition of the two main clans with their subclans and after that this we see the people and the areas where they live:

Amac Clans	Logo Clans
1. Nyiwara	1.0 Pinywil
2. Pumol	2.0 Thure
3. Pukuai	3.0 Puthudhi
4. Aduollum	4.0 Pajol
5. Tubo	5.0 Bungo
6. Pujuadhi	6.0 Akaa
7. Aciik	7.0 Kuela
8. Payang	8.0 Pugol
9. Pingor	9.0 Nyipuoc
10. Pabur	10.0 Pibuolo
11. Puthudhi	11.0 Nyipapo
12. Putong	12.0 Pabal
13. Pinkec	13.0 Pingeyo
14. Panluou	14.0 Papieu
15. Pukuno	15.0 Pujieth
16. Nyimai	16.0 Pinybelo
17. Filer	17.0 Pijulo
18. Pingodho	18.0 Pacam
19. Pukono	19.0 Pirama
20. Pinyburo	20.0 Pinykuai
21. Pijua	
22. Atongo	
23. Puding	
24. Puduno	

Source: (Apai Gabriel, 2005, P. 101)

Jo-Luo People tribal sections and areas:-

1.0 Pingor	-Alur
2.0 Athiro/Uthiro	-Athiro
3.0 Redom	-Abat
4.0 Pinigiem	-Pinigiem
5.0 Yau	-Gony
6.0 Ahuer	-Gony
7.0 Abiim	-Amac
8.0 Kuac (Minor)	-Logo
9.0 Kuac (Major)	-Logo
10.0 Thure	-Yabul
11.0 Bodho	-Bodho
12.0 Demo	-Demo

Source: (Apai Gabriel, 2005, P.12).

He went on to say that the above sections are in turn composed of many small clans to form allieds or divisions. Now let us compare the various sections:

Section of Pingor is composed of Piechero, Pinyikuawa and Piya

Section of Piechero is composed of Pajango, Panmachar, Panmakur and Pakan

Section of Pinyikuawa is composed of Pinyiyuo, Pibuolo, Puguak, Pakano, Demou, Utung and Piyua; all these clans have migrated from Alm to Logo some centuries ago. The Jo-Luo oral history tells us that these people are from Gir descendant, except for the Pakano people who are adopted nationals of Puguak. Traditionally, we can say that all the people from the above sections could trace their ancestor back to Utango —when boasting.

Section of Piya is composed of composed of Pinigilo, Pinyikumo and Puding. According to oral history of the Jo-Luo of Bhasr El Ghazal, the four sub-clans in the section of Piya, they are adopted nationals of Alm because Nyigilo came first from Amac and settled in Alm. The oral history tells us that Ayaga, the grand-daughter of Nyikumo was given to Nyigilo as genetic husband. Subsequently, the children he produced with Ayaga are known as Pinyikumo, while the children he produced with other wives are known as Pinyigilo. Uding, his maternal cousin, who came and settled with him (Nyigilo) in Alm, and his (Uding) descendants are now known as Puding clan. Therefore, all the clans in the section of Piya, when they are boasting, they refer to Unhadho of Piya.

Today, we have discovered that the Jo-Luo nationals are not solely the Luo descendants because a good number of the people from different tribes immigrated to Jo-Luo land —forming new clans and sections (Apai Gabriel, 2005)

Installation of the King (Jaa or Jago)

In the past when there were no place to protect people, the Jo-Luo of Bhar El Ghazal used to live in protective villages fenced with thorns. The chief of the village usually has his house built in the middle of the village. Homes in a village are built according to families and each family had a homestead and a footpath passing by the other houses. Settlement was based around family (*kith*) and kin groups to make mobilisation easy during the time of emergency –that is when there is a call to arms for collective protection. But today, because the South Sudan Government is there to protect people, the people felt secure and they protective settlements are abandoned together with the homesteads (Apai Gabriel 2005, P. 6).

As regards to the installation of the King in the past and present, there is little known about it. However, in this book, I would like to know and narrate step by step how the king is been selected or nominated through installation and ceremony in the Jo-luo society but we lack the information. I could not do this due to lack of written documents and unavailability of Jo-Luo elders to tell me more about the Jur Chol group. However, we know that at the installation of the new king and sub-chief, the head of the Dedom Luo group used to come and preside over the ceremony. A skin of hartebeest (thil or til) was the official dress of a king (Jaa) and a skin of teltel, dikdik, yin-nyikwol or what the Acholi called Amur) of a sub-chief (Rwoth/Rwot).

Jumjum

Demography and Geography

The Jumjum people live in Northern Upper Nile Province, in South Sudan; they occupied the areas along River Jujum, in the places called Jebel Tunga, Terta and Wadega. One of my reporter said that Jumjum people have no town or capital; the Jumjum in the region are scattered everywhere in the bush; they do not live in towns. However, the Jumjum has small town called Mayak. The informer further reported that in the region of Upper Nile, Jumjum people are perceived as famous thieves among the tribes of the Blue Nile. Many writers, **like.......** refer to the Jumjum as Nilotic people rather than one of the Luo groups. In my research for the Northern Luo of South Sudan, I discovered that Jumjum are part of the larger Northern Luo groups that include Acholi, Anyuak, Balanda, Jur-Chol, Mabaan, Shilluk, Thuri (Shatt) and Pari (Lokoro). Before, this, I thought Jumjum were either Dinka or Nuer people.

Surprisingly, Jumjum like the Jo-Luo Wau, know very little about their origin, migration and settlement in their current place. Thus, this resulted in having less written documents about them. Worst still, the Government of South Sudan (GoSS) even made no mention of the Jumjum tribe in her famous website called Gurtong.

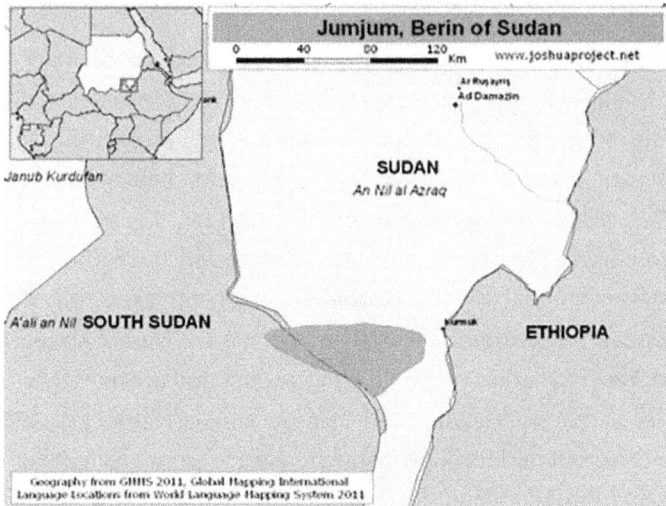

Source: Joshua Project
Figure: 39, Location of Jumjum in South Sudan

However, in this book, I am considering Jumjum as part of the Luo groups, because of the findings of some writers, like Bethany World Prayer Centre (1997) who argue, "The Jumjum are sub-group of the larger race called the Shilluk. Their original homeland is believed to be somewhere east of the African Great Lakes. When the population of the Shilluk and the sizes of their herds increased, they immigrated northward, reaching their present territory in the late fifteen century. As they settled into the region, they were often raided for slave traders. Although slave stations were established in many areas, the Shilluk were well organised and able to deter the raiders; nevertheless, due to this slave trade and constant attacks, the life of the Shilluk were greatly affected" (p.1). Thus from this simple fact we could easily deduce that the Jumjum separated from the Shilluk during these invasions and formed their independent group.

The Jumjum live around the Upper Nile valley, the area is flat with some rocky hills. They reside at the foot of these hills. In my research, I was a bit confused about the real name of Jumjum people. After reading some articles, I realised the people are known in four names —but still they are the same people. They are known as Jumjum, Berin, Olga and Wadega people.

When the Jumjum settled in their area, they have been raided by the Dink, Nuer and Shilluk (from where they have separated) —raiding it for slaves.

As discussed above, Jumjum is an ethnic group in South Sudan; they are approximately numbering between 50,000-80,000. They are closely related to the Mabaan and Burun people. However, many people do not consider them separate from Burun, they think, they are one. Moreover, we think they are closely related, because geographically and linguistically they are one. According to the Concordis International (2010) in the year, 2007 groups of Uduk and Jumjum entered and settled in Mabaan.

Jumjum people are quite tall, thin and long legged; they have broad noses, thin lips and frizzy hair. However, there are the typical characteristics of the Nilotic people. They also have lighter skin colour than most of the Nilotic people. Majority of men have long heads, while majority of women have rounded heads.

Interestingly, the Jumjum homesteads consist of two or three huts, a granary and several huts for sheltering animals. The huts are round with mud walls and thatched grass roofs. Each of the Jumjum village has a number of homesteads separated from each other by about one hundred yards. Each settlement represents one family's lineage and headed by a chief. The chief settles disputes that arise among the villagers and maintain order in his settlement.

In Jumjum tribe, polygamy is accepted and practiced, thus a man can have two or three wives (some even more), each wife has her own hut where she raises her children and conducts all other activities. To some African tribes, the birth of twins is considered a curse to the family members, contrarily, to the Luo group in general, and to Jumjum in particular, birth of twins is considered a blessing from God –they are children of Juok (God).

Jumjum people can easily be distinguished from neighbouring groups through the clothing and tribal markings worn by them. Both men and women do not wear much clothing, the men wear bracelets of ivory or wood on their wrists or upper arms; the women wear long dress with cloth covering their heads.

There is common practice of removal of the lower teeth by Jumjum young people, according to traditional laws, the must remove six of their lower teeth when they are below eighteen years of age. Traditional laws only forbid young people from royal family not to remove their lower teeth. There is no reason given why the young people from royal family are not allowed to remove their lower teeth. Both men and women have three to five rows of dots or scars on their foreheads, this is an indication of tribal identity. People without such dots or scars on their faces, are considered straight away as foreigners.

Mabaan

Demography and Geography

The Mabaan people are a Nilotic people; they are part of the Luo groups that is said to have separated from the Shilluk in the Sobat River. Their separation from Shilluk was not due to fights or quarrels but rather due to the growth of their population and herds. Consequently, the land at Sobat River was not enough to accommodate the growing populations and herds. They then migrated northward and settled in the plains, their current territory of today, which is between the Nile east of Renk, up to the Ethiopia Highlands. Seligman

C.G & Brenda Z. Seligman (1932) noted that this territory was formerly known as central and southern Kordofan. It lies between the White and Blue Niles, consisting of a flat open plain diversified by rocky hills. In the wet season the plain is covered with vegetation, in the dry season water is scare, water holes are far apart and south of Goz Abu Guma it is even harder to find water. The stone implements on these hills are found date back to immemorial antiquity, to the Old Stone Age. The population in the hills between the White and Blue Niles are as much an area of linguistic confusion. (P.413). They settled here around late fifteen century and today they are found near the Yabus rine, that runs through Mabaan County.

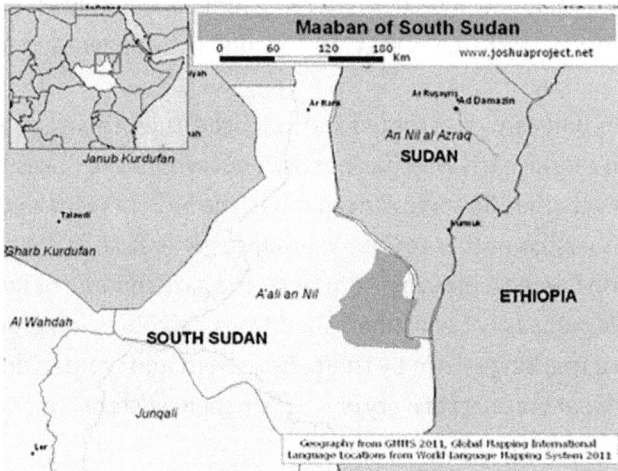

Source: Joshua Project
Figure: 40, Location of Mabaan in South Sudan

The people who live between White and Blue Niles are of a mix population, they are said less arabized Negroids. They speak their own dialect, which is non-Arabic languages, and traditionally they are pagan. Although these people speak Arabic and some of them profess Islam, many of them also speak their own language; they still keep customs, which they admit, came from their pagan ancestors. In many men of Mabaan breasts are very well developed. Women make cicatrices on the stomach, and both sexes practice circumcision. Mabaan are also tall people measuring between 1.80m and 1.92m, they have broader noses than they are long (Seligman C.G. & Brenda Z. Seligman, 1932).

According to Connordis (2012), Mabaan County borders four counties in Upper Nile States: Renk County to the North, Melut and Baliet counties to the west, and Longechuk County to the south. Blue Nile State of the Republic of Sudan borders Mabaan from the east.

Mabaan County possesses huge grazing land and water for livestock. Fellata pastoralists from the Blue Nile State annually migrated into the County to graze their cattle on these pastures and water during dry season. They are said to have peaceful co-existence with the local communities, nevertheless, periodically they fight with the farmers (Mabaano) over grazing land and competition over resources.

Thus, in June 2012, conference was held to address the problems between the Fallata pastoralists and indigenous community. The conference aimed to address fighting among them as well as refugee populations and issues in the areas. Both communities expressed a desire to create a mechanism for dealing with these incidents and strengthen relations between the groups.

However, with the influx of refugees from Blue Nile State into Mabaan County, tensions restated to rise over limited resources. The refugee communities want shelters and money for their living. They have to build shelters, sell poles and charcoal for making a living. The forest serves the same purposes for the host communities –this has resulted in noticeable deforestation of the areas. Mabaan initially requested that refugees only cut certain trees and to try to use branches for firewood instead of the destroying the entire plant. However, the refugees failed to stick to the agreement and in July 2012, Minister of forest passed order restricting the cutting of all tree (Denish Demining Group, 2013). Surprisingly, refugees even failed to obey the order of the Minister of Forestry. They continued construction of shelters from natural materials relying on grass for roofs. When the indigenous people realised that, refugees do not listen to any body, even to the Government authority, they took tough measures by taking away the tools from the hands of refugees if found in the bush and in one instance they went as far as burning the grass so that refugees could not cut it.

The South Sudan independence on 9th July 2011 has created suspicions and fears on the Mabaan people as well as on Fallatta who come periodically from Blue Nile State. On one hand the pastoralists from Blue Nile state fear they will not be given access to traditional grazing lands, that security forces will not protect them by new South Sudan Law. On the other hand, the local communities fear that the Pastoralists might now resort to carrying weapons and once disputes break up, they will start killing of people, animals as well as cutting and burning trees. Some even fear that pastoralists are acting on behalf of the Government of Sudan in obtaining information about South Sudan.

As Concordis (2012) noted, *"The militarization of the border by both Governments has inten-sified these problems, creating threats to the security of people in the borders areas. Insurgencies on both sides of the border had made the areas even less secure and caused tensions between cross-bor-der communities who may be perceived to be aligned with one side or another. Conflict could easily break out due to these tensions"* (P.2)

Mabaan consists of five Payams with population numbering 50,000 (SIL, 1987); however, some informants of Mabaan argued that now with the growing of population since 1987 the population should be between 60,000-70,000.

As oppose to other Luo groups, Mabaan societies is organised into matrilineal clans, while in other Luo groups like Shilluk, Anyuak, Acholi, Pari, Balanda and Jur Chol, are organised into patrilineal clans. Denish Demining Group (2013) noted that Mabaan society is more loosely structured than some of the Luo groups. Scattered patterns of family sett-lement are common with married couples, sometimes building houses outside of family compounds. Although there is a system of village chiefs, this is not hereditary position and is based on community selections, which is usually based on status measured in livestock or number of children. Worst still, the power of the chief seems limited and the ability to make binding decisions is restricted both by the geographical spread of the community and the lack of clear election procedures that provide for broad legitimacy.

However, Mabaan consists of several independent groups each speak different dialects closely to Shilluk, Anyuak, Dinka and Nuer.

The Mabaan have two important annual events, the first one is the feast commonly known as "Kornga" which is performed usually in October, when communities confess their wrongdoings for the year, and ask God for forgiveness. They go out early in the morning to a nearby river or stream to wash their bodies from the bad things they have done in the past one year. As they are washing their bodies, they appeal at the same time for tolerance and good health for both the people and the animals. After this, the people return home, and there they slaughtered animals (goats, sheep) and drink sorghum beer and will be followed by dance or "dukka-conkon". Traditional history of Mabaan tells us that, when they people are eating, drinking and dancing the Mabaan women and men put on the best clothes and decorate themselves with beads (burngo or bu-juok).

The second event is the harvest feast (gatti) which is usually held in December of every year. This is the month when matured boys and girls are preparing for marriage, animals are slaughtered, food and beer served during the gathering. At this time both boys and girls appear in their smartest dress 9apparel), wearing necklaces or linyan (Tim Mckulka, 2012).

The Mabaan like the Jumjum we have discussed in this chapter P. 122??), are been known with different names such as Barga, Gura, Meban, South burun, Tonko, Tungan and Ulu.

Pari (Lokoro)

Demography and Geography

The Pari are a Nilotic people, they are part of the Luo groups. Oral tradition has it that all Luo groups used to live together at *Wi-Pac* somewhere in eastern Bhar El Ghazal. They then dispersed because of the quarrel between the two brothers: Nyikang and Dimo. After separation from Wi-Pac, it said, there was another story of fighting between two brothers, Uthienho and Giilo, sons of Rwoth Eno, in which Uthienho killed Giilo because of jealousy.

Dimo, who became the founder of Pugeri Village, led the first Luo group that settled at the Lipul hill. Pari oral tradition says the Pugeri people migrated from Terekeka on the bank of White Nile, and through Lulubo-land and Lokoya-land to Lipul hill. They found the land was unoccupied. It is not clear as to whether Dimo and his group (the Pugeri) were the Mundari or they were Luo who left Wi-Pac and went to Terekeka, and from there, they found their way to Lipul hill. The second Luo group also migrated from the north, Wi-Pari. These include Wiatuo, Bura and Angulumere. The third and the last Luo group came as well from the north (Wi-Pari) and became founder of Kor Village. They are said to have been lead by Giilo to Lipul hill. Some informants told me that Pari are multi-ethnic in their origin. I think these informants said Pari is multi-ethnic in origin, because there is an allegation that when they Pari settled in their region, some people like Lopit, Lokoya, Bari and Lotuko migrated to Lipul hill, and were absorbed to the Pari community.

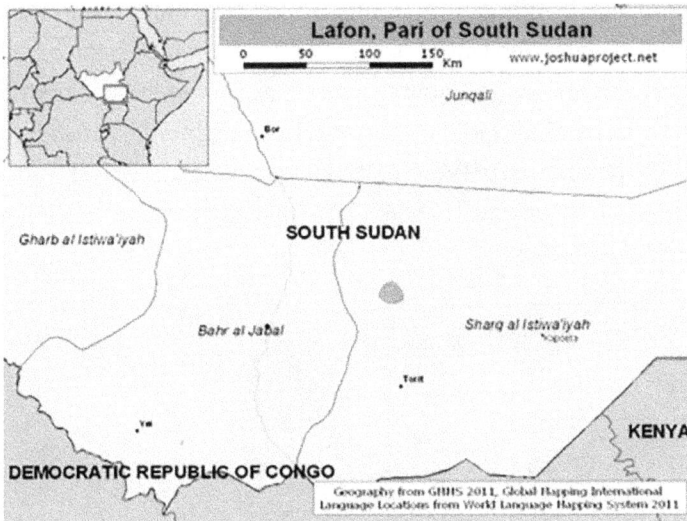

Source: Joshua Project
Figure: 41, Location of Pari in South Sudan

The Pari live in the six villages (mieci Pari) surrounding the Lipul hill in the eastern part of Eastern Equatoria, South Sudan. Each village is subdivided into various assembly yards commonly known as *baalo*—with cluster of huts. The *baalo* is properly the place for multi-social services. It may serve for assembly of elders for serious discussions, as a public council hall, or for entertainment such as dancing. Interesting, each village has *baalo*, so the council of a *Baalo* are held in the *Baalo* of the respective chief. The six villages are:

(1) Wiatuo and the following communities assemble in one *baalo* under the chief of Wiatuo: - the Dyier, Gari, Mol-Nyang, Nyiikwanya, Bupi and Wia-tuo.

(2) Bura and the following communities assemble in one *baalo* under the chief of Bura: - the Pukurjo, Bura, Paraao, and Ademac (Liluro is the proper name for *baalo* or assembly place of the Bura).

(3) Pucwa and the following communities assemble in one *baalo* under the chief of Pucwa: - Geri, Wiira, Alwori and Pucwa.

(4) Pugeri and the following communities assemble in one baalo unber the chief of Pugeri: - Lebaalo, Pokwaari, Laali, P'Ojwom and Pugari.

(5) Kor and and the following communities assemble in one baalo under the chief of Kor: - Lijaalo, Atwoo, Ajiba, Liboongi, Payweeri, Pokaal, Thwoor Abac and Loding. However, in the community of Loding they have a separate assembly place called Likidomok

(6) Angulumere and the following communities assemble in one baalo under the chief of Angulumere: - Adeeba, Adula, Pukwenyi and Angolomere.

Source: Crazzolara (1951) PP. 163-164

The place Lipul usually is known as Mt Lafon or Lafon hill is a solitary granitic rocky hill in the savannah plain. The Lipul is the name of the hill as well as the name of Juok of Pari. This hill is also known as a holy sacrifice place, where they at certain times offer solemn sacrifices to their Juok Lipul (Ajobi). It is just hour's walk around the hill through the six villages. The population of Wiatuo, the largest village, is almost 3,600, while that of Angulumere, the smallest village, is about 800. Thus, the total population is a little more than 11,000 (see below table).

Villages	Populations
Wiatuo	3,598
Bura	2,271
Kor	1,883
Pugeri	1,451
Pucwa	975
Angulumere	839
Total population	11,017

Figure 43
Source: Government Census 1983

According to Crazzolara (1951) in the old time, and before European intervention, the Pari lived under four districts Rwot or chiefs. This four district political units, are separated only by a very narrow path, threading its way between the court-yards heading to the "*baalo*" or "town hall". The four district units include Pugeri, Wiatuo, Pocwaa and the Kor.

However, today, Pari-land is divided into fan-like territories, the Lipul hill being the centre. Each of the six villages is composed of patrilineal clans *(tungi)*. A clan is localized and all Pari knows its affiliation to a village, although not all members live in the village because of uxorilocal and matrilocal residence. A clan has a gathering place called *(baali)*, in its own territory. Members of a clan are recognised as descendants of a founding ancestor, and share a history of migration to, and settlement at, the Lipul hill. Each village has its own chief (rwath) whose office is patrilineal inherited. His clan is called *"tung rwath"* or royasl clan. Among the six chiefs, only the chief of Wiatuo is a "rain-chief" *(rwadhi koth)* whose main role is to bring enough rain for the whole community. His power rests on an inherited special ability and the possession of rain-stones *(gwii)* and rain medicines *(yendi-koth)*. Apart from him there is a bird-chief (Rwodhi-winyo), whose job is rather specific that is to get rid of weaver birds that may destroy sorghum. The bird-chief is from the village of Pucwa, but he is responsible for the birds in the whole of Pari land. Therefore, on one hand the segmentary territorial structure is based on clan-village configuration, on the other hand it is related to the chieftainship (Kurimoto, 2012, P. 270).

The segmentary structure of Pari society is not pyramidal like that of the other Nilotic groups, because the structural weight of each village is not equal and there is little genea-logical relation among clans.

The six villages are divided into two midsections called Boi and Kor. The Kor section is comprised of only Kor village, while the Boi section comprised of five villages: Wiatuo, Bura, Angulumere, Pucwa and Pugeri. A village is a political and territorial unit and each has its own chief or Rwoth. But the chief of Wiatua, the largest village, is recognized as the chief for the entire Pari. Both in terms of population and political power, Wiatuo is prominent in Boi section. The two main divisions reflect the history of migration and settlement of the clans.

Boi is the name of a clan including the royal clans of Wiatuo, Bura and Angulumere. These three clans trace their ancestry to Jogi who lead a group of people to the Lipul hill. His three sons namely Uyu, Uthienho and Mucugo became the founders of the royal clans of the three villages. Thus, Uyu became the founder of Angulumere royal clan; Uthienho became the founder of Wiatuo royal clan and Mucuga became the founder of Bura royal clan.

Pari traditional history says Jogi first came to Lipul hill with his grandfather, Libala and mother Abongo. At that time, Jogi was very small he was carried on the back of a man called Nyoga. They spent some weeks in Lipul and went back to Wi-Pari (Anyuak). The Pari people believed they have all come from Wi-Pari, although there are some clans, which claim their origins in different places. Wi-Pari is allegedly located between the present Pari-land and Anyuak-land.

When Jogi grew to manhood, he led groups of Pari mentioned above to Lipul hill. It is said Jogi and his people found the Pugeri people already living at Lipul hill. This also tells us that the first people to go to Lipul hill were the Pugeri people. The chief of Pugeri submitted to Boi section, although he remains the priest of Lipul, which is the most important Jwok/Jok –god. Jogi came to Lipul hill with Nyoga, a person who carried him on the back during his first visit to Lipul; eventually Nyoga became the founder of Pucwa clan. Therefore, Pucwa is included in Boi section because of this historical episode. Today, all the people of Boi section accept the chief of Wiatuo as the chief of Boi in general. Not only Wiatuo, but also Bura, Angulumere, Pucwa and Pugeri villages pay tribute to him, cultivate his field, build, and repair his homestead. On the other hand, they also pay tribute to or construct the homesteads of their own chiefs.

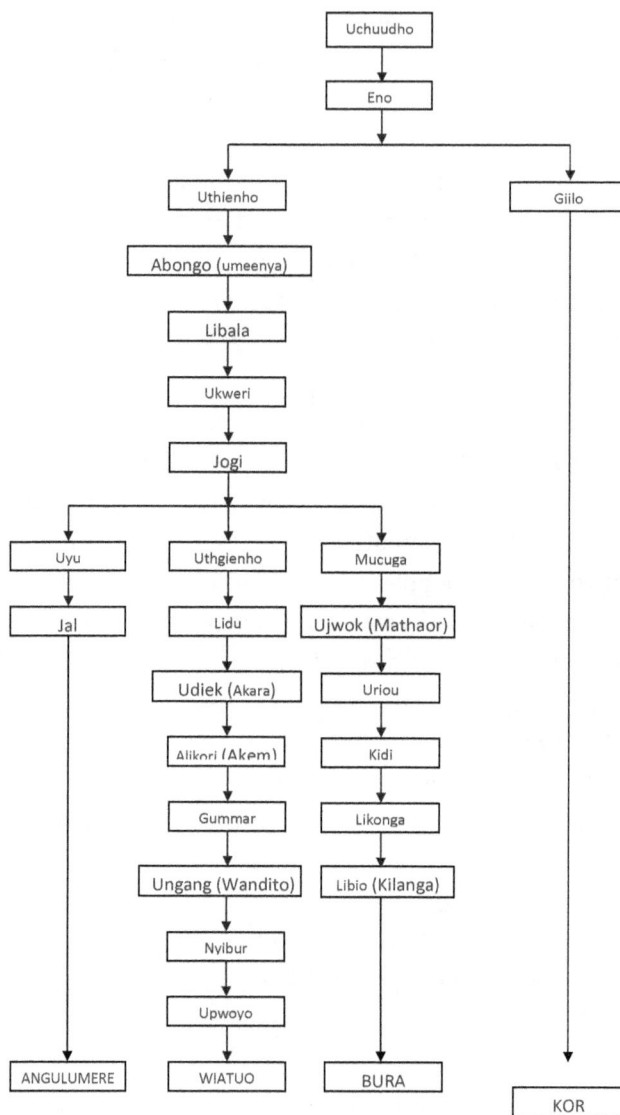

Figure: 42,
Genealogy of the Boi and Kor Royal Clans

Pari Oral tradition says Kor clan is the last to migrate to Lipul hill. Moreover, the royal clans of Kor and Boi recognize common ancestor patrilineal descendants of Uchuudho, a mystic figure who is said to have been discovered in the river. Uchuudho's grandson Uthienho and Giilo are respectively the founders of the Boi and Kor royal clans. Therefore, in genealogical order, the unity of Kor and Boi is guaranteed. This shared history, however, also refers to a primordial rivalry, because Giilo was said to have been killed by Uthienho (Kurimoto, 2012), as we have seen behind.

Crazzolara (1951) noted that the daughter of Cuai or Ocua, who is considered the first sovereign of Anyuak kingdom, gave his daughter called Koori or Koori-Nyairu to Ocuudho, the river man and the ancestors of Pari and Anyuak people. (Please see Chapter 10 under Anyuak). From Koori was born Eno and Giilo. Nevertheless, some informant said Eno was the son of Giilo. Thus, from Eno's wife were born Uthienho and Giilo (PP. 47-148).

After the settlement at the Lipul hill, the Kor did not submit to the authority of the Boi. They have been under their own chief. They do not even pay any tribute to the chief of Wiatuo and instead, cultivated the field of their chief, constructed and repaired his homestead. However, when chief Anywaa died and was succeeded by chief Kwara, situation changed. The Mojomiji of Kor started to pay tribute to the rain-chief of Wiatuo, while continued to cultivate the field of their own chief. Thus, in 1983, the chief of Wiatuo introduced centralization of political power, and stopped the Mojomiji of Kor from cultivating to their chief (Kurimoto, 2012).

Now, it is important to notice that rivalry and conflict is a recurring theme between Kor and Boi in the history of Pari. In 1819, it culminated in a clash, which led the entire population of Kor to leave home and take refuge at Pajok, in Acholi-South-Sudan-land for a period of at least ten years (1880-1890).

Pari traditional history says in 1819, the Wiatuo involved in mass buildings of their houses. They dug the soil for building from the same area, which eventually left deep one hole that acts as water catchment. In the dry season it is difficult to get drinking water, thus both animals and Pari people drink water from the same opened deep hole.

In the year 1880, the Wiatuo people became angry seeing their animals and Kor people as well as their animals are drinking from the same water. One day the chief of Wiatuo called his speaker, known locally as *Liyimme*, and asked him to tell Mojomiji to go and fight Kor. The Wiatua chief also requested chief of Lowathia village to support his people in fighting against Kor. The chief of Lowathia did not refuse the request of the chief of Wiatuo. Shortly, the two groups gathered and fought Kor and defeated them —since D.C. Mr Owen has collected all guns from the people of Pari. It is assumed, that if the Kor were to have guns, they would have not been defeated. The Kor then took refuge at Pajok in the eastern Equatoria. It is said, only small group of Kor, and from the sub-clan of Libongi, were left at Lipul hill, but the rest of the people of Kor went to Pajok (Onyala, 2014).

Thus, Crazzolara (1951) noted that under Rwoth Olum of Kor and Rwoth Alikori of Wiatua a grave difference arose between the two groups. The people from both sides clashed-fighting with sticks only, since they were brother; and no one were killed. Consequently, Olum with the great majority of his Kor group departed from Lipul in about 1880 and migrated to Pajok. The Kor lived in Pajok for a period of at least ten years (1880-1890). The departures of the Kor group subsequently weaken the Pari people. After ten years, the Pari decided to send a delegation to Pajok to make peace with Olum and invite him to go back and forget what happened in the past. Olum and his people made peace and agreed to go back to Lipul hill. When they were in Pajok, the Pari men married Acholi (Pajok) women and brought them to Lipul Hill, as they were going back to their land. Rwoth Olum had only daughters and no sons. At Pajok the young and only son of Okwon, Angwe, died, so the "keer" remained definitely in the family of Olum. When Rwot Olum and his group returned home from Pajok, he told Rwot Alikori of Wia-tuo, how the Obbo people used to suppress Kor people as they passed through their land, on their way to visit their brothers at Lipul, while they were still in Pajok, exciting his wrath. Thus, Rwot Alikori of Wiatuo led the expedition; they arrived Obbo Village unexpectedly by night, and had the village at their mercy. They slaughtered all the men and women they came across, they collected a good number of children and young women and drove them home with cattle, sheep and goats. It was on that occasion that a boy called Okumu was captured in 1884, and brought to Lipul as a slave of Rwot Olum he added him to his family. Okumu was only very much loved by Rwoth Olum, who chose and pointed him as his successor before he died. After the death of Rwoth Olum, Okumu, the adopted slave-son, became the Rwoth of Kor people in Lipul Hill. They even intended renewing the expedition, but the arrival of white men prevented them, and they had to forget all about it (pp. 159, 162).

There are two traditional systems among the Pari society that is chieftainship and mojo-miji, –a graded age set system. We shall discuss about Age Set System later in this chapter. The dual division of Pari society has another aspect. Next to the rain-chief of Wiatuo, the chief of Kor village has political significance. This is because Kor, as the last settlers at Lipul Hill, has remained as semi-independent. Both the chiefs of Wiatuo and Kor play the role as peacemakers (likweri) this is particularly in case of homicide. When a person kills somebody and takes refuge at the chief's homestead, no one can attempt to kill him; his life is secured. After a day or two, the chief as peacemaker negotiates with the two parties. The case is eventually settled by the payment of compensation (in what form?) in order to to make up for the lost of a human life. As a symbol of this status he has a forked stick (likweri) by which he is referred to, when the ceremony of compensation payment is held, the stick is laid down between the two parties, separating them and preventing a further fight. However, when a murder takes place between Wiatuo and Kor, neither of the chiefs is able to act as a peacemaker since he is not neutral. In such a case, a chief of another Village does the work. The two chiefs only act as peacemakers when a murder takes place between Bura, Angulumere, Pucwa and Pugeri (See the figure 41 below).

The chief of Pugeri, who is the direct descendant of Dimo, is the priest of Lipul, the owner of Juok Lipul. He is the one who offers to Lipul the first sorghum beer made of the first harvest, and the first wild animal hunted at the beginning of a new year. The celebration of the New Year is called "nyalam", which annually takes place in early December when the *star of nyalam* appears on the horizon. Men of the six villages go to the bush and hunt. The first animal killed is brought to the top of Lipul Hill, at the entrance of a cave, where it is cooked and offered to Juok Lipul by the chief of Pugeri village. Thereafter, a dance and beer feast follows for two to three days. It marks the beginning of a New Year and hunting season, and is the biggest festivity among the Pari people. After examining various villages in Pari, we will then discuss about "nyalam" in more details

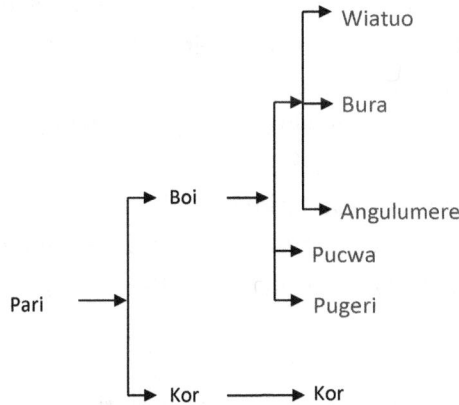

Figure: 43, Divisions of Pari Villages

The features of the territorial divisional structure discussed above are directly reflected in the age system that we will discuss later in this chapter. There is good side and bad side of the territorial divisional structure; in one hand it may work to bring unity among the villagers, on the other hand it may work to destroy the good relationship that exist among the Pari people, as a result of inter-village rivalry. According to Kurimoto (2012), the mojomiji of Wiatuo take a leading role, they are feared by the people of other villages because of they are often oppressive and with violent attitude. Their relations with the rain-chief are ambiguous. This is particularly witnessed when the rain-chief cannot make enough rain or stop too much rain. When it rains regularly, relationships are harmonious. Once the regular rainfall is disturbed, they urge, threaten and even beat him to bring the order back to normal.

The Pari oral history tells us that the Peri people of today are not all from Luo back ground. For this reason, I thought it is wise to revisit the migration of the six villages one by one.

Pugeri
As mentioned above, the Pugeri were the first group of people to settle at the Lipul Hill. It is assumed that people of Pugeri have migrated from Madi race in the west and came to south. They were already at Lipul when the Jo-Luo arrived from the north (Wi-Pari or Anyuak land).

As we have seen above the Pugeri were not from Luo race; they spoke originally Madi. As soon as the Luo group arrived at Lipul, the Pugeri dropped their own language and took up the Luo tongue. They have done this for two main reasons: firstly because their numbers were few as compare to the Luo groups who settled at Lipul. Secondly, they felt it was better to speak Luo language to enable the Pugeri people assimilated easily to the Luo groups.

Pari traditional history reveals the Pugeri people were, in origin, of Madi race. They were split from the other Madi tribal group, because of Lango's invasion of the Madi land. Some of the Puger went straight to Lipul, while parts of the group remain behind and they are the Pageri people of Madi of today. The Pugeri are also called "Minde Athum", which means female bow. This eventually points to their particular association with the bow arm. Moreover, the Luo, as a rule, use the spear as weapon, the Pugeri use bow and arrows only. This implies that the Pugeri belonged in origin to a race whose customary weapons are bow and arrows and these are the Madi.

The first Rwot of Pugeri was Rwot Dimmo and they went to Lipul under his leadership. Chief Dimmo was succeeded by Rwot Aleei, Rwot Gari, Rwot Caai, Rwot Onam 1, Rwot Oryau and Rwot Onam 11 in 1937 (Crazzolara, 1951, PP. 154-156).

Bura

The second group of people, who arrived at Lipul, were the group of Ademac and Bupi. It is said, that the Ademac group came from Bari, and the Bupi group came from Lolubo (who are said to be also Madi).

At that time the chief of Bari at home, when the Ademac group left was called chief Ocomer. The two groups (Ademac and Bupi) met at Lulubo village and continued moving together south, through Okaru Mountains and Lopit land to Lipul hill. The two groups seem to have stayed for some years in Lopit land, before they proceeded to Lipul. As a result, the Ademac women and men do intermarriage with Lopit. Crazzolara (1951) pointed out that the Ademac groups were and are related to the Lomia or Lomiaga of Lopit (p.156). The history of Pari confirmed that the Lopit are related to the Pari (Bura) as well as to the Murulee of the Pibor. However, it is remains unclear, as to whether this relationship of Pari (Bura) with Murulee, is due to migration or intermarriage between them. Thus, when the two groups of Ademac and Bupi arrived Lipul, they formed one village and called themselves Bura. Chief Ademac and chief Bupi respectively led the two groups to Lipul. When the four groups settled in Lipul, they soon joined the main group of Wiatuo.

Pucwa (Ajook group)

The third group that arrived at Lipul Hill after the Pugeri and Bura were the group led by chief Ajuk or Ajook. They are also said, to have come from Madi. According to Crazzolara (1951) Rwot Ajuk was the brother of Dimmo, who led the Pugeri group to Lipul. Dimmo left his brother in Madi land but sooner the brother followed the Kiir River, in search of his brother Dimmo, who came and settled at Lipul where Rwot Ajuk found them. Rwot Dimmo and his brother Rwot Ajuk formed small group that they called Pokwaari that became the sub-clan of the Pugeri.

Gari

Many years later, small fourth group, under Rwot Gari came from Anyuak (Wi-Pari) and joined the Pugeri, which had occupied the place earlier. No much detail is known about Gari group. Pari oral tradition said, after the Gari settled at Lipul, they also joined the Wiatuo

Wiatuo

After the Gari group, then came the main group of Pari, we say they are group of Pari because they brought this same name with them from Wi-Pari. The Pari oral history states, the Wiatuo left Wi-Pari under the leadership of Rwot Cogo. They came together with the group of Kor of Pajok, who proceeded southwards. It is said, at the time, the Wiatuo departed from Wi-Pari, general uncertainty prevailed among the people left behind at Wi-Pari.

The first chief of Wi-tuo was Omenya; Luo traditional history says Rwot Omenya led the Wiatuoa groups to Lipul —that is why we say there was general uncertainty because the few people left behind at Wi-Pari had no chief. The people could behave the ways they wanted without any disciplines or peacemaker.

The wife of Chief Omenya was called Abongo, who born Lebala. It is said that when Rwot Omenya led the groups of Wiatuo and the Kor to Lipul, after that he returned to Anyuak land at Wi-Pari, where he died. Crazzolara (1951) confirmed that Omenya was the chief of Wia-tuo, who died at Wi-Pari (p.157).

After the death of Rwot Omenya, his group came under the leadership of his son, Lebala, who we could say he is the second chief of Wia-tuo. Some of the informants said that Rwot Lebala was given cows to marry a wife with; contrarily the Wia-tuo people insisted that Lebala married at the cost of his own cattle. Rwot Lebala then married Ayu, who became the mother of Okwiri, who late on succeeded his father. Okwiri married Athoo, who became the mother of Jogi. When jogi grew to manhood, he married Akelo who became the mother of Othienho, Muucuga and uyu. From the three children, othienho succeeded his father, Jogi.

As soon as Rwoth Othienho was crowned in his father's place, he decided and went back to Wi-Pari, convinced the Kor and brought them to Lipul. Rwot Othienho married Apiyo, who became the mother of Ledo. When Ledo reached manhood, he married his wife Yege, who became the mother of udyiek and his brother Othienho. From the two brothers, Uthienho succeeded his father. Udyiek married a wife whose name is not recorded in the history of Pari; the wife became the mother of Alikori. When Alikori reached adulthood, he married a wife whose name not even recorded in the history of Pari. The wife without a name became the mother of Kidi, who later on succeeded his father Alikori. Kidi then became Chief of Jo-Wia-tuo. As Rwot Kidi became too old to administer Pari people, his son called Ongang replaced him.

Angulumere

This is a distinct clan group who came from Anyuak; they came together with the Wiatuo group and now became part of Wia-tuo.

Pucwa

According to oral tradition of the Pari, the Lucwa people came from Anyuak. They came together with the Wiatuo group, but more interestingly, they came under their own chief, and remained independent since then. But we do not know the name of the chief that brought them to their present place.

Jo-Kor

After the departures of other groups from Wi-Pari, the Kor people had dwindled down to a dangerously small group. In the old days, there were tribal wars, as such the Kor found that they could not face their enemies, even though they do, they could not have much resistance. As such, they thought of migrating to a safer place. However, they were in dilemma, they did not know where to go, south or north.

The Pari people at Lipul were similarly dissatisfied with the smallness of their number, which might well be a temptation for a stronger neighbours and a danger to their existence. Subsequently, the people at Lipul resolved to visit their brothers in Anyuak (Wi-Paajo) and to try to induce them to come and join them at Lioul. Shortly, Rwoth Othienho of Wia-tuo and Rwot Gari of Pugeri led delegations from Pari. On the arrival, Rwot Othienho met chief Giilo of Kor and invited them to move to Lipul; after a long persuasion, Chief Giilo and his group agreed to join Chief Othienho and his groups at Lipul.

Chief Giilo led his group to Lipul, but before long, his son Raanga (Amulanye) succeeded him. Thus, Kor group became the last group to leave Wi-Pari for Lipul. Raanga had a son called Cool, who succeeded him. Chief Cool was the father of two brothers Ajoori and Okwo. From the two brothers, Ajoori succeeded his father. According to Pari history, chief Ajoori had a son called Okwom, who succeeded him after his death but he did not survive long, he died. At the death of chief Okwom, he left a very small son called Angwe under the care of his brother Okwoor. Since Ongwe was too young, Okwoor succeeded as chief for a time being until Angwe grow up, he would then handover the chieftainship to him.

However, there is no record revealing that Chief Okwoor had handed over chieftainship to Ongwe.

Interestingly, Olum also known by another name as Okwom succeeded chief Okwoor. After the death of Rwot Olum, his slave son Okumu from Obbo succeeded him.

Pari oral tradition mentions that after the departure of the Kor group to Lipul, the people left at Wi-Pari moved north to join the Anyuak groups. They have done this for the sake of protection. Some of my informants said that since Kor group departed to Lipul, there had never been any communication between the Pari and the Anyuak. However, nobody could tell me the reason why there had never been any communication between these two groups.

Nyalam

Nyalam is one of the largest annual ceremonies among the Pari communities. This ceremony is held in November when stars of Nyalam, that is Arturo's and umpired of the herdsman, are seen early in the morning above the north-eastern horizon. This usually marks the beginning of the New Year as well as the beginning of dry season, indication of hunting season. This is also the time when mojomiji starts talking about wild animals. It is forbidden to talk about wild animals in the rainy season, because the people believe that once this is breached, the wild animals will come to destroy the cultivated fields.

Nyalam is always organised by the priest of Lipul who is also the chief of Pugeri village. Thus, when stars are seen, the priest of Lipul immediately sends nyi-dhok (massager) to spread the good news saying, "The day after tomorrow, we shall go under the tree in Wiatuo village and from there to the Adimac clan of Bura village". In this context, the tree refers to gathering place where collective invocation would be held the following day. They inform the Ademac after the formation to Wiatuo because their ancestors are said to be the second settlers around Lipul Hill; they are the people who followed the Pugeri. The chief of Wiatuo then send two messengers, the following morning. These two massagers move from house to house beating drums except the houses of the Pugeri people, the reason has been that, they are the descendants of Dimo who should not be disturbed. As the two messengers from Wiatuo go from door to door, at the same time they tell households, "Tomorrow we shall pray for the New Year".

Early morning, Nyalam starts; it consists of three parts: the first part is when hundreds of elders and mojomiji gather under big wild fig tree that is situated in opened space between Pucwa and Pugeri villages. They sit clan by clan and village by village. The people of each village occupy a fan-shaped position, the fig tree being the centre. Wiatuo, the most popular village, shares the largest space to the north of the tree. If one stand under the tree and look clockwise direction, from Wiatuo comes Bura, Pucwa Kor, Pugeri and Angulumere. The young people are allowed to sit around while looking at elders and *mojomiji* performing Nyalam.

Here, sixty elders from various clans will be given a chance to pray for the blessing of the New Year. The elders from Wiatuo will be given the first chance to pray, follow by elders from Pugeri, Pucwa —a part of Bura, Angulumere and this will be followed by elders from the rest of Bura, Wiatuo and Kor. The chief of Kor is the last to pray at this occasion. When and elder is asked to pray, he stands up with a spear or forked stick in his right hand facing the Lipul hill, and waving the spear towards the hill. Many of the spears are rather ulawi/ulau they are tied with magical charms called *daawi* which are used for invocation and blessing. Their shafts are black and glossy because os soot and smeared butter. Traditionally, the Pari people believe the spear *(the ulau)* of Pugeri chief is from Dimo, this is to be kept and pass down to generation and generations. Moreover the spear (ulau) of the chief of Wiatuo is called Ubel and is believed to have the power of making rain and producing good harvest of sorghum. The invocations are addressed to Jwok Lipul, in which the purpose of Nyalam is clearly stated. The content of the invocations are divided into two categories which is to bring good things into the villages and to expel all the evil things from the village. The good things according to the Pari people include rain, food, success and safety in *dwar or dwar nyicweny*, reproduction of domestic animals and human beings. On the other hand the bad things include evil-eyes (ci-jwok), ghostly vengeance (cien), disease, and thieves like the Toposa people.

Thus, the invocations in the first gathering will carry the following words; the elder citing the prayer will say the following words (and the words in brackets are the responses by all the attendants at the ceremony):

(1)   *Appealing to jwok*
   ▶   Lipul, we have come here.
   ▶   We brought Anywaa [*the senior most age-set of the mojomiji*] here
   ▶   Other Anywaa are also here for you
   ▶   Hay, may it [an animal] be killed at *kwaro* [*a place just outside of Pucwa village!*] be killed!
   ▶   Lipul people will not walk continuously
   ▶   Please do not catch people, people bring your things to you [*Lipul*]

(2)   *Making a deal with Lipul Jwok*
   ▶   Lipul, Lipul, I have come to give the spear to the hand of Anywaa so that the animal may be brought to you
   ▶   Tomorrow when people go, the animal at Wiatuo [a place about 4 km away from Bura village].
   ▶   May it come to Kwaro here! [Come!]
   ▶   The animal] at Amiru there, may it come to Kwaro here! [Come!]
   ▶   Look, it is you [Lipul] for whom the sacrifice is made first

- ▶ Those animals with long ears such as you used to give during the time of our grand-fathers, do we have them?
- ▶ Why don't you bring the grey one [male waterbuck]?
- ▶ If you give it, then it will be cooked and sacrificed to you
- ▶ Now, tomorrow when people go, the animal should be killed
- ▶ May the spear hit it! [Hit it!]
- ▶ May the spear hit really! [Hit!]
- ▶ Now the animal which will come
- ▶ May it stop here! [Stop!]
- ▶ Spear, spear, even though it were a stolen spear, may it hit the animal! [Hit!]
- ▶ May it hit the animal! [Hit!]
- ▶ The rain of this year, may it come from the east! [Come!]
- ▶ May it come from the east! [Come!]
- ▶ Oh, Jwok, why don't you help Anywaa?
- ▶ Don't you help Anywaa?
- ▶ Will the food come this year?
- ▶ Here are chase away god , here we pay to you
- ▶ And *jwok* agreed , *jwok* agreed
- ▶ Oh, may it [*jwok*] go immediately forever! [Go!]
- ▶ May it go immediately forever! [Go!]

(3) *Blessing spears: Calling animals to come to the nearer place*
- ▶ Alright, villages, our end now, the place to stop is now, let us stop.
- ▶ Lipul, we brought Anywaa, we brought Anywaa
- ▶ Now, [the animal] there at Amiru, may it come here to Kwaro [Come!]
- ▶ [The animal] at Wi-ith, may it come here to Kwaro [Come!]
- ▶ Hey, may the spear go straight to it! [Go straight to it!]
- ▶ May the spear go straight to it! [Go straight to it!]
- ▶ Alright, villagers, now all spears have been brought by boys.
- ▶ This is the place where we talk
- ▶ About the spear, we bless the spear for the animal there, it is coming there, we bless there, it is coming there, and the reedbuck is coming there.
- ▶ As we said, let [the animal] go and surround the hill
- ▶ Now another year has come
- ▶ Hey, may the spear go straight to the animal! [Go straight to the animal!]
- ▶ May the knife go straight to the meat! [Go straight to the meat!]
- ▶ May the knife go straight to the meat1 [Go straight to the meat!]

Sources: Kurimooto, 1996, pp. 29-30

As we discussed before, from this texts one can easily see that prayers are addressed to *jwok* Lipul. Thus, in the interval between invocation one, two and three, one elder or mojomiji who is present in the ceremony can stand up and state his opinion about good governance.

However, when the invocations are over, the young people who were standing watching the ceremony stand up and gather on the eastern side of the fig tree, forming a single line with each boy putting his spearhead on the ground. The several elders, including the priest of Lipul, dig some holes in the roots of the fig tree and take out some soil. They then spit on this soil and distribute it to other elders. They also sprinkle the soil on the spears while saying, "May you go straight to the animal".

After the elders have blessed the spears, young people pick their spears and run at full speed with spears on the hands towards the bush, which are usually outside the village. After this, boys come back home marching zigzag in compact groups, stamping on the ground and singing songs. This marching style, as we have seen before, is called *ipuura*

The first part of New Year celebration usually starts from 8.00 a.m. and lasts for three and half hours.

The second parts of Nyalam consist of asking the *mojomiji* to go for *Atondi* (a ritual hunt in the rain forest round the Hoss River) and the place is not far away from the village. The hunt will only come to an end when the *mojomiji* kill the first animal. Then the animal is brought to Lipul and sacrificed there.

After sacrifice, the third part of *Nyalam* follows, which is usually, consists of eating, drinking and dancing and this will continue for four days. In the evening of the ceremony day, all the fire in the villages is distinguished and a new fire is made with fire making sticks in Pucwa village; this is done by an elder from Alwari clan. In this same evening, the fire is distributed to all houses in the six villages of Pari.

(4)   Ritual Hunt and Sacrifice

The ritual hunt takes place on the following day after making deal with jwok Lipul and calling for animals to come nearer to the hunting ground. Thus, elders, mojomiji and young people gather in the bush that is found on the eastern bank of Hoss River. The gathering usually takes place in the early hours of the morning that is from 8.00 a.m. onwards.

Interestingly, majority of these elders, mojomiji and young people, who gather at the bank of the river, are mostly from the villages of Pu8geri and Pucwa. Not every elders, mojomiji and young people in the six villages attend ritual hunt and sacrifice. Kurimoto (1996) noted, *"It is usual that the people of Kor, Angulumere and Wiatuo villages do not attend. From Bura, only the Adimac clan participate"* (p.31).

Before the mojomiji and boys go to hunt, seventeen elders and two *monyomiji* make invocations. After the invocations, all mojomiji would stand round putting their heads of spears on the ground. The elders and mojomiji then come one by one to bless the spears, while a member brushing/pulling a branch of thorn tree called "*Ajiga*" over the spears.

After this, people divide themselves into two groups and they cross the river. The left wing will have to go along Hoss River, while the right wing has to enter the river-rain forest. Shortly, when the two groups start their hunting, it does not take long, in about one hour or so they kill dik-dik/duikers (mur/amur). As the duikers are being speared, the people shout at the top of their voices, blowing horns, which is hereditary of the priest of Lipul. The horn consists of two conjugated parts, usually with a cow horn at the base and a "*kudu*" horn at the mouth.

Each duiker (amur) is tied to a log and carried to Lipul hill, while men march in crowded procession stamping and singing —this is also called *ipuura*. The most senior age-set of youngsters, Madan, repeat the action of a mock charge towards the Pugeri village. When they reach Pugeri village, they are welcomed by old women who yeals (ululating) at them. The men passed through the dancing ground in the centre of the village and stop at the cattle enclosure (Kaalo) behind the dancing ground, where cattle are kept at night. The youngsters stand in a long line with their spearheads on the ground.

The priest of Lipul then take some soil from the root of a desert date tree called "thou" (balanites aegiptica). This is the place where chief Dimo (ancient chief of Pugeri) was buried. Then ten elders including the priest of Lipul bless the spears by sprinkling the soil on the spears, while making invocations. After this, the duikers is brought by men of Madan (age-set) to the entrance of the cave of Lipul, they leave the duikers at the entrance of Lipul, they draw and sit on the rock terrace, while singing songs. Before long the priest and his wife come and make invocations, while they are also sprinkling some soil on the men of Madan age sets, addressing them as *mojomiji*.

The priest of Lipul makes invocation without mentioning the name of Lipul or the victim and sacrifice. The contents of invocation consist of the following phrases:

- ► Alright, *mojomiji*
- ► The things of the village say that even though *jwok* came and spoil us, what about in the future that you are heading for,
- ► Today you are only watching the animal
- ► Now today, from a part of the Boma plateau may the animal come directly to you! {Come direct!}
- ► May it sleep {die}! {Sleep!}
- ► May it die! {Die!}
- ► You are helping the invocation, not for something else
- ► Now it is coming from Boma, may it come! {Come!}
- ► May it sleep {die}! {Sleep!}
- ► May the evil thing that is in your home, may the evil thing go out! {Go out!}
- ► May the evil thing go out! {Go out!}

▶   May this village be cool! {Be cool!}
Source: Kurimoto, 1996, p. 32

After invocation, the men of Madan age-set go down and join the people who have being dancing in the dancing ground. As the Madan join the dance, the priest, four elders from his clan and *mojomiji* remain sitting in Lipul. The *mojomiji* who remain sitting at Lipul, are usually the representatives of the *mojomiji* of Pugeri village. The main reason why they have to remain behind is to make sacrifice of the two animals to the Jwok Lipul. They then skin the animals and cut the meat into parts. Thus, the stomach contents called "wee" is smeared on the rock wall at the entrance of the cave. Stomach (Akaya), liver, heart, lungs, kidneys and pancreases are all cooked in one large pot. When the organs are cooked, they are cut into small pieces. Therefore, the liver, heart and kidneys are put into a small calabash bowl (awal) with some soup. This is then taken into the cave, there the contents are divided into three small pots, and they are left in the cave. Traditionally, this food is said to been offered to the snakes of Lipul.

However, the rest of the cooked organs are then eaten on the sport by the six participants. Uncooked meats are also distributed to the six participants, and one elder of the priest's clan, who is too old to climb up to Lipul hill.

Moreover, from the skinning of the animals through cooking and division of the cooked organs into small calabash, no one is allowed to speak. All the processes are done in silence, no invocation is done.

The animals that the people go to hunt for ritual ceremony are only hare, duikers (amur) and waterbuck (ruda). No other animals besides the three animals mentioned above are accepted for ritual ceremony and sacrifice to *jwok* Lipul. The three types of animals have significant in Pari norms and believes. It is generally believe that if waterbucks are killed during ritual hunt, this means that the New Year will be prosperous because waterbucks are bigger than hares and duikers. Contrarily, if hares are killed, it is believed that the New Year would not be prosperous as when waterbucks are killed. The bigger the animals sacrificed to *jwok* the better services *jwok* will offer to the people of Pari. In other words, the fortune of the New Year is measured by the size of the animals killed during ritual hunt.

Before taking the killed animals to Lipul hill, hunters stuff grass into its wounds to stop bleeding. This is only perform on the killed animals for the ritual hunts but in the ordinary hunts the hunters leave the wounds of the animals to bleed. We have just discussed how the animals in ritual hunts are cut, so in ordinary hunt a killed animals are usually cut into six or seven parts and the meat is divided at spot.

As we have discussed under Acholi, the animals in ritual hunts of the Acholi are captured alive, they are not killed. Contrarily, among the Pari, the animals in ritual hunts are rather killed. This is because the Pari people do not use nets and traps in hunting as the Acholi do, therefore, it is difficult for the Pari people to capture animals alive. However, the act of stopping bleeding may be seen as an alternative of capturing animals alive.

According to Kurimoto (1996) there is no much different between Nyalam and other rituals offered to other *jwok* in Pari cultures. There are many rituals performed for different purposes. For example, there is *Koor* ritual for women (collective rituals for women) or *Kongga Waato*, which is performed in May of every year before farmers put their seeds into the soil. In this ritual, Lipul is invoked for the fertility, welfare of community and prosperity of the New Year.

The other ritual is called *"Libangga"*, which is performed in August of every year. During the ceremony of Libangga milk porridge of a new and unripe sorghum are offered to Jwok. The other ritual is called "beer sacrifice" or "Konggi Waato" and this ritual is performed in September when the first fruits and new sorghum come from the fields. The fruits and new sorghum then offered to Jwok Lipul by the Lipul priest. The priest takes a small amount of porridge/beer, puts it in a small calabash pot, takes it to Lipul hill and leaves it at the entrance of the cave. The priest then withdraws from the cave, and sits to eat and drink the rest of porridge and beer. In both cases of *Libangga* and *Konggi Waato*, people make their first offerings to *jwok* Lipul, prior to making offering to other *jwok*.

After this, the priest tastes the porridge and beer prepared for other *jwok*; he then leaves the elders to continue with the offerings to other *jwok* found in different parts in Pariland. The elders take small porridge and beer and offer them to each *jwok*; when they finish offering, they then eat the balance of porridge and drink balance of beer. In this stage, people of the villages are allowed to join the participants in eating and drink.

Before, we forget, we should know that wild animals are sacrificed only in *Nyalam* ceremony, in other sacrificial, domestic animals, that is, he-goats are sacrificed. The person who has been intervened by the *jwok* usually offers porridge and beer in this ceremony to *jwok*.

However, it is important to know that the he-goats, which are to be sacrificed at "beer sacrifice", are usually killed by stabbing the heart with spear, after which the meat is roasted and eaten. It is only in the case of funerals that the meat of he-goats is boiled and eaten. The blood of the he-goat is then collected in a pot and its organs, such as tongue, nose, ear, eyelid, hoof and anus are cooked in one large pot and eaten by the participants. When participants finish eating, a mixture of blood, water and beer is then smeared on the foreheads of the mothers and babies and their hair shaved. Their bodies are also smeared with red ochre (pala)

Pari Age System

"One of the most important fundamental approaches to the study of African societies was the unfolding of classificatory patterns of social relatedness known as kinship. Thus, African societies (which their tribes and their villages) were perceived by anthropologists as made of

different families, lineages, and kin that operated in a structured manner through different governments, and were visibly seen as political systems that, in turn, were also reinforced and given authority through organised systems of religion" (Aguilar, 2011, p.17).

In the six villages of Pari, all men from age six or seven years old and above are organised into age-group which is locally called *"lange"*. However, there are two systems in Pari, one system is called age-set and the other system is called age group. We may not discuss much about age-set in this book, rather we are going to discuss more about age group. Moreover, an age group and age-set are called *"lange"* in Pari.

Interestingly, any non-Pari, who stays in the village for a certain period is accepted in his age group and will be given the name of that age group that he belongs e.g. Lidit of Pugeri or Lidit of Wia-tuo etc. This is an indication that kinship and age group affiliations are two key factors for the identification of a person.

It is important to study the age group system among the Pari that enrols most men except small children, because it is fundamental in understanding how the society itself operates. It bears a variety of functions, as we will discuss later, and these include political, legal, military, ritual, and economic and the list could go on and on.

Thus, in the book we are going to discuss briefly, about how Pari age system is organized, how it works and how young men are socialized in the system until they become the masters of the system itself.

However, we shall put more emphasis on the rolls of the *monyomiji*, which is the ruling age group and how they are coping with internal and external enemies. Further, we are going to emphasise on "warrior feelings" among the young people. Then, at the end of this article/topic, we are going to account for the system in the historical process of militarization in the region since the birth of Pari Villages. However, I understand , it is difficult to find written records to reconstruct the past history, nevertheless, we will hypothesis that the present graded age system of the Pari is an invented tradition after they migrated from Anyuak (Wi-Pari); this might have developed around when the entire Pari land was experiencing turmoil and upheaval.

It is common knowledge among the Pari that the present grade age system was adopted from the Lopit. The actual word mojomiji itself is a loan word from the neighbours. It derives from the word monyomiji, which means "father of the village" in Lopit, Lotuko and Lokoyo tribes.

However, it is commonly agreed among the Pari that the first age-set is Liborceri is borrowed from Lotuko; thus, ubor means "white" and ceri means "white cattle skin" in Lotuko language. The Pari people named the first age set Liborceri because its members used to dance with white cattle skin tied round their ankles. During the time of Liborceri, it is said, the Pari lived on the Lipul hill and cultivated only at its foot. It is also generally agreed, among the Pari people that chief Alikori of Wiatuo and chief Ujwok of Bura were the first persons to introduce the first age-set Akem and Muura (Jo-geedo) in mojomiji system. The introduction of Akem and Muura was done around 1887-1897. After Akem, there was a gradual transfer of power from Jo-geedo (Akem and Muura) to jo-tengo (Amukwonyin and Merithigo) that happened around 1897-1907.

Social Life of the Awope (youngsters)

(1)   How the Age System Organized

The age groups are formed in each village, with their distinct names; and once they are initiated into the *mojomiji*, they are no longer called by the former name. Rather they are now called by the new name, which they take up from Wiatuo age group that is considered to be over all umbrellas of various age groups in the six villages. Kurimoto (2012) observed that an age group is a village-based organization, while an age-set is formed based on Pari society as a whole. Thus, and age set is a general category consisting of the six age groups of the same age. However, both an age group and age-set are called *"lange"* in Pari. To differentiate between the two groups one needs to examine and clearly reflect on way they are formed and operate. To avoid any confusion, let us first examine age group. Each age group has its own name, as mentioned above, but when they are initiated into the ruling age group (*Mojomiji*), they take up the new name from Wiatuo age group as the general name for the six age groups. Former names are no longer used. The age group are strong and well organised government, that always act in unity. While relationship between age groups in an age set is rather lose and characterized by a potential rivalry, which often develop into stick fights. This rivalry can be seen as part of inter-village politics (pp. 264-265).

Age groups are formed according to biological or physical age. The age range within an age group is three to four years. In addition, once the group has been formed, the membership remains the same. In the age group, the junior people consist of boys who have less than ten years old.

As mentioned before, each village has age groups; this implies that there are more than twenty age groups in the whole Pari Villages. Particular age group is considered non-existence when its last member dies.

The term *mojomiji*, literarily means the "fathers of the village", they are also known as "Wigi-paac", which means pretty much the same "fathers of the village".

The restructure of the age groups start from the first age grade (*the Awope*) and they climb up the ladder of age grades gradually. The Awope/Awobe have their age set that include Ithilero, Yibidhieng, Agweno, Acac, Firik, Lidit, Madir, Morumaafi and Madan. For a young man to be considered in the age group of Awope/Awobe he must be between the ages of six years to thirty nine years old.

The second age group is called mojomiji, as we have mentioned above, they are the fathers of the village and their age set include Adeo, Maridi, Akeo and Anuak. People who are enrolled in the age group of Mojomiji are between the age of fouty one and fifty two years old.

The third and top age group is called *Cidonge*, they are elders of the village, and they have age set that include: Thangakwo, Kilang, Yuulu, Ithiyio, Bondipala, Kwara, Lilalo and Ukwer —see figure 46 below:

Group Levels	Ages in years	Age-Grades	Age-Sets
3	53-75+	Cidonge (Elders)	Ukwer Lilalo Kwara Bondipala Ithiyio Yuulu Kilang Tangakwo
2	41-52	Mojomiji (fathers of village)	Anyuak Akeo Maridi Adeo
1	6-39	Awope/Awobe (Youngsters)	Madan Morumaafi Madir Lidit Firk Acac Agweno Yibidhieng Ithilero

Source: Kurimoto, 2012, p.266

Figure: 44, Pari Age Systems

A man and his age-set pass climb up the ladder of age groups through this system. Nobody could reach the third age group without going through steps one and two discussed above.

However, in Pari customs, the numbers of age-sets in the ruling age group are principally four, that is, Adeo, Maridi, Akeo and Anyuak (going from bottom upwards). When the ruling groups retire to the elders grade (*Cidonge*), the junior age-sets of youngster (Awope/Awobe), that is, Madan are promoted to the elder groups to fill the gap left in *Mojomiji* by Adeo who have been promoted to the rank of Maridi.

Surprisingly, the flow up of age-sets from ruling group (mojomiji) is not simple, a person moves up the ladder to elders (Cidonge) from Adeo to Mridi, and from marid to Akeo and from Akeo to Anyuak to age-set of elders called Thangakwo, kilang and so on, it is a gradual process.

The four age-sets of the ruling groups may not retire at the same time. In one situation, they do, but in another, only the two senior ones retire and the rest remain, been joined by the two senior age-set of the *Awope*.

Moreover, we could argue that the age-set systems in Pari communities have four distinctive features that make them unique in comparison with other age-sets in their neighbours such as Lotuko, Lopit and Lokoya; interestingly, their age-set systems are also different from the age-set in other Luo groups.

Firstly, there is no initiation when boys are enrolled into the system and it has nothing to do with the marriage regulations. Thus, the youngest brother is free to get married before his elder brother or brothers. He can establish his own independent homestead if he can afford to do so.

Secondly, its organizing principle is physical or biological age and no generational principle in action at all. In other words, there is no generation sets in Pari society. Although we have seen that mojomiji are called 'fathers of the villages", this does not necessarily implies that genealogically all youngsters in the villages are their sons. Similarly, it does not mean that all elders in the villages are their fathers.

Thirdly, the age-set system among the Pari people is more of a politico-military system. Therefore, it is not required for the system to maintain the supernatural order by ritual or magical means or to keep and pass sacred knowledge from one generation to the next generation.

Fourthly, Pari age-set system is not a traditional institution but a relatively new one. Many of my informants told me that the age-set system in Pari society came into existence when chief Alikori was ruling.

Taking over

Let us now examine actual taking over. According to Kurimoto (2012), the age-set of *Cidonge* called Ukwer and Lilalo succeeded in 1947. After a year or so, Kwara and Bondipala joined them. These four age-sets held power together for about ten years. In 1958, Ukwer and Lilalo retired, but Kwara and Bondipala remained and were joined by the Ithiyio and Yuulu age-sets, which were promoted from the grade of youngsters. Then in 1967, all of these four retired and the next two age sets Kilang and Thangakwo took over; and after a year, Anyuak and Akeo joined them. This is one full cycle, whose duration is between twenty and twenty two years (p.267).

In other words, the three pairs of age sets namely Ukwer, Lilalo, Kwara and Bondipala,

Ithiyio, Yuulu they all comprise one class, although they may never be in the ruling grade at the time. Thus, in every pair of age-sets, the senior ones are always right and superior, while in the junior they are always not to be heard and inferior. The pair is usually referred to by the name of the senior age-set. For example, Ukwer referred to Ukwer and Lilalo age-sets. In case of the second pairs the name of the senior one refers to the four age-sets. The clear example is the Anyuak in the second age-sets include Akeo, Meridi and Adeo.

Moreover, when the first and second pairs take over, the first will be either called *Jo-Wic* (head people or *Jo-Geedo* (builders); and the second pair will be called *Jo-tengo* (rethatchers). The Jo-Geedo builds a new house where they live as Jo-Wic.

However, there is no specific name for the first pair, which is still in the youngsters' grade.

According to Kurimoto (2012) the *Jo-tengo* take over the house, but they only rethatch the old roof, that is, repair and maintain it. This rethatching usually refers to the hut on the Lipul hill in which the skull of a bull sacrificed on the initiation of the new *mojomiji* is kept. The *Jo-Geedo* built the hut, on the Lipul hill, and keeps the skull in it. Then the Jo-tengo rethatches it and keeps the skull in it together with that of Jo-Geedo. The Pari tradition says that *Jo-Wic* are the true rulers, although Jo-*tengo* stay in the ruling grade twice as long as the Jo-Wic, they are considered as a kind of deputy rulers. The first pair (Awope/Awobe) never enjoy the leading position in the ruling grade (pp. 268-269) (see figure.....)

Figure: 45, Hut on Lipul hill for sacrifices

The social order of the age system is symbolically expressed in the distribution of a scarified bull or ox. Thus, in Pari, when people want to commemorate the repair of the *"kabore"* and the purchase of new drums, a bull is usually scarified and the meats are divided among the *mojomiji* and elders (cidonge). They usually follow strict rules when dividing the meat.

Age-sets	Parts of bull receive by each age-set
Ukwer and Lilalo	Left hind leg & liver
Kwara, Bondipala, Ithiyio & Yuulu	Right hind rump
Kilang & Tangakwo	Left foreleg & chest
Anyuak	Right ear and eye, brain, right shoulder, waist, right stomach, halves right testacle
Akeo	Left ear and eye, mouth and nose, right foreleg
Maridi	Front neck & haves left testicle
Adeo	Back neck, testicle and heart
	The hump is distributed to the chief

Figure: 46, Distribution of Scarified Bull/ox
Source: Kurimoto, 2012, P. 269.

The figure 48 below, show the rule of distribution when Anyuak was in the *mojomiji*: The head always goes to the *mojomiji*, since they are the head of the village and head of people. It is interesting to note that, in relation to the head, the two junior age-sets of the mojomiji (i.e. Meridi & Adeo) are given the neck. The elders' share is substantial as the hind legs, the left foreleg, chest and rump given to them; however, these parts do not have any significance. While the liver, whose taste is highly appreciated, is distributed to the most senior age-sets, that is, to Ukwer. More importantly, the hump, which taste is highly valued as well, is given to the chief of Wiatuo.

Figure: 47, Mojomiji of Pari distributing meat

Features of Initiation of the Mojomiji

(A) *Initiation* The initiation of the new *mojomiji* takes place about every ten years and is very colourful events in the Pari society. This occasion marks the retirement of the old *mojomiji* into the elders grade (cidonge); it also provides an opportunity for the promotion of the youngsters (Awope/Awobe) to the grade of *Mojomiji* to become the head of the village.

In Pari, this occasion is called 'taking out the head" (*kanno wok ki wic*) and "taking out the village" (*kanno wok ki paac*). In other words, they called "seizing the village" (mak paac). According to Kurimoto (2012), the last succession of Anyuak by Madan took place in October 1988, in the midst of the south Sudan civil war. As we go backward, we discovered the age-set of Anyuak had taken over from Kilang in 1977. Kilang succeeded Kwara in 1967. Kwara succeeded Ukwer in 1958 and Ukwer took over Kalang in 1947. However, the exact years of successions before the year 1947 are unknown (p. 298).

The mojomiji as discussed before, are also known as "head people" (*Jo-wic*). In Pari tradition, there are two types of *mojomiji*, one is called the "builders" (Jo-geedo) and the other is called the "rethatchers" (Jo-tengo). As these names imply, the former is considered real "Jo-wic" who construct a new house to live in, while the latter are a kind of renewals or assistant builders who inherit the old house and only repair the roof instead of building a new one. Although the ritual process of initiation is long and complex and the different between the succession by "*Jo-geedo*" and by "*Jo-tengo*" are even more complex, nevertheless there are crucial stages that should be followed to avoid any confusion in the community. The stages are as follows:

(i) A proposal of succession the next *mojomiji* is made and the consent of the current *mojomiji* is obtained.

(ii) The next *mojomiji* drink beer spat out by the outgoing *mojomiji*.

(iii) The Tong clan makes a new fire after all the fire in the villages has been extinguished in order to mark a new start. All the mojomiji go out of the village discuss policies they will implement during the periods of their rule, and make invocations throughout the night. They also take the bull's skull. Which was scarified on the former *mojomiji*, initiation, and throw it (skull) into the river —this is an indication that they have taken over the head.

(iv) On the following morning, they go to the *thworo* of Wiatuo and capture the drums by force after a mock fight against the old *mojomiji*, which marks the succession; the capture of drums means the seizure of the village by the new *mojomiji*.

(v) After the capture of the drums, they dance in the *thworo*. This is followed by speeches and invocations by the new mojomiji. The new elders also make speeches. Then a dance or party continue throughout the day.

(vi)   The next day, the rain-chief offers a bull and blesses it by smearing sorghum flour and crop seeds on its back. By this act, the bull becomes tamed like an ox. It is then sacrificed by the *mojomiji*. The head of the bull is cut off, and each members of former mojomiji come one by one to spit saliva into its mouth. When this is finished, the new *mojomiji* go out of the village with the head, roast it and eat it without washing the saliva spate into its mouth by the former *mojomiji*. The act of taking the head of the sacrificed bull is a clear explanation of 'taking out the head". The rest of the animal's parts, except for the lungs is roasted by mojomiji and eaten by the elders. The lungs are given to the wives of the new mojomiji. After that the bull's skull is kept in a special hut on Lipul Hill.

The ritual process has a political aspect as well. The mojomiji in power do not hand it over easily and with pleasure. They do so unwillingly. They try to stay in power as long as possible, and therefore, it takes a few months before they finally agree; after the first demand of succession is made by the next *mojomiji*. During that period the negotiations and harassment, become the events of the day between the two parties. Thus, the next *mojomiji* demonstrate their physical strength and power; and challenge the current *mojomiji*, claiming that their rule is weak, inefficient and they have no plan for the future. The initiation usually takes place in a year of good harvest, so if initiation is to take place in a certain year and is a year of poor harvest, the *mojomiji* has the power to postpone the succession. When this is done, the next *mojomiji* could think that this was done deliberately by preventing rain, to cause hunger, so that they could stay in power longer (Kurimoto, 2012, pp.298-299).

For example between the periods of 1984-1986 the succession of mojomiji was postponed, when Pari were stricken by drought, hunger and cattle raids by the Toposa and the turbulent of the civil war in South Sudan. Due to all these events, the year was considered the year of poor harvest —there was no succession.

Moreover, these disasters were attributed to the weakness of Anyuak. During the *Nyalam* celebration in 1985, the Lidit age-set underwent the ceremony of "smearing black clay". Therefore, members of Lidit smeared black clay on each other's upper body. After this, Lidit together and with Madir are entitled to put on the headdress of black ostrich feathers, locally known as "lifaango" and they go for traditional dance. This symbolizes warriors who usually fight in the front line. At this stage, the *Madan* and *Morumaati*, who had been using black ostrich feathers *(litaango)* are ready to take over from Anyuak at any time.

However, this take over, might sometimes take up to three years or so, before they finally succeeded. Kurimoto (2012) observed that it took three more years before Madan and Morumaafi finally succeeded in 1988 (p. 299).

We have just discussed how the former mojomiji have to spit saliva into the mouth of the bull. At this point one could ask, "Why should the old mojomiji spit saliva into the mouth of the bull?" The oral tradition of Pari, says even after the mojomiji officially approve the handing over power, the experience showed that some of them would not yet remain unhappy. They may become very critical of the administration by the new mojomiji. Not only this, they may even wish that disaster would fall on the new mojomiji. Thus, this explains the reason why the mouth of each member of former mojomiji is checked, by some body, before he is allowed to spit into the bull's mouth. If nothing were found in the mouth of a former mojomiji, he would be allowed to spit the saliva into the mouth of the sacrificed bull. It is said, if no proper checking is done, those with malice (bad heart) may spit the saliva while having some wild edible fruits in his mouth, so that the new mojomiji would suffer from hunger. In addition, those with ghostly vengeance (cien) would bring misfortune on the new mojomiji. As a result, the relationships of confrontation and antagonism between the old and new mojomiji start well before the actual initiation and continue for years even after the succession.

Succession of Kwara by Kilang

(1) The Kilang at the time of succession are called *jo-geedo* (builders) and true jo-wic (leaders of people). The negotiation usually starts at some point and time between Kilang, Tangakwo and Ukwer, Libalo of a village. Kwara are to be divided because they are jo-tengo. Ten representatives will be selected from Kilang and Ukwer, these are normally sons of advisors to the age-group of Rwath (sons of atieli). The process of the succession is always expressed in the idiom like marriage negotiations as follows:

In the first meeting:

Kilang: "I want to marry your daughter".

Ukwer: "Do you want to marry? No, it is not possible. Do you have bridge wealth (jami or lim)?"

Kilang: "Yes, I do".

Ukwer: "Alright. If you want to marry, there is no problem.. Let me go and think about it well, I will come to you".

Second meeting, which normally take place after 2-3 months is in September, when sorghum is already ripe.

Ukwer agreed to hand over power:

Ukwer: "You, do you want to marry my daughter?"

Kilang: "Yes".

Ukwer: "Is it possible for you?"

Kilang: "Yes".

Ukwer: "Nowl leave it to you. You can marry her. But remember that if you don't catch her properly, the village will be spoiled".

Kilang: "We are able to do that".

Ukwer: "Ok. Now you need to *bring the head out*, as Alangore, Bongcut and Akem did in the past. Therefore, from today if you go to the *thworo*, you are the *Mojomiji*. And when you talk with someone, you young boys, and he speaks bad words and spoils your village, then you come together and discuss. Then let him die. Tomorrow at five in the morning, we shall come together".

As we have discussed before, negotiation between the *old mojomiji* and *new Mojomiji* usually last for a month or months. It is important to notice that in every village, Kilang discuss with Ukwer following same pattern or process described above. Thus, Ukwer of the six villages in Pari should agree. When the talk is finally finished, local beer has to be brought, and all the Ukwer members who are present in the final discussion, spit saliva into the gourd of beer (this is seen as sign of blessing).

The Kilang, who are to take over then drink the beer; drinking usually last throughout the night until the following early morning. As the sun is rising at the horizon, Kilang members go fishing in the Hoss River. When they caught fish, they bring home, cook and eat together with Ukwer members. They only eat fish and not animal meat, because according to Pari tradition, they are not to see blood soon after negotiation and succession. However, if fish is not available, customarily, the Kilangare allowed to cook dried or smoked meat or gettable. After the meal, Ukwer will call Kawara members and advise them that they have agreed to hand over power.

After some days, following the meal, all fire in the six villages of Pari are put out at midnight and the Tong clan make a new fire from sticks (yat-lapii) in the *thworo* of Wiatua. The new fire will first be brought to the *Kabore* and from there it will be distributed to all homesteads. With this fire, the six age groups of Wiatuo, from Kilang down to Adeo will have to go to Limungole, a place on the Hoss River, which is outside Pucwa village. They have to take the skulls of the bulls of Ukwer and Kwara that were kept in the hut. They not only take the skulls from the hut, they also break down the hut and throw the skulls into the river.

Shortly, the six age groups of other villages follow them and they all gather at Limungole. They have to move in full dress of mojomiji, which consists of leopard and black-and-white colours skins on the upper body and head dresses of different feathers. The group of Kilang and Thangakwo put on white ostrich feather (ligira or licaaro). While the Anyuak age-set put on white ostrich feathers (aleero) in front and their back. The group of Akeo put on black ostrich feathers (litango). Maridi group put on brown ostrich feathers (agiira); moreover, the Adeo age-set put on red feathers of the red bishop bird (uthwonh). As leopard skins symbolise power, pride and royalty it is essential that a leopard skin is slung across their naked chests. Traditionally, Adeo age-set are people who are actually fought in the front during the war. Thus, Ligira is a sign for people who organise wars.

At Limungole they discuss how to rule and make invocation (lam), so that their rule would be successful. Normally, the first speaker of this group is Akala, the *liloo* of Kilang of Wiatuo. Akala has special speech to talk; according to Kurimoto (2012), the speech of Akala would include the following:

"The village has already been given to us. We want food to continue to be in plenty during our rule. We from the six villages, we have taken out the village from Ukwer. Today, when we go back home, the village is ours. Now we do not want each village to go its own way. We should be united as one. There should be food. Our ideas should be one. We should cry only for food. Ukwer, who have become old, are in the village. They should have enough food to eat. Ukwer should not die because of lack of food. Now the sun is going down, it will rise tomorrow morning. Let the sun rise with you. Never go back again" (p.302).

Interestingly, each village in Pari haz a speaker who is the liloo of Kilang; thus, during the speech to the new mojomiji, the liloo of Kilang of Wiatuo come first then follow by speaker from Bura, Pucwa, Pugeri and Angolomere. The speaker from Kor village always comes last; this is in according to traditional protocol. The speaker from Wiatuo gives speech in the beginning of the gather, and after some hours the speaker from Bura gives, his speech and speeches from other speakers continue throughout night until morning when the people can see the star of "Limeth" on the horizon.

When the discussions and invocation is over, the mojomiji marched as a single group making "ipuura", while singing boasting songs (*twaaro*) of self-praise and blowing animal horns. They march around the villages of Pucwa, Bura and Wiatuo. At the end of the march, they then rush to the thworo of Wiatuo where the enemies (jur), that is, Kwara are waiting for them. At this stage and time, the Kwara are armed with sticks, while the new *mojomiji* are not armed at all. However, the new *mojomiji* are to make their way into the *thworo* by force through Kwara while being severely beaten. Finally, they reach the *thworo* and the *liloo* of Kilang of Wiatuo beat the drums as new *mojomiji* enter the *thworo*. This sound of the drums is a demonstration to the entire community of Pari that the new *mojomiji* have captured the drums from the enemies (i.e. Kwara).

Figure: 48, Dance to show that new Mojomiji are captured

Soon after this, the Kwara retreat to the Kabore where Ukwer will be sitting to wait for the news. The new *mojomiji* then dance in the thworo with three selected songs; after this they sit down in their respective place and in accordance to age-sets. However, this means, Kilang sit on the right side of *Thworo* and Tangakwo sit on the left side of Thworo and sit as well in the middle of *Thworo*. The Annual age-set sit next to Kilang, while Akeo sit next to Tangakwo. The liloo of each village then make an invocation, as described before, the *liloo* of Kilang of Wiatuo make invocation first, then come the liloo of Kilang of Bura, Pucwa, Pugeri, Angulumere and Kor.

According to Kurimoto (2012), the speech of liloo of Kilang of each village covers the followings:

My brother Ukwer and Kwara
Today, when you sleep, you sleep because of me
The land is now mine
Cows that will be going for grazing, they are mine
A woman going to get firewood, she is mine
Today, I have changed the land from the old form to the new form.
You just take off your hair, shave it
If you have cows, I will allow you to marry
If the girl does not want you, I can make it for you by force
May Sorghum be ripe!
May cows give birth!
May women give birth!
May sick people get up! (p.303)

Once the speakers from different villages have spoken, Ukwer speaker will be asked to jump on the *thworo* and his speech will cover the followings:

My brother, will you really be able to manage?
There are so many problems in the villages
I put down my hoe!
I put down my spear!
I cannot carry them now!
I give them to you!
Now Kilng, today, I take off the thread of my necklace
I put it round your neck
Today, you are an evil eye (ci juok)
You are witchdoctor (lilwo)
You are someone who stops rain
You are a thief
I, Ukwer, my place is far from you
Are you able to do that?
Enemy is coming, coming from Toposaland, from anywhere

I cannot run

That is your job

Are you able to do that?

After the speech from Ukwer, Kilang memebrs sing the song of head *(wic)*. They they dance together with Ukwer members. After a short time, Kilang members get out of thworo, leaving Ukwer members still dancing, Kilang go to drink beer for a warm up. When Kilang age-set are still out drinking beer, Kwara age-set enter the thworo and dance side by side with Ukwer age-set.

As soon as Kilang age-set feel that they have drank enough, they again enter thworo to continue dancing. When the group of Kilang enter a dancing place, the group of Kwara age-set then leave dancing ground (thworo) for them. Interestingly, when Kilang again leave a dance place (thworo) to drink more beer, Kwara again come back to thworo –these acts go on and on till morning.

In the morning, the chief of Wiatuo, take a bull and offer it to the new *mojomiji* as sacrifice (wal or tyier). Eight men are chosen to take the bull: four of them from Kilang age-set and four others are from Tangakwo age-set. They hold the horns of the bull and drive it by force to show their physical strength. When the bull is brought to *thworo* of Wiatuo, the chief takes some sorghum, beer four *(moko kongo)* sorghum seeds, cowpeas, groundnuts, simsim and other edible crops. The chief smears them on bull's back so that there would be enough food during the period of the new *mojomiji*. When the chief is smearing these food items on the bull, it becomes tame like a castrated bull (Ox). The *mojomiji* then drive the bull around the six villages, walking unti-clockwise direction and end up at the *thworo* of Bura.

When the *mojomiji* are driving the bull around the villages, all Kilang age-sets from various villages gather and walk after the bull while making *ipuura*. When they reach at *thworo* of Bura, and before the mojomiji and the bull enter the *thworo*, both Kilang and Tangakwo rush ahead into the *thworo*. When they are in thworo, the Kilang sit on the right hand side and the Tangakwo sit on the left hand side. As the bull comes between the Kilang and Tangakwo, four men from Kilang and four men from Tangakwo spear the bull at the same time. The bull runs with spears on it and falls down. When the bull is dead thye take it to kar-wic, the place of the wic, which is located at the back of the thworo. The door of the har-wic is then opened and stone place on it. Shortly, one man from Ukwer age-set come, he opens his mouth for another person to see if his mouth is free of objects. The man checking the mouth has to make sure that there is nothing in it. After checking the mouth, the man from Ukwer is then allowed to spit saliva into the bull's mouth for blessing. However, if the man does not check the mouth of a man from Ukwer well, and he spits the silava into the bull's mouth with a wild fruits in his mouth, then hunger will come. That is why his mouth is checked before spiting saliva into the mouth of the bull, to make sure that there is nothing in his mouth.

Shortly when spiting process is done, the head of the bull will be cut off. The head will be divided into upper and lower jaws. Traditionally, the lower jaw is given to Tangakwo age-sets and the upper jaw is given to Kilang age-sets. The age-sets of Kilang and Tangakwo the run out of the village, in different direction with the head of the bull, after that both parties meet at a place called *Ajwa*, which is located outside of Bura village. After this, the four age-sets from Anyuak to Adeo joined them at *Ajwa*. Here, they roast the head of the bull.

Interestingly, as the new *mojomiji* are eating head of the bull, the Ukwer member inivite the wives of *mojomiji* to join them. In reality, women should not accept the invitation, as their portion is usually the lungs. Nevertheless, the Ukwer will continue persuading the wives until they join their husbands. The bottom line is that the age-sets of Ukwer want Kilang age-sets to suffer.

Since the new *mojomiji* eat the head of the bull together with their wives, their heads also become hard (tek-wic). This simply means that they become stubborn. While the new *mojomiji* are busy eating the head of the bull, the Ukwer and Kwara age groups roast the other meat from the bull, and when the meat is roasted, both the Ukwer and Kwara go to eat at *Kar wic*.

When they finish eating the head of the bull, they take the skull, put it in the new built hut on the Lipul Hill and from there they scatter to their respected villages while making *ipuura*.

However, now, it is important to understand that the succession by *Jo-tengo*, we have described before, is slightly different from the succession done by *Jo-geedo* discussed before.

Thus, *Jo-tengo* proposes this matter of succession to *Jo-geedo* who are directly senior to them: Anyuak to Kilang, not to Ukwer or Kwara. They normally do not say, "I want to marry your daughter" –as we have seen in the succession of Kwara by Kilang. They instead say:

Every house has an owner!
Now you cannot stay in this house forever!
After sometimes, someone will come and take it over!
Kurimoto (2012, p. 304).

After the invocation, the *Jo-tengo* go at Ger to make a new fire. The place where the new fire is usual made is outside of Bura village and not at Limungole as we have seen behind. The jo-tengo sacrifices their bull in the thworo of Wiatua and not in *kar wice* and the skull of scarified bull is kept in the hut together with the previous skull of the new mojomiji. Subsequently, the jo-tengo do not build a new hut, but rather they rethatch (*tengo or roco wi ot macon*) the old hut.

2. *How does the system works in reality*

To avoid generalisation of how the age-set system works in Pari society, we are going to analyse and examine the system in its smallest units or components. These components include but not limited to playing, practical hunting game, communal labour and dance, stick fighting, wrestling and *mojomiji* community base governments (MCBG), which are divided into two categories, that is decision making on village affairs and war-fare (both internal and external. It is only by doing this that we would be able to understand how the systems work in age set-system.

(a) Children playing together:

The children of Pari society are not exception in Luo groups, in particular, and in East and Horn of Africa in general, the children play together according to their age groups. Thus, in Pari, boys of the same age in the neighbourhood start playing together when they are between the ages of two to four years old. The children playing in Pari communities start from inside the villages to outskirt playgrounds. Thus, when the children are still young too young, they play inside the villages, but as they reach the age four and half to seven years old, they take the playgrounds to the outskirts of the villages. From the age of seven and half to nine years old, children take their playgrounds along the rivers and finally to savannah.

Figure: 49, Children playground on the land and in water

(b) Practical hunting games

Hunting games are the favourite of Pari children *(why?)*. When the boys of the same age groups gather under a tree or in shades, they make figure of wild animals found in their locality, using clay. Thus, when the boys finish making figures of animals, they then sit around and put the clay figures in the middle. Each of the boys will then throw toy spears in turn at the clay figures, if he misses the clay figures, then he is said to be a bad hunter; but when he spears the clay figures, he is said to be a good hunter.

This is where the boys demonstrate whether they will be good hunters and warriors in future. In this way, young boys begin to think about responsibilities and obligations. In Luo traditional customs, it is the responsibility of the husband to feed members of family with meat from wild animals and to do this he should develop good skills in spearing animals.

Figure: 50, Children hunting games

Practical hunting game among the young boys in Pari communities starts when they boys reach the age of ten. They learn how to spear animals as one boy spins a vine of the sausage tree (yaa or yago), with a fruit at the end of the string, while other boys surround him spearing the fruit with toy spears (sticks). By the age of ten, the boys are already organised into an age group. Each age group is given name; and each age group has its own gathering place (baali), which is usually located at the periphery of the village.

According to (Kurimoto, 2012), a gathering place for the ten years old boys are usually located under the shade of trees and has a wooden plat-form (peedheo with a roof or a bench of wooden logs put on the ground. It is a place where they sit, have a snap, and discuss their own issues (pp.275-276).

As the boys grow older, the location of *bali* (gathering place) will be moved from the periphery to the centre of the village. The *bali* of the senior boys is built facing the *thworo*. It is only when the senior boys are promoted to *mojomiji* and elders; this is when they are allowed to enter the *Kabore*, which is a gathering place for *mojomiji* and elders.

There is no initiation to be member of the group and no ceremony when age group is formed. The reality is that, it is formed naturally an inevitably as a result of the physical growth of boys in villages. As the boys grow, they tend to spend more time with their age-mates than with family members, although they continue to eat and sleep at home. However, age groups in other African societies do not follow the same pattern of the Pari. According to Aguilar (2011) the Masaai and Nyakyusa boys that join the same age sets leave their fathers' households and go and live together, apart from their relatives, in order to prepare for their ritual path into adulthood. During the meat feast in the forest, the boys of Masaai behave as a group of classmate or unmarried friends sharing meals and feasts together and inviting unmarried girls to their celebrations (p. 18).

Interestingly, during daytime boys stay at the *bali* to discuss their affairs; and when they are tired of sitting they move together around the village. This implies that group activities by age mates expand as boys grow. Thus, dancing, hunting and stick fights become their first activities and by the time boys reach their late teens, communal labour and home constructions become their second activities to be added to their roles and responsibilities. Usually at the age of 12-15 years old, boys in Pari communities own their own cultivated fields, while they are also planning for marriage.

According to Kurimoto (20123) the age group of the young people are divided into three categories, this is done according to seniority. They use body metaphors to show these divisions. Thus, the top senior boys are called "head" or "wic" and the middle groups of boys are called "chest" or "kor" and the bottom age group are called "the lower body" or "thaar. It is important to understand that when the members of the chest (kor) become many they would be further divided into two sub-categories namely: the upper chest or "Ucula" and the lower chest or "Paar".

In the age groups there are many tittles given to the young people holding various positions in their age groups. According to Pari local tradition, the leaders of the various age groups, must come from the royal clan and he will be given the status or position of the chief (Rwoth or Rwot) for the boy age group. Like the village chief, he exercises no political power over his age mates, but rather acts as a figurehead who gives the final word when the consensus is reached in the issues affecting the boys. The chief of the boy groups usually has numbers of advisors commonly known as *"atieli"*. These *atieli* are considered wise enough to be consulted by the chief of the boy age group. They (atieli) are selected according to their past achievements, personal and physical characteristics. The advisors usually consist of two persons. The word *"atieli"* is actually derived from the nounce *"tielo"*, which means *"a leg"* –thus, they are the legs to the chief.

Among the boys, the bravest person is called *"alul"* and his bravery is demonstrated in hunting and stick fights as we shall discuss later. The *"alul"* carries the black flag known locally as *"beero or bero"* on his grass woven helmet *(likuluk)*.

The second bravest person in boy group is called *"atiep" or atiebi"*. He also carries red flag on his grass woven helmet. Other members of the boy age groups may also carry flags but their flags should have different colours from the ones mentioned above. They are allowed to carry white, yellow, blue and green but never carry black and red flags.

While the fastest runner among the boy groups is called *"dwero or dweri"*, the *dwero* plays an important role in hunting.

The most physical strong person among the boy group is called "lithir or lithire pl". The ability of lithir is usually shown in wrestling.

While the most serious dishonour person among the boy age grup is called *"lwar"*, which means a coward person.

Finally, each of the boy age group in a village has a speaker known locally as "liloo". He is the one who leads the discussions and plays a crucial role in the forming a consensus, His words should be powerful and convincing. He must be a wise man and acknowledgeable.

Thus, all these tittles indicate, through the activities of the age mates in an age group, physical skills, art of oratory and discretion are cultivated and expressed. The age group is a fundamental social framework in which the manhood of the Pari is fostered. As we shall discuss later in this chapter, this manhood is deeply related to the "warrior ethos".

(c) Eating system, Communal Labour and Dance

As we discussed before, age mates spend most of their time together in *baali*, rather than with their family members. They also share food and local brewed beer that is made up from sorghum. What we want to look and examine more critically here is that how do the go around sharing food?

Interestingly, the boys of the same age go around eat and drink at each and every homestead of the group members. Thus, it is the work of the mothers or the wives to make sure they make enough food and drink for each member of the age mate to have a taste. However, this does not mean that the first mother or wife has to provide enough food and beer to satisfy the wants of the group on arrival at that homestead. The reality is that the age mates want to eat and drink bit of food and beer as they go around the village. More importantly, the boys in whose house the age group are coming to eat and drink does not eat nor drink in his house. However, the mother or the wife keeps his portion that he will eat later.

Beer is the main source of family income in Pari villages. Apart from women brewing beer for age groups, they also brew beer for cash. When women brew beer at the homestead of a member, all age mates are invited to drink. At this time, although age groups gather according to their ages, the system for services will be completely different there is no free beer. The age mates have to contribute cash and purchase beer for their consumption. Nevertheless, not all the beer will be sold for cash; it is Pari custom that the close relatives and age mates are given free beer. As Kurimoto (2012) argues, "Not all the beer will be sold for cash, the close relatives and age-mates, have the privilege of taking it for free" (p. 279).

Kurimoto (2012) observed that in Pari society, age groups are units of communal labour (Koc) in agriculture and homestead constructions. In the case of agricultural work, the age groups organize themselves to clear, weed and harvest the crops. Thus, in the beginning of the raining season the age groups clear the lands and sow the seeds. When the time for weeding comes, they also gather and do communal weeding. Similarly, when the crops are ripening, age mates go to the fields and harvest. They harvest the crops and put them into granaries or pen (a sorghum plate-form built with four legs in the compound and with wooden top floor. These works are all been done on communal basis.

During the construction of a homestead, the age groups use the grass for thatching and rethatching; they build and repair the fences on communal basis.

Figure 51

A member of homestead fixes day and date for any activities the age groups have to do and call them on the fixed date. Here essential question arises, "How many members of the age group do actually participate in a given activity? According to Pari traditions, in each occasion of the activity, the number of age groups varies from 10-20 participants. The work usually starts early in the morning, this is the time the blood is considered to be cool and fresh, and then work finishes at midday. After the completion of the work, age mates gather around to eat and drink beer. The amount of beer per person is fixed: that is, half a tin of beer is allocated for an individual; the capacity of a tin in the village is about 20 litters, which implies that, two persons per tin of beer. In other words, this means if there are twenty participants, the women must make ten tins of beer (200 litters). This is a huge amount and requires a lot of sorghum and time for the women to labour. Although some beer is served during the work, the main portion is consumed afterwards in the organizer's house. This local law is never broken by the women.

When cultivating in the age group, peace and respect is of paramount important. All the activities are carried out in a very cheerful and friendly atmosphere. While the age groups are working; they usually chat, laugh and sing songs. On their way home they usually march zigzag, this may be because they want to talk to some of their friends in the front or they are under influence of beer. They keep on stamping on the ground due to alcohol and singing songs of the age group. Every age group in Pari society has its own "age group song". This marching with song is called *ipuura*, which is also performed on the way back from hunting and stick fighting.

When age group drink together, they become more cheerful than ever. The quality of beer is recommended to make them drunk. Once the age group are at home from the field, other people can join them for the beer feast.

Figure 52 Ipuura (marching with song)

In Pari society, there are several types of dances to be performed by the age groups. They are always the organizing units. However, the biggest dance is normally organized by the *mojomiji* and this dance is called "myiel" and is held in the *thworo* with drums.

When age group dance, after receiving their reward for labour, all men and women in a village are allowed to participate in the dance. This is one of the occasions, which age groups become visible to the society. During the dance, members of the same age group dance as one group and each age group puts on especial decorations on their faces or backs to distinguish them from others.

Sometimes an age group also organizes dances for fun. The dances for fun are usually held in the outskirt of the village and not in *thworo*; these dances are usually without drums. Participations by girls in this dance is indispensable.

(d) Hunting

In Pari society, there are two types of hunts, one is called *"dwar"* which is organised by the *mojomiji*; and the other is called dawn-lange, which is performed by the age group (boys). However, in both cases spears are the only means to catch and kill animals. In the case of the boys, as we have discussed before, they do not use true spears, they use sticks for *dawn-lange*; but in case of *mojomiji* they use true spears for *dwar*.

It is said, a good hunter at *dwar* throws spears just some 5-8 meters in front of running animal; if he throws the spear direct on the animal, he will miss the animal, because the spear reaches the place of the animal when it (animal) actually changes the position. Therefore, it is recommended that a hunter throws spear just in front of a running animal so that it bumps on it as it runs. In case of standing animals, (this also applies to clay figures) spears are thrown right at the body without much trust but at a target.

The head of a spear is short and small, but its shaft *(bol)* is long around two meters long. Interestingly, in other Luo group like the Acholi besides spears, they also use bows and arrows, traps and net, but unfortunately the Pari people do not any of these weapons.

Among the Pari communities, hunting is carried out in the dry season that is from December through March of each year, when herds of herbivorous animals migrate, and concentrate around, permanent watering places. In Pari-land, there are no South Sudan National Parks. Therefore, this explains why Pari people freely hunt in around their land and kill any of wild animals, whose meat constitutes an important part of the diet. Tanaka (1996) observed, "Hunting is not unique to the hunter-gatherers....there are many agricul-turalists and pastoralists who obtain a part of their diet through hunting" (p. 12).

As discussed before, hunting games are a favourite past time for young boys. When they reach the age of approximately fifteen years old, they start the age group hunts called *dwar-nyicweng*. Boys of an age group leave home early in the morning with spears and go to wooded savannah not very far away from the village; they come home in the afternoon. The main games are hunting smaller animals such as genet, mongooses and cane rates *(anyeri)*; all these animals are eaten.

The term *Nyicweng* is used to su-categorize classes of wild animals and these comprises animals with claws *(cweny)*, which is the opposite of 'lityiendu-ubongi" those are animals with hoofs.

In the *"dwar"* organized by the *mojomiji*, men hunt only four types of animals and these are elephants, buffaloes, lions and leopards, and they are known as big animals or "lai-ci-dongo". Nevertheless, in the late seventies, Pari people used to kill five animals that include rhinoceroses, but today rhinoceroses disappeared from the face of Pari-land. Interestingly, when Pari hunters come across other animals such as giraffe, antelopes and the like, during the expedition, they pay no attention to them because customarily they are not counted among targeted animals called big animals. Kurimoto (1996) correctly argues, "For the Pari hunting is not only a means to obtain meat. It is an occasion, especially for youngsters, who are engaged in specific hunts of four big animals (lions, leopards, buffaloes and elephants) when they are able to prove their manhood; by killing the wild animals, the group then establish their identity as a man" (pp.26-27).

Men take pride in numbers of animals they killed in a year or years. The same thing apply to the age group (age mates), they also take pride in the numbers of small animals they killed during dwa-nyicweny. In one-expedition lots of animals, we could say ten elephants and twenty buffaloes may be killed. Unfortunately, people do not bring all the meat of these animals at home, some of the meats are left to rot or to be eaten by vultures and other animals like wild dogs. Kurimoto (2012) noted, "They eat as much as possible on the sport and smoke some to be carried home, while the rest of the met is just abandoned; no one is called from the village to take the meat" (p. 282).

However, lion's meat is not eaten by the people of Pari, but they eat the meat of leopards, although the taste of leopards' meat not gorgeous; they also eat the meat of elephants and buffaloes. Moreover, both lions and leopards are hunted firstly because they are be counted and secondly because both animals are dangerous to both human beings and domestic animals. In addition, lions and leopards are killed because their skins are highly valued. The name of big animal could also be use by the people in certain occasion. Hurimoto (2012) rightly argues, "They (people) use animals' names such as elephants and buffaloes in particular, as interjections…..when one is astonished, he may utter "Liec" (elephant) or "joobi" (buffalo). However, this is allowed only to those who have actually hit elephant or buffalo with his first spear during the hunt" (pp. 282-283).

Unlike Acholi, the Pari people do not hunt in large community; they go hunting in small group consisting of twenty to thirty men. Each man carries three to four spears, some sorghum flour and water. A hunting expedition usually takes several days. In Pari-land, there are only six hunting grounds, and these men go in each of them looking for elephants, lions, leopards and buffaloes.

When they find herds of animals (a solitary male) they secretly approach it. There usually two positions of assaults on the animal: one is to surround and the other is just to attack it in one compact group. They come close to the animals so that they are within the range of throwing spears.

As the hunting method is very risky, when the group of hunters arrived at the hunting ground, the first thing they have to do, in the first night is to make ritual ceremony by calling birds *(cwondi-winyo or olik)* that is follows by taking an oath *(aiyeme or lamo dog kwaro)*. This is always done by a member of a clan, which owes the hunting ground. Kurimoto (2012) rightly says, "These ceremonies are performed on the previous night to the hunt in order to cheer up the group and enhance its spirit" (p. 284).

(1) Calling the birds
A member of a clan, which owes the hunting ground stands up in the open and cites the following prayer verses like:
- We have come here to count animals
- May animals come out to drink water here!
- You, the birds, if you are somewhere, come and inform us
- May we kill animals without any problem!

Thus, if birds (bats) come, flying over their heads, this is considered a good omen; the hunt in the following morning will be a success.

(2) Taking an oath
This then, is followed by taking of an oath by the bravest and fastest running people. At

the time of taking an oath, everybody sit down and just watch the best runners and a pair of the bravest ones stand up and say something to one another something like this:

- ► One runner (dwero): Says, if animals are far, it is I that will run in front of them, and stop them, while you are coming.
- ► The other runner (dwero): says, No, it is not you but I that will run in front of them and stop them, while you are coming.
- ► One bravest man (alul): It is me that will go closest to the animals and spear them
- ► The other bravest man (alul): No, it is not you but me that will go closest to the animals and spear them.

(e) Stick fights

A stick fights in Pari is occasion in which the members of an age groups cultivate and show their power. In other words is an appropriate occasion for members of age groups to try to show off their fighting skills and bravery. It is also an occasion in which the authority of *mojomiji* over youngsters is demonstrated. They reserve the rights to approve the fights among age groups. No groups of youngster can start fight among themselves without an approval from the *mojomiji*. Although the age group pick up quarrels durance dance, drinking place, crossing the border during harvest and crossing in front of other age group with their flags, they must first seek approval for stick fighting from the mojomiji.

However, in some case, *mojomiji* may not approve the fighting; these usually happen when fight is to take place in September, the month in which Nyalam is performed. No fight is accepted during the *Nyalam* or in the month *Nyalam* to be held. If the age groups persist to fight, members of mojomiji will tell them, "If you want to fight (x) age group, first start fighting us". Once the two parties give in *Kabore* then will announce to the entire village that the fight has stopped.

The stick fights usually happen between individuals of the same age-set or between boys of same age-set, which could be from one village or from adjacent village.

There are two types of stick fights in Pari, one is called formal fights and the other is called informal fights. The formal stick fight is called "goy-nyiponde"; the term *"goy-nyiponde"* comes from the Pari word *"goy-iwathinh"*, which means young boys. The group of *Goy-nyiponde* is composed of fifteen years old boys. The stick fights of *goy-nyiponde* usually take place in *thworo*, with the permission of the *mojomiji*. This stick fight is guided by strict rules.

The informal stick fights take place outside the village or in any open space, at the skirts of a village. The interesting thing about informal stick fight is that it is not regulated by any rule. More importantly, in the two types of stick fights, fight may be triggered off even by slight provocation of one age group.

However, it is strictly prohibited for individuals or communal stick fights to take place during the raining seasons, when sorghum ears are up and ripen. When anybody break this rule, he/she would be fined by the mojomiji. The fines are consisting of goats or sorghum. This is also another means, for the *mojomiji* to control young people and their violence. As we have discussed above, there are many factors that could triggered stick fights among the small boys. Thus, a stick fight of small boys (informal fight) may be caused by a verbal insult. For example, when the Ithiloro boys of Pugeri are playing and the Ithilero of Kor come and say to them, "*Tee wek-wu*, which literally means "The anus of your father". You are more coward than us". This insult could easily lead to stick fight between the two age groups on the same day. When both sides make up their mind, they can fight in open space, which is just outside the village.

Each members of the age groups takes stick called locally as "abeela/ abele or odoo" in their right hands and shields (ukwak or Okwot) in their left hands. A stick is a soft branch of any tree available at hand. And a shield is a branch with a hand-protector or a used plastic tank.

The confrontation continues for some minutes, while they insult and provoke one another. This is done, so that both sides could build mountains of angers in their hearts, which will then soon follow by actual fights.

Although the small boys clash in groups, nevertheless, the actual fight is held face to face. When someone is beaten and falls down, the fight is considered to have come to the end because one side has won and the other side has lost. Kurimoto (1996) pointed out that the losers run away in disarray and the winners, in a group, leave the place in high spirit, marching and making sounds by beating their sticks and shields. Later in the day, both groups meet in their *bali*, celebrating the victory or alternatively reflecting on the defeat. In this stick fight, usually no fighter should seriously be hurt. As we have mentioned before, this type of stick fight is usually between young boys (same age group) from the same village.

Stick fights among the age group from different villages follow the same patterns we have just described. However, before they actual involved in the fight, they have to raise the case to *mojomiji* —telling the *mojomiji* that group "A" and group "B" want to fight tomorrow morning. In Pari, the two groups, each may consist of no more than thirty members bringing the total number to sixty members.

Once the requests have been approved by the *mojomiji*, the *Kabore* make the announcement to the public that group "A" and group "B" will fight tomorrow morning. Early in the morning, many people will gather in the *thworo* to watch the fight; these include men and women both old and young, boys and girls.

Before the actual fight, the two groups then come and site in opposite corners of *thworo* facing each other. Interestingly, they are not allowed to wear shirts or trousers, all must come in short pants or come completely naked. Some of the smear white ash on their bodies; this is a sign to show to their opponent that they do not fear death. Everybody put on a helmet woven of grass fibber (*likuluk*) and holding a hard stick measuring about 120cm long and a shield with a hand protector of buffalo skin. Then the two parties stand up and start fighting in the middle of *thworo*. As the boys fight, they produce huge dust that could be seen from a distance. They also make terrific noise of shouting, creaming and beating of sticks and shields. Thus, if the fight is between age group of different villages it often becomes difficult to control. However, once a member is beaten and has fallen down, the spectators immediately rush into *thworo* to help him and to stop the fight. The spectators act as a buffer of violence. If a person bleeds because have been hit by the stick, and the hit part become swollen, it is cut out to remove congested blood. The scar remains as scarification, which serves as a mark of stick fight of the past, engraved on the body (Kurimoto, 1996, p.287).

Figure 53 Stick fight

The defeated group will the run in disarray while trying to counter attack as they run. In the afternoon of the same day, boys' parties come back to thworo for reconciliation ritual. During the reconciliation, each member of mojomiji soaks grass known in Pari language as *manhnho* in water that is in a gourd bowl and sprinkles the water with the grass on each member of the two age groups. The two age groups sit as the *mojomiji* sprinkle them with water. When sprinkling, each member of the monyomiji will say a word of blessing and pleading to the two parties so that they live in peace. This ritual is called "Kuk". The *mojomiji* also give each group a fine of a goat for having fought. The two goats will later on, be slaughtered and eaten by *mojomiji* and elders of Pucwa village.

After this, all the mojomiji and other participants including the members from the two parties sit down to drink beer (kongo-kwete); this is the end of reconciliation.

3. *Local Government by Mojomiji*
There are two types of the local government in Pari society, these are collective and Village

based government. Although there are two types of local government in Pari, nevertheless, each of the local government is administered by the *mojomiji*. In the collective government mojomiji from the six villages take care of the entire Pari society; while in case of village based government a particular mojomiji in a given village take charge of its own affairs. However, there is no any conflict of interest as these two groups work. For us to understand how these two governments work, we are going to going to discuss into depth how each one work. Thus for the purpose of simplicity and clarity, we are going to examine first how collective government work and secondly we shall discuss how village base government works.

(A) Mojomiji as a collective Government
1.0 Dealing with the internal problems
The mojomiji become a collective government when the mojomiji from the six villages are charged with taking collective care of the entire Pari society. They meet regularly to discuss different issues, and each member of the mojomiji is expected to participate and give his opinion on the varioues issues put on the table. The meetings of the *mojomiji* usually last between one to one and half hours. The meetings are usually held in an open place outside one of the villages because *Kabore* cannot accommodate the big number of the *mojomiji*. Some of my informants told me that during the discussions, there would be heated exchanges of opinions. For example the issue could be to counter attach raiders; interestingly, some people may oppose mobilisation of every man in the villages to pursue the raiders. While others could insist that it would not be good to pursue the raiders, better rather wait for police investigation –although there is no police search in Pariland. Moreover, the rest of people call the *mojomiji*, who proposed the above opinions, as coward.

The main legation function of the mojomiji is to fine wrong doers among the youngsters; this always happen when the youngsters fight informally. The mojomiji can summon them, fine them, even beat them with a stick, or whip. As we have discussed, the fines for the youngsters consist of sorghum grains or a goat. However, the case of misbehaviour among the boys is usually heard and the judgement is passed by at a court held in the *kabore*. Thus, before the court is held, mojomiji choose, from among them, one person to be a judge, and he is called *mukungu* and the court is called *lokigo or lokiko*. The word mukungu is found across all the Northern Luo of South Sudan; the word is borrowed from Baganda in Uganda. According to Kurimoto (1996), the Anglo-Egyptian Sudan established native court in the region during the 1920's, and native court system of administration was introduced based on the Baganda model –on terminological level. Each mukungu is under the government chief who is in charge of the Pari as a whole. Therefore, it is highly probable that the present village court is a colonial legacy (p. 291).

This then bring us to another question, "What sort of spheres of life are under control of *mojomiji*?" We have already seen that in the case of stick fights by the age groups of youngsters, the *mojomiji* put them under control, prevent further escalation of violence and secure peaceful relations between the two groups.

The other function of the mojomiji is to wage-war (many or mwony), and to secure enough food. However, due to the civil war in south Sudan, in the 1980's, the military role of the *mojomiji* has been reinforced. The Pari youngsters, who were recruited in SPLA, usually return home with weapons of their own and they fight side by side with mojomiji in any military operations. It is also believe that when these young soldiers decided to go back to SPLA camps, they usually leave behind the acquired guns.

2.0 Dealing with External Enemies

The reality is that from the year 1979 onwards, there were many guns among the cattle owners in Sudan. Presumably the Toposa got their guns from the Karimojong, Jie and Dodoth of Uganda and Kenya, with whom they have close trade link. Historically, the Toposa share a common ancestry with these people. According to Kurimoto (1996), after the fall of Indi Amin's regime in Uganda in 1979, many automatic rifles found their ways to Sudan. The Toposa were the first people to obtain these automatic rifles and to exchange them for cattle to other people in the South Sudan including the Pari. Unfortunately, the Toposa did not exchange enough guns for cattle to the Pari, subsequently; they started to dispatch cattle raiding on Pari because they were less armed at the time.

However, cattle raiding in Pari are not considered in the level of warfare, they called cattle raid rather "aliipa or kuu", which is totally different from warfare. The 1982 attack, by Toposa on the Pari, was the first attack, following attack that was conducted in 1930. Thus, when the Toposa attacked the Pari people in 1982, interestingly, the raids continued at on and off Basisa till January 1983.

In Pariland, the news of forthcoming raids usually reach villagers late evening between 7.00 p.m. and 9.00 p.m. Once the news is received, *mojomiji* beat drum to pass the news to other *mojomiji* in the six villages. Shortly, hundreds of men from the six villages (including the three age grade) gather in the open place of a village and rush to the direction of the cattle camps with spears in their hands. Each person carrying between 2-5 spears; other people carry rifles. As they reach cattle camp, where raids happened, the *mojomiji* track the foot prints of raiders and cattle throughout the night until in the afternoon. If they get the raiders, they usually fight them and recapture some or all of their cattle. However, if they do not catch up with the raiders, they go back home. Oral history tells us that Pari are always the losers in cattle raids. Kurimoto (1996) notes, "The Pari, however, remains losers; they could neither recover the loss of cattle, nor take revenge on the eternal enemies e.g. Toposa; to their disappointment, investigation by police did not bear any fruit" (p. 293).

In the meantime, as the men are returning home, the *mojomiji* of each village order women to go with food and water to receive the returning chasers.

In all the above cases, we can see clearly that the *mojomiji* are responsible for military action and other counter measures in the entire Pari society. Moreover, we should understand that even though the victim is only one village, it is considered an attack on the Pari as a whole, so all the mojomiji of the six villages must be involved and take action.

4. Dealing with Internal Enemies

The people of Pari believe that, in a community, there are some people who have magic power, they can control the rain, birds and insects. Therefore, if these people are hurt by an individual or individuals, even though good rains and good harvests are expected, they can still send out birds and insects to destroy the crops. To the people of Pari, enough food means enough harvest of sorghum.

However, due to irregularity of rainfall, sorghum production is highly unsustainable from year to year, and from village to village within the same year. Not only drought but also access of rain may result in poor harvest. Consequently, one of the functions of the *mojomiji* is to find out such kind of people, with magic power, and take necessary measures to remove the course. According to Kurimoto (1996), he observes, "The efforts of the *mojomiji* to fulfil this responsibility often result in violent conflicts. He took it further by saying that between the year 1982 and 1984, four people were killed, three people expelled from the village, two people fled in fear of being killed, while the graves of three men were also exhumed. In 1980, the former rain-chief, Pidele, died and his senior wife called Nyibur, because their eldest son was still small, succeeded the office; she was killed in 1984" (p. 294).

Let us follow the scenario that led to the death of rain chief Nyibur. However, even though we are going to look into only one scenario, we should be informed that there were series of rain conflict between the rain chiefs and *mojomiji* in the years 1982, 1983 through 1984. Thus, the narrative of how rain chief Nyibur was killed is taken from Kurimoto (1996) and it reads as follows:

"In 1982, after the sowing in April there was spell of dry weather, although the fields of Bura and Pucwa eventually received enough rain. At first, the *mojomiji* of Wiatuo met Nyibur, the rain-chief, and asked her to make rain. In May as the situation worsen; Nyibur and Anyala, an elderly member of the Wiatuo dominant clan who was believed to have rain-stones, were summoned by the *mojomiji* of wiatuo. They, together with the wife of Anyala, were beaten with sticks and whips. Nyibur repeated an exercise that it was not her but Akudu, a female diviner-healer (ajwa or ajwaka) in Bura that had been preventing rain. The plea was accepted and Akudu was brought to Wiatuo. She was beaten and finally confessed that she had been asked by five people from Anywaa of Bura to stop rain in the Wiatuo fields. The mojomiji of Wiatuo ordered the Madan age group to go and arrest these five men. Two of them were captured. On their way to Wiatuo they were seriously beaten by Madan and died. The mojomiji of Bura, who had been categorically denying the allegation, reported the case to the police. Nine Wiatuo men were arrested and brought into police custody in Torit. Later they escaped and came back home. The case was never pursued. On hearing of the murder, an elder brothers of one of the victims killed Akudu in a rage for having revealed the name. This murder was not considered a case either by the mojomiji of Bura or by the police.

In June, the drought even deteriorated. The mojomiji of Wiatuo repeatedly threatened Nyibur and Anyala, as a result they fled from the village. At the beginning of August, the *mojomiji* invited an Acholi rain-maker to Lafon –we do not know the name of the rain chief of Acholi that was invited to Pari; however, we assumed that this rain chief might be Rwot Aburi of Obbo. It was said that he came with very powerful rain-stones and as soon as he arrived Lafon, rain started to fall. The rain, however, continued until September when no more rain was needed. The mojomiji asked the Acholi rainmaker to stop the rain, but it continued. They decided to expel him. He left Lafon in the middle of September. The harvest of 1982 was one of the poorest for the last ten years. Not only the drought but also the excessive rain in August and September had a negative effect on the harvest. Furthermore, in the month August, a flock of weaver-birds caused more damage in the fields of Pugeri and Kor" (Kurimoto, 1996, p. 295).

Shulluk

Demography and Geography

"The Shilluk believe the probably originated from the land call Karoo, in Lake Albert, Uganda, and their ancestors were led by Nyikang after he had a disagreement with his brother, Dimo" (Saturnino Onyala, 2014) "Movement of people and reinforcement of settlements from south to north along the banks of the White Nile, and also from west to east along the Sobat, are indicated in the historical traditions of the Lwo-speaking Shulluk and Anyuak.......The Collo (blacks) were created along the banks of the White Nile from an amalgamation of different people brought into a particular relationship with the person and institution of the King (Reth).......It is difficult to disentangle the original Nilotic sources from their presentation by ethnographers, but the assimilation of existing and neighbouring populations into the Kingdom, often represented as the result of s specific action by Nyikang or subsequent kings, is one of the dominant themes of Shulluk historical traditions......the Shulluk clan traditions represent only one process of incorporation; large numbers of individuals were adopted into the main settlements of Shulluk people, coming originally as captives, marriage-partners or voluntary migrants.....Fashoda became the permanent resident of the king and the centre of focus for the kingdom, restraining devisive tendencies and facilitating the assimilation of new people" (Douglas H. Johnson, 1990, PP. 42-43).

When Nyikang arrived in what is now called Shillukland, the land was inhabited by Funj and Baggara Selim Arabs. This land was originally called "the land of Otango Dirim". The Funj have dark skin, Nyikang fought them, they were defeated, some were expelled and others subdued and then incorporated into the Shilluk nation. Shilluk oral tradition says Nyikang also found some red strangers in the land of Otango Dirim. These were the Baggara Selim Arabs, they have light skin; Nyikang also fought and defeated them, some were expelled, while others were subdued and incorporated into Shilluk nation. Thus, through war and diplomacy Nyikang conquered the people, and through course of time, assimilated Otango Dirim, giving every tribe therein a name and ritual to perform.

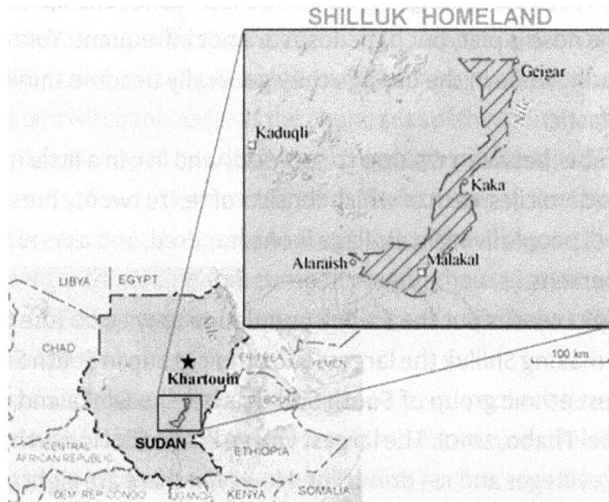

Figure: 54, Shulluk Kingdom (Source: Google)

By the second half of the century, what became known today as the Shilluk nation has emerged; in the old days they were not called Shilluk, rather they were generally known as Luo, the name Shilluk was given to them by Arabs later on. They originally occupied the area that now lies to the east of Bahr El Ghazal River—known as Sui River. The reason why Shilluk dispersed from Sui River remains a mystery to many of the scholars, which provides further challenging opportunity to the young Luo future researchers. However, Luo generally believe that they came from Wau, in South Sudan, and migrated to nearly all the neighbouring African countries that include Uganda, Kenya, Tanzania, South Sudan, Congo and Chad. According to Bethwel Ogot (1968), these Luo groups migrated from South Sudan from 1730-1760 A.D. The migration of the Luo to different parts of Africa resulted in the fragmentation of this large group into smaller but separate tribal groups that composed of many clans, and, in fact, the further they moved away from the next of kin, the lightly different accent they spoke and slightly different culture and traditions they practiced. The Shilluk, Anyuak and Pari have many of the words in common.

years and eventually Shulluk chiefs' umbrage at have been ruled by a woman; they then demanded that she step down. With all the pressures from the chiefs, Queen Abudok could no longer resist, she stepped down and a young man called Tugo was elected as her successor. It is said, before long, Abudok appeared in Fashoda with a bag of lily seeds, she strewed them about saying, "The royal lineage will grow larger and larger like this scattered lily seeds and spread across the country and fill the entire country".

As we shall see in this article, a divine spirit is passed from mortal body to another through traditional law and ceremony. According to Shulluk traditional law, King was to be killed after a fixed term of his services. The killing is usually done when the King shows some weakness into his body –that is when his weaken condition indicates that he is going to die.

In Shulluk tradition, King or Reth is not to die a natural death. As Crazzolara (1951) pointed out that the law of succession into divine kingship stipulate that the King should not die a natural death (from disease or wounds) but when he became weakened he must be suffocated (p. 134). The period of Chollo reign was asserted to be generally short (p. 130). The Chollo seem to share the view that a man in authority, like the Reth, should not live too long, but should make a room for successors (p. 134). In short, neither Nyikang the first and famous leader, nor his scarcely less famous son Daak, finished their days in the home country (P.131).

Death of Kings in Shulluk Country: Various ways King (Reth) been killed in Shulluk Kingdom
Thus, there are many ways that could lead to the death of the King:

(a) Waging war against the enemies:
The first way that leads to the death of the King is by him waging war against his enemies such as Dinka, Nuer, Nuba and Baggara Selim Arabs. In this battle, the King must take the lead so that he could be exposed to more danger and possibly death.

(b) Rival prince demanding a duel (Kingdom):
The second way the King could be killed by rival prince demanding a duel (kingdom), which they have the right to do so. However, when King has been killed with spear of prince, nobody should talk about it –it must be quite in keeping with the spirit of the law. Thus, if few persons become aware of the killing of the King, the matter will not be spoken publicly; if a person by mistake publicly brings out the news, this will bring danger to him. He might even be killed (Crazzolara, 1951).

(c) King kill by his own wives:
The third way that the King could be killed is through his own wives, if the King shows

sign of sickness and they see his is becoming weaker and weaker, they gather around him and suffocate him to death. It is important to remember that Shilluk traditional law and custom does not allow the death of King to be spread in the community. If the rival prince or wives kill the king, nobody is allowed to tell out to the community that the King has been. Crazzolara (1951) noted that if the King is killed by his rival prince, wives, children or relatives there is no legal case that could be compiled against them (p. 134). According to Evans Pritchard (1948) Shulluk people believe that when the King become physically weak the whole people might suffer and furthermore if a King becomes sick he should be killed to avoid some grave national misfortune, such as defeated in war, epidemic, or famine. Thus, the King must be killed to save the kingship and not only this, but to save the entire people of Shulluk. He went further by saying that there is no convincing evidence that any Shulluk King was put to death in either circumstance (P.414). As David Graeber (1996) cited Seligman 1911:222; Howell and Thomson 1946:10) that a Reth is said to be put to death when his physical powers began to fade –purportedly, when his wives announced that he is no longer capable of satisfying them sexually (P. 14).

The Shulluk customs seem to be changing with modern civilisation. In the old day, the King should not be sick, as we have just seen above that when the King become sick and his wives see him getting weaker and weaker they gather around him and suffocate him to death. In 2011 King (Reth) Kwongo Dak was seriously sick, instead of suffocating him to death, he was taken to Nairobi-Kenya for medical treatment. On his return to Juba, the King was received by Salva Kiir Mayardit, President of South Sudan, and King's followers. Not only this in the old day, King was not allowed to attend the Church, but today King attends Churches.

The three of my informants, Stephen Deng Lual, Fashak Deng and John Kuldit argue that in Collo custom, King (Reth) cannot be killed because of his weakness in sexual intercourse. He can only be killed (he can only disappear, as the Collo say it, when he sustains injuries on his body. When the King sustains injuries on his body it means that is the end of the Kingdom. When Akwoch Dok Kwanyiyek Reth (King) of the Shulluk people, who is regarded as the owner of the Shulluk country dies, he is believed to have disappeared together with Shulluk world (Piny away) he is believed has gone across the River Nile (*Reth log naam*).

However, writers like Evans-Patritchard (1948) claims that ritual king-killing was simply not real, it is only a matter of ideology; it is not something that ever really happened. There is no evidence to show that Kings of Shulluk were and are being killed in this fashion, subsequently, there seems doubt to claim that Shulluk Kings generally meet a violent death.

In my opinion Evans like any other western scholars, is not correct because he sees the history of Shulluk Kingdom through western lens, which emphasises that structure about African Kingdoms are not always true and accurate. However, the reality is that traditionally, foreigners are not allowed to enter the hut where the King dies, and not to monitor things that could have led to the death of the King. If foreigners were allowed to monitor and enter hut where King dies, this could enable him/her get better information. This is my conclusion i have drawn from reading reports on Shulluk Kingdoms by Evans Pritchard (2011).

In many ways when the King dies, a royal wife and her children retire to her father's home (pa-wigon/gol-gen). The children will then grow there; a royal wife seldom has more than two to three children. Traditionally she is not allowed to re-marry after the death of the King. The royal wife and her children are given cattle and other necessities.

However, according to my informant, Mr John Kuldit, who is an elder of Collo and lives in Melbourne-Australia, after the death (disappearance) of Reth (King), the royal wife and her children are not taken to her father's home (gol-gen), she and the children remain in Fashoda, the appointed King (Reth) will take over the wife, he also will take responsibilities for the children.

Moreover, other wives, who are considered not royal wives, and are called "baa-reth", they are also given cattle and other necessities and are allowed to remarry, and can live wherever they choose. Thus, soon after the death of the King they start recruiting men. As years pass by these *baa-reth* with their children, would form a village of their own. According to many of the Collo elders I have interviewed in Australia, they pointed out that King usually has many wives (between 20 and 50+ wives) and he (King) cannot have sexual intercourse with all of them. As such wives who the King does not satisfy them sexually developed secrete interests in some men to enable them produce children; customarily, when the wives conceived and delivered children from such relationships, these children would still be known as King's children. The elders argued that men from Shulluk tribe often fear to have sexual relationship with King's wives, even though they are been asked, for the fear that they might be killed. However, some of the workers and *askari* (guards) do take risk of having sexual relationship with King's wives, since they are closer to the wives than any other men in the society. Because King's wives find it difficult to have sexual relationship with Shulluk men, they resort to Dinka and Nuer men—yet the children produce from these sexual relations belong to the King and not to the Dinka or Nuer men.

Crazzolara (1951) pointed out that there are controversial reports about the death of Nyikang. The ordinary Shulluk are been always told, and have always believed that Nyikang has been carried away by wind to stay with God (Juok) in heaven. (Do you and the new generation of Shulluk also believe this?) Moreover, competent elders know better and differently that Nyikang one evening took a walk and never returned home. The people even never the direction in which Nyikang went and the place he died. The same thing happened to his son, Daak, he is said to have accompanied his father in exile—which means he followed him to heaven (pp. 130-131).

The explanation of the mysterious disappearance of Nyikang could be said to have come about because of the "never-ending disputes and quarrels among his people" in which Nyikang himself was unable to quench the problems and this made him decide to withdraw from his people. Interestingly, the most obvious disputes that led to the disappearance of Nyikang from home were rather disputes that were touching his royal position. This means, he was directly challenged by other elders or clan-leaders; it is said he was his royal position was even challenged by his own descendants (princes). All these acts were down played, since at that time, there was no particular traditional right to the throne. In all these challenges, they were saying Nyikang was not the first King, nor is he the founder of a royal dynasty of Shulluk. When some ancestors learned that Nyikang left home, they said, "He left home because he was the first person to disobey his father and quarrelled with his brother Duwaat".

We have seen above that the village where the royal children or princes grow up is called *paare*, and this is their village, although in reality this is the village of the uncles. If in the long run one of the princes becomes a King, he will still refer to *pare* as his true village. Although in Luo traditions, children generally refer to their father's village as their village, they do not normally refer to their mother or uncles' village as their village.

Traditional King spends few days of his leisure at *paare*, while normally living at Fashoda, his royal residence. When, he dies the body will be taken from Fashoda to *paare* where he will be buried. People then build Shrine (*Kengo*) for him at *pare*. More importantly, when the shrine becomes dilapidated (ruined), the people from the village (*paare*) rebuild it to preserve Shrine (*Kengo*) and heritage. This practice goes on in as long as the descendants of this prince (a person who became a King); continue to live in Shullukland (Crazzolara, 1951, p. 133).

Moreover, some studies show that in some case in Africa, Kings are killed –and this include Kings of Chollo, where a rival prince can pretend to kill his father or brother when taking over the throne. Thus, in the fact no killing takes place. In contrast, Frazer (1890) in his book entitled "The Dying god made it abundantly clear that the Kings been consecrated, their physical strength is tied to the property of nature, and that is why they could not be allowed to grow sickly, frail and old.

5.0 *Mystery of Kings two bodies: Spiritual and Moral Body*
Soon after the death of the King, it is said Nyikang's spirit leaves his mortal body and enters in a wooden effigy. Thus, when a new King is elected, he is to raise his army and fight a mock battle against the effigy's army and usually he defeats effigy's army and captures spirit of Nyikang from effigy, the sprit then directly enters into the mortal body of the new King.

In addition, Graeber (1977) cited Evans-Pritchard (1948) as arguing that there was no such king as a divine king, which means that Shulluk Kings were probably never ritually executed. Therefore, the installation ritual was not about transferring a soul, but rather about resolving the tension between the offices of Nyikang, that was set above everyone, including the current individual holding a sacred office. Therefore, it is the kingship, and not the King who is divine. Having followed the arguments of Graeber (1977), I would like to suggest that this argument offers an opportunity for readers and future researchers to detect the doctrine of the King's two bodies: political body or eternal office of the King and the natural body, which is the physical person.

More importantly, among the Africans, King is simply a sacred object through which profane society constitutes itself. This means that when we are looking for the King's sacred object, we should start examining it instead from the principle of his sovereignty rather than beginning examining from the idea of moral order and reality of violence. This from the aspect of African Kingship can be legitimately labelled as a "divine". Such creatures (men) transcend all ordinary limitations, whether they were said to embody a god is not the issue, but the point is that they act like God, and get away with it.

6.0 Selection of the King: The law of succession

To many of us, it is not clear if the Collo, in the old time, have established any regulation to save guide them during the selection of the King. In addition, there are no mentioned of how the law of succession operate in the past. As we have seen above, the rules and law of succession came after the reign of Nyikang, his son Daak, and other Kings down the line. According to Crazzolara (1951), Nyikang and his son Daak never cared to be officially installed (*reny/rony*) as King, which is an important ceremonial rite among the different Luo groups. Due to lack of royal traditions, Nyikang and his son and several of their succeeding princes (*nyi-reth*) were just headmen or Jaago of their villages. Therefore, lacks of fixed traditional previsions on kingships, bound to have repercussions on the history of the Shulluk. The first King who has been affected by this lack of tractional rules was Nyikang himself and his immediate descendants.

The most prominent among the sons of Nyikang was Daak; he therefore, might be counted, correctly, as successor of Nyikang in an official list of successors of Collo's Kings, while others, who are said to have lacked ability, were considered village Kings. Interestingly, such village kings occur at various times and places in the history of the Collo (Crazzolara, 1951).

The law of succession was also introduced by King Tugo, in which the law says, "When the King dies, his sons have right to lobe for the throne. The children of the previous Kings have also right to lobe for the same position". The rules fooled in choosing the new King are as follows:

- ▸ He should not be left-handed
- ▸ He should not be injured in any form or shape
- ▸ He should be born during his father reign
- ▸ He should not be the elder son
- ▸ He should not be condemned in any criminal of murder
- ▸ He should be well disciplined

Each village under chief (Jaago) always participate in the selection of the new King from among the princes, who are direct sons of the late King or the previous deceased Kings. Under the law of succession, many elderly princes aspire to the throne. Only one prince will be chosen; after the selection, the new King is to be installed at *Debalo-Kwom*, a place near Fashoda, according to the set rules, as we shall discuss under installation of the King.

In Collo oral traditional history, the only son who succeeded his father without lobbing was Okon wad Tugo, the first son of King Tugo. Oral history tells us that when King Tugo died, his son was unanimously elected to the throne. However, the only obstacle for all the princes aspiring to the throne, is the ruling King. If the ruling King does not die soon, the period for succession may be prolonged. As discussed before, a rival prince often finds ways to kill the ruling King. Sometimes, a rival may reach an agreement with the closest relatives, and these closest relatives may organise to kill the King. Mr Stephen Deng, a Collo elder who lives in Melbourne-Australia, argued that only the family who wants to lead will try to kill the ruling King.

According to Collo oral tradition, the mother of nominated King, prior to taking over the throne, trains the inspiring prince. The mother teaches him all the requirements of a good King; and these include the ability and characters of good King (be a good person, be responsible, be brave, be wise and be friendly to entire Shulluk people). Moreover, as the prince waits for the position of the King, he usually holds the position of a village chief – in *paar*. As *paar* chief, the prince decides on various cases, receiving gifts from the local people, very much like the King. The prince does all these to make him popular, kind and generous to everybody. Therefore, if he does not do these things, he would hardly stand a good chance to become a future King.

The neighbours usually call him and address him as King with woo (*pak*) of the King; this is done to flatter him. The true King at Fashoda never struck against such prince, if he does, he would hurt the feelings of the people, and as a result, the King becomes automatically unpopular. It is important to be born in mind, that chiefs and elders, who are the natural advisor of the King, would not allow the tensions to continue, they would suppress these tendencies. Yet, we can say without hesitation that oral histories of Shulluk acknowledge such cases (Crazzolara, 1951).

Looking through the various answers given by the Collo elders in regards to the election of the King, I am made to categorise the elections of the King into three theories as we shall discuss below:

Election of a King

It is surprising to many to learn that people of the same ethnic group and cultures, speak different language about the same thing. Although there are similarities in the answers given by the two-different groups of Collo, the question remains, "Which of the two responses, is the true answer to the question of election of King in Shulluk Kingdom?"

When I was doing my research in the election system of a King in Shullukland, on one hand I was informed by Collo elders that in Shulluk a King is elected by the magic of flint stones; and on the other hand, I was informed that the election of a King is entirely done by the Council of Chiefs (Jaagi Pa-Diwaad) from among the Nyiredhs.

According to the three men, John Kudit, Fashak Deng and Stephen Deng Lual that I have interviewed in Melbourne-Australia they stated that there is no election of the King in Shulluk society, rather the King is usually appointed by the family of rivals. And chiefs are there to facilitate the appointment. It is reported that some of the candidates to Kingship sometimes bride council of chiefs by giving them some cows; so that they (council of chiefs) give him the opportunity to become a King (Reth). After the selection of the candidate, by members of family, the council of chiefs consult the chief where this candidate has come from; the council of chiefs want to know about the characters and behaviour of the candidate. If the candidate is found to be a coward person, he would be disqualified because they people of Shulluk want a brave King to face intruders.

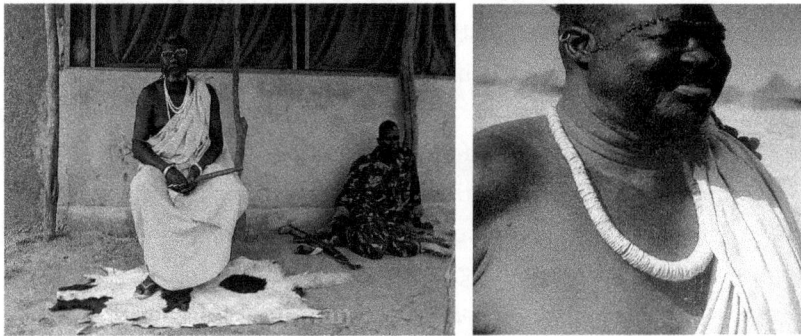

Figure: 55, His Majesty Kur Papiti left and King Aney wa Kur Nyidok right

However, the bottom line is that, for a prince to be chosen, he should have been born during the reign of his late father. He should not have scars of whatsoever on his body. He should not be known to be coward as he grew up under the supervision of the chief of the village in which he was brought up. Dr. Daniel Thabo (2010) pointed out that according to the Collo system of succession, any prince with any abnormal wound in his body cannot become a King or tone nominated as a candidate.

Nevertheless, previously, it was not uncommon that a powerful son of a King would occupy Fashoda or eliminate the reigning King and assume kingship; thus, circumventing the election process. Today, the kingship rotates in turn between the three remaining families and these are Papiti wad Yor, Aney wad Kur and Dak wad Padyied families (Simon Deng, 2013).

The competition between sons of King for the kingship is normal but these been pushed to bloody heights in some episodes in Shulluk history. The perfect example is of King

Dekwor wad Tugo, who is the twelfth King in a line; he made it a point to eliminate all sons of Kings (Nyiradh) he could lay hands on, except his own sons. Subsequently, the three of his sons called Muko, Waak and Dyielguth, and in this order, they became Kings, making succession after succession without any competition.

Yet, when Kudit wad Okon, who managed to escape the spears of King Dekwor wad Tugo, came to power in 1770, he decreed that Dekwor wad Tugo's family will no longer be entitled to become King in Shullukland. Interestingly, oral history of Shulluk tells us that since the installation of King Kudit wad Okon, no member from the family of Dekwor wad Tugo became a King in Shullukland. It is reported that, more still, the family right to throne by Dekwor wad Tugo's sons, were even eliminated by King Yor wad Kudit, the son of King Kudit wad Okon, who was elected in the position of his father. By the end of King Yor wad Kudit's reign, there simply were no sons of Muko, Waak and Dyielguth (Nyatho) still alive to compete for the position of the King. Another interesting case is the case of King Aney wad Yor, who also attempted to deny his half-brothers the chance of becoming King by removing their lower front teeth. Needless to say, some Shulluk traditions and customs seem to have lost their values or originality. As a result most of the archaic traditions have been dropped, while some have lingered on although are transforming under the pressure of modernity (Simon Deng, 2013).

According to Evans Pritchard (2012), in the kingship of Shulluk society, it is not the individual King at any time and point that reign, but rather it is Nyikang, who is resigning as the King. He is believed to be a mediator between men and God (Juok), and furthermore, it is believed, in some way that Nyikang is God himself, since he works in any King-elect. This fact is derived from Chollo logic that says, "Nyikang is the Reth (King) but the Reth (King) is not Nyikang". Thus, the participation of Nyikang in the Kings' reign, raises the kingship to a plane above all sectional interest, whether local or of descent (p. 414).

It is important to notice that the entire Collo share in the kingship; nevertheless, their loyalties may sometimes pull them apart, when they are faced with some serious cultural matters. Because in Nyikang are centred all those interests, which are common to all the people, which include success in war against enemies (foreigners), the fertility and health of men, safety of cattle, growth of crops and the life of wild animals which are to services men. Nobody has a saying these things. This implies that everything Collo value most in their national and private life has its origin in Nyikang. Because of the mystical values associated with the kingship, and centred in the person of the King, he (the King) must keep himself in a state of ritual purity by performing certain actions and observing certain prescriptions shall be discussed later.

For this reason, I find it difficult to reconcile the two answers we have mentioned before, that is about the election of a King. Therefore, I think that for the important of Shulluk traditional history and culture, and more importantly not to miss out elements in the election system, I decided to put the two views and answers under two separate articles as, "First Theory and Second Theory of Election".

(i) First Theory of Election of a King

According to the oral history of Collo, when Nyikang brought his people from Wi-pac (in Bahr el Ghazal) into the present Shullukland, he also brought with him some of the Nubians. However, it remains unclear as to how and when the Nubians came and settled with the people of Nyikang in Wi-pac that led them to move along with Nyikang when he was leaving Wi-pac. Nonetheless, it appears the Shulluk have been in some political connection with the Nubians.

It is reported that when the Nubian arrived in Shullukland, they were scattered and live in several villages among the Collo up to today. This fact is known by the Collo, although the people do not talk about it publicly. In their appearance, the Nubians of today in Shullukland do not differ from the other Collo.

Historically, when a King is to be selected, Shulluk chiefs summon the descendants of Nubians, because they play an important role in the constitution of the Shulluk dynasty. They are then sent to Nuba Mountains to bring some flint stones to be used for selection of the new King. It is generally believed that flint stones determine which inspiring prince is eligible for the election.

As soon as the Nubians come back from Nuba Mountains, the chiefs and elders soon organise a day for the election. The inspiring princes are then summoned and each is asked to bring a stick. On the day of election, people gather around to watch the mystery of flint stones. Fire is lead and flint stones put into the fire. The first inspiring prince is then called and one of the chiefs asks him to throw his stick into the burning fire. At this time, the people will be looking the colour of smoke. Shortly, fire is blown off, and if the smoke remains black, this means that a prince who threw his stick into the fire cannot become a King.

The chiefs blow up the fire again, and the second inspiring prince is asked to throw his stick into the fire. He then throws his stick into the burning fire. At this time, the people would be looking at the flames, if the fire flames up a little, the prince that threw his stick into the fire cannot become a King.

Again, the chiefs blow up the fire, and the third inspiring prince is asked to throw his stick into the fire. The inspiring prince throws his stick into the burning fire. At this time, the people would be looking at the height of the flames, if the fire burns up a little high, the prince who threw stick into the fire cannot become a King.

The Chiefs will again blow up the fire, the fourth inspiring prince would be asked to throw his stick into the fire. The inspiring prince throws stick into the burning fire. At this time, people would be looking at whether the fire will continue to burn or it will quench.

The inspiring prince then throws his stick into the fire; if the fire quenches, the prince who threw his stick into the fire cannot become a King.

The fire again blown up, another inspiring prince asked to throw his stick into the fire, and the prince throws his stick into the fire. At this time the people will be looking at whether the fire burns with a big flame and blazes up, if the fire burns up with a big flame and blazes up, this is indication that flint stones have selected the prince to be elected to throne of a King. When people see fire with big flame, they will shout with one voice, "Wauuuu! This is the King".

According to Hofmeyer (1910), before King Nyikang, Shulluk used to throw a number of little stones into the fire, each stone beared name of an inspiring prince. The stones are thrown into the fire one by one. The people would be watching at whether that stone cracks or not. The stone that cracks reveals to the chiefs and elders that those princes cannot become King. And one whose stone remains in the fire without cracking becomes a King. Moreover, it is important to know that this test is repeated until only one stone is left in the hand. Since other stones cracked up in the fire, the prince whose name was allocated in this stone automatically become a King without throwing the stone in the fire.

After this, all the chiefs return to their respective villages; and the King now continue to rule the people according to the Shulluk traditional laws.

(ii) Second Theory for Election of a King

The Second theory of election of a King among the Shulluk states that council of chiefs elects new King. Thus, when the day for election comes, the great chief summons chiefs and elders. They gather to discuss the procedures of election and possibly examine the life history of each candidate (prince). At the end of the discussions, the chiefs (Jaagi-Pa-Diwaad) then elect the new King from among the inspiring princes who are sons and daughters of the previous Kings. It is a condition that the father of the candidate must have been installed King according to the tradition (Simon Deng, 2013). On the contrarily, Evans Pritchard (2912) argues that when people from various parts of the country gather, prince is then seized and taken to his on village, there he is placed in a big open space, where people publicly elect him. But as we have seen above the King is never elected but only appointed by members of royal family.

Moreover, Evans Pritchard (2011) revealed that Mr P. Munro of the Sudan Political Services, who was an eyewitness of the investiture of King Fafiti Yor, the 29th King, in 1918, described the true phases of the investiture of a new King. Further, more, he pointed out that the accounts by Mr However (1946) and Mr. Thomson (1945) detailed the observations on the death of King Fafiti Yor in 1943, as well as the election and investiture of his successor, King Anei, in 1944.Therefore, their descriptions accommodate the Second Theory of election of a King. Evans went on to say that, on the death of a King his corpse is walled for some months in a hut and his bones are then buried in his natal hamlet, and not in the royal capital.

The remains are disposed of, and the mortuary ceremonies conducted by the royal clients (*ororo*) and number of the royal clan, the head of which is the chief of Fadiang. It is more a clan, than a national affair. On the contrary, the election of the new King, which takes place few days after the death of the King, is an affair of the whole Shulluk nation, who participates in the election through the chiefs of Mwomo and Tung.

From this review, I wish to emphasize that the procedure in the election of a King must ensures that, the prince to be selected to the throne, must have the backing of the whole country chiefs. Although some writers like Mr Thomson (1945), asserted that "the choice of the new King rests entirely with the chiefs of Gol Dhiang and the chiefs of Gol Nyikang.

It is important to notice that the election of the new king cannot take place unless the northern and southern chiefs agree; yet, this agreement must be reached after convincing arguments from both speakers. Mr Thomson (1945) and Mr Howell 91946) call this process a "electoral college", in which only conscious survival of the traditional structure of Shulluk tribe only listen to the decision of the chiefs starting from Mwom to Tung. However, in the investiture, all sections of the population are presented, which include royal clan, the disposed branch of royal clan, the commoner clans, especially those whose ancestors were among the original followers of Nyikang and the client clans (i.e. people from the clan of the new King). All these people have essential roles in the drama of investiture (Evans Pritchard, 2011).

According to the three informants that is Stephen Deng Lual, Fashak Deng and John Kuldit the Electoral College, as mentioned by Evans Pritchard is a system where every Reth (King) must go through; and all chiefs starting from Tung to Mwom must attend.

After the election or appointment of the new King, people reassemble in the court that is marked for Nyikang. Here, the people lift their hearts in prayer to both god and Nyikang. While people are praying, the newly elected King is held fast by the spirit of Nyikang. When people finish prayer, each person will sprinkle the body of the King-elect with holy water. Mr Fashak Deng said, *"Some people would be beating drums and holding the wing of the bird called "Owang" on the head of the Reth (King). While other people would be singing Nyikang's song, in order that the spirit will install a new King. This process will continue until the spirit of Nyikang accepts and appoints the new King"*. The King elect is then taken to another yard, there, the wives wash him and the body is covered with royal clothes, The King-elect is then given a royal spear and asked to enter into a royal hut, where he stays for one day and one night.

The next day, chief and elders gather again and the people of Nyikang are summoned. The question is then, where do the people of Nyikang live? The people of Nyikang live in Akurwa Village, this is the only village in Shullukland recognised to be the village of Nyikang. Shulluk people say Nyikang disappeared in a dust storm at Kurwa; they also believe that in the last day, Nyikang will return to Akurwa. More importantly, there are many shrines of Nyikang around Shullukland, but Akurwa is by far the most important place. It is in Akurwa where the effigies of Nyikang and his son Daak are kept.

It is from Akurwa that the people bring with them a woody statue of Nyikang and a wooden statue of Daak. Evans Pritchard (2011) observed that effigy of Nyikang and Daak are kept at Akurwa in the most northerly district of Shullukland, the effigies then brought out by priests and accompanied by army of Mwomo, who will hold mock fight with the clan of King-elect, for the possession of the capital. When effigies pass through each district, the people gather to pay their respects to Nyikang, for it appears that during the interregnum the effigy is believed to contain the spirit of Nyikang (the effigy is Nyikang himself).

On arrival, Nyikang, army of the north meets in mock combat with an army of the south, who are supporting King-elect outside the royal capital. This symbolizes the ceremonial division of the country into two moieties (parts of Shullukland). The balance between them is that the army of the King-elect is defeated and the King-elect himself is captured and taken to the capital by Nyikang —this is exactly what they mean by their saying, "The kingship captures the King". At this time, the two wooden statues are beautified with ostrich feathers. The great chief then welcome the people of Nyikang, as they are been welcome, the people cry out to the great chief with one voice: "Nyikang has come!" there, the wooden effigy of Nyikang is placed on the royal stool. Shortly, the wooden effigy is then removed and the King —elect is given stool to sit on. As the King sits on stool, the spirit of Nyikang enters his body, causing him to tremble. This is then the time when people would say Nyikang has possessed the newly elected King; and thereto he becomes a King (Evans Pritchard, 2011).

Soon after this, people bring a cow to be slaughtered for the welcome of the new King and Nyikang to a royal capital. Interestingly, when people bring a cow they also bring a man; and the man is stripped off his clothes, wrapped and put to lay on his stomach in the middle of the road. The cow is killed the same road, where the wrapped man is made to lye, the people put the new King in the middle and start marching towards the wrapped man. Each person walks pass over the back of the man —they step over him.

When everybody finishes stepping over the wrapped man, people of Nyikang come with a whip and imagining that Nyikang has caught a girl from the clan of the Kwa-okal.

"Shulluk oral history tells us that the clan of Kwa-okal came from Bahr El Ghazal, their ancestor was a relative of Nyikang. Chollo classified Kwa-Okal clan as ordinary Shulluk, because in the past they had committed crime against Nyikang and now, every time a new King is to be elected, Kwa-Okal clan has to pay the price by giving one of their girls. Traditionally, this girl has to stand beside the new King throughout election ceremonies" Crazzolara, 1951, P. 138).

In turn, the elders and chiefs of Kwa-Okal clan are supplied with clothes, beads, bells, lances (a long spear use by soldiers in war). While the relatives of the girl are given cows and oxen; and the girl now is called "Nya-kwer", which means "a child belonging to the authority of the King".

The cow is then skinned and the meat taken to women to cook; when the meat is cooked, women called a selected person to say a word. The same person also prays on food before it is distributed to the people. It is important to note that such kind of food is distributed by selected persons.

The people of Nyikang then approach the Kwa-Okal clan with a whip and start to beat them. Customary, the people from Kwa-Okal clan are usually not allowed to resist the mock fight, they have to run away for fear that the people of Nyikang may beat and besiege them. Not only this, traditionally, everyone who is beaten would be confined in a particular place and later on asked to pay fine to the people of Nyikang. The fine usually consists of 10 cows and occasionally with twenty sheep. However, the people from Kwa-Okal clan do not want this; because they believe that they have already paid a big fine —with their girl.

More importantly, when the people of Nyikang are performing this customary cere-mony with the clan of the new King, other tribe take privilege to perform different cere-monies. One clan may be chosen to fane the King with ostrich feathers; another clan may be chosen to secure dura sticks, which are symbols of defeat by the King-elect to other enemies.

According to Evans Pritchard (2011), no girl is seized from Kwa-Okal clan for the ritual of the newly elected King. It is rather a newly elected King who marries a girl from the clan of Kwa-Okal, because this girl has an important role in the ceremonies of investiture. Thus, when the newly elected King is put on the throne, Nyikang immediately seizes the girl from his side and refuses to surrender her to the new King on the ground that she was married with cattle from the royal herd, which is Nyikang's herd, and, therefore, the girl becomes Nyikang's wife (Evans Pritchard, 2011). As we have discussed before, when a new King is put on stool, the spirit of Nyikang would possess him —this means the new King is no longer himself but rather he is now Nyikang. This explains why the girl now becomes the wife of Nyikang. Nyikang is always King of Shulluk and when a King dies his spirit is conceived of departing in some manner from the dead King's body to take up its abode in the new effigy specially made for it accommodation at the shrine of Akurwa. Thus, by entering anew into the body of a new King, Nyikang is said to be once again ruling the Shulluk country from his capital at Fashoda.

Soon after Nyikang seizes the girl from the new King, the supporters from the side of Nyikang and the supporters from the side of the newly elected King will be summoned for a second mock battle. During this mock battle, the newly elected King captures the girl from Nyikang, and the girl now becomes his wife. Thereafter, Nyikang pays a visit to the new King to make peace with him; since they have been in war over the girl. When Nyikang is going to make peace with the new King, the chiefs always accompany him. After peace has been reached between Nyikang and the new King, in the morning the chiefs then give homage and exhortations to the new King, he is then given the green light to rule as a good

King. From this moment onward, Nyikang does not again contest the King's authority and some weeks later on the effigies are sent back to the shrine at Akurwa.

The most adequate interpretation of the succession of rites of investiture, would therefore, seem to be that, when the effigy and the new King make the first mock battle for the possession of the capital, the army of the effigy is victorious because Nyikang is said to be in the effigy. Again, when Nyikang and the new King make the second mock battle over the bridge, the army of the new King is victorious because Nyikang is now in the new King. The power has already passed from the Nyikang of the shrine at Akurwa to the Nyikang of the new King at Fashoda (Evans Pritchard, 2011, p. 417).

After this a bull is killed and its meat eaten by the men of particular clan called Ororo' who are said to have been degraded from the Reth clan or lineage. Then the Okurwa men carry the image of Nyikang into the shrine, and Ororo men place the King elect on the sacred stool, where he remains seated for some time, apparently till sunset. When the King gets up, the Okurwa men carry a stool back into the shrine, and the King is escorted to three new huts, where he stays in seclusion for three days. On the fourth night, he is taken quietly to his royal residence at Fashoda, and the next day the King can then publicly show himself to the people. The three new huts in which he has spent the days of his seclusion are then broken up and their fragments cast into the river.

According to Shulluk tradition, the new King must prepare for the last funeral rites (*wowo*) of the late King (*Reth*) before he is installs. It is only after the last funeral rites of the late King is held, that the new King would then embark on his own installation process. Once the new King has organised the last funeral rite of the late King, he would the starts the process of his installations by procurement of ritual items (*jami-kwer*). If the ritual items are all collected, and Nyikang accents to the choice of the council of chiefs (*jaagi-wi-pa-diwaad*), then the day for the coronation (*kwer rony*) begins.

How Collo View Their Kingdom

Before we continue with our discussions on the installation of the King, I think it is appropriate for us first to examine the perception of Collo about the disappearance of Nyikang and what impacts it has on the ritual ceremonies. The transitional period when Nyikang disappeared (died) is known as "wang yomo", which literarily means *period without authority*. The Collo described this period as "dangerous period", because if a person kills another person during this period, the offender must be ready to protect himself/herself at all time, reason been that the relatives of the victim can attack and kill the offender any time.

According to Shulluk customs, when Nyikang or a King dies, he is not said to have died but rather he is said to have disappeared across the River Nile, or he had made his final refuge across the river.

This belief originated from the time of Nyikang's death, which is believed, by all the

Shulluk people, to have disappeared in a storm on eastern bank of the River Nile, a place opposite Moroo clan in Mwom area. Some of the Collo elders asserted that Nyikang decided to disappear because he had failed to manage the Shulluk tribal affairs (Collo Adhalla Nyikang). Crazzolara (1951) noted that after the disappearance of Nyikang and Daak, the leadership of Shulluk tribal group remained with his descendants. The first seven successors of Nyikang followed each other in direct line, a son succeeding his father, but other successors were not, from indirect line, they could be cousin brothers (P. 131).

For us to understand better the immorality and eternity of Nyikang and what views the Shulluk have about their Kingdom, let us compare these two statements from Evans –Pritchard (1962) and Akwoch Dok Kwanyiyer (2008). The two statements show that Nyikang has a divine spirit and borrowing the body of elected king gives him a continuous existence.

The Shulluk live under the king. The king is believed to drive everything from Nyikang, the leader of the Shilluk in their heroic age. Nyikang's spirit dwells in the king and passes down the line of successors from king to king. The rule of succession is that only a son of a king can be invested with the kingship. The election of a new king is something that concerns the whole tribe, and tribal members participate through the chiefs from the two parts into which they are divided, northern and southern. To be elected king, a person must receive the backing of the whole tribe. In addition, for the enthronement ceremony that takes place approximately one year after the person is chosen, the collective participation of both halves of the tribe, the north and the south, is necessary. Rather than fulfilling legislative or administrative functions, the Shilluk king engages in ritual duties. On important occasions, especially when praying for rain or for victory in battle, the king performs sacrifices. Nyikang being immoral, the kingship is an abiding institution binding the past, present, and future. For the enthronement ceremony the effigy of Nyikang is carried forth from its shrine and carried to all regions of the tribe, after which it is placed on the royal stool. After some time it is removed from the stool and the king-elect sits upon it. At this point *Nyikang's spirit enters the king's body and makes him tremble*. This shows *that he has been possessed by Nyikang*, and he officially becomes king. When the king dies, his spirit (i.e. Nyikang) leaves his body and takes up residence in a new statue especially made for the purpose, whence once again it will enter into the body of the new king and performs its functions. In this way Nyikang is an eternal, immutable being, a divine spirit who borrows the body of the king to continue in existence; this is what gives the kingship a permanent, sacred place in that society (Evans-Pritchard, 1962, PP. 117-118),

The Collo believed that Nyikang, who has special gifts and wisdom from God, as witnessed in his special physical appearance, founded their Kingdom. He managed to form his Kingdom by confirming the special practices as roles to the individual members of Shulluk tribes or clans who had joined him. These assignments or practices, continued after the disappearance of Nyikang; and are renewed every time the Shulluk elect their new King (Akwoch Dok Kwanyiyer, 2008).

(1V)The coronation of the new King: Installation
"Ronyi Reth mi Collo"

We cannot emphasize in this book that there are many Chollo young people, as well as majority of elders do not know what take place during the installation or coronation of the new King (Reth). However, it is my hope that this book will help inspire many of the Shulluk people as well as those who are interested to know about Coronation of Collo's King (including foreigners).

The installation of the Reth of Shulluk attracted the attention of many contemporary scholars, both from within and outside The Federal Republic of South Sudan. As a result, few articles about coronation of Shulluk King started to appear in various newspapers since 1918.

The coronation procedures are usually done with very strict regulations by Akurwa clan who are assigned to take responsibilities in the coronation process (Akwoch Dok Kwanyiyek, 2008). But some people say the coronation is usually done by Ororo clan.

Shulluk tradition requires many things for Nyikang to coronate the King-elect. The people of Mwomo (Gol Dhiang at Abwor) and the people of Tango (Gol Nyikang at Akwabayo in Kwom) collect these materials; they collect the materials according to assignments given to them, as discussed below under the sub-title "step by step in the installation process". It takes months to collect ritual things, since the actual coronation of the King-elect takes place one year after the nomination or election. These materials include silver, cloths, ostrich feathers, bamboo, sacred ropes, skin of female antelope, ritual stool, palm leaves, sorghum sticks and royal spears; they are to be collected prior to the day of coronation.

Step by step installation processes

The Shulluk Kingdom is well set from Reth (King) to chiefs (jag); there are 15 chieftains in the society —each area with a chief. During the installation of the new King, chiefs order particular clans to do particular thing. E.g. collect silver, bring palm leaves etc. Now let us examine how these installation processes are carried on from step No 1 to Step No 20.

1. *Preparation for installation*: According to Shulluk oral history, the preparation for installation usually starts during the harvest of a new sorghum to enable people of Biw to collect special sticks from the new sorghum (dura) for the mock fight or battle at Fashoda over the prince (Nyikwer). The mock fight usually takes two days before speech day (as we shall see later). However, sometimes mock fight takes place two days after the enthronement of the new King. The special sticks of sorghum are collected by the descendants of the princess of Biw whose representatives take

all the gifts and cows, which are brought from Lwak (Southern Shullukland) for the installation of the new King (Akwoch Dok Kwanyiyek, 2008).

However, according to Okwom Nyikang, one of the Shulluk elders and my informant who lives in Melbourne-Australia disputed that the mock fight is only on the day of coronation (i.e. one day). Okwom said that in the old day the mock fighters used to aim at the bodies of people, which eventually caused harms as some of the sorghum sticks felt in the eyes and the bodies of people. He went on to say that today the culture has changed; thus, instead of aiming at the bodies, the mock fighters from both sides count together: one, two, three, four, five and they throw the sorghum sticks on the ground together.

2. *Collection of silver and cloths*: As part of the preparation for the installation, the chief sends the Kwa-Dikwor of Akwabayo of *kwom* to the northern frontiers of Maanyo, and they proceed as far as North Kosti to collect silver, spear handles for making rings for the new King. The two rings (*Ateg-Nyikang*) are to be worn on left and right hands; more importantly, by bringing such things (silver & spear handle), it marks the starting point of installation (rony). In the old time, the people of Mwomo used to collect silver and cloths from the local people, the work is done by selected people who go from village to village and collect these things. Today, as there are no silver and cloths in Mwomo district, men are sent to buy silver and cloths from North Kosti for the new King (Dr. Daniel Thabo, 2010).

    According to Okwom Nyikang, the case of silver and clothes started at the time of Reth (King)...........when Mr. Mougo learned that his brother was going to be coronate as a King, he bought silver and spotted (black and white) clothes as a gift for him. Soon after that, it becomes a custom to Collo that every time a King is to be coronate he should be presented with silver and spotted (black and white) clothes.

3. *Cutting palm leaves*: As soon as the people of Kwanyikwom return from Kosti, the Akurwa chief again sends the first comers from among the Nyikang's army (Kwa-nyikwom) of Akurwa (Anongo) with spears (*alot*) to Wod-Jouk-Awa (Omulka dhok-them) to cut the palm leaves (*bach can*) from Adodo with their special spears. The palm leaves will be used by the people of Kwanyikwom of Akurwa to make whips, with which they protect the new King and to keep the security at the time the King-elect is walking through the "*chor-paabo*", and on his way to the "kwom" (throne), this the important period of coronation –especially to protect the infiltration of unwanted princes.

4. *Leopards skin, elephant task, Rings and spears:* When silver, cloths and palm leaves have been collected by Kwa-Dikwor and Kwa-nyikwom, the chief will then ask the King-elect to look for the leopards skins and elephants tasks. King-elect will wear these leopards' skins after his coronation, some of the leopards' skin would be used

to cover the secrete stool (kwom) of Kwanyiyek. At the time, the King-elect is looking for leopards' skins, blacksmith from Kwa-Nyikang (descendants of Nyikang) of Golbanyo clan, at the northern side of Fashoda, will be busy making royal spears and rings for the new King. The work of blacksmiths usually follows the return of Kwa-Dikwor from Maanyo –where irons and shells are found. The work of making spears and rings, for the King-elect, usually start at Gotbanyo and finishes at Fashoda. While in Diballo, people will be busy carving ivory rings (*gial*). As people carve ivories, some people fix branches of thorn trees in front of their villages and on the road leading to the village. This group of people are been authorised by tradition to stop any man passing along that road without giving them gifts and offerings. In the same day the people of Tonga, who live far south, enter and paddle their canoes, they go to Lake No or Lai Shoropabo (stream) to look for *Lai Nyikang* (hippopotamus).

5. *Hunting of water buck (Gyiek)*: When the works of blacksmiths are almost going to finish, the people from Kwa-Ocwako of Nyilwak would be given permission to go hunting water buck in the country side. According to Okwom Nyikang, an informant from Melbourne Australia, the water buck is not hunted by the people of Kwa-Ocwako of Nyilwak, rather the water buck is hunted by people of Panyikang. And before they go hunting the water buck, the people of Kwany-nyiroo must give them green light to go ahead. Nobody in the society is allowed to kill water buck, and if somebody kills it, the King will give him heavy fines that range from 40 cows to 50 cows and sometimes even more.

6. *Bamboo (Bol Kau)*: As part of the preparation for the coronation of the King-elect, and before the coronation the people of Phading (Akwabayo clan), north-west of Fashoda, are sent to Fungor, in eastern Nuba Mountains to collect holy bamboo that will be used in the coronation. The bamboo is used by the Kwanyikwom of Akurwa as a symbol of Daak wod Nyikang.

7. *Sacred ropes*: The collection of the sacred ropes is the responsibility of the people from Daag clan. When the time of coronation draws nearer and nearer, the people from Daag clan leave for Panyilang to collect sacred ropes. According to Dr Daniel Thabo (2010), it takes two months for Daag clan to go to Panyilang because the place is far from the village of Daag clan. The sacred ropes are made up from the fibre of Dom palms (*bach kan*).

8. *Skins of female antelope*: The female antelopes are hunted by the people of Panyikang and before they go hunting, drum is beaten, men gather and dance. After the dance men cross River Lola, using canoe, each man would be carrying two to three spears. Customarily, men have to kill only one female antelope; its meat is eaten by hunters and only its dried skin is taken to Fashoda. Often men also killed

some male antelopes. The skin of female antelopes would be worn by the new King, the skin of female antelope is important for the ritual ceremony for the new King; Interestingly, all the skins of female and male antelopes are usually produced by chiefs of Panyikang, who live in the island of Panyilang. In Shulluk customs, female antelopes are royal animals, they must be treated with great care. Nobody is allowed to kill female antelopes in Panyilang Island. The female antelopes are only killed during the investiture of the King-elect. However, there are no customary attached to male antelopes, as a result people are allowed to kill them for food at any time.

9. *Ostrich feathers:* When nearly all the preparations are over, the individual adults of Chollo in Mwomo district (Gol Dhiang) including foreigners from other tribes will be asked to go hunting for ostrich feathers (*kona*). In the past, the people collect the feathers of the dead ostriches. But with the coming of the Arabs traders into the region, they killed ostriches and the people purchased ostrich feathers from them. All the feathers that are bought from the Arab traders are to be sent to Akuruwa village. In many cases the Arab traders took ostrich feathers to El Obei, in North Sudan, where they have market centre for ostrich feathers. Today, the King has right to ask government officials (police offers) in Kodok or Malakal to hunt ostrich feathers for him, at the time of need. It is reported that Tung people from Gol Nyikang (in the south) also help in the hunt for the ostriches. The black ostrich feathers (*kona*) will be used for the beautification of the wooden effigy of Nyikang and Daak, and the brown ostrich feathers are used for "Anonogo". Mr John Kudit, an informant form Melbourne-Australia, said that the effigy of Nyikang is normally short and is decorated with more ostrich feathers, while effigy of Dak is normally tall and is decorated with fewer ostrich feathers.

As we have just discussed above, the ostrich feathers are used for the beautification of effigy of Nyikang and Daak. Both of the effigies are beautified with black feathers, and the people of Akurwa, who accompany the effigies to the ritual ceremony, wear brown feathers. Interestingly, the effigy of Nyikang is beautified with short but blacker ostrich feathers, while the effigy of Daak is beautified with tall and less black ostrich feathers.

The ostrich feathers are also used for the blessing and empowerment of the King-elect by Nyikang's army (Kwa-nyikwom). The King-elect cannot become a complete King without possessing the Nyikang's spiritual strength -as will be discussed later.

The people usually go to hunt for ostriches in the region of Kaka. Besides collection of ostrich feathers, the people of Mwomo would be asked to collect ivory for bracelets to be worn by different functionaries during the coronation days. According to Evans Pritchard (2011) it is not the Chollo who collect ostrich feathers for the coronation of the King-elect, but rather it is the Arabs, who live in Mwomo district, who collect these ritual items. In addition to ostrich feathers, they also provide silver and cloths for the effigies of Nyikang and his son, Daak. The Arab in the northern part of Shullukland also collects skins of antelope from Panyilang Island. They also collect sacred ropes, sacred spears, royal drums, fibre of the Dom palm, cowries' shells. Besides all these, they also build a new house for the new King and provide other animals for sacrifice (P. 416).

10. *The search of Nyikang's spirit and his son, Daak:* As we have discussed, Nyikang is believed to have gone across the river, therefore, the only way to get him out of the river is for Nyikang's army from Morro Mwomo (Akurwa, Morro and Oring) to search for him. According to Shulluk tradition, the three groups of people (Akurwa, Morro and Oring) are the only people authorised to search for Nyikang's spirit. No any other clans or tribes from Shullukland can do the search for Nyikang's spirit. Joshua Ojwok Yor (2014) pointed out that one of the most important rituals in the coronation is the search of Nyikang and his son Daak. The people of Akurwa, Morro and Oriang clans are responsible for the search; they have to bring Nyikang's spirit from the river to constitute official induction of the King-elect (P.4).

On the day of the search, Nyikang's army (Kwanyikwom) of Akurwa and people from Oriang meet in Morro village. There, they gather in front of Nyikang's shrine to pray and dance to persuade him, so that when the searchers reach Jamel and Akoko areas, he would easily get out of the river. After the prayers, the people then proceed towards the north to Jamel and Akoko. People search for Nyikang's spirit in these two selected places, because it is generally believed by the Chollo that Nyikang's spirit usually come out at night, in form of white light like silver, in these two places to give light to the people.

When time has come, the Kwanyikwom of Akurwa go to meet the Kwanyikwom of Oriang and they travel together to meet the Kwanyikwom of Lo Morro. From there, they proceed northwards to Jamel and Akoko areas where they search the spirit if Nyikang. It is generally believed that Nyikang's spirit appears in any of these two villages at night in form of white thing (light) like silver. Thus, in the following morning, Nyikwer (a man offered to the spirit of Nyikang) is dressed with beads and special white dress known as "lau" and he is taken to the middle of the river with canoe, he is being accompanied by selected men, who will be beating special drums. There, in the river, Nyikwer is jumps out of canoe into the river. The men continue to beat the drum until Nyikang re-appears in the middle of the river. Shulluk people believe that when people are searching for Nyikang, his son Daak, is there participating in the search side by side with the delegates.

Interestingly, Nyikang's spirit does not necessarily accept gifts and offerings from every hands of *Nyikwer*. For example if Nyikwer is taken by selected men into the middle of the River Nile, a place where Nyikang's spirit is believed to appear to certain gifts and offerings, there Nyikwer jumps from the canoe and dives into the river. If Nyikang refuses to accept his offer, the blood of the man (Nyikwer) appears on the surface of the water, this is indication that Nyikang did not accept the offer, as such he (Nyikang) will not come out of the river. When this happens, the people dress the second *Nyikwer* and take him in the middle of Nile water. Nyikwer usually is accompanying to the river by beating of drums. The second Nyikwer then jumps from canoe and dives into the river, if Nyikang accept his gifts and offerings, he Nyikwer then appears on water surface with something in his hand, this is an indication that Nyikang's spirit has been found and he (Nyikang) has accepts the coronation of King-elect. When the delegations from Akurwa, Morro and Oriang, who are watching from the bank of the river see Nyikwer comes out with something in his hands, they immediately say, "Nyikang's spirit has been found". Nyikang is then taken into the canoe up to Akurwa and being accompanied by his son Dak. After two days, Anongo, the son of Nyikang, whose duty is to get "Ongero" (young growing palm tree) from Bol-Awar of Dhothim, appears. Joshua Ojwok Yor (2014) observed that on arrival at the river, people get down on their knees to pray again before Nyikang re-appears. At the river, people continue to beat drums until Nyikang re-appears in the form of white light in the middle of the River Nile.

11.   *Installation of Nyikang at Akurwa*: Nyikang arrives at Akurwa in a canoe, and from there, he is taken direct to Akurwa village. It is a common belief that as people take Nyikang's spirit home to Akurwa, Daak usually accompanies his father until the gate of Akurwa village. After that, Daak goes back to the river. Furthermore, it is generally believe that when Nyikang's spirit is brought home, after two days, Nyikang's son called Anongo appears in the place where the spirit of his father is put to rest. As we have just discussed above, the main reason of Anongo going Akurwa Village is to get Ongero (yung growing palm tree) from Bol-Akwar of Dhothim. According to Joshua Ojwok Yor (2014) when Nyikang's spirit is brought at Akurwa, the people rest for two days before they leave for installation in Fashoda.

12.   *Nyikang's spirit and Daak on the move to Fashoda*: After two days of rest and refreshments, Daak returns to Akurwa village and Kwanyikwom (the army of Nyikango) of Akurwa are asked to accompany Nyikang's spirit and Daak to Fashoda for installation (*rony*). Mr John Kudit, Collo elder from Melbourne, pointed out that before departure from Akurwa, King-elect usually bribes Kwanyikwom by giving them over one hundred cows, so that they give him proper protection at Arepa-Jur

and furthermore. This march could take ten days if all the needs of Kwanyikwom are met. Thus, on their ways to Fashoda, the people pass through the following stations where they spend one day and one night in each station:

i.   From Akurwa, delegates come to Adodo-noon; here they spend one day and one night before they leave for the next station.

ii. From Adodo-noon, delegates go to Delal dwong; here they spend one day and one night before they leave for the next station.

iii. From Delal dwong, they go to Okwath-yo-thuro, here they spend one day and one night before they move to the next station

iv. From Okwath-yo-thuro, they move to Panyikang- Otiego, here they spend one day and one night before they move to the next station.

v.   From Panyikang-Otiego, they go to Otiego-ogon; here they spend one day and one night before they move to the next station.

vi. From Otiego-ogon, the go to Otub-abyenyayo, here they spend one day and one night before they move to the next station.

vii. From Otub-abyenyayo, the move to Dethwork-dwong, here they spend one day and one night before they move to the next station.

viii. From Dethwork-dwong, the go to Delal-golo, here they spend one day and one night before they move to the next station.

ix. From Delal-golo, the move to Debwor-kodhok, here they spend one day and one night before they move to the next station.

x.   From Debwor-kodhok, they go to Pagach-padhyiang; here they spend one day and one night before they move to the next station.

xi. From Pagach-padhyiang they move to Adado-padhami Nyigir, here, they also spend one night and one day.

xii. And finally from Adado-padhami the Kwanyikwom travel and enter Fashoda with Nyikang's spirit at around mid-morning.

Source: Akwoch Dok Kwanyiyek, 2009

13   *The nominated King leaves Fashoda for Dibal*: When the people of Gol Dhiang and Gol Nyikang are in the process of collecting materials for ritual ceremony, at this time King-elect is waiting at Fashoda. However, before Nyikang army enter Fashoda (with Nyikang's spirit), when they are in the last station called "Adado-Padhami", the King-elect leaves Fashoda for Diballo village, to allow Nyikang to enter Fashoda. He (King-elect) first goes to Bankwago, where he spends a night and leaves for Diballo at dawn, reaching Diballo in the early morning at about seven O'clock. When the King-elect enters Diballo, the people there pretend to prevent Reth elect from entering the village but later they evacuate the village and King-elect

enter. The new King's flight to Diballo, Chollo simply say, "The Reth has fled from Nyikang". The army of Nyikang then say, "The march of Nyikang from Akurwa to Adado-paadhami is now over".

The appointed King is usually accompanied to Dibal by the people from the clan of Wichdiwad and group of chiefs. As King-elect enters Dibal, all fires in the whole village are put off, this is done to clear the way for the rituals that will follow the next day. Then new fire is made by the Kwa Nyutho (descendants of Nyutho). Thus, early morning, chiefs gather to pay their respect to the nominated King. Bulls and sheep are killed and the nominated King either puts his hands over the blood of these animals or simply put his hands over the dead animals.

In Diballo the King-elect waits for the completion of certain practices such as the ordering of the Nyikang of Ocoro (special beads) to go to Fashoda. It is believed that Nyikang of Ocoro determines the years, which the King-elect should take ruling the Shulluk This Nyikang of Ocoro is brought in form of special beads. It is supposed that all the chiefs of Shulluk could be in Diballo at this time, to determine the completion of the requirements for the installation of the King-elect. In the evening of the third day, following the coming of King-elect to Diballo, King-elect is dressed with ostrich-egg beads (*rek*) and the chiefs publicly confirm his becoming Reth of Shulluk in front of the people of Diballo; this takes place usually at around 4.00 p.m. South Sudan time. The chiefs confirm the would be King of Shulluk by saying four times the following phrases in one voice:

1. Wuoh! Wuoh! Wuoh! 1. Your Lordship! Your Lordship! Your Lordship!
2. Wuoh! Wuoh! Wuoh! 2. Your Lordship! Your Lordship! Your Lordship!
3. Wuoh! Wuoh! Wuoh! 3. Your Lordship! Your Lordship! Your Lordship!
4. Wuoh! Wuoh! Wuoh! 4. Your Lordship! Your Lordship! Your Lordship!

After these words of blessing and confirmation, the Nyiroro of Abuki-bal then kills a bull. On the fourth day, at about 12.00 noon, the Reth's army leave Dibal in a special march, which is done after exchanging special messages from "message carriers" (Kwabwol). They continue marching until Nyikang meet the King-elect army after Reth crosses the Dhok-Arepajur (Khoa Arepajur). At Dhok-Arepajur, a bull is killed and the King-elect jumps over it to meet Nyikang army at the northern side of Arepajur. Here, *Kwar-mal* of Makai beat their whips (*ongoro*) and *Kwanyidhiang* of Fashoda raise the hen. This then is followed by the first mock battle, using new sorghum stalks. At this point, it is important to mention that in this mock fight, the King-elect might sometimes be killed, if his army (*Kwanyikwom of Akurwa*) do not keep tied security; and if this happen, the people will have to restart the process by electing another prince to take over the position. Many of the Shulluk elders told me that the killing of King-elect usually happen in the first mock battle if that prince mismanaged Chollo affairs during his provisional period (*Reth ba Darooro*). When this killing takes place, the Shulluk people believe Nyikang has rejected King-elect. However, if the King-elect is not killed during this time, the marches continue until the army of King-elect enter Fashoda.

14. Nominated *King is back to Fashoda*: After blessing with the blood of bulls and sheep, people from the clan of Kwakel accompany the prince to join the procession of coronation in Fashoda. The prince (Nyikwer) leaves Diballo for River Arebjur, while Kwa-Julo (who are they?) are walking at his right hand and the Kwa-kel (who are they?) walking on his left hand. According to oral history of the Shulluk, when the nominated King leaves Dibal for River Arepjur, at the same time Nyikang leaves Fashoda for River Arepjur. King-elect usually reach River Arepjur before Nyikang. It is said, as Nyikang's spirit approaches River Arepjur, King-elect cross the river at certain point (or spot) to capture Nyikang. There the King-elect capture Nyikang's spirit and he continue to move towards a place called Pabur; here King-elect continue holding Nyikang's spirit, with his two hands, until they reach Fashoda (Joshua Ojwok Yor, 2014).

When King-elect and his sub-chief arrives Fashoda, he is then taken into the hut that is built near Nyikang's Shrine; here King-elect expected to take care of small herd of cattle which are brought from Diballo.

15. Nominated King performs Funeral Right of the dead King:

According to Chollo customs, before King-elect proceed with his coronation, he must first perform funeral right (mourning dance) of the dead King. This cultural practice is done to separate the coming King from the dead King. There is no option for this, it is a must. This traditional practice is known locally as *"woo Nyikang"*. However, it is important to notice that funeral right of the dead King cannot be performed at Fashoda; it has to be performed in his own village —where he grew. Therefore, this implies that before the installation, King-elect have to leave Fashoda for the village of the dead King to conduct the funeral right.

On the day of the mourning dance, the Kwanyiker dig holes into the ground and put in the drums (*bul*) leaving their mouths upwards. Unlike the Acholi, Shulluk people do not hang their drums on branches of tree. Kwanyiker clan then beat the drums to indicate the first signal of the ceremony. When people hear the sounds of the drums, they gather to pay their last respect for the dead King. This important event is marked by the sacrifices of bulls and sheep. When the King-elect finishes the performances of the funeral right, he returns to Fashoda.

16 Nominated King back to Fashoda and he is raised onto the throne:
After the completion of the funeral right, the nominated King returns to Fashoda to begin

preparation for his coronation process. The nominated King is called *Owooro* and not yet called King. And the wife of nominated King is at this stage is called *Nyirooro*. Thus, when the prince returns to Fashoda, the following morning his wife (Nyirooro) gets water wash him and take him in front of Nyikang's Shrine, where he is placed on silver coronation seat (*kwom Jago*) the Kwanyiyek covered the throne with white cloths. The wooden images of Nyikang and Daak are held over his head, so that he could be possessed by their powers. As a prince seats on the royal seat, his sub-chief holds his two legs with his two hands –the right hand holding left leg and left hand holding right leg. King-elect continue sitting on the royal stool for some minutes, while the people also continue holding the white cloths over his head. Here, Reth-elect is asked to nominate his deputy, who should hold his throne, while the chief is holding his two legs, he then nominates his deputy.

As nominated King is still sitting on the royal seat, Nyikang army start to beat their drums, while singing Nyikang songs. Shortly, King-elect is taken by the power (spirit) of Nyikang as the body starts trembling (shivering). According to Chollo customs, trembling body is the sign that the King-elect is now taking the power of Nyikang, as he posses-ses him. This then confirms the completion of installation process {Rony} (Akwoch Dok Kwanyiyek, 2008).

17. Clans Responsible for Installation

As many sons of Kings have never succeeded to the throne, there are today numerous and widely diffused branches of the royal clans whose members are ineligible for the royal office and lack authority, unless they are also chiefs of settlements, although they are treated with difference by commoners in virtue of their descent. With this view in mind it is worth mentioning here that not all clans in Shullukland can install King-elect. There are only nineteen special clans who can perform such responsibilities and these are:

Clans Responsible for Installation: (Tyeng Kwer Mo Rony)

1. Kwa-Nyikango of Golbanyo, Mooro etc
2. Kwa-Nyiyek of Kwom-Obuyo
3. Kwa-Nyikwom of Akura
4. Kwa-Mal of Makal
5. Kwa-Kell of Kwom, Nyingaro and wuobo.
6. Kwa-Nyimongo of Padhyang.
7. Kwa-Gyel of Nyigir
8. Nyiroro of Kodhok, Lul etc
9. Kwa-Wang of Dethwork or Acago
10. Kwa-Nyicu of Golbanyo.
11. Kwa-Owango of Lul
12. Kwa-Ocwako of Nyilwak

13. Kwa-Nyirow of Nyilwak
14. Kwa-Owenyi of Kwom-Dibalo
15. Kwa-Ojullo of Bol, Nyilwak
16. Kwa-Dik-wor of Kwom-Akwacabay
17. Kwa-Dikaki of Fashoda
18. Kwa-kit of Kwom-Pamuko
19. Kwa-Twot of Pabur

18. Reth is sent to adull:

Ather been seated on the throne, the new king is then sent into slavery huts of (*adull*) to labour there by collecting the cow-dung and cleaning the sleeping places of public treasury cows (Bwori-mach) which are kept there temporarily for that occasion. According to traditional laws and regulations, the new King is to spend four nights at the *adull*, sleeping in temporarily hut, built there, with grass for the purpose.

19. King-elect is sent to High Temples (Athurwic):

The installation ceremonies usually take two days; on his second day at Fashoda after the installation, the second mock battle takes place over the small virgin girl (Nyikwer), there, the side of King-elect (Gol Dhiang) takes over the small young virgin girl from the side of Nyikang (Gol Nyikang). From there, the small young girl and the King-elect are then taken to temporary huts (*adull*) at Athurawic (the headquarters of Fashoda). The King-elect and the young girl have to stay at Athurwic for another two days and two nights. When King-elect and the young girl are still in Athurwic, Kwanyimongo play with their shield around Fashoda, by doing this, they are wishing the new Reth a good fortune of having many cows that should pour to Fashoda during his rule.

At the end of the two days, senior elders and chiefs gather again in Fashoda for a final Advisory Council (Loko); thereafter, the King-elect is given the right to take up his new position (Joshua Ojwok Yor, 2014).

Figure 56 Chiefs and elders meeting the new King at Fashoda

This usually follows by speech day (*cang lok*), at around 8.00 a.m. Reth appears to his audience and the head chief from Mwomo District (northern) and head chief from Tungo District (south) give their speeches. The head chief from Mwomo district usually starts the speech; he stands up and asks if the head-chief of Tungo district is present in the ceremony. The head-chief of Tungo responses by getting up; head-chief of Mwomo then spears his spear into the ground just in front of him. The head-chief of Tungo, after his response, sits down to wait for his turn to deliver the speech that comes before Reth addresses his people. However, it is worth mentioning here that when the head-chief of Tungo is going to deliver his speech, he cannot spear his spear on the ground, as does the head-chief of Mwomo; but places it on the ground. When chiefs are giving speeches, everybody sit on the ground, with the exception of Reth, who sits on the throne (kwom) with an umbrella of red flamingo's feathers (Ober Owango) held over his head by the Kwa-Owango of Lul-agwoch. In their speeches, the chiefs would be advising the new King to perform his professional task justly and timely.

Interestingly, after each speech by a chief, drum is beaten as a sign of praise but only for good speech, they do not beat drum for a bad speech. Drum beating means that the speech contains useful advice to the Reth and this is followed by shield play by Kwa-Nyicu of Golbanyo who goes around the head-chief, who gives that good speech, playing the shield. Kwa-wang follows along them beating his drum. At the end of the speeches, chiefs give two silver bracelets to the new King —for him to put around his wrists, which designates responsibilities for the new office.

Finally, the new King gets up and concludes by addressing senior elders, chiefs and everybody present at the ceremony at this point and time. He is well dressed with Gyiel and silver-waist-rings on his both waists, the Kwanyikang of Golbange dressed him with these things in *adull* (Athurwic) before he appears in front of the Athurwic to deliver his policy statement. When the new King finishes his speech, he blesses the white bull, which is tied in front of the Temples (Athurwic), and there after the bull is killed. This marked the end of coronation (*rony*) and everybody is expected to go back to his or her respective areas.

20. King exercises his full power

Thereafter, King resumes his normal duties of looking after the Shulluk people and seeing cases among his people. He is always well dressed with "*gyel*" and silver-waist-rings (*atek*) at both waists; and these are dressed to him by the Kwanyikango (descendants of Nyikang) blacksmiths of Golbange in Athurwic before he appears in public right in front of Athurci, to deliver his policy statement.

The Shulluk traditional history says that when Kwanyikwom of Akurwa accompany Nyikang's spirit to Fashoda, they usually stay in Fashoda seven days and seven nights —they play trumpets (Thom) every night and people dance. The people who live in the nearby villages also join the dance, and sometimes the King also joins the dance.

It is reported by Collo elders that as the King-elect would be prepared for delivering speech to his people, the Kwanyikwom of Akurwa would be waiting for their offer of cattle before they return to Akurwa with effigies of both Nyikang and Dak. Soon after the Kwanyikwom of Akurwa receive their cattle, one of the residents of Fashoda brings the effigies of Nyikang and Dak with some money donated by the villagers and give them to Kwanyikwom of Akurwa.

Interestingly, before the King gives his speech, the head chiefs of extreme north (Mwomo) and south (Tungo) must give their welcoming speeches. Customarily, the head chief of Mwomo stands up first and asks, "Is the head chiefs of Tungo here?"

The head chief of Tungo responses by getting up, and this is to show that he is presence. He then sits down waiting for his turn, which usually comes after the speech of the head chief of Mwomo. The head chief of Mwomo then spears his spear right into the ground just in front of him and delivers his speech. If he gives good message, the Kwa-wang beats drum that signifies that the speech contains useful advice to the new King. This is followed by shield play by Kwa-Nyicu of Golbanyo, they go dancing around the head chiefs, while the Kwawang beat drum. Interestingly, if nothing useful is contained in the speech of speaker no drum will be beaten.

When the head chief of Mwomo district finishes his speech, the head chiefs of Tungo stands up. At this time, he does not inquire the presence of the head chiefs of Mwomo and does not spear his spear into the ground. He instead places his spear on the ground as he delivers his speech. Thus, if the head chief, in his speech, says a useful advice to the new King, the Kwa-wang also beats drum that signifies the speech contains useful advice to the new King. This is followed by shield play by Kwa-Nyicu of Golbanyo, they go dancing around the head chiefs, while the Kwawang beat drum. As I mentioned above, if nothing useful is contained in the speech of speaker no drum will be beaten.

According to Mr John Kudit, after the head chief of Tonga have delivered his speech, Reth takes over and he speaks with loud and clear voice to the Shulluk people. Whenever, he gives useful information to his people, this is followed by beating of drum, which means the citizens have accepted the policy statement delivered —but there is no shield play around him. When the King finishes his speech, he blesses the white bull that was tied in front of Athurwic and thereafter, the bull is killed. Collo traditional laws say that the killing of white bull after the speech of the King signifies the end of the rony (installation).

When the nomination or election processes are completed, chiefs and elders summon people from various villages around the country. These people, when they arrive in the place, they are not allowed to sleep in the village where the nomination or election of a King took place, but they have to sleep in the bush surrounding that village. Early morning, the people then enter the village where election was held. Shortly, the Nubians who brought the flint stones from Nuba Mountains are asked to accompany the selected prince from village to village. In each village that a King arrives, a cow is sacrificed and when a King visited all the villages, the Nubians and other people take him to Fashoda.

On the arrival at Fashoda, the people are met by chief, and the chief would ask them, "Who has been elected by the flint stones?"

One from the people accompanying the King elect answers, "….calling the name of newly elected King…is the person been elected by flint stones".

The following morning, a great chief summons all the chiefs starting from extreme north (Mwom) to the south (Tung) to come to Fashoda. When all the chiefs arrive at Fashoda, a big meeting would be held, chaired by a great chief. In the meeting they discuss important issues related to the investiture of the new King. These include the plan on how to grow the new King, and who is to do what and when. After this, the process for the gowning of the new King would go ahead as planned.

According to Evans Pritchard (1912), the burial of a King is more a clan thing under the chief of Fadiang. On the contrary, the election of the new King is a national affair. In the ritual configuration the two divisions of the Shulluk (Mwom and Tung) is presented by the ceremonial division of country into Gol Dhiang –the northern division and Gol Nyikang –the southern division, which represent the political division of the Kingdom in Ger, the northern half and Luak, the southern half (P.415).

After the gowning of nominated King, the chiefs and elders from Mwom and Tung accompany the King-elect to his own village (paar). On arrival at Paar, chiefs and elders do not enter the village; when night comes; they withdraw from the entrance of the village (paar) and sleep in a place called "adout" which is not far away from the village of the King. The nominated King is then given a hut to sleep in and the door is shut with door sticks. However, it is important to notice that the association of the princes with settlements and districts are related to the custom by which they are brought up, that is away from capital Fashoda. There it is easy to determine because the shrines of dead Kings are still maintained today where they were born and brought up (in paar-Village).

Early next morning chiefs get up; they wash their faces and enter the village (paar). There, they cut the door sticks with their spears, where the King-elect sleeps. The chiefs then enter the hut together with some descendants. They remove the King-elect from the hut, then a cow brought, slaughtered and eaten.

Soon after the meal, the chiefs take the newly elected King to Diballo, here, all the Shulluk chiefs are supposed to be present to determine the completion of the requirements for the installation of the Reth. Mr John Kudit, one of my informants and a Collo elder, who lives in Melbourne-Australia says, "And on the third day, the new King-elect is adorned with beads, clothes, royal stick, royal spear, silver bracelets (*ateg*), sacred ropes, skin of female antelope, giraffe male (yar wir), ostrich feathers and beads made from ostrich shells (rek) and a throne (*kwom*) –which is covered with tiger or leopard skins-and the chiefs confirm his especial ceremony in front of his residents at Diballo at about 4.00 p.m. In this ceremony all the chiefs would agree to confirm him as the would be King (Reth) of the Shulluk people or the father of the Shulluk-land. The chiefs confirm him by saying together four times the following words: -

1.   Wuoh! Wuoh! Wuoh! 1. Your Lordship! Your Lordship! Your Lordship!
2.   Wuoh! Wuoh! Wuoh! 2. Your Lordship! Your Lordship! Your Lordship!
3.   Wuoh! Wuoh! Wuoh! 3. Your Lordship! Your Lordship! Your Lordship!
4.   Wuoh! Wuoh! Wuoh! 4. Your Lordship! Your Lordship! Your Lordship

A bull is then killed by the Nyiroro of Abuki-bal soon after the confirmation of the King-elect". The King then sit with the chiefs and elders who advise him on what to do in future and the King responds in every advice, "Yes I will do all that you advised me to do".

According to Mr John Kudit, my informant, "On the fourth day, at about 12.00 p.m., the day of installation (*cang bony*), the Reth's army leave Diballo in a special march, soon after the delivering of special messages. The delivered messages are normally carried by Kwabwol (massages carriers). The Reth-elect and his army go and cross the Dhok-Arepajur/Khoa Arepajur where they meet with Nyikang. At Dhok Arepajur a bull is killed and the King-elect steps over it, the Reth's army then meet Nyikang's army in the place known as "Gar". Here, Kwar-mal of Makal beat their whips (*ongoro*) and the Kwanyidhyang of Fashoda raises the *hen* (protect the new King). This is then followed by a mock battle with stalks of new dura".

However, it is important to notice that sometimes King-elect might get eliminated or killed in this process, if security is not tightened by the Kwanyikwom of Akurwa –this often happen when the King elect mismanaged the Collo affairs during his provisional period (Reth ba garooro). And when this happens, the Shulluk people would say Nyikang has rejected the newly-elect King. Subsequently, the prince that killed King-elect would be installed as a new King.

The Shulluk people believe that during this mock battle at Arepajur, the newly King-elect would be taken over by the spirit of Nyikang when Nyikang's symbol (Nyikang-effigy) is being placed on his shoulders. This is a very important period for installation of the King-elect where security is tightened; no other people are allowed to come close to the place, only people from Kwa-kell, Nyiroro, Kwa-Ojulllo and other people who usually participate with Kwanyikwom in Chor-paabo's ceremony are allowed. Any unwilled stranger, who tries to come nearer to the place, is beaten with the whip. Customarily if the whip of Kwanyikwom's army touches the body of unwilled stranger, that stranger would have to surrender to them and he (stranger) would be asked to give a sheep or a goat, for otherwise, the wounds caused by whip on that stranger would not heal.

When Nyikang-symbol (effigy) is placed on the shoulders of the King-elect, the King then walks through "*chor-paabo* –road leading through to throne (*kwom*)", while the women suffocate to death Kwa-ojullo's cow that was brought through Diballo.

After marching on the "chor-paabo" the King-elect is raised onto the throne (*kwom-jago*) by the Nyirooro. At this very time, the Kwanyiyek and their chief would be holding white *domorria* clothes around the throne. Here, the King-elect would be shivering as Nyakang's spirit overtakes him on the "*kwom-jago*".

At the end of the celebration chiefs and elders give their speeches. The coronation ritual is always closed with special Nyikang's song. It is said that all the Nyikang songs that are sung to the nominated King, must be related to his father and his grandfather Kings. It is an offense to bring song which is not related to the current nominated King's father.

Thus the most common Nyikang's song that people sing after coronation is as below:-

Nyikang hero of the land! Nyikang yeda tek ki piny chol

When Nyiker pass away! Awen atou i Nyikeyo!

You let our Mother Atong be our Regent. Piny mayo wan Atong awii yin obeda piny wan

Now the Ceremony is over! Rony aum!

And the drums are heard from Akurwa. Bul aling ka Akurwa

We shall return home to give praise to God. Jwok apac odhwok pwoc wa

For our Summer Coronation! Rony ki dhdin

The Queen Mother has given the cow. Mayo Reth amuch ki dhyiang

Anei (X name) is installed at Pacod! Anei arony ki Pacod!

The Royal Village of Gwang! Path Reth Gwang!

We Praise Nyikang! Wa kwach Nyikang!

And the Praise of Akurwa! Wa pwoc Akurwa!

Go on beating their drums. Keth un ka bul goch wun

As soon as the King is grown, a particular clan give a virgin girl to him; in reality the virgin girl is not a gift to the new King, but he pays ten cows, twenty sheep, twenty goats and ten spears to the clan that gave him the girl.

And on his second day at Fashoda, after the installation, another mock fight takes place over the Nyikwer (a young virgin girl given to the King-elect). During this mock fight, a group of men called "*Gol Dhyang*", who are from one side of the King, takes the young girl (Nyikwer) from the group of men called "*Gol Nyikang*", who are closer to King. The Gol Dgyang takes the girl (Nyikwer) to Athurawic —it is headquarter of Fashoda, she waits here for the King.

After the second mock fight, the King (Reth) is sent into slavery huts of "*adull*", where he stays there for four days and four nights; while the King is in "*adull*", during these four days, he would be collecting cow dung and cleaning the place of cows of Bwori-mach (public treasurer cow).The cow dung are kept in the place for sometimes. In the morning of the fourth day, following labour in "*adull*", the Kwanyimongo come and play around with their spears and shields around Fashoda, wishing the new King good fortune and that many cows should come to him in Fashoda.

According to Mr John Kudit, an informant from Collo elders, in Melbourne-Australia, he stated that in the morning of the fourth day, at about 8.00 a.m., the King appears to his audiences. The King then gives speech to the chiefs and ordinary Collo, who are there at the time. He (King) sits in front of Athurawic, on the throne (kwom) with an umbrella of red flamingo's feathers (*ober Owango*), held over his head by the Kwa-Owango of Lul-Agwach to deliver policy statement to Shulluk people (for further information about the speech, please, see article "*Step by step installation processes* 20.0: *King exercises his full power*" below).

21. Ministry of the spiritual King:
The Shulluk people observe Reth as their inclusive representative of the spirit of Nyikang on the earth. His services to Chollo include carrying a divine message, which support national unity. One of my Shulluk friends in Melbourne told me, "We believe that *Reth* represents the spirit of Nyikang among us. He is responsible for everything in our institution, both in Diaspora and back home in South Sudan. His duties also include conducting constant prayers to God, so that the land in Malakal produces sufficient crop yields, ensuring health for the people and the animals (cattle). He prays to keep insects away from the crops and secure the rainfall needed for the natural growth of crops (Kuel Maluil Jok, 2010).

The new King thereafter resumes his normal duties as described above. Other people return to their respective villages while the *Kwanyikwom* wait for their cattle before they return to Akurwa with wooden image of Nyikang and that of Daak. One side, the residence of Fashoda expected to contribute some cows and money to be given to *Kwanyikwom*. On the other side, the new King give to *Kwanyikwom* ten cows, twenty sheep, twenty goats and spears and other things I cannot mention here. When the *Kwanyikwom* received the gifts, they remain in Fashoda for another seven days and seven nights and in all nights they playing trumpets (*Thom or tum*). They play *Thom* during the night and in the day; they eat, drink and sleep (Akwoch Doh Kwanyiyek, 2008).

22. Queen Abudok the only female ruler in Shilluk Kingdom
In the 17th century, around 1652, Shulluk Kingdom was ruled by a female called Abudok. Queen Abudok was the daughter of King Bwoch Alal, who was a rich King and had many cattle. Shulluk oral history tells us that King Bwoch had many children but never specified the number.

When King Bwoch came into power near Malakal, he did not stay long in this place, thus, he decided and moved to the northern part of Shulluk land. I was been informed that King Bwoch was a hostile King, as a result he has killed all of his sons, with exception of the two young sons who were hidden by Abudok.

When King Bwoch died in 1652, some writers like David Graeber (1998) argue that he was succeeded by Queen Abudok, who shortly abdicated the power because she was not like by the Shulluk people, and Tokot became a King (P. 23).

However, when Queen Abudok was ruling, many Shulluk people did not like her because she is a woman, and the people of Shulluk do not want to be ruled by a woman. This brought a lot of confusions among the people; they did not know what to do. One day elders went to Queen Abudok and said, "We are confused as we have no King and do not know what to do now?"

When Queen Abudok heard this message, she was angered, and realised she was not like by her people. She then called her two younger brothers and presented them to the people saying, "Take one of them to be your King". I believe the son that the people have taken to be a King, from the two brothers, was called Tokot, who David Graeber (1998) was referring to. It is said that soon after Queen Abudok presented to the people the two younger brothers, she walked away into Savannah (bush). She went into the bush and collected some seeds of water-lily brought the seeds home, spread them out in the sun to dry. When the seeds dried, Queen Abudok ground them, put them into a big bag. The bag was not full, she then asked for a smaller bag; she took small bag and poured the seeds in it and the small bag was filled up with seeds.

The following morning, Queen Abudok took the small bag to Fashoda and left it there without saying a word. She went back to her own village (what is the name of this village?), gathered people and told them, "One day, Shulluk land will be shallowed by the descendants of the Reth (Kings). When the descendants of the King build near your village, your village will start becoming smaller and smaller in the same way the small bag I put seeds in became small. The royal descendants will become many; just as the branch of the calabash plant become many in the bush" (Diedrich Westernmann, 1912, P. 150).

From this short history of the Queen, it remains unclear to me as to whether the people have taken one of the younger brothers and coronate him to Kingship or not.

More importantly, it is good to note that both King Bwoch Alal and Queen Abudok ruled the people of Shulluk before creations of the institutions of sacred Kingship; many people believe that during the reign of King Bwoch and Queen Abudok, there were continues debate about royal power.

23. Not all the Kings of Shulluk are installed: Its Impacts

Having seen the systematic process of installation of Reth in Shulluk Kingdom, this does not necessarily means that all the past Kings, who reigned in the Kingdom, were install in order described above. According to Akwoch Dok Kwanyiyek (2008), there are eight Kings (Reth) out of the 35 Kings, who ruled the people, without been installed onto the throne. These include (1) Muko wad Nyadway (1745-1750), (2) Waak wad Nyadway (1750-1760), (3) Dyelguth wad Nyadway (1760-1770), (4) Aney wad Yor Nyakwaci (1820-1825), (5) Awan wad Yor Nyakwaci (1835-1840), (6) Ajang wad Nyidhok (1869-1875), (7) Kuckon wad Kwathker

(1875-1882), (8) Dhanho Akol wad Kwathker(1903?) —who pretended to be the ruler, during the reign of Reth Kur (1) wad Nyidhok (1892-1903).

Since these eight Kings failed to be installed in the manner described before, they all faced the following challenges:

- They were regarded as princes (*nyireth*) who did not possessed Nyikang's spiritual leadership for the Shulluk people.
- They had and will face endless opposition from the princes who had and might have competed with him at the time of election or selection.
- Their sons shall not be entitled for the kingship position, when they die without been installed.
- They had and would not have full control over the Shulluk affairs.
- They had not and will not enjoy full respect as Nyikang's representatives.

Source: (Akwoch Dok Kwanyiyek, 2008)

24. Can a Prince refuge to be a King?

It is not uncommon, among the Shulluk that a prince can refuse to be a King. I assume that there were number of princes who have refused to be Kings but no records were taken. To demonstrate this, we are going to cite one example given by Diedrich Westernmann.

Diedrich Westernmann (1912) noted that Aleki, elder son of a King, was brought to be selected as a King but he refused. Aleki was informed secretly that chiefs planned to elect him by force. When Aleki learned of this, he fled with his two brothers to the Nubian country. What remain unclear to me up to this time is, "who was the father of Aleki?" When Aleki and his brothers were in Nubian country, they work in the fields of the Nubians farmers and they were doing the irrigation of the fields. They did this for food to maintain their lives; they learned Nubian language for communication with the employers and community at large. Shortly, the news for departure of Aleki spread among the people, and one of the princes on hearing this news said, "Let them stay in Nubian country". In the period Aleki was in Nubian country, council of chiefs elected another prince in the position of a King. After coronation, the new King moved to Fashoda and it seems that he even did not know that Aleki has left Shulluk land for Nubian country.

Although Aleki and his two brothers got jobs in Nubian country, Aleki was still feeling home sickness. Before long, he made a decision to return home; he came and built himself a village at a place that he called "*Pwot*" —which literarily means "beaten". I do not understand whether with the word "*pwot*" he meant he has beaten council of chiefs or he was going to beat the elected King.

When people learned that Aleki has returned home, they insisted that he should be a King but he continuously rejected the demand of his people. As Aleki continues to reject the demand, the people then elected his son he reigned. The son of Aleki then carved himself bracelets out of elephant-tusks. Since in Shulluk traditions, only Kings are autho-rised to wear ivory bracelets. When the King at Fashoda learned that a new King has been elected, he was angered and sent his army to fight the new King. The newly elected King was not ready for battle, so he gave fourteen cattle and ten men to quench anger of the King at Fashoda. The King at Fashoda received these fourteen cattle and ten men but he was not satisfied with the offer, so he then asked for more cattle and men as compensa-tion. The King at Fashoda also took ivory bracelets away from him. After this, the former King continues to reign (P. 148).

25. Parents of Nyikang: His Legends

The Shulluk people believed that in the beginning there was Juok (God), the creator, and he created heaven and earth. According to Kuel Maluil Jok (2010), although the Shulluk believe in God as the creator of everything, nearly all the Shulluk people unanimously claim that their spiritual leader, Nyikang, was and is exceptional. Every Shulluk believes that he (Nyikang) had a spirit, which he used to disperse the enemies of Shulluk with; and this spirit guided him to reach many dangerous places with his people, until they settled in their current home.

According to Saturnino Onyala (2014) Juok created the white cow, which came out of the Great Lake (This is probably at Lake Albert in Uganda or Red Sea in Sudan). This white cow was called Deung Adok or Deari Aduk. The white cow later on was married to Ud Diijil and gave birth to a man-child whom she nursed and named Kola (or Kolo). When Kolo grew to manhood, he married a woman Shulluk people do not know her name and later the woman gave birth to a son called Umak Ra (Omara). When Omara grew up to manhood, he also married a wife Shulluk people do not know her name, she later on gave birth to a son called Makwa (or Wat Mol). When Wat Mol grew to manhood, he married a wife, Shulluk people do not still know her name, and later on she gave birth to a son called Ukwa (or Okwa), and Okwo was the father of Nyikang. All these people live in a far off country called Kero near Lake Albert in Uganda. This is original land of the Shulluk people (For more information, read *Tekwaro Pa Acholi Me South Sudan* by Saturnino Onyala (2014).

Other episode, narrated that Nyikang was the son of Okwa, and Okwa was the son of Wat Mol, and Wat Mol was the son of Omara, and Omara was the son of Kolo, and Kolo was the son of Odyiang (Nyadhiang Aduk/ Deung Adok). Angelo Othow takes the genealogy further and he says that Odhiang was the son of Diwad and Diwad was the son of Oyel, and Oyel was the son of Cacre, and Cacre was the son of Lwor. It remains unclear on whether the Shulluk people beleive that Lwor was the grand-grand-grand-grand-grandson of Adam. However, from the two narratives above, we could conclude that the true genealogy of Nyikang could be summarised as follows:

GOD → Adam & Eve

Lwor

Cacre

Oyel

Diwad

Deuny Adok "Dyiang Kulu"

Omara

Kolo

Wat-mol

Okwa

Dumo | Okil | Gilo | Luo | Nyikang | Otin | Anyuak | Ju | Achol | Moi

Onongo | Odak | Bur | Chal | Dak

Nyidoro | Odak

The Thoro | Duwat

The Royal Line

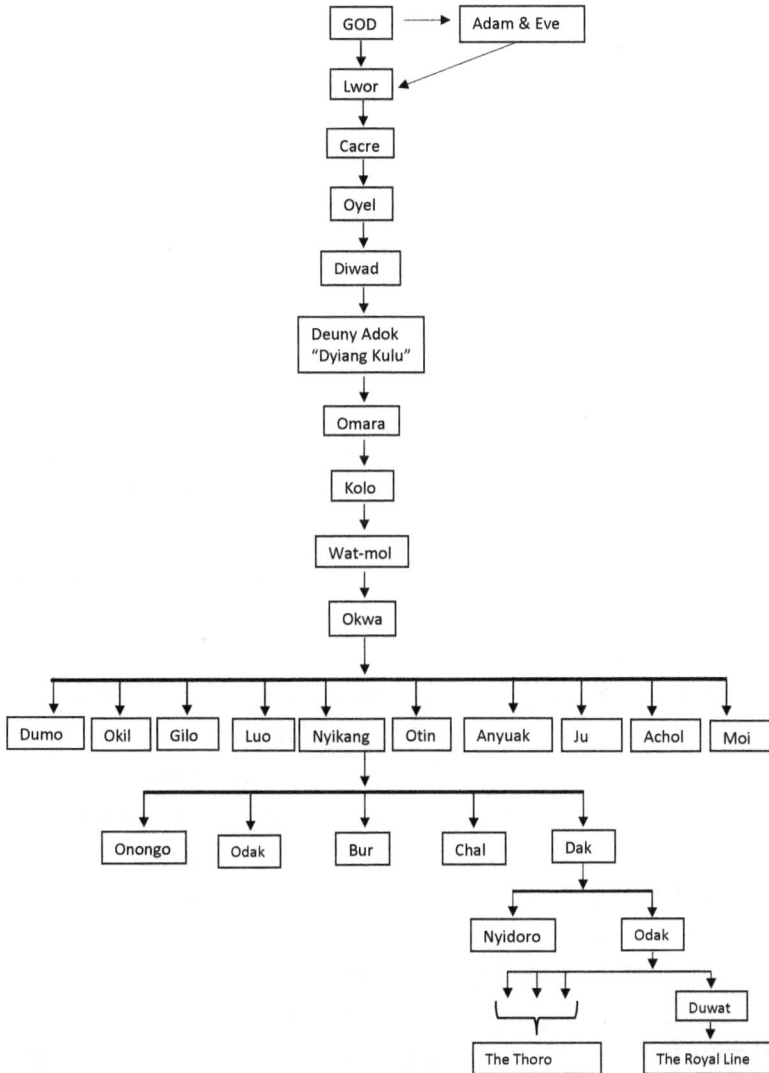

According to Shulluk oral history, one day, Okwa was sitting near the river (Lake), he then saw two beautiful and lovely young girls, with long hair, and they came out of the deep water and were playing about in shallow water. At this time, Okwa was still not yet married, so when he continues seeing the two girls coming out of the water many times, his heart beats fast with love for them. However, from the side of the two girls, they have nothing to do with Okwa, when they see him steering at them, they just laughed at him. Shulluk oral history tells us that the two girls were very beautiful and they had long hair, but their lower bodies were like the form of a crocodile, while the upper bodies look like women.

Thus, one afternoon, Okwa went as usual by the river side and found the two girls sitting on river banks; he then went slowly and quietly behind them and seized them by the necks. The two girls were fritted, they screamed on the top of their voices. When their father, Ud Diijil, heard of their screams from the bottom of the river, he came out of the water to see what was happening with his daughters. He then saw Okwa holding the necks of his two daughters, he protested mildly, yet allow Okwa to take the daughters for wives. Ud Diijil knew that Okwa held the two daughters because he wanted them for wives, he then gave in. It is said, Ud Diijil, like his daughters, had his right side had green colour and was made in form of crocodile, while his left side and his face was made up in human form (Diederich Westernmann, 1912).

It was unfortunate that the two sisters took a long time without having children. Thus, according to Luo cultures, since the two sisters failed to give birth within the expected time (one to two years) Okwa has right to marry another wife to test his fertility. He then married three wives, the first wife, whose name remains unknown, and this woman gave birth to Duwat (or Dumo), who is the elder son in the family. The second wife called Nyakayo, the daughter of Ke, gave birth to Nyikang plus three brothers and three sisters. The other wife was called Ogwat (Ungwet), who gave birth to only one son namely JU or Bworo (Bor). According to Diedrich Wassermann (1912) and Diedrick Westernmann (1912) Nyikayo gave birth to Nyikang plus three of his brothers namely: Okil, Otin, Moi and three sisters namely: Ad Dui, Ari Umker and Bunyung. Thus, Nyikang inherited the pleasing crocodilian attributes of his mother (Nyakaya) and Grandfather Ud Diijil.

The account of Nyikang's birth varies from person to person and from period to period. Thus, according to Godfred Lienhardt (1999) there are various accounts of the birth of Nyikang:

(a) In one account, it is said that Nyikang was the son of Okwa, who was the son of Omara, who came from heaven

(b) In another account, it is stated that a white or rather greyish cow (*dyiang aduk*), came out of the river; she brought forth a gourd, when this gourd was split, a man and animal came forth out of it. The man became known as Kolo; Kolo became the father of Omara, Omara became the father of Wat Mol and Wat Mol became the father of Okwa and Okwa became the father of Nyikang. As we can see Okwa is traced through several descents and forefathers to a great white cow, which was created by God (Juok) in the river. Nyikang's mother, Nyakaya, was either a crocodile, or partly crocodile, and now the mother is associated with the crocodile in Chollo thoughts, though she is also a woman.

(c) In other versions of the story, Nyikang's ancestor, his descent is traced through his father to a man who came from heaven, in the form of a cow, and his mother is a creature of the river with the attributes of the crocodile.

(d) Another account said, one day God was watching over His creatures and He saw, in a river, a woman called Nyakaya, with stars in her eyes. His glance made Nyakaya pregnant that led to the birth of Nyikang, the ancestor of the Shulluk people. The conception of Nyakaya conveys the message that Juok (God) is powerful and complete God, He can create any living thing. Therefore, He created Nyikang with exceptional powers. There is nothing that is difficult or impossible for Him; since He is omnipotent and a creator of rivers, animals, mountains, earth, heaven and everything on earth and in heaven. His sight penetrates all who exist and sees what is beyond oceans, rivers and forests. It was this metaphysical power of sight that caused the creation and birth on Nyikang (Kuel Maluil Jok, 2010, p.105).

According to some writers like Diederich (1912) Okwa married the woman of the river called Nyakaya, the mother of Nyikang and daughter of Ke. The brother of Nyikang was crocodile, it lived with the men; and the people used to play on its back saying, "Our grandmother, Eh!" One day Daak went to visit crocodile, there, he took some children of crocodile, he killed, roasted and ate them.

In the evening when the brother of Nyakaya was looking for other children of crocodile, Daak told him, "I have roasted them".

The brother of Nyakaya called Nyikang and told him this sad news, and said, "How is this?"

Daak replied, "I have eaten them".

The mother of the crocodile then said, "Nyikang, my grandchildren have been eaten by your son, Daak ".

Nyikang replied, "Have they really been eaten by my son, Daak?" –He asked, "Where will you go now?"

The woman, mother of crocodile replied, I will remain in the river".

Nyikang said, "No, because you and your children will in turn also eat the children of Daak when they go to wash and fetch water"

The mother of crocodile said, "You (men) can never pass a river where we live with my children, and you will never drink water from the river where we live".

The Nyikang said, "All right, if ever I and my children find you lying outside the river, we shall surely stab you. You shall never sleep outside the river; you shall only have sufficient time to lay eggs on the banks of the river". And a harpoon was made between crocodile and human being in Chollo's land (Diederich Westernmann, 1912, P. 155).

When the cows of Shulluk people swim across the river in search of grass, they are often seized by crocodile. This is the beginning of enmity between men and crocodile. So the men are now caught by crocodile.

According to Shulluk oral history, on Okwa's death there was a ferrous quarrel between Nyikang and his half-brother (elder bother) Dumo about who should succeed their father's throne. Interestingly, when we follow Luo culture, Dumo, been an elder son in a family, should succeed his father. Since Nyikang and Dumo could not reach common agreement,

Nyikang decided to leave original land, Kero, with his sisters Ad Dui, Ari Umker and Eun Yang; he also took along his brother Umoi and his half-brother Ju. He went his way till he reached Bhar El Ghazal and later on proceeded to Malakal (at the Conner of around Sobat River and Blue River Nile); the present Shulluk country. Here, they found the land was occupied by wicked Arabs (Funj). King Nyikang and his people fought the Funj and drove them out and Nyikang found a most successful Kingdom. This happened around 1545 A.D. (Diederich Westernmann, 1912)

However, writers like Godfrey Lienhardt, (1999) argues that Nyikang is not thought to have been the elder of the brothers, and there is no account of any superior claim on his part, except that he also had a following……. There is no emphasis in the myth on any customary order of succession, and so today any son of a King, who had once been installed is theoretically eligible to become a King if he can press his claim through council of chiefs, and elders concern. In the past, at least a man could retain the kingship only if he was able to defend himself against rival claimants (P. 97).

According to writers like Kuel Maluil Jok (2010) he observed that some people claim that parents of the mystical Nyikang had quite distinct identities. The mother, Nyakaya, was believed to be a mysterious creature, probably a crocodile and the father was believed to be a spiritual leader. During Nyikang's adulthood, a conflict arose between him and his brother Dumo, who swore to kill him. The circumstances of the conflict compelled Nyikang to seek sanctuary to avoid violence. He set off to another country, where he married the daughter of a local chief (King) and they have a son called Daak. Subsequently, more violence broke out, Nyikang was forced to leave this country, and he went looking for another place to settle. After moving for some days, they came to the present country, where eventually Nyikang took over as a new King in his land of exile (Kuel Maluil Jok, 2010, p. 104).

I want to draw the ideas of the readers that when we look closely at this narrative of Nyikang's legends, from Kero to Malakal, we could be forced to conclude that this legend is obviously a copy from the Christian bible. It related to the story of Moses leading people from Egypt to the Promised Land, by crossing the Red Sea. During their crossing God split the river for the purpose of allowing the Jews to cross (Exodus Chapter 2:1-10 and Chapter 12:31-39).

It seems certain at some points of this history that the Christian religious leaders might have influenced this legend changing its original form to some extent. Yet, the interesting part in the history is that Nyikang is not a hero but rather Ubago (Obago) is a hero because his head was cut and his blood sacrificed to God and River Sue (River Jur Chol) has opened for the Shulluk people to cross the other side. Therefore, in this context, we can say that Obago is more spiritual leader than Nyikang himself.

26 Division of Shilluk People and the four clans in the society
The Chollo nation comprises of about hundred different ethnic communities and clans that

we shall not discuss about them in this book because of their huge numbers. Talking about 100 Shilluk communities or clans need a book by itself. The previous indigenous, the Funj and Baggara Selim, who are descendants of the assimilated Otango Dirim, were the inhabitants of Tungu and Mwomo. The descendants of the assimilated Otango Dirim include Kwa-Nyidwai, Kwa-dway, kwa-nyidhiang, Kwa-mal, kwa-man, kwa-nyudho, kwa-mang, kwa-mwoy and kwa-jango. Interestingly, all the clans in Shilluk intermarry among each other without distinction; as we shall discuss more in depth under article "marriage in chapter 18".

Interestingly, in the past, particular during the migration period, the clan was not associated with any particular tract of country, but the members moved from one place to another as a body and the intense feeling of mutual dependence was kept active by proximity.

Today there are four clans found in Shilluk society. However, some writers like R. D. Oyler of Doleih Hill (2011) put the number of Shulluk clans between seventy four to hundreds. I personally think that this confusion come the translation of Shulluk word "*Kwar*". According to P.P. Howell (1941) the word *Kwar* may be translated to mean "Clan" that refers a group of people who would trace their descent unilaterally and patrilineal to a common ancestors. These people are bound by distinct rules of behaviour and mutual observance. He went on to say that word *Kwar* also means "descents of" which often has no wider reference than the descendants of a relatively recent ancestor, however, the exact meaning of the word "*Kwar*" vary according to the context. Shulluk tribe is divided into a number of traditional recognised clans, the numbers of which consider themselves bound by ties of kinship, however remote, and although their feeling of corporate unity is much modified by the dissolution of territorial ties, they are still vaguely clan conscious (p.47).

As P.P. Howell (1941) pointed out in his argument that the exact meaning of the word *Kwar* vary according to the context of the sentence; I would like to emphasize that there has been some confusions with the writers. The same tribe cannot have four clans and hundreds clans at the same time. From my research I have come to believe that Shulluk society has four clans and has between 74-100 lineages or kinships. The word *Kwar* is used in lineages or kingships to enable the people of that lineage to trace their descendants to a particular of the four clans in Shulluk society.

Therefore, let us first examine the four clans in Shulluk society and later on we shall examine the 75 so called clans. In this book we shall consider the 75 so called clans by Re. D. Oyler of Doleih Hill as lineages or kinship groups rather than considering them as actual clans.

Today, there are only four clans in Shulluk society any other traditional recognised groups are mere lineages or kinship groups. The four clans include the followings: *Kwar Reth*, Ororo, *Collo* and *Bang Reth*. The first two clans, that is Kwar Reth and Ororo, they together form the patrilineal descendants of Nyikang; and the last two clans that is *Collo* and *Bang Reth*, and they together form the general body of Shulluk people.

However, there is no outward difference between the four main clans, although the

Kwar Reth and Ororo are usually treated with certain degree of respect and they are served by Collo, who are considered by the people of Shulluk as the survivals or conquered.

The Kwar Reth is simply an enlarged clan and account for approximately fifteen lineages or family clans in Shulluk Society. The Ororo are relatively small in number as compare to Kwar Reth. On the other hand the Bang Reth is the majority in Shulluk society. We shall discuss later why the Bang Reth has majority numbers in Shilluk society as compare to Kwar Reth, Ororo and the Collo. Before we discuss the lineages or family clans, let us first examine one by one the four clans of Shulluk to give us the inside on how they are formed or how they have come to be.

(A) *The Kwar Reth Clan*: is the class of people owned by Nyikang and they are composed of:

-*Reth*, are family of the reigning King.

-*Nyireth* and *Nyareth* are sons and daughters of a reigning King or dead Kings.

-Nyinyireth, are grandchildren of King

-Kwanyireth, are great grandchildren of King or children of Kwanyireth themselves
The four generations above are regarded as a royal blood in descending degrees. The descendants of the King form a privileged group.

-Kwar-Nyireth, these are sons, grandsons, and their children.

-Kwa-jullo, are direct descendants of Nyikang

-Kwa-Jwok, are direct descendants of Jwok (who is Nyikang's cousin)

-Kwa-Obugo, are direct descendants of Obugo a man who sacrificed himself in order to open the weeds in the river to enable the groups of Nyikang to cross the river and continue with their journey. This includes the Ororo, who are not direct descendants of Nyikang but they are given privileges and right to become members of the royal clan. Nyikang had also established other historical sites called Nyilwal, Didigo, Akurwa, wau Chollo and Papwojo.

Interestingly, when a person from *Nyireth* or *Kwar-Nyireth* group is ask, "From which lineage are you?" They often do not refer to the four higher titles given to Kwar Reth clan, they simple answer, "I am from the clan of Kwar Reth". It is true that every member of the *Kwar-Reth* can claim to be *Kwar-Nyireth* but the reality is that the term *Kwar-Nyireth* is restricted to only to the grandsons of a *Nyireth*.

According to the Shulluk oral history, Kwar-Reth is bigger than any other clan in the society. There are three main reasons that can explain why the numbers of Kwar-Reth become bigger and bigger (grow abnormally):

The first reason been that the *Reth* are rich people, they have many cattle, and as such they married many wives. Their wives are locally known as *"bareth"* (which means the wife of the King), most of them live with the King at Fashoda. These wives delivered many children, which mean an increase in Kwar-Reth population.

The second reason been that people who refused to obey the order are punished and their penalty are usually paid inform of cows. The victims are always ready to save their lives at the cost of their cattle. Thus, these cattle add to the wealth of Kwar-Reth.

The third reason has been that when the wives of the King(s) become pregnant and ready to give birth, they are usually taken away from Fashoda. Customary, pregnant wives of Reth are not allowed to deliver at Fashoda; subsequently, these wives are sent back to their own villages and there the Nyireths are born. As soon as the Nyireth is weaned from mother's breast, he is planted out (trusted) to Collo chief, who is often the brother of King's wife; he then become responsible for his upbringing to manhood. The chief cannot refuse this trust as it will bring certain temporary benefits such as milk cows for Nyireth's nurture, not only this, the entrust of Nyireth to him, eventually will offer him an opportunity for his descendants to be made lineage of the Nyireth.

As Nyireth grows at the home of the uncle, he also looks after the cattle and sheep like any other children in the village. However, when Nyireth reaches the age of marriage he builds his home near the chief and moves to settle there with his wife or wives. The home or village that Nyireth builds is known as *"Pa Nyireth"* and in long run this village become the Nyireth's descendants. When the descendants of Nyireth give birth and multiply, the original village (Pa Nyireth) expand to immediate neighbourhood. As a result of this, the Nyireth's descendants are now rooted in every part of the Shulluk Kingdom and eventually growing in the area formerly owned by the Collo lineage. Each branch of the *Kwar-Reth* once planted, grows alongside the local Collo lineage and often tends to eclipse it (M.E.C. Pumphrey, 1941).

When the numbers of Nyireth's descendants multiply, these eventually generate quarrel or rivalry between them and the Collo lineage. However, more often the Kwar Reth are respected because of their superiority in the society, due to this, the Collo lineage will have to make a choose of between moving out of their area to a new location or remaining in the same area but become subject under the Nyireth descendants. In most cases, the Collo lineage moves to new area and settle there. At this point, I recall into my mind the prophesy of Queen Abudok when she said, "One day the *Kwar-Reth* will eat up the Collo". Today, this process is still taking place in Shulluk land. As a result, we now find more numbers of Kwar-Reth descendants at Detwok and Atoadwoi than the number of Collo, who are the aboriginal of the areas. Nevertheless, it is equally important to acknowledge that there are great areas in Shulluk land, where there are no Kwar-Reth descendants.

The Kwar-Reth descendants are scattered all over the Shulluk land, they have no political authority but sometime they are given the position of village chief although in most cases they do not. According to Shulluk traditions, the *Kwar-Reth* descendants intermarry with the Collo clans, with the exception of the King's daughters who cannot marry at all. Consequently, most, although not all of King's wives are women from Collo clan.

(B) *The Ororo Clan*

The term *Ororo* is used in Shulluk Society to describe any lineage of the *Kwar-Reth* which has been degraded by the act of the King, when he dubbed some sons of the King to lower rank of none-descendants of *Kwar-Reth* (Collo) and eventually these people formed what we call today, in Shulluk society, as Ororo, they are exclusive from the use of Kwar-Reth; thus, the term Kwar-Reth is restricted to the descendants of that King.

For us to understand more clearly what I am trying to explain above, let us take the case of King Duwadh. Shulluk oral history tells us that in the old time, sons of Kings were not going to war, but when King Odak became the ruler, he summoned Council of War, as the Nuer were preparing to wage war against them at Detang village. In the meeting the Council of War resolved that all sons of current King must join the fight against Nuer at Detang. After the meeting everybody men went across the river to fight Nuer. In this fight all the sons of the King, who joined the fight, were killed, it is said only one son, Duwadh, was not killed and he returned home with other men.

When King Odak died, his son, Duwadh, became a King and he degraded all the descendants of those Nyireth killed in war to Collo. Thus, in this way King Duwadh restricted the term *Kwar-Reth* only to his descendants. Subsequently, the descendants of the other sons from King Odak were dubbed to what we called "Ororo" –which literarily means the sons of group of girls from Kwar-Reth descendants (M.E.C. Pumphrey, 1941).

The Shulluk believe in the past and even today that the King possess the power to prune the royal family tree by distancing Nyireth from his own family. As such some of the Kings have been exercising these practices in the old day and today in Shulluk Kingdom; but other Kings did not succeed to distance Nyireth. And where the Kings succeeded to distance the Nyireth, then the distant Nyireth become the *Ororo*. M.E.C. Pumphrey (1941) observed that the *Ororo* play the part of masters of ceremonies at the induction of the new King in Fashoda (P.12).

For the purpose of the rules of exogamy the *Ororo* are considered as a Collo clan and can intermarry with the Kwar-Reth as well as any Collo man or woman other than their own. Shulluk oral history tells us that the King always has some Ororo who smother the King when his powers of procreation are failing.

(C) *The Collo*

Chollo group –these are the rest of the people who are not original descendants of Nyikang these people include Funj and Baggara Selim who are assimilated into Nyikang's Chollo community. According to Shulluk traditions the Collo are the aggregate of clans whose origins remain obscure to many. However, some of them claim that they are descendants of men who accompanied King Nyikang from Wau in Bhar el Ghazal to Malakal. These people include the clan of Kwar Abogo, Kwar Muol to mention a few. Others claim to trace their descendants to collateral relatives of King Nyikang, for example Kwar-Ojul. Others claim to trace their descendants to the prior inhabitants of Shulluk country for example Kwar-Jung (who originated from Dinka). Yet many people in Collo clan do not know where they have exactly come from.

(D) *The Bang Reth*

Bang Reth –these are servants of King, they include people who were slaves and refugees to Shilluk-land. Moreover, because of their royalties and faithfulness to the King they are awarded and allowed to be selected in the positions of chieftainships

We have just described how the Kings keep order among the Shulluk communities and how they enrich themselves through penalties. One of the ways the Kings enrich themselves with cattle is through the deployment of a large body called *Bang Reth*. These people are usually employed to the position of retainers in various ways.

In the old days, these people composed of captured people during the wars. But today, the Bang Reth are people who voluntarily recruit themselves to the position of workers or retainers because they are unable to raise the necessary cattle to marry their wives. For this reason, they join the position of retainers (of the King). M. E. C. Pumphrey (1941) noticed that such individuals from the moment they become *Bang Reth*, they blocked any outside communication with their own clan and focus on serving the King by building houses, digging and weeding the gardens of the King, cleaning the compound of the King, tending animals and some odd executive duties. In turn they receive cattle for marriage. The other way an individual can become a retainer is when he kills a person who is considered uncaptured, like member of *Ororo*, the King then compel him to be his retainer to avoid revenge and for the fear of been killed.

Since the establishment of Shulluk Kingdom, the relationships between the King and his retainers have been good and friendly; the retainers are treated equally regardless of whether a person is enslaved or voluntarily recruited to the position of retainer. However, the general custom is for the retainers of each King to build a village for themselves and these villages should be built near Fashoda where the King lives. The retainers will live in the village until the ruling King dies. When the ruling King dies, the body will be taken to his own village and sealed up in a house where it remains until the entire body is decomposed (for further information, please, read, read "Death and burial of a King"). According to M.E.C. Pumphrey (1941) after the decomposition of the body, the bones will then remove by his retainers and buried in his original village. (But we have seen under the article "the death and burial of a King" that the bone will be taken to a sacred place in the river). Some of the dead King's retainers will move to the village where the bones of the King are buried; their descendants remain in this village to attend to the King's Shrine. However, some informants told me that the grave shrines of the Kings are tended by certain old men or women, who correspond to the guardians of the shrines of Nyikang. They are usually widows or old men-servants of the deceased King.

Figure: 57, Shulluk Grave-Shrine

While others retainers will remain in their villages near Fashoda and they will have no further obligation to the memory of their dead patron except that their descendants will be known forever as *Bang Reth Kwadhker* or *Bang Reth Nyakwac*.

Thus, the retainers of a particular King, having collected enough cattle from their own clan, they become fictitious clan in themselves. According to Shulluk customs, the descendants of retainers cannot intermarry, though they have no blood relationships, in addition, a King cannot marry the daughters or sisters of retainers. A King usually addresses a daughter of a retainer as Kwara —which means my grand-daughter. By such calling it indicates the relationship subsisting between them (M.E. C. Pumphrey, 1941).

Today, the South Sudan government have disapproved of the practices we have just described above; the King of Shulluk is now deprived of his previous power and ways of enrichment of himself through retainers. As a result today, the King has no traditional retainers. Nevertheless, the *Bang Reth* still consist of descendants of retainers of previous Kings. These descendants have now moved themselves to different parts of Shulluk land; wherever they go or settle, they still remain *Bang Reth* in accordance with the names of their ancestors, and they will never join their original clan, where they were recruited from. P.P. Howell (1941) pointed out that today the clan is widely scattered and there is no periodic reunion and the clan never meets or acts as an entity (P. 47).

In the old time, when a Nyireth is entrusted to a chief (uncle), the King used to send some of his own retainers to live with him, eventually these retainers become *Bang Nyireth*, the retainers continue serving *Nyireth* in the same way *Bang Reth* serve the King.

Thus, after the dead of the *Nyireth, bang Nyireth* form a fictitious clan. Consequently the descendants from this clan continue to multiply and later they form what we called *Kwar-Reth*. I feel that using local terminologies often cause confusion to the readers. After having said this, I want to inform readers that *bang-reth* and *bang-nyireth* are all people who originated from Collo as we have described above. Yet, their status is considered inferior to those uncaptured Collo (or free Collo). The *bang-reth* and the *bang-nyireth* are considered inferior because they do not have right on the land or areas they live. For this reason, they are considered newcomers to the land. At this point and time, it is important for us to note that priority of occupation of land is something of great important to the Shulluk people.,

27. Lineages or family-clans in Shulluk Society
Lineage can be defined as direct descent from a particular ancestor or family. Every ancestor is at once a point of unity and division, thus through his sons his descendants are divided into separate descendant groups, yet in him unity is seen as one lineage group.

The Shulluk people are divided into 75 exogamous clans, kinships or lineages but some people even put the number up to hundred matrilineal and exogamous clans. Evans Pritchard (1948) noted that there has been some obscurities abut Shilluk descent groups. There are said to be in Shillukland about 100 groups designated by the word "Kwa" (descendants) followed by the name of the ancestor of the group (p. 409); as we shall discussed below.

Clans are not localized and have no specific territorials; they are scattered widely through different hamlets as we shall see below. The original clans are now been repre-sented by a number of off-shoots resident in widely separated parts of the country. These off-shoots are therefore referred to as lineages. It is natural that the numbers of one lineage are more closely bound to each other for they are not only more closely akin, but also in daily communication with each other. The Shulluk man is always conscious of his relationship to his widely scattered kinsmen, even though he often thinks in terms of his localized lineage. Thus, in using the word *"Kwa"* this often makes him easily refer to some comparatively recent ancestor (P.P. Howell, 1941 PP.47-48). Each of which is traced back to a common ancestors; and in most cases this ancestor is a man. However, some of the clans claim descendancy from an animal such as crocodile and cow.

Studies showed that many of the founders of Chollo clans have come from outside the country. One of the ways in which stranger lineages are grafted into the genealogical structure of the dominant lineage is a settlement. Thus, the descendant of a man, who has settled with his wife's people trace their descent through the wife to the lineage in whose home they live (Evans P. 1948, P. 409). Even this is so, it might be difficult for us to say with certainty what political significance the man play in the genealogical structure or group, unless we get to know the man more and more.

In Shullukland individual lineage members are grouped to form hamlets of agnostically related kin. Thus, a hamlet of this type may include as many as fifty homesteads and they usually form a settlement with a clear defined territory in a given region or area. In each hamlet there is an original or owner lineage called in Shulluk language *"diel"*.

The Chollo, like other Northern Luo groups, have a system of relationship using particular terminology, and this descriptive is used to approve the marriage of a boy and a girl (legal marital unions). If a girl and a boy have the same descriptive, for example if a girl or a boy says, "I am from the clan that does not eat meat from the knee-joint of animals", this means that both the girl and the boy are from the Kwa-Nyimol clan as such they cannot marry each other. Similarly, if a girl or a boy says, "I am from the clan that does not eat meat the meat of an animal that is eaten by eagle (*gu*)", this means that both the girl and the boy are from the Kwa-gu clan as such they cannot marry each other. Furthermore, if a girl or a boy says, "I am from the clan that does not eat meat of warrall (*ngwec*)", this means that both the girl and the boy are from the Kwa-mwoch clan as such they cannot marry each other. Another good example is if a girl or a boy says, "I am from the clan that does not touch piton (*nyelo*)", this means that both the girl and the boy are from the Kwa-Lap clan as such they cannot marry each other. There are many legal descriptive terminologies that I cannot mention them all in this book.

According to Chollo custom, a boy or a girl can get married when they reach the age of 18+; it is a criminal act to for a boy of 18 years old to marry a girl under 18 years of age. Fashak Deng, one of the informants, said that such illegal marriage, nobody will recognise in Collo society. We shall discuss this more under article "Marriage" in Chapter 18.

The following clans or kinships have been procured by Re. D. Oyler of Doleih Hill **(??)** and were examined by Chollo elders who are known for being well versed in the history and traditions of their people. As I mentioned before some people put the clans of Chollo up to one hundred but in this book we are going with the finding of Rev. D. Oyler that says Shulluk country consist of 75 main clans. These main clans include the following:

1. *Kwa-Ajal* (descendants of Ajal), this clan was founded by Ajal, one of the men who came with Nyikang from Kero. Today, this clan live at Nyelwak.
2. *Kwa-Mal* (descendants of Mal), this clan was founded by a man and a woman which Chollo believed they have come from heaven. Their names remain unknown up to the time I am writing this book. This man and a woman stayed for some years on earth and gave birth to children; after this they again ascended into heaven leaving the children behind. These children now form a clan called Kwa-Mal in Chollo country.
3. *Kwa-Lek* (descendants of Lek), this clan was founded by two celestial beings, a man and his wife. This clan was without a name but later got its name from the large wooden pestle that the Chollo use in rushing their sorghum (dura). The husband and the wife quarrelled over a "lek"; the husband wanted to use it to stir the cow

dung, while the wife wanted to use it for crushing sorghum. This man and his wife were in living in heaven, as they could not reach agreement, they continue fighting over the "lek" and eventually they felt down on earth. Before long, Nyikang captured them and told them to settle in Malakal. The husband and the wife did not refuse invitation of Nyikang. Shortly, the wife and her husband taught Chollo women how to brew beer (Ombuki). After staying in Malakal (on earth) for some years, the two people ascended into heaven leaving behind the "lek". This is how the clan got its clan name. Today, this clan live at Ogot.

4. *Kwa-Oman* (descendants of Oman), this clan was founded by Oman, who was found by Nyikang in the land of Malakal. Today, this clan live at Ogot.

5. *Kwa-Mon* (descendants of Mon), this clan was founded by Mon; like Oman, Omon is one of the people fond by Nyikang in present Shullukland (Malakal).late on Mon became the servant of Nyikang. Thus, Omon and Oman helped build the house of Nyikang at Upper Nile Wau. Today, this clan live at Ogot.

6. *Kwa-Ju* (descendants of Ju) this clan was founded by a man called Ju, who is a half-brother of Nyikang. Ju helped build houses for Daak; they built three houses for him at Filo on the White Nile. Today, this clan live at Mainam.

7. *Kwa-Nyadwai* (descendants of Nyadwai), this clan was founded by Nyadwai, who was one of the Kings of the Collo (please see the list of the King (see page 245). Nyadwai was the son of King Tugo and he was the servant of Abudok. Today, this clan live at Apio and Adit-deang.

8. *Kwa-Gwar* (descendants of Gwar), this clan was founded by Gwar, who was a servant of King Dokot. Today, this clan live Chet-Gwok.

9. *Kwa-Nyikang* (descendants of Nyikang), this clan was founded by Nyikang, since this group was a servant of Nyikang. Today, this clan live at Fakang.

10. *Kwa-Nwon* (descendants of Nwon), this clan was founded by Nwon, who was a hippo-hunter. Nwon was fond by King Abudok at Doleib hill when he was hunting hippos. He was the captured and brought home. Later on, the man became the servant of King Abudok. Today, this clan live in Twara.

11. *Kwa-Reth* (descendants of Reth), this clan was founded by Nyikang. Today when there is coronation of the new King, people of this clan all go for it. Today, the people of this clan live at Filo. Evans Pritchard (1948) pointed out that the royal clan is the largest single clan in the whole of the country. In some areas, their numbers are more than the commoners and they have supplanted comer lineages in the chieftainship of settlements. This process has been going on for a long time, this prompted Abudok, the only Queen and the eight ruler of Shulluk Kingdom, to prophesise that one day the royal clan would eat up the rest of the Chollo. However, this results from the custom of sending pregnant wives of a King from royal capital to bear their children

in the village where they were born and brought up. After the wives have children in villages where they were born; daughters grow up in these villages and when they reach womanhood they are not allowed to marry, but they can have children, and these children are counted as Reth's children. Moreover sons are brought up by settlement chiefs, often their maternal uncles, and when they reach manhood they are allowed to marry but after the marriages they are not allowed to move to royal capital with their wives. Thus, they settle in uncles' home. As a result of this a prince (nyireth) that marries builds a separate hamlet near the settlement of the chief that brought him up; and thereafter his descendants continue to live in this area.

12. *Kwa-Tuki* (descendants of Tuki) this clan was founded by a man that Nyikang discovered sitting by the river. Nyikang captured him and brought him home. This man later on taught the Chollo to build hearth-stones (Tuki) or (Keno), which is made up of three small pillars of mud built in a triangular shape. It is said that before, the Shulluk people learn about building Tuki/Keno for daily cooking, they used to dig a little hole in the ground for cooking. In the old time and possibly today, the main duty of Kwa-Tuki clan is to help look after the King's cattle. Today, this clan live at Didigo.

13. *Kwa-Chwal* (descendants of Chwal), this clan was founded by Chwal. According to Chollo oral history, this man called Chwal was found in Malakal by Nyikang, when he arrived with his group in the new country.

14. *Kwa-Jang* Nyikang (descendants of Jang-Nyikang), this clan was founded by Jang-Nyikang, who is a Chollo but married to a Dinka woman. They have children, and these children became descendants of the clan. Subsequently, the descendants from this clan call the Dinka people their uncles. Jang-Nyikang was a servant of Daak. Simon Kur, one of the informant from Brisbane Australia, said that Daak was chief of staff of Nyikang and his army captured Dinka men and women –the women are later on married to Collo men and the children they born formed Kwa-Jang Nyikang. Today, this clan live at Ojodo.

15. *Kwa-Tuga* (descendants of Tuga), this clan was founded by Tuga, who was believed to be an Arab. According to Chollo oral history, Nyikang married the sister of Tuga, and this man (Tuga) moved and settled in Malakal as brother-in-law (Juga), after that he found Kwa-Tuga clan. Today, this clan live Malakal.

16. *Kwa-Kelo* (descendants of Kelo), this clan was founded by Kelo; who was a servant of Nyikang. The oral history of Chollo tells us that it was Kelo who taught Shulluk people about building "Tuki" or "Keno" from the mud. Today, this clan live at Fone Nyikang.

17. *Kwa-Oguti* (descendants of Oguti), this clan was founded by Oguti, who was the servant of Nyikang. It is not clear whether Oguti came with Nyikang from Kero or whether he (Oguti) came from outside and settled in Shullukland. And if Oguti

came from outside, it remains a mystery from which country he came from. Today, this clan live in Twara.

18. *Kwa-Dak* (descendants of Dak), this clan was founded by Dak; this Dak is not the son of Nyikang but rather his servant. The main duty of this clan is to cut the new sorghum, from various fields, for the use in the house of Nyikang or King-elect. Today, this clan live at Owichi.

19. *Kwa-Oshodo* (descendants of Oshodo), this clan was founded by Oshodo, who was the servant of Dak. This Dak was the son of Nyikang; however, some elders from Shulluk claim that Oshodo was the son of Dak. Today, this clan live in Malakal.

20. *Kwa-Nebodo* (descendant of Nebodo) this clan was founded by Bodo, who was a blacksmith of Nyikang. The main duty of this clan is to build houses for the King at Fone Nyikang, and every year they have to offer dried meat of hippo to the King. Today, this clan live in Nyelwak.

21. *Kwa-Guga* (descendants of Guga), this clan was founded by Guga; this was a man who used to sit near Nyikang like a buzzard watching for meat. This clan today live in Nyilwal.

22. *Kwa-Obogo* (descendant of Obogo), this clan was founded by Obogo, the man who was killed by Nyikang when Nyikang's group arrived at the river, and they could not cross the river because it was blocked up with Sudd. Obogo asked Nyikang to take care of his children, and let him be killed, so that his blood could open the way for the group. He was killed, and when his blood touched the sudd, the sudd opened Nyikang and his group then crossed the river. Today, this clan live at Fone Nyikang.

23. *Kwa-Ogeko* (descendants of Ogeko), this clan was founded by Ogeko, who was also a servant of Nyikang. This clan got their name from what they are assigned to do; they were assigned to care for the sacred cow of Nyikang that he got from the river. Today, this clan live at Wau of Upper Nile.

24. *Kwa-Nemwal* (descendants of Nemwal), this clan is said, to be part of Kwa-Reth clan; that Nyikang was angry with them (what reasons?) and decided that the clan should no longer belong to the Kwa-Reth clan. Today, this clan live at Tung.

25. *Kwa-Okel* (descendants of Okel), this clan was founded by Nyikang himself. The primary duty of this clan is to give a girl to every King-elect. Thus, when the time of coronation comes, their chief seizes one of his daughters and gives to the King-elect. Moreover, Okel, the founder of this clan seems to have come from Bhar- El-Ghazal, they should have been related to Dimo, the father-in-law of Nyikang. Because of the crime they have committed against Nyikang, caused them to become common Chollo and as a punishment for their crime they have to pay a girl to every King-elect. Thus, the girl been paid for the purpose is called "Nya-kwer", which literarily means girl of taxes.

26. *Kwa-Oshu* (descendants of Oshu), this clan was founded by Oshu, who was the son of

Lobo and servant of King Abudok. According to Chollo oral history, Nyikang found a man called Lobo in the present Shullukland. Lobo was a woman and has a husband called Okola, they gave birth to a son called Oshu, who later on found Kwa-Oshu clan. Today, this clan live in Owichi.

27. *Kwa-Bungo* (descendants of Bungo) this clan was founded by Bungo, who is believed to come from a foreign country. The exact place this man came from remains unknown up to the present day. History tells us that Bungo later married a Chollo woman, that allowed him settle in Chollo country. For the clan members to become a good Chollo's family, they gave one of their daughters to the King. This payment was done once and for all. Today, this clan live at Nyigir.

28. *Kwa-Ororo* (descendants of Ororo) this clan was founded by Ororo. Like Kwa-Nemwal, Kwa-Ororo are said formerly they were part of Kwa-Reth. Today, this clan live at Yonj. Evans Pritchard (1948) noted that some lineages of the royal clan, the ororo, have been formerly deprived of their noble status and can now intermarry with their parent clan. They are said to be descendants of Ocolo, who was the fifth King, who was degraded by his successor. The ororo clan are few in number but they play a leading part in the royal funeral and investiture rites. The daughters from ororo clan always help King's wives in smoothing King's body.

29. *Kwa-Dokot* (descendants of Dokot), this clan was founded by Dokot, who was the servant of Dak. At the time Nyikang arrived in Malakal, the people from the clan of Dokot were around Sobat Region. Today, this clan live at Gur.

30. *Kwa-Nimono* (descendants of Nimono) this clan was founded by Nimono, This man (Nimono) was found in Malakal when Nyikang enter the country with his groups. The Chollo history tells us that later on, Nyikang married the daughter of Nimono. Subsequently, Nimono, as a brother-in=law of Nyikang became the founder of Kwa-Nimono clan. Today, this clan live in Gur.

31. *Kwa-Owen* (descendants of Owen), this clan was founded by Owen. Literarily the word "Owen" means deceivers. It is generally stated that Owen often tried to deceive Nyikang. However, we are informed that Owen and his people were brought by Nyikang form a distance, and later they became servants of Nyikang. Today, this clan live in Fone Nyikang.

32. *Kwa-Oreto* (descendants of Oreto), this clan was founded by Oreto, whom Nyikang found in Malakal on their arrival. Today, this clan live at Nigu and Wubo.

33. *Kwa-Wang* (descendant of Wung), this clan was founded by Wung. This is a man who tried to hide all the *shur* fish from Nyikang. It is said, one day Nyikang asked Wung, "Where are *shur* fish?" "I have not seen this fish for a long time" –Nyikang said. Wung answered, "There are none left", but Nyikang's Treachery found out that *shur* fish were just been hidden from the rivers in Tung. According to Chollo tradition de-

scendants of Wung were hiding shur fish from Nyikang first to protect them and secondly they did this because shur fish is a clan of Kwa-wang. That is why they do not want Nyikang to kill shur (fish). The Chollo oral history never indicated where the people Kwa-wang came from. Today, this clan live in Tung.

34. *Kwa-Nishine* (descendant s of Nishine), this clan was founded by Nishine. This man was found by Nyikang, on his arrival at Malakal, in the South country (Tonga). Today, this clan live at Tonga.

35. *Kwa-Nail* (descendants of Nail), this clan was founded by Nail, who was a servant of Dak. We are not sure whether Nail came with the groups of Nyikang or he was found in Malakal and he and his people were assimilated into Shulluk Kingdom. Today, the people of the clan live in two areas, some of the live at Obai and other live at Abijop.

36. *Kwa-Dwai* (descendants of Dwai), this clan was founded by Dwai, who was the servant of Dak. According to Chollo oral history, Dwai came from Nuba Mountains. Thus, we can conclude that the descendants of Dwai are the Nubian Chollo; this is probably right as the Nubians are generally addressed by the Shulluk people as "Nya-Dwai". As I mentioned this somewhere in this book, Shulluk people do not talk about this, to avoid separation among the entire Chollo population. Today, people from this clan live in a village of "Palo" of Panyidway chieftaincy. Some of them live in Agwoj of "Lulo" chieftaincy , while others live at "Agugo" of Obay chieftaincy.

37. *Kwa-Agodo* (descendants of Agodo), this clan was founded by Agodo, who was a servant of Nyikang. According to Chollo oral history, Agodo and his people are the aboriginal of Malakal —they were already there, when Nyikang came with his group. Today, the people of this clan live in Obuwa.

38. *Kwa-Nideang* (descendants of Nideang), this clan was founded by a man from Dinka tribe, his name was Nideang. He was a servant of Dak. This man moved from Dinka area and settled among the Shulluk people that led to his assimilation. It is not clear whether this man came with his wife from Dinka land or he came without a wife and later married a Shulluk woman. Today, the people from this clan live in Obai.

39. *Kwa-Nikogo* (descendants of Nikogo), this clan was founded by Nikogo, who was a servant of Nyikang. Nikogo is said to be aboriginal of Shullukland —he might be from Funj people, who are left behind, when Chollo drove the Funj out from the present Chollo country. Today, the people from this clan live in Didigo.

40. *Kwa-Dung* (descendants of Dung), this clan was founded by Adung, who was a Dinka and a servant of Abudok. It remains unclear whether Adung came from Dinka land with a wife or he married a Shulluk woman after his settlement in Shullukland. Today, the people from this clan live at Owichi.

41. *Kwa-Okwai* (descendants of Okwai), this clan was founded by Okwai, who was aboriginal of the land and a fisherman. According to Shulluk oral history, Okwai was a

man from Dinka tribe; he was found fishing in the river and brought home by Du-waat. He later on married a Shulluk woman; they have children that led to the find-ing of Kwa-Okwai clan. Today, the people from this clan live in Adodo.

42. *Kwa-Jalo* (descendants of Jalo), this clan was founded by Jalo, who was a servant of Odak. Today, the people of this clan live at Adit-deang.

43. *Kwa-Ogwat* (descendants of Ogwat), this clan was founded by Ogwat, who was a ser-vant of Odak. Today, this clan live in the south at Tonga.

44. *Kwa-Omal* (descendants of Omal), this clan was founded by Omal, who was a servant of Odak. Today, the people of this clan live in Malakal; however, the people from this clan have relationship with the people of Kwa-Mal clan. According to Chollo oral tra-dition, Omal and Mal both came from heaven to dwell among the Shulluk people.

45. *Kwa-Wang* (descendants of Wang), this clan was founded by Wang. According to Chollo oral history, it was this man (Wang) who crowned Nyikang as a new and the first King of Shulluk Kingdom. Wang is said to be an aboriginal of the Shulluk land. It remains a mystery, as to why Nyikang allowed a foreigner to crown him instead of to be crown by his own people. Today, the people from this clan live at Okun and some of them live at Dur. I have been informed by one of the Shulluk elders that as of today, this clan has important part to play in the coronation of the King-elect in Shulluk land.

46. *Kwa-Okongo* (descendants of Okongo), this clan was founded by Okongo, who was a servant of Nyikang. According to Chollo oral history, Okongo is an aboriginal of Shulluk land —he is one of the people left behind. Today, the people from this clan live at Kakugo.

47. *Kwa-Duwaat* (descendants of Duwaat), this clan was founded by Duwaat, who was a servant of Dak. Toda, this clan live in Filo.

48. *Kwa-Ku* (descendants of Ku), this clan was founded by Oku, who was a servant of Nyikang. According to Shulluk tradition, this man called Oku is an aboriginal; he was found by Nyikang by the river and was brought home. Today, the people from this clan live in Arumbwut.

49. *Kwa-Yodo* (descendants of Yodo), this clan was founded by Oyodo, who was the ser-vant of Nyikang and he was an aboriginal of Shulluk land. Today, this clan live in Fone Nyikang.

50. *Kwa-Okogi* (descendants of Okogo), this clan was founded by Okogo, who was the servant of Nyikang. According to oral tradition of the Chollo, Okogo was a man who was brought from Nuba Mountains and settled in Malakal. Today, the people from this clan live at Det-wuk.

51. *Kwa-Mui* (descendants of Mui) this clan was founded by Omui, who was a Nuer by tribe and a servant of Nyikang. Today, the people of this clan live at Adit-deang.

52. *Kwa-Obon* (descendants of Obon), this clan was founded by Obon, who was a servant

of Nyikang. The main duty of Obon at that time was to eat the meat off the skins of Nyikang's cattle. It is believed that today; the people of this clan are still eating meat off the skins of Reth's (King's) cattle. According to Chollo oral history, Obon was an aboriginal of Shulluk land. Today, the people of this clan live at Nyelwal.

53. *Kwa-Chwai* (descendants of Chwai), this clan was founded by Chwai, who was a servant of Nyikang. The word "Chwai" in Shulluk language literarily means "soup". Therefore, the Kwa-Chwai was founded by a man who used to prepare soup for Nyikang and he was an aboriginal. Today, this clan live in a village of "Nyilwal" of Panyikango chieftaincy.

54. *Kwa-Ringo* (descendants of Ringo), this clan was founded by man called Ringo, who was the servant of Nyikang and an aboriginal of Shulluk land. The man duty of this man (Ringo) was to collect meat of ox that was killed for Nyikang, thus, from his action the man was named "Ringo". We understand that this man had a name before, but we do not know how he was called previously. And the area where the people of Kwa-Ringo clan live in Shulluk land today remains unknown to the writer of this book.

55. *Kwa-Fyien* (descendants of Fyien), this clan was founded by Ofyien; his main duty was to get skins of Nyikang's cattle. Ofyien was a servant of Nyikang and he was an aboriginal of Shulluk land. This clan found a name of Ofyien from the word "Fyien" which means "skin" because they used to gather skins of Nyikang's cattle. Today, the people from this clan live at Nyelwal.

56. *Kwa-wich* (descendants of Owich), this clan was founded by Owich, who was the servant of Nyikang. According to Chollo oral history Owich came from a place called Liri in Kordofan. The main duty of this clan was to collect the heads of Nyikang's cattle. This is how they found their name "Wich" or "wic", which literarily mean head(s). According to Chollo oral history this man call Owich came from Dinka land and settled in Shulluk land. Today, the people from this clan live at Nyelwal.

57. *Kwa-Shin* (descendants of Shin or Cin), this clan was founded by "Shin" or "Cin"; he was a servant of Nyikang. The main duty of this clan was and is to remove intestines from Nyikang's cattle, when they are killed. This also explains how the clan found their name from the word "Shin" or "Cin". In Luo language the word "shin" or "cin" literarily means intestines. Today, the people from this clan live at Nyelwal.

1. *Kwa-Nilongo* (descendants of Oleng or Oling), this clan was founded by Oleng,; this man came from Nuer land and settled in the southern region (Tonga). Today, this clan live in Tonga.

58. *Kwa-Nyidok* (descendants of Nyidok), this clan was founded by Odok, who was a servant of Dak. Today, the people from this clan some of them live in Dur and others live in Obai.

59. *Kwa-Ayado* (descendants of Ayado), this clan was founded by Ayado, who was a servant of Dak. The main duty of ethic clan was and is to prepare bean leaves, gives it to the King, who puts it on his body. Ayado was a servant of Nyikang. Today, this clan live at Dur.

60. *Kwa-Anut* (descendants of Anut), this clan was founded by Anut, who was a servant of Nyikang. According to Chollo oral history this man (Anut) was an aboriginal of Shulluk land. It is said that this clan has taught the Shulluk people to make fire from stick of tree through friction. As we have seen before, that at certain point and time, fires in the whole village have to be quenched. When this happened, it provided an opportunity for the people of Kwa-Anut clan to teach the Shulluk people about how to make fire from two sticks of tree. Today, when a new King is elected and people reach a decision to quench fires from the entire village, it is Kwa-Anut clan that will produce the new fire, from sticks of a tree. Today, the people from this clan live at Fotou.

61. *Kwa-Nyerit* (descendants of Nyerit), this clan was founded by Nyerit who was a descendant of Nyikang. The people from this clan are also called Kwa-Reth (No. 11). The people from this clan belong to the royal clans. Usually, Kings are selected from Kwa-Nerit clan. Today, the people from this clan live in Yoyin (Yogin).

62. *Kwa-Dong* (descendants of Odong), this clan is founded by Odong. This man (Odongo) came from Nuba Mountains and settled in Malakal. Odongo was a servant of Nyikang. Today, the people of this clan live in a village called Twor that is near to Tonga.

63. *Kwa-Odengo* (descendants of Odengo), this clan was founded by Odengo, who was the servant of Abudok. The country where Odengo came from remains unknown up to the present moment. Today, the people of this clan live in Twara.

64. *Kwa-Wubo* (descendants of Wubo), this clan was founded by Wubo, who was the servant of Nyikang. This man had no respect for the cows of Nyikang; he used to kill them when they got into his fields. For this reason, Nyikang cursed them and wished them death. As a result of this one day the village of Wubo was attacked by the Nuer, many people from the village were killed. This explains why today the Kwa-Wubo clan is the smallest in Shulluk land. Today, this clan live at Ajwogo.

65. *Kwa-Nikai* (descendants of Nikai), this clan was founded by Kir, who was a servant of Nyikang. This man was found by the river and was captured by Nyikang. Today, the people from the clan are charged with responsibilities of beating drum when the King dies (disappears). Today, the people from this clan live in Gur.

66. *Kwa-Yo* (Descendants of Yo) this clan was founded by Yo, who was a servant of Odak. But other people said he was a servant of Nyikang. Today, the people of this clan live in Obwo.

67. *Kwa-Gau* (descendants of Ogau), this clan was founded by Ogau, who was the ser-

vant of Odak. This man (Ogau) was from Anyuak, he came and settled in Malakal. Possibly he married a Shulluk woman. Today, the people of this clan live in the southern country (Tonga).

68. *Kwa-Mwal* (descendants of Mwal), this clan was founded by Mwal, who was a servant of Nyikang. According to Chollo tradition, the people of Kwa-Mwal clan are coward, when they are in the battle field; they crawl away from the battle. This people are also known of not eating meat (flesh) from the knee-joints of animals. Today, the people of this clan live at Ogot; they are related to Kwa-Nemwal clan

69. *Kwa-Okati* (descendants of Okati), this clan was founded by Okati, who was the son of Dokot. However, other people said Okati was from Arabic descent. Today, this clan live at Fone Dwai.

70. *Kwa-Kam* (descendants of Kam), this clan was founded by Kam, who was the servant of Nyikang. According to Chollo oral history, this man (Kam) was caught as a fish by Nyikang and he was changed to human being later on. Today, the people of this clan live in Fone Dwai.

71. *Kwa-Bel* (descendants of Bel), this clan was founded by Bel, who was a servant of Nyikang. Bel came from the land of Anyuak and settled with the people of Nyikang in Malakal. Today, this clan live at Mainam.

72. *Kwa-Niyok* (descendants of Oyok), this clan was founded by Oyok, who was a servant of Nyikang. Today, this clan is charged with the responsibility of ringing bells during the coronation of the new King. Today, the people of this clan live in Pashodo.

73. *Kwa-Neyok* (descendants of Neyok), this clan was founded by Oyok, who was the servant of Nyikang. Like Kwa-Niyok clan, Kwa-Neyok clan are also charged with the responsibility of ringing the bells during the coronation of the new King. Today, the people of this clan live in Fashoda.

74. *Kwa-Netyien* (descendants of Netyien), this clan was founded by Otyien, who was a servant of Nyikang. Today, the people of this clan live in Fakan.
    Source: Diederich Westernmann, 1912, pp. 128-134

75. Bang-Reth, these are descendants of retainers of past Kings, they were captured and now they become possessed by the spirit of Nyikang, the people from this clan attach themselves to the court and they are chains of exogamous lineages. In the past when the body of the King used to be buried in his natal settlement some men from Bang Reth move with the elderly widows of the King, remarried them and their descendants remain there to attend his shrine.

76. Bang Nyireth, when a prince is "planted out" in a settlement, his father sent some of his retainers to live there and these people become bang-nyireth. They will serve the prince during his lifetime and after his death their descendants continue to live near the prince's descendants as a fictitious lineage (Evans Pritchard, 1948, p. 411)

28. The Fashoda Incident between Britain and France

Fashoda Crisis was the climatic event caused by years of territorial disputes in Africa, between France and Great Britain that started in the 1670s, which ended in the partition of the entire continent of Africa by 1895. Since the colonies of both the French and the British were widespread over the continent, both countries wanted to link their respective colonies with a system of railroads.

Thus, Fashoda, which is a royal capital of Shulluk Kingdom, was a place where British and French almost came to blow in 1898. This act was promoted by the fact that in the late 19th century, Britain wanted to link Uganda to Egypt by building railway-lines from Cape Town, (formerly known as Cape of Good Hope) in South Africa, through Fashoda to Cairo in Egypt. What the British did not realised was that the French were also thinking of spanning their territories in Africa by building railway-lines from Senegal, in West Africa, through Central Africa and Fashoda to Djibouti in the horn of African coast. The French wanted to hold stretch of lands from the east coast of Africa to the west coast, controlling the all-important head-waters of the Congo River and the River Nile in Upper Nile region. Subsequently, both the France and England were very busy colonizing as much as possible of African land.

Therefore, the two axes of the railway-lines could meet at Fashoda, which necessary would mean that the two armies could meet at Fashoda to protect their railway workers.

Figure: 58, British & France Railway Crossing

The French foreign minister, Gabriel Hanotaux, in his attempts to continue French impe-rialism, sent an expedition of some 200 men east to Gabon under Jean-Baptiste Marchand. British forces, too, sent its troops under Sir Herbert Kitchener some 400 kilometres south of Khartoum. Eventually, both Marchand and Kitchener and their respective forces reached Fashoda. However, Marchand arrived at Fashoda on 10th July 1898.

When Lord Kitchener learned that Marchand and his forces were on their ways to Fashoda, he advised Captain Marchand of France not to attempt to enter Fashoda. Unfortunately, Captain Marchand downplayed the warning on the ground that Fashoda was abandoned since 1882 by the Anglo-Egyptians during Mahdist war in the Sudan. Nevertheless, Lord Kitchener told the head of French army that he would not loose Fashoda at any cost.

As the head of French downplayed the warning, Lord Kitchener then sent army to Fashoda. Kitchener and his forces arrived Fashoda on 18th September 1898, two months following the arrival of French troops to the area. The army were backed by navy on the Nile since the British gained a fresh victory over Omdurman (Khartoum). The British army outnumbered the French force. Subsequently, the two parties resolved to diplomatic solutions. France was ready to fight British forces, hopefully with the support of Russian who showed no interest in the war. The newly formed French Government resorted to diplomatic solution.

29. Background to Confrontation
Sir Henry Percy Anderson (1831-1896) was put in charge of African policy for the two decades of the 19th century. He made decisions about the fate of the strategic areas of the Congo River and the Nile valley. In 1885, he helped to establish the Congo Free State, and this has frustrated the French. The British African policy was focused on preventing other European countries, especially France, from gaining control of the Nile Valley, which regarded as the key and strategic post in Africa. In the same way, East African presented important route from South Africa to India, when it comes in the case of defence. Whereas, the coast of Western Africa, was considered the region for commercial interests. Despite all these, British never imagined to go to war with France due to the importance of Upper Nile.

The Nile Valley drew attention of many European countries such as France and Belgium that resulted in the administration of Egypt in former manner rather than "de facto". Thus, in 1885, The Egyptian Government sensed the danger of the Mahdists and advised Emin Pasha to withdraw from his garrison in Cairo through Uganda to the cost of East Africa. However, Emin Pasha did not take necessary measures to withdraw from Cairo due to passivity of Mahdists in the south of the Sudan.

Meanwhile King Leopold of Belgium was of the opinion that the withdrawal of the Egyptian Government would enable him to extend his Congo Free State to Equatoria and Bahr-el-Ghazal regions in South Sudan.

James Rennel Rodd (1858-1941) was quick to realise the potential conflict of interests that would result from uncertain frontiers and borders of the Congo Free State, which started in 1892. But, with regards to the maximalist claims of the French colonial groups, it was not possible to conclude the negotiations successfully. Subsequently, a British diplomat, James

Rennel Rodd arrived in Brussels (Belgium) at the end of April 1894 and he started sacred negotiations. He very quickly agreed on Anglo-Congolese agreement, which was a paramount. This agreement put the left bank of the Nile of Leopold's disposal and the narrow strip of territory between Lake Albert and Lake Tanganyika at the disposal of the British.

Therefore, in return, Great Britain leased Leopold 11 the left bank of River Nile as far north as Fashoda and as far west as longitude 30 degree east, but only for his lifetime. This implies that the Belgian King and his successors could own larger area in Bahr-el-Ghazal between longitude 25 degree to 30 degree east. By this, the British made up a "buffer zone", which was to protect the Nile Valley from French penetration. The Belgians did the same thing; they enabled the British to gain similar area, cutting the Congolese territory from the north to the south for a permanent lease of the corridor between the Congo and the Nile.

In the years 1894-1895, the Upper Nile was the object of negotiations between Paris and London. Shortly after, when the Anglo-Congolese Agreement was made public, France and Germany raised an objection to its contents. The Secretary of State for Foreign Affairs, Lord Kimberley, was ready to negotiate with the French counterpart, Gabriel Hanotaux, on all African matters, but Sir Henry Percy Anderson, who was the head of the African Department of Foreign office, did not want French friendship, which Britain would be bound to compensate the French with the piece of African territory. On one hand, the French disagreed with the fact that the territory based in the Congo State, for the British, would prevent them from the advance towards the Nile; on the other hand Germans were embarrassed about the north-south corridor in proximity of the German East Africa frontiers. Due to this pressure, King Leopold 11 asked Great Britain to avoid its former agreement concerning the 25 kilometre wide corridor for the British between the northern point of Lake Tanganyika (Tanzania) and the southern shore of Lake Albert. Thus, on 22nd June 1894 the British and Leopold signed this modified agreement.

In the years 1894-1896, the Congolese Sate was in financial crisis, subsequently forcing King Leopold of Belgium to *cede* his private African empire to Belgium. So the France succeeded in effectively eliminating the barrier between its possessions and the Upper Nile. Yet, the rout to Fashoda was still open.

As we have just discussed above in our introduction, in the year 1894, Egypt and Upper Nile became object for discussion between British and French, at the same time (period) there were unreliable information and various rumours that French was planning to control more of the African territories. These matters were tabled in London but diplomatic solution was not successful. In September 1895, the War Office and the Foreign Office asked France's Reginald Wingate (1861-1953), who was the head of the Intelligence Office of the Egyptian army, if it was possible to reach Fashoda through the swamps of Bahr-el-Ghazal. Wingate told the War Office and Foreign Office that it was not possible to reach Fashoda through the swamps of Bahr-el-Ghazal because of the power of the

Mahdists and the unpredictable demeanour of local people. The Mahdist ruled a country without borders in a part of Africa without countries. British Prime Minister, Salisbury also did not believed that numerous French expedition would penetrate to the heart of Africa. While the British were blundering, the French in 1895 began to make expedition to the Nile Valley. The French commander ordered three small forces to march to Fashoda, one force was from Senegal and the other forces from Congo. The expedition of the French to the Nile Valley still came to the British as a rumour and surprise. Thus, in response to these rumours, Edward Grey, the Permanent Under-Secretary of State of Foreign Affairs, made speech in the House of Commons, and in his speech he declared that French expedition to the Nile valley is unfriendly and unacceptable.

Moreover, Major Marchand met Captain Gabriel Hanotaux to discuss about prospective action on the Upper Nile. Major Marchand assured Captain Gabriel that with the assistance of 200 men and 6000,000 francs they would reach Bahr-el-Ghazal without difficulty; for a time being Marchand made no mention of occupying Fashoda. Soon after the meeting, with Captain Gabriel, Major Marchand submitted a memorandum in which he already made a plan to reach the Nile. He thought the best way to avoid a conflict of interests is by quickly advancing the Nile Valley and that would force the authorities to call for international confe-rence that would help solve the problem of the Sudan and the Nile Valley questions. The memorandum was approved by French Minister for Foreign Affairs, Marcellin Berthelot —the Minister also approved finance to assist Major Marchand in his plan of action.

According to Marchand, the expedition should have been informal visit of European travellers to local inhabitants with whom the French would maintain commercial and friendly contacts in the territory bordering on the Upper Ubanghi River.

British Prime Minister, Salisbury, in November 1895 foresaw that British movement in the north would stimulate French advance towards Fashoda, which also means that British could only made effective response to French expedition after the completion of the costly railway-lines from Cairo to Fashoda to Mombasa in Kenya, according to him this would be too late. For this reason, Prime Minister Salisbury ordered for British expedition from Uganda, under command of Colonel James MacDonald, since Lord Kitchener in Cairo did not act faster.

The Prime Minister Salisbury supposed that MacDonald would easily advance to Fashoda and in this way he would overtake the French expedition that was heading towards Nile Valley. Thus, on 9th June 1897, MacDonald received sacred instructions from Salisbury, on how he should officially explore territories on the east shore of the White Nile. He was also authorised to offer agreements on the protection to the local people. Unfortunately, before MacDonald could carry out the order from the Prime Minister, Salisbury, in September 1897, his Sudanese troops in Uganda had mutiny, and this prevented him to go ahead with the plan. However, in April 1898, when the situation calmed down in Uganda-barracks, his mission was aborted.

30. The Encounter in Fashoda

By the end of 1897, news reached Paris that Marchand and his companions were massacred in Upper Nile (Fashoda) but this was not true. Thus, in January 1898 Sirdar, a comman-der-in-chief of Egyptian army and Sir Horatio Herbert Kitchener heard that the French expedition was successfully moving into Bahr-el-Ghazal but not yet reached Upper Nile. In August 1898 Wingate wrote a letter to his wife saying, *"It seems almost certain from our information that Europeans-French probably- are at Fashoda and no doubt we shall make our way south very soon after the Omdurman business if all goes well. It should be an interesting trip"*.

On the 2nd Sepember1898, the Egyptian army with assistance of British army won the Battle of Omdurman and this finalised its long lasting campaign. The moment the world learned about the battle of Omdurman, the French Government in Paris found out that it was only a question of time that the British would meet with major Marchand. However, due to poor communication between Fashoda and Paris, where a letters sent from Paris took between eight and nine months and vice versa, French expedition in Fashoda could not be warned on time and effectively.

Soon after the Battle of Omdurman, the new French Minister for Foreign Affairs congratulated Sir Edward Monson, the British Ambassador in Paris, on the victory in the Sudan. He also mentioned to him the French expedition that was located on the Upper Nile. Several days later, French Minister for Foreign Affairs tried to find out the attitude of the British to Marchand's possible in the region of the Nile Valley. Ambassador Monson answered that Great Britain had earlier openly warned France that a penetration into the heart of Africa (Fashoda) was unfriendly and unacceptable. The French Minister for Foreign Affairs maintained that French could lay claims to Barh-el-Ghazal and Fashoda since these areas were believed to have been abandoned by the British following their fight with the Mahdists. Since then, the French has not recognised them as been part of the British colony.

Thus, in August 1898, the Khalifa in Khartoum sent out two steamers with 500 men to Fashoda to collect grains from farmers. One steamer was called the Twefkich and the other called Sajja. On arrival at Fashoda, the pilot of Twefkich, Soghier, was informed by the local people that eight groups of European soldiers, accompanied by some black troops have arrived Fashoda. There were assisted by the Shulluk warriors and they drove out the British army that was based at Dervish garrison, and now they installed themselves at Fashoda. It is said that Soghier with his 500 men fought the European troops and withdraw since the number of the European soldiers outnumbered them. Soghier withdraw at Dem Zeki, a camp near Fashoda. He then sent streamer Twefkich down to Omdurman for reinforcement.

When steamer Twefkich arrived in Omdurman, Lord Kitchener sent a runner from Babiu with a letter written in French and directed to whoever might be the chief of the party. In the letter Kitchener demonstrated how they have captured Omdurman and about their little fight at Renk. This is simply to advise the French army that British are capable to recapture Fashoda. In other words, this is to show Major Marchand, a French Commander at Fashoda, that it is just a matter of time, British will recapture Fashoda.

In September, Kitchener, Wingate and four gunboats carrying battalions of infantry, set off on a voyage to the south –towards Fashoda. Eight days later they arrived at Fashoda; Kitchener assumed that French could open fire on his steamers and gunboats. Thus, before he enters the city of Fashoda, Lord Kitchener informed the commander of European in Fashoda by writing him a letter telling him about his arrival and entrance into the City. In the letter he also talked about the destruction of the Mahdists and how they reconquered the Sudan.

In the morning on 19th September 1898, Kitchener and his troops entered the City of Fashoda. As Kitchener was entering the City, a black man dressed in red cap, came out from small boat to meet Sirdar's expedition. The man was carrying an answer to the letter of Kitchener that he sent to the commander of European in Fashoda. Major Marchand welcomed British troops on behalf of France Government, and congratulated Lord Kitchener for his victory over Omdurman. Major Marchand also informed Lord Kitchener that Bahr-el-Ghazal and Upper Nile region are now under the protection of France following agreement with the local chiefs.

Subsequently, the Frenchman agreed to an invitation, he (Major Marchand) personally visited Kitchener and Wingate on the board of the Egyptian steamers. They dined together and in a friendly atmosphere over a cup of coffee and a glass of beer, they discussed the situation.

After a short speech (conversation), Sir Horatio strongly protested against the direct infringement of the rights of the Egyptian Government and Her Majesty's Government by the presence of the French forces near Fashoda. Major Marchand was of the opinion that he has occupied the old Egyptian fortress (Fashoda) through order from his Government and for this reason he and his troops had to stay in Fashoda and wait for instructions from Paris.

Therefore, he declined the offer of the Prime Minister, Sirdar, to transport all European expedition members to Cairo. Furthermore, Major Marchand declared to Lord Kitchener that he is prepared to defend the French positions by force of arms. However, Kitchener asked him, "What would happen if the French resist against hoisting the Egyptian flag?" Captain Marchand paused for a while and after a short moment he agreed, because he understood that he wouldn't have means to prevent Sirdar from recapturing Fashoda from French and he might take protection of the entire Upper Nile region.

Thus, in afternoon on 19th September 1898, Kitchener decided to erect his camp approximately 600 yards south of French post near the village they called today Kodok; and on the same day the British-Egyptian army solemnly hoisted the Egyptian flag in Kodok, which symbolised the Khedive's claim to Fashoda.

The British Government left a strong garrison at Fashoda, which Reginald Wingate described in a letter to his wife as follows, *"Thus, our poor Froggies are virtually our prisons, they cannot budge a step, and they have only three small row boats"* (DALY, p. 31).

Furthermore, Wingate talked to himself, *"Is not the whole situation too absurd? Here is this little expedition of 120 men and 8 Europeans shut up in a position hundreds of miles from their nearest support"*.

On the same day (19/09/1898) in the afternoon, Kitchener repaid a visit to Marchand. There, he watched a parade of the French forces in clean uniforms, and he accepted presents —which consists of vegetables and flowers.

Shortly after this, Sir Horatio left Fashoda for Omdurman. Before he left for Omdurman, he wrote two protest letters to Marchand; in one letter, he formerly protested against the French occupation of any part of the Nile Valley, and in the second letter, he announced that all transport of war materials on the River Nile was absolutely prohibited, since the country is under military law.

When Major Marchand received these letters, he was not happy about the contents of the letters, subsequently he raised his complains and wanted Major Jackson, the British Commander in Fashoda, to submit his complains to the authorities in Omdurman —but Major Jackson refused to submit his complains to the authorities in Omdurman, where they could telegraphed the message to French Consul-General in Cairo. In doing this, Kitchener had managed to cut off Marchand from the rest of French authorities as well as from the world. Lord Kitchener then left for Cairo; several days after he arrived in Cairo the chiefs and elders of Shulluk went to Egyptian camp in Fashoda, and told the commander that they have not signed any treaty with the French and they welcome the British protection.

31.0 At the Edge of War

In the month of September 1898, there were exchange of newspapers coming from London and Paris. The British newspaper described Commander Marchand of French, as a band of "irregular marauders" abandoned in the heart of Africa without any outside help. On the other hand the French newspaper keep on telling British people that Fashoda must be retained by French at all cost — even though it means war.

During the meeting of Kitchener and Marchand, he (Kitchener) repeatedly told Marchand that the situation in which they are must be solved diplomatically. With both armies wishing to occupy the fort at Fashoda, Kitchener and Marchand agreed that they did not want military solutions.

On 27th September 1898, Delcasse informed Monson that French Government did not received reports from Marchand, who was in Fashoda. As such French Government could not decide way forward. For this reason, French Government asked for direct communication from Paris, with Major Marchand in Fashoda via Cairo and Khartoum. One day later, British Prime Minister, Salisbury, answered that Her Majesty's Government (French Government) could not at any circumstance, assume responsibility for Marchand's health or safety. Meanwhile Salisbury did not raise any objection for French Government sending message to Marchand through Cairo and Khartoum. However, British Prime Minister predicted that Marchand and his troops would leave Fashoda soon due to lack of food supplies.

Before long, the Fashoda crisis started to escalate, as on 30th September 1898, Delcasse added oil on the fire, when he unofficially declared that he reject on strong term the proposal of the British commander, which offered the French to withdraw its troops from Fashoda to Cairo. And by occupation of Sobat, the Sirdar had already committed to what could be summarised as act of war —which is opposed to unfriendly act on the part of the Great Britain. He also stated that France understood the Fashoda affairs as a matter of national honour and France is ready to accept war rather than submission to the British.

When Prime Minister, Salisbury, received this message, he was not happy either. He immediately ordered Kitchener to prevent Marchand from acquiring any kind of reinforcement from French territories in Africa or in Paris; and to stop supplies of food for the French troops in Upper Nile. Several days later, the British Premier even ordered Major Jackson to block the French at Fashoda by force —which he did but there was no war.

In early October 1898, Delcasse started to persuade his colleagues in the cabinet to surrender Fashoda, as to avoid war with the British. He went further by saying in return for the French withdrawal from Fashoda, he demanded territorial concessions from the British and these territories include Bahr-el-Ghazal and Upper Nile. At this time, British French negotiations were moved from Paris to London.

Thus, in the second half of October 1898, the insular press began to mention more and more about the war, emphasizing that war is the only point for departure from recurring Fashoda crisis.

On 22nd October 1898, French Government received report from Major Marchand at Fashoda; and in the report Major Marchand stated in strong term that the only way forward with Fashoda crisis is for France and Britain to go for war. The winner r would then control Fashoda and Upper Nile. The report was so perplexed that made it more difficult for the French to make a decision on how to avoid war with Britain.

Three days, following the receipt of the report from Marchand, French domestic crisis emerged; consequently the situation of French Government became very complicated. Thus, in the morning of 25th October 1898, more than five thousands of nationalistic and anti-Dreyfus sympathizers crowed the "dela Concorde"

As a result, General Charles Chanoine, French Minister of War, resigned same day from their positions, this brought great tension to the Government of Henri Brisson that felt on that same evening.

By the end of October 1898, three British fleets namely Channel Fleet, House Fleet and Mediterranean Fleet were all on alert but a group of people that include George Joachim and Joseph Chamberlain started to openly require a preventive war against France.

In the second half of October 1898, Marchand left Fashoda to Cairo to communicate with the Parisian Government. But Delcasse was angered by the move taken by Major Marchand, he did not approve the going of Marchand to Cairo, thus he considered Marchand abandoned his troops at Fashoda, leaving Captain Germain in command. But, on the other hand, obviously Marchand could only return to Fashoda by the grace of the British Government. Captain Baratier was sent from Paris to meet Marchand in Cairo and ask him to return to Fashoda immediately. Marchand spent a week in Cairo. He then went back to Fashoda with Captain Baratier to arrange for the evacuation. Furthermore, the relationship between the garrisons at Fashoda worsened due to the French effort to contact the Ethiopian Government.

On the 2nd November 1898, the new government under Charles Dupuy was formed in Paris. The new government had no mood to fight the war. Subsequently, the new French Government dropped the idea of fighting for two main reasons: firstly the French wants British to be an alliance to give support against German. Secondly, the French gave up the war because of the Dreyfus affair, which was the source of great division among the French. Eventually, Theophile Delcasse, the new French foreign minister, on 4th November 1898, ordered Major Jean Baptiste Marchand to withdraw from Fashoda through Djibouti or Obok. This was good news for the British. However, there was still fear that Paris Government would seek alliance with the German and continue with Fashoda incident. Surprisingly, the incident at Fashoda did not eventuate in war because of the large measure taken by Kitchener's act as we have discussion behind and refrain from physical war. . In the end, despite the bellicose talk in London and Paris, there was no fighting between French and Britain at Fashoda. This was a positive sign.

In December 1898, Marchand's expedition withdrew from Fashoda as directed; he hauled down the French flag. Then he set off for his long journey from the Atlantic Ocean to the Red Sea Coast through Addis Ababa and Djibouti to Paris. He finally arrive Paris on 1st June 1899, and was received with immense enthusiasm.

"However, Delcasse did continue to push for a string of smaller posts that would have allowed the French to control a corridor to the White Nile. Lord Salisbury, the British foreign secretary, and the prime minister of Britain, rejected France's idea of occupying forts, in hopes of controlling the headwaters of the Nile. Eventually, on March 21st 1899 the French and British governments agreed that the watershed of the Nile and of Congo, respectively, should mark the boundaries between their spheres of influence" (Guatam Bahl, 1997, p. 2).

32.0 *Fashoda Royal Residence and Capital*
How Fashoda became the royal residence and capital of Shulluk King remain unknown to many young and elders Chollo. Thus, the story of the establishment of royal residence and capital (Fashoda) could not be explained apart from the general belief system that

exist among the Shulluk people, which attributes all the phenomena of the universe to the power of God and His will.

According to Shulluk oral history, Fashoda was established by Reth Tugo wa Dakoth (1690-1710). Like the former Kings, Reth Tugo wa Dakoth ruled Shulluk Kingdom from his village called Badiang or Nyewajo, which is located south of the present Kodok. Chollo traditions stated that Reth Tugo wa Dakoth had two bulls without horns. The two bulls used to isolate themselves from the other cattle when they are taken for grazing. The story said, the two bulls without horns, before they reach the grazing land, they always took different direction. The story said the bulls used to go to the side of the present Fashoda; there they used to dig hole into the ground with their legs and heads. The bulls were called "*Chod*", which means without horns. This happened many times, when the King saw that the two bulls repeatedly dig hole in the same site, he then said, "My God! Why are these oxen always dig the ground?" The repeated actions of the oxen attracted the attention of King Tugo and all the chiefs of the tribes of Shulluk country.

The King and his chiefs then held a meeting to discuss the matter. In the meeting, it was resolved that the repeated digging of the ground in the same place by the bulls is an indication that God has chosen the site as a sacred place —huts must be built there.

The following morning King Tugo wa Dakoth and his chiefs mobilised people to offer their services. The people gathered and the hole that was dug by bulls was filled with earth till it became higher than the rest of the ground around.

Figure: 59, House of King of Shilluk at Fashoda

The King then order the people from the clan of **.......** to build four huts on elevated ground, This was done within short period of time —four huts were built and they gave a collective name as "Athurwic" in Shulluk language. The four huts were considered as the sacred huts. Johnson D. (1990) observed that Athurwic is the sacred enclosure of four huts built on a mound which dominates the capital. These huts are rebuilt at the installation ceremonies of each new King, and the mound on which they rest consists entirely of the debris of previous huts. Thus, Athurwic becomes periodically the focal point of the royal rites and the royal capital itself.

The Village where the four huts were built became known as "Pachodo". When the constructions were completed, Reth Tugo moved from his village Badiang or Nyewajo to live in his new place (Pachodo). King Tugo wa Dakoth then declared, "This village shall always remain the village of election and the village of the coming Kings". Eventually, from that time Pachodo remains a place of election and coronation of each King-elect.

The word "Pachodo" or "Pachot" —means bulls without horns. With the changing of time and European colonisation in Sudan, the place then was called "Fashoda".

According to Shulluk traditions, they can be renovated by the descendants of the clan who first built the huts at the time of King Tugo rule in 1690.

The history of the political system of the Shulluk Kingdom revealed the fact that the Shulluk land had been divided into two provinces or districts known as Luak that constitute the southern part of the Kingdom and offend referred to as Gol Nyikang —which means areas occupied by Nyikang and his descendants; and the Ger that constitute the northern part of the Kingdom, and often referred to as Gol Dhiang —which means the areas occupied by non-descendants of Nyikang and slaves.

Interestingly, in the old days, Reth used to appoint two main chiefs to keep order in the northern and southern parts of the Kingdom. The two main chiefs, together with the chiefs of Mwomo (Northern Province), and chiefs of Tango (Southern Province), all were considered the four high ranks in Shulluk Kingdom. The high chiefs were and are considered as the King's representatives in their administrative units. Their role is almost administrative rather than judicial. They are restricted to local disputes. The rest of cases are usually referred to Fashoda to be settled by the King. In this way, Fashoda became the focal point of the indigenous systems of the Shulluk government.

Today, Fashoda is considered important cultural space because of cultural activities that take place within its domain. These cultural activities are related to simple fact that Shulluk people believe in the ancestors and their power as mediators between God and man. In this point, it is important to notice that these mediators cannot be performed outside Fashoda —all must be performed in Fashoda. However, this fact is clearly reflected in the coronation ceremonies from day one of election to the last day of crowning the King, when two silver bracelets are put around his wrists to indicate that he is authorized by his people to take up his responsibilities as a King.

33.0 *The Importance of Fashoda*
According to Shulluk belief systems every step in the coronation process of the King must be done in a certain way, by certain clan or people and on the fixed date and month. Some

of the cultural activities that are performed in Fashoda as regards to the coronation of the King could be summarized as follows:

(1) Fashoda is a c capital of Shulluk Kingdom where all senior chiefs of Shulluk tribes meet to attend the election of the ne King.

(2) Fashoda is a place where the newly King-elect (Reth) is taken to spend a quiet life before he performs mourning dance for the dead King.

(3) Fashoda is selected by the will of God, as such all the rituals and coronation of each King-elect must reach their climax in Fashoda.

(4) The sacred building (Athrwic) and the shrine of Nyikang are found in Fashoda. This is also a place where the traditional documents and traditional artefacts of Shulluk Kingdom are kept. These places are visited by many of local people as well as tourists.

(5) Fashoda is then cultural court to which major disputes are referred to be settled by the Reth himself.

(6) Fashoda is a place where the Advisory Council (Loko) of the Shulluk meet to discuss matters of election and coronation of any newly King-elect

34.0 *Building houses among the Shulluk people*

The hamlets of the Shulluk are almost continuous, like beads on a string, along the west of the Nile from near Lake No to about latitude 12n. With a number of settlements on the east bank and along the lower reaches of the Sobat. The hamlets (*myier* or *pac*) are built from 100 yards to a mile or so apart on the high ground parallel to the river, vary in size from one to fifty homesteads. A homestead is a residence of a family (*goh*), consisting usually of two huts encircled by a fence. Each hamlet is occupied by members of an extended family with their wives Evan Pritchard, 1948, p. 408).

Shulluk people have well developed systems of building houses in a community. They have a clear labour distributions, each adult member of the family has a role to play in the construction of a family house.

The women are charged with responsibility of cutting and carrying grass and corn stalks to the building site. They are also charged with the responsibility of fetching water for mixing mud for the foundation and wall during the construction.

Men are charged with the responsibility of cutting and bringing poles to the building site for roofing. They are also charged with responsibility of cutting grass and making it into a rope or ropes; mixing and carrying mud for filling the foundation.

Men make ropes from the tall grass that is not easily destroyed. They cut the grass, soak it in water and bruise with clubs until the fibber comes out. The fibbers are then dampened and twisted into ropes. The rope maker sits on the ground while holding the rope between his toes and forms the rope by constantly adding new fibber and rolling the rope between the palms of his hands.

Generally, it is required that all building materials should be collected in the building site before the actual construction begins.

Customarily, Shulluk people start building of their houses by digging foundation. According to Luo definition, foundation is a shallow circular trench where the wall of the house is to stand.

According to R. W. Tidrick, house building among the Shilluk is a trade to be learned, where only few men learn (p. 97). The reason why only few men learn the trade for building needs future research, But as I know, Luo young people learn building skills from youth hood.

While the circular trench is being dug, men mix the mud with manure, ashes, broken grass and sand; they carry the mixed mud with their hands and drop it into the trench. In addition, the men pour certain amount of beer (Ombuki) together with the mud into the foundation trench. The builder forms the mud into the desired shape —using his hands. Interestingly, builder put only two to three layers of mud a day. A layer of about six inches deep is put on top of one another until the foundation is complete.

After this, the foundation is left to dry between three to four days. In the fourth day, the builder continues with his work. When the foundation reaches a required door, an elliptical band of grass about three feet in depth is then put in place to form the door.

As the wall is being built, the mud is added against the door, to keep the desired shape of the door. The walls are then built and left for another four days. After the fourth day, the builder opens a round-door in the wall and continues with the construction in the fifth day.

In the six and seven days, the thatcher starts work on the roof, as men cut poles burn them so that the poles become hard and later on they cut these poles to equal length to form roof-sticks. Soon after this the roof-sticks are arranged and tied up in a circle. Thereafter, circular ditch is dug and the ends of the roof-sticks are placed in this circular ditch and buried with soil.

Now the roof is built by twisting and tying the grass on the poles. When this is finished, the roof-sticks are removed and the roof is lifted upon the wall. At the same time, the junction between the wall and the thatch poles are made tight with mud and the lower roof-ends are tied together to the wall (Diedrick Westernmann, 1912).

According to R.W. Tidrick, there are two types of grass used for roofing in Shulluk country (P. 97). However, the family class dictates which type of grass to be used for the thatching of family house.

The first work in thatching is to put on what the Chollo call "*cienot*" (the "apron of the house)". This is when a short layer of grass is put around the top part of the wall and is tied strongly to the poles. Thatcher climbs on the roof and bundles of grass is cut and thrown to him. This is the time we say with certainty that thatcher begins his work, where grass is tossed up in small bundles. The craftsman then removes the bands and ties them down tightly.

The thatcher continues to work until he reaches the point that he has to leave small opening on the top of the roof known locally as *"wang-kac"*. Thus, when thatcher reaches this stage, he asks the proprietor to give him a hoe; he takes the hoe and covers it with *"wang-kac"* -the small opening that was left on the top of the roof.

Figure: 60, Shulluk Houses

The grass lies on the roof from six inches to a foot thick, and if the grass were kept free from white ants, the house could last between six to ten years.

Before the owner of the house move into the new house, he is to make sure that that surface of the roof is smoothen, dripping —eaves cut, door is opened to a required shape, the floor is made smooth and hard, a fence is constructed. When all these are in place, certain ceremony is required before the family moves into the new house. The man (owner of the house) takes red-rooster (a Cock) and leaves it in front of the new house as a sacrifice. He usually takes this red-rooster to the new house early morning or late evening when the sun is almost setting. The red-rooster is left alive in front of the new house. What happen with the red-rooster after that is not explained to me and this required for future research. After this, the man and his members of family then move into the new house.

Moreover, in the old days, Shulluk people did not used money to pay for labour of the builder of thatcher, the rather used to give a sheep for such labour. However, today the system has changed; a builder of a thatcher of the house is given money instead of a sheep. What is not clear to me is whether a thatcher of today is paid with money that is equivalent to the value of one sheep.

35. Hunting among the Collo

Hunting is an important economic activity to the Shulluk people. Hunting is usually done during the dry season after people have gathered in their crops. When the villagers want to go for hunting, the hunting must be well coordinated for otherwise the animals might migrate to a far places and this often cause disputes among the people. The Collo are allowed to kill gazelles, antelopes, hippos and water buck in particular circumstances. However, the Collo are not allowed to kill leopards, water bucks and giraffes, as well they are not allowed to kill ostrich and the red flamingo. If by mistake one kills one of these forbidden animals, he must report to the chief of the clan that is in charge of that animal, and must give him the animal's skin. The chief will then report the killing to Pachodo, taking with him the animal's skin. In case of birds, the flamingo wings, and ostrich feathers must be taken to Pachodo.

36. List of Shulluk Kings 1545-1992

In the administrative of South Sudan Government, the Collo of South Sudan is composed of 100 clans (which are divided into sub-clans), 4 Payams and twenty-five Bomas.

The Shilluk Reth (Kings)

S/N	Names	Period of	Reign	No. Of yrs	Name of villages
		From	To		
1	Nyikang wad Okwa	1545	1575	30	Nilwal-Panyikang (1537-67)
2	Caalo wad Nyikang	1575	1590	15	Denyo-Tungo (1567-1582)
3	Dak wad Nyikang	1590	1605	15	Paalo-Panyidway (1582-97)
4	Nyidoro wad Nyikang	1605	1615	15	Nyilyiec-Gaar (1597-1607)
5	Ocollo wad Dak	1615	1635	20	Detang-Obwa (1607-1627)
6	Diwaad wad Ocollo	1635	1650	15	Obudhyang-Malakal (1627-42)
7	Bwoch wad Diwaad	1650	1660	10	Paroro-Abyienayo (1642-52)
8	Abudhok Nyabwoch	1660	1670	10	Thworo-Dhothim (1652-62)
9	Dhokoth wad Bwoch	1670	1690	20	Thwonh Kong-Awajwok (1662-1682)
10	Tugo wad Dhkoth	1690	1710	20	Nyimongo-Padhiang (1682-1702)
11	Okon wad Tugo	1710	1715	5	Palabo-Pabur (1702-1707)
12	Nyadwai wad Tugo	1715	1745	25	Dibwor-Kodhok (1707-1737)
13	Muko wad Nyadwai	1745	1750	5	Pabo-Nyigir (1737-1742)
14	Waak wad Nyadwai	1750	1760	10	Burkyenyi-Obwa(1742-1752)
15	Dyetguth wad Nyadwai	1760	1770	10	Panyatho-Fashoda (1752-1762)
16	Kudit wad Okon	1770	1780	10	Palabo-Lul (1762-1772)
17	Yor wad Kudit	1780	1820	40	Agwoch-Lul (1772-1812)
18	Aney wad Yor Nyakwaci	1820	1825	5	Nyiwudo-Gar(1812-1817)
19	Akwot wad Yor Nyakwaci	1825	1835	10	Diballo-Ogot (1817-1827)
20	Awan wad Yor Nyakwaci	1835	1840	5	Opat-Nyigir (1827-1831)
21	Akoch wad Akwot	1840	1845	5	Anyiago-Agot (1831-1836)
22	Nyidhok wad Yor Nyakwaci	1845	1863	18	Dur-Dhothim (1836-1859)
23	Kwathker wad Akwot	1863	1869	6	Opathiwan-Ogot (1859-1870)
24	Ajang wad Nyidhok	1869	1875	6	Radilep-Fashoda (1869-1875)
25	Kuckon wad Kwathker	1875	1882	7	Apodho-Ogot (1875-1881)
26	Yor Adodit wad Akoch	1882	1892	10	Bapi-Kodhok (1882-1892)

S/N	Names	Period of		Reign	No. Of yrs	Name of villages
		From	To			
27	Kut(1) wad Nyidhok	1892	1903		11	Pareth-Golo (1892-1903)
28	Padiet wad Kwathker	1903	1917		14	Tumier-Wau (1903-1917)
29	Papiti Gwang wad Yor	1917	1944		27	Abwogatho Golbanyo (1917-1944)
30	Aney wad Kur Nyidhok	1944	1945		1	Ganawat-Golbanyo (1944-1945)
31	John Dak wad Padiet	1945	1951		6	Kujo-Panyidwai (1945-1951)
32	Kur(11) wad Papiti	1951	1974		24	Ywodo-Kodhok (1951-1974)
33	Ayaang wad Anei Kur	1974	1992		18	Owiykyiel-Fashoda(1974-1992)
34	Kwongo wad Dak	1992				Panyi-dway(i)(1992-

Geographical Distribution of Shulluk Clans and their Title Sets

Clan Hero (Kwey)	Title used in addressing a respectable clan member (Dhwom)	Title used in replying to a respectable clan member (Paag)	Clan Settlement (or Village)	Chieftaincy	Great Chieftaincy
Ababo	Kwababo	Kal	Olelo	Warajwok	Makal
Aboro(Ayang)	Kwaboro	Nyigol	-	Oriang	Muomo
Acolo	Kwacolo	Wedi	Ayidhajo	-	Tungo
Acoolo	Kocolo	Diang	-	Nyibanyo	Tungo
Acoolo	Kwacoolo	Godo	Dhekoodt	Thworo	Dethin
-	-	-	Pacoolo	Odwojo	Penyikango
Acuag	Kwacuaag	Dhiang	Pekier	Nyiluag	Penyikango
Acu	Kwacu	Godo	-	Nyiluag	-
Adhog	Kwadhog	Dway	-	-	Tungo
Adong	Kwadong	Dhiang	Pakwadong	Nyibanyo	Tungo
Adung	Kwadung	-	Pebag	Oweci	Dethin
-	-	-	Agwoj	Lulo	Wireg
-	-	-	Ongacien	Ogoon	Luag-Bolbeth
Adway	Kwadway	-	-	Odong	Penyikango
Agudo	Kwagudo	Godo	Tegothwonh	Pedhing	Wireg
Agwad	Kwagwad	Nyigol	Akoonre	Gol-Banyo	Wireg
-	-	Godo	Lelo	Obwar	Makal
-	-	Dhiang	Ayidhajo	-	Tungo
-	-	-	-	Nyilual	Penyikango
Ajulo	Kwajulo	Dway	-	Bol	Wireg

245

Clan Hero (Kwey)	Title used in addressing a respectable clan member (Dhwom)	Title used in replying to a respectable clan member (Paag)	Clan Settlement (or Village)	Chietaincy	Great Chieftaincy
-	-	-	Dod	Warajwog	Makal
-	-	-	Dubo	Kodhog	Kodhog
-	-	-	-	Palo	Penyikango
-	-	-	-	Nyiluag	Penyikango
-	-	-	Pekano	Thworo	Dhethin
-	-	-	Pelidho	Pedhiang	Wireg
-	-	-	Pepuojo	Kodhog	Kodhog
-	-	-	Balowudo	Odhucogo	Dithwog
Akaj	Kwakaj	Colo	Pebool	Dinyo	Tungo
Akaji	Kwakaji	Tongo	Adhekoong	Warajwok	Makal
-	-	-	Oriemcang	Pekaang	Pekaang
Akajo	Kwakajo	Kogo	Dibalo	-	-
Akeg	Kwakeg	-	Lelo	Oweci	Dhethin
Akoogi	Kwsakoogi	Dway	-	Obwar	Makal
-	-	Godo	-	Nyiyar	Penykango
Akoong	Kwakoong	Godo	Anyiago	Adidhiang	Penyidway
-	-	Nyigol	Dhekoodh-Paj	Warajwok	Makal
Akway	Kwakway	Godo	-	Adodo	Dhethin
-	-	-	Pakway	Agworo	Muomo

Clan Hero (Kwey)	Title used in addressing a respectable clan member (Dhwom)	Tetle used in replying to a respectable clan member (Paag)	Clan Settlement (or Village)	Chieftaincy	Great Chieftaincy
-	-	-	Nywudo	Dilal	Kwojo
-	-	-	Pebag	Abienyayo	Dithwog
Aku	Kwaku	Dway	Wibur	Adidhiang	Penyidway
-	-	-	Paku	Arobuodh	Kwojo
Alabo	Kwalabo	-	Lelo	Oweci	Dhethin
-	-	-	Nyaper	Godo	Dithwog
Alami	Kwalami	Godo	Dilal	Golbanyo	Wireg
-	-	-	Obuyo	Obwar	Makal
Aloongo	Kwaloong	-	Ngoor	Pedhiang	Wireg
Luudh	Kwaluudh	Dway	Agwoj	Lulo	-
Amadh	Kwamadh	Gudo	-	Nyilual	Penyikango
Amoong	Kwamoong	Kwojo	Alaal	Lulo	Wireg
Anag	Kwanag	Paan	Petwaro	Dur	Dhethin
-	-	-	Punag	Ajogo	Pekang
Anay	Kwanay	Kelo	Burkag	Akurwa	Muomo
-	-	-	-	Penykango	Luag-Bolbeth
-	-	-	Abujob	Obay	Pekang
-	-	-	Agunjwok	Niluang	Panyikango
Areedi	Kwareedi	-	Otongog	Lulo	Wireg
Atengthooj	Kwatengthooj	Dway	Ameng	Oriang	Muomo
Atibil	Kwaatibil	Dhiang	Patibil	Agworo	-
Atugi	Kwatugi	Godo	Didigo	Peju	Penyikango

Shatt or Thuri

Demographic and Geographic

The Thuri people live around Wau town, Tonj and Aweil in Western Bahr el Ghazal, and some live around Warap and Raja in Northern Bahr el Ghazal. They are found between Wau town and Aweil and some are found between Jur and Lol Rivers, on Raja Nyamlell road up to Wau-Deim Zubeir road. Shatt are part of the larger Luo family, made up of the Shulluk, Anyuak, Acholi, Jumjum, Mabaan, Balanda Boor, Pari and Jur Chol, as well as the Luo in Kenya, Tanzania, Congo Republic and Acholi in Uganda. The Shatt society is divided into three main groups the Yabulu, who live around Yabulu located between Wau and Raga; the Achana who occupy the territory Mariasl Bai and Raga; and the Chekou who occupy the land between Aweil and Raga; and the fourth group of Shaat live in the crest of the Nile-Congo watershed. The four groups are also divided into agnatic lineages that make up to more than thirty clans. The people elaborate traditions and customs for nearly everything in their life.

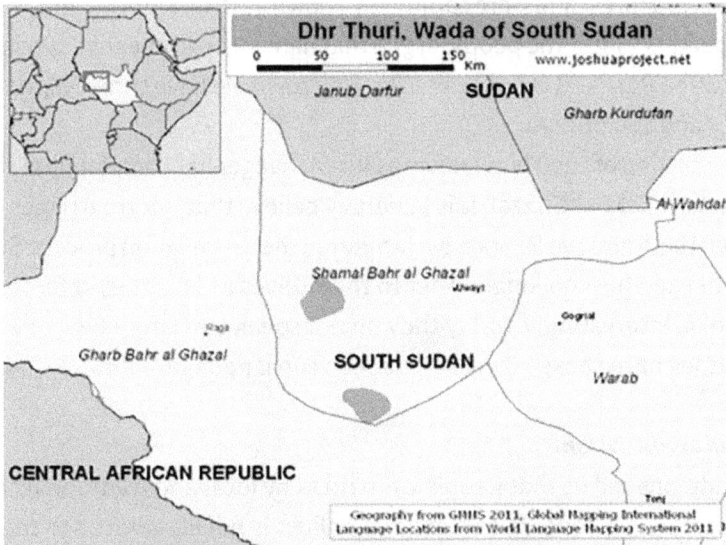

Figure: 61, Locations of Shatt/Thuri in South Sudan

More specifically, the Thuri live in the following districts or villages: Dhiama, Umal, Nyinbwoli, Yaalo, Achaano, Par-Amaado, Maper Gir Faama, Areia, Jod-Kul, Apuojo, Wuj-yaa, Aguli, Awoda, Thikidi, Thikou, Aguarbel, War-mon, Abul, Kuuru, Doi-Leela, Doi-Kou, Aweil, Nyalath, Umang Kiwang, Dagajim, Wadhilang, Umoora and Wau. They are said to have come to settle in these districts in 1600s. The Shatt live in plains with isolated hills cut by deep valley in which drain several perennial streams.

The real population of the Shatt remains unknown to the present day. Writers are estimating the population, for example Angelo Ugwaang estimated the population of the Thuri at 8,000 (in Joshua Project 2016); on the other hand Madut (2008) put the population to 16,720.

In their creation myth, the Thuri believe they are the descendants of a legendary ancestor, Dimo, whose brothers Nyikang and Gilo left Bahr el Ghazal for Upper Nile after a family dispute over the beads; and Nyikang ended up finding the Shulluk Kingdom. They believe Dimo was the father of their ancestor, Othuru (or Utho). Subsequently, the Shatt are referred to as Shulluk of Bahr el Ghazal, as St Andrea (1938) thinks that they are part of Nyikang group immigrated to Upper Nile Region where they currently settled.

The Thuri are also called Luo Thuri, Jur-Shatt, Dhe Thuri, Jo-Thuri, Shatt, Boodho, Jur Chol and Wada Thuri. So whenever you hear or read these names do not think that these are different people –they are the same people. Edward Thomas (2010) observed that Fertit are part of Thuri, they were separated by the British. This was confirmed by Ireneo Kunda, a Balanda chief, who says, "The name Fertit was given by the English to some people of Thuri who lived in a forest and were commonly known as "fruit eaters". Thus, literary "Fertit" means "Fruit eaters" but some people argue that name "Fertit" means associate of people who live between Raja and Deim Zubeir. In Wau town we have three Nilotic tribes Shatt, Balanda Boor and Luo of Wau.

However, a section of Luo Thuri known as Jur-Mananger has completely associated with Dinka of Western Bahr el Ghazal. It is generally believe that intermarriages, which have been between them and the Dinka have facilitated the assimilation process. Subsequently, this section of Luo Thuri no longer refer to themselves as Luo, they refer to themselves rather as Dinka. Interestingly, today they do not speak Luo language (Dhe Boodho), in other words they have chosen Dink language to Luo language.

Socio-political organisation

The Thuri traditions tell us that people were ruled by local and traditional chiefs known locally as "Rwot". Their concept of State and politics is not elaborated to the level of the Shulluk as we have discussed in this chapter. It is worth mentioning that today; the position of traditional Rwot is eroded by the government chiefs who do little for the people but more the government in term of collecting taxes.

Hunting

Hunting is one of the social activities that occupy the live of the Shatt. Hunting is usually carried out during the dry season and is performed by large groups of people and in accordance with certain local rules. Rules are strictly applied when hunting big animals like elephants, buffalos, rhinos and giraffe. This is done to avoid any conflict over the distribution of tusks, skins, horns and tail hairs. Thus, the person who stabs the animal first is called the killer or owner of that animal. More importantly, the person who stabs the elephant first according to local rules he is to receive the right tusk while the second person who stabs the elephant receives left tusk. The meat of an elephant can be taken by anybody who in the hunting games. The people do not keep the tusks, they sell the trunks of elephants to Arab traders and often they exchange trunks with cattle from the Dinka.

Chapter 11:
Saturnino Obwoya Opio: The Hero and Initiator of Liberation of South Sudan Independent

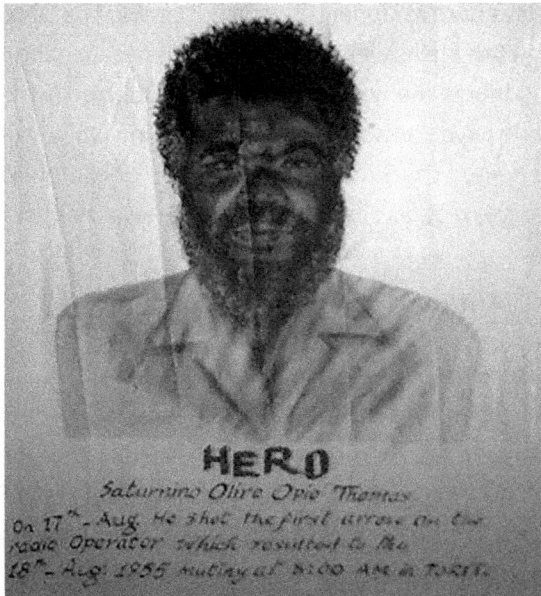

Figure: 62, Saturnino Obwoya Opio

McCall Storrs (1970) *took the argument further as he says, "On August 6th, 1955, at Torit, a southern soldier named Saturnino Obwoya shot arrow in the darkness at a man near the post office who he mistook for his northern commanding officer. He was arrested, his house was searched, and a number of documents discovered. The documents revealed that Obwoya, who styled himself "President of South Corps" was involved in a plot to murder all Northern officers in the South; and 24 other officers and men of the Equatoria Corps were also involved, including soldiers at Wau and Malakal.....It appeared that Obwoya had already tried to get Loleya to head an army uprising in Juba on August 4th, 1955, with the aim of capturing the ferry and the aerodrome, but that Loleya had urged him to be patient and wait. Obwoya was of too impetuous character and took the matter into his hands with the shooting of the arrow" (McCall Storrs, 1970, p. 7). "Today in South Sudan history: 18/08/1955 the members of the British administered Sudan Defence Force Equatoria Corps mutinied in Torit by using primitive tools and the first man who used his bow and arrows wisely to light the candle of the liberation war was ladit Saturnino Olire Opio Elias Lanyok. Unfortunate his name is kept behind curtain in the history of the Sudan and South Sudan......" Oluku Andre' Holt (2016)*

Birth of Saturnino Opio

The birth date of Saturnino Obwoya remains a mystery to Obbo elders; although some elders assume he was born in 1910s. His father is called Thomas Obwoya Adwat Alias Aboyii from Abong clan (sub-clan Abong Kure) and the mother is called Ayaa from Abong pa Rwot sub-clan. He was born at Upper Lerwa or got Lerwa. The Obbo elders do not know much about his early youth life as well as his education because the existing Obbo elders were born after his childhood. Many of the elders from Obbo do not know anything about his education, yet many have known that when Saturnino Obwoya grew up in Lerwa to manhood, he later on joined the Sudan military force. For example, Justice John Ongee puts it clearly, "I do not know of his early youth because I was not yet born".

Root Cause of Torit Mutiny

a) Arabs occupying southern positions left by British officials

People like Justice John Ongee observe that as independence of the Sudan was approaching, tensions arose between Northerners and Southerners. These tensions were brought about because the British Government was silent on the future of the South Sudan. People started to hear rumour that Northerners (Arabs) were rushing to the south to occupy positions been left by the British officials. Before long, the position of the Governor, Deputy Governor, District Commissioners and their Assistants as well as the positions of Mamur were all occupied by Arabs. Therefore, Southerners saw these as a challenge of masters from outgoing British officials. The takeover of the positions started in May 1955 and ended in JULY 1955. The Arabs did these because they knew the British were leaving the country but the Southerners were kept in darkness.

Writer like McCall Storrs (1970) took the argument further, and pointed out that in May 1955, on the day of the Muslim Bairain Celebration, Prime Minister, Ismail el Ashari, publicly announced that his government would abandon its policy of linking Egypt and Sudan —they rather aim for complete independence. Consequently, the Northerners started to send Arab Administrators to the South Sudan. Eventually the Prime Minister asked his Administrators in Yambio to let the local chiefs sign telegram supporting the government and disassociated them-selves from Liberal Party ideology. As a result, thirteen Azande chiefs signed the telegram in support for the government.

Prime Minister, Ismail el Ashari was totally against the Southerners. Mc Call Storrs (1970) notes he has written a letter to Administrators in the South saying, "To all my administra-tors in the three Southern Provinces: I have just signed a document for self-determination. Do not listen to the childish complaints of the Southerners. Persecute them, oppress them, and ill-treat them according to my orders. Any Administrator who failed s to comply with my orders will be liable of prosecution. In three months times all of you will come around and enjoy the fruits of the work you have done" (Mc Call Storrs, 1970, p. 2).

It is said that this telegram circulated in the South Sudan like wild fire; and undoubt-edly, it has contributed to the outbreak of Torit mutiny. It is believed that a Southern clerk, Ruben Yakobo, who was on duty, received this telegram in Juba. It is reported, after Ruben finished his duty, he handed the telegram to the post-master as usual. It is not clear whether the post-master read the telegram or not but Ruben put it in his drawer and went home. When Ruben reached home, his heart was not at peace because of the contents of the telegram.

The following morning when Ruben went to work, he decided and stole the key of poster-master's drawer. It is asserted that when everybody left the office, Ruben came back to the office and obtained the original Arabic version that was translated into English by a man called Francis Ako. They then made many copies of the translated telegram and circulated the message in the three Southern Provinces. These copies were distributed by secrete committee headed by Marko Rume and Daniel Jumi (McCall Storrs, 1970).

b) The arrest of Elia Bkuze

On the 7th July 1955, when Elia Kuze, the Liberal MP of western Equatoria in the parlia-ment learned that thirteen chiefs signed the telegram, and the telegram has been sent to Juba, he organised a public meeting in Yambio. The meeting resolved that they do not want to be ruled by Arabs and all the chiefs who have signed false declaration in support of Khartoum government must be removed from their offices.

On 25th July 1955, MP Elia Kuze was taken to court chaired by chief Soro. The court claimed MP. Kuze was found guilty and was sentenced in prison for two years. After the chief announced the verdict, a crowd of approximately 700 staged a demonstration outside the court; the police broke the crowd with tear gas. The crowd split into small groups and they started to attach some Arabs on the streets and raided their shops. This affairs seriously damaged North-South relations in Zande land (McCall Storrs, 1970).

c) Deployment of No. 2 Company to Khartoum

On July 9th, 1955, the new Governor and his Deputy, the army commander and his Deputy and six other Arabs met in Juba. The meeting agreed on three important points: Firstly, it was agreed that No. 2 Company army unit in Torit must be taken to Khartoum on 18th August 1955 without fail. Secondly, all the ammunition must be removed from Torit garrison before the departure of the unit. Thirdly, troops that will be brought from Khartoum should not be taken to Nzara but must be deployed three miles away from Torit to monitor the departure of the unit. It was reported that many Southerners did not attend this meeting. As tension grew among the Southerners, British officials were already leaving the South Sudan one by one and quietly.

On July 10th, 1955, approximately four platoons arrived in Juba from Khartoum. The southerners did not know why these troops were brought to the South; this eventually increased the tension among the Equatorians

d) Nzara incident

According to Justice John Ongee, on 26th July 1955, before breakfast, a collection of workers of the Nzara Board went to meet the chairman demanding for equal pay with the Northerners working in the factory. Nevertheless, the chairperson told them to select representatives from among them to come after breakfast and discuss with him about their demand. The crowd listened to the suggestions offered by the chairperson and soon withdraw. Thereafter, the chairperson sent message to Yambio calling for police and army to come to Nzara to calm the situation. Ahmed Hassan, Acting District Commissioner of Yambio, received the message from the chairperson. He then informed other officers such as Nyang Diu, an army officer, a junior police officer, assistant inspector of police, leaving Placido Laboke, who was the senior police inspector in darkness. The army and police officers then sent their forces to Nzara. When Placido Laboke learned about this, he rushed to Nzara, with a hope to interfere with the plan, but unfortunately, he found that the army and police forces have already taken their positions. They shot and killed seven workers; this act created tension particularly among the people of equatorial. The office of District Commissioner (D.C.) of Yambio immediately sent message to Khartoum demanding army Headquarters to send northern troops to Nzara.

McCall Storrs argue that on the 26th July 1955, following Kuze's public meeting and demonstration, the workers at the Equatoria Projects Board cotton mill in Nzara rioted in demand for pay rise but the crowd was dispersed by the army using machine-guns. Both Northerners and Southerners were working at Nzara cotton factory. Thus, following the riot, 300 Southern textiles workers were dismissed from their positions and the Arabs took over the administrative position in the industry from the British officials. The impression was inevitably created that Southerners were losing jobs to Northerners. As a result, the Arab administrators were abused and told to go back to their country (where they came from). When the situation was almost getting out of hand, the chairman of the Board sent a telegram to Yambio requesting District Commissioner to send in five police men and eleven soldiers to go down to Nzara to help control the situation. The army came; they were armed with rifles, brand-guns and stand guns. Both police forces and soldiers came from Yambio under the command of Mutassim Abdel Rehman. When the soldiers and police officers arrived in Nzara, they found the crowd were continuing with their riots. The Commander Mutassim ordered soldiers and the five police officers to open fire on the crowd. It is reported that at least six people were shot and killed and two other people died in the resultant stampede of the crowd that bring the total number of death to eight people. Thus, by the end of july1955, the tension reached full scale of rebellion for Southerners that triggered army mutiny in Torit on 18th August 1955 (McCall Storrs, 1970).

Hero Saturnino Obwoya in Prison

Although not much is known about the level of Saturnino's education, the reality is that he was a hero, strong fighter and a politician. It is assumed that he joined Sudan Arm Forces (SAF) in the 1940s and later on worker in Torit as alliance corporal (clerical) in mechanical depot of the Equatoria Corps. Justice John Ongee notes that in those days, politics were unknown to Southerners, as a result, people had no interest in politics but Saturnino was the only exception person in eastern equatoria to develop interest in politics. He developed his political skills through discussions with his colleagues. He liked chatting about politics with Jeremiah Lotiang (the father of John Ongee). To demonstrate that he was a politician he often use big words such as "Governor General" "Senior Police Inspector", District Commissioner" the "Prime Minister of the Sudan".

Elder Marcello Odiyo went further and emphasised that Saturnino was a leading military personnel and had connections with southern politicians of the time. He was also a liberal party leader in the army; most of the meetings of both politicians and military leaders were done in his house.

Saturnino was well informed about Arab occupation of the positions supposed to be filled by the Southerners, the motives that led to the arrest of Elia Kuze, the reason behind the deployment of Company No 2 to Khartoum and what led to Nzara incidents. Been a politician Saturnino also knew the motives and content of the telegram sent by Prime Minister Ismail el Ashari to his administrators in the three southern provinces.

The telegram sent by Prime Minister Ismail al Azhari reads, "To all my administrators in the three South provinces, I have just signed a document for self-determination. Do not listen to the childish complaints of Southerners. Persecute them, oppress them, ill-treat them according to my orders. Any administrator who fails to comply with my orders will be liable to prosecution. In three month time all of you will come round and enjoy the work you have done" (Arop Madut-Arop, 2012, P. 40)

As the South was already politicized by the activities of the Liberal Party following the Juba Dance Hall Conference of 1954, the telegram was distributed and circulated all over the Southern provinces, towns, in the schools, in the army, police, and prisons, up to the village levels. It would then be a matter of time and South Sudan would take action to fend for themselves, as the people who had been defending them were about to leave forever.

After having received the telegram from the prime minister, Saturnino changed the content of the telegram to suit his own aspiration, and reworded the telegram to read, *"To all my Northern officers in the three South provinces, I have just signed a document for self-determination. Do not listen to the childish complaints of Southerners. Persecute them, oppress them, and ill-treat them according to my orders. Any officer who fails to comply with my orders will be liable to prosecution. In three month time all of you will come round and enjoy the work you have done"*

Saturnino Obwoya was even aware of Arab's evil plan about Southerners. Not only had these, Peter Debong pointed out that Saturnino had a secret military group who used to meet regularly by night in his house. The meeting had only one agenda: *"Arabs are the worst people; we should never live with them. We should stand together and fight them and chase them away from our regions"*.

Saturnino Obwoya was the president of the Equatorial Corps based in Torit and same time he was the chairman of underground cell in the Corps. Obwoya succeeded to establish underground cell branches among the Liberal Party within Wau and Malakal military garrisons. The leaders of these underground cells were all sergeants with Corporal Saturnino Obwoya as the coordinator. Thus on July 20, 1955, Saturnino Obwoya called for an urgent meeting with the leading members of his underground cells. The following members attended the meeting (1) Lieutenant Emilio Tafeng the apparent chief organizer, from Torit (2) Sergeant Akech (3) Sergeant Musa from Malakal (4) Sergeant Luboyo, (5) Sergeant Lumanyia and Corporal Abeniko Latoya from Terrekeka. In the meeting Obwoya told the members that the telegram was true and that the army in the south Sudan must do something about it. All the members in the meeting agreed and swore that if the government send troops to the South, and the troops mistreated Southerners, they would retaliate to defend their people (Arop-Madut-Arop, 2012)

When the message to fight Arabs were disseminated among underground cells in various parts of South Sudan, it was reported that the secrete group was divided in their visions and ideologies, some of the agreed to fight Arabs and chase them away from Southern Sudan but others were hesitating. Having been confronted with these divisions, Saturnino asked members of the secrete group, "What should the people do to sympathise with the innocent citizens of Anzara killed by the Arabs?" Majority of the members answer, "We must kill the Arabs to pay for the cold blood of our innocent brothers". Nevertheless, the manor numbers still opposed the idea of killing the Arabs. He then praised those whom supported the idea of killing the Arabs and called that minority of people, who refused to fight Arabs, as *cowards and people who practically know nothing.*

More importantly, Saturnino Obwoya, apparently was in touch throughout his underground organisation with the leading members of the Liberal Party, such as Daniel Juma, Marko Rume in Juba; Buth Diu, chairman of the Liberal Party and all other members of parliament. He was also in touch with Lieutenant Renaldo Loleya, who was based in Juba. According to Saturnino, the uprising should start earlier, on August 4th, 1955, to pre-empt any move that the North may have taken to bring Northern Troops to Juba. In line with his war plan, Saturnino Obwoya sent a telegram to Lieutenant Renaldo Oleya on August 3rd, 1955, alerting all the underground cells to start the war without delay. Obwoya telegram reads: "There is war tonight at five o'clock in the morning.....You must treat this at the same time....Do not be late...Send two platoons to Juba Airport and Mangalla Post tomorrow.... Put this in mind..." (Arop Madut-Arop, 2012, P.41)

Unfortunately Lt Renaldo Loleya did not response to telegram of Saturnino Obwoya. When he saw that there was no response to his telegram, Saturnino sent another telegram to Lt Ranald Loleya on August 4th, 1955 giving instructions about the progress of the war. This second telegram reads: "You must capture Juba airport and the ferry boat. Wait for my letter. I am telling you the truth. You must take two platoons to Mangalla post and you must send some lorry to collect all the company that is in Yambio"

Interestingly, after having received the two telegrams from Saturnino Obwoya, Lt Renaldo Loleya refused to carry out the order. Instead he sent the following telegram to Obwoya, advising him to delay the action until the leader of Liberal Party, Buth Diu return from Khartoum. The telegram reads: "There is no matter. Do not do anything now. Wait for my letter. The message will reach you early tomorrow. Do not think of anything. I am telling you the truth; do not do anything..."

When Saturnino Obwoya realised that his orders were not been carried out by Lt Renaldo Loleya, who preferred to wait for Hon. MP Buth Diu and other members of Liberal Party to return from Khartoum, Saturnino Obwoya became so angry and on August 5th, 1955 he resigned from his position of underground cell coordinator as well as his position of President of Liberal Party branch in Torit. Saturnino Obwoya then turned his order to senior Sothern Sudanese armies in Torit, instigating them to kill their Northern officers. For unknown reason, the orders of Saturnino have not been listened to by the army officers. This made Saturnino Obwoya to boil more in anger, and he decided to kill Colonel Salah, who was the Officer Commanding (OC) in Torit. He then borrowed a bow and arrows from his friend, Mr. Giovani; and went round and round looking for Colonel Salah but did not find him. When he failed to find Officer Commanding of Torit, he came across another Northerner, the post master of Torit and shot him (Arop Madur-Arop, 2012)

However, according to Justice John Ongee and Peter Debong on August 6th, 1955 between 7.30 p.m. and 9.00 p.m. Saturnino went to the neighbour and borrowed arrows and bow on the pretext that there was a cat which use to disturbed him at night, he wants to kill this cat. The neighbour did not hesitate he gave him arrows and bow, and he went straight to the post office, that was located within the army quarters. He was looking for Commander Salah he wanted to kill him. Unfortunately, Commander Salah escaped to Juba two to three days ago. The postmaster on duty was an Arab civilian. As Saturnino was passing by, the man saw his shadow and became frightened. The man then cried out with loud voice: Arami! Arami! Arami! (Thief! Thief! Thief!). Saturnino took the bow and arrow and shot him, the man felt down shouting: Sadu an! Sadu ana! (Help me! Help me!).

When Saturnino shot the postmaster, he ran away. There was a southern soldier nearby, when he saw him running, he ran after him with intension to remove the bow and arrows away from his hands, for the fear that he might shoot more people. Saturnino looked behind, saw a man running after him, and told the man, "Do not follow me, I will leave you with the arrow". Stupidly, the man thought he was joking; he continued to run after him. Shortly, Saturnino left the man with the arrow. The man cried for help, people came and took him to the hospital, he did not die after a while he was cured.

Saturnino Obwoya was immediately arrested and detained pending investigation. Following his arrest, the list containing the names of the would-be conspirators was found in his possession. The company's number two would have come under the direct command of Lieutenant Emilio Tafeng the organiser of uprising. Upon discovery of his name on the second list of conspirators as the organizer of conspiracy; Lieutenant Emilio Tafeng, who was on his way to Wau for duty, was also ordered to go back to Juba on the pretext that he was going to collect the salaries of the army of Torit. On arrival, Emilio Tafeng and his bodyguards were placed under arrest. The following was the second list of more suspects on the plot, they are:

1) Lt Emilio Tafeng, Number 6 Coy. Torit
2) Sergeant Major Lobulho Lohia, HQ Coy. Torit
3) Corporal Silvio olweny, Number 1, Torit
4) Sergeant major Mutek Ingong, Nuber 2 Coy. Torit
5) Sergeant Major Akiyo Lopiaramoi, Signals Platoon, Torit
6) Sergeant Latar Lelong, Number 2 Coy, Torit
7) Sergeant Solong, Engineer Coy, Torit
8) Sergeant Samusa, Number 1 Coy, Torit
9) Sergeant Lomanya Lomerek, Numebr 2 Coy, Torit
10) Corporal Yesiya Yingki, Number 6 Coy, Torit
11) Corporal Lofotir Ikille, Number 4 Coy, Torit
12) Sergeant Nyombe Mallwa, Terrekeka
13) Sergeant Nimaya Boramilai Number 3, Coy, Torit
14) Sergeant Akim Geliba, HQ Coy, Terrekeka

15) Sergeant Yakobo, Number 1 Coy, Juba
16) Warrant Sergeant Major (Sol) Lobuho (on leave in Juba) HQ
17) Sergeant major Mario Okello, Signals, Torit
18) Samuel Okello, Number 3 Coy, Kapoeta
19) Sergeant Simone Mufuta, Number 3 Coy, Wau
20) Sergeant major Tertaliano Number, 5 Coy, Kapoeta
21) Sergeant Ejidio Okwir, Number 5 Coy, Kapota
22) Sergeant major Mizan 4 Coy, Malakal
23) Sergeant Odong Oto, Number 4 Coy, Malakal
24) Corporal Abdenigo Latoya, 3 Coy, Terrekeka
Sources: Arop Madut-Arop

According to Nathaniel Polo when Saturnino shot the postmaster, soldiers quickly rushed to his house, he was arrested and they took number of documents. He was taken to Torit prison where Regimental Sergeant Major Nicola Obwoya and Nathaniel Oyet Polo, both from Pajok, came to investigate him. He told them, "I will not speak to you. If you want to hear words from my heart, bring my best friend Esbon Mundir". He then abused and called them "useless and cowards people". Esbon Mundir is from Muru tribe, from western equatorial, he was the man who always shares ideas with him. It is reported that Esbon Mundir was not in Torit, he was believed to have gone to Juba or Mundri District. Many people told me that Saturnino refused to speak to Nicola Obwoya and Nathaniel Oyet Polo because both of them were strong supporters of the Arabs.

As I mentioned before, Nathaniel Oyet was best friend of the Arabs, he then took all the documents (with the above list) found in Saturnino's house to Arab military officers. The documents were dangerously incriminating. The document revealed that Saturnino has asked the Southern MPS in Khartoum to support the secret group in their endeavour to liberate the three southern provinces from the Arabs. Unfortunately the Southern MPs turned deft ears to the requested made by him. Moreover, the specific date to start the work was not even agreed upon.

Following the discovery of the plot of the impending mutiny, the commanders of the Southern Command and the Governors of the three South provinces held an urgent meeting in Juba and made the following resolution:
a) It was essential that Northern troops be brought to the South immediately. An urgent message was sent to that effect
b) Until preparations were made, no more arrest would be made
c) The sergeants (Shawishia) from Wau and Malakal, whose names appeared in the conspiracy, were to be lured to Juba under the pretext that they were coming on courses, and they were to arrested and detained on arrival in order to discover the extent of the conspiracy.

d) The civilian authorities would proceed to arrest those civilians who were suspected to being involved in the mutiny conspiracy.

The members of the underground movement in the Equatorial Corps, in the list found with Saturnino Obwoya, who would have led the uprising, were lured to Juba. They were arrested on arrival and detained, accused of conspiracy. The suspects were Sergeant major Luboyo, Sergeant Akec, Sergeant Latayo, sergeant Musa, sergeant Lumanya and Corporal (ombashi) Lavota and, Sergeant Mizan.

From the above narratives, we have seen that Saturnino Obwoya was the Coordinator and secretary of the secret military group that use to meet at night in his house. The secret military group had very strong link with Liberal Party Committee in Juba because Saturnino was also the President of Liberal Party. They had planned to liberate South Sudan from the Arabs. Therefore, I find it difficult to comprehend when I hear Dinka people and their alliances called John Garang de Mabior as the founder of South Sudan Liberation or Independent of the New Nation. The true hero and founder of liberation of South Sudan from the Arab is Hero Saturnino Obwoya Opio, (a man from Abong clan in Obbo Village), South Sudan; John Gang de Mabior and others, are just his students.

Torit Mutiny 18th August 1955

> *"Because of an event which took place on August 6th 1955 we are able to know something*
> *about the origin of this mutiny and the planning which preceded it (McCall Storrs, 1970, p. 6)*
> *"During Saturnino Obwoya was in Prison during Torit mutiny. —said Peter Debong"*

When the Prime Minister Ismail al Azhari, whose National Unionist Party had won the pre-independence election in August 1955, foresaw possibility of rebellion in South Sudan, he immediately gave orders that all Southern troops be transferred to the Northern posts. The transfer was to take place before the formal granting of independence to Sudan.

Thus, on August 14th, 1955, all soldiers in No 2 Company army unit were called in the marching ground and were ordered to prepare for the departure to Khartoum on 18th August 1955 in accordance to what was resolved in Juba meeting between new Governor, his Deputy, the army commander and others. Almost all the southern soldiers were disturbed with the order but none of them could say a word. After the order, soldiers were asked to return to their respective places of work.

Again, on the 16/08/1955, the Commander Lt Colonel Banaga Abdel Hafiz called the soldiers in the marching ground and informed them that the departure to Khartoum will commence in two days' time; Captain Salah Abdel Magid will lead the unit. While the soldiers were still standing, one of them asked Lt Colonel Banaga, "Why are the soldiers from No. 2 company army unit going Khartoum?"

Lt Colonel Banaga answered, "The unit are going to Khartoum to take part in the parade to see off foreign troops". By the term "foreign troops", Lt Colonel Banaga meant British soldiers.

Figure: 63, Saturnino Opio middle

Again, the same soldier asked, "How long will the soldiers from this unit remain in Khartoum?"

Le Colonel Banaga answered, "I do not know; but it is better, for all the soldiers going to Khartoum, to send their families home".

It was already the middle of the month and soldiers have not received their salaries. The soldiers were left with only two days to depart to Khartoum, and these two days are not even enough for soldiers to send their families home. The northern soldiers started to mistrust the southern soldiers and police; they started to be unfriendly to them.

It is reported that more than half of the soldiers from No. 2 and No. 5 Company Unit were men from Otuho tribe. Joining army became one of Otuho cultures (customs). These Otuho men said they could not depart to Khartoum without saying welfare to "mwonyo-muji" (warriors or ruling generation), families, relatives and friends. It is the custom of Otuho people to say welfare through traditional dance. Consequently, on August 17th, 1955 they staged a big farewell dance that was followed by speeches from some "mwonyo-miji". After the welfare dance, all Otuho soldiers went back to the barrack.

On August 18th, 1955, all the soldiers from company No. 2 were ordered to assembly in the marching ground. By 7.00 a.m., all the southern soldiers assembled in the open ground within the barrack and were divided into four platoons. The first platoon was under the command of sergeant Mathiang, and at 7.15 a.m. they were ordered to go and pick their guns from deports about six hundred metres west of the marching ground. They went and found a northern officer was in charge of the store. The man then issued them with rifles without ammunition and ordered them to get into the car and travel to Juba. The southern soldiers became suspicious and asked, "Where are the ammunitions?"

The soldiers then decided not to enter the cars; they went back to the open ground and occupied a rear position. Shortly Captain Salah, an officer supposed to take the unit to Khartoum, drove behind the returning soldiers. He then jumped out of the car and asked, "Why did you come back here?" One of the soldiers answered, "We have returned back because we were issued with rifles without ammunitions. Captain Salah was not happy with the soldiers; he then went into argument with the soldiers to the extent that he became so angry and asked Bina, his driver who is from Ifotu village, to drive off. He was assigned to drive Captain Salaha to Juba and bring the car back to Torit. Bina responded negatively to the order of Captain Salah, who then took his revolver and shot Bina on his hand, and shot more bullets into the air—it is reported that Bina did not die. Captain Salah jumped into his car leaving the wounded Bina bleeding; it is reported that he drove straight to Juba for the fear that southern soldiers might kill him. This was the second shootings in Torit; following Saturnino shooting of the postmaster.

The commissioner of Inquiry did not even bother to look for Bina nor did he examine Captain Salah the accused. Soon after the incident of Bina, the No. 2 Company army unit went into rampage, they broke into the store sand took rifles and ammunitions. Within a short period, shooting spread in Torit town like wild fire. It is reported that six northern officers were killed and many escaped by lorry to Juba, they reached Mongalla where they were rescued by army who came from Juba.

It is reported that as soldiers went into rampage, the area Commandant, Ismail Salah, Lt Colonel Mahgoub Taha, Inspector of police, Major Nicola Obwoya took to their heals from the open ground and reached Uganda after a day of walking. Many northern traders and southern civilians lost their lives. It is said that many of the civilians were drawn as they tried to cross-river Kinaite that had a narrow wooden bridge. The bridge was full as a result many people were pushed into the water, and because they did not know how to swim, they then drawn.

After three days, following the mutiny in Torit, it was reported that many civilians were massacred in Juba. The southern soldiers then took this chance and massacred northern traders and their families. They broke Torit prison, killed all the Arabs (who were in prison) and realised all the southerners which include Saturnino Opio Obwoya. However, according to people like Peter Debong, Saturnino was realised from Torit prison by his colleagues, and he ran into the bush with untied chain on his hands. There he untied the chain and came back into the barrack

When Saturnino came into the barrack, he took a car, branded a bran-gun on top of the car, and started touring Torit town. Life was no longer favourable in Torit town; after some of his colleagues surrendered Saturnino refused to surrender. He then decided to leave Torit town and hide himself in the bush not far away from the town. The military officers knew that Saturnino has not gone far, so they sent security to search for him. When Saturnino learned that security personnel were looking for him, he decided to go and live in a place Marcello Odiyo described as "a no-man-land" that lies between South Sudan and Uganda. From there, he proceeded to Lokung Village in Uganda. He built his house at Lokung in a place called "Tee-Agu". News reaching people in South Sudan was that the Arabs went looking for Saturnino up to Uganda; subsequently, his pictures were put on the walls in the shopping centre of Lokung, Kitgum and Kampala. The Sudan Government wanted that any person who knew where he lived should report to Sudan security. Good enough, at that time there was strong link between the people of Acholi Lukung and Acholi Obbo. For this reason, Saturnino became under the protection of Lokung people and was given the name "Tadeo Okeny". Saturnino remained in Lokung until 1964 then joined Anya-nya one movement.

e) Saturnino Joined Anya-nya One Movement
He joined Anya-nya One Movement in 1964; not much information been recorded about his roles and achievements in the Movement.

f) Saturnino Obwoya Joined the SPLA
It remains unexplained whether Saturnino Obwoya actually joined SPLA but the reality is that he was the first man in Obbo village to accommodate SPLA soldiers in his house at Labato. At this time, most people in South Sudan and Obbo in particularly fear SPLA. Peter Debong, one of my informant said people were very scared of the SPLA, when they come in a village people usually run away from them. People hide in the bush or forest from SPLA but Saturnino has no fear of them, when SPLA come in Obbo village, he usually go and meet and invite them into his house. Eventually he became popular among the SPLA and he continuously encourage Obbo men, women, girls and boys to join the SPLA.

To accomplish his mission, on May 8th, 1986, Saturnino organised a big meeting at Loudo shopping centre. At this time almost all, the population of Obbo were sleeping in "abak" or house built in the bush, for emergency during the time of insecurity. Prior to the meeting, he went from *abak* to *abak*, accompanied by his bodyguard from SPLA, mobilising people to attend this important meeting. He was an influential person among the community, as a result, he succeeded to convince population and on the day of the meeting, many people turned up for the meeting. The meeting was conducted under a bid tree at the centre commonly known as "Tee Iduro". The chief speaker was Isaa Obuto Mamur and assisted by Saturnino Obwoya. The agenda of the meeting was "Recruitment into SPLA soldiers". Some 200 SPLA soldiers that include William Okeny, from Agoro village, and Isidoro from Magwi, also attended the meeting. After Mamur delivered his speech, chance was given to Saturnino. In his speech, he called upon the people of Obbo from all walks of life to join the SPLA. Kalisto Okun, said messages from Saturnino were stronger than that given by Mamur.

It is reported that soon after the meeting, seventeen Obbo boys included Kalisto Okun (the informant) and six Obbo girls registered into the "Red Army". It is the policy of SPLA that all children under 18th years of age must be registered under "Red Armies". In other words, Red Armies" are groups of children under the age of 18, who voluntarily or forced to joined the SPLA. The six girls from Obbo village became the first girls, from Equatoria, to join SPLA.

After the meeting in Loudo centre, the group of 200 SPLA soldiers and the twenty-three new recruits to Red Army, travelled to Galario Modi's camp in Imurok area. Galario and his high officers warmly received the group. Galario was particularly surprised to see that the Obbo girls were recruited into Red Army. He said, "The Obbo people have taken the right step. They have demonstrated that girls can serve in the Red Army". It is reported that Galario was disturbed by the thought taken by the Obbo people to recruit the girls into Red Army. He went and sat under the tree and asked himself, "Why should the Acholi recruit their girls into Red Army? Why Otuho girls remain for marriage as house wife?" Immediately, he recruited two of his daughters into Red Army. From here, the SPLA group proceeded to Nakichot where the twenty-five Red Army soldiers were praised for their thoughts and courage to join the movement.

Before long, the whole of Obbo land was under control of SPLA and many men boys and girls joined the SPLA. This made the SPLA administration to think that Saturnino is not a good friend but he has contributed very much in the recruitments of the young men and women into SPLA ranks. Because of this, he was given the rank of Commander and at the same time, he became the chief of Obbo Village. The SPLA appointed Saturnino as chief of Obbo because they know and understand the population of Obbo listen more easily to him. It is reported that, in his capacity as commander of the SPLA in Obbo, one day he ordered the SPLA soldiers to attack Pajok military garrison where the Sudan Army Force are stationed. According to Saturnino, all the SAF soldiers in Pajok, including their Commandant Lt Colonel Charles Ogeno (an Acholi from Palwar) are Arabs

The death of Saturnino: Who killed him?

"Saturnino was moving under the escort of SPLA soldier, he was always kept in the middle with one SPLA soldier in front and one behind him," Peter Debong said

In August 1986, the people of Obbo and Owiny-ki-bul (Ngaya) were in the funeral of mother Elias Lapoko and Ester, daughter of Musa, in Obbo Village. On the same day, the

SAF soldiers and militias from Pajok and Palwar (Korean Army), led by Lt Colonel Charles Ogeno, were on their routine tour to Magwi. They came, surrounded the people at the funeral of Elias' mother, arrested, and took them to Loudo centre. It is said that Loboyi, the son of Isuma and the bodyguard of Saturnino, and Lowi, the son of Saturnino, were among the people arrested at the funeral. Obbo elders reported that a Lotuko soldier called Mathew killed Loboyi Katula at Loudo centre, without ordered from Charles Ogeno. He (Loboyi) was found with a clan-scope rifle from SPLA.

Soldiers and militia spent night at Loudo police station. In the evening that day, the news of those arrested Ogeno's soldiers, reached Saturnino at his home in Labato. He then thought that one of his sons called Loci Abraham, who is an SPLA soldier and did not spent night in at home, might have been among the people arrested in the funeral place. Therefore, in the morning Saturnino walked to Loudo centre. He did not only think of his son, but would like to know the names of those arrested by the SAF. When Saturnino arrived at Loudo police station, he found the convoy was already on move to Magwi, taking with them the civilians arrested at the funeral. Because of the talk grass in Obbo, Saturnino could not see the trucks but he could only see ground soldiers heading towards Magwi. The SAL soldiers were dressed in uniform, commonly known among the Acholi people as, "lakaca-kaca" that is similar to SPLA uniforms. Thus, Saturnino became confused; he thought those were the SPLA soldiers pursuing the SAF on their way to Magwi. The SAF were together with Militias from Pajok and Palwar. He then followed them, and started to command, shouting at the top of his voice and saying, "Comrades, follow them one by one until you kill them. They are Arabs and you should spare no one. Leave no one to go back on his foot to Pajok".

Unfortunately, among the SAF soldiers there were some Acholi people. One of them understood what Saturnino was saying and turned to his fellow Acholi and whispered to their ears saying, "Jal! -which means Hello! The old man is here". The Acholi soldiers, whose names were not disclosed, started to withdraw tactically backwards to allow him enter in the middle of the SAF. With confidence Saturnino found himself in the middle of the soldiers; he was captured. As soon as they caught him, the soldiers rang Commander Lt Colonel Charles Ogeno and told him, "We have caught the old man". On hearing this message, Charles Ogeno quickly rushed with his jeep, he came back at rear convoy. Ogeno found Lanyok in the hands of his soldiers, picked him and told him "Yaba! I told you many time and I have sent you many messages requesting you not to continue with your old thinking of hating the Arabs and siding with the SPLA. I also advised you to move away from Obbo village and come to me at Pajok to save you from the hands of SPLA. You are too old to continue to fight against the Arabs; this will never help you but rather put your life at risk. Now you have risked your own life, which I cannot help".

Saturnino answered to Ogeno with a man voice, "Have you known me today? You know me very well that I and Arabs we cannot put our elbows together, and yet you want me to come to Pajok and leave with Arab? I will never accept to go to Pajok; if somebody wants to kill me he should kill me here in Obbo land".

Charles Ogeno seated him in the front seater for the fear that soldiers and Militias might kill him as they killed Loboyi Katula.

Reaching at Pajok military barrack, Saturnino and other people who were caught at funeral in Loudo were thrown into military custody. They were interrogated, and forced to accept that they are supporters of the SPLA. The people who were under investigation together with Saturnino were fearful and spoke out with fear but Saturnino never fear any army-man and never opened his mouth when asked questions by military investigator. When he was forced to asked questions, Saturnino did not answer the questions but criticising soldiers by telling them, "What do you know? You know nothing. I cannot speak with people like you—because you know nothing. When soldiers insisted to question him, he spat saliva into their faces.

After two to three days, some SAF and Koreans militias, who are from Palwar village, the homeland of Lt Connell Charles Ogeno, removed him out from the custody. The soldiers and militias took Saturnino and killed him at Lerwa. The body was left in the open and people were instructed never to bury him. His black jumpers were taken and hang on the tree at Ayaci. This was to demonstrate that Saturnino is now dead. The Acholi elders told me, the body of Saturnino was not rotten or eaten by birds or while animals. It is commonly believe that the body of Saturnino was later burned by wild fire—neither the bone nor the skull was seen or found. This is a brief life history of Lanyok and how was killed by his own people.

Chapter 12:
Religion Beliefs and Cultures

Abstract: "In this Chapter a depiction of the ethnographic landscape is offered as an attempt to capture the spirit of a people who have turned inwardly to focus on the spiritual mainte-nance of their community. The majority of the Northern Luo are fervent Christians and their Churchgoing and parochial activities are a focus for empowerment in their social and material environment. About 80% of South Sudanese population professes Christianity, they tend to be the most educated and about 5% of population professes Islam. However, this Chapter explores further on the concept of the Northern Luo connecting the past with the present. The Northern Luo Communities have many rituals directed at ensuring good health, protecting their people and their animals and crops. Subsequently, health, illness, hunting, harvesting are religious-cultural matters which have to be dealt with in religious-cultural ways"

Christian Denomination in the South Sudan

Religion is a fundamental, perhaps the most important influence in the life of the Northern Luo, yet its essential principles and practices remain a mystery to the foreigners, especially people from the Western World. They often misunderstand the Northern Luo Christianity and traditional beliefs; without understanding that the religion enters in every aspect of the life of the Northern Luo and it cannot be studied in isolation. Its study has to go hand-in-hand with the study of the people who practice it.

Among the Northern Luo, there are two types of religions: one of the religion I called it "Western Religion" because the ideology of this religion was imported from the Western World; and the other religion I called it "African Traditional Religion" or "Luo Traditional Religion".

The word "traditional" means indigenous, that which is handed down from generation to generation, upheld and practice by Northern Luo today. This is a heritage from the past, but treated not as a thing of the past, rather it is something that connects the past with the present and the present with eternity. This is not a dead religion as some people might think, it is a living religion – it is a religion practice by Luo men and women, boys and girls. It is in people's hearts, minds, oral history, rituals, shrines and religious functions. It is founded by grand-grand-grandfathers and not by one person like Christ, Asoka, Gautama the Buddha or Muhammed. It is not a religion of one hero; it is a religion of ancestors.

I quite agreed with J.O. Awolalu (1976) when he argues that the Western scholars attempted to write off African as a spiritual desert. While some scholars admitted that the whole of Africa could not be a spiritual vacuum, they, however, raised doubt as to whether the God that the Africans believed in was the real God of the Bible or their own God. To undermine God that Africans worship, they started to coin expressions such as a "High God" or "Supreme God". In this way they pronounced many supreme Beings existence in order to place the African God (Luo God) at a lower level than the Deity in the Jesus Christ.

As a result of this, there came a number of investigators including theologians and trained researchers who did their investigations among the people they do not unders-tand their language. Although interpreters were used, one could not be sure that the interpretation would be accurate. The noticeable fault among the Missionaries was that they were particularly subjective, and they would not see anything good in African Traditional Religion. The impression they had of it was that it was not worth knowing it (African Traditional Religion) at all and they expected that the religion would soon perish – but they were proved wrong.

Despite the fact that the foreign researchers were committed to writing their investi-gations about African Traditional Religions, we need to point out that a great number of them used misleading term in describing the Africans' beliefs, using words such as *primitive* (old fashion), *savage* (wild, uncultured, uncivilized), *fetishism* (emotional attachment to inanimate objects e.g. charms and a mullets), *paganism* and *heathenism* (one who worship idols and does not acknowledge the true God – he is neither a Christian, a Jew nor Muslim), *Animism* (attributing a living soul to inanimate objects and natural phenomena), *Idolatry* (worshiping false god) and *polytheism* (worshipers of variety of gods).

In this book I want the readers to understand that the Northern Luo of South Sudan practice Christianity, traditional religion as well as Muslims. However, due to the short time I have dedicated to research materials for this book, I am not going to discuss or examine Islamic and Christian practices among the Northern Luo. However, I would rather prefer to demonstrate first the Christian history in the country (South Sudan) without even entering deep into the Christian worshiping among the Luo groups. Therefore, I prefer to discuss more into depth about the Luo Traditional Religion – which is the focus of this Chapter. Everywhere in the world, there is concept of the God of the bible; beside this every locality has its own local deities, its own festivals, its own names of the Supreme Being, but in essence the pattern is the same. There are about one thousand African tribes and sixty-four tribes in South Sudan; each has its own religious systems.

As we shall see later in this chapter, the Northern Luo of South Sudan, have special regard for the ancestors. A belief in ancestors is largely based on the notion that life continues after death and that communion and communication are possible between the living dead. Fundamentally, ancestors are the spirits of the deceased relatives, and for many individuals are the primary intermediaries. Ancestors are believed to take interests in the day to day affairs of the community and have powers of influence the affairs of the living either for good or for bad. They play the role of guardians of public and private morality, and unseen presides at family and community social gatherings. They also act as intermediaries between the living and Divine Being (God). They are believed to be involved in the practice of medicine and can, upon request, send special cures to the members of family or community who are seriously sick. The ancestors, by virtues of being part of the extended family and living in the presence of God, are endowed with special power. The ancestors continue to be part of the family and community and are included in all the events of the family, clan or community. Although we cannot see ancestors, they reside in the family and community like everyone else. They have special places where they live such as shrines, cattle byres, houses to mention a few. Thus, in Northern Luo Groups' belief, offerings to the dead are the expression that the living this world and going to the next world has not destroyed life in the family or community. Ancestors have exceptional powers, and can make their presence known through fortunes, misfortunes, or materialize in different forms such as tortoise or certain type of snakes. Let us now examine briefly the Christian history in South Sudan.

Christian history in South Sudan

The Christian history of the present South Sudan can be backed to the 1860s. The Roman Catholic Church had much interest in expanding to central Africa. As a result Pope Gregory the sixteenth, in 1846, created what they called "The Vicariate Apostolic of Central Africa", which embraced South Sudan. The Catholic Church had been attracted to South Sudan for two main reasons: firstly because the colonial Government suddenly opened a virgin filed in Equatoria African that provides an opportunity to expand Catholicism. Secondly, the Roman Catholic Church wanted to ensure that they reach the South Sudanese population before the Protestants and the Muslims. The Congreatio de Propaganda Fide, then prepared and sent the first Christian Missionaries to Upper Nile in 1850. The mission was to make no attempt to proselytize among the Muslims (if any was there) but to seek converts among the Negroid people beyond the sudd (Robert O. Collins (1971).

The social context of mission was largely created by the British colonial Government to meet their interests. The colonial administration fostered a mission of its own, on secular rather than sacred ground. Despite this, each mission felt short of intended goals. The secular process of conversion met with little success in the first four decades of British rule in the South Sudan due to the ill-conceived, poorly defined, and superficially administered system of indirect rule. As a matter of fact, colonial officials created traditions in order to impose a regime that was intended to work on the basis of traditional usage. The British consistently relied upon missionaries to advance the interests of secular rule; ultimately administration directives frustrated their evangelical work (Burton, 1985). Therefore, if the success of Christianity in the colonial period is to be measured with the number of individuals converted, we can conclude that ten Missionaries did not make great achievement.

The mission was led by Ignaz Knoblecher and they travelled to Gondokoro in 1850, where they were met with tension from the Bari. Their position became so untenable that they temporarily withdrew to establish the Holy Cross station 150 miles to the north. When Ignaz Knoblecher died, there was no priest to replace him, consequently in 1860 Gondokoro were abandoned. In 1861 the Franciscans attempted to found a mission among the Shulluk, but soon they withdrew. Protestant interest in the Upper Nile was first aroused in 1878 by General Gordon. He provided support for a party of the Church Mission Society (CMS) passing up the Nile to Uganda and urged them to establish mission stations along the River Nile. The support of General Gordon to CMS was seen by the Mahdist as a threat to Islam, as s result he was killed in Khartoum –that followed by establishment of Gordon Memorial Mission. The Roman Catholic Church was the first to enter south Sudan and they were followed by the American Presbyterian Church. The first Catholic Church was opened at Lul in Shulluk country in 1900; the second Roman Catholic Church was opened at Attigo in 1904. It is said that Reth of the Shulluk was not happy with the missionaries at Lul as a result he castigated the people who attended the Mission School. Time and again the Missionaries among the Shulluk appealed for Government to intervention on their behalf for, having no means to intimidate the Shulluk, they depended increasingly on the moral support of the government. In 1904 Bishop Geyer founded the first eight Mission stations in Wau that include Kayango, Mbili and Dud Akot Missions. In December 1919 the relationship between Missionaries and Reth (King) of Shulluk improved but the hostility did not disappear. In the same year (1919) King Agot attached the mission station and drove off its members (Robert O. Collins, 1971).

However, some people argue that the Christian Church in Sudan began with the coming of the Coptic Christians, of the black African, to Nuba. Their influence spread throughout Egypt and as far south as to Khartoum, the capital of Sudan. Before long, there was split in the Churches; nobody could explain to me the reason that led to the split of the Churches in Egypt. Despite all this division, Christianity continued around Aswan among the Nubians until the 14th century, when the arrival of Islam extinguished all Christian spirits among the Nubians, all were converted to Islam. However, the Luo groups, who were living in the Northern Sudan, bordering Egypt, continued to practice their Christianity even in their new settlements. Christianity has had a major impact on the Northern Luo religious beliefs and practices. Today, religious communities draw on beliefs both from indigenous practices and from Christianity. The Roman Catholic Church and Anglican Church are very significant among the Luo groups.

The missionary spheres of influence delineated by the British Government in the Sudan during 1905, allocated approximately 40 percent of the South Sudan to the Verona Fathers, who are the Roman Catholic Church. These spheres of influence were allocated in Thuri/Shatt region, Jur Chol (in Wau and Tonyj), Malakal area, in and around Juba City, Madi area and in Acholi region. The Church Missionary Society were granted an equivalent area, stretching from central Dinka in the west, across the main channel of the Nile, to the border with Ethiopia in the east, including the traditional homesteads of the Atuot, the Mandair, the Anyuak, the Bari, and a part of Nuer speaking population as well as Yei and Kajokaji Districts. The American Presbyterian Churches were given the smallest territory, being restricted among the Nuer communities between the Nile and the Akobo Rivers. These territorial divisions were created to meet the needs and interests of European powers, not to the people in residence. Why more than half of the Equatoria and Bahr el Ghazal regions should be exposed to an Italian form of Christianity and the remaining areas plus Upper Nile were exposed to an Anglican Churches –this was not clear to many people. It is reasonable to presume, therefore, that missionary spheres were intended to correspond with administrative boundaries (Burton, 1985).

There are many Christian denominations in the world, and all these Christian denominations cannot be found in the South Sudan. To the best of my knowledge, there are currently only nine Christian denominations operating in the South Sudan and they are as follows:

The Roman Catholic Church:
The Roman Catholic Church was the first Christian denomination that enters Sudan from the north in 1842, and they built their first Church in Khartoum in 1846; this Church was destroyed by an insurrection in 1881 and was reconstructed in 1898. The Roma Catholic Church is currently the largest Christian denomination in the South Sudan. According to Saturnino Onyala (2014) the Roman Catholic Church entered South Sudan in 1905 through Uganda in East Africa. In the South Sudan today, there are approximately 80 catholic Congregations in existence, comprising some 2,000,000 members.

The Anglican Missionaries:
Anglican Missionaries began working in Sudan in 1899, they were commonly known as Church Missionary society (CMS). By the year 1916, the Anglican Church has grown to tens of thousands Christians. It is now the second largest Church in the country with approximately 1,300 Congregations and with some 630,000 members. Later on they became known collectively as the Episcopal Church of Sudan.

The Presbyterian Church:
The Presbyterian Church emerged in South Sudan in the early 1900s. Currently, it has around 620 congregations (they are found mainly in Upper Nile among the Nuer and now in Juba City) with some 132,000 members.

The Sudanese Church of Christ:
The Sudanese Church of Christ was founded by Sudan United Mission (SUM) workers in the 20th century; they have 560 Congregations with some 80,000 members; they are confined within the major towns of South Sudan.

The African Inland Mission
Soon after the establishment of Sudanese Church of Christ, the African Inland Mission team was founded in 1949. Shortly, the name became known as African Inland Church, today they have approximately 55 Congregations with some 12,000 members.

The Serving in Mission:
The Serving in Mission (SIM) entered Sudan in 1937 and after the Addis Ababa Agreement, it spread to South Sudan, where it became known as Sudan Interior Mission (SIM). Today, they have approximately 130 Congregations with some 20,000 members.

Assembly of God Church:
Assembly of God Church emerged in 1983, which currently has 50 Congregations and around 7,000 members.

The Sudan Pentecostal Church:
Sudan Pentecostal Church was established in 1986 and today it has around 65 Congregations with some 5, 200 members.

Jehovah Witness:
The Jehovah Witness emerged in South Sudan around 1890 and today it has around 10 Concretion with some 1,000 members.

The Muslims
There are few Muslims in South Sudan particularly among the Northern Luo Group. The Southern Sudanese Muslims are more concentrated among the Dinka and Nuer. They tend to follow Sunni Islam and are mostly of the Melkite rite although there are some Shafiites. Although there are some Muslims in South Sudan, they do not strictly follow Islamic teaching, some of the do not even attend prayer in Mosque all their lives; what can show them, that they Muslims, is only the long dress called "jalabia" , which is worn like a normal dress.

Source: Missionary Atlas Project Africa-Sudan

In South Sudan we have no Hindus, Bahayi and Budish. When civil war broke out in 1964 many Missionaries were forced to flee south Sudan, with the signing of Addis Ababa Agreement in 1972 many Missionaries returned to the country. Again their work was halted following the formation of Sudan People Liberation Movement (SPLM) in 1983. At this point and time it is good to remember that as from 1983-2017, the work of the Missionaries have been run by the native priests and pastors.

Christianity has had a major impact on the Northern Luo religious beliefs and practices. Today, religious people draw on beliefs both from indigenous practices and from Christianity. The Roman Catholic Church and Anglican Church are very significant among the Northern Luo of South Sudan. Many Luo, however, do not draw sharp distinction between religious practices with European origins and those with African origins. Many Churches with European origins, however, draw on a rich Luo traditional music

For many Christians, the ancestors continue to play a significant role in their lives. According to traditional beliefs, the ancestors reside in the sky, on earth in places such rocks in the rivers, mountains, forest, caves and underground. The spirits of ancestors are believed to communicate with the living in their dreams.

The Luo generally values their ancestral and belief in life after death, the practice that has facilitated their acceptance of Christianity that was imported from the West. Ocholla-Ayayo (1976) says, "The religious beliefs of the Luo are centred on a few questions which seem to be common in all religious beliefs elsewhere. These are the questions of origin of man and of his destiny; who created man and what will become of him after his death? And what are the causes of man suffering and remedies to them? These questions are crucial to religion, and there are several ways in which they have been answered in different societies. These complex questions call for answers, which are based on man himself, since it is only man among all animals that can imitate the arts of making, or try to cure, try to change nature. The one who created so must in some way be in the form of man himself (Gen. 1:26-27). In English, we call this person "God", in Luo language this person is known as "*Juok*", "*Jok*" or *Nyasaye*" (P.166).

The Northern Luo of South Sudan holds the following fundamental premises that the entire universe was created by and continues to be sustained by the Supreme Being (Juok/Jok/God). Everything that happens does so because it was willed to be so by Juok/Jok. Man is the centre of the creation and all things were given to him by the creator. This is also confirm in today bible. According to Holly Bible, New International Version (2011): Gen. 1: 26 Then God said, "Let us make mankind in our image, in our likeness, so that they may rule over the fish in the seas and birds in the sky, over the livestock and all the creatures that may move along the ground".

Ocholla-Ayayo (1976) noted that in Luo mythological version that Juok made man first and then animals for men to hunt. In the mythological context Juok made things by his own hands, moulding them like a potter. It should be emphasized that the concept of the Supreme Being, the divine creator, has been among the Northern Luo, from the origin of the Luo tribe and this has been transmitted orally in forms of proverbs, songs, prayers of their daily life. Juok is conceived as a dominant universal power that knows everything, sees everything and hears every words coming from our mouths.

The Northern Luo believe that Jouk may strike men individually or collectively, he can do this by bringing epidemics, sickness, locusts, draught and flood.

In the old days, old Luo man when he gets up in the morning, he met the rising sun by spitting gently to him (sun) while saying, "Thank you. You have given me good night. Give my village a good day!" In the evening when the sun is setting the same man says, "Thank you. You have given me a good day. Give my village a good night!" These acts are all the manifestation that God is in control.

According to Kamwaria (2012) the Luo general believe that there are two worlds, that is, the world of the living and the world of the dead. The two worlds are alike in many respects, and there are stories told by people who visited the world of the dead either in dreams or in short period of death, and returned to tell what they have seen. The Luo groups believe that after life everyone goes to the ancestors in the land of the dead below the sea or in the stomach of the earth. In the underworld, the life of the comer is similar to the one he lived in the upper world. The people in underworld do not suffer anything but just carry on with their favourite activities such as herding, hunting, fishing, dancing and so on.

Similar views are reported by Crazzolara (1954) from the Nuer side that the land of the dead is situated in the bowels of the earth. There are villages, steppes, fishing waters and so on, in which there is of course a cattle-byre (*mura*); only the spirits of men killed by lightening can go to the sky the rest of the spirits go in underworld.

Acholi Religion and Expressive Culture

According to Latigo (2008), prior to colonialism the Acholi people maintained a traditional government that was rooted firmly in their religious beliefs, norms and customs, which demanded peace and stability in Acholi land at all times, based on their philosophy of life. This structure was maintained by the real anointed chiefs of the Acholi, what they called the "*Rwodi moo*".

The old Acholi word for God was Juok or Jok, but the missionaries from Bunyoro introduced the name Rubanga or Lubanga. The Juok of Acholi with other Luo group's Juok became a devil. Juok is associated with the creation of universe and sending rains to the people where the rains gather above through the power of rain-makers (Rwot-kot). C.G. Seligman (1976) observed that Acholi like Shilluk, Jur-Chol and other Luo groups believe that Juok (Jok) is God, and the word *jok* to certain extend also connected to beings living in particular places, especially in big trees called in Acholi language *Kitoba* and *Olua*; these beings exist in snake form. A man breaking off a branch of such tress would die unless the *ajwaka* perform a ceremony to save him. Among the Acholi when a man sees spirit in form of snake or large-headed dwarfs, he immediately tell members of family about it, and in the following morning people take what is called *"moko-tobi"* mixed with water in a small gourd (*agwa-deku or abit*) and is taken to the place where the person saw the snake. An elder person in the family pour small mixture of thin substance (*moko-tobi*) in the place and the small gourd is left there. This will then avert the future illness from him; but if on the contrary the spirit (snake) sees the man and the man did not see the snake (which is difficult to justify), no sacrifice will avert illness from him and the unfortunate man will die. According to Acholi traditions, illness is not attributed through neglecting of ancestral spirits, it is more considered to be due to the curse from God (Juok).

Many Acholi like other Luo groups, do not draw sharp distinctions between religious practices with European origins and those with African origins. Most of the Acholi people of South Sudan are Catholic, with a few hands of Protestants and one or two Muslim. Nevertheless, the traditional beliefs in guardian and ancestors' spirits remain strong, though it is now often described in Christian or Islamic terms. Thus, to many Acholi Christians, the ancestors continue to play a significant role in their lives. They believe life continues after death and that communications are possible between the living and the living death. The spirits of the dead communicate with the living in dreams, which they are apparently heard rather than seen. It is this spiritual dimension of the Acholi, which is little understood by non-Acholi, especially foreigners from Western World.

Interestingly, from the narrative above, I can say without certainty that there is no much different between Acholi and Dinka beliefs. As Kamwaria (2012) puts it, that the Dinka have special regard for the ancestors; their belief in ancestors is largely based on the notion that life continues after death and that communications and communion are possible between the living and the living dead. They believe ancestors are the spirits of the deceased relatives, they are believed to take interest in the day to day affairs of the community and have powers to influence the affairs of the living either for good or for worse. They play the role of guardians of public and private morality, and unseen presides at family meetings. They also act as intermediatory between the living and the divine beings (P.51).

The Acholi religious beliefs focus on three types of spirits: The first been the spirits of grandparents called *"tipo pa kwaro"* and the second been the spirits of the dead father called '*tipo pa wora* (singular)" or *"tipo pa wego* (plural)". The spirits of both of these are approached in regard to good health, fertility, and appeals or thanks for good harvests. The third been the spirit of the, "The Almighty God of the bible", called *"Jok" or "Juok"*. It should be remembered that *"Jok"* is not "evil spirit" as assumed by the Western religious leaders. To the Acholi in particular and the Luo groups in general, the term "*Jok*" or "*Juok*" is the true God of today bible.

Jok is the creator of the world and he has established the order of things, it is he who sends the rain from the place where rains gathered (above), which is his especial dwelling home. The Acholi attribute anything of an unusual nature and unusual occurrences to some aspect of *Jok's* power. When lightning struck a house or when rain failed or hail, locust destroyed crops, prayers are offered to Jok and sacrifices made to ancestral ghosts, just as other troubles occurred. C.G. Seligman (1976) pointed out rightly that the old Acholi word for God was "*Jok*" but the Missionaries from Bunyoro introduced the name "*Rubanga*" or "*Lubanga*" —and "*Jok*" turned to Satan or evil. Acholi corresponds closely with the Shilluk "*Juok*". She went further to elaborate that "*Juok*" is associated with the creation of universe and sending in rains to the people when approached in prayer by the rain-maker -"Rwot-kot" (P.122).

The Christians worship their God in Churches, School buildings, Halls, Community Centres, under big trees where there is no facility. In Acholi land of South Sudan there is only one Catholic Parish that is Palotaka Mission. However, I am not going to discuss how Christians worship their God of the bible, because this is already taken care of by the different religious leaders in the South Sudan and around the world. For this reason, I would rather take the readers through, how traditional sacrifices are made to the ancestors, and why such sacrifices are made. Among the Acholi these sacrifices are usually done in shrines (*Kac* or *Abila*). Let us now see how *Kac* works.

A) Kac or Abila

Kac is believed to be the God of Acholi sub-clan. As a result each sub-clan has its own *Kac* placed at the door side of the elder who holds it at his home. This ritual is performed in order to give blessing and fortune and to take away misfortune and calamity whenever it happened within the sub-clan; thus, the family who is responsible for *Kac* shall perform the ritual to prevent such calamity.

Before elders perform any rituals at the *Kac* they perform a form of prayer, this prayer could be in form of killing a goat or spiting saliva in the small gourd with water and sprinkle the *Kac* with it; and pour the remaining water under *Kac*. This is a way to call the ancestors and the spirit world in order for them to accept elders' rituals or ceremonies. If this prayer is not done, *Jok* will not be notified and the ritual will fail as a result.

The "*Kac* (Shrine) is built opposite the door of the hut, this is done to enable spirit (*tipo*) to watch what is going on in the house and around the house. However, in some cases, Shrine is built four yards from the hut entrance.

The shrine consists of built table supported by four uprights wood at a height of around. It is built about three to four feet from the ground. On the "*Kac*" one can find objects, such as a stake supporting the skulls of animals formerly sacrificed to God on this alter. C.G.

Seligman (1976) argues, "There stand the "*Kac*", on which the horns of the beats sacrificed or killed in game are hung as a token of gratitude to the death people for favours they have given to their children" (PP. 125-126).

Figure: 64,
Kac/Abila of the Acholi

The Acholi believe that spirits of the dead grandparents exist in the earth below the shrine (*Kac*). I can recall when I was a young boy, my paternal uncle told me, "When you build a "*Kac*" (shrine) you call the spirit of your father to help you in hunting, forming, provision of good health and so on; and spirit of your father emerges from the grave and comes to live in the shrine (*Kac*) and will never go back to the grave. The home of the dead is in the cave, shrine (*Kac*), forest, and mountains, under big trees such as *Kituba* and *Olua* trees. C. G. Seligman (1976) rightly observes that the use of shrine (*Kac*) is in everyday life of the Luo, it must be remembered that spirits grandparents and father (*tipo pa kwaro/kwar and tipo pa wora*) are in most intimate association with the shrine and the people can only approach spirits through *Kac* or *Abila* (shrine) (P.124).

The shrine (*Kac*) is renovated once years, at the time family members want to offer their prayers to the ancestors. This is usually done during the month of December and January, when the people finished harvests of the crops. It is not a surprise that one could find broken or fallen shrines in Acholi villages. However, this does not necessary mean that the people do not care about the shrine –they do.

The *Kac* (shrine) is never built by youngest son nor *can Kac* be transferred to the wife if the husband dies. When the husband dies the second son can erect Kac; and if there is no second son in this family, the shrine (*Kac*) can go to the door of cousin brother. Writer like Mr Driberg misunderstood the information, as a result he asserted that the "*Kac*" (shrine) is built by youngest son and that young son once he is marries, he transfer the "*Kac*" to his wife's court-yard, where the elders brothers come to pray and sacrifice (P.125).

When a man moves from his current location to another location he transfers the *Kac* with him. Before a man settle in the new location, he builds the new *Kac* (shrine) and sacrifices a goat and uses the wood of old shrine as fuel (*yien*) for the fire at which meat is cooked and the new *Kac* being anointed with blood and gut contents (we). The meat of the goat is eaten by the elder folk of family and the beast is avowedly sacrificed so that the spirit of the dead father can see it. The blood of a comer killed in war is left unburied, but the dead man's son makes a shrine (*Kac*), sacrifices and prays the sense of his prayer being, "I have made Sacrifice for you and brought your spirit to your shrine (*kac*), do me no harm".

Moreover, for foreigners whenever they see the broken or fallen shrines, they immediately assume that people do not care about them. A good example for this is C.G. Seligman (1976) when she was conducting a research among the Acholi of South Sudan, she came across a number of fallen shrines and immediately concluded, "No particular care seems to be taken of the *Kac*, for in many instances it was falling to pieces and was sometimes represented by single decaying upright wood" (P.124). The bottom line is that people care about *kac*.

Having examined the views of foreigners about Acholi's shrines (*Kac*), let us now show how sacrifices are conducted to the ancestors among the Acholi. There are many religious practices that could be conducted at shrine (*Kac*) as we shall see below.

Between the Acholi people and the souls or spirits, incorporated in memorial "*Kac*" to which man offers prayers; there is a link of kinship, it is to this type of cult, in which the living and the dead are the kin to one another, therefore, the term "ancestor worship" is best reserved. So, as long as a man dies leaving a son, he becomes an ancestor of equal standing with any other ancestors, it is believed he has power to intervene in the life and affairs of his descendants in exactly the same way as any other ancestor. Subsequently, ancestor worship is an extension of the authority component in the jurally relations of successive generation. It is a long-lasting and intimate, between kinsmen, neighbours and relatives.

The spirit of the father, mother, and grandparents may at any time ask for food in a dream. And if their dreams are disregarded, they may send sickness, misfortune and death —and these matters can only be remedied by sacrifices. C. G. Seligman (1976) noted that although there might not be preliminary dreams or request from the ancestors but yet the reality is that routine treatment of all sickness is to make offerings to the ancestors in hope that they will remove the sickness for which they are held responsible (P.187).

When a person is ill, black goat or black sheep is killed as a sacrifice to the spirits of ancestors. The Acholi choose black colour for such occasion, because they believed that the black colour chase away illness, misfortune and presumably death. The black colours also close the eyes of the people with bad eyes, which the spirits sometime operate through them.

The goat or the sheep killed is left lying on the ground for sometimes while elders pray, "Let the bodies of your children be healthy. We pray when we go hunting, we may kill animals for meat; may we recognize your help (power) at this time of the year".

The goat or sheep killed at shrine is usually provided by member of the family and it should be killed by old men who are married with children. Some of the cooked meat is left over night in the house of the sick person for ancestors to eat. However, experience tells us that there has never been evidence we have seen or have been shown that ancestors have eaten part of the meat offered to them. Or could we assume that the eaten part of the meat, by the spirit of ancestors, cannot be seen with the necked human eyes?

Subsequently, in the morning elders bring out the food (meat), the clan's folk eat the food and bones from the meat are collected and put in a pot and again left in the house for the ancestors to continue eating.

The blood of the sacrificed animal is left to dry on the ground and afterwards buried in front of the house near the place where the animal was killed.

Still at the shrine (*Kac*) at the harvest time, after hunting, meat and porridge (kwon) are placed on the *Kac*/Abila, it is left there for sometimes, and after a short period the food is taken away and eaten by elders; a small portion will be left for the spirits to eat at the tree-peg shrine.

Yet in the shrine (*Kac*) if a man has been in danger when hunting he may come to his *Kac* and pray to the spirit of his father (and if the father is still alive, the man can pray to the spirits of his grandparents). This is especially occurring when a person escapes from danger without been harmed or injured.

Furthermore, when there is outbreak of epidemic diseases, family elders can pray to the spirit of the *father to cool the skins of children. And when an animal is sacrificed a notched stake (lot-dyiel* or *uno dyiel*) the gut contents (we) and blood is smeared on the grave of the father.

When a man kill animal such as antelope, the horns will be placed on the four-peg shrine to obtain success of hunting. But sometimes if a man is sick relatives go to witchdoctor (ajwaka) and the witchdoctor may suggest that the sick person has neglected his father's shrine and this is why he is now sick. The witchdoctor might suggest that the sick person to go home and get a bowl of water spit saliva into it and spread the shrine with the water from the bowl. The sick person will then go home and do as instructed by the witchdoctor. It is commonly believed that after doing this, the sick person will be cured.

Ancestors are believed to be involved in the practice of medicine and can, upon request send special cures to their relatives, who are seriously sick. The ancestors, by virtue of being part of the extended family and living in the presence of the living God, the creator, are endowed with special powers. The ancestors continue to be part of the family and are included in all the events of the family or clan. As I mentioned early, traditionally, elders are the people who are involved with religious practices.

All the above information concerning the *kac* of platform shape applies to both Acholi of South Sudan and Acholi of Uganda. But Acholi Uganda preferred to offer their sacrifices to the bad spirit of the dead father in Abila rather than *Kac*. The abila usually consists of minimum hut and sometimes a medium sized building, big enough for a person to enter; it is erected in the courtyard. This is briefly, how the sacrifices are conducted among the Acholi of South Sudan.

Father Crazzolara (1954), from whom many writers learn about the information concerning *abila*, suggests that true *abila* are perhaps built only for the "chiefs". The importance of "*abila*" is not in the hut but the three-pegs it has in the ground and this is not from the common wood but it is from the fig-tree wood pieces. From the account given by Father Crazzolara we may infer that the northern Acholi of South Sudan, do not commonly build an "*abila*" but have retained the essentials, the peg-shrine and the "*kac*", for presenting offerings. Writing of the central Luo (Acholi of Uganda), Father Crazzolara suggests that kac and *abila* are more prevalent among the Alur and Jopaluo, on the other hand the "*okango*" (a tree of God) at which they invoke God, performing a waive offering (*buku*) and praying for the good health. Any members of family can pray at "*okango*" for example a woman can take her child in front of "*Okango*" holding a hen by the legs and can waive the hen over the child and towards the "*okango*", while birds are signing in the surrounding bush or trees. After this the woman throws the hen down –it is not always killed on such occasions.

B) Ajwaka(Healer)

Ajwaka (diviner or healer) is a person (Male or Female) who is said to have the spirit of his/her father and as a result of the dreams, *ajwaka* uses his/her power to cure illness that come upon the victim. Because *ajwaka* feels confident that he/she is possess by the spirit of the father, *ajwaka* puts up in front of his/her hut a wooden stake with some traditional medicines tied to it. And as he/she produces his/her power, this will eventually prove whether or not he/she has the power that he/she claims.

In other words, we say that *ajwaka* are men and women who are able to see and communicate with the spirits. They have the powers that can be directed to the diagnosis and treatment of sickness; *ajwaka* also gives advice on good fortune –how to get it.

The amount influence exerted by the *ajwaka* varies from person to person. Some ajwaka are said to be effective and strong, while other *ajwaka* are considered to be weak and not reliable. The good example of a weak *ajwaka* is Mr Ogili, a man form Panyikwara village and who claims to be the *ajwaka* for women who are childless. I have been informed that Mr Ogili from his boyhood to manhood, he has been wearing women dresses, never a day he worn man dresses. More importantly, he likes staying in the company of women.

Thus, when a woman come seeking for fertility, Ogili tells that her childlessness has been attributed by the spirit of the dead father or dead grandfather. He then gives her some native medicines to drink and at the same time asks her to spend at least a night or two with him in the house. The reason has been that Ogili believes that the cure of childlessness is only on the tip of his penis. This means that the woman can only be cure when he has sexual intercourse with her. Because the woman is in need of a child of children she will not refuse. The woman then will send message to her husband and relatives she will not come until after a day or two –because Ogili will be working on her.

When the sun sets the woman enters the house and Ogili come and closed door behind him. They share one bed whole the night. The other woman, when they see the woman enters Ogili's house, they do not talk about it.

During the one or two nights that a client spend with *ajwaka* Ogili, many women suspected that Ogili also advises the woman to try herself with other man, which means to have sexual intercourse with other men, when he perceives that the husband might be impotence. The woman in need take this advice very seriously, thus, when she goes home, she would be using medications given by Ogili and at the same time seek for secret sexual intercourse with a man. I was told that about five to ten women became pregnant after visiting Ogili. This made me raise a question, "Were these five or ten women conceived by Ogili or by other men she was advised to have sexual intercourse with or by the husband?" Given this fact if the husbands of the women were impotence, then the women properly might have been conceived by *ajwaka* Ogili or by the other men.

According to Acholi culture a woman will never tell if the husband is impotence unless the husband acknowledges his impotence to his wife or to his relatives (mostly to the blood brother).

However, example of good *ajwaka* may tell the husband to pay dowry to the father or relatives of the wife if he/she perceives that childlessness might have been attributed by lack of dowry to wife's family –who *ajwaka* perceives they are angry. A good *ajwaka* also in certain cases may ask the relatives of the wife to make offer to the husband's family –if he/she perceives the uncles might have been angry for no reason that led to childlessness of their girl.

In many occasions, when the source of *cen* (bad spirit) is not readily apparent (following the consultations among the elders to determine the circumstances that must be addressed) elders may call upon diviner/healer (*ajwaka*). Ajwaka believes that when he/she is consulted by elders, he/she in turn consults with the spirits. The spirits will then provide him/her visions to assist him/her with the work; the same spirits also show diviner which herbs are good for healing and give instructions on the types of herbs to use for certain sickness.

Ajwaka (healer) are able to look into the future and frequently consult the ancestors on pressing matters. Using powerful rituals, healer (*ajwaka*) are adept in interpreting whether or not a person has *cen* or *ayweya*, and is so, where it was derived from. Ajwaka is believed to have strong communication link to the spirits worlds, and ability to heal those affected. He uses traditional instruments such as a shaker constructed out of a gourd (*ajaa*) to conduct the supernatural world and with their guidance establish the truth about particular incident, for example, thief, illness, death, childlessness and so.

More importantly, *ajwaka* (healers) are believed to be able to perform good and evil deeds. For example, should the power of *ajwaka* be denied, he/she is able to cast a curse on the family of a person in question (that is the person who abused or denied the power).

2.0 Acholi Rituals and Ceremonies

Despite the degradation of culture by the religious leaders and government offices, traditional rituals continue to hold an important value to the people of Acholi in the villages and those inside the towns of South Sudan, as opposed to the Acholi in Diaspora, who are been brain washed by Western Culture.

(2.1) Cien (Bad spirit)

The *cien* (bad spirit) and the *tipo* (soul) are distinct entities, the *cien* is more linked to the physical remain of the dead person but *tipo* is linked to *kac* or *abila*. According to the Acholi cultural elders, *cien* (bad spirit) is sent when a wrong against the dead has been committed. The phenomenon of *cien* illustrates the centrality of relationships between the natural and supernatural worlds in Acholi, the living and the dead are linked in one way or the other.

Cien can be described as the entrance of an angry spirit into the physical body of a living person that seeks appeasement. As C. G. Seligman (1976) puts it, "If a man or a woman dies angry, something called in Acholi language *cen* comes into existence and continue to act against those whom he/she was angry with at the time the dead person was still on earth —causing their children to die. Thus, if a man, whose father died with anger, is experiencing continuous illness, members of the family go to *ajwaka* and *ajwaka* might attribute this illness to the sick person's father" (P. 126).

The appeasement of the spirit is usually done in form of a sacrifice or compensation and reconciliation between the offended and the offender. The spirit of *cien* always haunts the wrong doers by entering their minds or bodies in form of visions or nightmares that may result in mental illness until the wrong committed is corrected.

Cien can also send illness and nightmares to the rest of the family, especially to those who were involved in the wrongdoing either directly or indirectly. Therefore, treating the dead or dying person in ill manner brings abomination (*kii*) to the family. Likewise committing murder or manslaughter cause abomination (kii); and further still inheriting the wrong doing from a person who committed suicide is also susceptible to *cen*. According to Acholi tradition, a person who committed suicide always brings *cien* to the family.

This is the reason why the Acholi treat dying person with respect and dignity. They would not like the person to die in a bad way (with anger), or to die in the stage when he/she is been neglected by family members or been sent out of home by force, worst still if a person die of hunger because the relative refused to give him/her food. The dying person would then say, "Let no food be given to any man or woman from this family or clan when he/she is at the point of dying with hunger". And if a member of the family neglected the word of the dead person and give food to a person suffering from hunger, that act alone will bring *cien* to the family members.

It is generally believed that if a person dies in a bad way, his/her spirit will not rest in peace, and the spirit will seek appeasement and correction of the wrong committed. As a result, elaborate burial and funeral rituals and ceremonies have evolved to show respect to the dead; this is considered vital to maintaining the well-being of the family or clan.

Cien haunts those who disrespect the dead, either by failing to provide proper burial or failing to show adequate respect as we have discussed above. Cien could also haunt people who refused to pay "*paneyo*" (the demands of the relatives of the dead wife). These could be balance of the dowry or the whole dowry might have not been paid and wife's relatives are asking for before their daughter is buried. And if the relatives of the husband turn debt ears to the demand of the in-laws, this brings *cien* in the family. When this happens Acholi usually say, "Wutwee cen pa dako-wu". Which literally means, "You must appease the spirit/soul of your wife"? However, *cien* cannot affect a person who is not related to the dead person.

Interestingly, *cien* does not manifest itself immediately after the wrong done, it takes months, years, decades or generation before people would realize a person is affected by cien but typically attacks the family or sub-clan of a person who treated dead or dying in an ill-manner; a person who committed murder or manslaughter and a person who inherited the wrong-doing from a parent or grandparent. This is the reason why elders are concerned about future of generations. If *cien* is not ritualized, it may follow the family lineage of the killer. According to Akena Ponse Otto (2016) people who are forced out from their family home in anger, and they die outside home, their spirits will not rest, they will actively seek correction to the wrong committed. Likewise, the people who die at family's home but their burials were not respected, in term of providing a proper burial (uncles or "*paneyo*" were not paid their due compensation) the spirits of such persons would also not rest; they would as well seek correction to be done for wrong-done.

There are many types of rituals used to chase away cen. In most cases, the sacrifice of an animal is used to chase away the bad spirit (cien). The animal is killed, cooked and eaten by elders of the clan. After eating, everybody washes his hands in one container of water (calabash). One elder then takes this water sprinkles on the body of the affected person and balance of the water first he pours it towards the east saying, "*Maber ne obin ki yoo tung nyango*". Which literarily means, "Let the good things (good spirits) come from the east" and he pours the remaining water towards the setting sun, saying, *"Marac ne odoki/ociti yoo tung piny/poto ceng"*. This literarily means, "Let the bad things (bad spirits) go to the west".

C.G. Seligman (1976) observed that among the Acholi, one of the ways to stop cien, is by digging up the bones of the dead man and burning. After this sheep is killed in the place where the bone were burnt, and it is believed that after this cien will not be harmful. But if cien continues to be active, ajwaka will be consulted, who normally instruct that a female goat with its young one to be brought (tethered) and place between the house and the shrine)kaco; thus, if both die, the cien is said to be still active and further sacrifices of a sheep is required. On the other hand if the cien is of an old woman, a gravid goat is tethered near the four-peg shrine; thus, if the goat produces male offspring it is a good news, but if she produces female offspring, it is a bad news, which means that Juok has not accepted the former sacrifice and further sacrifice is required. This sacrifice will continue until the ajwaka and elders find a solution to the wrong done.

2.2 Ryiemo Gemo or Cien

Ryiemo gemo or ryiemo *cen* is a ritual organised by elders and senior women and this ritual is perform at night. The elders and senior women beat calabashes, saucepans, drums, empty tins, door and other noise-making objects. Acholi people believed that *cien* is afraid of the noise, as such when it hears this noise; it runs out from the area towards the west where the sun sets. When the elders and senior women believe beyond doubt that the bad spirit has be driven away, the women then prepare sheas butter (moo yaa) and put on their bodies. Shortly, the women perform Acholi traditional dance called *acut* (by Acholi Uganda but Acholi of South Sudan just call traditional dance for chasing out bad spirit). The occasion always end up by heavy drinking of local beer called *"kongo-kwete"*. Nevertheless, Acholi believe that although *cien* is been driven out of the area, it continue to linger in places such as forest and mountains.

The other way the Acholi do to drive out *cien* is by transferring it from a person to an animal. There are many ways of doing this but the common way is by swinging a chicken around affected person or by giving a goat to the affected to spear. The person then walks on the blood of the goat so that the bad spirit could transfer to this sacrifice goat. If this is successful, the bad spirit is then said to have been chased out.

Interestingly, after performing all these necessary rituals to chase out the bad spirit from a person, and if abnormal continues in a person, an ajwaka will then be consulted and asked to perform the same rituals over as we have just discussed above. At this time, ajwaka will be responsible for the entire ritual; he/she slaughters a goat or chicken and throws it into the site where people believe the bad spirit (*cien*) is coming from to harm people in the village. Such places could be rivers, wells or rocks.

2.3 Other illness among the Acholi and their causes

The Acholi of South Sudan also believed that sickness like epilepsy (*twoo-yamo*) is caused by smelling certain object that the spirit of the father does not want the son or daughter to smell. C.G. Seligman (1976) pointed out that when she was doing her research among the Acholi of Obbo village, one man from Kitaka clan informed her that he was affected with epilepsy every time he smells meat of elephant being cook (P.127).

It is said in the old days, once a person is diagnosis with epilepsy, the person wears part of skin of antelope (*abur or pura*), a bell is hang above his ankle, these are all done following command of *ajwaka* —who diagnosis epilepsy in patients. *Cien* manifests in persons who purposely committed an act of murder that is unresolved. In some cases, the responsibility for setting things right falls not only on the individual, but on the family and entire community of the village.

However, something bad or good is always associated with the word "jok" for example there is a common saying, "Jok peri ber". It signifies "good luck" and "jok peri rac" signifies "bad luck". The same apply to the word "*nyima gum* and *nyima kec*". That literarily means "I have good luck and I have a bad luck".

Among the Acholi tribe the monor child always killed at the time of delivering as soon as the women realise that the baby has only one tentacle (monorchild (longelere) and this child is regarded as extremely dangerous. When they kill the child, they tell other people that the baby has breathed in the blood.

Apart from disease sent by God or ancestral spirits, illness may be caused by somebody who is said to have "evil eye". This person is called in Acholi language "*latal*" or "*lajok*" —means wizard, the "*latal*" or the "*lajok*" is not the same a healer (*ajwaka*). *Lajok* is considered responsible for the introduction of some foreign substances into the body of the patient. He/she is said to be a person with "evil eye" and possesses supernatural powers and he/she can cast a spell on someone just by gazing upon them. The evil eye is a specific type of magical curse. It is believed to cause harm, illness and even death. This is why in Acholi land, parents are very careful, they make sure that children do not to eat meat in the open where *lajok* could see them with evil eye.

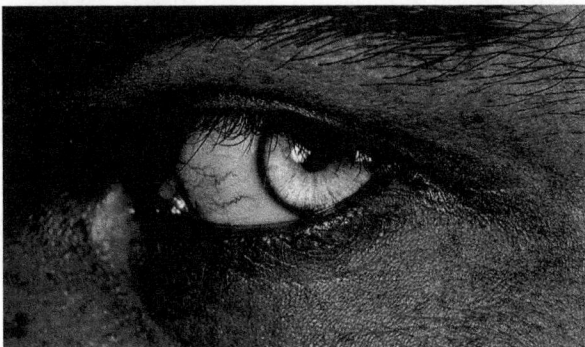

Figure: 65, Evil Eye

The "*latal*" is believed to kill a person by just bad wishes or sometimes by poison. The person who is possessed by "tal" run necked at night behind the house of a person he wants to die. In Acholi land there are no women who are witchers (*lotal*) this is the profession of men only. C. G. Seligman (1976) cited Captain Grove, who we owed debt of thanks to be saying, "The *latal* or *jatal* uses supernatural drug called "anya-nya', which when pointed at its victim, the person can be killed a hundred yards away from the latal/jatal and some-times latal/jatal puts "anya-nya" on the house of a victim —this will cause the victim to be stroked by lightning. Grove observed that many diseases are attributed to foreign bodies having been introduced into the body, and a class of doctors known as the "*Jotago*" (Plural) and "*Jatago*" (singular) make a speciality of extracting bits of woods, stones, charcoal and so on in large quantities from the body of the sick person (P.128). When I was ten years old, and living in Obbo Village, I saw the "*jatago*" extracting charcoal and stones from the thigh of a sick person. The *Jatago* applied oil on the thigh of the sick person and suck bits of woods, charcoal and stones from the thigh. It is difficult for a person, who has not seen this with his or her own eye, to believe that this doctor is really extracting charcoal or pieces of wood from the sick person. But the reality is that it is true.

C) Taking Oath

In Acholi land, disputes are common among the population, this could be due to stealing somebody property or having sexual intercourse with the wife of somebody. When this dispute arises between two persons, elders try always to resolve it through talks but if the resolution is not reached, the oath is then taken. This often happen when both men believe they are both right. But in dispute resolution there is no way that both could be right. The reality is that either one could be wrong or both could share the blame.

When this happen the two persons take spear or knife or rifle or axe and swear while saying, "If am mistaken in this case let the spear/knife that I am going to lick kill me". He then licks the spear or knife three times. In some cases oath are taken on a special spear kept by rain-maker, but in most cases people take any spear or knife from the home and swear with it.

Thu, if it is true that one person or both men were wrong, the person who lying and licked the spear or knife will become ill and eventually die if the elders do not reverse then oath. C. G. Seligman (1976) cited Captain Grove as saying, "Oaths are a good deal employed in disputed cases and are really much the same as the ordeals. The ordinary form of oath is then taken on a weapon such as spear, knife, rifle and so on. The man taking the oath calls on the spear to kill him if he is lying and then licks it three times. In a few days the man accused (if guilty) or his accuser (if innocent) will become ill and die, unless the proper measures are taken. The spear said to have caught him. As soon as he gets ill, he calls in the other party to the oath and pays him an appropriate fine, this usually in form of 15 sheep. A sheep is then brought to the sick man and killed with the weapon (spear or knife) on which the oath was taken on the spot where the oath was made. The sick man is the smeared with the food from its stomach, and the anger of the weapon is thus diverted into the body of the sheep" (P.129).

More importantly, for oath to be made accurately and with high degree of success, it should be taken on the weapon which has previously killed a man, is there is no spear in the home that has killed a man, people can then use chief's spear (*tong-ker-pa-Rwot*).

However, there are many ways oaths could be performed, sometimes if two people are in disputes and cannot reached a solution, they could step over a door that is put down on the ground. In other cases the parties may step over spear, knife, rifle or axe instead of licking it. They step over it three times and the outcomes are always the same. Sometimes if a man accused the other man of having sexual intercourse with his wife and that person denies, the ajwaka then be approached. The ajwaka prepare two gourds of water (with some local medicines in it) he/she then dips chief's spear into this water. The two principals each then drink off the water from two gourds. If the accusation is false, the husband of the woman will immediately be sick. The body will start to swell from face down to the legs. People who are watching the event will the say among themselves, "See how he has accused this man wrongly, and see how the body is swelling now?"

When this happen a sheep is then hastily brought and killed and the undigested food from its stomach smeared over the man with swollen body. After his recovery, he is to pay a fine to the man he wrongly accused as compensation for spoiling his name in the public. But if the accusation is true, the reverse process takes place. Furthermore in some case, the ajwaka could struck the two persons with a red hot spear and the person that gets burnt is said to be guilty and the other person who is not burnt by red hot spear is said to be innocent.

D) Kwir (Abomination)

As we have discussed above, there are many ways oaths could be taken. In some cases individuals from both party might not lick the spear, knife or axe and they may not even step over the weapon such as spear, knife, rifle or axe, but the ordeal is carried out by proxy each person bringing his own chicken who are made to drink the water that has been swear upon by the ajwaka (diviner) to kill the wrong doer. This type of water, that is put aside to kill, is called "*Kwir*" in Acholi language and when the water is drunk it causes illness the Acholi called "*Kii*". After the two chicken have drank the water, people will keep eyes on which chicken will die. The one who has been lying, his chicken will definitely die. This is known in Acholi traditional as drinking "*kwir*" and the person, whose chicken has died, acquires "*Kii*".

When this has happened, and the people know who is really wrong, elders call the two parties to "*Tumu Kii*", which is a sacrifice to appease the God (Jok) for the abomination that has occurred. There are different types of sacrifices that can be offered in the appeasement of God during "*tumu kii*", these could include killing goat, sheep or chicken depending on the requests of the elders.

E) Kwong (swearing)

Another interesting form of conditional curse among the Acholi is called "*Kwong*" (swearing). This usually happens, when two people are disputing and cannot reach a solution, each person accusing the other for either stealing potatoes or chicken. When the other party is not agreeing of stealing either potatoes or chicken from his neighbour, they will ask if they could swear "make *kwong*". When making *kwong* the accused will go first he/she stands upright, looking into the sky and says the following phrases, "If I have really stolen the potatoes or chicken of this person, let the lightening strike me!" After this, the accuser gets up, stands and looking into the sky saying this phrases, "If I am telling lying and this person has not stolen my potatoes or chicken, let the lightening strike me!"

When happy rain comes and the accuser knows that he accused the second party wrongly, he runs to him and asks for forgiveness before the rains falls on the ground. Alternatively, if the accused knows that he was lying to the owner of the potatoes or chicken, when heavy rain is about to fall, he runs to the accuser and asks for forgiveness for the rains touches the ground. But if one of them knows that he was telling lying and does not ask for forgiveness, when the rains comes, the first thunder or second thunder will strike him/her.

In Nilotic languages "*Lam*" is a core religious and ritual notion. It means prayer, blessing, invocation, taking oath, sacrifice and so on. What lies underneath the term "*lam*" is the power of word of words uttered from the mouth as we have seen above, to determine the present and the future state of being. Therefore, narrating history or events among the Acholi is more than simply narrating events of the past. Subsequently, this usually follows by "*lam*". This is simple to reinforce that what they tell is the truth from their own point of view (Cressida Marcus, 1991). If somebody says, "*Adaa pa Rubanga*" this becomes a "lam" (swearing) among the Acholi.

Once the parties have swear and would like to forget what was in the past, the do what Acholi called "Goyoo-pii", this exercise involves elders taking calabash of water and this calabash of waster is passed from each of the party to pray saying, "*Dano ducu timo bal, tin dong wa-gure kany ka lwoko cwing-wa. Runabga lwok cwing-wa pi bal-wa ma okato angec*". Which means, "Everybody makes a mistake; today, we have gathered here, to wash our hearts. God cleanse our hearts from the past mistakes". The water is then poured on each person to provide forgiveness and purity.

F) Kwero-Merok (Cleansing the killer)

When clan fight against one another or a tribe fight against the other in Acholiland, people are usually been killed. The spirits of these people killed, if not appeased could be dangerous in the community, they could cause illness and death. Thus, when returning soldiers kill foreigners, on arrival at home, ceremony has to be performed to appease the blood of the foreigners. To perform this ritual perfectly, men dig ants-hill big enough for a person to lie in. The killers are then put into the opening one by one for a period of three to five minutes; this period is to allow ants with read heads to bit the killer. During this period, a killer should not cry or to make any fearful signs. If he does the blood of the foreigner will not be washed away from his hands and members of family will continue to suffer from various illness.

G) Stepping on Egg (Nyono-tong-gweno)

In Acholi land, if a person had been out of home for extended period of time, and when he decided to come home, before he enter the com[pound he is to step on the egg. This is a ceremony to welcome a person who has been out for extended period to family home. However, there are many reasons why a person could be out of home for extended period, these could include studying overseas, been in prison, coming from the rebel groups, running from home due to disputes, coming home after two to three years of working in the towns and so on. In either case, there is a perceived need to receive a person back home in order to reconcile any problem or feeling of alienation that might have resulted from the person's extended absence, and to ensure that the person feels once again a full member of the family.

This ceremony has been adopted to facilitate the reintegration process of returnees from wherever he/she might have been during the period. In Acholi of South Sudan, this ceremony involves the returnee stepping on an egg that is placed in the middle of the road, just before entrance into the compound. But for the Acholi of Uganda, it is a little bit different; the ceremony involves a returnee stepping on an egg that is placed on a slippery branch of tree called "*Opobo*" and a stick with a fork called "*layebi*" used to open granaries.

The egg is said to symbolize purification of heart, body and mind. The fact that an egg has no mouth to speak against others and it is fragile, it signifies a restoration of innocence. The inner egg is slippery, which helps to cleanse returnee from any external influence he or she might have encountered in the towns, schools or bush. Stepping on egg can be individually or communal if there are two or three persons are returning home.

Anyuak Religion and Expressive cultures

The Anyuak are strongly religious and have strong beliefs in spirits to which one returns when one dies. They believe that one could communicate with the dead through a medium or when one becomes possessed by the spirit. Tghe people attach important to "cien" as we shall discuss below. Anyuak community in South Sudan has had some contacts with

Christianity for a long time but the most intimate contact came around 1961. It was mostly active in Akobo, Pinyudo and Akado. American Missionaries who were active in the villages and towns complained repeatedly about not being to convert many Anyuak into Christianity as there were only a very small percentage of the Anyuak became Christians. Interestingly, the Anyuak of Gambella have never heard about the Gospel of Jesus Christ, as a result they are not Christians. One of my informants, Mrs Man Diel, from Gambella in Ethiopia, who currently lives in Victoria, confirmed this to me when she said, "We have no churches in Gambella. The Anyuak of Gambella started to hear and know about Christian churches when the Anyuak of South Sudan took refuge in Ethiopia, following the outbreak of civil war in South Sudan in 1983".

But in the past fifteen years, many Anyuak in South Sudan and in Gambella Ethiopia, have turned to Christianity and Islam, in favour of their traditional religious customs, although we cannot verify this as there are no records. The Anyuak mass conversions to Christianity have enormous positive impact on the family; it creates a sense of responsibility in the family, such as having more responsible parents and a strong community. Thus, these traits intend to increase the lives of those who believe in religious lives than those who do not believe in it—therefore, I am forced to call this as incentives to the family.

When we look in Anyuak society, we see elders and some young people who spend much of their lives following religion, on the other side of the equation we see elders who follow traditional religion and some young people who do not care about religion at all.

The Anyuak believe in Supreme Being called "*Juok*", who is regarded as the creator of all things. Jok is the creator of the world and he has established the order of things, it is he who sends the rain from the place where rains gathered (above), which is his especial dwelling home. They also believe that bad spirits brings misfortunes to the people; their explanations and interpretations of personal misfortune call attention to a variety of spiritual agents, who are thought to bring illness and ultimately death to human beings. The Anyuak believed that troublesome spirits may cause misfortune if they are not remembered or respected. Anyuak are especially concerned with the capability of ghosts of the recently deceased to seek vengeance (an act of inflicting troubles or injuries) among the living.

Many Anyuak, like the Acholi and other Luo groups, do not draw sharp distinctions between religious practices with European origins and those with African origins. Most of the Anyuak people of South Sudan are Protestants, with a few hands of Catholics and hardly any Muslim. Nevertheless, the traditional beliefs in guardian and ancestors' spirits remain strong, though it is now often described in Christian terms. Thus, to many Anyuak Christians, the ancestors continue to play a significant role in their lives. They believe life continues after death and that communications are possible between the living and the living death. The spirits of the dead communicate with the living in dreams, which they are apparently heard rather than seen.

The Anyuak sacrifice animals to "Juok" when they want help from him, especially when somebody is sick or when someone wants to take revenge. Unlike the Acholi, the Anyuak pray directly to Juok. But when one wants to communicate with the spirits of ancestors, they do it through a medium that is a person who is possessed by the spirit. This man who is possessed by spirit is called in Anyuak language *"cijor"* (Sorcerer).

The Anyuak, like the Acholi, attach important to the *"cien"* (bad spirit) and *"gieth"* (blessing).

When we look critically into the beliefs of the Anyuak, we shall discover that *"cien"* and *"gieth"* create order in their society. Thu, for better understanding of the culture of the Anyuak, we shall discuss the *"cien"* and the *"gieth"* as separate article, and see what are involve in the ceremonies.

a) Cien (Bad spirit)

Cen can be described as the entrance of an angry spirit into the physical body of a living person that seeks appeasement. For example, in Anyuak Society, before a man dies, he confides his will to somebody, who will declare himself a trustee after the death of the man. Traditionally, nobody can change or disobey the will of the death person. Should anybody changes or disobeys the will of the death person, something called in Anyuak language *"cien"* comes into existence and continue to act against those who disregard the will –causing their children as well as adults to die.

b) Cijor (Sorcerer)

The Anyuak practice divination and magic in certain events; for example is there is a dispute between two persons, one person could go to sorcerer (*cijor*) and ash him or her to curse the other party. In Anyuak village, *"cijor"* are mainly used by elderly people who cannot avenge when someone done them wrong. But for the young people, if somebody done them wrong, they take revenge immediately.

Balanda Religion and Expressive Cultures

The Balanda, like the Acholi, Anyuak and other Northern Luo Groups in South Sudan, they are strong religious and have strong beliefs in spirits to which when one returns when one dies. The Roman Catholic Church, Anglican Church and Islam have significant impact on the life of the people. It is unfortunate that we do not have reliable records that quantified the number of believers in each of the religion. However, it is assumed that Balanda people have a very small percentage of Christianity and Islam. Majority of the population practice traditional religion. They practice all kinds of witchcrafts, sorcery and magic. It is said; Balanda borrowed a witchcraft known as *"mapiang"* from the Dinka.

The Balanda believed in Supreme Being they called "Juok", who is regarded as the creator of all the living and the non-living things on earth; they also believe in the spirits of their departed ancestors. Juok is the creator of the world and he has established the order of things, it is he who sends the rain from the place where rains gathered (above), which is his especial dwelling home. They also believe that bad spirits brings misfortunes to the people. The Balanda, like the Anyuak, believed that troublesome spirits may cause misfortune if they are not remembered or respected. Balanda are especially concerned with the capability of ghosts of the recently deceased to seek vengeance (injuries) among the living.

Of the immortality of the soul, the Balanda have no defined notion; and their only approach to, knowledge of a beneficent deity consists in a vague idea of luck. They have, however, a most intense belief in a great variety of petty goblins and witches, which are essentially malignant.

The Balanda sacrifice animals to Juok once a disaster have befallen a homestead; they sacrifice animals at the beginning of the cultivation season as well as at the harvest time. They believe the spirits of the departed relatives stay with God (Juok) up in the sky, and therefore, act as intermediatory between the living and God. They also believe that spirits live in the rivers and hence a sick person would be taken to the stream to be healed. They also believe in the power of the witch-doctors. More importantly, much of Balanda cultural heritage is contained in music and facial markings. Many have converted to Christianity and a few to Islam.

Jo-Luo of Bhar El Ghazal Religion and Expressive Cultures

The Jo-Luo of Bahr- el-Ghazal, like any other Luo groups in South Sudan, believe in God, they called God in their language as "*Jwok*". They make sacrifices to *Jwok* once a disaster has befallen a homestead, at the beginning of cultivation season and at the harvest of crops. Jwok is the creator of the world and he has established the order of things, it is he who sends the rain from the place where rains gathered (above) through the power of rain-makers (*Rwot-kot*). *Jwok* lives in the sky above mankind.

The Jo-Luo people believe that the spirits of their dead relatives live with God (*Jwok*), and therefore, they act as intermediatory between the living and God. They also believe that spirits live in rivers, which customarily called "*Iwok naam*" and hence when a person is sick, he is taken to the stream to be healed. The Jo-Luo people also believe in the power of the witch-doctors (*kwir*) and other spiritual leaders.

The ancestral linkages are of particular importance to the Luo of Barh El Ghazal cultures and customs. The understanding of the Jo-Luo about ancestry through oral history, which has been passed from mouth to mouth and from generation to generation, made the *Arace* (*Apace*) and linkages among themselves and to the other Luo groups much easier. The ancestry is of particular importance as they maintain the wellbeing of the family and groups even after death. The Jo-Luo of Bhar El Ghazal values their ancestral and belief in the life after death; therefore, with this view in mind, we could say that this practice has facilitated their acceptance to Christianity during the colonial period.

However, there are no written records about Jo-Luo religion and expressive customs; this made it, somehow, difficult for me to collect the necessary information. I found it difficult to get well informed Jo-Luo elders, in Australia, who have knowledge about their region and cultures. Nevertheless, many individuals in the community are well informed about *Jwok* and what ceremonies are involved.

About fifty percent of Jo-Luo of Bhar El Ghazal population are Christians, with about ten percent of them are Muslims; the remaining individuals practice traditional religion. Many of the Jo-Luo of Bhar El Ghazal, like other Luo groups, however, do not draw sharp distinctions between religious practices with European origins and those with African origins. Most of the Jo-Luo people, of South Sudan are Catholic, with a few hands of Protestants and few Muslims -if any. Nevertheless, the traditional beliefs in guardian and ancestors' spirits remain strong, though it is now often described in Christian or Islamic terms. Thus, to many Jo-Luo Christians, the ancestors continue to play a significant role in their lives. They believe life continues after death and that communications are possible between the living and the living death. The spirits of the dead communicate with the living in dreams, which they are apparently heard rather than seen.

Apai Gabriel (2005) argues that according to Jo-Luo philosophy, there is a supreme God (known as Master God) who created everything on earth and in heaven. The dead people will appear before Him for judgement. So, when the Master God finds a dead person with some wrongdoings, while on earth, He would pass a devastating judgement on that person, that would cover individual wrong-doers down to his off springs and this judgement may even be extended descendants from generation to generation. However, this often depends on the severity of the crime the dead person committed while still on earth. One of the worse crimes is "*wanton*" killing. Today, the Jo-Luo of Bhar el Ghazal of South Sudan still believe in Master God that curves and direct their behaviour in daily life (P. 71).

Needless to say, the Jo-Luo people also worship other gods, which are thought to have supernatural powers. They believe that these gods can use their powers to cause death to a person or to destroy completely their human interests. In Jo-Luo society, these gods have different names, and they are honoured and revered by individual families, clans or tribes. As a sign of good worship, people rear animals in the names of these gods and food containers are put up in the homes in which people throw delicious food into them for gods to eat. It is very unfortunate that, during my research, I did not find the origins of these gods and their relationship to human beings. Given the fact that the relationship of these gods to human beings remains a mystery, yet, these gods are highly respected by members of community including the well-to-do families.

One of the interesting things in the customs and norms of the Jo-Luo people is that they do not call their gods with Jo-Luo names, they rather call them with Dinka names such as achiek, arop and mathiang-guk. Apai Gabriel (2005) pointed out that since these Jo-Luo gods have Dinka names; this implies that they have been adopted by the Jo-Luo from Dinka people.

The original Jo-Luo god (Koy)
However, from the many gods that the Jo-Luo people have, the only original and true god that the Jo-Luo people believe in is called "Koy", who always makes a woman his agent. The "Koy" always talks to his customers in Jo-Luo language. When talking, "Koy" keeps on murmuring in Jo-Luo language in special shakers, which are kept in a basket that is covered with cloth. The performance of "Koy" is always done at night; the two shakers are shaken by an agent (woman) while singing queer songs in which only the names of the dead people are mentioned. Shortly, this is followed by sounds of low voices which are murmuring in the shakers saying, "Why have you brought us here?" The agent answers, "We have called you here because we want you to help us bring the spirit of our dead person (they may even mention the name of that dead person here) to come to shakers so that we talk to him/her face-to-face. Koy then facilitates the coming of the dead person into the shaker by performing some magic actions. After the talks, the relatives and the spirit would agree on what should be done to take away any misfortune that is looming over the relatives due to anger of the dead person. In this way, the problem is said to be solved (Apai Gabriel, 2005).

More importantly, the adopted gods may attribute the cause of cases referred to them, to other home-gods who may be angry because no domestic animals have been sacrificed to them for quite a long time. For this reason, home-gods often threaten to kill people or to cause harm to them if nothing is offered to them. In such circumstances, "Koy" may be called to reveal the cause of illness. In the cause of revealing, when "Koy" realises that he is called o reveal the cause of certain illness, which is supposed to be brought by what Jo-Luo called "*cien*" or "*chien*", the "Koy" would avoid mentioning the name of the person whose spirit cursed the sick or caused illness.

At this point and time, the agent of "Koy"(god) will not be opened by beats around the bush by giving some clues of the names, to the complainant, so that he/she says mention the name of the suspected dead person. Once the complainant names the suspected dead person, his spirit will be called in the myth of "Koy". However, if the name mentioned, is a false name of the suspected dead person, the agent will interprets the murmuring in the shakers negatively. However, in all cases, when the name of a person whose spirit is believed to be causing harm is a accepted, the relative, who called Koy's agent, would then want to witness miraculous proof. In this case, the closest relative of the dead person would be requested to come with a Jude venerated in the name of that person and place it under the bed of the sick person. Thus, if the condition of the sick person worsened that night, the Jude would be removed and thrown away. But if the condition of the sick person improved that night, then this would be taken as a correct and right proof; then the relative, who brought the Jude, would be asked to perform a reconciliatory ritual between the dead person and the sick person (Apai Gabriel, 2005).

During the performance of this reconciliatory ritual, a relative of the sick person brings small pot and makes small hole underneath (the pot). After this, a relative who brought the Jude is asked to catch a special lizard, brings it alive to the house where the sick person is. As a relative is out, hunting for special lizard, the relative of the sick person dig a hole, right in the centre of the entrance of the door of the sick person. The hole has to be dug deep enough to shallow the small pot. So, when the relative found the special lizard and brought it to the house and elder male would chop off the head of that lizard as a sacrifice of reconciliation between the sick person and the angry spirit.

The male elder then takes the head of the lizard and throw it into the hole (that is dug in the entrance of the door) and a pot is lowered with the mouth facing down, so that the opening at the bottom of the pot remains above the ground. This reconciliation with the dead person is called in Jo-Luo language as "*adeed*" (which literarily means promising good thing to happen).However, it remains a mystery to many of us, why the Jo-Luo people have decided to choose only that type of lizard among all other animals to be sacrificed for reconciliation.

It is important to know that no money is paid in the reconciliation between the dead person and the sick person. If the case is not settled because the sick person refused to accept that fact in favour of the dead, he/she must equally agree to accept the conse-quences that might follow after refusal (Apai Gabriel, 2005).

Traditionally, if an agent of "Koy" (god) performs a reconciliation services between the sick person and the dead female, that agent of "Ko" will remain in the house of the sick person for four days; and during these four days, she will not drink nor eat in this house. Likewise, if an agent performs a reconciliation services between the sick person and the dead male, she must remain there for three days; and during these three days, she will not drink nor eat in the house.

After having discussed about "Koy" and spirit of the dead, let us now turn to two very important tales told about "*cien or chien*" (bad spirit) and "*adojo*". We shall first briefly examine about "*cien*" and later on we shall examine about "*adojo*".

Chien or Cien (Bad Spirit)
The *cien* (bad spirit) is linked to the physical remain of the dead person. According to the Jo-Luo cultural, *cien* (bad spirit) is sent when a wrong against the dead has been committed. The phenomenon of *cien* illustrates the centrality of relationships between the natural and

supernatural worlds in Jo-Luo society, the living and the dead are linked in one way or the other. However, Cien could be described as the entrance of an angry spirit into the physical body of a living person that seeks appeasement. Therefore, if a man or a woman dies angry, something called in Jo-Luo language *cien* comes into existence and continue to act against those whom he/she was angry with at the time the dead person was still on earth —causing their children to die. Thus, if a man, whose father died with anger, is experiencing continuous illness, members of the family go to agent of "koy" as we have discussed above.

Historically and culturally, we find that many people of the Jo-Luo society talk about "chien"; and the one commonly referred to is the concerns of a slave woman who was badly treated my her master that she told the people that on the day of her death, the people would see something they had never seen before. On the day the slave woman dies, many people were happy and feasting for her death in the village; they were dancing and enjoying themselves. When the people were in the centre of their happiness, suddenly the earth opened and shallowed all those who were dancing and feasting, leaving a high rock standing in the place. It is reported that some people continue to hear sounds of drum coming out from slave rock for days following the death of the slave woman. This rock is called "slave rock" up to today; both old and young people know about the slave rock. But the most interesting thing about this historical event is that, people talk widely about the slave rock but when one goes around the Jo-Luo region one cannot find or see slave rock in the region. The question is, where has "slave rock" migrated? —the answer to this question remains a mystery to many.

Adojo

In Jo-Luo society, when a man dies and he has an enemy left on earth, that living enemy would be worried about what the dead person is going to do to him/her in regards to a dead-curse. This worry usually forces living enemy to seek way of how to get away with the dead-curse. The enemy then calls somebody who is specialises in dead-curse diversion —which is commonly known as "*adojo*". This specialist normally uses herbs, shams and supernatural powers; and the specialist believes that when herbs are put together and directed to do a particular job, they can divert a spiritual curse. It is reported that this dead-curse diversion practice is adopted from the people of western Bhar el Ghazal tribes —mainly from the Azande.

So, when a specialist is called to perform dead-curse diversion, he/she brings along the herbs and shams. The specialist then mixes his/her herbs with sesame paste, cooks them until the oil appears on the surface inside the pot. He/she then takes the oil and puts it in a small bone and hangs it on the roof from the inside of the house to scare off the bad spirit. After this, a chicken is slaughtered and the head is cut off and taken to the outskirts of the house; there the mouth of the chicken is opened and placed between bars of a special tree and twisted together, and tied to face upwards. In this occasion a specialist has no talking god (Koy) but sometimes he is also a fortune teller (Apai Gabriel, 2005).

The work of specialist in performance of "*adojo*" is never understood by the local community as a commercial business, but the bottom line is that the specialist always wants to get more and more money from his/her customers; in this case he/she may tells his/her customers that the dead man is so angry that diversion alone does not work. He/she would then tell the customers that digging out the head of the dead person from the grave and hanging it on a tamarind tree may do better. However, digging and chopping off the head of a dead person is a shameful and disgraceful thing to do, and at the same time it is dangerous because if a relative of the dead person finds you performing this nasty thing, he may kill you or sue you to court. Because of the seriousness of the act, the specialist will demand large sum of money. It is commonly believe, by the contemporarily anthropologists, that the practice of diversion (*adojo*) is dying out among the modern Jo-Luo communities.

Patriarch Prayers

Despite the practices of the "*cien or chien*" and "*adojo*" as we have just discussed above, the Jo-Luo people also believe in their patrons who are the representatives of descendants of men consecrated in the past to pray to the Master God to relieve and protect people from any mischief. Theses patrons have a fishing spear (*dedho*) as a symbol of their belief, just as Christians and Muslims have the holy bible and Quoran. During the prayers, an old male calls on Master God, and the spirits of their ancestors, which they believe are with the Master God in His village in heaven. An old man prays to Master God so that He relieves the sick person from sickness. Every sentence an old man says in patriarch prayers is repeated by interpreter. And old man from time to time refers to the degree their patriarchs made for them to do in such situations.

These prayers always culminate in a blood offering by sacrificing a bull, a he-goat or a red rooster. However, if these domestic animals or birds are not available, a god-fruit called "*kuoljok*" is offered by cutting it into two halves and tossing it into the air. The part of the god-fruit (*kuoljok*) with the inside facing upward is taken as the accepted offer and it is smeared on people's foreheads, while the other part with the inside facing downward is thrown to the west to go with the sun and all the bad things. Today, many patron performers have emerged both in rural and urban areas. They are commercially minded hijackers who roam about in villages and towns looking to be paid money for performing patriarch prayers (Apai Gabriel, 2005).

Initiation to Manhood

In every Jo-Luo home, there is a house designated for guests, this house is locally known as "*bachelors' house*". This is a house where the boys in a village gather every evening to sing and talk freely about their youth affairs; it is also a house where boys from a given village spend their nights. This togetherness of the boys is perceived as an opportunity where young boys come together to practice their tribal culture.

In Jo-Luo society, both boys and girls remove their six lower teeth, boys go further for group circumcision –but girls do not practice "genital cutting". Culturally, boys who reach 18 years of age can decided to declare their entrance into manhood. For the boys to do this, they must contact the boys of the same age groups from surrounding villages about their intention to enter manhood. When the same age groups agree among them, they then fix a time for initiation according to coincide with the local events; this could be done after a social football, which might be in progress in the nearby village. Thus, when the time of event comes, all the boys who want to enter manhood, suddenly undress themselves completely, they take up sticks and defence shields from nearby neighbours' houses. They then emerge as one group forming what is called "warrior troopers" and go about annexing other boys from nearby villages. It is reported that sometimes boys from nearby villages come to the groups by themselves. Traditionally, the warriors trooping usually take place around the middle of December.

Apai Gabriel (2005) observes that when the warriors have assembled, armed with sticks and shields, a war game begins. The war game is guided by strict rule, which states *"nobody shall beat his opponent on the head"*. Here a boy who believes he is the strongest of all, stands up and claims leadership. When another strongest boy saw the claims of his age made, he also stands up and claims leadership. Shortly, the two boys undergo war game; nobody is allowed to interfere or to stop them, they must fight until one is defeated and gives in. The winner then becomes the leader of the young people. Once the leader of the young men has been identified, one boy stands up to contest for the position of "General". Another boy also stands up and contest for the same position. The two again undergo war game, they fight until one is defeated. The winner then becomes the "General" of the young people. The duty of the "General", in the young men groups, is to call or to stop a challenge for leadership at any time in any place in surrounding villages.

Thus, all the positions from front to the back of the line are contested in the same way. The collective fighting goes on daily from sunrise to sunset. Occasionally, the "General" allows it to go on longer than expected until he hears somebody crying or sees many of the young men surrendering to discontinue the fight, and then he commands them to stop.

War-game is a harsh training, and those who undergo it, know the result of force and will not advocate it later in their manhood. The exercise continue for about a fortnight, with warriors moving from village to village where they use passwords for their arrival and food is cooked for them. During this time, they do not sleep in their respective homes. They rather sleep in the forest on the way to other villages.

There are two reasons that explain the important of "togetherness of the boys"; the first reason being that the nature of their practice does not allow them to sleep in their respective homes, because they have early morning group fighting, which is known as "morning greetings". In such fights, many are often beaten until they cry and the warriors would not like women and girls to witness –because they will say, the crying young men are coward and they do not deserve to be married. In addition, a homestead is not an ideal place for fourty to fifty warriors to fight. The second reason being that the young men want to avoid appearing in their houses for the fear that evil eye men would bewitch them. Even food cooked for them is often eaten under big trees on the skirts of the house.

However, as days pass-by, the protocol keeps on changing, because those who won positions of leaderships before are now losing their positions. Therefore, when there is a change in leadership, the whole process of protocol is removed with bitter fighting among the contestants from top to down –new protocol is developed. The leaderships then moved from the former winners to the new winners.

Interestingly, the main individual fighting is done by midday in the forest at water pools and wells where they spend the day to avoid being seen in public. As the young men stay in such water locations, they send out teams to nearby villages to steal sweet potatoes and bring them to be roasted for the whole group. It is very important that team who are sent to steal sweet potatoes must go secretly without farmers know. But should anyone see the team in the fields and shout, "Warriors", the team would be forced to run away leaving sweet potatoes behind. When this happen, nobody would raise a case against them; because it is a communal traditional practice, which everybody went through it in the past. The strongest boys usually force weaker boys to roast sweet potatoes, but nobody roast sweet potatoes for the weak boys.

Stealing and running away from the fields is not a humiliating thing to do, but running away from troopers and going home is considered the worst thing to do. In fact, this is a lesson the practice of warrior-ship wants to communicate to the troopers, that confrontation by force has a degrading result. Everybody in the process should realise and avoid it in future.

When the troopers have spent two weeks and are exhausted from fighting, they go for the final competition. It is reported that, when the young men are going for their final, all young men in the surrounding villages go to witness it. The final competition, like other fights, is done under a palm tree far away from home. The young man, who come from different villages bring with them hats made out of palm leaves and badges that symbolised the group, and put on them.

In as soon as the hates are put on their heads, someone among the troopers orders another to surrender his badge to him. The other worrier, always reject the command –here onward the contest begins. One after another, warriors get up to demand that others give them their badges, and again and again a fierce fight ensues over the badge.

Sometimes the trooper who demands for a badge from somebody is defeated, and the one he under-rated takes his badge. Other warriors surrender their badges without a fight when they know that they cannot match their opponents. In the end, one warrior gets victory and the warrior he defeated last becomes second in command in the young people organisational systems (Apai Gabriel, 2005).

However, it is important to note that on rare occasion, the last two contestants may fight it out nearly to death until the group from the young men intervene. As eyewitnesses, they will judge according to how they see the fight progressing and advise the weak warrior to take to take the second position in the protocol. Subsequently, the new protocol, after the final competition, will be recorded in the book of the year.

While the warriors are preparing to go for the final competition, one elder, who wants to name the troopers, gets a bull ready. He also chooses a tattoo and calls one man to mark the tattoo on the young men. After the final competition, the new leader and his team produce a new protocol; all the warriors would then follow this protocol as they walk or march to a traditional pool, where they dump their shields. They then take their shields and run away without looking behind; they go straight to their respective homes.

The next morning, all the young men who entered the manhood assembled, the bull is slaughtered in a traditional ritual way to bless them, after that their heads are marked as a sign of maturity and manhood. They are given new dresses and new spears; they are now no longer boys but rather men.

The elder then names the troopers. Traditionally, the group, in many cases, is given the name of the bull that was slaughtered –this may be Marial or Mathiang. Nevertheless, when no name is given to the group, the troops automatically become the second badge of previous warriors named. Soon after this the warriors then retire to their villages.

Mysterious facts among the Jo-Luo

Among the Jo-Luo people of Bhar el Ghazal, there is strong belief that there exists a rarely seen creature –perceived as a Master of the reptile family. This creature is described as a creeping reptile with numerous iron-tipped feet, which grind even stony ground into powder as it passes over it. This animal is believed to have being living in a bushy mountainous area with big tall trees. It is reported that the creature does not live outside but rather it lives inside a tree and eats the pith.

It is alleged that when the creature sees any living thing, that living thing will dies instantly. As the same way, if the living thing sees the creature first, the creature would then dies instantly. According to Apai Gabriel (2005) this creature spends many years without seeing or being seen by a living thing. When the pith, in the tree that the creature lives, get finished, the creature opens up a passage, so that it would look out at night. Using the lights, the creature surveys the area where there are big, tall trees in order to come down from the old home in a tree and go to a new home in another tree. The great-grandfathers who saw its track when moving from place to place, they said it was like track of a lorry over sand. And those who saw its bones, they said the bones were as heavy as pieces of iron.

Due to lack of written documents, the Jo-Luo elders from Alur decided to pass this useful information of mysterious creature to the coming generations through songs —they refer it to as the miracle of Abilek. Thus the people from Alur composed the song as follows:

During the seventeenth century,
When the slave trade was at its peak in South Sudan,
The slave traders advanced on Abilek from Ting Aguar,
Where they have set up their headquarters,
They fought for days trying to occupy Abitlek,
Where many Jo-Luo people had run after the battles of Agota and Ting Aguar.
One morning a certain man sat quietly
And emotionally near his hut.
The sun went up and the man did not moved to look for shade or talk;
People gathered around him asking
What had happened!
But he did not talk;
Suddenly, some footsteps were heard
Coming from above!
Then a big venerated fruit called "Kuol Jok" (fruit of Master God) dropped from heavens.
The silent man got up and took it
Blessed the people with it in a traditional way,
And says,
Slave traders are coming tomorrow,
Call people and tell them not to run,
Because their guns will not work;
They will flow with water, not fire
The next day the slave traders came and fired at people,
Their guns flowed with water
And they were all killed like fish
In a dry pool!

Jumjum Religion and Expressive Cultures

Most of the Jumjum people, as oppose to other Luo groups, are animists, although the primary religion practice in the country is Islam, which some people called it a monotheistic religion built on around the teaching of Qur'an and of the prophet Muhammad.

Nevertheless, majority of the people believe in God, which they called in their language "Juok" or "*Dyong*", who lives in the sky and suits on a horse day and night.

The Jumjum believed in Supreme Being God that they called "Juok" or "*Dyong*", who lives in the sky and suits on a horse day and night. They regard Him as the creator of all the living and the non-living things on earth. *Dyong* is the creator of the world and he has established the order of things, it is he who sends the rain from the place where rains gathered (above), which is his especial dwelling home. They make sacrifices to *Dyong*, like other Luo groups, once a disaster has befallen upon a homestead; they also make sacrifices to "*Dyong*" at the beginning of the cultivation season and at the harvest times.

They also believe in the power of the "evil eye". They believe a person with evil eye is envy of something or angry for certain reasons, he/she may bring misfortune upon other parties.

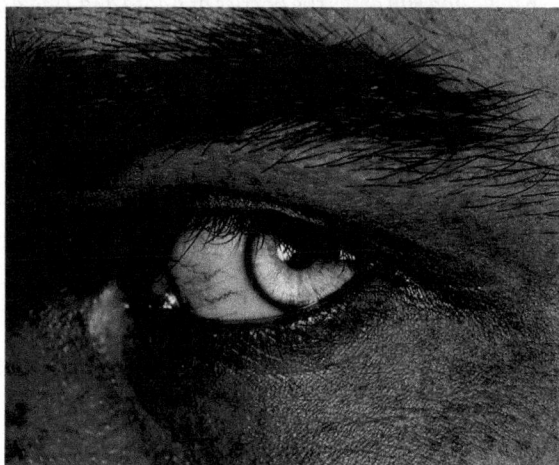

Figure: 66, Evil Eye

To chase away the curse or misfortune brought by a person with evil eye, and to attempt to safeguard themselves from this power, they use prayer, sacrifices and magic. How the sacrifices are made, remains unclear to many of us due to lack of written information from the scholars and oral information from the elders.

The Jumjum also believe in the spirits of the ancestors; the belief in ancestors is based on the notion that life continues after death and that communication and communion are possible between the living and the living dead. The spirits of ancestors communicate with the living in dreams; they are believed to take interest in the day to day affairs of the living either for good or for worse. They also act as mediator between the living and God (*Dyong*).

Islam was the first religion to enter Jumjum country in around 1800s; despite the fact that Islam was the first religion to be preached in the country, not many inhabitants were converted to Islam. Although I could not find the statistic to prove my preposition or assumption, it is believed that only 12% (8,040) of the population are Muslims; and according to Joshua Project research centre U.S. (2012) on 2% (1,340) of the population are Christians – they are Catholic and Orthodox. Thus, the remaining 96% (64,320) of the population practice African religion or some people would prefer to call it traditional religion.

Mabaan Religion and Expressive Culture

The Mabaan believe in a supreme being (God) with whom they communicate through a medium that could be elders or witch-doctors. The people have two important annual events; the first event is the feast of "Kornga" which is held in October of each year. This is the time when the community come together to confess their wrongdoings in the past year, and ask God for forgiveness. They appeal for tolerance and good heath for adults, children and their animals. When the time comes to do this event, they walk out early in the morning to a nearby stream to wash bad things away from their bodies, and on returning home, they slaughter animals, drink local beer made from sorghum and dance local dance called "Dukka-conkon". During the dance, they put on their best clothes and decorate themselves with beads (burngo).

The second event takes place after the harvest and is commonly known as *"harvest feast"* (*gatti*), which is usually held in the month of December. This second event is very important to the people of Mabaan, especially to the boys and girls who want to marry. This is the month when boys and girls can date and prepare for the marriage in March. Thus, during the dance both boys and girls appear in their smartest dresses, wearing neck-laces (*Linyan*).

Pari Religion and Expressive Culture

Nearly all the Pari people are animists who follow their traditional ethnic religion. They believe in an all-powerful spirit (God) they called "Jwok", who is regarded as the creator of the universe. The "Jwok" with the capital "J" is the true living God and the "jwok" with small letter "j" that they believed in is the spirit(s) of the ancestors. There are many places of "jwok" (spirits) where offerings and sacrifices are made to the spirits of ancestors but there is only one place they offer sacrifice to only one true God "Jwok", and this place is called Lipul (see article under Pari as Luo group). The people believed that God is like a wind, he has nobody, and therefore, he is everywhere. However, the Pari believed that the central dwelling place of "Jwok" is in the sky where he lives with the spirits of the dead ancestors.

Unlike many African tribes in general, and the Northern Luo groups of South Sudan in particular, the Pari people do not have mediators to intercede with God on the behalf of the people, but they communicate direct to "Jwok" (God).

The Pari also sacrifice to Jwok once a disaster has befallen upon a family member; they also make sacrifices to Jwok at the beginning of the cultivation season and at the harvest times. Interestingly, when someone is sick or when someone wants revenge against the enemy, animal is sacrifice to Jwok for help –so that the mission is successful.

a) Cien (Bad spirit/Curse)

A dying person makes either a blessing (*gweth*) or curse (cien). The power of a curse is very much feared among the people, as it may bring disaster not only to individual but to the whole community.

In the Pari Society, it is believed that ghostly vengeance (*cien*) is very powerful and destructive. An effective ghostly vengeance may cause misfortunes to the community, such as draught, fire and cattle raids. Some of the Pari elders informed me that when "monyomiji", who take the responsibility in maintaining the welfare and prosperity of the society, come to know that these disasters come from ghostly vengeance of a particular person, they usually do and dig the grave of that person and remove the born and throw them away in the bush or throw them into the nearby river so that ghostly vengeance becomes ineffective.

b) Ajwa (Traditional healer)

There are traditional healers or witch-doctors among the people of Pari. The traditional healer could be a man or a woman.

c) Cijot (Sorcerer)

The Pari practice magic and the cijor is a type of sorcerer who exercises magical powers against those he envies. He puts curses on those who have done wrong to someone else. Elderly people who are unable to avenge often resorted to using these curses.

Shulluk Region and Expressive Culture

In Shulluk country there are four components of religion, which are clearly distinguishable. The first component is the religious practices to "Juok" (God); the second component is the religious practices to Nyikang, who is the progenitor and national hero of Shulluk; the third component is the religious practices by witch-doctors or sorcerer (ajwogo); and the last but not least component is the imported religion what we call today as "Christianity". These four religious practices do not exist separately from each other, but have many relations among one another. However, when talking about the four components of religious practices in Shulluk land, we should not forget that there are still other forms of religious services but they are not as important as the four mentioned above. To some of us, these types of religious worships might look complex and confusing. I want to make clear, before we enter deep into the discussions about the various religious practices in Shulluk country

that Shulluk believed in various non-human beings living in bush and river. For the importance of our subject matter, we are not going to explore all other non-human beings; we
rather shall focus on the four types of religions mentioned above. And to avoid any confusion and misunderstanding, we are going to discuss them separately.

(a) *Religious practice to Juok or Jwok (God)*

Shulluk like other Luo groups believed in Supreme Being they called "Jwok", who resides
in sky together with the spirits of the ancestors. But report from "People and Language"
(1993) in contrast argues that although Shulluk believe in a supreme being who is a creator,
they also believe that this creator, called *Juok*, does not live in the sky, he lives under the
ground. They believed *Jwok* is a creator of all living and non-living things in the universe.
But writers like Diedrick Westernmann (1912) argues that Shulluk people do not regard
"*Jwok*" as a creator, although some natives told him that "surely *Jwok* is a creator" (P.39).

Although Shulluk people believe that God is the Creator of everything on earth and
above the earth, yet they believe that God Created the Cow and cow is their grandmother
(Adam and Eve are not their grandparents). They believe that this cow did not conceived
by a bull but rather through the power of the spirit of God, and bore two cows, one white
and the other black. This is the reason why Shulluk people call God their Father.

Shulluk oral tradition says that the black cow was loved by the mother, while the black
cow was hated by the mother. One day God went in the home of the mother cow, to see
the baby cows; she showed him the white one and hidden the black baby cow.

God asked mother cow, "Why do you hide the baby black cow from me?"

She answered, "For nothing".

God told her, "Well done! As you have hidden the black baby cow from me, I like the
white one. I also want to tell you that the black people shall be ruled by the white people
because of your action".

When the mother cow heard the curse from God on her black baby cow, she went and
brought the black baby cow out too.

God then asked her, "Why do you bring this baby black cow out?"

The mother cow answered, "Oh, I just brought him out without any special reason".

From the above narrative the Shulluk people believe that today, the white people are
given books, guns, swords and all kinds of goods by God because he loves them. And the
black people are not given knowledge and all kinds of goods because God hate them. This
also explains why today the black people are being governed by the white people.

Shulluk people also believe that *Jwok (God)* is formless and invisible, and, like the air,
is everywhere at once time; he is above Nyikang as well as above witch-doctor (*ajwogo*)
and ancestor spirits. Though Nyikang is considered inferior to Jwok, but in the political,
religious and personal life Nyikang takes a far more important place thank *Jwok*. He is

the national hero, on whom each Shulluk feels proud, everything they value most in their national and private life, has its origin in Nyikang. While Nyikang is their father, who only does them well, but *Jwok* is great with uncontrollable power, which is to be propitiated in order to avoid is inflictions of evil (Diedrick Westernmann, 1912).

God
|
Nyikang
|
Dak
|
human beings

Prayers and petitions are offered to god (Juok) through the mediator of dead people. It is believed that the dead intercede for the living and so it is good to appeal to dead relatives and friends to approach God on their behalf –through Nyikang.

According to Shulluk traditional belief, no one can approach *Juok* (God) directly, unless through Nyikang. Juok is so linked with Nyikang in the Shulluk mind that he is seldom approach in prayer without the mentioning of Nyikang.

When Shulluk man prays, one can easily understand which of the four components of religion he is really citing/praying; because prayers through these four components of religions carry different expressions. Thus, when a man, chief or an elder wants to pray to *Juok*, they begin the first word with *"lam"* –which means to pray. And when a chief or elder wants to pray to Nyikang, they begin the first word with *"Kwacho"* or *"kawayo"* –which means to ask for or to beg on certain occasion an ox is killed as a sacrifice to Juok, but the reality is that more sacrifices are frequently offered to Nyikang; prayers are also offered to Juok probably once a year (C.G Seligman, 1976).

Example of Shulluk prayer to Juok (God):

"We praise you, you who is God. Protect us, we are in your hands, people are in your hands. It is through Nyikang that you save your people and give them rains. The sun is yours, and the river is yours, everything belongs to you.

Our father Nyikang, you have come from under the sun, you, and our God (*Juok*) have saved the earth and we believed Dak is above all people. Here, today, we are scarifying to you this cow, and let its blood go to our God (*Juok*), through you as our mediator".

Source: C.G Seligman, 1976, P. 75

Other example of prayer to Juok in Shulluk society also goes as follows:
Juok has no visible symbols or temples, nor are the prayers to him offered by priest or sorcerer, but prayers to *Juok* are always offered by the chief of the village or a village elder, where there is no chief. According to Diedrick Westernmann (1912) the Shulluk people do

not think that Juok possesses deep-rooted rank. Many a times, they attribute good and evil to Juok, especially when a person become ill, they quickly conclude that the sickness is brought on the person Juok. Furthermore, when a person dies suddenly and violently, people regard this as being caused by God (*Juok*).

(b) *Religious Practice to Nyikang*

The second factor of religious practices among the Shulluk, is that which is practiced to Nyikang. However, the tradition on the origin of man or rather of the Shulluk leads to the second and most important part of the religious practice, which is the worship of Nyikang.

The majority of Shulluk think of Nyikang as a divine King or semi-divine being, human in form and physical qualities, though unlike his recent successor he did not die but disappeared in a great storm of wind during a festival held at Akurwa around the year 1575 AD.

Interestingly, to many of his worshippers, Nyikang is essentially a spirit being, who manifests himself to human being by way of incarnation and often he comes in dreams as bright light. Diedrick Westernmann (1912) notes that Nyikang at times appears in the forms of certain animals, as ichneumons, rats, snakes, lizards or in form of a bird. And when Nyikang appears in form of a bird, the tree on which a bird stands, is considered holy, and is henceforth dedicated to him (Nyikang); beads and pieces of cloth are suspended on its branches, sacrifices and prayers are offered below this tree. Such a tree is not to be cut down.

But the objects kept in the shrines of Nyikang are only objects used for a dead body with human body, subsequently. This casts doubt into the minds of many people, especially the foreigners. Therefore, if the objects in his shrines, which are claimed to be his, are only objects used for a dead body with human body, and Nyikang is said to have disappeared in the storm.

The shrines or tombs of Nyikang are known locally as "*Kengo* Nyikang" or "*Kengo* Reth". There are ten Kengo-Reth (tombs) in Shulluk country and all are in the names of Nyikang, but according to the worshippers he was not separated/divided and buried in these tombs. I asked some of my informants, "Why do Shulluk elders build ten tombs in different parts of Shulluk land instead of building one tomb at Akurwa, which is the home town of Nyikang?"

In according to Bunydhuru Oyay, Shulluk intellectual, who lives in Melbourne, he said the ten villages where the tombs of Nyikang have been built, are the villages where Nyikango is believed to have stationed during his movements on arrival in Upper Nile region. As a result, Shulluk built temples in these villages and they become worship places of Nyikango. Mrs Tarizah Otor, who also lives in Melbourne, agreed with Bunydhuru Oyay and emphasizes that out of the ten tombs built in Shulluk land, the most important one tomb built at Akurwa, this special tomb is known as (Kengo-Reth).

The spirit of Nyikang is regarded as being present at any time in his wooden effigy that is kept in Akurwa. However, out of these ten shrines of Nyikang, the most important ones been at Akurwa and Fenyikang villages. Moreover, some writers like Dietrich Westernmann

(1912) argue that *"Kengo Reth"* are found in most villages because these villages are devoted to Nyikang. These sacred villages include: Akurwa, Wau, Fenyikang, Nyibodo, Otonyo, Nyelwal, Osharo, Otigo and Didigo. In each of this shrine, there is a statue of Nyikang made of ambach-wood, the holy spear, drum, shield of Nyikang, a digging stick, ancient metal ornaments and cloths. The contents of the *"Kengo"* vary, but they always include sacred spear called *"alodo"*, representing those used by Nyikang and his companions.

Thus, in Shulluk land each shrine consists of a group of two to five huts (structure), which are of the same circular form but rather larger than dwelling house. Each shrine also has a small circular hut or shade called *"Ludi"*, which is neatly fenced with millet stalks. The *ludi* is stated to be place of the King where he may urinate and perhaps take bathe when visiting shrine (C. G. Seligman, 1967). Each hut in the shrine has a name as follows:

(i)   *Nyikayo* –this is a hut ear marked for the mother of Nyikang. She is supposed to live in; hut Nyikayo is also where millets are donated by villagers, during the harvest, they are kept on one side of the hut or leaving space for the mother of Nyikang to sleep and sit.

(ii) *Kwayo (Duwad)* –this hut is devoted to Nyikang for him to sleep. This is known as a royal sleeping room. This is the most sacred hut of the shrine, with the horns of sacrificed animals fixed into the ground just at the entrance (door). According to Shulluk traditional history, all the previous or former belongings of Nyikang such as spear (alodo), royal stool, drum, his lyre (*thom*), eight gourds, seven red and white plates, leopard skin, cooper cooking vessels, clay pots, four elephants tusks and sacred water are kept in this hut called *"Kwayo"*. It is said, Nyikang himself visits hut *kwayo* at night, and he comes like a wind. People then know that Nyikang has arrived in *"Kwayo"* only when they hear sound of music coming out of hut *kwayo* as he plays his lyre (*thom*). No any other clan in Shulluk Kingdom is allowed to build, repair and clean hut kwayo, this hut is only build, repaired and cleaned by chief of Nyelwal clan in Fenyikang District. Dietrich Westernmann (1912) observed that on the huts of Nyikang (kengo-Reth), drawings are made with different colours: white, red, and black, consisting simply in line-ornaments or representing animals. These drawings are made by woman and are renewed every year before the chief prayers are offered.

(iii) *Wad Mac* –this is a hut built to be used as kitchen by Nyikang's cook or his wife. The hut contains cooking pots, salt, oil, flour and other food items.

(iv) *Dag* or *Kwayo* –this hut has two names that is "Dag" and "Kwayo". In this same, hut number two also name "Kwayo". Why is this? The reason been that from time to time Nyikang uses hut *Dag* as his sleeping place, this is why the people have given two names to this hut, another possible explanation is that when Nyikang is in the hut *Dag*, the hut could no longer be called *Dag*, but must be called "Kwayo" and when Nyikang is outside the hut you can call it any name you want (*Dag* or *Kwayo*). It is said that when Nyikang is not sleeping in hut Dag, his son, Dak uses the hut as his sleeping room.

(v) Duwol—this is a hut used by Nyikang as a court room, when he hears cases of his peo-ple. Today, when *Reth* visits Fenyikang, and there disputes that needed to be settled, he hears the cases of his people in this room.

The huts are particularly well thatched, and the apex of the roof ends in ostrich egg—from which the blade of spear protrudes/projects outward, while the surrounding fence is kept in notably good repair.

The five huts and their enclosed area, area considered sacred, nobody is allowed to enter the compound, if a person is not qualified to do so. The only people authorised to enter shrines of Nyikang are the *bareths*, who traditionally are considered attendants and priests of the royal grave shrines of Nyikang. The term "*bareth*" in this context means "king-wife" and this include certain old men and women who are said to have hereditary connection with the shrine.

However, the old women could be the ex-wives of the kings and the old men could be the blood relatives of Kings. Moreover, for a woman to be a member of *bareth*, it is said, she must have born a reasonable number of children in the clan. *Bareths* are been charged with the responsibility of carrying for the shrines of Nyikang, their main duty include cleaning the tombs, receiving donations from the population, sacrifice and facilitate ceremony during the coronation of the new King.

It is easy to distinguish a member of *bareth* from other ordinary people, from what they usually wear. They usually wear necklace of disc beads of ostrich egg-shell, they also wear green and white ankles glass beads. This was even noted by C. G. Seligman (1976) when she cited Hofmaye as saying, "The distinctive mark of the *bareth* is a necklace of disc beads of ostrich egg-shell, but they also wear ankles of glass beads, white and green" (P. 78).

The clans of Golobogu and Pameiti are closely associated with the up keep of the shrine at Fenyikang. People who keep shrine must be sexually pure for otherwise they will not enter the shrine. Traditionally, the children from the two clans mentioned above, before they reach their puberty age, they can enter the shrine but after they have already reached their puberty, they are forbidden to enter the shrine. Furthermore, a man from the two clans may enter the shrine if he lives in chastity, but once he sleeps with a woman a day before, he is forbidden to enter the shrine; but if he forces himself and enters, he will die. A married man whose wife is pregnant, will in all possibility be living chastity, such man can enter the shrine without any comment from the villagers.

When there is sacrifice to be made to the shrine, it is the chief of Golobogu who always takes the bull from his cattle to be sacrificed at the shrine.

Interestingly, in 1921, it was reported that shrine of Nyikang at Fenyikang was entrusted to an elderly Dinka woman called Aker, who was the widow of King Kur and Fadiet. When Aker became the priestess of shrine, it was asserted that she did not know how to care for the shrine and a man called Anyurial helped her by giving instructions on how to care for the shrine. It is difficult to believe this assertion because Anyurial is junior to Aker and he has no right to instruct her. On the other side of the coin, it become extremely difficult to believe that a royal wife could be sent to a village to look after the shrine, where she might find it difficult to accommodate her cattle and those of the shrine attendants.

The worship of Nyikang at Fenyikang and other shrines consist of solemn ritual of sacrifices and prayers. All the ten shrines in Shulluk Kingdom are the places where annual sacrifices are made to *Juok* (God), through Nyikang when the country is expecting to be threatening by national and natural disasters such as drought, outbreak of cholera, and other contagious diseases. When a man is sick he may bring or send a sheep to the nearest shrine of Nyikang, there the attendants sprinkle it with water and spear the animal. After this *bareth* Nyikang pray for the sick person; and when the meat of the animal is cooked, they give him/her part of the meat to eat and drink some of soup (the water in which the meat has been cooked). If the sick person just send the animal to the shrine and never appears there, part of the cooked meat and soup will be taken to the sick person to eat and drink. The rest of the meat is then eaten by the guardians of the shrine and their friends without any ceremony; but after the meal, the bones and perhaps the intestines will be taken and thrown into the river. However, due to lack of time and space in this book, I am not going to cover all the annual ritual ceremonies at various shrines in the country. Therefore, I feel oblige to give at least two most important annual ceremonies.

1.3 People Prayer for the sick

The Shulluk believed it is God (*Juok*) who causes illness and it is he who can also cured a sick person. Thus, when a person falls sick, that person is to be taken to shrine of Nyikang, to ask for blessing and curing. There a sick person is sited in front of hut *Kwayo* and a member of *bareth* or chief or elder gets up and recites a generational prayer (a prayer that Nyikang taught his people to pray to God (*Juok*), when a person is sick and before killing a sacrificial cow). This is more like the Roman Catholic Church prayers that are written long time ago and are recited by the priests in every day of their prayers. A member of *bareth* or a chief or an elder gets up, lifts up his heart to God and says, "I implore you God, I pray to you during the night. How are people kept by you all days! And you walk in the midst of grass. I walk with you; when I sleep in the house, I sleep with you. To you I pray for food and you give it to the people; and water to drink; and the soul is kept alive by you. There is no one above you, you God. You become the grandfather of Nyikang; it is you (Nyikang) who walks with God; you become the grandfather of men and by son Dak. If a famine comes, is it not given by you? Oh God, to who shall we pray, is it not to you? Oh God, and you who becomes Nyikang, and your son Dak! But the soul of man is it not from you? It is you who lift up the sick" (Diedrick Westernmann, 1912, P. 171).

After the prayer, the cow is then speared and the contents of the stomach are taken out, and thrown on the body of sick person; and he is sprinkled with holy water. One ear of the cow is cut off and cut into strips, tie together and tied round the leg of sick person. The right forelimb of the cow is cut off, cooked and left to God (*Juok*) nobody is allowed to eat it. The celebrant takes a soup and pours on the ground –this is considered a property of God (*Juok*). In few days to come, the sick person is expected to get cured.

1. Swearing Oaths by Nyikang

One of the important aspects of Shulluk cultures is swearing in the name of Nyikang. This happens when two parties have disagreement on certain vital thing such as thief, sleeping with somebody wife etc. In this circumstance each one claims he/she is right and the other party is guilty. According to Shulluk culture, such claim could only be solved by bringing a sheep that is killed and both the accuser and accused are then smeared with its blood; after this one of the person takes spear, holding it in his left hand and swearing by name of Nyikang saying, "If I am mistaken, I swear by the name of Nyikang of Akurwa, let this spear kill me".

When the first person finishes, the second person takes spear, holds it in his left hand and while repeating the same words saying, "If I am mistaken, I swear by the name of Nyikang of Fenyikang, let this spear kill me".

But if the matter is too complicated and he swearing is to be done in front of the court; the healer (*ajuago*), who keeps the holy spear of Nyikang, brings the spear to the court. The traditional healer then sacrifices a sheep and put the blood upon the accused and the accuser; the healer offers prays to Nyikang saying, "Nyikang indeed! Let Justice follow our sacrifice, let anyone, between these two persons, who is telling lies to die".

Firstly, a healer gives the holy spear to the accuser and asks him/her to swear by Nyikang. The accuser takes the holy spear, holds it in his left hand and swear using the following words, "Nyikang her! I swear by the name of Nyikang of Nyielwal, let this spear kill me, if I am not telling the truth".

Secondly, a healer gives the holy spear to the accused; he takes the holy spear, holds it in his left hand while saying, ""Nyikang her! I swear by the name of Nyikang of Nyielwal, let this spear kill me, if I am not telling the truth".

At this juncture, it is important to note that if a person perjures (deceives) himself, death is eminent to follow as a penalty. Nevertheless, the Shulluk always make promises under oath, which they, however, readily break without any fear of penalty. Sometimes, people also swear by the name of *Dak* or in any other name of the ancient Kings (Diedrick Westernmann, 1912). This is acceptable in Shulluk religious culture.

It is customary that when people are swearing, they have to name one of the villages that has a shrine. This means that a person could swear by naming for example Nyikang of Akurwa; Nyikang of Fashoda, Nyikang of Wau or Nyikang of Nyielwal. This is recommended because when a person swears in the name of Nyikang, which implies that Nyikang is being asked to come out from one of the ten shrines and witness the swearing. Not only this Nyikang in turn to ask God (*Juok*) to kill the person who is not telling the truth.

(c) *Religious practice by Ajuago, (healer) or witch-doctor(Sorcerer)*

In Shulluk land there are two kinds of healers one of whom is called "*Ajuago*" and the other is known as "*Jalyat*". The term "*ajuago*" literally means "The man of God (*Juok*)", that implies a medicine man who supplies good medicine to cure sick persons. The *Ajuago* always speaks about good work of the community, and also cure the sick persons.

The term "Jalyat" literally means "a man of herbs, who is dealing essentially in black magic; he/she satisfies the wants of an individual rather than the wants of the community. Therefore, to understand how these two healers perform their work in the community, we have to discuss them separately.

1. The Jalyat (Monorchid magic)

Typically, though not always, the "*jalyat*" is a monorchid; nevertheless, even though a man with double underscending testes or with small glands are regarded as "*Jalyat*" or "*okum-ajuok*". Traditionally and customary, in general Luo customs, babies (*okum-ajuok*) born with one testes or with double underscending testes are killed at birth, as we have seen among the Acholi.

In most cases, Shulluk women find it difficult to kill such baby (*okum-ajuok*) at birth; therefore, they prefer to wrap the baby and thrown it in the river. This usually happen after two or three days, following birth, when the baby is enclosed with a special woven basket (*dong*) and taken to the river by three women (usually aunty or grandmother), these women perform certain ceremony while at the river.

More importantly, these ladies must be ladies who no longer give birth or are barren. As they accompany the baby (*okum-ajuok*) to the river, they take with them a goat to sacrifice to Nyikang. When the ladies reach the river, they gather lilies (*otwoc*), tie them together to provide base where to put the woven basket (*dong*) later. Before the woven basket is put on the tied lilies leaves and push into the water, on woman takes one egg (*tong-gien*), maize grains (*abwok*) and dura (*biel*) and put them into the woven basket (*dong*), where the baby (*okum-ajuok*) is lying and place this woven basket on the lily leaves. One woman then pushes the baby into the river, while saying, "Dok nyakali", which literarily means "Go back where you came from".

The woman does this with a hope that the child will sink under water and dies. And if baby does not sink under water, but dies in the woven basket on top of the water, it can be eaten by birds. However, it is reported that as the baby is being push into the river, one woman would jumps into the water; she swims toward the woven basket (*dong*) to save the baby. As one woman swims to save the baby, the other two women call name of Nyikang and kill the goat, they then cook goat meat for Nyikang. When the women are performing this ceremony, customarily, children are not allowed to see the processes.

Figure: 67 Lily leaves (otwoc)

In contrast, it is said that a woman pushes the baby into the water for the sake of custom; the bottom line is that she pushes the baby (*okum-juok*) into the water with intention to save it, but if she cannot reach the basket in time, the child will then be pronounced dead. But more often, the infant is saved and survives.

When it happens that the baby is saved and it survives, the women take the baby (*okum-juok*) and return home, on arrival at home they do not speak about what happened at the river to any man; they rather prefer to talk about it only to their closed relatives. This is an act that must be treated with high confidentiality.

But on the contrarily, if the basket sinks with the baby, the women will return home with tears (crying); and tell lies to men and relatives that the child has slipped from the hands of one of their colleagues and was taken by the river (died). When people hear this message, they immediately understand that the baby is born abnormal (e.g. with one testis or bisexual things not accepted by the customs.) and that is why it is to be taken to the river.

It is generally believed that when the survived baby grows up, he becomes the *"Jalyat"*. A common practice of *Jalyat* is to work on parts of the body of the person to be injured. The man or woman, who wants Jalyat to injure his /her enemy, usually tells the *Jalyat* which part of the body (of that person) he/she wants the *Jalyat* to destroy or injure. The parts of the body usually targeted include legs, hairs, hands, stomach and eyes.

The process required for *Jalyat* to injure this person requires that person to be injured has hobbies such as club, petticoat or in some cases effigy representing the person could be made, and this object is heated or pierced to destroy the person in absentia (C. G. Seligman, 1976).

In Shullukland, the practice of destroying somebody is not only practice by man, there are certain women known as "Daiyat", who are recognized by their actions and they do speak enviously of other, are regarded as the feminine equivalent of the *"Jalyat"*. The magic of *daiyat* is associated with her body; one mode of working evil is to urinate into a gourd and place this in the house of her victim.

Some people said, a female healer (*daiyat*) has no difficulty of getting married, or does she brings misfortune on her husband. But the *"Jalyat"* has difficulty of getting married, because the parents of the girl would think he might bring misfortune to their daughter as well as to their family in general.

According to C.G. Seligman (1976) the female healer should not marry but might have lovers, and if she marries, the husband accesses her only during the dark half of the month (the period she should not get pregnant). Seligman argues that this matter was confirmed by Aker (a Dinka woman who was trusted to care for Nyikang's shrine in 1921). Aker said, "It is impossible for a female healer to marry and if she decides to marry, the husband cannot approach her sexually, if he does the female healer will die. The female healer is not even allowed to have sexual intercourse with other men. As regard to male healer he is allowed to marry but when he wants children he is to arrange with some men (preferably his brothers) to have sexual intercourse with his wife" (P. 101)

This then means that all traditional healers in Shulluk land live in celibate lives, more importantly; a male healer may give his cattle to a relative avowedly because he no longer requires them to procure a wife. Not only this, a male healer will be physically afraid to approach any woman.

It is generally believed that the power of traditional healer is due to the immanence of the spirits of the early Shulluk Kings. The guardian of Nyikang shrine could or could not be a healer; they do not become a healer because of their connection with the shrine, nor does their being healer alters their official position. C. G. Seligman (1976) notes, "Only the spirits of Nyikang, Dag (Dak) and Boc (the first three and the seventh of Shulluk Kings) become immanent in men to make them traditional healer" (P.100).

It is assumed that the spirits of Dag (Dak) possesses the traditional healers the most. Therefore, for a man or woman to become a "healer" he or she must first become ill, perhaps waking up at night from time to time, trembling and agitated in a dream.

If this happen, this is the time the person affected could consult the one of the existing traditional healers and tell him/her all what he has been experiencing. The "healer" may tell the person, "No, you are not ill; you may have the spirit of one of the early kings (which could be the spirit of Nyikang, Dag/Dak or Boc)".

After this, a long and complicated ceremony is then performed in order that the spirit may not affect him too severely, for without this ceremony, the spirit will remain strong in his body and continue to affect him severely.

The ancestral spirit of any one King (*Reth*) may be imminent in many healers at the same time, often healer passes this spirit into one of his children at the death or afterwards and that child become a healer himself; the power is frequently hereditary. And once a person is convinced that the spirit of the King is in him, he builds the spirit hut.

Deidrick Westernmann (1912) observed that Sorcerer heals the sick person by administering charms. Sick people give two sheep, goat or even ox as a price or cost of the medications to be administered on him. The animal is killed, and the contents of its stomach are laid on the sick person's body; the person wears head anklets, and skin of the animal is cut into strips and fastened below the knee of the patient (P. 44). The ankles are considered protection against future dangers.

However, the reality of the power is practically tested afterwards by the success of the would-be healer in effecting cures. If he heals, especially in his early cases of cures, it will be recognized that he is not truly possessed by the spirit of any of the three powerful Kings mentioned above.

1. Ajuago (Sorcerer or witch-doctor)

The Shulluk people believe in the sorcerers, because they said sorcerers have visions – and the visions work. Let us take two good examples of sorcerers with working visions in Shulluk society:

(A)   In the first example it was said there was a certain man in Shulluk land, he was called Wet Kwa Oket, and also known as "Agweratyep", he was a very strong man and a famous sorcerer.

One day Wet Kwa Oket had a vision, and said to his people, "The white people came and destroy the society!" Meaning: *I have a vision that white people came in our land and they destroy the society*. And it was true that the white people later came to Shulluk land and destroyed Shulluk society by introducing in the society what are not culturally acceptable. They destroyed the society.

And before Wet Kwa Oket died, he told his people again, "Ha, the chieftainship will be taken over by "Ajalong" after my death. I am not dying a natural death but this man "Ajalong" is killing me with his witch-craft, hover, he will also die soon after me". Shortly, he died and he was mourned. And before long, "Ajalong" who had bewitched him, was struck by lightning and he died –the people believe this man died because he was cursed by the sorcerer (Wet Kwa Oket). When Ajalong died, all the people believed in him, saying, "Agweratyep is a strong man indeed". Subsequently, the medicine men in the society became afraid to bewitch sorcerers, as a result for a period the villages lived in a peaceful condition (Diedrich Westermann, 1912).

(B)   The second perfect example of sorcerer with vision was "*ajwoga*' called Agok; he was tested by one of the wizards in the village. One day a wizard bewitched a cow and it felt down. One of the men from the village then ran to Agok (*ajwoga*) and told him, "Ah! Agok, a cow has died in our village".

Agok asked, "Who killed the cow?"

The man answered, "I do not know?"

Agok then ordered all the people in the village to assemble. Everybody gathered, including the wizard, and Agok told the wizard, pointing his figure to his eyes and saying, "Man, is it not you who bewitched the cow?"

The wizard answered, "Yes, it is I".

Agok asked him, "Why did you kill this cow?"

The wizard replied, "Because I want to try you, as I want to know whether you are really able to find out the killer or not".

Agok said a wizard, "Ha, you are a cursed man! You black-eye man, you are cursed! Why are you always bewitching the cattle of the people?"

A wizard answered, "I have been doing these only to try you whether you are really strong in your wisdom".

Agok said, "Well, we have met. Now go and correct your thoughts and bad actions".

Having seen the examples of the visions of sorcerers, I want to inform our readers that the main business of sorcerer is to satisfy the wants of community as opposed to the *Jalyat*. Shulluk people prefer to take their cattle across the river during dry season. When the time comes, and before they people move their cattle across the River Nile, the sorcerer prepares charms to protect them from being seized by crocodile and other enemies. In contrast the Shulluk believed that the mother of Nyikang, Nyakayo, lives in the junction of the Sobat and White Nile and occasionally appears out of the river in the form of a crocodile or other forms. No worship or sacrifices are offered to the mother of Nyikang, but when a man or animal is taken by a crocodile, while crossing the river for grazing, people do not complain because they believe that Nyakayo has taken the man or the anima. It is rather an hour, when she takes her sacrifice from a village. What would this mean to many of us? It means that when sorcerer offers charms, this is only to protect cattle from wild animals such as lions and leopard but not crocodiles.

It is generally believe that *ajwogo* (witch-doctor) also performs miracles to kill a man by witchcraft, prevent rain or even cause cows to be barren. To avert this act, another witchcraft must be consulted to nullify the curses of the former witch-doctor (*ajwojo*). Many people believe that nobody, in the society, has power to avert the curse of the witch-doctor. Two of my informants, Terizah Chol Otor and Banydhuru Oyay, pointed out that people can avert the curses of witch-doctor by sending old men to the river, there, they assemble to pray to Nyikang asking him to avert what is happening in the village. When the old men completed their prayers, they take a ram or goat and offer it as a sacrifice to Nyikang; they spit into a coconut dish while praying against the witch-doctor or perpetrator. It is believed that after this performances (prayers and sacrifice) Nyikang always hear people's prayers and averts the actions of the witch-doctor from befalling on his people.

(C) *The Sorcerer prepare warriors for war*

When the enemies are drawing nearer and nearer, the people send for sorcerer and when he comes, they give him a cow; the young and middle age men then gather before the sorcerer, in open space near the village. After this, sorcerer asks two men to take his spear and stick it on the ground; when they finish, sorcerer asks for a rope and tells them (two men) to fasten it on top of the magic spear. The two men fasten the robe on the magic spear, living it hanging one side of the spear; and sorcerer asks the warriors to pass below the rope one by one. Interestingly, all men touched by this hanging rope, he will be place in a separate area and they will not be allowed to go to war, because they will be killed.

When the exercise is completed, sorcerer then tells people, who passed untouched below the rope, to sit down. A he-goat is then brought, thrown into the ground; the he-goat is cut up and its head is cut off; the contents of the stomach are removed and thrown among the people. The head of the he-goat is taken and given by sorcerer, he takes it and throws it towards the country where the enemies are expected to come – everybody there will be watching what sorcerer is doing. Thus, if the fore-head of the goat points in the direction of enemies' country, the sorcerer will tells the warriors, "The enemies will be defeated". But if the head of the goat falls pointing towards Shulluk army, Sorcerer tells warriors, "It is a bad war".

Thus, if the forehead of the goat points towards Shulluk armies, then in this case, sorcerer makes his witchery once more; this is to minimize death among Shulluk armies. At this time, grass is brought and tied on a rope, sorcerer then asks for another he-goat to be sacrificed. He-goat is killed and the head again is cut and thrown towards the country where the enemies are expected to come. At this time the forehead of the goat will points in the direction of the country of the enemies, and sorcerer says to warriors, "All right! Let all the people come!" The warriors gather, the contents from stomach of the goat is removed and thrown on their bodies; the sorcerer leaves the place, the head of the he-goat is then buried in the ground in the open space and water is put on fire, and sprinkled on the people who are standing around.

Shortly, the armies take their spears and proceed to the battle field. There, the warriors will kill as many enemies as possible and defeat them. Even though many enemies are killed, nevertheless, some warriors from Shulluk side would also lose their lives. When returning home, Shulluk armies bring along the bodies of their dead ones. On arrival home, they send for a different sorcerer, he is given cattle and asked to perform another witchery.

It is generally believed that the second witch-doctor that warriors send for is the most powerful of all other sorcerers. This sorcerer then performs his witchery, and the armies of Shulluk would be commanded to go back to the battle ground. It is also believed that in the second fight, Shulluk armies will kill many of their enemies and defeat them without a single person been killed from their side (Shulluk side). After the fight warriors return home with full satisfaction. On arrival home, the warriors report to the King (Reth), and the King calls his royal ambassador (usually one of his chiefs) and sends him to the chief of the enemies with strong message. The message is, "The Shulluk people want enemies to compensate for the men they killed with twenty cows". When the chief receives such strong message, he usually acts quickly to avoid further confrontation. These twenty cows are then loosening from the cattle camp (*mura*) and given to the ambassador, who drive them home. When the warriors see that the twenty cows enter their village, they sit and relax waiting for another provocative war.

The Evil Eye: Meaning and Protection

In human history around the world, man looks for assistance of magic objects called "talismans" to protect him and members of his family from evil eyes. We cannot over emphasized that the case of evil eye came into existence thousands of years age and one can find it in almost every culture around the world. People around the world have different names for evil eye for example in Jewish culture it is called "hamsa" or "the hand of Mariam", in Arabic it is called "Ayin hasad" (eye of envy); in Hungarian it is called "szemmel veres" (beating with eyes); in Greek it is called "Matiasma" or "Mati" (cursing with eye); in French it is called "Mauvais Oeil"; in German it is called "Blick"; in Turkey it is called "Nazar"; and in English it is called "Evil Eye" (Evil Eye Store, 2016).

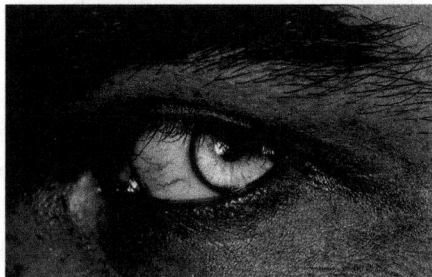

Figure: 68, Evil Eye

Therefore, when we are going to discuss about evil eye, I want readers to have broad mind that this evil eye is something practice world wild. It is stronger in the Middle East, East and West Africa, South Asia, Central Asia and Europe.

What does the term "Evil Eye" really means in our modern day? The term evil eye could be defined as a curse cast by a person (with evil eye) on the other person by looking or glaring at him or her with bad wish. Thus, many cultures, including the Luo believe that receiving the evil eye cause one misfortune or injury. The common illness caused by evil eye include but not limited to loss of appetite, excessive yawning, hiccups, vomiting, headache, stomach ache and fever.

It is believed that there are three types of evil eyes that exist in our modern world today. The first type is said to be unconscious eye –which means that the person does not actually intend to cause illness to you, but yet when he looks at you unintended you will get a mild illness. This is why it is called "Unconscious evil eye" and it is not dangerous. The second type of evil eye is caused when a person possessed by evil eye intend to harm you to certain extend but not severely. The second type of evil eye is also said not to be dangerous and cannot kill –although the person intended to cause you harm. Moreover, the third type of evil eye is the most scared one and can kill if the action of a man with evil eye is not averted. The third type of evil eye is our main focus in this book.

Some of the Shulluk people asserted that "*Jalyat*" and "*daiyat*" are associated with evil eyes. But I am not sure of this if it is true, there is need to make more research on this. It is generally believed that the power of evil eye is something hereditary (although not always necessary so) and it is the most common cause of illness among the Shulluk in particular and among the Luo groups in general.

The illness caused by evil eye (*youp*) is cured only by sympathetic magic, known locally as medicine man.

Thus, when a medicine man undertakes to cure a case of the evil eye, he aims to cure the person suffering, and to put that particular charmer out of business. In this occasion, offering is always presented or animal is sacrificed in order to remove illness imposed in the body of affected person.

Frequently, a sheep is brought and the medicine man heats a nail till it becomes red-hot and blind with the eyes of the sheep. By doing this, the medicine man wants to demonstrate that as he blinds the eyes of the sheep from this end, the eyes of the person, with evil eyes, are also blind from the other end regardless wherever he/she may live. The eyes of the person waste away. If the eye of the person, who cast the spell, do not become inflamed, the cure is said not to have taken place (Seligman, 1976). After all these performances, the medicine man will then tell the sick person to go home –if he is cure, and if he is not cure another performance is to be done till the eyes of the person who cast spell waste away.

The Age Set System

The age-grade organisation is a system of compulsory associations based on classifications of age and sex; outwardly, the organisation looks tribal but internally it is based on democracy. In Shulluk society, the age-sets are always recruited from within the local group. As we shall see later, the role of the age-grade organisation in community life is confined practically on specific military purposes.

Some authorities stated that the age-set-system in entire Shulluk society is now broken down beyond recognition. Where every man is a soldier of himself and his family members, they only unite against common and powerful enemies when it deems necessary. Henderson (1941) in Sudan Notes and Records Vol. 24 cited Holmahyer (1923) as saying that every man is a soldier and literally he is always under arms; and every Shilluk who has ceremonially danced is warrior as long as he is able to bear arms. Every village is independent with its own principle singer and president (*Bany*) but they only unite when there are common enemies coming to invade the country. Thus, different villages combine their efforts under one leadership of the eldest or senior singer and president of war (*Bany*). Customarily, the *Bany* is only advisory powers and he is chosen by the chiefs of the villages; and the unmarried youths form what we called permanent army of the country, and they are divided according to ages with various names (P.56).

However, this could not be a surprising if the age-set-system in Shulluk society is based on purely political ground, but the reality is that it is not, the organisation is rather based on social welfare to unite clans; some of the readers will agree with me that social and political functions cannot exist in a society side by side where authority is held by the virtue of blood. Furthermore, if by bad luck it exists, it is likely to raise ties and mutual obligations which often create conflict with the patterns of behaviour that aims to unite members of the clan. Therefore, in Shulluk society, social functions are more important than political functions. Nevertheless, the head chief (*Jago*) and elders of the original lineage (*Dyiel or Diel*) have limited political power in the villages and over them is the King (*Reth*) –although some people argue that the King has no political authority in the community.

In Shulluk society, the group solidarity of the village springs from the power and prerogative of the original lineage (*Dyiel or Diel clan*) and from common interests. The compulsory association of the age-grade organisation and the stratification of all male members in the village put more emphasis on the we-feeling of solidarity, but does not create any conflict with the political functions of the *Dyiel* clan, because its purpose is limited only to welfare of the community. Thus, when there is dispute between a lineage of one village and the other village, the age-grade organisation does not interfere, but when there is dispute between tribe and tribe, this is when the organisation works hard to unite all lineages to fight as one body, this is when the young people and warriors are mobilized into age-sets to face the common enemies in the battle field (Henderson, 1941).

According to Shulluk customs, every male who lives in a village, when he reaches the age of puberty, he is automatically grouped with his age-groups and together they are recruited to the new age-set. Each age-set, known in Shulluk language as *Ric*, is made up of series of sub-divisions, which are in turn grouped together temporarily until they are move into the general structure of the organisation. However, there are two grades that each age-set must pass through before they retire from active military services. The recruitments to these two grades are compulsory and are done collectively and not individually.

For us to understand the explanation of the two grades, I think it is better to categorize grades into "A", "B", and "C" (see below figure 69-70). Where grad "A" consists of three sub-divisions and grade "B" also consists of three sub-divisions. Whereas the three sub-divisions in grade "A" and grade "B" are made up of young men of approximately same age, and often they know themselves because they might have herded cattle together.

The members of age-set "A" are usually married men and probably have children although not necessary so. The age-set "A" is static and new sub-divisions do not join them until a new and formal division between age-set "A" and age-set "B" is made.

Meanwhile, age-set "B" is continually increasing in number as new sub-divisions are formed from age-set "C" which is the main supply chains. The number of individuals in each sub-division varies as well as the rapidity with which the new sub-divisions are formed, it is not always constant. As a result, after a certain period age-set "B" will become too many and very much larger than age-set "A".

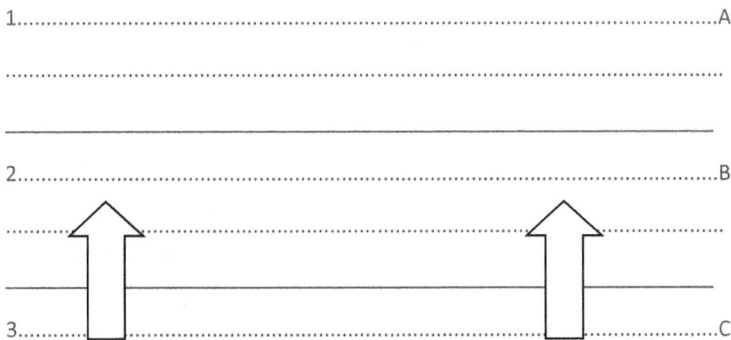

Figure: 69, Shulluk Age Set
Source: Sudan Notes and Records, 1941, P. 69

Moreover, the older sub-divisions of grade "A" will be getting too old for active military services. Many of whom, from the grade, might even be dead causing the organisation to reach unbalance and unwieldy stage; so to sustain age-set in grade "A", new divisions from age-set "B" must be formed and promoted to age-set "A".

Traditionally, it is the younger age-set who initiate the suggestions that they take over the duty of the old and dead men –in age-set "A". It is acceptable that a young person in grade "B" can nominate himself as a new war-leader (*Ban*) and based on this request a division would be made. Therefore, no division can be made without a war-leader.

When sub-divisions are formed from the age-set "B", ceremony then follows; and in this ceremony, all the young men promoted from age-set "B" to age-set "A" have to cut their faces, this is known as "*Ngol*" (cutting ceremony). When everybody has finished cutting his face, they would now be organised into two age-sets, while the members of sage-sets "A" pass out into retirement. When we talk of age-set "B" being promoted to age-set "A", we should not forget that the young men in grade "B" are actually been divided into two, so that half of them move up to age-set "A" and the other half remain within age-set "B" –which means they are approximately equal in numbers (See figure, 70 below)

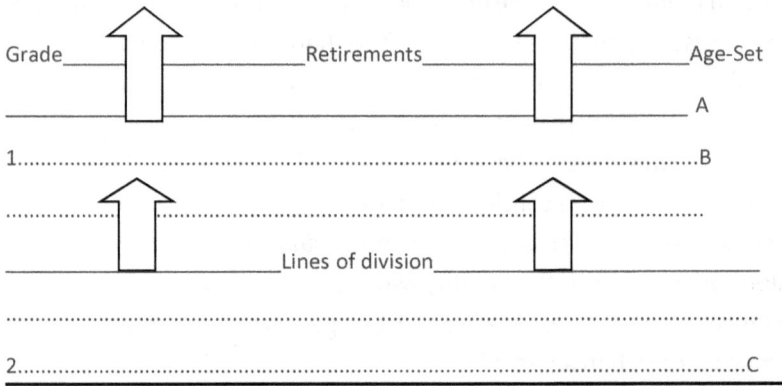

Figure: 70, *Shulluk Subdivision in Age Set*
Source: Sudan Notes and Records, 1941, P.59

The regrouping of sub-divisions age-set "C" is often the subject of heated controversy. This is because the young men left in age-set "C"; each one wants to be in the highest sub-division 1 of the newly formed lower age-set. The young men feel that when they are regrouped in sub-division 1, they acquire special privileges and duties. Thus, if this is not handled with care, the ceremony might come to blows. When the age-set "A" retired from their positions, they are no longer allowed to participate in the war nor are they allowed participating in the ceremonies associated with the organisation.

Henderson (1941) notes, "The sub-divisions of any age-set are known in Shilluk as "Head", "Neck" and "Middle" respectively. Thus, while the "head", "neck" and "middle" sub-divisions of age-set "B" occupy the same position in a new status, the two or three sub-divisions below them are "head", "neck" and "middle" in "C" age-set. On ceremonial occasions when the age-sets function as such, the highest age-set is privileged to eat the back and chest of any animal provided, while the lower sub-division will eat of the hump and legs only. This division of meat is rigidly observed and any infringement is severely punished. This rule is particularly enforced at the cutting ceremony (Ngol) when a bull is slaughtered and ceremonially divided and eaten —to emphasise the newly acquired privileges of the age-set in the higher status" (Henderson, 1941, P. 60)

After the promotion of sub-divisions from age-set "B" to age-set "A", the new groups have to appoint their war-leader —they are now connected as senior age-set. The newly elected war-leader would be raised and carried around the villages on the shoulders of his age-mates. In many cases this procession goes on until the group enter the territory of another village. When this happens, often the younger warriors fight with the young men of this neighbour village.

The war-leader has no administrative or political functions; he has no connection even with the *Jago* of the village. The main duty of war-leader is to organise and mobilise warriors in time of war and to chair the meeting of the age-sets in that grade. In certain cases, his age-mates go as far as cultivating fields of war-leader (*Ban*).

When the ceremony is over, each age-set is given a new name; these names are usually taken from some current events or from the colour of the bull slaughtered at the *Ngol* ceremony. It is a custom for the age-sets to use this name when entering battle fields –they always shout on top of their voices, each calling the name given to their age-set.

However, in Shulluk society, there are no elaborate and prolonged *rites de passages* through which the youths acquire manhood status and there are no separate ceremonies associated with their recruitments to a given age-set. Interestingly, for the young men to be recognised as a grown up ma (reach manhood) they have to perform dance locally known as "*Cong Bull*". The "Cong Bull" is purely a social dance which is arranged from time to time by permission of the *Jago* of the village; this dance has no religious act associated with it. The dance is considered as the final initiation of young men to manhood. It is not in itself a ceremony of initiation. For the young men, dancing publicly for the first time, they take it as a qualification rite by which they are now recognised as adults.

Before this first dance, a young man is subject to certain restrictions such as not to fight with spear prior to initiation, not to take part in large-scale warfare and more importantly not to take part in tribal raids; not to take part in the performance of dance in the bright moonlight; not to wear cloths and ornaments and so on. The young Shulluk men fear to break these rules, if one person breaches, he might not get a girl for marriage. Therefore, no youth would dare to disregard the warning of his elders. All these restrictions and many more are removed from young men at the time of the first "*Cong Bull*". At this time, the initiates are then allowed to wear the forbidden cloths and ornaments and they are allowed to carry spears (Henderson, 1941).

The initiation dance takes place about four O'clock in the afternoon, when the heat cools down and the population of the village have returned home from the fields. In the morning of initiation dance (*Cong*), each candidate is accompanied by his friends to traditional healer (*Ajuogo* or *ajwaka*) who performs certain rites on them before they continue on their way to the dancing ground.

Before the warriors arrive in the dancing ground, two drums called "*Bull Dwong*" (big drum) and "Bull thin" (small drum) are to be brought out and placed on a wooden structure in the centre of the village. The drummers furiously beat the drums, at various intervals, to call local inhabitants. Soon, the old women, who are too old to dance, assemble round the huts, gossiping and drinking local beer (*ombugi*) with the occupants of the huts. The important men of the village known as "*Jak*" and their assistants, and few of the most respected elders assemble near the hut of the *Jago*, who offers them local beer and some tobacco.

Normally three to four boys from the same village are initiated together. Their participations in this first "*Cong*" usually occurs when the boys are age between sixteen and seventeen, but sometimes some boys may undergo the rites at an earlier age (between fourteen and fifteen), this often happen when the fathers of the boys are dead and it may become necessary to celebrate initiation to gain status and to present the family. But for the young men who are between fourteen and fifteen years of age and their fathers are still alive, they might obtain permissions from their fathers to undergo initiation. Each candidate then acquires from his father the necessary ostrich-egg beads and cat-skin, and presented with two spears. The sons of the rich people are presented with bull (*wath owono*) as well. The sons of rich people usually take names from one of the bulls of their fathers because "*wath owono*" is a source of great pride to its owner and normally sons are composed about it, which are sung for the benefit of girls. This bull will be treated with great care and affection by the new and young warrior. Traditionally, this bull is kept throughout the life of a young warrior, but if the bull dies, new bull with similar colour will have to be bought to replace it. The "*wath owono*" is never given out as bride-wealth. The candidate takes all of these gifts to a man who is appointed to be his teacher or sponsor. Positions of sponsor require that a person must be so popular with girls. This teacher then ceremonially ties on the young man dancing ornaments before he actually enters the dancing ground and after dressing he (teacher) accompanies him to the dancing ground.

Meanwhile, the warriors will be arriving in the dancing ground in small groups; each candidate remains under the supervision of his sponsor until the actual dancing begins. The warriors at this time would be wearing clot known as "*lau*" and fasten skins on their waits; put on ornaments with special dancing beads. They will also be carrying with them their best spears but without the fighting shields, because the final initiation dance (*Cong*) has no military associations. However, some warriors would be carrying small shields of hippo or crocodile skins, which are made in such a way to represent the heavy war shields, but majority of the warriors will be carrying parrying-sticks or sticks-shields. Some of them are wearing strings of small beds round their legs (Henderson, 1941).

When warriors finally gathered in the centre of the village (which is a dancing ground), they get up and start to shout at their top voices. They then shamble and jumps around the drums, and before long they separate in small groups as other warriors outside the village prepare themselves to enter dancing ground. Thus, when every warrior has enter the dancing ground, they salute the drums and move quickly towards the Jago's house until in a wild rush they fling themselves as one body before Jago . They salute Jago, while girls who are sitting around the house of Jago admire their handsomeness, good behaviour and strengths. As a matter of fact, these girls are there to look and select from among the candidates their partners. When the new warrior finish saluting the head chief (*Jago*), they retire for further rearrangement of their dresses, and at this particular time, they put away their spears.

Thus, in the beginning of the dance, candidates join the ranks of warriors and continue showing respect to the head chief, before the girls are allowed to make their choices of husbands. At this time the candidates will be standing around the drums armed with sticks and small shields waiting for girls to approach them. Soon the girls appear in the dancing ground, they come two by two or three by three wandering round the circle of the waiting and love hungry men. They choose their partners by show of hands or a discreet nod of the heads.

The choice of the husband is entirely the prerogative of the girl and the man has no say in the matter. Therefore, if a man is chosen by a girl, he cannot refuse under any circumstance. It is a honour and a privilege that the new warrior is chosen by a girl; and once he is chosen they have to dance with the girl. This is considered as part of the ceremony, but if he refuses or ignores the girl (which is impossible) the man would be the subject of criticism (Henderson, 1941).

After this, the warriors with their girls dance in groups of eight or ten, the men dancing facing the girls, while the girls would be dancing backwards. In the actual dance, these groups advance anti-clockwise, hopping on both legs in time to rhythm. The head is held erect, the men with their arms flexed, holding in one hand a dancing stick, while the girls dance with their hands at their sides.

As the dance is going on, the audiences are entertained further by various individuals who greatly to their own enjoyment imitate animals or some individuals from the neighbouring tribes. The Shulluk are born mimers (Mimes) and every trick is possible for example they can amuse people by acting like foreigners arriving with wild bulls, in the country where tame bulls, accompanied by two herdsmen are and they lose the bulls to run around so that the crowd scattered.

As we have seen before, the final initiation dance (*cong*) starts at four O'clock and usually last for about two hours. Shortly, before sun set, they stop beating drums and the exhausted dancers leave dancing ground and in just a few minutes the whole village would be clear as all guests return to their houses. Dancers would be visiting their relatives on their way home. Henderson quoted some writers like Hofmayer (1923) as saying, "After the dance, it is the custom for the Shulluk initiates to lead a nomadic life for a time, moving across the country from one end to the other" (P.64).

Shatt (Thuri) Religion and Expressive Cultures

Although I have not been in touch with people from Shatt community here in Australia, to learn more about their religion and beliefs, I will attempt to explore their religious and expressive cultures through different scholars, who also had some difficulties getting in touch with the Shatt in South Sudan.

The Shatt people believe in Supreme Being they called "*Jwok*" to whom they make sacrifices once a disaster has fallen a homestead. They believed "*Jwok*" is the creator of human race and other things we see in the world today. "*Jwok*" is also regarded as ruler of the universe. The people offer sacrifices to God (*Jwok*) at the beginning of the cultivation season so that God gives them rain; they also offer sacrifices to him at the harvest of crops as a token of thanks-giving.

Like other Luo Groups, the Shatt believe that the spirits of their dead relatives stay with God in the sky and they act as inter-mediatory between the living and Supreme Being (God). However, although the Shatt believe that the spirits of their dead relatives stay with God in the sky, nevertheless, they also believe in good and bad spirits; some of spirits stay in the river, and hence, when a person is sick, patient is taken to the stream to be cleansed.

The Shatt have confessed Christianity as well as Muslims; unrealistic source tells us that majority of the Shatt are Muslims forms about 60 percent, Christians forms about 30 percent and the traditional religion forms about 10 percent of the population. I do not understand why the majority of the population of Shatt are Muslims; there were no elders to explain to me the reasons behind this. However, this is an area that required further research.

Although I have not come across scholastic reports and statistic to support my arguments, I assumed majority of the population of Shatt are Muslims, but traditional history tells us that they do not follow Islamic teachings daily. For example, Friday is marked for all the faithful Muslims to attend prayers at the Mosques, but the so called Shatt Muslims, many of them do not attend the Mosques, and the restriction of alcohol is often ignored. In addition, many of their traditional beliefs are retained and mixed with Christian and Muslims beliefs. It is important to notice that all the Shatt who practice Islam are Sunni Muslims.

The Shatt also believe in magic and charms. As such, each Shatt family or clan have own totem; for example one clan or family may own crocodile, another own hippo and other own certain snakes. They also believe in the power of witch-doctors and other spiritual leaders. Let us now examine witchcraft practices in Shatt society.

(a) *Witchcraft or Sorcerer*

People of Shatt believe in the power of the witchcraft. They believed that witchcraft has some knowledge of the properties of herbs that could be used for treatment for the sick or to defeat the magic of enemies. The witch-doctor usually tells his client that his illness is caused by the presence of foreign bodies such as root or stone charmed into him by a witch. To treat a person sick of these foreign objects, the witchdoctor removes the root or stone into his cheek, applying his lips over the seat of pain vigorously. Finally he produces the root or stone and persuades the patient that this having been removed, the cure is complete. At this point and time, patient is relieved by just seeing root or stone been removed from his body, because he believes this is the cause of his sufferings. –therefore, the treatment is beneficial to the patient.

However, the people believed that the witchcraft can injure a person or animal by just looking at them, this act makes him so unpopular in the community; not only this, he also finds it very expensive –if not difficult –to buy a wife. Furthermore, it would be difficult for his daughters and sons to find husbands and wives respectively.

(b) *Evil-Eye*

The Shatt people believed that there are people in the community with evil-eyes. They believed a man with evil-eye can dry up cow's milk or goat's milk; he can also make a fat healthy child lean; he can destroy the beauty of a girl or of a woman.

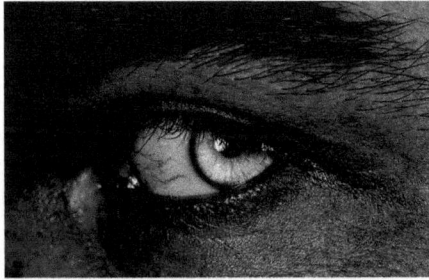

Figure: 68, Evil Eye

A man with evil-eye does these by charming roots or stones into the bodies of the victims or by casting spells over the path where the enemy is likely to pass. Therefore, a man with evil-eye in Shatt society is regarded as a criminal person and is often punished by death.

Chapter 13:
Language

. . . .

In general, we can say that Language is the ability to acquire and use complex systems of communication especially where the human has the ability to do so. Thinkers such as Rousseau have argued that language originated from emotions while others like Kant have held that it originated from rational and logical thought. However, the 20th-century philosophers such as Wittgenstein argued that philosophy is really the study of language. Languages evolve and diversify over time, and the history of their evolution can be revisit and to be compare with modern languages to determine which traits their ancestral languages must have had in order to come out with comprehensive conclusion how it is has developed and been preserved up to the present day. A group of languages that descend from a common ancestor is known as a language family. Thus, the Northern Luo language has descended from their common ancestor (Okwa or Kolo) as such; Luo language has now become a "language family".

Of all the cultures in a given society, appears to be the question of language for communication –it boils down to the mother-tongue. Thus, it is the culture of language that a person or people could be identified in a country or a society. For the Luo groups to have totality in their social environments, they must speak Luo language. Without mother-tongues, human beings would have no countries of societies. A language, over and above its value as a means of communication, is an integral part of the individuality of people, initially connected with every aspect of its social life.

The Luo languages in South Sudan are spoken by the nine Luo groups in the country. Thus, Northern Luo languages are spoken in the South Sudan and ranging from Eastern Equatoria to Bhar-El-Ghazal, and extended into Upper Nile region. The Northern Luo groups are from one of the branches of the Western Nilotic family. Although the northern Luo groups are branches from Western Nilotic family, we cannot emphasise that they have small variation in as long as the spoken language is concern. Unlike the Southern Luo, which has one single language, the Northern Luo languages varies in pronunciations from one group to the other, but the meanings of the words are the same. It is believed that before mass migration, the Northern Luo had one single language like the Southern Luo.

The Northern Luo of South Sudan include the following people: Acholi, Anyuak/Anywaa, Balanda Boor, Jur-Chol (Luo Wau), Jumjum, Mabaan, Pari/Lokoro, Shatt/Thuri and Shulluk. The level of historical separation between these groups, as I indicated elsewhere, is estimated some 500-800 years ago. The Luo groups of South Sudan are allocated in various regions of The Democratic Republic of South Sudan as we can see from the below map.

Locations of Luo Language in South Sudan

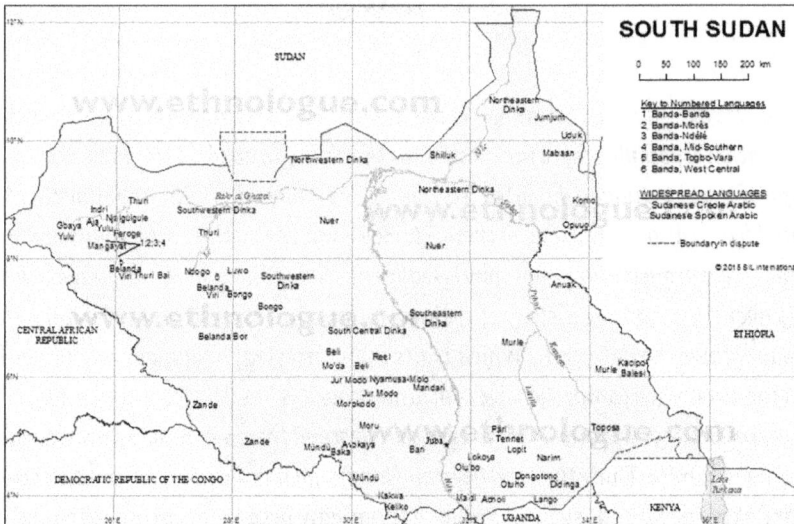

Figure: 72, Location of Luo Language

Luo language of the South Sudan remains unwritten for many years or centuries. When the Missionaries entered South Sudan, they became the pioneers in rendering Luo language into literate expression. However, the concern of the Missionaries at that time was not been the achievement of Luo Literacy as an end itself, but rather literacy as a facilitator for evangelization and the amelioration of the human condition.

It is unfortunate that the Luo alphabet were not developed equally by the Missionaries. As I already indicated elsewhere in this book that the Luo language undergone number of changes as people move further away from each other. As such, a vowel may carry a contour pitch; nevertheless as we can see the examples below, a monosyllabic word which has pitch (2) like *"neen"* in Pari and *"Nen"* in Acholi, this does not affect the relative values and meaning of the word across Luo groups and their dialects.

Let us now examine each of the Northern Luo Language one by one and see their similarities; please see below the examples of Northern Luo Words, starting from the Acholi through to Shulluk.

The Acholi Language

The Acholi mothers are ministers and directors of education of the Acholi language, because from day one, mothers are the initial teachers of children to speak mother tongue. Acholi is a language primarily spoken by the Acholi people in the seven villages in Magwi county, Eastern Equatoria, South Sudan, These villages include Agoro, magwi, Obbo, Omeo, Pajok, Palwar and Panyikwara. Acholi is one of the sixty four major languages in south Sudan in term of number of speakers.

It is a language that serves as a means of expressing the Self, as a dedium of art and self-actualization, and often a medium of writing. It is spoken at homes, towns and public places by the Acholi members who identify themselves to other Luo groups.

Acholi is also spoken by the Acholi people in the districts of Gulu, Kitgum and Pader, a region known as Acholiland in northern Uganda. Historically, the Acholi of Uganda moved from Bahr el Ghazal, in South Sudan between 1400 and 1800 A.D. They were in search of pastures for their domestic animals. Approximately 2.5 million people speak Acholi language, of these 1.17 million (2002, Uganda Census) live in Northern Uganda and 0.45 million (2000 SIL) live in South Sudan.

The Acholi language has 24 alphabets (see below); all vowels in a word have to belong to a single class. However, there are two sets of five vowels. Acholi language and pronunciation is more close to Anyuak especially in syntax and structure. Acholi is a tonal language, given the fact that tones are not normally written. Thus, two seemingly identical words can actually means different things depending on the tone of their vowels. For example word "be'l" with "low tone" means "wrinkle"; yet same word "be'l" with "high tone" means "corn". Likewise the word "ka'l" with "low tone" means a compound or a place, still the same word "ka'l" with "high tone" means "millet". Furthermore, word "ka'ny?" with "high tone and a question mark at the end means "where?"; yet same word "ka'ny" with "low tone" and without a question mark at the end means "here".

Therefore, we can conclude that if we want fluency in the Acholi language, tone is very important to reduce any ambiguity in the meaning of a given word.

Leb Acholi	English
Lacuc	"right" or "north"
Lacam	"left" or "south"
Remo	"blood"
Ogwang	"wild cat"
Pii	"water"
Ringo	"meat"
Poto	"field"

Leb Acholi	English
Rec	"fish"
ot or udi (pl)	"House" or "houses" pl.
jubi	"buffalo"
apwoyo matek	"Thank you very much"
but maber	"Good night"
icoo maber?	"Good morning"
irii maber?	"Good afternoon or Good evening"
atyie maber	"I am good or I am O.K."
bin ka cam	"Come and eat"
cawa adii?	"What is the time?"
kec tyie ka neka	"I am feeling hungry"
amito pii	"I want some water"
amito nino	"I want to sleep"
piny lyiet	"It is hot"
piny ngic	"It is cold"
kot tyie ka cwee	"It is raining"
wa citu	"Let us go".

Acholi alphabets and pronunciations

a	b	bw	c	d	e	g	i	j	k	l	m
[a/ɒ]	[b]	[bw]	[tʃ]	[d]	[e/ɛ]	[g]	[i/ɪ]	[ʤ]	[k]	[l]	[m]

n	ŋ/ng	ny	o	p	pw	r	t	u	w	y
[n]	[ŋ]	[ɲ]	[o/ɔ]	[p]	[pw]	[r]	[t]	[u/ʊ]	[w]	[j]

Source Omniglot (12/08/2016): Online writing systems and languages

The Anyuak Language

The Anyuak mothers are ministers and directors of education of the Anyuak language, because from day one, mothers are the initial teachers of children to speak mother tongue. This language sometime is known as Dho Anywaa or Dha Anywaa. Anyuak language is primarily spoken by the Anyuak members in two main counties: Akobo and Pachalla. Anyuak language is more similar to Acholi language in south Sudan and northern Uganda than to Shulluk and other Luo groups.

Anyuak language is one of the sixty four major languages in South Sudan, in term of number of speakers. It is a language that serves as a means of expressing the Self, as a medium of art and self-actualization and often as medium of writing. It is spoken in the homes, towns and public places by the Anyuak members who identify themselves to other Luo groups.

Approximately 144,710 people speak Anyuak language; of these 52,000 (1991 UBS) live in South Sudan and the remaining 92,710 people live in Gambella (Ethiopia). As I am writing this book, I believe the population of Anyuak in South Sudan might have grown 52,000 in 1991 to 62,000 in 2016.

The Anyuak language, like Acholi language, has 24 alphabets. More importantly, the Anyuak language is more close to Acholi language especially in syntax and structure. Since the Anyuak language is more similar to the Acholi, I am forced to share the alphabets between two of the groups as you can see below.

But what remains a mystery for me is whether Anyuak Language is a tonal language as we have seen in the Acholi. I am unable to confirm this because there was no woman or man from Anyuak community who could give me the example and to confirm that Anyual language is a tonal language.

Dho Anywaa	English
Re'mo	"Blood"
Gwang	"Wild cat".
Pii	"Water"
Ringo	"Meat"
Puodho (Single) Poudh (pl)	"Field" or Fields
Reeo (single) or re'c (pl)	"Fish"
Otho or Udi	"House" or "Houses
Jobi (single) or jobe (pl)	"Buffalo" or "Buffaloes"
Ina pwoch (single)	"I thank you very much"
Unu apwoch (pl)	"We thank you very much"
Ina mada kiwar	"I wish you good night"
Una math kiiwar	"We wish you good night"
Ina mad kamula (Single)	"Good morning"(single)
Una mada kamula (pl)	"Good morning" (Pl)
Ina mada kidiceng (single)	"Good afternoon" (single)
Una mada kamula (pl)	"Good afternoon" (pl)
Ina mada abwoya (single)	"Good evening" (Single)
Una mada abwoyo (pl)	"Good evening" (pl)

Dho Anywaa	English
Ani bong gin murac (single)	"I am good" or "I am OK." (single)
Wani bong gin murac (pl)	"We are doing well" or "We are OK" (pl)
Wubedo nidi(pl) and (single)	"How are you?" (pl), "How are you?"(sing)``
Oi nocemo (single)	"Come and eat"(single)
Ou nocemo (pl)	"Come and eat" (pl)
Caa adio?	"What is the time?"
Ani deri dekec or Dara dekec (single)	"I am feeling hungry" (single)
Unu deri dekec (pl)	"We are feeling hungry" (pl)
Amanya cam or nacama (single)	"I want to eat" (single)
Ou nocamo (pl)	"We want to eat" (pl)
Omal nocio nocamo (pl)	"Let us go to eat" (pl)
Ani ocoo nabuta (single)	"I want to sleep" (single)
Oni oia no buto (pl)	"We want to sleep" (pl)
Piny liet or Diceng piny liet	"It is hot"
Piny ngic or Diwa piny ngic	"It is cold"
Koth opotho	"It is raining"
Anai (single)	"I am going"
Onai (pl)	"Let us go".

Anyuak alphabets and pronunciations

a	b	bw	c	d	e	g	i	j	k	l	m
[a/ɒ]	[b]	[bw]	[ʧ]	[d]	[e/ɛ]	[g]	[i/ɪ]	[ʤ]	[k]	[l]	[m]

n	ŋ/ng	ny	o	p	pw	r	t	u	w	y
[n]	[ŋ]	[ɲ]	[o/ɔ]	[p]	[pw]	[r]	[t]	[u/ʊ]	[w]	[j]

Source Omniglot (12/08/2016): Online writing systems and languages

The Balanda Language

The Balanda Boor mothers are ministers and directors of education of Balanda Boor language, because from day one mothers are the initial teachers of children to speak mother tongue. Balanda Boor language is a language primarily spoken by the Balanda people in Western Bahr El Gahazal and Western Equatoria. In Western Bahr-el-Ghazal the language is spoken in two counties: Wau and Jur River. The language is spoken in the following districts or villages Raffili Tirga, Bazia, Ayo, Gitten and taban. While in Western Equatoria the language is spoken in Nagero County that covers the following villages or districts Komai, Nagero, Bangazegino and Tambura town.

Although the arrival of the Boor in Bahr-el-Ghazal and Western Equatoria of South Sudan is believed to have taken place approximately between 300 and 500 years ago, this area is perceived by the speakers as their traditional home-land. The Boor language is classified by Storch (2005) as Northern Luo language. It is smallest language of the Northern Luo Groups and is more similar to Shulluk, Anyuak, Jur-Chol, Thuri/Shatt and Pari than to Acholi.

It is said that balanda Boor language has been influenced by Ubangi language of Ubangian Bviri also known as Balanda Bviri in Western Equatoria. In my opinion many students of linguistic study of Balanda Boor are driven by two main things. Firstly, Boor language was nearly undocumented language and secondly they found it interested to examine the insights and the complexity of this language itself. Although gathering data consumes times, these students bear it because of the two things I just mentioned above.

The Balanda language, like the Colo language, has 29 alphabets (see below). Their language and pronunciation is more close to Shulluk, this is especially in syntax and structure. What remains a mystery to me is whether Balanda Boor language is tonal language or no. I could not confirm this because there was nobody from Balanda Boor community that could confirm.

Dho Balanda	English
Kaur c'am	"Left" or "North"
K'ur k'uc	"Right" or "South"
Re'm or re'mo (pl)	"Blood"
Gwa'ng or gwang' (pl)	"Wild cat (s)".
Fii or nyi-fii (pl)	"Water".
Ringo or ring (pl)	"Meat".
f'odo or ka-f'odo (pl)	"Field".
Reeyo or ka-reeo	"Fish".
Kot or ka-kot (pl)	"House".
Jobi or ka-jobi (pl)	"Buffalo".
	"Thank you very much"
	"Good night"
	"Good morning"
	"Good afternoon or Good evening"
	"I am good" or "I am O.K."
	"Come and eat"
	"What is the time?"
	"I am feeling hungry".
	"I want some water"
	"I want to sleep"
	"It is hot"
	"It is cold"
	"It is raining"
	"Let us go".

Balanda alphabets and pronunciation

The words in this dictionary are listed according to the order of the Belanda Bor letters. Belanda Bor has 37 letters, and they are in the following order.

Balanda Boor Alphabet			
A a	[a]	amuga	rhino
B b	[b]	bul	drum
C c	[c, s]	cɛlɔ	leg
D d	[ɖ]	dungo	basket
'D 'd	[d]	'dübor	lion
E e	[e]	ceŋ	sun
Ɛ ɛ	[ɛ]	lɛt	finger
F f	[f]	for	hippo
G g	[g]	guk	dog
Gb gb	[g͡b]	gbada	bed
Gw gw	[gʷ]	gwaŋ	cat
I i	[i]	winy	bird
Ï ï	[ɪ]	wïr	giraffe
J j	[ɟ]	jobi	buffalo
K k	[k]	kit	scorpion
Kp kp	[k͡p]	kpɛndu	anteater
Kw kw	[kʷ]	kwɛr	hoe
L l	[l]	lɛc	elephant
M m	[m]	mac	fire
Mb mb	[ᵐb]	mburmbur	butterfly
N n	[n]	nati	child
Nd nd	[ⁿd]	ndɔt	door
Ng ng	[ᵑg]	ngabu	jaw
Ngb ngb	[ᵑg͡b]	ngbom	okra
Nj nj	[ⁿɟ]	njɛngɛrɛ	twig
Ny ny	[ɲ]	nyaŋ	crocodile
Ŋ ŋ	[ŋ]	ŋu	leopard
Ŋw ŋw	[ŋʷ]	ŋwɛn	termite
O o	[o]	combo	snail
Ɔ ɔ	[ɔ]	jɔt	cloud
R r	[ɾ]	reyo	fish
T t	[t̪]	to	fox
'T 't	[t]	'tula	owl
U u	[u]	tuŋ	horn
Ü ü	[ʊ]	übit	hook
W w	[w]	wara	shoes
Y y	[j]	yey	canoe

Source: Belanda Bor Dictionary © 2014 Belanda Bor Language Committee and SIL - South Sudan

The Jo-Luo of Bhar El Ghazal Language

The Jur-Chol mothers are ministers and directors of education of the Jur-chol language, because from day one, mothers are the initial teachers of children to speak mother tongue. The language is locally known as either Dhe Luwo or Dhe Lwo. Jur-Chol language is primarily spoken by the Jur-Chol people who live in two counties Jur River and Wau in Western Bahr-el-Ghazal, South Sudan. The language is spoken in Warrap State that cover Gorgrial east, Gogrial west and Tonj North counties.

The Jur-Luo of Bhar El Ghazl language is one of the sixty four languages spoken in South Sudan in term of number of speakers. It is a language that serves as means of expressing the Self, as medium of art and self-actualization and often as medium of writing. It is spoken in the homes, towns and public places by the Jo-Luo members who identify themselves to other Luo groups.

Approximately 80,000 (1983 Sudan Census) people speak Jur-Chol language. However, as of today the population of Jo-Luo of Bhar El Ghazal might have grown from 80,000 in 1983 to approximately 100,000 in the year 2016.

The Jo-Luo language, like the Shulluk language, has 29 alphabets (see below). The studies revealed that Jo-Luo language and pronunciation is more close to Shulluk especially in syntax and structure.

What remains a mystery to me is whether Jur-Chol language is a tonal language as we have seen in Acholi. I could not justify this because there was no man or woman from the community that could explain to me. Let us now see some few words from Jo-Luo to give us the insight and similarity of the language to Luo groups of South Sudan.

Dhe Luwo	English
Tha'r p'iny'	"Buttocks of the earth" or "South"
rem or remo	"Blood"
dom –pago or gwang or gwange' (pl)	"Wild cat"
Pii	"Water"
ringo or ring	"Meat"
Puotho	"Field"
pajo	"House"
jobo or jobe (pl)	"Buffalo"
Khori Krray	"Thank you very much"
Uru kiyom rogu	"Good night"
Uru kiyom rogu?	"Good morning"
Wuriya	"Good afternoon or Good evening"
Riya yom or rokwa yom	"I am good or I am O.K."

Dhe Luwo	English
Beni cam or beni mathu ki pii	"Come and eat"
Caa adii? Or ceng niker	"What is the time?"
Ana keje	"I am feeling hungry"
Amdak ki pii	"I want some water"
-	"I want to sleep"
Ngo aledho	"It is hot"
Ngo nginya	"It is cold"
Ngo avii	"It is raining"
Hayo	"Let us go".

Jur-Chol alphabets and pronunciations

A a	Ä ä	Ɛ ɛ	E e	I i	Ï ï	Ɔ ɔ	Ö ö
[a]	[ɐ]	[ɛ]	[e]	[i]	[ɪ]	[ɔ]	[ʌ]
O o	U u	W w	Y y	B b	C c	D d	Dh dh
[o]	[u]	[w]	[j]	[b]	[ʧ]	[d]	[ð̞]
G g	J j	K k	L l	M m	N n	Ŋ ŋ	Nh nh
[g]	[ʤ]	[k]	[l]	[m]	[n]	[ŋ]	[n̪]
Ny ny	P p	R r	T t	Th th			
[ɲ]	[p]	[ɾ]	[t]	[t̪]			

Sources: Omniglot (12/08/2016): Online writing systems and languages

The Jumjum Language

The Jumjum mothers are ministers and directors of education of Jumjum language, because from day one, mothers are the initial teachers of children to speak mother tongue. The language is primarily spoken by the Jumjum who are sometime known as Berin and they live at Jebel Tunga, Terta and Wadega, in Upper Nile, South Sudan.

Jumjum language is one of the sixty four major languages spoken in South Sudan in term of number of speakers. It is a language that serves as means of expressing the Self, as a medium of art and self-actualization and often as medium of writing. It is spoken in homes, towns and public places by the Jumjum people who have mixed feeling about their identity—some of them identify themselves with other Luo groups, while others identify themselves with Dinka.

Approximately-------people speak Jumjum language. I am not sure whether Jumjum people are using their alphabets that might be similar to Shulluk or they are using alphabets which are similar to Dinka alphabets. This area needs further research. But for the sake of this book, and we have considered Jumjum people as part of the Northern Luo

Groups, we assume they use same alphabets like the Shulluk (see below). Like Shulluk, their language has 29 alphabets and is more close to Mabaan especially in syntax and structure.

Another thing which remains a mystery to us is whether Jumjum language is tonal language as we have seen in Acholi or not. There was no man or woman from Jumjum community to explain and confirm to me. Thus, this area also requires further research.

Dho Jumjum	English
rem or remo	"Blood"
gwang or gwange' (pl)	"Wild cat"
Piu/Pii	"Water"
ringo or ring (pl)	"Meat"
pwod	"Field"
re'jo or re'c	"Fish"
ot or (pl)	"House"??
jobi or ka-jobi (pl)	"Buffalo"???
	"Thank you very much"
	"Good night"
	"Good morning"
	"Good afternoon or Good evening"
	"I am good" or "I am O.K."
	"Come and eat"
	"What is the time?"
	"I am feeling hungry"
	"I want some water"
	"I want to sleep"
	"It is hot"
	"It is cold"
	"It is raining"
	"Let us go".

Jumjum alphabets and pronunciations

A a	Ä ä	Ɛ ɛ	E e	I i	Ï ï	Ɔ ɔ	Ö ö
[a]	[ɐ]	[ɛ]	[e]	[i]	[ɪ]	[ɔ]	[ʌ]
O o	U u	W w	Y y	B b	C c	D d	Dh dh
[o]	[u]	[w]	[j]	[b]	[ʧ]	[d]	[ð]
G g	J j	K k	L l	M m	N n	Ŋ ŋ	Nh nh
[g]	[ʤ]	[k]	[l]	[m]	[n]	[ŋ]	[n̪]
Ny ny	P p	R r	T t	Th th			
[ɲ]	[p]	[ɾ]	[t]	[t̪]			

The Mabaan Language

The Mabaan mothers are the ministers and directors of education of Mabaan language, because from day one, mothers are the initial teachers of children to speak mother tongue. The language is primarily spoken by the Mabaan people in Mabaan County in Upper Nile Region of South Sudan. The tribe called Mabaan is known by many different names. They are also sometimes called Barga, Gura, Meban, South Burun, Tonko, Tungan and Ulu, yet they speak one language called Mabaan. The language is one of the sixty-four major languages spoken in South Sudan in tern of number of speakers. It is a language that serves as means of expressing the Self, as a medium of art and self-actualization, and often a medium of writing. It is spoken in the homes, towns and public places by the Mabaan people who hardly identify themselves with other Luo groups of South Sudan. According to them, their language is a modified language and they prefer to identify it with Dinka rather with other Northern Luo Groups of South Sudan. In this respect, I wonder if the Mabaan people really know their origin; the question is, "If the Mabaan people know they are part of Northern Luo of South Sudan, why, then, should they identify themselves with the Dinka?

Mabaan language is also spoken in South Western Ethiopia. Approximately 50,000 to 80,000 (1987 SIL, 2006 Joshua Project) speak the language. The Mabaan language, like Shulluk language, has 29 alphabets (see below). Mabaan language and pronunciations are much further away from Luo group languages. As I have mentioned elsewhere in this book, Mabaan people were pure Luo people, they were even speaking Luo language but in course of time and marriages with neighbour Dinka, they became emerged or assimilated completely to Dinka. This might be one thing that could explain why the Mabaan people identify themselves more with the Dinka rather than other Luo groups.

Another thing which remains a mystery is that we do not know if Mabaan language is a tonal language as we have seen with other Luo groups or not. This is an area that needs further research.

Dho Mabaan	English
Rem or remo (pl)	"Blood"
Gwang or gwange' (pl)	"Wild cat (s)"
Fii/Pii	"Water"
Ringo or ring (pl)	"Meat"
P'wod	"Field"
Re'jo or re'c	"Fish"
Ot or (pl)	"House"???
Jobi or ka-jobi (pl)	"Buffalo "???
	"Thank you very much"
	"Good night"
	"Good morning"
	"Good afternoon or Good evening"
	"I am good or I am O.K."
	"Come and eat"
	"What is the time?"
	"I am feeling hungry"
	"I want some water"
	"I want to sleep"
	"It is hot"
	"It is cold"
	"It is raining"
	"Let us go"

Mabaan Alphabets

A a	Ä ä	Ɛ ɛ	E e	I i	Ï ï	Ɔ ɔ	Ö ö
[a]	[ɐ]	[ɛ]	[e]	[i]	[ɪ]	[ɔ]	[ʌ]

O o	U u	W w	Y y	B b	C c	D d	Dh dh
[o]	[u]	[w]	[j]	[b]	[ɟ]	[d]	[ð̧]

G g	J j	K k	L l	M m	N n	Ŋ ŋ	Nh nh
[g]	[ʤ]	[k]	[l]	[m]	[n]	[ŋ]	[ṋ]

Ny ny	P p	R r	T t	Th th			
[ɲ]	[p]	[ɾ]	[t]	[t̪]			

The Pari (Lokoro) Language

Pari, a Western Nilotic Language, has a terraced-level tone system with total downstep. Although Pari could be analysed as having three basic tone levels and automatic downstep, there is morphological evidence that it has two basic tone levels and non-automatic downstep.

Furthermore, there is evidence that downstep is the manifestation of a floating high tone. Floating tones thus behave differently from tones of deleted vowels. In spite of many surface differences between Pari and Luo, a related language, a single tone change accounts for their underlying differences (Torben Andersen, 1988, p. 261).

The mothers of Pari are ministers and directors of education in Pari Language commonly known as "Dhi Pari", because from day one, mothers are the initial teachers of children to speak mother tongue. The language is primarily spoken by Pari (or Lokoro people as some neighbours called them) who live around Lafon Hill in Torit District of Eastern Equatoria Region in South Sudan.

Pari Language is more similar to Anyuak; the two languages are mutually intelligible, they are similar in term of syntax and structure. In terms of linguistic affinity, next to Anyuak come Jur-Chol in Bahr-el-Ghazal and Shulluk in Upper Nile Region.

Dhi Pari is one of the sixty four major languages in South Sudan in term of number of speakers. In Kohler's (1955) generic classification, Pari language belongs to the Northern Luo sub-branch of Western Nilotic along with Shulluk among other Luo groups. The language serves as means of expressing the Self, as a medium of art and self-actualisation, and often as a medium of writing. It is spoken at homes, towns and public places by the Pari who identify themselves to other Northern Luo Groups.

Approximately 10,000 (1988 Torben Andersen) people speak Dhi Pari. However, as there has not been up-to-date census in the Republic of South Sudan, it is estimated that today (2016) the population of Pari has grown from 10,000 in 1988 to 15,000 in 2016. The Pari language, like the Acholi, has 24 alphabets (see below).

Pari is a tonal language with terraced-level ton systems, although the tones are not normally written. Thus, one word or a sentence could have one pitch or two pitches. Pari has a less tonal system as compare to other Luo groups of South Sudan. The tonal system is even less straightforward than for example the discrete-level systems of the neighbouring Moru and Madi languages, which belong to the central Sudanic language family (Torben Andersen, 1988, P. 262). Because of the complexity b and unusual character of the tone in Pari, we are not going to details the tonal systems. However, we shall briefly give example of the complex and unusual tonal systems as follows:

1. Pitch Level
The Pari language is known for its two tones or pitches, that is what we called high and low pitches. At this point, I would like to inform readers to note that the absolute pitch level of a vowel, with a given integer value, is perceptually and exactly has the same value within an utterance, especially when the word is whistled.

The monosyllabic word like *"nee'n"*, that means "see" , is pronounced with "low pitch;

yet the same word "nee'n" when pronounce with "high pitch" means "see there!" and yet it does not affect the relative value of the word "Nee'n".

Thus, the following are example of words with only "low pitch" in Pari Language:

-Pal'a means "knife"

-Gu'ok means "dog"

-Li'ec means "elephant"

Dhi'ang means "cow"

To'ng means "spear"

The following words are pronounced with both 'low" and "high" pitches.

-Wiin'jo' means "bird". This means you start the pronunciation of the word with "low pitch" and ends it with "high pitch"

-Dhaa'go' means "woman". This means you start the pronunciation of the word with "high pitch" and ends it with "low pitch".

-Ngee'do' means "rib". This word has low pitch throughout. This means you start the pronunciation of the word with "low pitch" and end it with "low pitch".

More importantly, there are words or sentences in Pari language that require you to start with "low pitch" and go to "high pitch" and come down to "low pitch".

Example: Nee'ni kan'jo (Neeni Kanyjo) means " Look at_____there!

"L" "H" "L"

"Neen yire" means "Look at that thing"

This means you start the pronunciation of the" first" word with "Low pitch" (L) and in the beginning of the "second" word you pronounce the word with "high pitch" (H) and in the last word, you pronounce word it with "low pitch" (L).

a-neend-a means "I am looking at___!"

"H" "H" "H"

This means that you start pronouncing a word with "high pitch", and continue the middle word with "high pitch" and the last word you pronounce with "high pitch"

Neen guok kanjo means "See a dog there!"

"L" "L" "H" "L"

This means you start pronouncing the first word with "low pitch", the second word also with "low pitch" but the beginning letters of the third word pronounce with "high pitch" and the last letter pronounce with "low pitch".

Neen dhaag-o kanjo means "Look at a woman there!"

"L" "H" "L" "H""L"

This means you start pronouncing the first word with "Low pitch" and the beginning of the middle word pronounce with "High pitch" and the last letter of middle word pronounce with "Low pitch" and start of the third word pronounce with "High pitch" and the last letters of third word pronounce with "Low pitch".

For further information about Pari tone systems, I recommend you read a research note entitled *"Downstep in Pari: The tone system of Western Nilotic Language by Torben Andersen, 1988, University of Aalborg"*.

Dhi Pari	English
Rimo	"blood"
gwang or *gwange'* (pl)	"wild cat"
Pii	"water"
Rino	"meat"
Pwodho or *pwoth* (pl)	"field"
Reo (single) or *ric* (pl)	"fish"
Hotto (single) or *Hudi* (pl)	"house"
Jobi (single) or *jobe* (pl)	"buffalo"
Mara na puch dwong	"Thank you very much"
Buttu mubeer	"Good night"
Unaa buttu?	"Good morning"
Unaa riyoo?	"Good afternoon" or "Good evening"
Abedo mubeer	"I am good" or "I am O.K."
Oyii chamii	"Come and eat"
Wang ceng naadio?	"What is the time?"
Anni anekechii	"I am feeling hungry"
Amitaa pii	"I want some water"
Amitaa nine	"I want to sleep"
Piny lieth	"It is hot"
Piny ngic	"It is cold"
Koth cwier	"It is raining"
Onaa choo	"Let us go"
Neeni!	"You see!"

CONTEMPORARY DHøG Pari ALPHABETICAL ORDERS

1	2	3	4	5	6	7	8
Aa	Ää	Ee(ε)	Ëë	Ii	Ïï	Øø (Ɔ)	Öö
9	**10**	**11**	**12**	**13**	**14**	**15**	**16**
Oo	Uu	Ww	Yy	Bb	Cc	Dd	Dh/dh
17	**18**	**19**	**20**	**21**	**22**	**23**	**24**
Gg	Jj	Kk	Ll	Mm	Nn	Ng/ng (ŋ)	Nh/nh
25	**26**	**27**	**28**	**29**			
Ny/ny	Pp	Rr	Tt	Th/th			

Aa Ba Ca Da Ee Fa Ga Ii Ja Ka La Ma Na Oo Pa Ra Ta Uu Wa Ya

The Colo Language (Dhocollo)

The Collo mothers are ministers and directors of education in Collo language, because from day one, mothers are the initial teachers of children to speak mother tongue. The Collo language is locally known as Dhog Colo. Thus, Dhog Colo is a language primarily spoken by the Collo in approximately 100 clans in four counties of Upper Nile region, South Sudan. The four Shulluk counties include Fashoda, Malakal, Manyo and Panyikang. The language is one of the sixty four major languages in South Sudan in term of number of speakers. It is a language that serves as a means of expressing the Self, as a medium of art and self-actualization and often as a medium of writing. It is spoken in the homes, towns and public places by the Collo members who identify themselves to other Luo groups.

Approximately 175,000 (1982 SIL) people peak Collo language but I expect the number to have grown from 175,000 in 11982 to 200,000 in 2016 (or even 1 million). Unfortunately there is no census conducted in South Sudan to confirm this.

Shulluk, like the other Luo tribes, speak directly or indirectly same dialect as others. Their differences are only in alphabets which have some miner crosses here and there. The Collo language has 29 alphabets (see below). The language is written with an orthography developed by Missionaries during the early 20th century.

I understand Collo language like other Luo language is a tonal language. The Collo language, like Pari language, is known for its two tones or pitches, that is what we called high and low pitches. At this point, I would like readers to note that the absolute pitch level of a vowel, with a given integer value, is perceptually and exactly has the same value within an utterance, especially when the word is whistled. Here below are the examples of pitches or tones:

Shulluk Alphabets and pronunciations

H L

Ngëëdø	Rib

L L L

Ngandhajø	Woman

L L H

Winyø, Winy pl	Bird/Birds

H L L H

Lidh winyø a cineel	Look at bird

H L L L H

Ya nena ri ngandhajø!	I am looking at Woman

H L L L

Lidh gwök (g) a cinee!	See a dog there

H L L H L H

Lidh ngandhajø a cinee!	See a woman there!
L H H L	
Ya dwada nenø	I want to sleep.
H H L	
Bï yï cäm	Come and eat.

OLD DHøG CøLLø ALPHABETICAL ORDERS

A a	Ä ä	Ɛ ɛ	E e	I i	Ï ï	Ɔ ɔ	Ö ö
[a]	[ɐ]	[ɛ]	[e]	[i]	[ɪ]	[ɔ]	[ʌ]
O o	U u	W w	Y y	B b	C c	D d	Dh dh
[o]	[u]	[w]	[j]	[b]	[ʧ]	[d]	[ð]
G g	J j	K k	L l	M m	N n	Ŋ ŋ	Nh nh
[g]	[ʤ]	[k]	[l]	[m]	[n]	[ŋ]	[ɲ̊]
Ny ny	P p	R r	T t	Th th			
[ɲ]	[p]	[ɾ]	[t]	[ṯ]			

Sources: Omniglot (12/08/2016): Online writing systems and languages

NEW DHøG CøLLø ALPHABETICAL ORDERS

UPPERCASE LOWERCASE EXAMPLE UPPERCASE LOWERCASE EXAMPLE

A a Agagø Ä ä Älø
B b Bulø C c Cängø
D d Dagø DH dh Dhøgø
E e Erø Ë ë Ëërø
G g Gwelø I i Idhø
Ï ï Ïdhdh J j Jaldwøngø
K k Këëlø L l Lebø
M m Malø N n Namø
NG ng Ngerø NH nh Nhuunhø
NY ny Nyäälø Ø ø Ødø
Ö ö Ölø O o Olidhø
P p Palø R r Rëëjø
T t Tulø TH th Thølø
U u Wunø W w Wangø
Y y Yadhø

CONTEMPORARY DHøG CøLLø ALPHABETICAL ORDERS

1	2	3	4	5	6	7	8
Aa	Ää	Ee(ɛ)	Ëë	Ii	Ïï	Øø (Ɔ)	Öö
9	10	11	12	13	14	15	16
Oo	Uu	Ww	Yy	Bb	Cc	Dd	Dh/dh
17	18	19	20	21	22	23	24
Gg	Jj	Kk	Ll	Mm	Nn	Ng/ng (ɪ)	Nh/nh
25	26	27	28	29			
Ny/ny	Pp	Rr	Tt	Th/th			

Dhog Cøllø	English
Lidh winyø a cinee!	< Look at bird there!
Ya nena ri ngandhajø!	< I am looking at woman!
Lidh gwök (g) a cinee!	< see a dog there!
Lidh ngandhajø a cinee!	< see a woman there!
Ya dwada nenø	< I want to sleep!
Bï yï cäm	< come and eat
Rëjø (singular) and rij is (pl)	< fish
Kal	< house > with many rooms
Ød	< room > a hut with one room
Ogïd(sglr & plr), Ogïïd(plr)	< buffalo
Pwøc me gir carø	< thank you very much
Ji aciew	< good night > not common in Cøllø
Ji ariic or ariij	< good morning
Ya døøc(j) or dëlla yöd	< good afternoon
	< "I am good or I am Ok.", depends on the
Bï yï cäm	question posed
Tha ba adi?	< come and eat
Ya da käj	< what is the time
Ya dwada pï	< I am feeling hungry
Ya dwada nenø	< I want some water
Piny lëdh	< I want to sleep
Piny ca libø	< it is hot
	< it is cold

Dhog Cøllø	English
Køth ca mägø	< it is raining
Wa keth	< let us go
Acol gweda wänyø	< Acol is writing a book
Acol agweda wänyø	< Acol has written a book
Acol ogwedø ki wänyø	< Acol will write a book
Ajag (k) ba jaldwøng	< Ajak is an elder or old man
Cwøl yan!	< call me!
Telepön ca ywøk (g), tïng mal.	< Telephone is ringing, pick it up.
Yømø wang rudø	< North wind
Yømø wang wurlwal	< South wind
Yømø wang cängmwøl	< East wind
Yømø wang cängbörø	< West wind
Remø / rem	< Blood
Gwäng	< wild cat
Pï	< water
Wøk	< field
Pwödhø	< farm
Rïngø (sglr), rïng(plr)	< meat and different types of meat (pl)

OLD DHØG CØLLØ NUMERICAL SYMBOLS

o Bugɔnɔ 0 Zero
I Akyɛlɔ 1 One
II Aryɛwɔ 2 Two
III Adägɔ 3 Three
IV Aŋwɛnɔ 4 Four
V Abïjɔ 5 Five
VI Abïkyɛlɔ 6 Six
VII Abïryɛwɔ 7 Seven
VIII Abïdägɔ 8 Eight
IV Abïŋwɛnɔ 9 Nine
X Pyärɔ 10 Ten
XI Pyärɔ wïj da akyɛlɔ 11 Eleven
XII Pyärɔ wïj da aryɛwɔ 12 Twelve
XIII Pyärɔ wïjɛ da Adägɔ 13 Thirteen
XIV Pyärɔ wïjɛ da aŋwɛnɔ 14 Fourteen
XV Pyärɔ wïjɛ da abïjɔ 15 Fifteen
XVI Pyärɔ wïj da abïkyɛlɔ 16 Sixteen

XVII Pyärɔ wïj da abïryɛwɔ 17 Seventeen

XVII Pyärɔ wïj da abïdägɔ 18 Eighteen

XIX Pyärɔ wïj da abïgwɛnɔ 19 Nineteen

XX Pyärri aryɛwɔ 20 Twenty

Γ Pyärri Abïjɔ 50

XØIV Pyärri abïŋwɛn wädhi gɛn da abïŋwɛnɔ 99 ninety nine

Ø Miyɔ 100

ØØ Miy Aryɛwɔ 200

ØØØ Miy Adägɔ 300

ØZ Miy Aŋwɛnɔ 400

Z Miy abïjɔ 500

S Alïbɔ 1,000

SS Alïb aryɛwɔ 2,000

SSXIV Alïb aryɛw ki pyär wädhi gɛn da akyɛlɔ 2,014

SSS Alïb adägɔ 3,000

SΛ Alïb aŋwɛnɔ 4,000

Λ Alïb abïjɔ 5,000

S¢ØSXØIV Alïb abïŋwɛn ki miy abïŋwɛn ki pyärri abïŋwɛn wädhi gɛn da abïŋwɛnɔ 9,999

¢ Alïb pyärɔ 10,000

ЖН Alïb pyärri abïŋwɛnɔ 90,000

Н Alïb miyɔ 100,000

SЮ Alïb miy abïŋwɛnɔ 900,000

SЮØS Alïb miy abïŋwɛn wädhi gɛn da miy abïŋwɛnɔ 900,900

The Thuri/Shatt Language

The Thuri mothers are the ministers and directors of education in Thuri language, because from day one, mothers are the initial teachers of children to speak mother tongue. The language is primarily spoken by the Thuri or Shatt people in two counties and these are Aweil central and Raga Counties. The Thuri people are also known as Shatt, Dhe Thuri, Jo Thuri and Wada Thuri. Whenever you hear of these names, do not think that they represent different people but rather they represent the same Thuri people.

In Aweil Central County, Thuri language is spoken in the following villages or districts: Bar-Mayen and Nyabulo; Lol river west of Marial Bai. While in Raga County the language is spoken only in one area that is the area covering east of Deim Zubeir on road from Raga Town to Wau.

The language is one of the sixty-four major languages in South Sudan in term of number of speakers. It is a language that serves as a means of expressing the Self, as a medium of art and self-actualization and often as a medium of writing. It is spoken in the homes, towns and public places by Thuri people who identify themselves to other Luo groups. The Thuri language is very close to the Shulluk and Jur-Chol languages especially in syntax and structure.

Approximately 8,000 (Joshua Project 5/6/2016) people speak Thuri language. Due to lack of census in the South Sudan for the last thirteen years, it is assumed that the population of the people might have grown from 8,000 to 10,000.

The Thuri language, like Shulluk language, has 29 alphabets (see below). What remains a mystery is whether Thuri is a tonal language. This is because when I was doing my research about the Northern Luo of South Sudan, I did not find men or women from Shatt Community that could have been consulted.

Dhe Thuri	English
W'an w'uudo'	"North" or "Western wind"
W'an ud'ulual	"South" or "Eastern wind"
Rem or remo (pl)	"Blood"
Gwang or gwange' (pl)	"Wild cat (s)"
Pii/Fii	"Water"
Ringo or ring (pl)	"Meat"
Pwo'do or pwoth (pl)	"Field"
Re'jo or re'c	"Fish"
Kot or ka-kot	"House"???
Jobi or ka-jobi (pl)	"Buffalo"
	"Thank you very much"
	"Good night"
	"Good morning"
	"Good afternoon or Good evening"
	"I am good" or "I am O.K."
	"Come and eat"
	"What is the time?"
	"I am feeling hungry"
	"I want some water"
	"It is hot"
	"It is cold"
	"It is raining"
	"Let us go"

Thuri Alphabets

A a	Ä ä	Ɛ ɛ	E e	I i	Ï ï	Ɔ ɔ	Ö ö
[a]	[ɐ]	[e]	[e]	[i]	[ɪ]	[ɔ]	[ʌ]

O o	U u	W w	Y y	B b	C c	D d	Dh dh
[o]	[u]	[w]	[j]	[b]	[ʧ]	[d]	[ð̪]

G g	J j	K k	L l	M m	N n	Ŋ ŋ	Nh nh
[g]	[ʤ]	[k]	[l]	[m]	[n]	[ŋ]	[n̪]

Ny ny	P p	R r	T t	Th th
[ɲ]	[p]	[ɾ]	[t]	[t̪]

Chapter 14:
Reconciliation: Means to Healing the Inner Wounds

. . . .

"If one engages in war as a fighter, he/she learns how to kill others, and become afflicted by the evil. The spirits of those he/she killed count him/her, family members and even the entire community.....he/she is not normal (ill) because during the war, many other things also happened to him/her such as seeing or coming into contact with the blood of others, carrying or touching dead bodies besides killing others....When such person comes back to the village, he/she can hardly fit into the normal life....those angry spirits will haunt him throughout his/her life....He/she has to be treated to become his/her own self again (Field Research Interview, December, 2009)"

Since the mid-1990s, there has been a proliferation of attempts to adapt and institutionalise forms of traditional justice as part of conflict policy. This has occurred in many parts of the world, which include Timor Leste, Sierra Leone, Rwanda, South Africa, Chile, Guatemala and Latin America (Tim Allen, 2013). Here the crucial question arises, "Will reconciliation and healing work among the Luo Groups of South Sudan and South Sudan Nationals?"

Traditional reconciliation is not something new among the Luo Groups of South Sudan; but it is a big question mark, if South Sudan National Commission would accept traditional reconciliation to be part of national reconciliation.

In my opinion, for the national reconciliation to succeed effectively, it is significantly important to combine traditional mechanism of reconciliation and justice with international mechanisms. The reason behind this logic is that in our world today, especially in developing countries like South Sudan, people are more incline towards local accountability mechanism and would appreciate and adopt the internal criminal law that is expanding from North to South. Needless to say, these mechanisms should be seen as being part of the same process in that they are all seeking forms of viable justice in its totality.

More importantly, the actual content of the traditional justice should be viewed through the lens of *"people-to-people peacemaking process"*. This is because the responsibility for peace rests with individuals and communities themselves, which means that the desire for the reconciliation and healing must come from within the affected people themselves. Once the affected people have decided and agreed to go ahead with reconciliation as a means of healing their inner wounds, the chiefs, community elders, international members and government officers would facilitate the dialogue (Agwanda & Harris, 2012).

South Sudan has been the scene of tribal and ethnic war, which is associated with power-struggle and aspiration of tribal domination of the country. Since 15th December 2013, conflict, war, organised violence and so on leave thousands and thousands of people dead, including the innocent people such as women, old people and children. Some people are uncounted for; others are physically and psychologically wounded. Each region of South Sudan has its share of violent conflict, but Upper Nile Region has a lion's share of this unnecessary two years conflict.

What remains a mystery to many of southern Sudanese people is that politicians talk that the root cause of the conflict has not been addressed, without pointed out *"what is the really root cause of this conflict"*. As a matter of fact, majority of South Sudanese, up to now, have not understood the root cause of a two years' war that broke out in the Capital Juba –the root cause of the conflict remains a hidden agenda.

On one hand many South Sudanese do not want to learn *"what to learn about the war"* for the fear of being killed. On the other hand the government of South Sudan prevents scholars from conducting research on the conflict because they will expose the weakness of the government and her soldiers.

In brief, President Salva Kiir dismissed his cabinets in July 2013 and on the 15th December 2013 conflict broke out in Capital Juba, South Sudan –soon the nation splits along ethnic lines. According to the African Union (AU) and Intergovernmental Authority on Development (IGAD), they characterised the situation in the Capital Juba as a "Crisis" but President of South Sudan, Salva Kiir viewed the situation as an attempted coup by The First Vice President of South Sudan, Riak Machar and his Nuer groups. This followed the arrest of Taban Deng, Oyayi Deng, John Luk Jok, Majak D'Agoot, Gier Chuang, Deng Alor, Pagan Amum, Peter Adwok, Afred Lado, Cirino Hiting, Kosti Manibe and Chol Tong.

I n the three days of killing in Capital Juba, over 20,000 people lost their lives, many of whom were Nuer. In the three days, gunfire and explosions continued to shake Juba City and residents were terrified and restricted to remain in-door under a curfew. It was reported that Tigard soldiers of President Salva Kiir went from door to door hunting and killing Nuer people. Only Nuer people who took refuge in UN Compound or in non-Nuer houses were saved. At night, bodies were collected and taken outside the city; the bodies were either burned or buried by some selected people possibly from Salva Kiir tribe.

The violence quickly spread like wild fire resulting in deep nation-wide political and security crisis. The following States were severely affected of the conflict: Central Equtoria, Jonglei, Lakes, Unity and Malakal.

Agreement was reached between the two warring parties on 15th August 2015; Riek Machar returned to Capital Juba, for the implementation of peace agreement, on ....April 2016. Riek Machar was sown in as First Vice President on ....2016. When Riak Machar on Sunday 22nd May 2016 went to Emmanuel Church, this is the Church for the Bor people only, and tried to talk about peace, reconciliation and healing, he was shut down by the faithful. Many people walked out of the Church in protest.

On 7th July 2016, when President Salva Kiir, First Vice President Riek Machar and Second Vice President Wani Igga were meeting in JR palace, war broke out in which more than 700 soldiers lost their lives. When the shooting subsided, Riek Machar and his body guards were escorted to his resident by Kiir's soldiers. First Vice President and armed opposition Leader, Riek Machar, fled South Sudan on 11th July 2016 and walked from Eastern Equatoria region to the Republic of Congo in 40 days and 40 nights. On Tuesday 23rd August, 2016, Sudan's Minister of Information Ahmed Belal Osman revealed that the former First Vice President of South Sudan and armed opposition Leader, Riek Machar has arrived Khartoum, the capital of Sudan, for treatment. The Minister did not say "how" and" when" Riek Machar arrived Khartoum.

Having discussed briefly about the back ground and development to the two years conflict, let us now return to our subject matter: *"Forgiveness, Reconciliation and Healing"* of the affected people in South Sudan. For the important of the subject, we shall attempt to explore the meanings of forgiveness, reconciliation and healing and how they have worked in general and how they should work in our situation. In addition, we shall spell out the ten top steps in reconciliation. After this, we shall discuss into depth the proposed South Sudan National Reconciliation and Traditional Reconciliation and Healing among the Northern Luo Groups. The ideas of discussing the Northern Luo traditional reconciliation and Justice procedures here, is for the Government of South Sudan to see the important of traditional reconciliation and justice and if possible they could use both International and traditional reconciliation mechanisms in peace building in the new broken nation.

What is forgiveness and why forgive others?

The word "forgiveness" means to wipe the slate clean, to pardon, and to cancel debt. When one did or does wrong thing to others in Juba, Bor, Malakal, Bentiu, Wau, to members of family or to the communities, he/she should seek forgiveness in order for the relationship to be restored. However, it is important to know that forgiveness is not granted because a person deserves to be forgiven. Instead, forgiveness is an act of love, mercy and grace.

Therefore, when we talk of forgiveness, it means that the way we towards our victim must change, we should think of not going back into a harmful situation (or think of doing revenge if we are the offended). It should rather mean to release the wrong doers from the wrong they have committed against us, or against our dead relatives or against members of family. Thus, we forgive them because God forgive us (Ephesians 4:31-32; Romans 5:8). Jesus Christ warned us that God will not forgive our sins if we do not forgive those who sin against us (Mathew 6:14-15: Mark 11:25: Roman 12:18). We forgive others to gain control of our lives from hurt emotions (Genesis 4:1-8). To us Christians, scripture reveals that forgiveness remains very important to God's heart, helps us combat darkness, and brings us blessing in ways we cannot imagine.

One reason some people resist forgiveness is that they do not really know and understand what forgiveness means and how it works. Many people claim to know forgiveness, but the reality is that *"they do not"*

The first step towards forgiveness is *to learn what forgiveness is and what it means*; secondly to give ourselves permission to forgive and to forget. We all know that it is easy to forgive but is difficult to forget—nevertheless, we should strive hard to forget and let it go. Instead of just requiring for forgiveness, it is equally important that reconciliation should create culture space where legitimacy is accorded for collective reactions and people are encouraged not only to forgive but they are also allowed to find other ways of dealing with their sorrow, anger and resentment id they are not willing to forgive.

In the correct situation, one should fix the issue and then forgive then because the offenders could be your country-man, members of family or members of community. However, forgiveness is always considered in the context of one's relationship with God without necessarily having contact with the offender. Meanwhile, reconciliation is more focussed on restoring broken relationship between a man and a man or between a man and God, to create an environment of tolerance and mutual co-existence. This is when trust is completely broken and the only means to repair and bring back the trust is through "reconciliation process" regardless how long it might take.

What is reconciliation and how it works?

Reconciliation means different things to different people, communities, institutions, and organisations. However, in this book, I want to define the word "reconciliation" to mean the restoration of peaceful or amicable relations between two individuals or groups of people, who were previously in conflict with one another. It is only through the process of reconciliation that the two individuals or group of people could be made friendly again after they have argued, fought, killed or sexually assault one another. Kriesberg (1998) defines the term "reconciliation" as *"The process of developing a mutual conciliatory accommodation between antagonistic or formerly antagonistic persons or groups. It often offers to a relatively amicable relationship, typically established after a rupture in the relationship involving one sided or mutual infliction of extreme injury"* And McGill (2015) defines "reconciliation" as *an ongoing process of establishing and maintaining respectful relationships".*

A critical part of this reconciliation involves repairing damaged truth by making apologies, providing individual and collective reparations, and following through with concrete actions that demonstrate real societal change. Establishing respectful relations also requires the revitalization of the traditional laws as we shall see later. It is important for Luo Groups in general and South Sudan Government with its population to understand how traditional justice and reconciliation work among the Dinka, Nuer, Murule, Madi, Acholi, Bari and Collo. Traditional elders and peacekeepers, many a time, dealt with conflict resolutions in their communities. Therefore, it is important to acknowledge that these traditional and practices are the foundation of native laws; they contain wisdom and practical guidance for moving towards reconciliations in South Sudan following the 21 months of tribal war in the New Nation.

The reconciliation is often depends on the attitude and actions of the offenders, the depth of the betrayal, and the patterns of offense. When an offended party works towards reconciliation, first and most important step is the confirmation of genuine repentance on the part of the offender (Luke 17:3). It is difficult to genuinely restore a broken relationship when the offender is unclear about his/her confession and repentance.

Anybody who killed or physically and emotionally disturbed somebody, he/she must be willing to recognize that reconciliation is the only way forward for returning back to friendship. Montville (1993) suggested that for the reconciliation to be successful, the parties must go through these three steps: There should be *acknowledgement* from and *contrition* from the perpetrators and the offended must be read for the *forgiveness.* These three steps are very important because they reassure the victims they will not suffer the same abuses in future and the relationship can move beyond a cycle of revenge and retaliation to something more positive. However, some writers like Charles Lerche (2014) argue that different people think differently about reconciliation. Some think that everyone who is confronted with conflict can achieved healing (peace and development) through only one form of reconciliation process; others think that reconciliation process has multiple forms.

For example Montville (1993) envisions reconciliation can take form of special designed "workshop", where participants from both sides feel secure and the trained third parties conduct various therapeutic exercises such as "walks through history". The trained third party would have undergone course exploring the richness of democratic governance and participatory decision-making. The trainees should be able to learn from indigenous knowledge systems how to strengthen community organisation and community peace building. Fisher (1999) supported the idea and said such workshops can work well among members, communities or groups in conflict who want to participate in intensive exchanges ideas and experiences in safe environments, provided that the workshop is directed by trained third party. He went on further to say those participants in Interactive Conflict Resolution (ICR) workshops often are opened to recognize and accept responsibility for their actions that caused harm to the other. In this way, participants can begin to see beyond their reciprocal feelings of victimization and more interestingly, they begin to experience the spirit of true reconciliation.

However, we cannot overemphasize that such workshop approach, described above, are more for the individuals thank for communities and ethnic groups. Nevertheless, if the individuals that participate in the workshop are selected by the people because of the influences they have in the community or society, their participation in ICR workshop will have positive impact on a broader 'communal peace process" between and among these same groups. According to Charles Lerche (2014) reconciliation must be proactive in seeking to create an encounter where people can focus on their relationships and share their perceptions, feelings and experiences with one another, with the goal to creating new perception and new shared ideas and experiences.

Every individuals or community always have the purpose of seeking reconciliation. Therefore, if the reconciliation has no purpose, then, it is useless and meaningless to talk about practicing reconciliation. Kelman (1996) made it clear that the primary purposes of IRC workshop are first to produce changes in the participants, and secondly to maximize the likelihood that the new insights, ideas and proposals developed in the course of the workshop are fed back into the political debate and the decision-making process within individuals, communities and societies. This means that the shared feelings and ideas must move from interpersonal to the collective and public domain —collectively and publicly sharing the resolutions reached thereto.

When people are in the process of reconciliation, they should avoid offering amnesty or impunity as a price for telling truth. If they do this, the logical progression of reconciliation would become muddled. Let us take one example: If a former torturer regrets for his act of torture and apologizes to the relatives of the victim or to a surviving victim, but yet he/she maintains that the act of wrong doing he/she has committed was necessary —although it was not entirely legitimate at the time; this makes the reconciliation process ineffective if not impossible.

In the two years of civil war in the south Sudan, the ethnic groups, mainly Dinka and Nuer, have developed "culture of violence" in which violence had been seen as the n normal and functional response to the meaningless conflict that broke out in Capital Juba on 15th December 2013. Thus, for the South Sudanese to reach peace building through reconciliation, all people involved in the criminal acts should honestly ask themselves, "What aspects of our societies and cultures got us where we are in today?" Curiously, a vital part of human experience —learning to live together- has been badly neglected throughout the world.

The seven signs of genuine repentance

There are seven signs that can clearly indicate that the offender is genuinely repentant. If two or three of these indicators are missing, it is advisable that this offender seeks the help of a wise counsellor, one who understands the difference between forgiveness and reconciliation. Such a counsellor can help the injured person establish boundaries and define steps toward reconciliation that are restorative rather than retaliatory. The seven signs include the following:

(1) Accept full responsibility for his or her actions. (Instead of saying since you think I have done something wrong...).

(2) Welcome accountability from others.

(3) Does not continue in the hurtful behaviour or anything associated with it.

(4) Does not have defensive attitude about being in the wrong.

(5) Does not dismiss or downplay the hurtful behaviour.

(6) Does not resent doubts about their sincerity or the need to demonstrate sincerity—especially in cases involving repeated offences.

(7) Makes restitution where necessary

Source: Steve Cornell, 2012, PP. 2-3

What does it mean to heal through reconciliation?

The word "healing" means to experience relief from emotional distress and give the grieving family the time for relief of inner wounds......it implies that the affected people think it is now time to revive and promote existing traditional methods of conflict resolution, conflict transformation and peace building. It is important to understand that the old and modern conventional means of settling conflicts through military means does not always achieve peace, let alone healing the inner wounds of conflict and violence. Reading through other book about healing, I am forced to suggest that *Gacaca Courts in Rwanda* and *Mato Oput among the Acholi in Uganda* are the best principles to follow. The guiding principles in these two healing processes are:

(a) The offender expectations include:

- ► There should be a voluntary confession
- ► They are going to disclose the truth by telling exactly what happened in the killing: removes suspicion among the communities, Societies and Nation.
- ► They must be ready to accept the responsibility and accountability.
- ► They must be ready to truly repent: establishes individual responsibility
- ► Where the cleaning rituals are required, they must accept to go through them.
- ► They must be ready to compensate the aggrieved party in any form or shape.

(b) The expectations of the offended or grieved party include:

- ► Coming to terms with the crime committed against him or her

- ▶ Be prepared to forgive (although he or she may not forget).
- ▶ Be ready to reach out to the offender to amend the broken relationship –this is
- ▶ what we call reconciliation.

(a) The roles of the mediating party include:

- ▶ Promoting dialogue between the offender and the offended
- ▶ Listen very carefully to the story of the offender
- ▶ Not taking side with any of the two parties –they must be neutral
- ▶ Give enough time for each person to tell his/her story
- ▶ The mediators must ensure that the reconciliation process is complete (which may involve certain rituals)
- ▶ Agree on appropriate compensation to be paid to the offended or grieved person –if necessary.
- ▶ To ensure that the person found guilty should be prosecuted or punished.
- ▶ Administer justice in accordance with community/national standards and often requires no money, no lawyers
- ▶ Integrate the guilty person into a society faster

Example Gacaca Courts in Rwanda:

The aftermath of the 1994 genocide in Rwanda offers an interesting comparison of modern and traditional techniques of achieving some element of justice for the 800,000 Tutsi and moderate Hutu killed during this wave of violence. Two processes have been created to deal with the many of Rwandans believed to have committed acts of genocide.

(1) In 1996 the United Nations established the International Criminal Tribunal for Rwanda in Arusha (Tanzania). In the same year, the formal court in Rwanda started prosecuting some 120,000 people jailed for suspicious of been involved in killing. By the year 2002 (six years later), the regular judicial system had tried less than 7,000 suspects. The Rwanda government then concluded that the case of reconciliation among such large number of suspects could only be handled by reviving the traditional court system known as *gacaca*. Rwanda villagers when solving conflicts between families, they sit on grass called *gacaca*, they operate on collective principles; the individual has no rights or duties other than within his or her group. In Rwanda, the word reconciliation is understood to mean "to mend ourselves". The idea is to make hold the offender accountable and require the person to make amends for the wrongs committed. The government created more than 12,000 *gacaca* courts and divided hearings into a pre-trial investigation phase and the actual trial. Rwandans elected approximately 225,000 persons of integrity, to act as judges of the *gacaca* courts (Linda James M. and David H. Shinn (2010).

Under the *gacaca* system prisoners are taken to their hometowns or villages, where local residents are invited to assemble, often in an open field, and give evidence against the accused. Local elected judges, few of whom have any formal education, receive six days of training in law, conflict resolution, and judicial ethics. These judges render the final verdict after hearing evidence from people in the local community. Villagers and judges are closely linked to the genocide, which tests the limits of fairness. However, some villagers are reluctant to offer evidence for fear of retribution. There have been a few cases in which witnesses were killed. The *gacaca* system looked at the cases of some 400,000 suspected perpetrators, which is better as compare to Internal Criminal Tribunal for Rwanda.

   The vast majority of those convicted of genocide were asked to do community service program, while those who are convicted of genocide when they were in prisons have to be reintegrated into the communities. AS of the end of 2009, more than 94,000 convicted suspects have worked in some form of community service program (Linda James M. and David H. Shinn, 2010).
(2) In the year 1994, what we called Rwanda's pre-1994 judicial system was established and it was comprised of 758 judges and 70 prosecutors, many of whom perished in the genocide. Others fled the country for fear of prosecution. By November 1994, there were only 244 judges, 12 prosecutors, and 137 support staff left in Rwanda. Subsequently, this system was clearly inadequate for the job.

The Ten Principles of Truth and Reconciliation

Each country or community may have their ten top principles of truth and reconciliation differently, created for different purposes; but in my opinion the top ten principles of truth and reconciliation should be as below. When we talk about conflict resolution, we should rather follow Unity Nations Declaration, which I am convinced that it provides the necessary principles, norms, and standards for reconciliation to flourish today in South Sudan. Although societies, communities and Nations have different ways of understanding reconciliation, yet, I believe in the case of South Sudan, the following principles of truth and reconciliation could be of help to various communities in the country to move forward:
(1) The United Nations Declaration on the Rights of the Natives is the framework for reconciliation at all levels and across different communities in South Sudan
(2) The indigenous people such as Collo, Dinka, Nuer, Murule, Bari, Acholi, to mention a few, they are the original of the places, and they have human rights that must be respected.
(3) Reconciliation is a process of healing relationships that requires public truth sharing, apology for the wrong done, and commemoration that acknowledge and redress past harms.

(4) Reconciliation requires constructive action on addressing the damages caused by the 21 months of civil war in South Sudan and the impacts it has on the people who have lost their fathers, mothers, sisters, brothers, relatives and properties.

(5) Reconciliation must create a more equitable and inclusive society by closing the gaps in social, mental and emotional damages that exist between the local people and the SPLA and SPLA-IO

(6) Both the SPLA and SPLA-IO share the responsibility for establishing and maintaining mutually respectful relationships.

(7) The traditional practices of reconciliation among the Dinka, Nuer, Collo, Bari, Murule, Acholi and Madi are vital to long-term reconciliation.

(8) Supported the offended people, and assist each tribe to move to their original land, all these are essential for reconciliation process.

(9) Reconciliation requires political will, joint leadership, trust building, accountability, and transparency, as well as a substantial investment of resources

(10) Reconciliation requires public education from the grass root to the towns and Juba City. Youth must be engaged; and general public must be told what exactly happened in Juba, Bor, Malakal, Nimule and Bentiu, in addition they must be educate of their Rights.

Together South Sudanese must do more than just talk about reconciliation and healing; each one involved must learn how to practice reconciliation in their everyday lives, one has to start with himself or herself, then with family members, and Community members, and in places of worship and government officials. For the relatives and survivors this is all about healing themselves, their communities and their nations.

Traditional Justice and Reconciliation

Due to lack of books and wise northern Luo elders, I will be discussing traditional justice and reconciliation of each of the seven Northern Luo Groups. However, we have enough information about Acholi and Shulluk traditional justice, for this reason, I will take only three examples (Acholi, Pari and Shulluk) as traditional justice among the Northern Luo. Thus, if God wishes, in the second edition of this book we may be able to cover traditional justice among all the seven Northern Luo Group. Let us first explore how traditional justice and reconciliation work among the Acholi tribe, and after that we shall explore how traditional justice and reconciliation work among the Pari and Shulluk societies.

According to Akena Ponse Otto (2016), prior to colonialism, crimes were handled in "open courts" held at different levels of social organisation –house hold, sub-clan, clan, inter-clan and inter-tribal, and in accordance to the nature of the conflict.

Acholi Traditional Justice and Reconciliation: Mato Oput Process

In order to understand how the Acholi traditional justice works in relation to conflict management, it is important to highlight the indigenous system of governance under which this tradition operates.

In pre-colonial and post-colonial times, the people of Acholi settle all the conflicts in the community through well-developed mechanism for the prompt resolution of conflicts as soon as they arose. More importantly, it is good to note that prior to the British rules in South Sudan, the Acholi maintain a traditional government and it is rooted strongly in their religious beliefs, norms and customs, which demand peace and stability in the community at all times, based on their philosophy of life. This structure was and is maintained by the anointed chiefs (*Rwodi*) assisted by influential elders in the communities. The chiefs, who head the Acholi local government, are believed to have been given ultimate authority by the people and their God.

In pre-colonial times, chiefs were enthroned and anointed with fat preserved from the carcasses of lions in solemn religious ceremonies. After these ceremonies they believed to be initiated into an esoteric (to the select few) relationship with the world of invisible God and spirits of ancestors. They are highly respected by their people and foreign visitors. These traditional chiefs have no executive power as such they cannot dictate. However, they governed strictly through the intercession of sorcerers known as "*ajwaki*"; and they are always under the guidance of the most powerful council of clan elders called "*lodito kaka*" or "*Lodito gang*".

The members who make up this council are democratically elected by the particular clan to sit on the Grand Council known as "*kacokke madit pa lotela*". In this gathering, the elders identify the problems those are urgent and need people to think together to find solutions to them. The primary aim of the Grand Council is to eliminate the vast and complex social issues of unhappiness in the society.

The Grand Council also acts as "supreme court" to try cases of killings and land disputes between different clans, by handling cases which are of both a criminal and a civil nature. They also make local laws and take decisions to be observed and implemented by members of community for their own good. They govern the society in accordance with Acholi traditional beliefs, norms and customs. These ensure that no person can commit a crime and go unpunished, although they do not have prison services in the society.

As the governing body, the council of Elders at all levels deal firmly with individuals and groups and ensure that everyone follow the Acholi traditional laws.

Although the British colonialists later on stripped traditional chiefs of their political power and replaced them with colonial administrators, they system of traditional justice and reconciliation continue to work up to the present day among the Acholi. The Acholi, like many other ethnic groups in South Sudan, greatly rely on traditional ways to solve disputes between and among people to uphold social harmony.

Since pre-colonial and post-colonial regimes, the recurring civil wars in the country, diminished status of cultural leaders, as people are constantly displaced within the country or in the neighbour countries.

Traditional justice and other rituals have been used throughout Africa to serve important elements of accountability during and after conflict. For example in Angola, where the traditional ceremony of *"Consuelo"* base on general encouragement given to people to abandon the thoughts and memories of war and losses was used as a healing process. Again in Angola and Mozambique, cleansing rituals attended by the family members and entire communities were carried out when welcoming ex-combatant child soldiers back into the community.

For example, in Mozambique, despite government's policy of "forgive and forget", survivors living in the former epicentres of the civil war in Gorongosa, inspired by their own cultural wisdom, developed their own socio-cultural mechanisms to create healing and attain justice and reconciliation in the aftermath of civil war. In Sierra Leone, healing ceremonies aimed to "cool the hearts" of former child combatants and encourage their reintegration in the community has been used. Traditional justice has also served a critical accountability role in Rwanda and Burundi.

In Northern Uganda, among the Acholi of Uganda, traditional rituals and justice ceremonies were performed way before the government and the international community took up the issue, even in the midst of the conflict. The rituals continue to be practiced today and will be practice at even larger scale when peace returns to northern Uganda, with or without the involvements of the government and international agencies.

All the above examples clearly show the important role that traditional justice plays in healing and reconciliation after conflict.

Therefore, traditional justice should not be overlooked and there is need to explore the potential of system as practiced by different communities in SPLA/SPLA-IO affected areas, to enhance access to justice, truth and reparations.

Traditional justice is more likely to play a positive role if it is applied at the individual and community levels and if its participation is voluntary and complementary to other accountability mechanisms. This inevitably requires an investment by the Government of South Sudan to take initiatives to enhance the capacity of traditional leaders but also preserve the autonomy of non-political character of the traditional institution.

The government will need to accord respect to these leaders and give them space to exercise visible influence within the communities, and to make traditional justice an alternative mechanism to condoning impunity for the grave crimes perpetrated in the SPLA/SPLA- IO conflicts.

The Rite of Reconciliation Called Mato Oput

The South Sudan government must agree in principle that the application of traditional rite, among others, should be one of the appropriate mechanisms to address the issue of accountability and reconciliation. The Acholi people believed firmly that man is a sacred being whose blood ought not to be spilled without just course. Killing of human beings is strictly forbidden by their traditional norms and customs. But if it took place, negotiations for blood money would have to be led by the victim's family, with agreement followed by rituals of reconciliation and ceremony to restore the killer to the community, and to bring peace between clans. In other words, we could say in traditional Acholi culture, justice is done to restore harmonious life.

Within Acholi society, if one person kills a person from the same or different clan, the killing will provoke the anger of God and ancestral spirits of the victim. It is believed up to the present day that the angered spirits of the ancestors will invite evil spirits to invade homesteads and harm the inhabitants. Moreover, such killings automatically create a supernatural barrier between the clan of the killer and the clan of the person who is killed (International IDEA, 2008).

Thus, as soon as the killing happens, the members of the two parties immediately stop eating and drinking from the same bowl or vessel (calabash), and engaging in social interaction of any form. Thus, supernatural barrier remains in force until the killing is atoned for and a religious rite of reconciliation has been performed to clean the taint. In the meantime the killer is treated by community as an outcast or unclean person. At this stage, a killer is prohibited from entering any homestead other than his own for fear that he is a companion of the evil spirits and will pollute the soil of the homestead with evil spirits. To clean the evil spirits from the body of the offender and to restore good relationship between the offender and offended families, they have to go through "*Mato Oput* Process", which consists of seven steps, which include personal testimony, putting the offender in confinement, beating a stick by offending clan, truth telling, payment of compensation, sacrificing animals, eating smell full green vegetable called "*boo mukwok*", drinking *oput*, and sharing cooked food from both sides of parries.

Mato Oput Process

Mato Oput literally means "drinking *oput root*". *Oput* is a tree found in Acholiland; its bitter root is ground and used to prepare a drink that is shared at the peak of the ceremony. *Mato oput* aims to re-establish relationship that break down between two individuals or clans, as a response to the killing done deliberately or accidentally. According to Latigo (2008) *Mato Oput* process might not begin until as many as ten or twenty years following the killing, after misfortunes have befallen the offender's clan and social pressure has motivated the perpetrator or the perpetrator's family to seek reconciliation (P. 362).

It is, however, important to stress that even for the traditional Acholi justice and reconciliation, *mato oput* does not cover all crimes, especially those committed with impunity. In other words, traditional *mato oput* ritual is never conceptualised as a process to address systematic cases of murder such as rape, kidnappings, mutilation, pillage, destruction of properties, among others, as the crimes committed by the SPLA/SPLA-IO in the conflict and it will be a great challenge for the mechanism to serve that purpose. But rather mato Oput can be used for crimes occurring between individuals, clans of the same regions that have friendly relationships with each other. As mentioned above, the "mato oput process" has seven steps to be accomplished for the totality of the "Forgiveness, Reconciliation and Healing".

(1) *Personal Testimony*

According to Rt Rev Ochola (2009) the process of *mato oput* only starts, when a perpetrator at the gate of his/her village or clan, is not allowed to enter into the village with the blood in his/her hands. He/she stands outside the gate and testifies himself or herself to the people of the village. The offender is to introduce himself by saying his/her name, mother's name, father's name and uncle's name. In Luo communities, mother, father and uncle are keys in the life of every child. The personal testimony includes confession of motive for the homicide committed. The offender must also give the name, gender and clan of the victim and tells why he/she has decided to kill that person. Cecily Rose (2008) pointed out that the affected families create an account of the facts which emphasizes the perpetrator's voluntary confession, including: the motives, the circumstances of the crime, an expression of remorse. She went on to say that in Acholi culture, until perpetrator confesses and seeks rectification the spirit of the dead may plague the perpetrator's family members through nightmares, sickness and death (P.362).

However, Latigo (2008) argues that central to mato *oput* process is that the perpetrator must be able to identify the victim to be able to initiate the process of reconciliation with the victim family and clans. However, to some perpetrators it might be difficult to identify any victim for his family to initiate the process of *mato oput* (P.210).

Once the offender's community or clan are convinced by the evidence in the personal testimony at the gate, the Elders of the village immediately take full collective responsibility on the behalf of the offender and on the behalf of offender's family. However, the Elders will hold offender responsible and accountable to bringing shame and disgrace to members of the family and entire clan (RT Rev. Ochola, 2009).

After the offender has testifies himself or herself and the Elder assume to take responsibility on the offender's behalf, ritual of stepping on the egg (*nyono tong-gweno*) is performed at the gate. According to Acholi norms and customs, stepping on the egg symbolises three things: Firstly it is a symbol that the offender has been accepted into the community despite the fact that the offender has committed serious crimes. Secondly, it is a symbol that the offender has accepted to take step forward for purity and sanctity of the human life he/she has destroyed. Thirdly, stepping on egg is an indication that the village accepted the offender's act of destruction of human life as an act of murder.

After the confession, elders move to begin the process of *mato oput*. Without confession the mechanism is of no use. Sometimes family may implores a suspected perpetrator who shows signs of being possessed by "abomination (*cen*) to confess but the final decision to commence lie with the offender. This is related to the practice used in central Mozambique where "Magamba spirits" are believed to possess the living causing misfortune to the individuals until they are followed to air out their grievances (truth) and are appeased by the perpetrator, for the victim to find rest. The offender must be willing to visit the elders and ask for assistance and members of the offended family must also be willing for the ritual to proceed, without which the ritual will not precede.

(2) *Offender is kept in confinement and isolation*

Soon after stepping on egg, the offender is allowed to enter the village but he/she is not allowed to go back to his/her family house; the offender is kept in confinement and isolation within the fence of the village for certain period of time. The offender is supposed to use this period as a time of reflection and re-examining of him-self or her-self.

While in confinement, offender is not allowed to communicate or interact with anybody in the community. However, the offender is provided with the basic necessities of life, such as drinking water, food and fire. Traditional law states that only an innocent young girl, who is between the age of eight and twelve, is allowed to deliver drinking water, food and fire to the offender –without any contact whatsoever with the offender.

The small girl takes food and water in separate containers and puts in other containers used by the offender at specific point and time. On the other side, the offender is expected to put his or her containers at the appointed point and time before food and water is brought. The offender is also expected to collect food and water from the same place when the girl is long gone. The whole ritual here is to emphasize the seriousness of murder.

According to Cecily Rose (2008) the separation of the offender and the affected clan serves as a "cooling off period" to prevent immediate revenge killing. This separation requires the complete suspension of relatives of both parties, during this time, the two clans are forbidden to intermarry, trade, socialize or share food and drink. Such separation is significant because of the communal nature of Acholi culture, where in families from various clans share food, water, land and social-relations (P.362).

(3) *Offender's clan beats a stick*

As people gathered a representative from the clan of the offender beats a stick and run away from the Assembly. Cecily Rose (2008) reveals that beating of a stick by the offender's clan constitute part of *mato oput* process. A representative from offender's clan beats a stick and runs away. Thus, beating of a stick in *mato oput* process symbolizes restorative purpose of the process, and running away signifies the acceptance of guilt as a murder from the side of the offender.

(4) Truth Telling

When both parties sit down to listen from the offender what exactly happened, the offender is expected to be sincere, straightforward and transparent. In Acholi traditional and cultural justice system, there are no denials, lies or deceptions. Thus, anyone who accepts to go through the process of truth-telling he/she must tell the whole truth and only whole truth without fear or favour. In other words, the offender must be open, sincere, transparent and truthful, not only to him-self or her-self but to the entire communities even to unborn members of the communities.

The offender voluntarily takes the lead in the process of confrontation of telling the whole truth in the spirit of love, mutual respect, and mutual understanding. For this reason, offender's clan have to publicly acknowledge and accept ownership of responsibility on the behalf of the offender and his/her family. Akena Ponse Otto (2016) observes that, during the hearing, elders listen to both parties, who are involved, before they decide what compensation is appropriate for the offence. Interestingly, the Acholi traditional justice is usually distinguished from formal justice in term of the voluntary and willingness of the perpetrator to confess. Part of the logic of Acholi cosmology is to illicit fear and shame if one break a social norm and to encourage people to take the appropriate steps towards restoration.

The whole truth helps both parties to own a common memory of their recent past history. The public acknowledgement is meant specifically to deal comprehensively with malicious intent of murder. The malicious intent to kill is called in Acholi language "*aneko nyong*".

Through public confession, the offender's clan becomes vulnerable and guilty for the crimes committed by one of its members. This is the fundamental basis for community-based collective responsibility of the clan of the offender.

RT Rev. Ochola (2009) argues, "The whole revealed truth functions like the naked electric wire, which is dangerous and deadly to both communities involved in the violent conflict. It can only be handled through the power of mercy and forgiveness by the offended community. This is the beauty of reconciliation of the heart of offended community" (P.28).

(5) Payment of Compensation: Blood Money

In all cases of deliberate or accidental killing, the clan of the offender is required to pay "blood money" to the offended clan. Traditional processes of paying compensation in Acholi are presided over by "Council of Elders", which include chiefs (*Rwodi*), Elders and clan leaders. Cecily Rose (2008) pointed out that in Acholi customs, when offender declares that he or she has committed a "wrong", the traditional conflict management system is triggered. However, the mato oput process takes place in the context of an intentional or accidental killing. It is a long and sophisticated process that may last for weeks, months, or even years; this involves separations of affected clans, mediation and payment of compensation.

Once the killer confesses his guilt, the entire community of the offender finds itself in state of vulnerability and guilt. The clan must become willing and ready to pay compensation without any question. In the past, a young girl from the offender's clan would be given as compensation for deliberate committed murder as a form of compensation; today, this has been replaced by a system where cattle (10-15 cattle) or money ($10,000 to 30,000) may be used as forms of compensation. Latigo (2008) said that cows, goats and sheep are central to the success of most traditional justice process. This will no doubt create immense financial constraint on the poverty-stricken communities that have been relying on humanitarian aid for survival for years. Therefore, this implies that where possible the government and donors should support such processes by providing items needed and putting aside funds for compensation to benefit persons who will undergo traditional justice process.

The payment of compensation is not meant, in any way, to replace the life loss because human value is invaluable. The compensation is just to demonstrate or show the sincerity and the purity of the heart as graphically as expressed in the public confession. The focus here is to address the conflict in order to bring the healing of the wounds in the heart of the deeply affected by the offense. Compensations after truth telling are mechanisms of conflict transformation (RT Rev. Ochola, 2009). Payment of compensations are made only after establishing the truth.

However, to many people the payment of compensation by the government or donors on behalf of perpetrators that undergo the process will inevitably cast doubt on the legitimacy of process. Some people will think that perpetrators are been rewarded.

To many of the Acholi people, struggling to provide item needed for the ceremony and payment of compensation is seen as a sign of atonement and the punishment for the wrong doing. They also think that traditional leaders should encourage, and engage community members to come up with novel ways of payments of compensation and atoning for crimes that does not require immediate exchange of money or property. Therefore, one of these ways could be requiring the perpetrator's clan to farm or herd animals for the offender's clan for a certain number of years.

In the process of blood money payment, the council of Elders appoint a leading man from a different clan to mediate between the two affected clans. The mediator, who is expected to be completely impartial, coordinates the arrangements for the payment of the blood money. It is strictly unacceptable to pay for the killing by killing the offender or his or her member of the family, since it is believed that this will only mean a loss of manpower to the society, without befitting either side. The payment of blood money is preferable since the money paid to the bereaved family can be used to marry another woman who, in turn, will produce children to replace the dead person –form of reparation.

After the payment of many (reparation), the Elders arrange for the customary rite of reconciliation to take place in order to bring the estranged clans together to resume their normal working relationship. The reconciliation ceremony always takes place in an uncultivated field which is usually somewhere between the villages or communal settlements of the two parties or clans, away from any foot path or any place frequently visited by women and children.

(6) *Animals are sacrificed and meat exchanged*

To provide the reconciliation ceremony, the killer provides a ram and a bull called *"dyiang me dog bur"* (cow for blessing the hole for burial), while the next of kin of the offended provides a goat. But according to some writers, like Latigo (2008) each party provides sheep which will be cut into half and exchanged by the parties' members. However, the common characteristic of the ceremony include slaughtering animals, which could be two sheep, or a sheep and a goat, or a sheep, goat and a bull. Unused new vessels are required and a large quantity of beer is brewed for the occasion.

On the appointed day, conciliators and elders from both sides assemble at the chosen site and stand facing westwards in solemn since. An invocation is then performed my two masters of ceremonies one from the side of the offender and the other from the side of the offended.

Figure: 73,
The two animals are cut into half

Master of ceremony, from the side of the offender, comes first, and prays to the spirits of ancestors saying, *"You, our ancestors and the children of the Supreme Being! I now plead with you and ask you to realize that sin is part of man's life. It was started by those who ever lived before us. This man, whose fault brought us here today, has merely repeated the perennial SIN which man has inherited from our forefathers. He killed his own brother. But since then, he has repented of his deed. He has paid blood money which may be used to marry a woman who will produce children who, in turn, will keep the name of his killed brother for our posterity. We now beseech you our ancestors to let the two families resume a brotherly relationship"*

All the assemble elders join Master of ceremony at the end of his petitions to the spirits of ancestors with one voice as they say,

"Let a man who will be given the blood money to marry a wife be sharp and pick on vicarious

woman....a virgin woman who will produce many and healthy children to grow well and to take over the empty home".

After this, Master of ceremony, from the side of the offended clan, gets up and responses to the solemn invocation sited by his mate, saying,

"We are not the first clan t suffer premature death of this kind. Offender has repented his misdeed. He has paid for it. We now supplicate you our ancestors to bless the blood money given to the family to marry a wife to produce a replacement to our brother"

All the assemble elders join Master of ceremony at the end of his petitions to the spirits of ancestors with one voice as they say,

"Let us accept the blood money and wash our hearts clean, and begin to live and work together as we have been doing in the past....Our enemies who have heard of this reconciliation are not happy that it will now bring peace and prosperity to out two clans....Let their ill will be carried away by the sun to the west, and sink with it down, deep and deep down".

Source: (International IDEA, 2008, PP. 104-105)

(7) *The two parties eat spoiled and smell-full green cooked vegetable (boo mukwok) together*

After the two Masters of ceremonies (from the offender and offended sides) finish their prayers or petitions to the spirits of ancestors, the next step is for the two clans to eat spoiled and smell full green cooked vegetable (boo mukwok) together. It is assumed that this vegetable might have been cooked and kept in a container outside the fridge for weeks or months so that germs containment it, that produces bad smell. The eating of smell full green cooked vegetable signifies that tension between the clans persisted long enough for food to spoil, it also symbolizes that clans readiness to reconcile after this long period of time (Cecily Rose, 2008). Some readers may argue that eating this smell full green cooked vegetable may cause problem in the stomach and make people sick. But the reality is that since from the beginning of this ritual in Acholi, no report have been received that somebody has fallen sick or died from eating smell full green vegetable during such ritual ceremony. Therefore, this indicates that God hears the prayers of the Acholi and protects them from falling sick or dying after eating such food.

(8) *Representative from both side drink oput*

After the petitions and drinking of smell full cooked vegetable, the two Masters of ceremonies then mix the pound extract from the roots of the *oput* tree with alcoholic drink in a new vessel; and then the killer and the next of kin of the offended kneel down and begin to drink *oput* (bitter root) from the same vessel or calabash simultaneously, while women from both sides of the clans make shrill cries and shout the war cries. The root of *oput* tree represents the bitterness between the clans, and drinking it symbolizes washing away the bitterness between them.

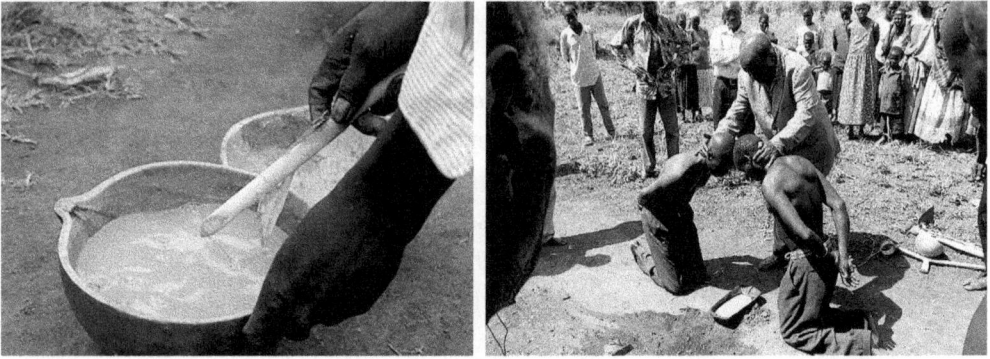

Figure: 74, Mixing pound extract from oput tree Figure: 75, Drinking Oput in Acholi Justice
Source: Trip Down Memory Lane, 2013

Mato oput is predicated on full acceptance of one's responsibility for the crime that has been committed of taboo. In its practice, healing of inner wounds become possible, but only through their voluntary admission of wrong doing, and their acceptance of responsibility, and the seeking forgiveness which then opens way for healing.

Tolerance and forgiveness is shrine in the principles of *mato oput* and other associated ritual. The rite embraces collective guilt as well as individual guilt. It is a process where the parties to conflict resolve to deal with the consequences of the conflict and its implications for the future in a collective, mutual and democratically acceptable manner.

After the representatives from both sides drink *oput* from the same vessel, the Master of ceremony, from the side of the offended clan, cuts off the head of the goat that they brought, and give it to the offended clan. The Master of ceremony, from the offender side, also cut off the head of the ram they brought, and give it to the offended clan. The Master of ceremony, from the offender side, then continue to cut the ram into half and gives one side to the offended clan to cook, while the offender clan cook the other remaining half of the ram. The ram's head is ceremoniously handed over to the next of kin of the dead man and the goat's head is also handed over ceremoniously to the kin of the killer. The bull is ritually speared to death and skinned and the meat is cooked and eaten together as we shall see later.

The act of slaughtering the goat and ram as well as exchanging heads remind the perpetrators and those witnessing the ceremony that there is a price to be paid for violating the agreed rules of co-existence. *Mato oput* embraces the principles that society and the perpetrators contribute to the extent possible to the emotional restoration and repair of the physical and material well-being of the victim. These principles, the Acholi people use them successfully in conflict management for generations and generations.

However, we cannot deny that traditional approaches to conflict resolution might not be meaningless to other people who are non-Acholi or non-Luo. This might also be true for the young Acholi boys and girls, who are born and grew in the Western World, as they have had no opportunity to experience and participate in such practices. Some fanatics Christians also reject traditional practices outright as being "Satanic".

The *mato oput* ceremony itself has various forms across different clans in Acholiland, but the common characteristics include the slaughter of two sheep which are cut in half and exchanged by both clans, and the drinking of the bitter herb *Oput* by both clans to 'wash away bitterness'. The ceremony continues to be practiced to date.

(9) *Ritual of sharing food*

After compensation has been paid, a daylong *mato oput* ceremony takes place. The local chief (*Rwot moo*) presides over the ceremony that brings the clans of the perpetrator and the victim in order to eat together and to re-establish harmony. Cecily Rose (2008) and Latigo (2008) noted that the process of *mato oput* is not about individual victim or perpetrator but rather involves the entire community; since the Acholi people believed that the perpetrator's offense affects the whole clan. This is in consonant with Desmond Tutu's view that African justice aims at the healing of breaches (or wounds), the redressing of imbalance, and the restoration of broken relationships, seeking to rehabilitate both the victim and the perpetrator. However, this is not to suggest that there are no challenges to the use of traditional justice, particular as an accountability measure for crimes committed in SPLA and SPLA-IO conflict in South Sudan.

Both parties cook and eat the livers (*cwiny*) of the sheep and goat (where applicable), to show that their blood has been mixed and unity in harmony has been achieved. This symbolically was away the bitterness within the blood of the human liver. One of that last rituals involves consuming "*deyo*" or "*odoyo*" (the remains of a saucepan), which is thought to free the parties to eat together again. Other cooked food, from the side of offender and the side of offended clans, are served to the elders, who are allowed at this time and moment to mingle freely. This means that from now onwards, the members of the two clans resume their normal social intercourse. In this way, the Acholi people make good the damage caused by the spilling of the sacred blood of human beings. The ceremony is not complete until the parties have eaten all of the food prepared for the day (occasion), thus, finishing food that is prepared for the occasion like this means no bitterness remains between the two clans.

Rituals to restore Evidence

Different rituals are undertaken when the facts of the crime are not all available, and there is no way to establish the truth. In one version, a spear is put in the fire flames until it turns red. Then all the suspects are called and made to hold the red spear or lick it. The Acholi believed that if a person is innocent, even though he/she holds red, hot spear or lick it, the spear will not burn the person. But if the person is a wrongdoer, the red hot spear will burn him or her and eventually lead to death. This type of ritual is only conducted by the *"ajwaka"* (sorcerer).

Another way to determine innocent or guilt is by digging a hole and ask all the suspects to jump over it. It is said that is the person is guilty that person will fall in the hole.

The third version to determine the wrongdoers states that to establish truth from the suspects, each of them would be asked to feed a hen using poisoned food. It is said that if a guilty person throws poisoned food to the hen, it will move towards the offender and eats the poisoned food straight away but if the person is not guilty, when he or she throws the poisoned food towards a hen, that hen will ignore the poisoned food and moves away from the person —leaving the innocent person with the poisoned food —the hen will not eat the food. From these narratives, one can note how the rituals play with the psychological conscience of the offender in the same way that the spiritual world does.

Last but not least, another means to establish the truth is to consult an "ajwaka" (divine healer). In doing this, a group of elders would go around consulting three different *"ajwaki"* (divine healers); who are living far away from way the place of incident. In this way, the ajwaka would not know exactly about the crime. When a group of elders enter the compound of each "ajwaka" (sorcerer) that "ajwaka" (sorcerer) would invite the spirit of the dead, the spirit then would reveal to *"ajwaka"* (divine healer) the perpetrator or perpetrators. Therefore, if the three *"ajwaka"* (divine healer) identified the same perpetrator (s) then truth would be established.

Having said this, we should note that establishing the truth will not necessarily lead to a mediation of the conflict. However, it is always up to the individual, the offender, to initiate reconciliation process.

The Ten Guiding Principles of True Reconciliation

Many writers in anthropology have written many and various guiding principles of true reconciliations. For the important of this book, and to avoid getting confuse with so many principles, we are going to focus on ten top guiding principles of reconciliation. Thus, when reading this article, I want us to put in the back of our minds that outside there, there are people who are hesitant about reconciling with their offenders. It is my hope that these ten top guiding principles of reconciliation would help such people.

People who are seeking for reconciliation should understand and accept that reconciliation offers a new way of living together. Therefore, for the offender and the offended to walk on a long path of reconciliation, they should recognize that the below ten top guiding principles are applied to all parties, whether be it in family, clan, society, individuals, or national conflict resolutions. These principles should be viewed as tools or framework in achieving true and lasting reconciliation and healing; they are as follows:-

(a) Be honest about your motives and inner feelings

If you are a Christian or a believer in God but does not practice Christian doctrines, you should make sure your desire for reconciliation is to do what pleases God and never think of taking revenge. Reconciliation is a process of healing relationships that requires public truth sharing, apology and commemoration that acknowledge and redress post harms.

(b) **De humble in your attitude**

Don't let pride ruin your attitude. Don't think that the other person earn you forgiveness but rather working towards true reconciliation. This is not easy; it demands humility, which means either lowering yourself with others or lifting up yourself with other. You should have shared responsibility for establishing and maintaining mutual respectful relationships.

(c) **Be prayerful about the person who hurt you**

All of us in this world pray to our God in one way or the other. Jesus Christ taught his disciples to pray for those who mistreat them. "Bless those who curse you, pray for those who abuse you" (Luke 6:28). You need to pray that God gives you strength to go through this long and harsh road to reconciliation and healing of inner emotional wounds. "Let us then with confidence draw near to the throne of grace, that we may receive mercy and find grace to help in time of need" (Hebrews 4:16)

(d) **Be willing to admit ways you might have contributed to the problem**

It is true that some of the perpetrators might have not started the dispute that led to the killing, nevertheless, we must understand that lack of knowledge, careless use of words, impatience or failure to respond in a loving manner might have aggravated the situation that lead to the killing of other person. All-in-all one must admits all aspects that led to the killing, if he/she wants a true reconciliation and healing between the two parties.

(e) **Be honest with the offender**

If we want to find out and absorbed the reality of what exactly happened at the killing, we should be honest and friendly to the offender. If we are not honest and friendly to the offender, we shall not find the facts of the matter from perpetrator. Do not be aggressive to the offender as he/she is telling his/her story. If you are aggressive, it will close the line of good communication between you and the perpetrator.

(f) **Be objective about your honesty**

The offender might have some good reasons for being hesitant to reconcile, but he/she must state clearly and objectively reasons behind his/her being resistant. We understand some times repeated confessions and offenses of the same nature may make it hard for the truth to be built. And if you are a mediator who doubt about the sincerity of the offender, for the reason best known to you, please, define your reasons for doubting; because reconciliation requires constructive action plan on addressing the identified issues.

Be clear about the guidelines for restoration
Reconciliation must create equitable and inclusive society by closing the gaps between the offender and the offended clans or families. Reconciliation requires sustained public education and dialogue, including young people agreement, about the history and legacy of residential schools and one's own rights.

(g) *Be alert to Satan's schemes*
In Ephesian 4:27 Paul warns about possibility of giving Satan an opportunity in our lives when he says," And give no opportunity to the devils". He went on to write, "Let no corrupting talk come out of your mouths, but only such as is good for building up, as fits the occasion, that it may give grace to those who hear. ......Be kind to one another, tender-hearted, forgiving one another, as God in Christ forgave you. Therefore,......walk in love, as Christ love us and gave himself up for us, a fragrant offering and sacrifice to God" (Ephesian 4:29-5:2).

(h) *Be mindful of God's control*
God will not allow you to be tempted beyond your strength. He is outside, walking by your side all the time long, to give you helping hands. As the apostle Paul wrote, "...and all ate the same spiritual food. And all drink the same spiritual drink (1 Corinthians 10: 3-4).

(i) *Be realistic about the process of reconciliation*
It is said, "Change often requires time and hard work". Repeated offences in our societies are often tending to replace a powerful grip on a person's life. The key indicator of change is the attitude of the offender. Don't give up too easily on the process of reconciliation. Be open to the goal of a fully resolved relationship.
   (Source: Pastor Steve Cornell, 2012).

Family conflict resolution among the Acholi (Dispora): CASE STUDY
This is a true case study done on a Sudanese couple and because of the policy; the names here do not reflect the true names used and the suburbs where the family live in Victoria (Australia) and the tribe are not mentioned.
   Mr. Peter Opoka age 32 years old and Grace Acayo age 27 years disclosed before and during the mediation session that they were happily married in refugee camp (in Africa)

in the Year 2000. Both the husband and the wife are from the same tribe. The relatives of the couple with the marriage and as part of Sudanese traditional culture, soon after the payment of dowry, the mother of the girl gave the couple a piece of advice by telling the daughter and the husband to cultivate a spirit of good understanding and always to work in harmony wherever problems arise in their marriage life.

Before the marriage Mr. Peter Opoka applied for humanitarian visa to Australia and successfully past the interview; and soon after the wedding ceremony, Peter went to Australia Embassy and included the name of his partner into the visa.

While the couple was waiting for their visa for traveling to Australia in the Year 2001, they have their first baby son named John Garang. After which the husband again went to Embassy and informed the staff about the new born and his name was included in the application. After three months, visas were granted to them.

The couples were then faced with another problem that is "Where to get money for the three air tickets?" However, messages were sent to the relatives of the husband who live in America and Canada and they willingly contributed fund that enabled Opoka's family to purchase air tickets. The couple came to Australia and settled in Melbourne, after three months of self-adjustment, Mr. Opoka was employed in one of the factories and he earns good wages but Mrs. Acayo remains unemployed. One of the reasons Mrs. Acayo could not get a job was because of her poor English language. However, during this period, John Garang was enrolled in one of the Child Care centers and she attended English classed at AMES. Nevertheless, both were happy with their incomes but before long, they started to have family problems. The wife wants the money to be sent to her relatives overseas while the husband wants money to be saved for family welfare as well as for the education of Garang and any child that might be born. It is important to understand that Sudanese people want to have as many children as they could, so if there is some money to be saved, potential children are always included in plan.

In 2003 the couple was again blessed with another baby son name Kevin Howard Oketayot. The couple fights almost every month and each person blamed the other of starting the fight and police have been in the house four times. The husband said during mediation session that the best thing for him to do in this situation was to leave the house. He also stated by doing so he would avoid being moved from the house through intervention by the police. Since then even no place in the house, life became difficult for both parties. Opoka confessed during the session that his life was difficult for him; he thought to leave Australia for America or to look for love, peace of heart and mind outside marriage. Consequently, the husband decided to move away from the house and live with his friend for six months. Mrs. Acayo reported to Centrelink that they have separated and Centrelink approved for her "Single Parent Payment" and child support being deducted from Opoka's income.

Before long Mr. Opoka made an affair with a single mother named Akwach Ajak who live next door to friend's house. Ms. Abalo learned of the affair and reported the case to her friend (Mrs. Acayo). Although Mrs Acayo live apart from her husband, she was very much hurt when her friend (Abalo) briefed her about the situation.

Our self-community in Australia is managed by elected committee members who hold different offices and welfare office is one of them. The officer of welfare is responsible for the wellbeing and welfare of the entire members of the Association. They have mandate to appoint an elder or elders to resolve disputes that might arise among families or individuals. I was then appointed to help resolve disputes that led to separation of Mr. Opoka and his wife. I was appointed a mediator because of the knowledge, age and trust that community has in me. Consequently, this implies that I was charged with the responsibility of "intake process".

I look deep into myself and realized that I am going to deal with family conflicts. I thought to myself that for effective mediation, I should understand what is meant by family here? Because the term family in our day and age takes different definitions: family can be defined as a social entity, a legal entity or biological entity but yet the structure varies from community to community. In this situation, I believed, I was going to dial with family as legal entity. And we understand that in many cases families come into conflict because their interdependence of dealing with children, money, housing, social and sexual life (FMC Relationships Services, 2005, P. 16)

At the end of six months, I thought some has come to look into the case of the couple meanwhile, I hope that at this time, angers from the couple might have come down that would provide room for better negotiation. It is also traditionally believed that during such period, both parties would recall their thoughts and reflect on their past words and events (actions).This reflection is very important, it provides a better ground during negotiation stage.

I approached husband and wife individually at different times and before I started the interview, I introduced myself (even though they already know me) and told them that I am been appointed by Community members to coordinate and facilitate the forth coming discussion between them. I informed each one that the discussions are meant for family benefit. And this individual session is confidential in the sense I will not tell the husband that you the wife told me and vice versa. We can say to certain extent that this is in conformity with the policy practice at Family Relationships Centers in Australia which stated "Each family member is then seen individually and is informed that this separate session is confidential This enables parties to talk about issues openly in the knowledge that this information will not be shared. It also enables family disputes resolution practioners to gather more information and assess carefully the appropriations of family disputes resolution (FMC Relationship Services, 2005 P.21).

I also have to make sure that both parties accept that I help them resolve their disputes. Thus before, I proceeded with mediation process, I asked them if they have any objection for me doing this mediation? Both parties said they have objection and this has given me mandate to go ahead with intake interviews. As I was going intake interviews, some of the questions I asked include the following:

- ▶ Can you please tell me exactly what happened in these disputes that led to separation?
- ▶ What do you think are the root cause of your family problems?
- ▶ To what extend do you think these issues affect the children?
- ▶ What do you need be done to help resolve the problem?
- ▶ Are you willing to come together and discuss the issues with your partner? – To reach common agreement.
- ▶ Are you confident to say what you want in the presence of your partner?
- ▶ What do you think children want from you as parents?
- ▶ How would you like things to be after family resolution?
- ▶ What changes would you like your partner to make after a successfully resolution?
- ▶ What are you ready to give out so that your partner rebuilds trust and confidence in you?
- ▶ What made you leave home and stay with your friend?
- ▶ How did you feel when your partner left home?

By asking such questions, I was really establishing foundation for effective assessment. It is at this stage that a mediator could predict whether mediation could be successful or not. The process enables me understand the intention and interests of the parties. And when they told me that they were willing to come around the table to discuss the issues that indicated they want family resolution. During my visits to the couple, I made it absolutely clear to them that "life and welfare of children are top priorities in any family life.

I spent one hour with each couple for three alternative days. I met Mrs. Acayo at her house and met Mr. Opoka at his friend's house. During my visits, I asked open and closed questions and some of the key questions are as outlined above. Meanwhile, during intake interview, I tried hard to make the couple appreciate mediation. In addition, I tried to find out whether party 1 and party 2 are willing and also able to sit around the fire and negotiate the issues in the dispute. I was amazed when by the end of the intake interview, the couple agreed to discuss about the family problems.

However, the next issue was the venue for the meeting; I know that culturally it is not acceptable to meet in the house of Opoka's friend. And on the other side, Mr. Opoka would feel uneasy meeting in the house where his wife stay with the children because culturally this is not acceptable secondly Opoka, like some Sudanese men, might feel that her has invested in a lot paying air tickets. So it was suggested that the meeting be conducted in my house and both the husband and the wife agree to this.

In African culture, and particularly Sudanese culture, family mediation is never done by one or two mediators – family mediation is rather done collectively and is known as "collective mediation" and this could include relatives from the sides of both parties (mothers, fathers, uncles, aunties, brothers and sometimes sisters). The number of people involve in collective mediation normally varies from 3-10 or even more. In the western model of family mediation, two people (male and female) do family mediation – Mediator 1 and Mediator 2. Mediator is somebody who meet the client first and do the intake and assessment process and brief Mediator 2 of his or her findings prior to actual mediation process. Mediator I also explain to the clients the voluntary nature of mediation, neutrality of mediators and the confidentiality of proceedings; while Mediator 2 is a person who never met clients before. And the role of the Mediator 2 is to explain to the clients' different steps in mediation as well as ground rules. The ground rules normally include: no interruption when somebody is speaking; it also states that everyone must listen to one another and finally each person must talk with respectfully during the session. In addition, Mediator 2 tells clients to write down what each wants to commend on while other person is speaking (FMC Relationship Services, 2005 PP 42-43).

At this point it is worth mentioning that Sudanese traditional model of family mediation has much in common with the western model of family mediation. After having said that we should not forget that although the western and Sudanese model of family mediation have much in common, nevertheless, there are still significant differences And these differences are found in the role of Mediators 1 and 2. In Sudan, Mediator 1 is commonly known as Chairperson that is a person who resides over the session and helps direct the discussions. According to Sudanese culture, woman will never be Chairperson in family mediation. The main reason for choosing man to be a chairperson is that men are usually considered, by the society, to have better knowledge and skills about problem solving. And women are considered to lack such knowledge and skills. On the contrary in Western World, Mediator 1 could be either of the sex that is female or male. This is right and true for the Western people because there is high assumption to "gender equality". Another different between African traditional mediation and Western model of mediation consist in that in the Western World clients are expected to take notes during mediation session. But in Africa and Sudan in particular, clients go in mediation room with their hearts and mind to remember issues that required commence - They have to remember everything they have to commend on. And last but not least, mediators in traditional model of mediation although they claim to be neutral but the fact remains that they are not - however they minimize degree of neutrality (i.e. they are particularly neutral).

In Sudanese traditional model of family mediation, mediator I combines his role as well as the role of mediator II. Therefore, he explains to the clients the purpose of the session (known as the purpose of the gathering), the voluntary nature of mediation, partial neutrality of mediators and the confidentiality of proceedings. In addition, chairperson explains to the clients the different steps that would be followed in the mediation process (that each party tells his/ her stories; each mediator will re –state the key issues he/ she thinks have been raised; allowing room for discussion and the key issues; drawing up binding agreements and finally check with the parties whether there is any objection on the agreement). If family mediation is successful this will be followed by shake of hands which are symbol of peaceful and genuine agreement. The chairperson also explains ground rules that will guide, not only party I and party II but everybody in the mediation room, during the session.

In this case, I become chairperson or Mediator 1 because of the trust and hope the community members have in me. However, it is a common practice that community members could appoint chairperson plus other members that formed Mediator 2 and these people could be from both sexes. The main role of Mediator 2 is to assist the chairperson in helping the couple to effectively discuss their issues and possibly reach binding agreement. But sometimes, community members appoint chairperson and give him mandate to nominate people he knows have knowledge and skills in problem solving. As in this case, community members gave me mandate to choose people I know have the knowledge, ability and skills to help solve family issues. In Africa, mediation session will never be done by one person (solo mediation) or by two persons (co-ordination) rather mediation is done collectively.

I then nominated 5 men and 5 women and invited them for the sessions. I understand sometimes one of the parties might not want one or two of the nominated members to be part of the mediators. In this case, before we proceed ahead with the discussions, I asked the parties if any have objection of any nominated members? There was no objection from the parties.

As mentioned above in this case study, Mediator 1 or chairperson has roles. As people gather in my house, I explain to party I and party II that we have gather here today because community members have noticed that there is dispute between parties that led to separation. I also informed the parties that traditional mediators are to facilitate communication but the final decision of the problem lies with the parties. But for the couple to achieve the goals and expectations, this required that each party must be open, frank and honest. This is to enable traditional mediators to detect and see the weak and strong points of each party. And hope that by the end of mediation session, part I and party II will reach genuine, commonly and acceptable agreement. I therefore ask each of the couple to put aside the past grievances and rather "to focus on the future".

Before we proceeded with mediation, I did inform the 5 men and the 5 women that I have met the wife and husband individually and each one was able to tell me his / her story. So this mediation session will allow each couple to tell again their stories to mediators. In addition, I explained ground rules to the clients by telling them when one person is talking, the other person should listen very carefully and never to interrupt. Above, all we want each person to speak respectfully. I also re-assured the parties that as we are from same sub tribe in Sudan, matters discussed here will remain confidential. In other words, the information about this session should not be talked outside community members. During the session time will be given to each person to talk about and commend.

I then asked the couple "Who do you want to talk first?" There was no answer to this question: the wife said that the husband should talk first and the husband said the wife should talk first. I then told them for smooth running of the session, it is important that one person accepts going first. There was a minute of silence, which indicates that the couples were reflecting on what to do? Soon the husband broke silence by saying "I am ready to go first".

I thanked the husband for his discussion and asked him to give his statement and invited traditional mediators to listen carefully and take notes of the key issues in the statement. This also the stage traditional mediators can identify whether the concerned person are telling the "truth" or "lies" or he is exaggerating the truth. The husband talked for about 50 minutes after which traditional mediators asked questions to clarify some of the points raised thereto. I can remember one of the question asked to Mr. Opoka was "Why did you decide to have affair with Ms. Ajak?" He answered "I am unhappy and depressed in my marriage with Mrs. Acayo. I am looking for love, consolation and breathing space that I believe I could only find in Ms. Akwach Ajak and not with Mrs. Acayo".

After having heard from Mr. Opoka, I gave time for Mrs. Acayo to tell her story. She was also given 50 minutes and after she finished her statement, floor was opened for traditional mediators to ask questions to clarify some points raised by her. I can also remember one of the questions put to Mrs. Acayo was "Why do you fight over money?" And she answered "Mr. Opoka is working and since then I have never seen his salaries and do not know on what he is spending the money. Opoka does not want me to ask him about his salaries. And when I do, he always get cross that follows by verbal abuses and often ended in fight".

When the couple finished giving their statements, I then checked with the local mediator whether there are areas that might still need clarification. However, it was generally agreed that all areas of conflict been explored. So the next step was to list down the key issues raised by two parties.

KEY ISSUES:

- ▶ Ms. Akwach Ajak
- ▶ Money
- ▶ Bills and education of children
- ▶ Bank card
- ▶ Listening to outside advisors

- Raw information to relatives
- Child care
- Domestic work (cleaning, washing and ironing)
- Application for humanitarian visa for Mr. John to come to Australia
- Salaries and centrelink money
- Visits to friends
- Coming late home

After having listed key issues, times were given to Party I and Party II to negotiate the key issues; because this is about a crucial part or stage where resolution of the issues in the disputes commences. The couple was also reminded that when one person is talking, the other should listen carefully and no interruption allowed; they were also reminded to talk respectfully. Negotiation over key issues took three hours. Chances were given to each party to talk and ask questions to the other and at the same time observing the ground rules. Thus after six hours of discussion, the couple realized their mistakes and weaknesses and made the following agreement:

- Mr. Opoka to stop an affair with Ms. Ajak
- Agreed to send money overseas once a month at an agreed amount
- Mrs. Acayo to stop listening to outside advisors
- Mr. Opoka to help in domestic work (cleaning, washing and ironing) every weekend from Saturday and Sunday.
- Mrs. Acayo and Mr. Opoka to have separate bank account and each person keeps his / her bank card.
- Mr. Opoka to inform Ms. Acayo every fortnight of his income; and Mrs. Acayo to be involved in decision making about the usage of income of the family.
- Mrs. Acayo to stop giving wrong information to parents and relatives – overseas.
- Mr. Opoka to assist Mrs. Acayo in lodging application for humanitarian visa for Brother John.
- Mrs. Acayo to visit friends whenever she chooses.
- Agreed to save $100 every forth night for children's education.

As regard to the late coming into the house, the couple realized that such agreement would be difficult to keep as the nature of Mr. Opoka's work sometimes required him to come home at 11.00p.m (with evening shift). Nevertheless, it was agreed that Mr. Opoka should be in the house in the earlier possible hour.

According to the Sudanese culture, hand shake on occasion such as this is a sign of "Unity" and "forgiveness". As a traditional practice, I invited the couple to shake their hands that was followed by deep smile from the parties and applause from traditional mediators. This is a symbol of appreciation to couple for their efforts and reconciliation. The couple re- united the same night.

To make sure the couple respect the agreement two men (which include myself) and one woman were asked by traditional mediators to make irregular and surprised visits to the house of the couple as part of their follow - up. After four months of irregular visits, delegated members discovered that the wife and the husband respect agreement – They have implemented all agreements and now they live a happy family life. Among Sudanese community, this couple is spoken pf as "model of family resolution".

The principbles of problem solving in Sudanese culture states, "Problems can be solved if both parties are opened, flexible and receptive to good advice". No behavioral change can be imposed on an individual but this change must come from within with support from traditional elders or worker". As a global citizen, I assume that the principle could be applicable to people in developing and developed countries - human issues are the same all over but ways of solving them varies from community to community and from individual to individual.

Anyuak Traditional Justice and Reconciliations

Anyuak is not exception among the Northern Luo groups, in the pre-colonial, tribal killings were very common among the people of Anyuak. However, it remains a mystery that nothing or little is written or spoken about Anyuak traditional jusctice proceedures.

It is believe that customary when there is problem offender family present the case first to village chief. If the offender is not satisfy with the judgement of the village chief and his elders, he/she can appeal to district chief known as the high-chief. When the offender is not satisfied with the judgement passed by the high chief (district chief) he/she can appeal direct to the King.

Source: Anyualjustice.org
Figure: 76 Anuak people for reconciliation

Interestingly, the Anyuak people in Dispora formed what they called "Anyuak Justice Cou7ncil" (AJC) in December 13th, 2016. The formation of AJC came as a result of massacre of the Anyuak people on 13th Decewmber 2003, by bthe Tigre People's Liberation Front (TPLF) regime in Ethiopia.

The intention of the formation of the AJC was good, but the reality is that they cannot meet the immediate needs of the people at the grass root in term of conflict resolutions. The Anyual Jusctice Council was formed to represent the voice of the Anyuak people worldwide, and they are based in Canada; they formed partnertships with international law groups to bring legal pressure to bear against the Ethiopian Governement in internal court. This prompt a question, "How do AJC help address conficts among the families and community in South Sudan and Gambella?" There is a wise saying, in Acholi that "If you want to kill a snake, use a short stick —you will never kill a snake with long stick".

In conclusision, I want to say that the AJC would never help locsal Anyuak people solve their family and tribal conflicts -as we have seen among the Acholi and other Northern Luo Groups.

Collo Traditional Justice and Reconciliations: Forms of Compensations

In order to understand the working of Collo traditions in relation to conflict management, it is important first to high-light the local system of governance or legal system under which these traditions operate.

The Collo like the Acholi, in pre-colonial and post-colonial times, they settled all conflicts in the community through well-developed mechanism for the prompt resolution of conflicts as soon as they arose. At this juncture, it is important to note that prior to the British rules in *South Sudan, the Collo maintained a traditional government that is strongly rooted in their* beliefs, norms and customs, which demands peace and stability in the community at all times. As we shall see later, in the past all cases of conflicts were directed to the King who deals with them. But today the structure has changed, where we have Reth (King) on the top of the hierarchy, follows by the elected provisional governors (Jangi *Podho*), these are followed by District Commissioners or high-chiefs (*Jangi Paj*) and comes Village chiefs (*Rwadhi*) who deal with native courts at village level, they are assisted by sub-chiefs and elders in the communities.

The provisional governors, high chiefs, chiefs and sub chiefs in Shillukland, are believed to have been given authority by the people and the Reth (King) to correct the misbehaviours of certain individuals or clans in the society. Fr. Bauholzer described chiefs as persons who correct ethnical behaviours and assisted by the sub-chiefs. Chiefs let no stranger go hungry, he does not use abusive language.

Like the Acholi, the Collo have Council, who look over affairs of the entire communities in Shillukland reigning from disputes, marriages to murders. We have seen in Acholi traditional justice and reconciliation that the Grand Council act like "Supreme Court" to try cases of killings and land disputes between individuals or clans involved.

Taking the provisional governors(*Jangi Podho*), district commissioners or high-chiefs (*Jangi Paj*) and village chiefs as the governing body at all levels, they refer cases among themselves (from village chiefs through provisional governors to the King who has the final judgement) as we shall see later. All the authorities take oath to observe the traditional laws and duties, in token of which each elected provisional governor, high chief and chief licks the blade of a spear, presses it to his forehead, and then waves it over the ox he sacrifices for the occasion. He remains in seclusion in his house for four days, only after this he "then becomes the chosen one by God".

The Collo like any other Luo groups in the South Sudan, greatly rely on traditional ways to solve disputes between and among people to uphold social harmony.

Rite of Reconciliation

In the pre-colonial, tribal killings were very common among the people of Shilluk. It is reported that when a man wanted to kill his opponent, (form inter-clan) he gets out of the village and hides in the bush, where the man in question often passes. Thus, when the man comes nearer and nearer, the potential offender jumps out from his hiding place and spears his enemy twice or three times –he does this to ensure that the man is really dead. When the offender is sure the victim is dead, he then runs home to inform his people of what he has done. When he reaches home, he first goes to village leader (or chief of the village) and tells him, "I have killed a man".

When a village leader hears this, he immediately orders for drum to be beaten; and people assemble when on hearing the sound of the drum. They assemble in open space, where drum is beaten; there they would be informed of what one of their men has done. It is not clear to me whether the killer in Shilluk society also narrate his/her story as we have seen among the Acholi. However, killing like this in Shilluk land always followed by blood revenge as a form of compensation.

So, when everybody is well informed about the killing (murder), elders then send message to women to carry away any valuable items and hide them in the bush –for fear that when the offended clan come to fight their people, they will seize everything they come across. The elders also send another massage to warriors to drive cattle into Dinka-County -for the same fear that the offended clan will seize the cattle. Eventually women hide valuable items and young men take the cattle to Dinka-county.

Moreover, when women and young people are hiding items and driving the cattle out of the village, warriors would be preparing to face their opponents, who are coming to revenge. They are ready to fight should they be provoked but more often no fight takes place. The offended clan just come into the village and rob the village of all the fowls and dura. After this the offended clan return to their village.

When offender clan see that their opponents have returned to their village, elders again ask women to bring valuable items out of the hiding place and also ask young men to bring back cattle from Dinka-county.

As a token for reconciliation, the chief from offended clan releases two cows and give them to the King as information fees. In Collo society, when such killing happens, all the compensations to the offended clan would be paid through the King. And according to Collo norms and customs, giving only two cows to the King, is an offense; not only to the King, but it is also an offense to the offended clan. When the two cows are brought to King, he gets angry and sends back the delegations with an instruction to village chief to ask villagers to contribute more cows. The villagers of the offender then contribute; one gives a cow, and another one also give a cow, till they collect ten cows the cows are then collected in one place (D. Westernmann, 1912, P. 115). Sudan Notes and Records (1941) notes that where a compensatory payment for killing is due the net is cast wider and not only the entire lineage but even members of other lineages of the same clan may be required to contribute (P. 23). The same principle is applied today in Collo society. "At any rate, compensations are in kind, because of their spiritual content, and are fixed by the King in a legislative council, or in a judicial decision. For a Shulluk man, the compensation fixed by the King is viewed as divine rules because the King is the only person in whom the spirit of Nyikang reincarnates, particularly after the enthronement ceremony" (Nyawelo, 1992, P. 7).

One of the song composers in Shulluk land, Thonh Akol Anyong, expressed this believe of divine rules, in the following song of King Ayang (1974-1992)

Cøllø	English
Ayang-dit Nyikango ya adwog wuo,	Ayangdit, Nyikango, I have returned, father
Ya ciet a yi gony oweni;	I have come for the judgement of last year.
mog a muj yi Nyikang i dhog	The judgement delivered by Nyikango for cattle
Di mag bung ngan lith gen	Attachment nobody could execute it.
Go pegi be madh ki mogo,	People are drunk, and
Gony a jiang piny. Mogo	Court work neglected. Why Shilluk judges
Madha ngo yo yi jagi colo,	Drink alcoholic drinks? There is liquor,
Keny da mog adhhadh ki mogo	Sherry, and what the wife prepares in the form of local beer
Cug ki men a gwog yi ciege	Being the third item.
Gon a athobobo gon a dag gen.	How would the judges settle the disputes
Gony di lidh adi con yi jagi,	When they are seriously drunk.
Ji ciathi ge thamo yi mogo.	

Source: Nyawelo, 1992, P. 7

As the ten cows are now ready, the village chief send message to Reth (King) advising him that the people have contributed more cows. When the King receives this message, he sends his ambassador, with few men, to offender village, to examine the ten cows if they are all in good health and fit to be given for compensation.

Figure: 77,
Cattle for compensation

On arrival, ambassador goes straight to the village chief and asks, "How far have you gone with the settlement of the matter?"

The Village chief answers, "The matter is settled".

After this, the ten cows are brought for ambassador to examine. Before, ambassador actually starts examination, traditionally, offender people give him some sheep to share with the people who accompany him in the mission —this is done purely to please them.

Ambassador takes the sheep and tells offender clan, "I am sorry, the gifts are not sufficient?" Because they are supposed to add two cows for King's information fees; this then could bring the total cows to twelve.

On hearing such comment from ambassador, offender clan try hard to please the ambassador with some drinks, and convince him to accept the gifts and take the ten cows to the King. In doing this, they bring one cow and kill for ambassador and his people to eat before their departure. So after eating and drinking, the ten cows are then brought to ambassador who takes them to the Reth (King) in Fashoda together with the family of the offender.

The King receives the ten cows and gives five cows to the offended family as a mourning-fund; and the other five cows remain for him (King). In this way, the case of murder is settled, justice done, reconciliation reached and healing in process.

However, there are many ways the Shilluk people practice justice and reconciliation. The three Kings namely Reth Kuijkon, Reth Padiet and Reth Papiti have contributed greatly into Collo traditional law, customs and social justice and reconciliation. Let us then examine some few examples of compensations and they are as follows:

It is customary that when there is problem offender family present the case first to village chief. If the offender is not satisfy with the judgement of the village chief and his elders, he/she can appeal to district chief known as the high-chief. When the offender is not satisfied with the judgement passed by the high chief (district chief) he/she can appeal direct to the King. To understand this better let us go through some examples.

For example in November 1904, when King Padit Kuathker, lost his hearing, some cows went into the field of an old, childless woman and trespassed the sorghum; the cows eat part of the crops before maturity. The woman raised the case to the village chief but there was no solution reached; she then went direct to the court that was chaired by King Padiet and she threw herself down and cried with loud voice, before the King and council of Shulluk provisional governors. The King felt a sharp pain in his head and suddenly his ears opened. The King calmed down the woman, and asked what her problem was! The woman told the court that her sorghum has been eaten by cows of a man unknown to her. The court looked for this man, and when he was found, he was summoned to court and he was asked to compensate the woman, with ten cows, after the assessment was done by the committee. As from that time, the court laid down the judicial precedent stating that should a domestic animal damage crops, the owner of the field must kill that animal - when he/she finds the animal still eating the crops. But if the animal has damaged the crops and has gone, the case must be brought to the court that will help persecute the owner of the accused animal.

Nyawelo (1992) observed that the court at that time resolved that if a bull (any domestic animal) goes astray and damages the crops that bull must be killed by the owner of the field, and on top of this the owner of the bull must pay equivalent of the dead bull. Similarly, if a dog kills a domestic animal or birds such as cow, sheep, goat or chicken the dog's owner must pay the value of that dead animal or bird (P. 10).

Homicide compensation under the Shulluk law is based on bride price. As we shall see later Shulluk people fixed bride price to ten cows. Therefore, this means that if a person kills another person he/she must pay the offended family ten cows. This is done regardless of the age of the deceased. Interestingly, if the offender and the offended are closed relatives, the offender still pays ten cows to the King, the later gives five cows to the offended family and he keeps five cows for himself. Besides the payment of the ten cows by the offender, both the offender and offended families each will pay one cow to the King. As we have seen behind, the two cows are taken as information fees to the King.

But where the unintended accident happen, say between Dinka and Shilluk, and a Dinka causes death of a Shilluk or other way round, Shilluk law states that the offender family or clan will pay only seven cows to the offended family. The reason has been that the act was not intended.

There could even be killing between Shulluk people and Dinka or Nuer. In such cases, inter-tribal precedent was laid down during the reign of Jal Nyiluaag, the chief of Dinka, and King Papiti (or Gwang) of Shulluk and Jogo Mayom, who was the chief of Panyikango province at that time. According to Nyawelo (1992) this precedent came as a result when a Shilluk man, from Tugo province, and some Dinka men, from Abek clan, went hunting. The groups went into the bush and came across an antelope (Taang); they started chasing it; and in the process of chasing the antelope, a Dinka man released his spear, the spear missed the animal and fixed itself firmly onto the ground. At the end of the spear's wooden pole, there was a sharp pointed iron spiral known as "kol tong" or "din". Unknowingly, a Shilluk man came running after the animal, he felt on this sharp end of the handle, he was then pierced in the abdomen and eventually he died. When the news reached King Papiti, he immediately went to Atar to settle the case with Dinka leaders. There, it was decided that the offender family be paid seven cows as a compensation. The reason is the same as we have seen before, because this incident is not intended by the dependent. Since then, this has remained a local law that is being applied between Shulluk and Dinka in case of accident like this.

Another good example of inter-tribal conflict resolution was even cited by Nyawelo (1992) when he said in 1973 a group of Dinka and Shulluk young people went fishing together along Oluudh River. Accidently, the fishing spear of a Shulluk man called Apom Kat Ayomo trucked a Dinka man and he died later. The matter was narrated to King Papiti, he and Jogo Oluk immediately went to meet Dinka chief called Winyiluaal, in his village, Pejur. They discussed the matter and it was agreed that seven cows be paid to Nyadeng, the father of the offended as compensation and it was done. The two families then restarted their good relationship again. Justice has been done and reconciliation reached (P.11).

In Shulluk customs, when inter-tribal killings occur, they do not always pay cows as compensation. In the other case, a lady or a girl is paid as compensation to the offended family or clan. Nyawelo (1992) also provides us another good example, when he says, in lieu of cattle; a young lady may be surrendered to the offended family or clan. When a lady is given, a closest male relative of the deceased takes over the lady. This is only feasible between clans or neighbours, who want to strengthen their cooperation, under the condition that there are no barriers of marriage between them. If they are not allowed to marry among themselves, then the usual ten cows are paid as compensation. The best example of lady been surrendered offended clan is found during the reign of King Papiti. At that time the Shulluk people crossed from Penyidaway to the eastern bank of the White Nile, they went and gathered in a place called Lual-Goker Acien. Here, the group of Shilluk left Jago Duyding and some elderly men in the shade. The Shulluk group then entered River Sobat for fishing. As the Collo were fishing, people from Anywag village, which is in the Awang Island, came and launched an attack on Jago Duyding and elderly men. They killed Jago Duyding, cut off his head and took his ruling silver wristlet. After this the leader of Anywag village ordered his people to swim across River Sobat, they then ran through Adidhiang (Apiew) being short cut to Didigo, where the main force was based.

When the Shulluk returned from fishing and learnt of the incident, they went straight home and mobilized large army. Very early morning, the army crossed the river and killed many people from Anywag village in revenge for their Jago Duyding. In order to solve the matter, the Anywag's clan, which caused trouble, gave a girl called Nyamuj to the offended from the clan of Shulluk, as a compensation for killing Jago Duyding and removing his silver wristlet. The girl was taken, as a wife, by Gokwej Aru Abangdit, who later succeeded his late brother, Jago Duyding, as Shulluk Provincial Governor of Penyidaway. Subsequently, the two factions resume normal relationship (Nyawelo, 1992).

According to Seligman (1948) in pre-colonial time, a girl could be taken in lieu of the cattle, and this girl would not be taken anywhere other than the house of the Reth (King). If a boy wants to marry the girl, the King would allow them to marry on the condition that both the girl and the boy be kept hostage in King's palace, until all cattle for dowry are paid. The King would keep the bride wealth for some months, and after this, the King would give one or two head of cattle to the family of the deceased as compensation (P. 6).

In Shulluk land, compensations are not only paid when a person kill a person. As we have seen somewhere that Shulluk people are peace loving people, they extend the compensation as far as to people who injured or hurt, people with the broken limbs or teeth. Thus, when a person aggressively hurts his colleague, that leads to total or partial paralysis, the person causing it must pay five cows and five sheep as compensation to the victim. Why should a person who just cause and injury pay compensation? Nyawelo (1992) observed there are two reasons for doing this. One reason has been that, according to Collo, a paralysed person is considered half-dead owing to loss of life amenities. The second reason been that the Collo considered a paralysed person to have been reduced to a non-productive consumer, he/she is reduced to the level of a child or old man or an old woman, who does not have sexual feelings even though he/she sees a naked woman or man. Therefore, this state of affairs adds additional burden to immediate family members, which is morally obliged to cater for the victim.

Interestingly, paralysis compensation of five cows and five sheep is due even if the offender is from the same family or clan as the victim. The five cows and five sheep represented compensation of half-death suffered by the victim. The five cows paid for causing paralysis or disfunctioning any one set of the paired human organs, obviously constitutes have of the full compensation payable for homicide. Furthermore, should the person already compensated for paralysis and this same person is killed by a member of same family he has paid the compensation, the person causing such death will also pay five cows and five sheep to the family of the new victim. Here, we can clearly see how the payment of compensations circulate whether it be in the family or outside the family.

In Shulluk land today, this rule has been extended to include the situation where any paired of the human body such as legs, eyes or ears have been disfunctioned by an injury wrongly inflicted by another person. However, where the infliction breaks any one of the limb bones or more ribs, one or more of the upper permanent teeth, the offender pays two cows and two sheep to the family of the victim. This same rule is also applied to the married woman, thus, if the husband breaks her ribs or upper permanent teeth in fight, the husband must pay two cows and two sheep to the family of the wife, because the marriage buys only the economic and sexual life of a woman, and not her bones.

In any other case, what the wrong doer pays as compensation to the injured party is left to the discretion of the court. For example if the offender has fought with his friend and he has dislocated his arm, in this situation the offender must be proved wrong in the process of the law in court. Even though the offender is sometimes proved wrong, the court may or may not ask the offender to pay compensation for causing injury during the exercise of the right of self-defence.

However, where there are no cows or a girl paid as compensations to the offended family or clan, often the people resort to blood revenge as a form of compensation.

Blood Revenge
In other cases, when the offender clan fail to pay compensation (the ten cows) to the offended relatives, people usually organise for blood revenge. This blood revenge usually happen, when crime is committed between Collo tribe and neighbours, who might be rigid and do not embrace compensation, for the reason best known to them. Such people normally include Dinka and Nuer, who have different systems and forms of compensations. Thus, when one of the Collo men is killed by either of the above two mentioned tribes, the only option to get compensation is for the Collo warriors to organise themselves for blood revenge.

However, experiences revealed that giving a girl as compensation did not often solved the matter – because one girl is not equal to ten (or more) dead people in the tribal war. Some Collo elders reported that in many case, Collo do not like to take a girl for compensation, they rather preferred blood revenge. When the offended tribe go for blood revenge, they usually fight until they are satisfied that the number of the dead from the offender tribe is equal to the dead from the offended tribe.

When the day arrives, Collo warriors move into the village of the offender with big force. They kill as many people as they could, and capture many people as they can. Oral tradition says that often as many as 50-60 captives are taken by Collo. They drive the captives to Shilluk land and detained them behind closed door.

The family of the offended then brew beer and they take small amount of the beer, mix it with flour. When they finish mixing the flour with beer, they then sift it; and one person would be asked to bring a drum out from the hut. As the man is carrying the drum outside the hut, one captive is brought out from the place where captives are kept, he is killed to bless a drum. Traditionally, the drum is to stay overnight in the open, and in the next morning, it is returned into the hut. As a man is returning a drum into the hut, another captive is again brought and killed. Eventually all the fifty or sixty captives will be killed one by one. As D. Westernmann (1912) observed, "Early next morning the drum was removed from house; one man was brought out and killed; and he was mourned. In the evening, a drum was been returned into the house, again another man was brought and killed. Thus, all the sixty men were brought out and killed in the place of the one man who had been killed" (P.115).

Tribal or Clan war

As we have just seen above in the preparation for the war; thus, when all the preparation has been completed, people move to battle field. William Oboc said, in the tribal war, before the warriors actually march to a battle field, a skilful magician is paid a bull to predetermine the casualties and evoke his magic toward spears. The magician bewitches some of the combatants, are like to be killed on the front line.

There are many causes of war in Shulluk land for this reason O. Deng (2016) quoted Ojwok Nyawello as saying, "There are many causes of war in the Shulluk community. There could be fight caused by cattle raid, blood feud, girl eloping, farm demarcation, unblessed fishing, friction at the youth dance, funeral rites dances, hunting or incursion on another village, forest or cow grazing land, foreign invasion and the punitive expedition" (P. 7). It is true that in Shulluk community, if some people from a particular village want to attempt to fish out of a pool or lagoon that belong to a certain village, without paying royalties, this would lead to war between inhabitants of the two villages. Likewise, eloping of girls could obviously case wars in the community, especially among the young people. We have seen somewhere in this book that if initiated youth in a village dance with a group of girls and certain youth group from a different village as the same girls to dance with them, this would cause a bloody fight. Furthermore, if a certain village fails to execute King's orders, the King can send punitive expedition to raid the cattle, goats and all the precious materials of the village as punishment.

Traditionally, tribal war, in Shilluk land, is fought in the open bush outside the village. It was observed by D. Westernmann (1912) that during tribal war one tribe comes and goes into the bush, another tribe also come and goes into the bush, shortly, they start spearing at each other. One man is speared and he falls down; the other man again is speared he falls down, until the battle field is full of dead bodies.

Collo are law-oriented community; they have defence system, which is under the command of War Commander (*Bany*). Bany must have courage because he should not run away during a pitched battle. If he runs away the warriors lose morale and they too will run in disarray. According to one of the Shulluk men called Along Owett, those who rum away from battle gain bad names and offensive song is composed against them.

In war, there is always the winner and the loser. Thus, when the losing side see that they are losing more of their warriors, they scattered running towards their village or county. As soon as the losing side withdraw from the battle field, the winning side also withdraw, making sure that no one is left out in the bush during night time. Because it is common among the Collo that the losing side often come back at night and wait in the bush near the village of the winning side. They do this with a hope that if somebody comes out to urinate or defecate (poo), he/she would be killed for revenge of their dead ones. For this reason, after the tribal war, Shilluk people do not come out at night to urinate or defecate. They urinate and defecate in houses, in a gourd, for fear of being killed.

The next day, early morning, women from both sides go to the battle ground to collect their dead ones and carry them home. There has never been a report that women fought in the battle ground, when they come to collect the dead bodies of their people.

As reported by O. Deng (2016) in Shulluk community there are two levels of reaching peace settlement: one is territorial and the other is inter-families as we shall discuss here below. However, territorial peace settlement is always done by Reth (King) between one clan and another or between one village and another. The two warring parties are asked to bring two bulls which will be exchanged, slaughtered, cooked and eaten at the site. This is called *Togi Cogo* (hammering of the bone till it breaks). Another type of peace settlement is of blood feud when the goat will be killed and its meat is cut up into pieces mixed with sorghum and pieces of charcoal fried together till it is cooked. The cooked meat is then distributed among the members of the two families. They then chew the piece of charcoal in the meat, together with cereal, and another separate dried meat. After chewing properly, each must swallow the piece of meat mixed with charcoal. This is called *"mwot ke cugol"* (chewing of the charcoal).

More importantly, the killing of a preson as sign of compensation is only ordered by the King (Reth) and for particular reason. However, in most case it is reported that when a Shulluk man kills a Shulluk man, the Reth (King) will immediately intervene. The King then take the offender to Fashoda and according to the traditional law the offender will stays in Fashoda for seven years. During these seven years a criminal would be sent to the other side of the Nile River, this is known locally as *"Lok-giang"*, which literary means living away from other people; and once the correction is completed (in 7 years), it is assumed that the family of the victim might have cold down, subsequently, a criminal is realised to go to his community. As the offender arrives in his community, a cow is killed it is skinned and the born is broken, the meat is cooked and eaten between the two parties; this is known locally as *"shiul"* (reconciliation). However, once the offender is realised to the community, no one would be allowed to touch him.

The Judicial Process

From our explanations above, we have seen that there are different circumstances where Collo fight and kill one another; we have seen that these fights include fighting with foreigners or neighbours, such as Dinka and Nuer; or fighting among themselves (a clan against another clan). Furthermore, we have also seen that there are families who experience family violence or family disputes. These are all considered to be ingredients of conflict resolutions. Shulluk are mindful of wrongs done to any person.

From a constitutional point of view, the Shulluk Kingdom exhibits a strong federal system of government as discussed below. A great deal of executive, legislative and judicial powers are delegated to the provincial governors (*Jangi Podho*), who in turn delegates powers to the district chief (High chief or district Commissioners). The district chiefs in their turn also delegate power to village chiefs and their elders. However, the residual powers are vested in the King and exercises such federal powers only with respect to matters referred to him from the provincial Governors (*Jangi Podho or Jagi*). Their arrangement as shall be discussed, simplifies the task of running the Kingdom at the federal level. A unique feature of the Shulluk royal system is that individuals, including the King, ascend to positions of authorities through democratic elections or nominations. And all cases are decided with reference to a judicial precedent by basing its decision on aspect of Shulluk law and customs.

Case heard by head of family

In Collo society, systems for hearing cases are well set up from the head of family to Reth (King), this mechanism is developed to guide the people in the community. Customarily, a Shulluk case begins when the accuser presents his/her oral petition to the head of family of which the dependent is a member. The head of family takes this case to other elders of the village and subsequently a quasi-judicial tribunal is set up that will consist only of the elders of the offended family. As the elder sit down, the summoned both the accuser and the accused to come before them and narrate exactly what had happened between them. The elders listen to the stories from both sides, and on the basis of the stories told, they make and pass decisions.

Appealed made to district chief

If the offended is not satisfied with the decision of the elders, because the offender is not penalized, he/she can appeal before the high-chief (district court or Buri Jangi Paj). In the same way, if the offended is not satisfied with the decision of the elders, he/she can also appeal before the high-chief (district court or Buri Jangi Paj). The high chief or District Commissioner presides over district court (Buri Jangi Paj). Before the high chief proceeds ahead with the court, he asks them whether this matter was dealt with by the village chief and elders. When the high chief is sure that the case was first seen by village chief and his elders, he then orders one of the party (which could be the accuser or the accused family) to appear before court on specific date, day and time. Nyiwelo (1992) says, on that appointed date and time, the court would consist of High-chief (District Commissioner), the Jurry (Jo Tog Buro) and the audience "Aneni" (P.22).

Once the people have gather, the high-chief (*Jangi paj*) orders head of the accused family to report to the court what their son or daughter told them in their family gathering. The head of family gets up and report to the court all their findings at the time of family hearing. When the head from the offender family finishes, the high chief then asks the offender to react on the reports. In some cases the offender may admit the report is correct or sometime the offender may deny the report. Thus, if the offender admits that the report is correct, the offended would be asked to comment on the report, comments of the offender and fact findings by elders.

After this floor is opened for the jury to comment and when they finish their comments, a floor is again opened to the audience to comments of the report, offender's comments and the comments of the offended. The debate goes on until everybody who has an opinion has been given the chance to voice his or her views.

When everybody voices his/her views, the district chief (high chief) summarises what have been said and pronounces the judgement.

But where the offender denies the case, the offended is asked to state his/her case and produce evidence to support the case. Nevertheless, the statement of the head of family tribunal is always taken as conclusive evidence with respect to admissions made before family tribunal by the offender.

Next the offender is given the chance to present his/her defence and provide evidence to support the case. As in the previous case, the floor is also given to the Jury and the audiences to debate out the issue. Such a debate reviews the Shulluk law and judicial precedents as far back as they can be remembered. As such, the debate guides the district chief (commissioner) to the proper legal principles to be applied to the case.

After all the views have been heard, the district chief summarises the case and then pronounces the judgments. In passing the judgement, the district chief does not need to follow the majority opinion; he may uphold the opinion of only one man if it, appeals more to him than the rest (Nyawelo, 1992).

According to Nyawelo (1992) when the case of murder is taken before the district chief, payment of court fees is left to the ability and conscience of the offended. The offended may or may not pay the court fees; yet the case must be heard by the district chief. Should the offended decide to pay court fees, this is usually considered as a contribution to the general fund for the district chief earmarked to welcome and support the guests. In other words, this fund is to be used for safe-guarding the security, peace and social welfare in the district. However, in the absence of such a fund, the district chief (Commissioner) will be expelled to make expenditures for his guests and the needed people from his own pocket. Since, in the first place the running of the district is generally his personal responsibility.

The Shulluk people believe that any court in session is guided by God's spirit. Therefore, any statement made to it in form of testimony, opinion or judgement is made direct to God. Thus, people making such testimony, opinion or judgement, they must do it factually, frankly for otherwise they will be accountable to it in front of God when they die; not only this, if they are not frank, God's wrath will descend upon their off-springs. As such for a Shulluk man, woman, boy or girl giving to court amounts to taking an oath in advance in regard to what a person is going to say. Although the people assume that a person going to court has already taken an oath in advance, nevertheless, many a times, one of the parties, for practical reason, could request his/her the other to make an oath before the court. In Shulluk customs, such request cannot easily be accepted by the court, the court only can accept such request on the condition that a person requesting for an oath has paid ceremony fee. This is the reason, why in Shulluk court, a witness is generally not required, because the accuser and the accused are supposed to speak the truth and only truth.

The divine nature of a Shulluk court is expressed by the following song, which is meant to be an advice to the King (Reth).

Collo	English
Atwolpiny, yagi nyar Akwey,	Atwolping, son of Akwey
Gong gonya Jwok, wuo mier.	A case is a disclosure to God, good father
Ya Kwaja Adiang nya kuathker	I beg Adiang, son of Kuathker
Bany Padiet, Adiang Nyar Akwey,	The sovereign Padiet, Adiang son of Akwey
Padiet Adiang gony nyi gonya Jwok.	Padiet Adiang cases are made to God
Collo lala mal o di cang a rony Padiet,	The Shulluk were lamenting the time Padiet took over
Adiang nya Kuathker labo Kwaja nini	Adiang, son of Kuathker, the nation is begging for peace

Source: Nyawelo, 1992, P.23

Another way of reaching conflict resolution between the warring parties could be done through women. According to William Oboc ayul Deng, one of the Shulluk elders, if the warring parties fail to listen to the chief, a woman takes away her clothes and marches in between the warring parties naked wailing and calling out to them. "See where you come from". The warring parties when they see the naked woman, they get scared by the scene and quickly break up the fight or argument and walk away.

Appeal made to provisional Governor (Jangi Phodho)

However, if the accuser is not satisfied with the judgement of the district chief's court, he or she may lodge an appeal to the Shulluk provisional Governor (Jangi Podho). In Shulluk custom there is no obligation in law to pay fees for an appeal lodged with the provincial Governor (Jangi Podho). Such payments are pure moral obligations and are taken as individual participation in alleviating the financial burdens from the provincial Governor. My informants in Melbourne stated that where fines are imposed on one of the party, the people from the other party would discuss and one person may agree to pay the fine on the behalf of the offender. Once, this person who sacrificed to pay the fine for the offender has committed a crime, and he is fined in future, the one member from the community would do the same by paying the fine on his behalf.

Final appeal to the King

Interestingly, if one of the parties is not yet satisfied with the resolutions reached thereto, appeal could be made to Reth (King) whose decision would be final. This means that when the plaintiff is not satisfied with the decision made by the commissioner (Jang-Podho), he or she may appeal to the King. In the appeal the plaintiff is expected to pay court fee based on one ability, but traditionally, there is no obligation to pay the court fee. Whether the plaintiff pays the fee or not his or her case would still be heard. Therefore, in Shulluk custom there is no obligation in law to pay fees for an appeal lodged with the King (Reth). Such payments are pure moral obligations and are taken as individual participation in alleviating the financial burdens from the King. Thus, such concept as bribes does not exist in the Shulluk legal philosophy.

Family Violence and conflict resolution among the Collo

When there are cases of family violent the plaintiff may present his or her case orally to the head of family of which the defendant is a member. Once the head of family receives the case from the accuser, a family judicial tribunal is then set up that consists of elderly family members.

The offended (Plaintiff) is then called, and questions put to him/ her based on the claims raised by the accuser. The elders look at the both sides of the stories from the accused and offender. This is done to justify whether the claims are true or false.

When elders are satisfied with the claims and answers, they pass their decisions and judgments based on the confronting arguments. When plaintiff is satisfied with the decision, fine or warning is passed to the offender. And this would be the end of conflict resolution. However, if the plaintiff is dissatisfied with the family tribunal, or the dependant failed to heel the judgement of the family tribunal, the plaintiff may proceed to the district court (Buri Jangi Paj).

At the district court (Buri Jangi Paj), the local chief resides over the meetings and assisted by influential elders (Jury). When the chief and Jury members are satisfied with the claims and answers, they pass their decisions and judgments that would be based on the confronting arguments. When the plaintiff is satisfied with the decision, the offender is asked to pay fine or he/she is given a strong warning -not to repeat such kind of offense again in future. Interestingly, if the plaintiff is dissatisfied with the district tribunal, or the district court failed to heel the judgement of the district tribunal, the plaintiff may make appeal to the King (Reth) as we have discussed above the final appeal to the King. This is step by step, on how conflict resolution is carried out in Shulluk society.

Traditional Nuer justice and reconciliation

Having explored how reconciliation works among the Acholi and the Collo, I feel that we should also briefly look at how reconciliation is done among the Nuer Community, who are the neighbours of the Shulluk people.

In the old days, is a person killed someone, another person from the offended family or clan would also kill another person from the offender family or clan in revenge. The killing from both sides could go on and on generating a spiral of violence. According to Nuer customs, people involved in the killing, to spear a person under this revenging circumstance is not considered a "crime"; revenge has to take place as a kind of compensation, for the life of those who were killed. Subsequently, violence and death between the two parties will not stop. For this reason, the chiefs and the elders of that time decided that the situation should not go on indefinite. So, around the year 1940s, the chiefs and the elders decided to change the system. They started calling conflicting parties to solve the issue peacefully. They then performed traditional ceremony rite for conflict resolution.

As one of the informants, Kerubino Nyuon Yar, chief from Panyijiar, said, "Our grandfathers had laws for life. If there was conflict between two persons or groups, they were brought under these traditional laws, and the person or groups found guilty was persecuted, and the problems and conflicts would disappear next day. Unfortunately, today, our traditional laws have been overlooked and their values are forgotten by the South Sudan authorities".

According to Nuer tradition, not everybody can make laws. Traditional laws are made only by these four groups of people: the chief of the land, the chief of cows, the chief of sticks and the chief of spears.

In Nuer traditional justice and reconciliation there are only three ways of organising reconciliation.

(1) *Reconciliation of broken relationship in fights.* If a person fights someone and wounded that person, he (offender) is required to kill a cow or a goat as form of reconciliation. The cow or goat is to be speared by the local chief for the ceremony; and the meat of the animal is shared by the two sides and this is considered or taken as total reconciliation has been reached. After eating meat of the animal, the chief determines the number cows to be paid to the family of the victim.

During the reconciliation ceremony, the chief of land gets up first and calls upon the parties for unity, since they all live in the same land.

After this, the chief of spears gets up and bends a spear in front of the two parties saying, "This is my spear, by bending this spear in front of you, I want to tell you that no one is again allowed to fight using spear against his/her brother or sister".

When the chief of spear finishes, the chief of stick gets up, breaks a stick in front of the of the two parties saying, "This is my stick, by breaking this stick, I want to tell you that as from today onward,, no one should again beat his/her brother or sister with stick". After this ceremony, no revenge is allowed between the two parties, because violence is believed to have finished.

One of the Nuer elders told me that the above processes are parts of the ancient ceremony in reconciliation among the Nuer society, which have been passed from generation to generation and today chiefs and elders are following the same foot step.

Thus, when the wounded person feels better, the victim is advised not to take any revenge against his/her offender's family or clan. Customarily, many people abide by this advice, but in some cases, some people fail to fulfil the traditional order. Person who fails to keep the advice, and later takes revenge, that person will die. For this reason, the Nuer people often abide the advice given to them by their chiefs and elders.

(2) *Reconciliation in case of murder* If a person kills the other person; he/she (offender) is required to pay number of cows. It is the chief who determines the number of cows to be paid to the offended family or clan. However, sometimes the chief may decide the offender to give a cow with a calf to the family of the offended. But in many cases many cows are paid for killing a person.

Thus, when killing happened the chief calls the two parties for negotiation. Once the negation has been successful, number of cows are decided by the chief are paid to victim's family; and reconciliation ceremony organised.

During ceremony, the meat of the thigh of the cow is cut and roasted. When the meat is already well roasted, the people mix meat with gall-bladder. The two sides are then asked to eat the roasted meat together with gall-bladder. After eating and sharing the meat and some milk, each part is asked to choose a representative and the two persons drink water from the same bow—possibly this is the water that the two parties have washed with their hands, before and after eating meat.

After this, one of the elders takes long bone from the killed animal and breaks it in front of the two parties. As the bone is broken, meat and milk shared and water drank from the same bow, this is a clear indication that reconciliation has been reached between the parties.

What remains unclear to me is where they take the broken bones after the reconciliation ceremony!

(3) *Reconciliation for the broken relationship:* There are many things that could cause relationships to be broken; and these happen mainly among the members of the same family or clan. Thus, if the relationship is broken between two members of groups of people from the same community or village, the person causing the conflict would feel pains and uneasiness in his/her heart. This person most of the time may think that the other person may retaliate.

With this view in mind, the offender then calls elders and tell them what happened between him/her and the other person and how he/she feels now. When the elders hear this report, they summon the other party and go for peaceful negotiations; sometimes these negotiations finish without killing a cow but in other cases, where the offended has his or her leg or hand is broken, in this case compensation depends on the demand of the offended family or clan. If the family or clan demand two or three cows to be paid they are usually paid.

This is briefly the traditional justice and reconciliation among the Nuer. Should the readers want to know more, about Nuer traditional justice and reconciliation, I refer you to read other books.

Jo-Luo Traditional Justice and Reconciliation

Tribal and Clan War: Compensation and Reconciliation

The Jo-Luo society of Bhar el Ghazal is the only society in South Sudan, who maintained their traditional justice and reconciliation, with no influence by the colonials and pre-colonial systems. Thus, among the Jo-Luo people, whenever there is inter-tribal, clan or family conflicts, elders assemble and mediators are involved to investigate the root cause of the conflict. They are also slow in declaring war with other side. Apai Gabriel (2005) observed that the Jo-Luo people would not easily declare war on others for emotional reasons. Only a repeated provocative action would make them declare war. But, to do so, elders would assemble to investigate and find out the root cause of the dispute (P. 80).

Therefore, if the dispute is among Jo-Luo sections, mediators step in and summon both sides together for a peaceful solution. Alternatively, if the dispute if between the Jo-Luo and another tribe, elders also come together to examine the root causes and put forward their suggested solution to the other tribe through an envoy. However, if the atrocities continue and there is no response from the other side, the elders would reassemble to review the situation, in which case declaration of war might be the first option.

Once the elders have reached a decision to declare war with that tribe, they call all the generals from various sections of their communities. They also call spiritual leaders to pray for the generals and warriors who are now preparing to wage war against a tribe. In their prayers, spiritual leaders would be recalling the aggression which the other tribe had been repeatedly making against the people. In doing this elders would get up one by one and say, "God of our grandfathers, who made our grandfathers successful in their wars against the enemy; God who judges the wrongs of the enemy, I pray you to help our generals and warriors to carry out a successful war, as they are now going to face our enemy". After prayers, spiritual leaders then spray the warriors and their generals with holy water so that God protect them, and above all, so that God gives them peaceful return from the war (Apai Gabriel, 2005).

When each elder recited the prayers, they then declared war against the other tribe. This is followed by mobilisation of the warriors. To mobilise warriors to take up arms, the Jo-Luo people blow a war hooding horn. The war horn is usually blown at dawn so that the nearby villages hear the sounds. As a man blows the war horn he would be sending to the people the following messages: "War war" "Danger danger" "Here here". This is usually interpreted as, "*You warriors, war is coming; it is dangerous, so prepare yourselves, and come and gather here*".

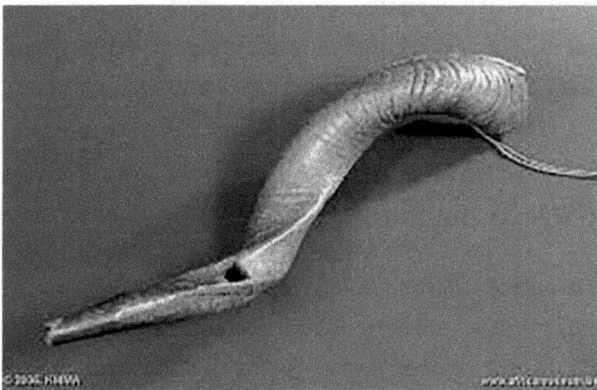

Figure: 78, War-hooding-horn

Figure: 79, Jo-Luo Warrior

The surrounding villages, in turn, reply with the same tune, setting people in the further villages on full alert to move down to where the sound of the war horn is coming from. In a very short time, people start to arrive in groups, armed to the teeth with spears and shields. If it is an alarm for a suspected attack from enemy, warriors take up positions according to sections and clans. But if it is an aggressive campaign against the enemy, generals gather in the area to make plan; they then send scouts to survey the area before attacking. The generals and warriors wait here until they hear from the scouts. Traditionally, the best time to attach an enemy is by night when everybody is sleeping, so as to take the enemy by surprise and to find men unprepared and at home; particularly those who perpetrated the dispute and those who killed.

When the first cocks crow, the warriors the attack the enemy killing any man they would come across. In conventional wars, women and children are not to be killed, they are taken as captive; and women will be divided among the men who have no wives to give birth to more warriors. However, boys who are found to be carrying spears along side with the men are likely to be killed.

When fighters come back victorious, war drums are sounded and the heroes are carried high on shoulders while people dance the "*gumo*". Today, people dance "*gumo*" only when one has killed aggressive animals such as a lion, an elephant, a leopard or a tiger but in the old days, people dance "*gumo*" only when a human being was killed.

According to Apai Gabriel (2005) when a clan or a tribe is at war with another, lonely visits to other areas are observed. People may pay visits to relatives in groups. But although they go in a group and the area is hostile one, the host area would find men to escort them out on their return journey for fear that they may be attacked on the way.

Once there is war among the tribes or clans, life becomes difficult on both sides, despite the fact that escort is done night and day, the risk remains the same. Because the Jo-Luo warriors use tactics of guerrilla war that make security for everybody look slim. This is also one of the reasons why the Jo-Luo people in the past used protective villages.

As times pass by, the people begin to realise that there is much war's disasters, the people then come to their senses, and elders from both sides assemble to find out solutions to end the recurrent wars. As the elders assemble, time would be given to elders from both sides to explain the root causes of the wars. In this gathering, mediators are also called to listen to the root causes of wars as raised by each side; and the mediators would address them accordingly. The offenders would be asked to pay compensation to the relatives of the victims. In Jo-Luo society, compensations for men killed are sheared by the cousin. Thus, when a cousin hears that his first, second or third cousin has killed a man; he would be alert, knowing that if he is found by the enemy of his cousin, he would be killed in revenge. Apai Gabriel (2005) pointed out that if the war has involved only a clan, the amount of any payment would be distributed among the three groups who would, in turn, break it down to the family level. But if the war involved the allied, their share would be given and the clan which is being assisted take a greater share of the burden of payment (P. 86).

When peace has been realised and compensation paid, the mediators will ask a relative of offenders to slaughter a bull as a sacrifice for a bond peace. The blood of the bull is then dropped into water that is in the calabash, and the people spray themselves with the water, and once more, they become socialised.

But in case when a relative has killed a relative, the blood reconciliation is somehow different. In this case both sides would be called and a bull is slaughtered and spear handle is held up and people from both sides catch the spear handle; at the same time pieces of charcoal would be distributed for people to chew. When everybody has chewed piece of charcoal, an axe would be thrown upwards and downwards by its handle three times. After the third round, the people drop the handle and each marks himself with charcoal. The compensation of a relative is symbolic, not the full amount of compensation is required by tribal organic law (Apai Gabriel, 2005).

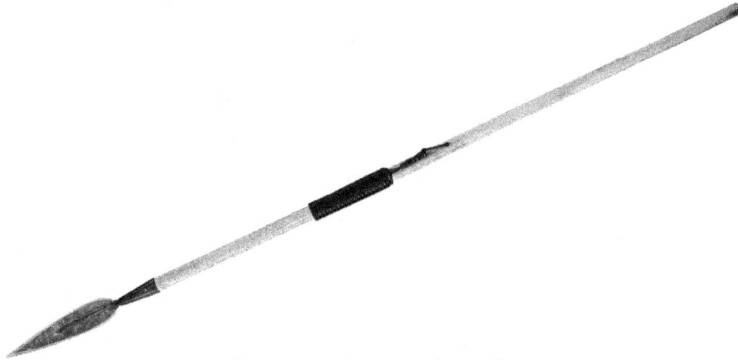

Figure: 80 Spear and handle
Media War (Achong)

The media war, commonly called *"achong"*, is very popular among the Jo-Luo of Bhar el Ghazal in South Sudan. This is also known as "cold war" it is fought by singing and disgracing somebody within opponent's clan. In the songs the singer calls the opponents bad names such as thieves, misers, cowards and wicked persons.

The cold war is often caused by simple problem, for example, when a girl is officially married and another man goes around trying to spoil her mind by disturbing the engagement and despising her partner in front of the girl; if the partner hears of it, he would react by composing song against that man without mentioning his name, but revealing, in his song, what he has told the girl. When the other man hears the song, and understands the meaning of the song, he would immediately realise that he has been mocked and returns the song with another bitter song, such as *why good girls are allowed to be engaged to certain people with no qualities*. Subsequently, the beating around the bush with songs goes on and on, until it comes to the point where names are mentioned.

When names started to be mentioned in songs, the composers from each side would contest and they sit with their elders to investigate the long cold war. They evaluate the meaning of the songs one by one; the greater singers, who are like war generals, will have their final saying on the songs. Thus, if the greater singer is related to both sides, therefore, he would have nobody to sing against. Eventually, they elders then tell him to stop the war of songs, at the same, time the elders would inform the other side that there will be no more media-war because of strong relationship that exists between the two parties. But, if there is no strong relationship, the media-war may continue indefinitely. Apai Gabriel (2005) pointed out that after the evaluation of the songs by the elders, if the elders find that reasons that lead to mockery songs are genuine, and the generals agree to it, the two sides would go ahead with composition of songs against one another.

Shortly, the great singers will start to compose songs against each other and these songs become popular among the young people in the villages. When singers from both sides have composed enough songs, they will appoint a time to meet in a particular place and time. The best time for the singers usually to meet for show of "war-words" is in the evening. When time comes, both sides gather in a given place, and as a rule, each side is allowed to sing four songs at a time, while the other people from the other side listen to then without interruption. And once the four songs are sung, they cannot be repeated in the sho within that day of the contest. And if it happens that one side ran short of songs, the other side will continue to sing against them, and if this is the final day, they will be declared winners. There are different ways in which a party could be considered as defeated. One of the ways a party would be considered defeated is when a person feels abused or despised by the songs, in such a way that he gets angry and try to fight —fighting in or during the 'war of words" is not acceptable. Generally, this media-war battle may take up to fourty eight hours, only with breaks to eat and come back. The young men thought that to make the event interested, after four songs the participants make the rhythmic music of the "kanga" and everyone dance.

This media-war custom has a profound impact on the Jo-Luo people's social behaviour, because the songs which children gear from childhood to adulthood are full of criticisms of the bad things alleged to have been committed by some individuals. Therefore, a child who grows up environment, hearing and sing these kinds of songs, automatically becomes cautious in his/her attitude towards such acts and avoids these criticised behaviours for fear of being publicised in a media-war in the future (Apai Gabriel, 2005).

In the Jo-Luo society, some of the songs are composed to remember what has befallen singer. The best examples for such songs is what Kuo or Kwa Lual Youk sang in early 1940 when his bride was abducted by another man with the full knowledge of the girl's father. Kuo Lual Youk, when he learned that his future wife has been abducted he approached the family of his bride and demand his money and animals for the dowry to be reimbursed but bother the father of the girl and the abductor failed to reimburse the dowry. Kuo Lual Youk then expressed his rejection song as follows:

My thing, my thing, how little it may be
Like seed of sesame
My heart will never leap
My heart will never claim

Things of a rich man
Who owns lions and elephants!
Has rhinos and hippos!
And rears buffalos as his herds!
The ducks of the river are his chickens
He eats with a spoon as big as a canoe
His sorghum is as plentiful as grass
Though he builds houses
As high as Twin Mountains —highest in the Jo-Luo area
And delivers a daughter like Dhalka (Apollo)
I, Makuei, will not consider her
Haa! Upieu, even if it is sweet like honey!
Though as good as gold!
Which outmatches all dresses in the market!
And an honourable bull in the touch
I, Makuei, will not look at her
Satisfaction brings mental disorder
One talks like a mad man
And spoils his own thing
Which would have not
With his own tongue
Alas! In-law, you are satisfied to your neck
You say no more, because you are hero of the world
Ho! On earth is for work and sweat
Satisfaction is a loan from God
One cannot find comfort without a walk
Bring my small thing
Source: (Apai Gabriel, 2005, PP. 90-91)

The Jo-Luo Customary Courts

Jo-Luo Courts are presided over by chiefs or male elders from communities. The court judgements are often for reparations and condemnations. For example, if a person is found guilty of committing adultery, with somebody's wife, he would be sentenced to pay reparation (*aruok*) to the husband.

Before the British conquered the Sudan, and established a modern legal system in the Sudan, The Jo-Luo of Bhar el Ghazal in South Sudan had their own systems. It is reported that the old Jo-Luo system was based on tribal organic laws, which dealt with crimes such as adultery, murder and thefts.

Before long, the British abolished most of the Jo-Luo tribal organic laws and replaced them with their own system; and in addition they introduced death penalty which was unknown to the Jo-Luo people. The death penalty made the Jo-Luo people more scare. Eventually the British removed or ignore the Jo-Luo traditional jurisdiction. One of the best examples is that in Jo-Luo organic law, sexual relationships between relatives were and are forbidden and considered disrespectful, but with the modern law, sexual relationships between relatives is considered a normal incident. This is because the British laws are derived from the reverse of a communal source, which does not recognise Jo-Luo traditional values (Apai Gabriel, 2005).

Swearing

The Jo-Luo people swear in many ways; some of them swear by snakebite, others swear by thunder or anything that can kill instantly. Today, the Jo-Luo people avoid swearing because they have heard and learned many miraculous things that happened in the past due to false swearing. Furthermore, the modern Jo-Luo discovered by themselves that some of their traditional organic laws are out-of-date; as such they have agreed to remove them out from the modern judicial practices. The good example of out-of-date judicial practices are swearing by fire (chut-mac) and drinking a poisonous juice from a tree, this is known locally as "*kuir*".

The people use swearing by fire when there is evidence that a man and a woman had committed adultery, but each of them denying the fact on the ground that there is no eyewitnesses. Thus, in such case, the court would ask the pair if they could swear by fire. If both agree, a consecrated stone master would be called to come and swear them. The master of ceremony comes with stone; he puts it on a pot full of water and boils the water before the court. When the water boils and gives hot steams, the man would lubricate the accused couple's hands and arms. The master of ceremony then asks the man and the woman to pick up the stone from the boiling water in the pot. It is generally believe that if a person had not committed crime, that person would take the stone from the boiling water and feels pains but his/her skin would not make blisters. However, the result of this type of swearing is always not clear because in the swearing process one of the accused person gets blisters, while the other may not get blisters —remains clean. Subsequently, this practice was seen by modern Jo-Luo as a torture rather than a solution —torture would not serve any purpose.

Needles to say, the truth testing habit called "*kuir*" is even was than swearing by fire (chut-mac). The "kuir" (juice swearing) is administered to a woman who has been accused of being a witch but she denies. In this case, the accused woman would ask for a bitter juice drink. She wants to drink this bitter juice to clear her name and to prove innocence. According to traditional law and procedure not any person from the street could be selected to administer the "*kuir*". Customarily, this person is selected from among the community and he must fast and abstain from certain food and acts, including sexual intercourse for a given period of time before administer the "*kuir* juice" to the accused. And to fulfil the demand of the woman, one man would be asked to press juice out from a "kuir tree" and makes it into a drink. The woman would then be taken to a nearby glade and given a drink. As a woman drinks this juice, she becomes much intoxicated to the point of losing her sense and makes her talk nonsense. It is generally believe that if the accused woman is not a witch, she would vomit out the "*kuir*", but if she is a witch she would drink more and more without vomiting until she becomes drunk and begins to confess to the people that she has killed people through her witchcraft. Eventually, relatives of those she admitted to have killed would even kill her. Therefore, calling a person a witch in Jo-Luo society is a major offence because the Jo-Luo believe that witchcraft is a generic heritage, meaning from mother's side, Hence, when a daughter is accused of been a witchcraft, this implies that her mother, grandmother and great-grandmother are all accused of been witch (Apai Gabriel, 2005).

Today, in the Jo-Luo society of Bhar el Ghazal, in South Sudan, traditional swearing by spear remain an ideal legal swearing-in procedure to lay before native courts.

Figure: 81, Spear for swearing

Therefore, swearing by spear has been taken as a model because it harms nobody but only the person who swears. Thus, when a person swear by a spear, he/she is telling Master God that he/she is innocent and that if he or she is lying, let Master God use the spear he or she holds in his/her hand to kill him/her in any way possible whether now or later on.

Pari Traditional Justice and Reconciliations

The traditional instruction of conflict resolution in the Pari society is the head chief. The chieftaincy in the Pari society came into existence through the evolution of the society in its contact with the neighbours. In my research, I have made to understand that the people of Pari practice conflict resolution similar to that of Lotuko and Lopit who are the immediate neighbours.

We have seen in the Luo group of South Sudan that Pari society is composed of six villages. Each village has its own chief (*rwath*). The Pari traditional justice and the conflict resolution centres on the court known as "Jo Likweeri". The chief of Wiatuo is entrusted with the rain administration and he is the most important chief among all other chiefs.

When it comes to the mechanisms of conflict resolution in Pari society, the chief of Kor and chief of Wiatuo are responsible for hearing cases and to bring justice among the population.

However, there are many types of disputes in Pariland that can be handled by respective chief of the village. Murderer is the extreme case in the society that could easily start a cycle of vengeances and violence between clans. There are traditional laws, regulations or customs which are used to manage murderer cases. The normal practice is that when a person committed murder, for the offender to escape revenge, he/she must seek a protection of the chiefs of Kor and Wiatuo. During the first few weeks, before the case is heard, the two chiefs selected judges to escort the offender wherever, he/she goes (Wassara, 2007).

The two chiefs will then decide the day and date for the court hearing; and both family members of the offender and family of offended would be summoned. It is customarily that the clansmen of both parties accompany their family members to the court. It is to be noted that the court is opened for any member of the villages who wants to attend. Elders in the community are encouraged to attend and they are expected to give their opinion on how the matter could be handled. The views or opinions given by the elders always help the two chiefs to reach conclusion and make decisions.

Hanging or killing a person, who killed another person, is usually prohibited by Pari traditional laws. The verdict always centres on compensation of the victim locally known as "kwoor". Compensation is paid in form of a girl or heads of cattle. This is usually determined the family of the offended (Victim).

In case where the family of the victim want a girl to be given as a compensation for the victim, in this case, the responsibility falls entirely with the family of the offender who is to give one of their girls as compensation. The two chiefs will then ask the offender to provide a goat that would be sacrificed to purify the compound from the evils brought about by the murderer. This practice (of giving a girls as form of compensation) is been condemned by the Catholic Church, which is predominant in the area.

Otherwise, in case where the family of the victim choose heads of cattle as form of compensation, here, the responsibility falls on the whole members of the clan of the murderer. Besides the cattle, the clansmen of the offender must provide a number of goats to the court. One of the goats has to be slaughter for the performance of the rite, while the rest of the goats will be divided among the panel of judges and the chiefs (Wassara, 2007).

However, when a Pari man killed a foreigner, like Toposa who come to steal cattle, no compensation of whatsoever is made, but the murderer would be affected with what they called "mira" (the power of the blood of the victim). "Mira" usually affects murderer regardless whether he kills a clan man or a foreigner. Mira is said to be contagious and when a man kill a foreigner, he should be cleaned as soon as possible for otherwise, the murderer could spread "mira" to other people. When a person gets "mira" from the murderer he/she becomes sick with leprosy. Kurimoto E. (1992) observed that when a man kills an enemy of the society, such kind of murderer is said to be good but when he kills a member of the clan such kind is said to be bad. Thus, when a man kills any person, his children would be in more danger of contracting leprosy, so the state of "mira" is contagious. In order to prevent the effect of this, a goat should be slaughtered and its stomach contents smeared on the boy: on his door steps, knees, middle of the chest, forehead and top of the head. The same stomach content is smeared on the bodies of murderer's family members: on their knees, middle of the chests, foreheads and top of the heads. The same thing is done to the relatives, sisters and brothers of the murder –and on all those concerned.

If a murder takes place in the bush, like the case of Toposa man, where a goat is not available, a cucumber or some soil of anthill would be put in to the calabash of water and smeared on the body (Kurimoto, 1986, P. 136). It is reported that these items are smeared on the body in order to remove "the smell of the blood". In other words sacrifices are made in order to restore the normal condition.

The removal of "the smell of blood" from the body of a murderer takes months if not years. Thus, this implies that whenever a murderer wants to visit friends and neighbours, he would be required to keep some of the animal stomach contents, so when he comes to the home of the friend or neighbours, he is to take a piece of stomach contents of goat and smear it on the body of the friend or neighbour before he greets him/her.

Therefore, when a murderer contacts a person without smearing goat stomach contents on him/her that person would be affected by "mira". Subsequently, if a murderer greets a person without smearing goat stomach contents, mira would be transmitted to that person. What would then a person do to remove "mira" from the body? The person has to follow the same processes we have described above –slaughtering a goat etc.

Riek Machar attempt for Reconciliation

Riek Machar has made many attempts for reconciliation and healings, of which all have failed. Riek Machar, after having achieved his goal of becoming the first Vice President of South Sudan, on 22th May 2016, he went into Emmanuel Parish Episcopal Church to recultivate reconciliation and healing with the people of Bor State. This sparked tension among the church congregation.

Figure: 82,
Riek Machar in Dinka Emmanuel Church Juba

Emmanuel Parish is the Church where the majority of Bor sub-ethnic Dinka people pray every Sunday. The Bor people experienced horrifying genocide from Riek Machar in 1991 and again in 2013 his forces massacred the people of Bor. As Machar was talking about peace, reconciliation and healing, and defending himself that he did nothing wrong, many faithful left the church, and those who decided to remain behind were lining their heads against the tables, and some slammed the Church, while the pastor kept looking through the window, this is a clear indication that he was not interested in his speech. The man on the right hand of Dr Riek Machar is not even listening to him; he puts his head against the table. All these were done in cool protest as many Bor Christians, who were at that time in Church, wondered as to why the Church leader has taken that drastic step.

After few days, a group of people, which referred to themselves as the relatives of Riek Massacres of 1991 and 2013 to 2015, organised a protest against the action taken by the top church leaders, they said the Church leaders should not allow Machar to polarize his political agenda under the pretext of mission reconciliation. The group condemn, in a strong term, the Church leader of Emmanuel Parish, they said church leaders should not permit politicians to discuss politics in the Church; this is against the Church doctrine and regulations. The Bor Christians believed that Machar was preaching false reconciliation and he was in the Church not to apologize to the relatives of the victims but rather to laugh at them. He (Machar) never acknowledged that his forces have raped and killed elderly women, who were praying in Emmanuel Parish, in Bor town. Machar never acknowledged that his forces have killed patients on their beds at Bor Civil Hospital. Therefore, Riek's tour of churches in Juba City is not a true reconciliation; he is promoting his political agenda. If the churches really want true reconciliation, they should organise a dialogue and reconciliation conference where both the relatives of the victims and the perpetrators could have dialogue and reconcile. Riek was doing false reconciliation that the people do not want. However, if the Churches cannot bring together the relatives of the victims and the perpetrators, they should rather focus on worship of the living God rather than mixing politics with religion and creating more hatred and division within the members of Congregation.

Reconciliation and Healing in South Sudan

"If you truly want to set things rights, you must walk the path of peace. Peace is denouncing violence and accepting the diversity in a society"..."To forgive others is in fact being kind to ourselves —Chinese Proverb"

Figure: 83, Army tank-car patrolling in Juba City

The South Sudan is composed of sixty four tribes, of these the Dinka, Nuer and Shulluk are considered major tribes in the country. Before the signing of Comprehensive Peace Agreement (CPA) in 2005, and yet before the ballot paper war referendum in 2010 that followed with independent on 9th July 2011, the South Sudanese viewed themselves as one entity with common enemy called "Arabs". Wassara (2007) observes that conflicts have become more complex in the aftermath of the CPA; new conflicts have emerged that have the bearing on people returning from refugee camps in to the country.

Interestingly, soon after the independent of South Sudan in 2011, life in the New Country was characterized by uneasy relationships between the two major tribes, the Dinka and Nuer, due to the political ambitions, where each group claim to be only right people to lead the New Nation. The Dinka and Nuer forgot that the achievement of independent of South Sudan were the collective efforts of the sixty-four tribes in the country. Secondly, they forgot that the New Nation can be led by any person, be he or she from minor tribe, provided that he/she has the ability and knowledge to lead the country. As sun rises and sets, some people are been misled to believe that the conflict was cultivated by the scare natural resources and how to use them in the diverse needed population.

Subsequently, this led to split and outbreak of civil war in December 2013 where many people lost their lives. The Nuer group formed what they called Sudan People Liberation Army –in opposition (SPLA-IO) on the basis of the word "*Wanathin*", which means let us just do it whether we succeed or not (no thinking, planning and strategies). The SPLA-IO eventually become a family movement with Riek Machar as its leader and Angelina, his wife, Minister of Security and Defence; Deng Taban Gai, the brother-in-law of Riek Machar, become supply officer and Minister of Finance and Ezekiel..........become Minister of Foreign Affairs and communication.

Figure: 84, Two members of the JCE reading Leaked list of their members

While the Dinka group formed what they called Sudan People Liberation Army –in government (SPLA-IG) on the basis of the word "*Konkoc*", which means let us think, plan and put our strategies and tactics before we execute them. The SPLA-IG remain under leadership of Salva Kiir Mayardit, with all the power of leadership in the hands of Jieng Council of Elders (JCE) that is composed of 45 members; and the below names of the 45 JCE show representatives coming from various States around the country.

1. Representatives from Upper Nile State.
   1) Hon. Deng Chol Deng, Abialang section.
   2) Hon. Akot Dau, Ager section.
   3) Hon. Gatwec Nyok, Nyiel section.
   4) Hon. Thon Mum, Dongyol section.
   5) Mr. Joseph Nyok Abiel, Ngok section.
   6) Dr. John Antipas Ayiei, marbek section.
   7) Mr. William Sunday, Paweny section

2. Jieng Representatives from Jonglei State

1) Hon. Joshua Dau Diu, Luac (Padang) section (Bor).
2) Hon. Anne Lino Wor Abyei, Hol section (Bor).
3) Hon. Daniel Deng Lual, Nyarweng section (Bor).
4) Hon. Deng Dau Deng, Twic section (Bor).
5) Hon. Deng-Tiel Ayuen, Athoc section (Bor).
6) Hon. Maker Thiong Maai, Gok section (Bor)

3. Jieng Representatives from Unity State:

1) Hon. Benjamin Majak Dau, Ruweng (Panaru) section.
2) Hon. Battaria Monyror Makuei, Alor section.

4. Jieng Representatives from Lakes State

1) Hon. David Deng Athorbe, Atuot (Yirol) section.
2) Hon. Gabriel Daniel Ayoal Makoi, Kiec section
3) Hon. Permana Awerial Aluong, Aliab section.
4) Hon. Chief Daniel Dhieu Matuet, Agar section (Rumbek).
5) Hon. Gabriel Daniel Deng Monydit, Pakam section (Rumbek)
6) Hon. General Daniel Awet Akot, Gok section (Rumbek).

5. Jieng Representatives from Warrap State

1) Ustaz Lewis Anei-Kuendit, Rek section (Tonj central).
2) Hon. Cauor Adong, Luacjang section (Tonj north).
3) Hon. Mawien Akol Aduol, Rek section (Tonj South).
4) Dr. Riak Gok Majok, Rek section (Tonj East).
5) General Agasio Akol Tong, Aguok & Awan section (Gokrial)
6) Just. Ambrose Riny Thiik, Apuk (Giir Thiik) section (Gokrial).
7) Hon. Joseph Lual Achuil, Nyok, (Kuac) section (Gokrial).
8) Hon. Mayom Kuoc Malek, Twic maryardit section (Twic West).
9) Hon. Charles Majak Al, Twic Mayardit section (Twic central).
10) Hon. Kuany Mayom Deng, Twic Mayardit section (Twic East).

6. Jieng Representatives Bahr El Ghazal State [Mading Aweil].

1). Joseph Aguer Alic, Akuang Ayat section.
2) Hon. General Albino Akol Akol, Malual Geernyang section
3) Mr. Cleto Akot Kuel, Paliopiny section.
4) Hon. Arthur Akuen Chol, Paliet section.
5) Hon. Pio Tem Kuac Ngor, Abiem west section.

6) Hon. Aldo Ajou Deng Akuei, Abiem central section.

7) Hon. Kom Kom Geng, Abiem East section

7. Jieng Representatives from Western Bahr El Ghazal State

1) Mr. Moris Yol Akol, Wau marialbai section.

Source: The Upper Nile Time, 7th April 2016

The JCE did not include minority Dinkas from Upper Nile Region; the organization is formed primarily to kill Nuer and all its alliances such as Equatorians and Shulluk. The above elders prior to December 15,2013 crisis according to reports, often meet in person or on phone with President Salva Kiir and Gen Malong Awan to strategize plans, recruit youths and to instill tribal agenda within their states. The agenda are often: 'Riek Machar should not be allowed to take leadership' and 'Nuer are our only enemies in South Sudan'. The Jieng people must unity to avoid what happened in Rwanda in 1994, (for otherwise the Nuer and its alliances ill massacre thousands and thousands of Jieng people)

This action brings us to a crucial question, "Were/Are there no people from other sixty-two tribes in the SPLA-IO or in the SPLA-IG?" The answer is simple and clear that there were and are people from other 62 tribes in both SPLA-IO and SPLA-IG. These consist of political players, senior military commanders, youth from different political backgrounds, as well as girls and women from different walk of life. However, the main issue is not the presence of these people in the Movement or in the Government, but the reality is that these people who are fighting sided by side in the two parties, are actually viewed as alliance with no voice and can be killed or removed any time the leader wishes.

Is Reconciliation Possible in South Sudan?

It is important to start this article by noting two things that might have influenced South Sudan's involvement in conflict resolution. The first thing most probably might have been that the South Sudan Government is convinced that conflicts deter development in the country through sheer destruction that includes massive death in Juba, Bor, Bentiu, Nimule, Yei, Yambio, Malakal town and Wau as well as the displacement of civilians and the destructions of the little infrastructure in existence. The leaders are convinced that development in South Sudan can only be achieved if there is stability and peaceful in the country. As a result we can see that there is political will and commitment to engage in conflict resolution and Traditional Authorities. The bottle neck is, as I said before, reconciliation does not mean the same to everybody in the country.

Nevertheless, given the fact that the country is better resourced than most of the bordering countries, this has raised expectation for both international communities and South Sudan Government to play a major role in resolving the conflicts in the new nation.

Secondly, the current political leaders in South Sudan are convinced that the only way forward is for the country to adapt Federalism and accept good governance. But this requires a strong unity among the States, for otherwise the negotiation on conflict resolution may not succeed. Katabara (1980) notes that any negation on conflict resolution could only succeed on the ground that the States must continue to be strong. But, if much of the States in the country are weaken and fragile conflict reconciliation will collapse.

However, as we have seen in South Sudan, there is a clear division on conflict resolution by two major tribes that is the Dinka and Nuer; without understanding that other tribes from Malakal and Equatoria are also affected by these brutal killings. Therefore, whenever we talk of conflict resolution, it should be more inclusive –including all tribes affected by civil war.

When considering South Sudan's role in the conflict resolutions in the country that is still under President Salva Kiir Mayardit and his opponent Riek Machar Teny, on one hand we should not forget that Riek Machar Teny has broken too many resolutions and on the other hand, we should also understand that Salva Kiiri is being driven by Jieng Council of Elders (JCE), who are not interested in peace in any form or shape. As a result he has also broken series of resolutions. We cannot talk about conflict resolution without reaching tangible Agreement.

Reconciliation looks to repair right relationships, it attempts to figure out how to repair damage inflicted to these relationships through injustices, and to heal the resulting wounds. Reconciliation practices must therefore identify these wounds, find a way to address them and employ practices to achieve their objective. The driving force behind reconciliation is that of mercy. In order to achieve this Llewelyn and Philpott identify six practices that can help bring about reconciliation.

The first is a *presence of just political and economical institutions*; so that people do not feel discriminated against. Second *acknowledging the past and acknowledging the suffering of victims* is a practice common to the theory of reconciliation. We see this reflected in the thinking behind truth-comissions, as well as museums, monuments and remembrance rituals. Third we have *reparations*; a practice where victims are given material compensation for their suffering. While material compensation is unlikely to ever be sufficient repression it does serve as a public recognition of victim's suffering, and reinforces the legitimacy of the victims' grievances. Fourth is the idea of *accountability*; those guilty of oppressing people and committing crimes must be held accountable in some way. This can be done through reparations, acknowledgement of wrongdoing or other means of holding perpetrators accountable. Fifth is *apology*; the need for those guilty of oppression and wrongdoing to apologise for their actions. Be it apologising on your own behalf and your own misdeeds, or apologising for the role of government or political leadership in the past. Finally we have the concept of *forgiveness*. Those who have been wronged look towards the future rather than dwell in the past. They decide to not take a path of revenge and anger against those who wronged them. (Llewellyn & Philpott, 2014, pp. 26-27)

In conflict resolution, somebody must be held responsible and he is to account for the crimes committed. Looking to our current leaders in South Sudan, nobody want to accept his mistakes, they tend to trade blames between them, and this cannot help in any form or shape. Daniel W. Van Ness (2014) reiterates the importance of accountability in his article on the issue. How you decide whom to hold accountable and how to do so has a significant impact on wider society and in extension any peace and reconciliation effort. As restorative justice and reconciliation looks to restoring relationships that have been harmed it is also important to consider how to hold individuals and groups accountable for what they have done in a manner that aids the restoration of these relationships. To achieve this Van Ness presents three fundamental elements in a restorative understanding of accountability, namely that of reparation, truth-telling and taking responsibility (pp. 131-132).

If the actors of peace-building and reconciliation want the conflict resolution in South Sudan to bear fruits, I think it would be advisable for them to revisit the writing by John Paul Lederach (1997), there he presents us with a clear theoretical framework for peace-building and reconciliation. He too sees relationship as the core element in any conflict. Conflict breaks relationships, and harms them. It becomes the goal of reconciliation to change these relationships into peace-building and reconciliatory relationships.

In order for reconciliation to have such a capacity to change relationships there has to be a way for people to express to one another feelings of trauma and anger for past events. However, while giving the past just credence it is also crucial that the focus of all reconciliation is towards the future. Past grievances and traumas must be acknowledged, stories must be shared and justice must be delivered. At the same time the objective of reconciliation is to provide a foundation for a shared peaceful future. Therefore reconciliation must provide room, or in the words of Lederach; a space. In this space people can share in and acknowledge the past without indulging and burying themselves in it. (Lederach, 2013, p. 30).

Truth is understood as the necessity to validate and acknowledge people's experiences and trauma. People's longing to have their losses, their hardship and their stories not only heard, but also validated as a powerful motivator. Further, the concept of *Mercy* is understood as the process of acceptance. Of not letting past grievances stand in the way of a peaceful future and *letting go*. While *Truth* is necessary, it must be tempered with *Mercy* to ensure a peaceful future. *Justice* speaks to the need and the desire to see injustices punished and restituted. While *peace* relates to the need for security, well-being and interdependence in society. (Lederach, 2013, pp. 28-29).

Lederach understands *reconciliation* as both a *focus* and a *locus*. As a perspective he believes *reconciliation* at its heart deals with the relational aspects of the conflict. As a social phenomenon Lederach argues that *reconcialtion* represents a social space in which parties in a conflict meet. It should be the goal of reconciliation to provide a setting in which the past can be discussed with the goal of achieving a better future. (Lederach, 2013, pp. 30-31)

In order to visualise how one can develop approaches to building peace, Lederach shows us a pyramid model of the different levels of social actors. He identifies three primary levels of social actors; top leadership, middle-range leadership and grassroots leadership. We can imagine a pyramid where the grassroots - the group with the highest population -makes the foundation, followed by middle-range leadership and topped off by the apex of top leaders (See Figure below).

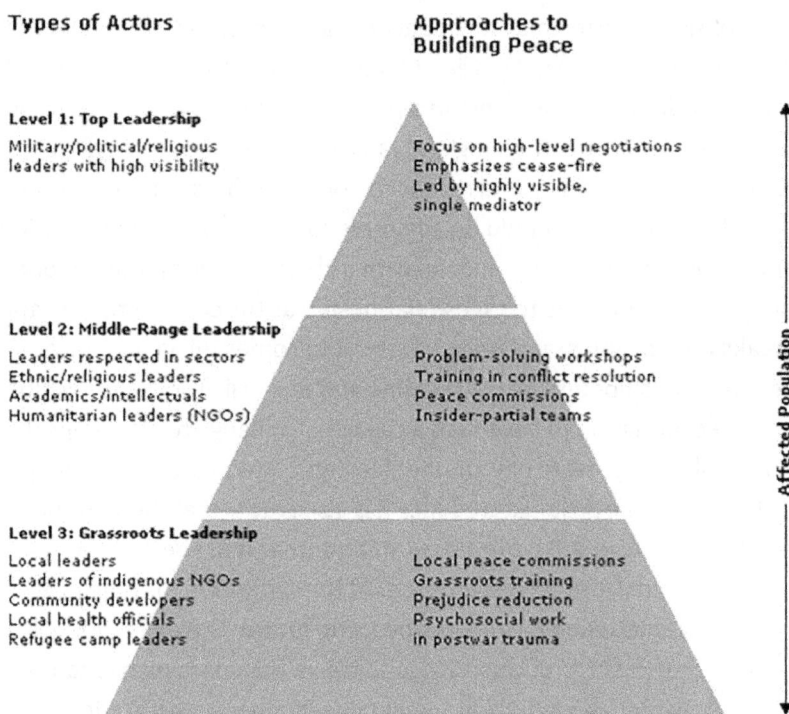

Types of Actors

Approaches to Building Peace

Level 1: Top Leadership
Military/political/religious
leaders with high visibility

Focus on high-level negotiations
Emphasizes cease-fire
Led by highly visible,
single mediator

Level 2: Middle-Range Leadership
Leaders respected in sectors
Ethnic/religious leaders
Academics/intellectuals
Humanitarian leaders (NGOs)

Problem-solving workshops
Training in conflict resolution
Peace commissions
Insider-partial teams

Level 3: Grassroots Leadership
Local leaders
Leaders of indigenous NGOs
Community developers
Local health officials
Refugee camp leaders

Local peace commissions
Grassroots training
Prejudice reduction
Psychosocial work
in postwar trauma

Affected Population

Derived from John Paul Lederach, *Building Peace: Sustainable Reconciliation in Divided Societies* (Washington, D.C.: United States Institute of Peace Press, 1997), 39.

Figure: 85
Ledearch's Pyramid of Actors and Approaches to peace

After having revisited Ledearch's philosophy, let us now briefly review on how Burundi and Democratic Republic of Congo (DRC) reached their conflict mediations. After a deep reflection on these examples, South Sudanese authorities might find it important to adapt the philosophy used by either Burundi or DRC to bring peace and stability in the New Country.

(A) *Conflict mediation in Burundi*: - The Hutu-Tutsi conflict in Burundi can be traced back to the struggles for political control that resulted in massacres in the 1960s. In 1993 Hutu president, who was elected under 1992 constitution, he was killed. Nelson Mandela was called upon to mediate; when Mandela was still in the process of mediation, his successor died in April 1994, when the plane he was travelling in with the president of Rwanda was shot down over Kigali. These incidents sparked a violent power struggle in Burundi that resulted in a new power-sharing arrangement known as "the Convention of Government in October 1994", which guaranteed the Tutsi to have 45 percent share of the government. The Hutu were not happy about this arrangement, because in their minds, this arrangement annihilated the Hutu's 1993 election victory, as well as setting aside the 1992 Constitution. The Hutu responded by forming the National Council for the Defence of Democracy (CNDD) and continue to fight Tutsi army. Mandela then stepped down from the position of mediator. By July 1996 the army took over the government.

Between 1994 and 1996 the Organisation of African Unity (OAU) and the United Nations (UN) attentions were focused on Rwanda following 1994 genocide. It was only in March 1996 that the OAU leaders appointed Julius Nyerere, the former president of Tanzania, as a facilitator for the all-party negotiations in Burundi. The conflicting regional and international interests in Burundi hamper Nyerere's effort at mediation (Katabaro, 1980). There was a coup in Burundi in 1996, and this undermined the regional efforts that resulted in the internal promulgation of transitional constitution in 1998, thus legitimizing the coup and Buyoya's presidency. Unfortunately, Julius Nyerere died in 1999 before he could finish his assignment.

After the death of Nyerere, Nelson Mandela took over as mediator for Burundi in January 2000. The expectation was that Mandela would use his political and iconic stature to force the parties to negotiate and adhere to the deadlines set for reaching an Agreement, and to garner international support for the negotiations. However, Mandela asked international communities for financial supports, which was granted. He used this money to host publicized Burundi Peace Summit in Arusha in February 2000. Six African presidents were involved in the three days negotiation meeting, while President Carter, of USA participated through televised appeal to Burundi leaders. Mandela also did succeed in pressurizing the parties to sign the Arusha Agreement in August 2000. Three times, Mandela made it abundantly and clear to the negotiators that if they do not accept his proposals, he would quit. Thus the negotiators did sign the accord as a mark of respect to Mandela, although there were still numerous reservations from various parties. Discussions were expected to go on and indeed did go on to accommodate the reservations raised by many signatories (Katabaro, 1980).

One of the tricky issues, for which Mandela was again to be involved, was the transitional leadership. Mandela proposed a split of the three years transitional period into two equal periods of 18 months. It was resolved that in the first period Buyoya (who has now taken over the government and he is from Tutsi) is to be a President, with Domitian Ndayizeye (who is a man from Hutu) to be vice President. In the second period, the Hutu to select their man for the position President, with a vice president from Tutsi; and this resolution was endorsed by the Heads of States of the Regional Initiative for Burundi. This was followed with the formation of a transitional government on 1st November 2001. Then after the end of 18 months of the second president, there came general election and it worked.

(B) *Conflict Mediation in Democratic Republic of Congo*:- The government of South Africa became first involve in the conflict mediation of Democratic Republic of Congo (DRC) in early 1997, when it attempted to broker a deal between Mobutu and Kabila. It was unfortunate that the South African government did not understand the root cause of the First Congo War. As a result, they ignored the fact that the DRC was invaded by Rwanda and its allies Uganda and Burundi with some supports from Angola. The Rwanda and its allies brought Kabila's Alliance of Democratic Force for the Liberation of Congo (AFDL) to lead the invasion of DRC—that is the Kabila's allies were put in front and the Rwanda and its allies fight behind them. So that Mobutu and his government think that they are being fought by the groups of Kabila. So the Alliance brought together four separate movements that include: (1) The Democratic People's Alliance (DPA), which consisted mainly of Zairean Tutsi, they were fighting for their right to citizenship and was lead by Deogratias Burgera; (2) the Revolution Movement for the Liberation of Zaire led by Anselme Masasu Nindanga; (3) The National Resistance Council for Democracy led by Andre Kisare Ngandu, who became the first leader and commander of the AFDL forces; (4) and the People's Revolutionary Party led by Laurent Kabila, who was referred to as spokesman of the alliance, he later took over the leadership after the mysterious death of Ngandu.

The four groups were held together by Rwanda alliance and by their common desire to out Mobutu's regime, although they have not enough military capacity to overrun the regime.

South African equally ignored the fact that Mobutu's army already been defeated by the Rwanda alliance. Despite all these, the multiplicity of internal parties allowed Mobutu to hang on to power. The arrival of Rwanda refugees into eastern Zaire had brought with its renewed external support for Mobutu, allowing him to postpone the election from 1994 to 1997. Thus, what allowed the invasion to succeed was the weakened support from Mobutu and the existence of a power vacuum in Kinshasa. This then allowed Kabila to declare himself president on 16th May 1998, pending referendum by December 1998, and legislative and presidential election by April

1999. However, none of these took place as war broke out in soon after Kabila became president.

The military defeat of Mobutu brought Kabila to power, but it did not result in an effective control of the country or the establishment of popularly supported government in Kinshasa. In fact, when Kabila came in power he completely ignored the internal opposition to Mobutu. So, Mobutu's soldiers were replaced by Rwanda soldiers in Kinshasa, and the Rwanda Commanders were installed in the villas of the fleeing Mobutu's Commanders, the Rwanda Commanders then drove around Kinshasa to show the people that they are now in control. This act sparked anti-Tutsi and anti-Rwanda sentiments in the capital; they began to say that Kabila is just a puppet or prisoner of Rwanda —with Rwanda soldiers guarding him. Subsequently, this threatened Kabila's survival as president and he immediately reacted calling on all foreign troops to leave the country by 27th July 1998 —it happened.

This was immediately followed by the departure of Kabila's three Tutsi associates from Kinshasa to Goma. The three Tutsi associates include Bazima Karaha, Minister of Foreign Affairs; Deogratias Bugera, Minster for Presidential Affairs, and Moise Nyarugabo, Kabila's presidential secretary.

On 2 August a second rebellion was launched in Goma and Bukavu with the sole aim of removing Kabila from the power. As a result, a new movement was quickly created on 16th August 1998 that became known as "The Congolese Rally for Democracy" (RCD). This movement brought together the DRC Tutsi, the old Mobutuists under one leadership of Alex Tambwe, assisted by Emil Ilunga, and other anti-Kabila forces who were wondering under Rwanda Alliance. Before long, this sparked off what become known as African's First World War" (AFWW). We called this African's First World War because Angola, Namibia, Zimbabwe, Chad and Sudan all came to the aid of Kabila (Katabaro, 1980).

By 1999 the war has reached its peak turning rebel leaders into warlords and external armies into resource extraction forces. It was under these recommendations that the Lusaka Ceasefire Agreement was signed on 10th July 1999 by six countries involved in the mediation in the Democratic Republic of Congo. This was followed with the rebel groups signing the Agreement in August 1999; that was followed by lack of implementation of the Agreement, in particular the aspect that called for a national dialogue between the rebels and the government, as well as unarmed opposition.

It was until December 1999, when the OAU appointed Sir Kitumile Masire, the former president of Botswana, to be a facilitator for mediation in Democratic Republic of Congo. Despite all these efforts, little progress was made, and it was only after the assassination of Laurent Kabila on January 16th, 2001, when he was replaced by his son Joseph Kabila, that is when things started to move slowly forward. Soon after taking over the position, the young Kabila accepted the deployment of United Nations Organization Mission in the country. Kabila also told his forces to recognize the African Unity (AU) facilitator when carrying out his mission for the Inter-Congolese Dialogue. Subsequently, the Security Council Resolution number 1341 was passed in February 2001, which demanded the withdrawal of foreign forces from the country and urged the parties to the Lusaka Ceasefire Agreement to adopt a precise plan that was scheduled to take place by May 2001. This was also the time Mbeki, president of South Africa, re-entered the mediation process in the Democratic Republic of Congo.

However, it is important to notice that the conflicts in DRC had been complicated by the varying regional and international interests they have in the country –although the regional and international authorities claimed to have no interests in the country, but their continue support for DRC was only due to the fact that they had security concerns of Rwanda and Uganda. Therefore, it became clear at that time that the regional and international authorities wanted to deal first with the mending of political relationship between Democratic Republic of Congo and hostile neighbours (Rwanda and Uganda). Sooner than later, Mbeki was able to broker an agreement between DRC and Rwanda following the signing of what became known as "the Pretoria Agreement" in 2002, which led to the withdrawal of Rwanda troops from Democratic Republic of Congo. After this, Angola also helped to broker an Agreement between the Democratic Republic of Congo and Uganda by signing what became known as "the Luanda Accord" in 2002 that also led to the withdrawal of Uganda forces from the country. With the two main backers of the rebels appeased, it was possible to go ahead with the Inter-Congolese Dialogue.

However, the USA, France, the EU and Belgium still maintained their interests in the Democratic Republic of Congo as the country is rich of mineral resources. Thus, Mbeki has to deal first with the interests of the actors before the parties concern go ahead with Inter-Congolese Dialogue (Katabaro, 1980). One critical element of Inter-Congolese Dialogue was the mobilization of funds to implement the agreements. In this, South Africa government agreed to meet 50 percent of the total cost of Inter-Congolese Dialogue; this opened doors for the Western donors to make their financial contributions. The other critical element of Inter-Congolese Dialogue was power-sharing arrangements in the transitional government. To break some of these deadlocks, Mbeki had to directly intervene in the dialogue and he succeeded in convincing the negotiators to retain the incumbent president Joseph Kabila. The main concern of Mbeki was to prevent a power vacuum as the process of transition began. With the incumbent in place there would be a more smooth transition.

Having secured an agreement for the formation of transitional government in the Democratic republic of Congo, South Africa had to become a kind of midwife by providing overall support for the implementation process of the Pretoria Agreement. The committee for the follow up of implementation of Pretoria Agreement was formed and this consisted of South Africa, United Sates of America, United Kingdom, France, Belgium, the European Union and Angola. The committee members have helped the parties in the Democratic Republic of Congo to reach a Transitional Government Agreement in April 2003, and the Agreement on Military Integration was reached in July 2003. The committee also became involved in the organisation of the election by providing logistical and technical supports through its Independent Electoral Commission. The supports include the provision of 128 electoral experts and 118 observers.

As there was no complete return to peace, in eastern part of the Democratic Republic of Congo, South Africa government and Mbeki in particular did help to bring to an end Africa's First World War (AFWW); he did this be creating Bi-National Commission in 2004, and it was charged with the responsibility of implementing various agreements reached. Among these agreements are: - those dealing with defence, health, economic cooperation and investments, public administration and diplomatic consultation.

The forgoing reviews of conflict mediations were presented to serve as background or a mirror to the conflict resolution in South Sudan. The process of conflict management and resolution is subject to cultural diversity in South Sudan. Each of the sixty-four tribes in the New Nation has rules and procedures of conflict resolution. The practices and rules to settle the disputes may, nevertheless, differ according to culture and customs. However, the different actors in conflict resolution, in the sixty four tribes play different roles in similar cases. The respective research and studies revealed that the Dinka, Nuer and Northern Luo Groups have represented key features of conflict resolution among the population of South Sudan.

Unfortunately, due to the scope of this book, we are not going to discuss about mechanisms of conflict resolution among the Dinka and Nuer tribes; in this book we shall rather focus on the Acholi, Pari (Lokoro) and the Shulluk mechanisms of conflict resolution. It is my hope that some of the readers might prefer to adapt one of the Northern Luo Groups systems of conflict resolution. However, these three groups, above, are selected to represent mechanisms of conflict resolution among the Northern Luo. If you want to know more about the Northern Luo Groups, please, see chapter 10.

However, without any prejudice, we can argue that the success of conflict resolution in South Sudan depends entirely on how the population define the term "Reconciliation" and the term "Healing". We my experience working with South Sudanese Communities back in the New Nation and in Australia, I have discovered that the term "reconciliation" and "healing" do not carry the same meaning, they mean different things to different people, tribe, region and family. There is no universal accepted definition of "reconciliation" and "healing" in South Sudan. Unless the people of South Sudan deny their perceptions of conflict resolution, it will remain a political game and financial source to the country.

Before we go ahead with our exploration to answer the questions, "Is Recompilation Possible in South Sudan?" let us first examine traditional systems of governance and how it would work in the New Nation.

Traditional Systems of Governance and the New Nation

The freedom fighters for independent of South Sudan, when they envisaged that peace agreement was just at the corner, they started to think on how modern and traditional institutions would work together for the development of the people in the country. They have realised and admitted that both the modern and traditional institutions of South Sudan are severely weakened by the 21 years of civil, which lead to a critical institutional vacuum in the country. For this reason, they thought that it was important to rescue whatever is left of legitimate institutions, and to rehabilitate and to adapt them into the new environment.

In this light, many attempts were made to re-organize the traditional systems of governance prior to the peace agreement. As a result, two workshops were organised, during the period 2000 and 2001, for Sudanese intellectuals in Khartoum. In addition, in November 2000, Ambassador Bucher organized a three days workshop in Aberdare, Kenya aimed at exploring how governance in South Sudan could be shaped; and the idea of the House of Nationalities (HoN) was in this workshop (Mason Simon A. 2007). Again in January 2003 many community leaders, from South Sudan Civil Society and representatives of various ethnic groups attended a three days meeting in capital Nairobi-Kenya.

From June 29th to July 10th 2004, the Sudan People Liberation Movement (SPLM) invited over three hundred kings, chiefs and spiritual leaders of South Sudan to Kapoeta, in Eastern Equatoria Region; they meet in small village called Kamuto. There, they have eleven days of meeting and the participants came out with the following resolutions that became known as Kamuto Declaration:

(1) The SPLM and the Government of Sudan to assist with establishing the country, States and traditional forums for Traditional Leaders and Chiefs. At the State level forums be organised every six months and at National level, the forums be organised every three months.

(2) Establish peace-building networks and institutions, devoted to popularising the sentiments and values of peaceful co-existence within and among communities.

(3) All disputes must be solved amicably and peacefully through established legitimate institutions.

(4) The role of Traditional Leaders and chiefs in the ownership of the lands and other resources belonging to the communities must be respected.

(4) The Traditional Leaders to conduct meeting among them or go out for study tours

to acquaint themselves with customary laws of each others both within and outside the country.

Source: (Modern Government and Traditional Structures (2005-Switzerland)

All the above meeting we have talked above were to explore the good systems of governance that could be used in the New Sudan. John Garang de Mabior, te leader of SPLA/SPLM had no dream of separation of the Sudan. He was purely for Unity and the change of name of the country from "Sudan" to "New Sudan".

However, it is assumed that in one those meetings the participants attempted to adapt the term "House of Nationalities". The term was invented by Peter Adwok in one of his leaflets; unfortunately the term "House of Nationalities" was misinterpreted to mean only for the young people. Riek Machar Teng, pointed in the meeting of Upper Nile Youth Association for Development (UNYAD) by telling the youth, "The House of Nationalities is not being addressed to youth alone, it is been addressed to women, whole tribes and the political parties like the SPLM" (Machar's Speech 2004). The idea of House of Nationalities was to do experiments with the youth and women, by various bodies and see if it could become the basis of viable political programme.

Thus, the signing of Comprehensive Peace Agreement in January 2015 brings with its new opportunities and new challenges to the people and government authorities. One of the challenges is the rebuilding of the damaged societies in South Sudan. Another challenge is to accommodate of the ethnic diversity in the country (a country that composed of 64 tribes). Thus, this implies that there will be no unity in South Sudan without respecting its diversity.

To overcome these differences and to explore way to bring unity among these diverse communities, the government of Switzerland in 2005 decided to support South Sudan project in establishment and training of Traditional Authorities. The aim of the project is to establish a forum for the representatives under the name of "House of Nationalities" –where people can meet and consult each other in matters of importance; and the forum could be organized on regular basis. According to the report from Switzerland (April 2005) numbers of youth and women, who have participated in the workshops, have emerged as the most active supporters of the "House of Nationalities". Bankie (2013) observed that Traditional Leaders need to undergo training in cultures, if they are to promote traditions, because many traditional leaders lack adequate knowledge of their own culture in conflict resolution. This is due to the simple fact that a significant part of the collective traditional memory had been eroded by decades of conflict and colonial rule. As such study tour should be organises for South Sudan's kings and chiefs to visit three African countries that is Ghana, South Africa and Botswana. We shall not discuss what South Sudanese delegations have found in South Africa and Botswana as their tours in these two countries were not documented. However, we shall rather examine their visit to Ghana.

South Sudan's Kings and Chiefs in South Africa, Botswana and Ghana

Swiss government (with the help of Josef Bucher, former Ambassador in Libya from 1994-1997) has been engaged in helping resolve the North-South conflict in Sudan since 1994 to 2006. Their attention was focused in three engagements: First, the House of Nationalities project, which aimed at supporting and developing traditional authority structures in the South Sudan and Nuba Mountains. Second, the Nuba Mountains cease-fire negotiations, was facilitated by a joint Swiss-US mediation team, and which was central in the preparing the IGAD North-South negations. Third, the IGAD negotiations between the government of Sudan and the Sudanese People's Liberation Army/Movement that led the Comprehensive Peace Agreement —ending a war that cost the lives of some two million people (Mason, Simon A. 2007)

In 2006, when the Swiss Government has seen that the South Sudan Government has interest in working with the Traditional Leaders, they set up to support the training for a group of Traditional Leaders, some government representatives who are dealing with local governments as well as some members from the National Constitutional Review Commission by organising study tours to three African countries that is South Africa, Botswana and Ghana. It was hoped that after the study tours, the participants would share their knowledge among colleagues and more importantly they would act as promoters of the inclusion of Traditional Authorities (TA) in South Sudan's institutions.

The first leg of the South Sudan's kings, chiefs and other government officials' tours in the three African countries started in South Africa (from 18th-22nd August 2006), where they started with the visit to the National House of Chiefs (National House of Traditional Leaders), in Pretoria. In the following days, the South Sudan delegations travelled to various parts of South African, which include Kwandebele (where they met King Makhosoke 11) and the coastal city of Durban (where they met King of the Zulus). The delegations also visited Barolong boo Rratshidi Trabal Authority at Montshioa-stad in Mahikeng. In one of these visits, a member of the South Sudan delegation admitted that the revolution in South Africa was almost similar to that of South Sudan, where the war destroyed almost everything, especial in rural areas. He went on to say that once South Sudanese Authorities have learned how the government of South Africa supported the traditional leaders, and how the traditional leaders can deliver services to the communities, resolving disputes and administering justice, them they (South Sudanese) would rebuild the ruined country. In these meetings, Anyuak King, Adongo Agada Akwai Cham, shared with participants the plight of his people across the border with Ethiopia, to where most of his subjects were cut off by colonial borders. In response, Moshe Mabe, Deputy Chairperson of the North West House, in South Africa, applaud the South Sudan delegations for their visit that shows how much they intend to pick up from the aftermath of the war in South Africa, and move into rebuilding the new country and establishing institutions of traditional leaders. He emphasized that the traditional authority must co-exist with the National Government and together endeavour to effect change in the lives of their communities.

The second leg of South Sudan's kings, chiefs and other government officials was in Botswana (from 23rd -27th August 2006). Here the delegations visited the House of Chiefs in Capital Gaboroni (where they met paramount chiefs and chiefs "Kgosi"); this was followed by the visit to Mosadi at her court "Kgotla" in Ramotswa.

The third leg of South Sudan's kings, chiefs and other government officials' tours was in Ghana (28th -31st August 2006). The delegations left Botswana capital Gaboroni to Ghanaian capital, Accra via Johannesburg. When the delegations arrived in Ghana, Tody Collins (2013) notes that during the visit the South Sudan kings and chiefs were met by media, politicians, academic; and in its first day of tour (28/08/2006) they held the debate in Alisa Hotel and the discussions were focused on the role of the Traditional Leaders in the context of developing the young Nation in Africa. During the meeting, Ghanaian Traditional Leaders, who are known as "Nananums" performed a ceremony involving pouring and splitting schnapps on the ground. This is done to invoke their ancestors to accept and welcome these guests.

Figure: 86,
Nananums performing traditional ceremony

In the conclusion of the tour, Bari Paramount Chief, Dennis Dar Amallo Kundi, promised to acquire a land in Juba capital where "a rest house for chiefs" would be built. These bring us to two questions, firstly, "Has a rest House for the chiefs been built in capital Juba, following the visits of South Sudanese delegations to those three African countries?" Secondly, "What types of services have been delivered to the communities by the chiefs after having learned from local authorities in the three African countries?" The answers are, "I am not sure".

In meeting in Ghana, a representative from South Sudan, Tor Deng Mawien Bak, who was also the leader of the delegation, informed the Ghanaians that the kings and chiefs and other members from South Sudan were in their country because they want to find out how Ghanaians have administered Justices, handled Spiritual affairs and coordinated the work of Traditional Leaders with the National Government. In other words, members of delegation to learn from the Ghanaians and take back the findings; he further suggested that when the South Sudanese delegations go back home, the government of South Sudan will contact the Carter Centre to establish a legal framework for the role of the Traditional Leaders. Tor Deng Mawien Bak also advised the Ghanaians that the government of South Sudan planned to call for general elections of the chiefs before the end of 2009, to implement the South Sudan Local Government Act that would be formulated in the same year 2006; the South Sudan Local Government Act 2009 that would come to effect by November 2011.

The South Sudan government talks of "inclusion of Traditional Authorities" but the selection of kings and chiefs for the study tours in these three African countries, revealed that "inclusions" is the word of mouth and not to be practiced in the true sense of the word; and this can be seen clearly from the members of South Sudan delegations below, who were selected for study tour in the three countries:

S/N	Names	Tittles	States	Tribe
1	Tor Deng Mawien Bak	Presidential Advisor on Decentralisation and International Linkages	Gogorial State	Dinka
2	Garang Aguer Akok	Office manager of the Presidential Advisor on Decentralisation and International Linkages (In Prison)	Awil State	Dinka
3	James Lual Deng Kuel	Chairman of Committee for Foreign Affairs and International Cooperation, National Legislative Assembly	Gogorial State	Dinka
4	John Masua Madanza	Chairperson of Committee for Decentralized Governance and State Affirs, National Legislative, Council of States	?????	?????
5	Agol Ayuel Adway Agol	Chairman of COTAL Upper Nile	Fashoda State	Chollo/ Shulluk

S/N	Names	Tittles	States	Tribe
6	Alphonse Legge Loku Tombe	Chairman of COTAL Central Equatoria	Guba State (Central Equatoria)	Bari
7	Wilson Peni Rikito Gbudue	Zande King	Gube State (Western Equatoria)	Zande
8	Akwai Agada Akwia	Anyuak King	Bumba State (Akobo)	Anyuak
9	Achien Achien Yor	Paramount Chief of Northern Bahr El Ghazal	Northern Bahr Ghazal State	Dinka
10	Angelo Bagari Ungbanga Ukungu	Paramount Chief of Western Bahr El Ghazal	Wau State (Western Bahr El Ghazal)	Dinka
11	David Mangar Nial Kon	Paramount Chief of Western States (Lakes)	Western States	Dinka
12	Manoon Ater Guot Chol	Executive Chief of Warrap	Gogoria State	Dinka
13	Del Rumdit Deng	Director General of Local Government and Traditional Authorities, Local Government Board, Office of President	Tuic State	Dinka
14	James Alala Deng	President of Court of Appeal Greater Bahr El Ghazal	Lakes State	Dinka
15	Adam Abwol Kiir Deng	Legal Counsel, Ministry of Justice	Yirol State	Dinka
16	Jackline Novello Nailock Tamot	Director for Gender, Ministry of Gender, Child and Welfare	???	Dinka/ Nuer?
17	Margaret Akon Isaiah Majok	Gender Focal Point, Ministry of Gender, Child and Social Welfare	Awil State	Dinka
18	Acuil Malith Banggol	Gurtong Trust Board Member(Secretariats SPLM Popular Organisation)	Tuic State	Dinka
19	Oliver Humbel	Human Security Advisor, Federal Department of Foreign Affairs, Switzerland	Switzerland	Expatriate
20	Kwesi Kwaa Prah	Professor, Centre for Advanced Studies of African Society (CASAS)	CASAS-Juba	????

S/N	Names	Tittles	States	Tribe
21	Victor Oduho Lomiluk	King of Latuko??	Imotong State	Otuho
22	Dennis Dar Amallo Kundi	Paramount Chief of Bari/Juba	Guba State	Bari
23	Madalina Tito Ohirong Ohire	Chief of Lopit	Imotong State	Lopit
24	Louis Lopua Naita Lobor	Paramount Chief of Toposa	Gurungang State?? Kapoeta	Toposa
25	Philip Manytong Awin Manytong	Representative of the Reth of Shilluk Kingdom	Fashoda State	Chollo/ Shulluk
26	Issa Ruot Lam War Kur	Paramount Chief of Nuer Luo	Akobo State??	Nuer
27	Ireneo Kunda Tabur Unango	Chief of Raja	Lol State	Cresh
28	Pio Tem Kuag	Chief of Awil North (In Assembly Juba)	Awil State	Dinka
29	Dut Malwal Arop Tong	Chief of Rumber	Rumbek State	Dinka
30	Jacob Madhol Lang Juk	Chief of Tuic	Tuic State	Dinka
31	Emir Kuol Deng Kuol	Chief Dinka Ngok (Dead)	Abyei	Dinka-Ngok
32	Emir Elamen A. Elgadir Dauod	Chief of Nuba Mountains	Nuba Mountains	Nuba
33	Emir Yagoub Gebril Abdalla Makki	Chief of Nuba Mountains	Nuba Mountains	Nuba
34	Nikodemo Arou Man Chot Ngot	Local Government Secretariat Representative	???	Dinka
35	Owar Ngot Ojang Odol	Assistant of Anyuak King	Bumba State	Anyuak
36	Edwin Baba	UNDP	Switzerland	Swiss
37	Dr Conradin Perner (Kwacakworo – Anyuak Nickname)	Swiss anthropologist	Switzerland	Swiss
38	Ms Amer Ajok (Amira Junub)	Gurtong Discussion Board	????	Dinka

The Ghanaian leader reassured the delegations that their forefathers fought for communities and established ancestors' rules that should not be forgotten by their successors in the roles they play in the modern communities. He went on to say that in Ghana whatever is more than 100m under the ground is said to be the property of the State, but the owner of the land must be compensated for its extraction. I am sure, these arguments made the South Sudan's delegations ask themselves, "What kind of compensation has the government of South Sudan given to the people of Bentiu for the extraction of oil from underground in their land?"

The Ghanaian leader was saying this in response to the statement from Tor Deng Mawien Bak, when he said, "In South Sudan the land ownership is still a problem as local population do not have voice and the companies, like Chinese and Arabs developers extract the oil in unsound way and without consideration of paying compensation to the land owners. Not only this, people are also being poisoned through the water supplies and the cattle being poisoned through the grass they graze upon".

Ghanaian leader concluded by saying that in the history of Africa when the State gains control of the land it is the beginning of the end.

Angelo Bagari Ungbanga Ukungu, a paramount chief of Western Bahr El Ghazal, pointed out that the new system the the government of South Sudan wants to introduce by electing the chiefs to replace Traditional Leaders in the communities would be a problematic. Because the communities do not like the new systems, and the elected chiefs would be seen as person the government would want to impose on them. The local people know themselves better, so it would be better if the government of South Sudan would allow the local people to continue with the traditional methods of elections, while they act as a bridge between the government and the people. In contrast, James Lual Deng Kuel argued that it is not the proposed systems of elections that would be problematic, but rather it is the respect for the traditional leaders, that was lost during colonial period and the war, that would be a problematic in the country. The Traditional Authority Systems were damaged and the powers of traditional chiefs have been diffused since the colonial era. In the old days, it was known that the Traditional Leaders Structures are closest to the hearts and minds of the people, especially in the rural areas, and can therefore play a significant role in the development of the regions or country

One of the Ghanaians leaders told the South Sudan delegations that in his country, Traditional Councils are based on the original clan systems with chiefs appointed by the paramount chiefs, where he appoints five chiefs in each of the ten regions in Ghana. He also told the South Sudanese delegations that there are also female chiefs been appointed. These female chiefs are known as "Queen Mothers" and they must be from the royal lines and they cannot substitute the King in any form or shape. Unlike South Sudan government, the government of Ghana always support the chiefs financially; chiefs and King are included in the budget. The Ghanaian Government does this because the role of the king and chiefs are very expensive, they are expected to receive and accommodate guests that come into the communities. While in South Sudan chiefs depend on the collections of taxes from the communities in their areas, yet they are not responsible for the money collected.

Figure: 87: South Sudanese delegations pictured with Ghannaian Leaders
Source: Toby Collins: Sudan Tribune, 17 March, 2013

The Ghanaian leaders told south Sudan delegations that when the country got independent in 1957, the administration was unsure how much power to instil in the chiefs as the government returned more power to themselves. At that time, Traditional Leaders were removed from active participation in governance and this allowed the Traditional Leaders to approach the future of their communities more far-sightedly.

Consequently, soon after independent in 1957, the local chiefs who were led with totalitarian rule were already leading their people into war. Their roles has adopted with the changing societal situation.

Today, the Ghana's customs and traditions make them unique from other African countries, which have gone through the civil war. Chiefs carry out legislative and judicial duties—mostly with administration of Traditional Authorities; and they are now focusing on developmental activities. The Ghana's government support chiefs and kings financially so that they remain financially strong.

Some chiefs have been in their positions for decades, seeing governments come and go, instilling them with a wealth of knowledge. This happened because chiefs are answerable to their ancestors as well as to their communities, they use alternative laws.

Formation of Councils of Traditional Authority Leaders

Due to increasing inter-ethnic conflict across the country, the government of South Sudan (GoSS) resolved to strengthen Traditional Leaders Structures with the signing of Local Government Act in 2009, setting out governmental structures at various levels including the Councils of Traditional Authority Leaders (COTALs). Thus, in October 2009, the representatives from the Kingdoms, States, and National Government as well as representatives from the Department of Justice drafted a draft law setting up a COTAL in each of the ten States in South Sudan. According to the Laws, the Councils of Traditional Authority Leaders are to function as an intermediary between traditional societies and the modern Sate Structures, they should advise and provide the national government with the matters related to community affairs; and from time to time organise forums for civil society groups.

The draft law consultation, were then conducted among some people of South Sudan between the month of July and October 2010. It is not clear to me why some people from South Sudan were excluded from this consultation; if it was for the national interest. After consultation and training for the Councils of Traditional Authority Leaders were the formed in the ten States with hope to form one Councils of Traditional Authority Leaders at the National level based in Capital Juba.

The Traditional Leaders were approached as significant, not only as a means to judicial and developmental ends, but as bastions of cultural heritage. In the implementation of the Local Government Act in November 2011, the South Sudan Local Government Board and the Human Security Division of the Swiss Confederation renewed their memorandum of understanding stating that the COTALs will be established in the country within the next two years. Therefore, the Swiss government agreed to support South Sudan by assisting in the establishment of the legal frameworks for the Councils of Traditional Authority Leaders; this mean that they would support the election processes of the ten proposed COTALs, provide support for the training of the members through supervision of technical staff to make sure the Councils of Traditional Authority Leaders function properly as expected.

More importantly, the 2011 Traditional Consultation of South Sudan also recognized the role of Traditional Authority as an institution at the local government level. As a result Central Equatoria was the first to elect their Councils of Traditional Authority Leaders that was followed by the Upper Nile in January 2012. The other eight States did not elect their Councils of Traditional Authority Leaders because their bills remained in the Legislative Council, although the first draft of the national Councils of Traditional Authority Leaders was at the same time under consultation in Legislative Council. Even though I am living some hundreds of thousands of kilometres away from South Sudan, I am forced to believe that no further elections for the Councils of Traditional Authority Leaders have been conducted in the country since civil war broke out in December 2013.

To what degree can reconciliation be implemented in South Sudan?

This is a difficult question to answer. According to Lars Christoffer Skogrand (2015), there are too many issues in South Sudan at the time for a reconciliation effort to be fruitful. In his research, he found in the empirical data that the state appears to be absent in much of people's lives, and for this reason abd due to lack of good-practices for reconciliation and restorative justice the reconciliation can hardly be implemented. The lack of any real system of governance on the local level makes it difficult to engage in fruitful reconciliation processes. Despite this, there is clearly a need for relationships between the various tribes and peoples to improve drastically. In his analysis, he presented South Sudan as a country that lacks a common understanding of what it means to be South Sudanese; that they lack a positive national-identity. It is here, he believes, the authorities could find way forward. It might possibly be helpful to foster a national identity within the country. The various peoples of South Sudan are split into numerous tribes and their identity is often tied to their tribe rather than their nation. In providing South Sudanese with the chance to get to know each other and each other's cultural and historical heritage, leaders enable their people to naturally come to an understanding of what it means to be South Sudanese. Having various peoples meet with the purpose of learning how the other tribes think, what they believe in and what they have in common with themselves could help manage the fear of the unknown.

It is here that the South Sudanese church can be highly effective and helpful. No other organisation in the country has such a large area of influence and trust among the various local communities.

The church has both the will and the ability to adapt their approach to various cultural and traditional necessities, a skill that would be very valuable in keeping people comfortable than meeting the *threatening other*. Negative-identity is reliant on defining oneself in relation to a threatening other, when the threat dissipates it creates a void where people define a new threatening other to fill that void, and a negative-identity spiral is born. The necessity to redefine "*we*" and "*them*" into an "*us*" is quintessential in this discussion. That process must be maintained locally; a job the church is ideal for.

Yet the creation of a national identity or common understanding does not remove the need for reconciliation, it is simply the first step. There needs to be a clear common ground between two sides for reconciliation to take place. In modern reconciliation this is arguably achieved by utilising clear systems to help guide the process. In a country where these processes aren't present, it is necessary to both help develop them as Norwegian Curch Association (NCA) are doing, as well as provide people with the opportunity to interact.

Positive interaction fosters understanding. It is easier to see how two groups are alike if you discuss what you have gone through together, rather than what pains you have inflicted upon each other (PP. 109-110)

From the analysis we can say that the biggest issues South Sudan is facing is the deteriorating relationships between tribes. While the war between Riek and Salva is destructive to the long term effects of hostile relationships between ethnic groups and this can be catastrophic. As people of peace advocate for war to end in South Sudan, I suggest that we should also be aware of the relationships between groups. It could be argued to be beneficial to all parties involved, if a concrete effort to improve relationships between people outside of a classic reconciliation process was undertaken. The natural evolution of a common ground, fostered between men and women of the various peoples in South Sudan could provide the necessary foundation to better handle reconciliation efforts in the future.

Bewitch (hex) and South Sudan Politics

We cannot overemphasise here that some African leaders use hex (*ajuogo/Juju*); they believe that hex plays part in National Politics and enables them to stay longer on power. However, it is important to notice that these leaders use hex for many reasons, the main reason being to protect them from harms; and more importantly hex enables them to maintain power for longer periods. They also believed that "hex" can be used to influence the reasoning capacities of the opponents, so that they accept leader's opinions and principles. South Sudan leader is not an exception, it is reported that President Salva Kiir, has purchased his hex ("black hat") from Nigeria; it is also assumed that when he was leaving Nigeria, he (Kiir) was instructed by the witch doctor (the seller of the black hat) that whenever he is going in public places or in the offices, he should wear the black hat at all time and never put it down. If he does, he would become weak and powerless and eventually he would lose his job. If this happens, he would be forced to flee to another country or he would be shot to death by any military officer who does not wants him to stay longer in power. If you live in Juba or you have ever met President Slava Kiir, I wonder if you would testify that one day you have seen President Salava Kiir been driven in the streets of Juba or entering any office without his "Juju hat".

Some readers may ask, "Where do African Presidents get their hex (Juju) materials from". It is reported that most of the materials use for hex, come from Nigeria. We should know that the materials use for hex take different forms, which varies from skulls, monkey's hair, dog's hair, cat's hair, sticks (e.g. use by Omer el Bashir), Jacket (used by President Mugabe of Zimbabwe) and "black hat" or "Juju hat" (used by President Salva Kiir).

President Museveni
(Uganda)

President Kiir (S. Sudan)

President Indi Amin (Uganda)

President Odasanjo (Nigeria)

Emperor Bokassa
(Central Africa)

President Koroma
(Sierra Leone)

President Mbutu
Sese Seko
(Congo)

President Omar el
Bashir (Sudan)

President Mugabe's Jacket hex
(Zimbabwe)

Figure: 88, African Leaders who use bewitch (hex) for national politics

The Republic of South Sudan, President Salva Kiir, like other African leaders, regardless of his education, social class and religion, still believes in hex and its power. He believes that political problems, of South Sudan are caused and supported by spiritual forces. He believes that, besides Jeng Elders Council (JEC), hex gives him instructions on how to run the government. Indi Amin was said to have deported all Asians in Uganda because he was instructed by his hex. (See Figure 88 above).

Although some African leaders believe in the power of hex, it has been reported in the past and in the present that by the end of the day, all the leaders in Africa, who have been using hex, have become weak and powerless in the mid of hex and eventually they lost their jobs. This implies that Salva Kiir, despit the usage of "Juju hat" he will onday lose his jobs. Guns do not know the power of hex – subsequently, it was reported that some leaders have been shot and killed in the mid of hex; it is also reported that other African leaders fled the country in the mid of their hex -for asylum seeking.

Chapter 15:
Kinship and Family Life

Kinship is reckoned in a number of different ways around the world, resulting in a variety of types of descent patterns and kin groups. In most cases descent is traced through a single line of ancestors, for example either through the father or through the mother. Although some Western anthropologists might argue that traditional Northern Luo family patterns are slowly being altered as a result of population movement from the rural areas into the urban areas, the reality is that the people's movements from rural areas into the urban settings has never trigger changes in family structure. The traditional norms and values remain unchanged. However, we cannot deny that the Northern Luo groups of South Sudan are increasing faced with the challenges and pressure from the people coming from Western World to work in South Sudan. They are trying to sell modern or Western family values to the people of South Sudan, which are met with high resistance.

In this article, the kinship and family life seeks to examine and explain the trends which are assumed by many to affect family patterns among the Northern Luo Communities. It aims to examine to two key issues connected with the Northern Luo family patterns that is nuclear family or a compound nuclear family and extended family (let us note that some people deliberately confused the meaning of a compound nuclear family with an extended family); and more importantly, we are going to examine how marriages are being conducted among the husband's clans or village and wife's or wives' clans or villages. Let us first look at patrilineal societies.

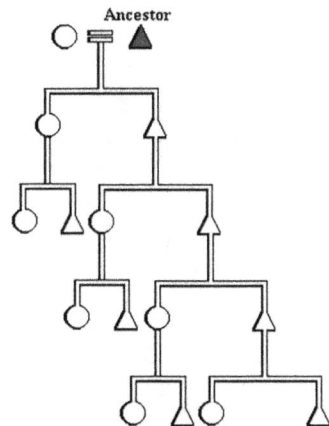

Figure: 89,
Structure of patrilineal societies

In patrilineal societies, children are considered to be part of their father's kin group and not of their mother's clan or kin group. This implies that only males pass on their family identity to their children. A woman's children are members of her husband's patrilineal line.

The black people in the diagram below are related to each other patrilineal. The diagrams below illustrate kinship relationships.

In Europe and America today people practice bilateral descent, which means the system traces descent from all biological ancestors regardless of their gender and side of the family. Whereby all children whether males or females are members of both their father's and their mother's families. This system does not work in South Sudan particularly, among the Northern Luo groups.

In Luo groups, kinship and family structures bind the Luo people together. It is the kinship ties which determine a person's rights, responsibilities and behaviour. In other words, we say it is the kinship that defines the obligation of the members of the Luo society; it defines privileges of each member in the family or clan.

Thus, from the Luo perspective, kinship has two levels: - one is known as nuclear family and the other is known as extended family. The nuclear family always is made up of wife, (wives), children, dog and husband as the head of the family, which literarily form a lineage. The extended family is composed of parents, grandparents, children, grandchildren, relatives, aunties, uncles, cousins, brothers and sisters, which literarily form a clan. In this book we are going to focus on patrilineal systems, since it is generally practice among the Northern Luo Groups of South Sudan.

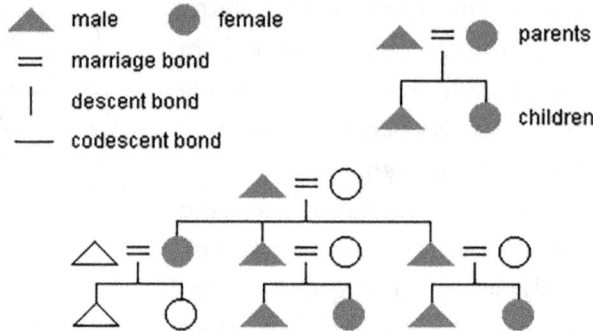

Figure 90

PATRILINEAL DESCENT

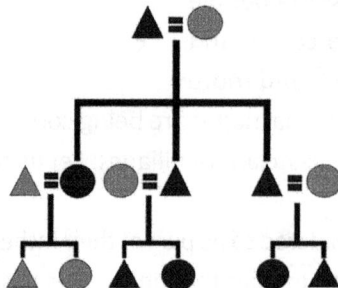

Figure: 91, Patrilineal Descents

The family is a universal group throughout Africa in general and throughout the Northern Luo in particular, with many different forms and functions. Everywhere, the basic family unit is what we called the elementary or nuclear family, a small domestic group made up of a husband, his wife or wives and their children. The nuclear family is formed by a marriage and supposed to end by either death of one of the spouses or with the divorce, following serious reason. Polygamy is permitted among the Northern Luo in the past and today; where a husband and his wives form a compound nuclear family. Unfortunately some people do not understand the different between a compound nuclear family and extended family. Both the nuclear and a compound nuclear family are found among the Northern Luo in the rural area and in urban settlements.

According to African traditions, customs and norms, in nuclear family, a husband is the head of the family; he has powers and roles that are unchangeable. However, forces migrations have made many men (husbands) have lost their positions, powers and roles, as they sought protection in the Western World: Australia, America, Canada and Europe.

In South Sudan, nuclear family is viewed as a hierarchy where husband, as head of the family, where he comes first, the wife (wives) comes second, child comes third, and dog comes fourth. But in the Western World, particularly in Australia, the family hierarchy has been turned upside down, where the child comes first, the wife comes second, the dog comes third and the husband comes fourth. Given this fact, some men believe that Australian culture and Australian Community Service Sectors always favour women and children over men. Due to this belief some men are sceptical of using community Services in Australia because they believe that Australian System deliberately wants to separate husbands and wives –this is something not acceptable in Southern Sudanese culture. It appears that men have less right over their wives and children.

More importantly, the new family structures in the host country has caused many divorces among the husbands and wives that led to many children ending up in foster cares as some the mothers are unable to care for their children without their fathers.

The nuclear family serves many functions and is important in the entire Luo societies; it is the centre of education where parents teach their children values of things, common practices and customs. Therefore, the husband, as head of family, plays important part in politics, economy and education in the family. He makes essential decisions for the family welfare and development. Interestingly, in the Luo groups, education of children are done on gender basis, where the females like mothers, wise grandmothers and aunties educate girls; and males like fathers, wise grandfathers and uncles educate boys.

Kinship today still plays important roles in the Luo societies. For example, when a person has a child, that child views not only his "biological father" as a "father", the child also views "father's brothers" as his/her "fathers". As a father's brother is identified by a child as "father", this implies that the "children of the father's brother" will be viewed by the child as "brothers" and "sisters" —rather than cousins as it is the case in the Western World.

On the other side, the mother's brothers are identified as "maternal uncles"; mother's sisters are identified as "mothers". So when Luo refer to their family, they invariably mean their extended family. Therefore, in Luo traditional customs, a person can have several fathers, several grandfathers, several mothers, several grandmothers and many brothers and sisters.

In the Western Societies the structures of social interaction and roles and obligations change as individuals move out from the immediate family circle, to the wider society. In contrast to this, in the Luo societies the family structures and the set of rights and obligations underlying them are extended to whole society. Thus, when individuals move out, with their wives from immediate family circle, to live in a nearby local group, this does not necessarily mean that those individuals are out from the memberships of the nuclear families. Not at all, they are still considered members of nuclear families. Because distance or changing locations does not affect membership of nuclear family. For example a person can live in America, Australia, London or Hongkong, but yet he/she is considered a member of nuclear family in the Luo society in South Sudan.

When speaking to, or about, another person in Northern Luo societies, family names and nickname are commonly used. In case of a man, a person is addressed using family name or nickname e.g. Okot (family name) or Dikomoi, Lutukumoi (nicknames); but for females they can be addressed only using their family names as they have no nicknames e.g. Lakot (family name), Nyakot (family name).

With this brief summary of kinships in Northern Luo Societies, I would then like to take us further to explore how kinship is viewed in each of the Northern Luo Society of South Sudan.

Acholi Kinships and Family Life

The Acholi kinships are organized into localized patrilineal lineages or brother lineages; and in other cases they are organized into temporary groupings of two to four lineages. These have long been the fundamental social and economic units in Acholi. These groups claim descent from Okwa, the father of King Divine Nyikang, the founder of Shulluk Kingdom. Nevertheless, we should not forget that fact that in the past and present the Acholi incorporated and continue to incorporate many types of outsiders, who in course of time have now become part of the kinship in extended family.

We have just seen in the summary above that kinship in Luo groups consist of two levels. In nuclear family, where there exist lineage males, here we find lineage males called grandfather, father, brother and son —depending upon generational relationship to the speaker.

All affine are known as "mother", and the relationship between the actual mother's child and the child of the mother's sister are clearly demonstrated in the time of nurse-child (*lapidi*). When the real mother asks the elder child to nurse her new born child, and that child will feels strong bond between him/her, and the child being nursed; because the nursed-child looks at the nursing child as a brother or sister. But on the contrary, if the mother sends one of her children to nurse the new born baby of her sister, the nurse-child (*lapidi*) will see small gap between him/her and the new born of his/her aunty.

To understand the kinship in Acholi society, let us look at how persons are related to one another. In Acholi a child calls his/her biological father as a "father" (*wora*) and biological mother as a "mother" (*ma*). Although the terms for father and mother are only used for the individual parents, and not for a class of persons as is commonly the case in the classificatory system, however, clan brotherhood is recognised.

Here are some few Acholi kinship terms:
- a child calls biological father as "my father" (*Wora*)
- a child calls biological mother as "my mother" (*Maa*)
- a child calls father's brother as "father" or 'brother of my father' (*Omin wora or Wora*)
- a child calls father's brother's son as "brother" (*Uwura or wowora*)
- a child calls father's sister's daughter as "sister" (*Nyawora*)
- a child calls father son as "my brother" (*Omera*)
- a child calls mother's sister's son as "brother" (*Omara*)
- a child calls mother's sister's daughter as "sister" (*Nyamara or Lamara*)
- a child calls father's sister as "aunty" (*Waya*)
- a child calls mother's brother as "uncle" (*Nera*)
- a child calls mother's brother's son as "Uncle" (*Nera or Latin pa nera*)
- a child calls mother's brother's daughter as "<u>aunty</u>" (
- a child calls mother's mother's brother as "uncle" (*Nera or Latin pa ner maa*)
- a child calls mother's sister as "mother" (*ma or Lamin ma or Nyamin ma*)
- a child calls sister's son as "nephew" (*Okeya*)
- a child calls sister daughter "niece" (*Lakeya*
- a chill calls father's sister's son as "nephew" (*Okeya*)
- a chill calls father's father as "grandfather" (*Kwara*)
- a child calls mother's father as "grandfather" (*Kwara*)
- a child calls father's mother as "grandmother" (*adaa*)
- a child calls mother's mother as "grandmother" (*adaa*)
- a man calls children's children as "grandsons or daughter" (*Nyikwaro or Nyakwara*)
- a man calls wife's mother as "mother-in-law" (*Mara*)
- a man calls wife's father as "father-in-law" (*Ora*)

-a man calls wife's brother as "brother-in-law" (*Ora*)

-a man calls wife's brother's wife as "mother-in-law" (*Mara*)

-a man calls wife's mother's sister as "grandmother or mother-in-law" (*Mara*)

-a woman calls husband's mother as "mother-in-law" (*Mara*)

-a wife calls husband's father as "father-in-law" (*Ora*)

-a wife calls daughter's husband as "brother-in-law" (*Ora*)

-a wife calls husband's sister's husband as "brother-in-law" (*Ora*)

- a husband calls wife's sister as "wife" (*Dakona* or *Ywera*)

-a wife calls sister's husband as "husband"(*Cwara* or *Ywera*)

-a husband calls brother's wife as "wife" (*Dakona* or *Ywera*)

-a wife calls husband's sister as "husband' (*Cwara* or *Ywera*)

-a wife calls brother's wife as "wife" (*Dakona* or *Ayena*)

-a husband calls wife's sister's husband as "brother-in-law" (*Obaa* or *Muko*)

The term "*Wora*" is used to include the biological father and his brother); and the fathers of all those whom the father calls them "father's sons" (although the term "*Omin wora*" is also used for father's brother. Thus, as descent is patrilineal, therefore, all the men of the father's clan and generation are called "*Wora*" (fathers); and all the men of the "grand-father's generation" are called "*Kwaro or Kwara*" (grandfathers), but not all men of the same generation of the grandfather are called "*Kwa*", they are rather called "*Omera*" (father's sons).

Unlike the other Luo groups, in Acholi, all those succeeding generation, a man's own sons, are called "sons", called "*Woda*" (sons); likewise, all those succeeding a man own blood, such as brother's sons, he calls them also "*Woda*" or *wod pa Omera*. The sister's sons are called "*Okeyu*"(Pl) and "*Okeya* (Single), while the sister's daughters are called "*Lakeya*"(Single) and "*Likeyu*" (Pl); they all belong to the clans of their fathers.

"*Maa*" is an individual term used for "biological mother" and in Acholi custom, all persons whom "mother" calls "*Omera*" her children called them "mother's brothers" (*Nera* "Singel" or *Nero*"Pl"); and those persons whom the mother calls "fathers" (*wora or wego*), her children call them "mother's fathers" (*Kwara or Kwaro*); the word "*Kwara*" in this sense indicating the generation where the child or children belong and does not indicate clan.

Interestingly, all the children of mother's brothers "*Nera or Nero*", her children call them "mother's brother's sons" (*Nera* or *Lotino pa Nera*) and if they are girls the children call them "mother's brother's daughters" (Nyina**).** Therefore, all the men and women who belong to the mother's clan are considered as the children of "mother's bother**".**

The Acholi girls and boys usually marry among themselves, in other words they marry in clans, or what they commonly called "*dog ot*". Thus if clan "A" marry from clan "X", "Y" and "Z", this does not mean that no boy from clan (*dog ot*) "A" will not marry again from clan (*dog ot*) "X", "Y" and "Z". Traditionally, marriage among the Acholi is a mixed game, there is no avoidance among them as we shall see later among other Luo groups. Thus the word "Oraa" or "Or" become reciprocal between village "A" and villages "X" "Y" and "Z". In Other words, the marriage in Figure 78, below is not a one way traffic but rather it is a two ways traffic: When a person calls the other person as *"Oraa"* or *"Maraa"* one needs to distinguish them.

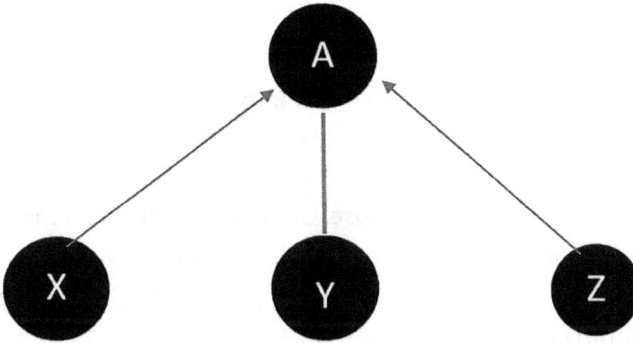

Figure: 92, Simple kinship relationship
Source: Sudan Notes and Records, 1941, P.30

But from a particular group or lineage there could be average of 30-40 married men. In this case, the number of in laws (*ore*) will be numerous. The example of the lineage marrying many wives that eventually create more in laws (*ore*) are demonstrated in the below diagram.

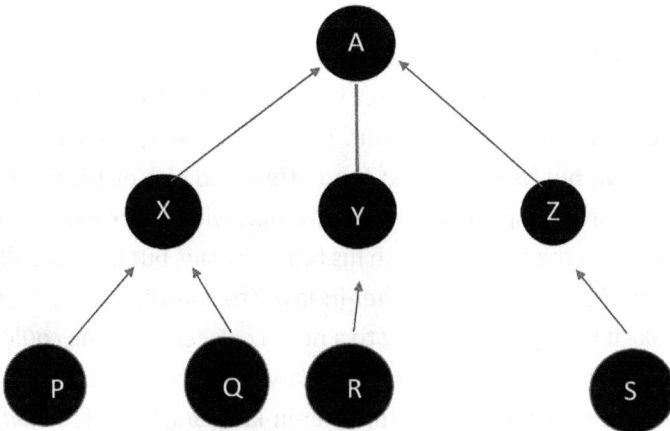

Figure, 93, Complex kinship relationship
Source: Sudan Notes and Records, 1941, P.30

We have seen in the first diagram that there are two ways traffic, that "X", "Y", and Z call

"A" their in-laws (*Ore*) because men from "A" having married women from them. In the same way people in village "A" call people in village "X", "Y" and "Z" their in-laws. However, in the second diagram we can see that men from village "X" have married women from village "P" and "Q"; and men in village "Y" have married women from village "R"; and men from village "Z" have married women from village "S". This creates numerous transactions of in-laws (*Ore*) in the society. Subsequently, this implies that other men from village "A" are allowed to marry other girls from village "P", "Q", "R" and "S" without any restriction. Similarly, men from village "X", "Y" and "Z" are allowed to marry girls from village "P", "Q", "R" and "S" and vice versa. In other words, this means that the entire memberships of Acholi society can ,marry "P", "Q", "R", "S", "X", "Y" , "Z" and "A" on the condition that the relationship between the bride and broom should not be too close. Unlike the Shulluk and the Anyuak and Jur Chol , men from Village (*dog ot*) "A" are not forbidden to marry women or to have sexual relationship with women from the four villages "P", "Q", "R" an "S", because they are seen as being distance from family relationships.

More importantly, the "term" *Ora* or *Oree*" is a reciprocal term that could be used by either the husband or the wife to address wife's parents (in case of the husband) and to address husband's parents (in case of the wife). The husband can also address the wife's brothers as "*Ora*" (in-laws); but the husband's sister can address the wife's brothers "*Oraa*" or "*dako na*" (my wife).

Furthermore, from the above narrative, we could see that the term for mother's brother (uncle or *nera*) is a common Nilotic terms; it is also used for the mother's mother's son, although Acholi often do prefer to address mother's brother's son as "*latin pa nera*" or "son of my uncle". Among the Nilotic, a large number of descriptive terms are employed like "the son of my uncle".

Both husband and wife call their mother-in-law "Mara"; customary a husband has to speak to his mother-in-law with due respect, but should they meet in the market or hunting place (bush) both will make a slight detour. The husband may drink beer offered to him by his mother-in-law, but he cannot eat food offered to him by his mother-in-law. The husband is also required to speak to the father-in-law with due respect; however, he may drink beer drawn from the same pot with his father-in-law, but he is not allowed to drink in the same room or place with the father-in-law. The man can draw his beer from the same pot and take it in a separate gourd to a nearby place, where he could drink; and on no account the man may eat with his father-in-law in the same pot.

Although the wife's brother is called brother-in-law (*Ora*) he is not treated with much respect as is done to the father-in-law. The husband of the wife may eat and drink together with his brother-in-law from the same pot and in the same room and talk familiarly.

The uncle from the side of the wife is feared as among the Shulluk than the father-in-law. The reasons have been that uncle has power to curse the children of her sister. He could make the children unproductive or children could produce but their children die one after another. The wife of the brother-in-law is also feared for the same reason, and she is treated with due respect more than the mother-in-law (the mother of the husband).

Acholi customs forbid the husband and wife from seeing their father-in-law or mother-in-law naked. Thus, if a son-in-law sees the mother-in-law nakedness, he must give her a sheep, and this sheep the mother-in-law will not kill but puts in her flock. Likewise, if the wife has seen her father-in-law nakedness, she will make enough beer for the whole group that her father-in-law used to hang with. When beer is ready, the father-in-law calls his group (friends) they will then sit under a tree or in a room chatting and drinking. The same rule applied to all who are called mother-in-law; and to the wives of the uncles, who are been addressed as mother-in-law (*Mara*). If they see their father-in-law nakedness, they have to make beer and give to them.

According to Seligman (1948) the man is extremely unlikely to see his mother-in-law nakedness, as the Acholi women in the old days wear a small fringed apron; and today, Acholi woman covers herself well with clothes from neck to knee and sometimes from neck to ankles, unless the husband comes upon her accidently when she is bathing –the same thing apply to wife, she might not see nakedness of her father-in-law, unless she comes up him accidently while bathing.

It must be noted that the wife's brother's wife is also called "mother-in-law" (Mara); this implies that her daughter (s) cannot be married by father's brother's son (s), because this girl is addressed as 'daughter of the brother-in-law" (*Nya pa Ora*); and the brother-in-law is not respected as the in-laws but rather treated with familiarity address towards sister's husband (*Ywera*).

For a man his wife's sister and his brother's wife are both called "Ywero" and the man may treat them with complete familiarity, he may even hold their breasts, and touch private part of the body, and there is no harm in seeing them naked. Due to this familiarity, weaken wife may get pregnant with the brother of her husband. This is why the Acholi say, "*Tuko ki ywero oketo latin i ic*", which literarily means, "familiarity and unnecessary touching of the private parts of the wife's body by husband's brother could lead to sexual intercourse and eventually to pregnancy".

If a woman becomes pregnant by husband's brother, she cannot be married by him. This is the same as the man cannot marry the sister of his brother's wife.

We have also seen briefly that in Acholi Land a person is addressed by his family name or nick name. It is important to notice that family name (*Nying* kwon) is given soon after three to four days following birth of the baby. The family name is often derived from some events connected with birth, situation or behaviour of the husband or the wife. In Acholi a boy born during famine is called "Okec" and a girl born during famine is called "Akec". And in Acholi if the wife does not want to have sex but the husband forces her and eventually the wife become pregnant as a result of that sexual contact, the boy is called "Odiya" and if it is a girl, she will be called "Adiya". The name Odiya and Adiyp might be confusing in Acholi, because if the wife does not want the marriage and she is forced, the child born, if a boy will be called Odiya and if a girl will be called Adiyo.

The nick-name (*nying mwoc*) is given when a man does something peculiar and his companions immediately acclaim his nick-name. For example nick name "*Dikomoi*" is given to a person who has killed many people during the war; and nick-name "*Lutukumoi*" is given to the person who caused fighting between two parties or is given to a person who has killed a Lutuko-man or Lotuku-woman in the bush or house. The nick-name "*Okwoto cet i gang pa maro*" is given to the person who breaks wind at his mother-in-law's house.

Anyuak Kinship and Family

Marriage is forbidden between all persons (males and females) to whom relationship can be traced directly, whether through the father or through the mother. This is correlated with custom governing inheritance and the transfer of bride-wealth, where the cross-cousins must stand together, depending on the mutual obligation that exists between the bother and the sister; on the other hand, the children of the sister and the children of the brother are supposed to stand together in case of any family issues, but yet all these depend on the mutual obligation that exist between sister's children and brother's children. Before we go further with our discussions on kinship and family life, let us see how kinship terms are used among the Anyuak.

The following are the kinship terms use among the Anyuak (Anywaa)
  -a person calls biological father as "father" (e.g. Wora)
  -a person calls biological mother as "mother" (Mera)
  -a person calls father's brother as "Uncle" (Wora)
  -a person calls father's brother's son as "son of my uncle" O'Wora or Wor Wora
  -a wife calls husband's mother as "mother-in-law" (Amang or Mango)
  -a father calls his boy as "son" (Woda)
  -a mother calls her boy as "son" (Woda)
  -a father calls his siblings as "children" "Nyara" (single), Nyaa (pl)
  -a mother calls her siblings as "children" (*Nyara (single) Nyaa (pl)*)
  -a child calls his/her father's son as "the son of my father" (O'wora)
  -a child calls mother's son as "son of my mother" (Choka)
  -a person calls father's daughter's as "the daughter of my father "or "Sister" (nyiwora)
  -a person calls wife's brother as "the wife of my brother" (Chi-coka or Chi-mada)
  -a person calls mother's daughter as "daughter of my mother" or "sister??" (Chura)

-a person calls wife's sister as "sister" (Nyimera) or (Chik-ciya)

-a person calls mother's sister as "mother" (*Mara*)

-a person calls mother's brother as "Uncle" (*Nara*)

-a person calls mother's paternal grandfather's grandson as (Kwa wur mia)

-a person calls mother brother's son as "uncle" (*O'nera*)

-a person calls mother's sister's son as "*O'mera*"

-a person calls mother's sister's daughter as "*Nyimera*"

-a person calls mother's brother's daughter as (Nyinera)

-a person calls father's sister as (Awaa)

-a wife calls husband's sister as (Awaa)

-a person calls father's sister's son as (O'naa)

-a person calls father's sister's daughter as (Nyi-waa)

-a person calls father's brother's wife as (Ci-woo)

-a person calls sister's son as (Choyo) or (O'Choka)

-a person calls sister's daughter as (Nyichoyo or Nyichoka)

-a person calls father's father as "grandfather" (Akwa or Kwa)

-a person calls mother's father as "grandfather" (Akwa or Kwa)

-a person calls father's father's brother as (Akwa or Kona)

-a wife calls husband's father as "parent-in-law" (*Akwa or Kuum*)

-a person calls father's mother as (Awango or Wango)

-a person calls mother's mother as (Awango or Wango)

-a wife calls husband's mother as (Aweng or Wanga)

-a person calls son's children as "Nyikwara" (single) "Nyikware" (Pl)

-a person calls daughter's children as "Nyikwara" (single) "Nyikware" (Pl)

-a husband calls wife's father as "parent-in-law" (*Ora* or Wora or Anuo)

 =a person calls wife's brother's son as *Ora*

-a wife calls husband's father's brother as (Akwaco or Kuik)

-a husband calls wife's mother as (Ayang)

-a husbands calls wife's mother's sister as (Amaro or Mara)

-a husband calls wife's brother as (Omera)

-a wife calls daughter's husband as (Wada)

-a wife calls sister's husband as (Omera or Yuera)

-a husband calls wife's mother's mother as (Awang)

-a wife calls husband's brother as (Cwara)

-a wife her co-wife as (Nyaka)

-a wife calls husband's brother's wife as (Nyaka or Chi Nyaka)

-a husband calls his brother's wife as (Chi madha or Chi madho)

-a wife calls her brother's wife as (Chi-madha or Chi-madho)

-a wife calls husband's sister's daughter as (Nyi-waa)

-a person calls father's sister's daughter as (Nyi-waa)

The term "*Wora*" is used to include the biological father and his brother; and the fathers of all those whom the father calls them "father's sons". Thus, as descent is patrilineal, therefore, all the men of the father's clan and generation are called "*Wora*" (fathers); and all the men of the "grandfather's generation" are called "*Kwa*" (grandfathers), but not all men of the same generation of the grandfather are called "*Kwa*", they are rather called "*Uwa*" (father's sons).

At this point it is important for us to remember that, in Anyuak, not all those succeeding generation are called "sons", only a man's own sons are called "*Wode*" (sons); the brother's sons are rather called "*Umade*" (sons of a friend –"*mat*"). The sister's sons are called "Oka", while the sister's daughters are called "*Nyaka*"; they all belong to the clans of their fathers.

"*Maiya*" is an individual term used for "biological mother" and in Anyuak custom, all persons whom "mother" calls "*Uwa*" her children called them "mother's brothers" (N*a*); and those persons whom the mother calls "fathers" (*wa*), her children call them "mother's fathers" (*Kwa*); the word "Kwa" in this sense indicating the generation where the child or children belong and does not indicate clan.

Interestingly, all the children of mother's brothers "*Uwa*", her children call them "mother's brother's sons" (*Una*) and if they are girls the children call them "mother's brother's daughters" (*Nyina*). Therefore, all the men and women who belong to the mother's clan are considered as the children of "mother's bother". Seligman (1948) notes, In Shulluk, clans are patrilineal and all men under one general are called "*Uwa*" because their mothers' clansmen are mother's brother's sons "*Una*". All men from a person mother's generation, the person call them "*Na*", while his mother calls them "*Uwa*. In all these lines, only mother's father's sons are important as "*Na*" or "Uncles" (P.53).

Sex taboos and in-laws

When there is intermarriage between village "A" on hand and villages "X", "Y" and "Z" on the other hand, the men in village "X", "Y" and "Z" call men and women in village "A" their 'in laws' (*ora*). The Sudan Notes and Records (1941) pointed out that the term "in-law" (*ora*) is not applied by the husband to only his wife's kinsmen, but also, strictly speaking, all kinsmen of the wives of all his own kinsmen.....thus, term "in-law" (*ora*) embrace a very large group of people. But the word "in-law" (*ora*) in Anyuak customs has two senses (or meaning): the wider sense and the narrower sense. In the wider sense the term embrace a large group of people but yet limited by the interpretation of the word "kinsmen', this is especially when we come to the sex taboo such as "*dwalo*". However, in the narrower sense of the word "in-law" (*ora*), it ties the husband to the homestead (*gol*) of his wives only. And throughout the earlier years of the marriage, the men from the village/*pac* ("A") of the husband must pay special respect to the women from villages/*pac* "Y" "X" "Z" of the wives. This is done mainly to avoid two things:

(1) To avoid "*dwalo* sickness", this is a form of sickness that resembles dropsy. Thus, when a man from village/*pac* "A" has sexual relationship with the woman from village/*pac* "X", "Y", or "Z", who are considered as in-laws (*ora*), both the woman and the man (if there is a child) all will be affected by "*dwalo* sickness". It is believed that if no sacrifice is done to serve them, both man and woman (and child) will die. According to Shul-luk customs, this requires that a man must kill a sheep and put the stomach contents on top of the gourd by the road side, so that people who are passing through the road may carry away the "*dwalo* sickness". The same sickness (*dwalo*) is said to be produced by a man having sexual relation with a woman to whom he is related.

(2) To avoid all possible cause of friction during what may be called "the period of tension". During the tension period the husband will not speak to his in-laws (*ora*) and if by any chance he meets them on the road, he must avoid them by going off the road to let them pass. After the in-laws (*ora*) pass, the husband then could return on road and continue his journey. This tension is usually the product of husband not paid all the required bride-wealth. He might have just paid ten head of cattle as a pledge to see the good behaviour of his wife. The full bride-wealth would then be paid as the wife produces children; the husband will be paying five head of cattle for every two baby girls born and two head of cattle for every five baby boys born. This will, bring the total cattle to thirty including the ten head of cattle paid for a pledge. Once the husband has paid this number of cattle, he is said to have completed the marriage and the tension is said to be resolved (P.29).

Village "A" being reciprocally in law (*ora*) o villages "X", "Y" and "Z". Thus, if men of village "A" have married women from villages "X", "Y" and "Z", no men from village "X", "Y" or village "Z" may marry women from village "A" because they call women from village "A" their in-laws (*Ora*).

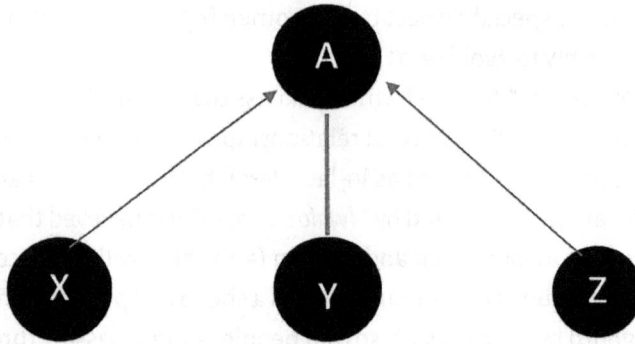

Figure: 92, Simple kinship relationship
Source: Sudan Notes and Records, 1941, P.30

But from a particular group or lineage there could be average of 50-70 married men. In this case, the number of in law (*ora*) will be numerous. The example of the lineage marrying many wives that eventually create more in laws (*ora*) can be seen from the below diagram.

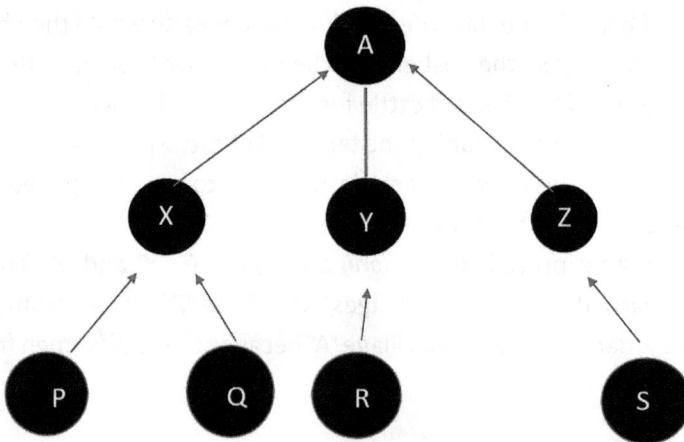

Figure: 95, Complex kinship relationship
Source: Sudan Notes and Records, 1941, P.30

We have seen in the first diagram that there is only one way traffic, that "X", "Y", and Z call "A" their in-laws (*Ora*) because men from "A" having married women from them. Thus, in the second diagram we can see that village/*pac* "X" have married women from village/*pac* "P" and "Q"; men in village/*pac* "Y" have married women from village/*pac* "R"; and men from village/*pac* "Z" have married women from village/*pac* "S". This also implies that other men from village/*pac* "A" are allowed to marry other girls from village/*pac* "X"; as similar with other men from village/*pac* "X" are allowed to marry other girls from village/*pac* "P" because the girls from village/*pac* "P" are said to be from the village/*pac* of the "in-law"(pa *ora*). In other words, this means that the entire membership of "P", "Q", "R" and "S" are look-ing at Village/*pac* "A" as the village of the wives of their "in-laws" (*ora*), which called locally as "*myer ora*". In other words the entire members of Village "P", "Q", "R" and "S" become the in-laws (*Ora*) of the people from village "A". Therefore, it is forbidden for any man from village "Village/*pac* "A" to marry or to have sexual relationship with women from the four villages "P", "Q", "R" an "S" above.

However, if for one reason or the other, another man from village/*pac* "A" seduces another girl from village/*pac* "P", people or men from village/*pac* "X" will be annoyed because they regard the unmarried girls in village/*pac* "P" as in some sense their prerogative (men in village/*pac* "A" should not jump over them). As such this case would be taken seriously, as something that would spoil the good relationships between the in-laws "*ora*" (Sudan Notes and Records, 1941).

For a man, all his wife's clansmen he calls them "*Ora*" (See kinship terms above) More importantly, the "term" *Ora* is a reciprocal term that could be used by either the husband or the wife to address wife's parents (in case of the husband) and to address husband's parents (in case of the wife). The husband can also address the wife's brothers as "*Ora*" (in-laws). The terms Cousin that is *Nyiwa, Nyimia, Nyima* or *Nywaja* are all reciprocal terms, which could be used by both husband and wife to address: wife's brother (in case of a husband) or to address sister's husband (in case of a wife).

In Shulluk when we talk of "*Ora*" to mean "mother-in-law", this would include wide range of relationships, in this sense the term "*Ora*" would mean the wives of all men, the married wife calls their fathers her fathers, as well as any woman who stands in that relationship (of mother-in-law).

However, in Anyuak customs, the wife's sister is called "*Nimeraa*"; and the wife's brother's wife is called "*Chi-madha* or *Chi-madho*". The Anyuak, like the Acholi, they also call the wife's brother's wife as "mother-in-law" (*Mara*). It must be seen that the term "*Ora*" cannot possibly belong to one clan but the terms mothers, fathers, and mother-in-law, mother, father and brother belongs to clans, they cannot go out of the clans. In other words, they are "copy right" of the clans.

The use of the terms have significant connotation, when a woman speaks of the "*Ora*" this specifically referred to her husband's brother and possibly to the husband's mother's sister. The term "*Ora*" implies respect and ceremonial behaviour when used by the opposite sex.

As we have seen from the terminologies use in Anyuak kinship, the term "*Mera*" is applied only to the biological mother but mother's sister is called "*Mara*". In Anyuak customs, like other Northern Luo Groups, daughters do not belong to the clan of their mother, they rather belong to the clan of their father, what we called "husband lineage", so a person calls the sons of his/her mother's sister as "*O'mera*" and calls the daughters of his/her mother's sister as "*Nyimera*", subsequently, differentiating them from brothers and sisters, with whom they are usually identified according to the classificatory system.

Anthropologist, Lewis Henry Morgan, defined classificatory kinship system as putting people on the basis of abstract relationship rules, sons of father, daughters of father, daughters of mother and sons of mother are put genealogically although the classes have no overall relation to genetic closeness. For example a stranger may marry from Anyuak family, when they have children with the wife; all these children will belong to the class or clan of his wife. In Anyuak the term "*Ci-woo*" is used for the father's brother's wife. Thus, a person calls his/her father's brother's wife as "*Ci-woo*". I am not sure if the term "*Ci-woo*" is used throughout the Anyuak society.

Therefore, from the above illustrations, it becomes clear to us that the Anyuak systems, whether we considered it as classificatory or not, yet it recognizes clan relationship among the people, it also recognizes family relationship and their differences can easily be seen according to the generation.

While the term "*Yuara*" is applied by the wife (woman) to the husband's brothers (*Uwora*). The same term "*Yuara*" is also applied by the husband (man) to the wife's sisters. It is important to understand that the term "*Yuara*" does not has the same meaning with the word "*Ora*", although generally it is assumed that the word "*Yuara*" might have been driven from the word "*Ora*". However, the term "*Ora*" implies that the man and the woman must show due respect to one another, any breach of the etiquette between them is usually taken as a serious matter.

On the contrarily, the term "*Yuara*" does not require respect from both sides of the man and the woman; subsequently, many a times one could see a woman target her husband's brothers and her father-in-law's brother's sons (*Uwora*). Seligman (1948) notes that in case of death among the Northern Luo the "*Uwora*" may inherit each other's wives, so that "*Yuara*" become legitimate spouses (P.54).

In Anyuak, like any other African groups, the biological father is the head of the household; and more importantly, a man belong to his father's clan and all of his father's clans are also his own clansmen. Thus, in a society, once a man is married, he sets up new relationships to his wife's relatives. Which implies that all the relatives of the wife become husband's brother-in-law (*Ora*)? These are the people the husband must show great respect, he must talk to them respectfully; in addition the husband is to avoid any misbehaviour that could lead to quarrels or fights before them; he will never become familiar with any of them. His intimate friends will be chosen from among his clansmen. However, the relationship to the *yuara* is more flexible between clan-brothers (*Uwora*), who stand a better chance to have access to each other's wives. They can play together with any restriction.

Traditionally, a woman who is not yet married can use kinship terms in the same way other people use, but a married woman must use kinship terms according to the manner and attitude expected of her. For example, a married woman may call her husband's sisters as "*Awaa*" and she may call the husband's father and mother as "*Ora or Wura*" and "*Aweng*" or "*Wanga*" respectively. So, it is not accepted that the wife addresses the father and mother of her husband as relatives by marriage, but should rather address them as her own relatives. The reason why the wife should now address her father-in-law and mother-in-law as her own relatives is because by virtue of marriage the wife considers herself as a member of husband's family.

According to Anyuak customs, when a father dies, his son or sons takeover his younger wife or wives; while one of the father's brothers can inherit their mother; customarily, no son can inherit his own mother as a wife. In addition, in Anyuak society, it is not considered disgraceful for a young unmarried son to sleep with his own father's wives.

Interestingly, the husband calls his wife's sister as "Nyimera" or "Chik-ciya"; it is not clear to me whether when wife's sister comes to visit them, the husband is not bound by any taboos (like in Shulluk Customs); he may have sexual intercourse with her, since custom dictates that in case of death a wife's husband could marry one the wife's sisters.

Among the Anyuak, people avoid all persons that they call them "in-laws" (*Ora*), especially the wife's mother. According to customs the man must avoid entering into the house of the mother-in-law, and above all he should take care of her properly; and when caring for the mother-in-law, the husband is not allowed to look directly into the eyes of the in-law, especially in public places. Thus, if a man sees his mother-in-law at a distance, she should take different direction to avoid them meeting faced to face.

Yet some informants told me that the husband is not only to avoid meeting his mother-in-law face to face, but he is also not allowed to enter unnecessarily into the village where the mother-in-law lives, until all the cattle for the dowry are paid all.

However, the husband cannot avoid entering into his brother's wife's house; he can enter into the house at any time to eat and drink on the condition that when he wants to enter the house, woman must leave the house. And when the man enters the house, he is not allowed to sit on the brother's wife's sleeping mate; he is to sit upon bare ground. However, if by mistake the man sits on his brother's wife's sleeping mate, this will be considered as breach of etiquette; and such an act might be taken as he intended to have sexual intercourse with the woman, although the woman is not in the house at that time.

The wife's father's brother's wife "*Ci-woo*" is also treated as "*Wanga*" (mother-in-law) and must be avoided and given due respect. However, if by any chance the father's brother's wife "*Ci-woo*" comes suddenly into the group, so, the person who calls her "*Wanga*" must quite the group and site on the other side of the building where the group are, so that he does not see the woman and the woman should not see him. Interestingly, if this woman (in-law) is the man's wife's mother, the man must not remain around the place, he must disappear completely from the surrounding, for otherwise, the mother-in-law might claim that the man has seen her nakedness.

In Anyuak society, like in Shulluk society, there are four aspects of the in-law (*Ora*) relationship as follows:-

(1) *The avoidance of the mother-in-law* is so common a custom among the Northern Luo of South Sudan. The aspects of respect to mother-in-law is so complicated that we cannot discuss all of them here but I want to let us know that there are three elements in the custom that we should always remember in relation to the culture. Firstly, it is a MUST that the mother-in-law be given due respect without any condition; Secondly, the in-law (*Ora*) MUST not show any sexual feeling towards her. Thirdly, both the mother-in-law and the son-in-law (Ora) MUST avoid seeing each other nakedness. Among the Anyuak men, who wear garments only knotted over the shoulders, leaving the genitals exposed, the mother-in-law must take a great care to avoid the presence of the in-law, since the eyes travel faster than light. The custom forbids the son-in-law (*Ora*) to have sexual intercourse with his mother-in-law. Thus, if the young man wants sexual satisfactions, he could approach the sister of his wife. However, if the young man in Anyuak land has not yet collected enough cattle to marry, and he wants to satisfy his sexual feeling, he can exercise his rights and consort with the wife of one of his "*Uwura*"; he is also expected to treat the mother of the woman with respect as though he is the actual husband, if he does not practice this attitude trouble may arise at the time he wants to inherit his brother's wife after death.

From all the above narratives, we can see clearly that avoidances of in-laws (*Ora*) are of paramount important in Anyuak society and in Shulluk society. Therefore, men are checked against avoidances, for example, if the husband of the wife dies and his brother (*Uwura*) wants to inherit the wife, he might meet challenges from the mother-in-law. The mother-in-law may make it difficult for him to inherit his brother's wife, if previously he did not avoid her. However, to enable this man to inherit his brother's wife, he must kill a sheep and ask some of old influential elders to act on his behalf as peacemakers. Because in Anyuak society, the inheritance of a brother's wife is always interpreted to how much the husband's brother has done to avoid the then mother-in-law when his brother was still alive (before his brother actually dies).

It is to be noted that the wife's brother's wife is addressed as "mother-in-law" (*Amang* or *Mango*), so the same avoidance applied to her. She must be looked upon as mother-in-law.

(2) *The attitude towards those relatives of the wife to whom she stands in some definite relationship.* Thus, among these relatives we may consider the wife's father and her brother. They are her natural protector, to whom she may complain if her husband is mistreating her. The bride-wealth is always given to the father of the brother of

the girl and if the girl dies childless, the father of brother will take this bride-wealth as a compensation for the loss of her children and not as a dowry. The wife's father's brother (*Ci-woo*) and his son receive parts of the bride-wealth, although they do not belong to her clan, however, they are called in-law (*Ora*)

(3) *The attitude towards the husbands of female relatives is the converse of the (1) and (2) above; the behaviours are reciprocal but not identical.* Although the man show respect towards husbands of the female relatives, yet he is more in favourable position. They pay bridge wealth of his sister to him and he can still demand anything he wants from these husbands.

(4) *A woman's attitude towards the relatives of her husband.* The wife looks upon her husband's blood brothers and his mother's sisters' sons as her in-law (*Yuara* and *Uwura*). By traditional norms she cannot avoid these people; she could be too familiar to them and this would not be considered as an offence. It is not a shame for the wife to have sexual relationships with any of these people; it is acceptable.

Although the conception of the "*Yuara*" and "*Uwura*" relationship is broad among the Anyuak society, the omissions from it are as striking as the inclusions. The husband does not call wife's sisters as "in-law" (*Ora*) in any form or shape but calls them "*Yuara*". As such, the husband always treats them with familiarity; and if he also wants he could take one of them as his second wife. Thus, if he so desires, priority would be given to him but he must give full bride-wealth.

Staying away from the in-laws is very important especially for the wives or women. Traditionally the wife is forbidden to call her husband's parents as "*Ora*"; she is to address them as her own grandparents (*Kwara or Kwaroo*); and treat them with respect. When she is handing something, like water, to her father-in-law, she must kneel in front of him to show respect; and when she is speaking with the father-in-law, she should not see him direct into the face, but must turn her face away as she speaks; however, there is no strict avoidance required.

Sometimes mother stays with the her son if her husband dies; in this case she is given small separate hut to sleep in but that hut must be closer to the son's hut in case of any emergency. At breakfast, lunch and dinner time the husband's mother is to eat separately but the wife and the husband could eat together if they so wish.

Young people are face to marry the partner of their choice provided that there is no exogamy and no genealogical relationship can be traced on the mother's side. Every Anyuak, young man wants his sister or his father's sister's daughter to be married properly, with enough cattle, because the bride-wealth from these girls are given to him to marry with, and without the cattle from his sisters or his father's sister's daughter he cannot marry a wife. The father, who uses to marry more wives for him, while his son remains unmarried, is often regarded as useless and irresponsible father. In the Sudan Notes and Records (1941) observes that the head of homestead is the sole legal owner of cattle in his homestead, he can dispose them in the ways he chooses. He holds them in fact however subject to a trust which equity will enforce. The terms of trust are that all cattle shall be applied solely towards the marriages of male members of the homestead in order of their seniority, that no cattle shall be sold except for the purpose of grain in time of famine, and that the homestead leader shall not himself marry a second time unless there is no unmarried male member of the homestead of marriageable age; a breach of these terms of trust would be universally censured but would probably not be interfered with by the courts (P.23).

In Anyuak society, there are no houses built particularly for boys and girls. The boys normally sleep in the cattle byres (Luak). In a village, where there are many unmarried girls, parents usually give them a house to sleep in. The girls sleep together in this one house and such house is known locally as *"Otho mar nyakwe"*. Thus, when the boys want the girls, they come at night into the house where the girls sleep, and ask the girls, if they could allow them into the house. It is not clear, whether customarily; boys are forbidden or allowed to share a house with the girls (Seligman, 1948, P. 63).

In contrast, Howfmayr states that among the Northern Luo Groups, meeting of boy-friends and girl-friends must take place outside the village and the parents are not supposed to know what is going at that appointed time; although parents are not allowed to meddle in the affairs of the boys and girls. He went further by saying that the girl chooses the boy she likes and often the girl breaks several engagements before finally deciding to marry one man, this is not considered disgrace except when the girl has done to excess (too much).

A girl who has more than ten lovers is looked upon as a feast woman (*Ngati cou*). It is said that men who do not have girl-friends often take opportunity to have sexual intercourse with such girl when she is on her way to fetch water from the stream or river.

It is reported that the girl's choice of a husband is often influenced by the thought of how wealthy is the boy-friend or the potential husband. The girl wants to know if the boy has enough cattle to marry her. The girl's thought is often supported by the father, brothers and father's brothers (Nara), they always try to dissuade her from getting married to a poor man.

Nevertheless, we cannot deny that the general opinions of the Anyuak communities are that, they are against forcing a girl to take a repugnant partner for the sake of cattle. Interestingly, the provision of bride-wealth prevents very early marriages; as a result many men appear not to marry when they are still under twenty years of age.

An Anyuak man may have sexual intercourse with the wife of any man whom he calls *"Uwura"*. He can also have sexual intercourse with the wife of his father's brother's son and his clansmen of his own generation. The boy may even ask the younger wives of his father to have sexual intercourse with any of them (with exception of his mother). The boy could even ask to have sexual intercourse with the wives of those he calls them "father" (Wura). Usually, the father knows that the son is consorting with one of his wives, and as this happen, the father leaves the son to sleep in the hut with the wife, while he goes to sleep in the hut of the other wife; he does nothing about it.

It is said that the Anyuak man, like the Shulluk man, supported their sons to have intercourse with one of father's wives, because this is better than going to have sex with another woman, where the father would be obliged to pay the fines for adultery.

Thus, in Anyuak cultures, it is acceptable that a grown up boy goes with the wife of his father; or goes with the wife of his father's son or father's brother's son (Uwura). It is reported that when a man has been out of his house for days or months or years, when he comes back home and learn that his "Uwura" has had sexual intercourse with his wife, as a human being he might feel very angry, he could even say some bad words to him, but that would be all. In such a case, local chief usually comes and tells the husband of the wife, "This man is your brother "Uwuru". It is alright for him to sleep with your wife while you are away". On hearing this, the man (husband) would accept the idea, however, if it was a stranger who has sexual intercourse with his wife, the man would try to spear that stranger at all cost. Thus, the young unmarried man, who was exercising his rights towards the wife of his father's brother's son "Uwura", would say to his relative that his father's brother's son "Uwura" knew before leaving the village that he was interested in the wife; the "Uwura' made no objection to it, therefore, such a union should not be a surprise to him. This implies that a child born from such a union always the child is considered to belong to the woman's husband and any idea of it belonging to the wife's temporary lover is ridiculous.

Anyuak traditional custom forbids a man from making an attempt to come closer and closer to the wives of his mother's sister's sons (O'mera) wife's brother's sons (Uwura), mother's brother's son (Unara), sister's son (Oka or ocowa) as well as the wife of his father's father (Kwa); if he does, this would be considered more worse than committing adultery, and it is believed that both the wife and the man who are guilty would die. Not only this, it is also believed that having sexual intercourse with the wife of in-law (Ora) is looked upon in the same way. A man who is involved in having sexual intercourse with the wife of his in-law is commonly called *ci-juok* (a bad man).

In many cases, married brothers and married sisters happen to live near one another and their children would address themselves as "Unara" (uncle's son), "Nyinara" (uncle's daughter), "Uwaya" (father's sister's son) and "Nyiwaya" (father's sister's daughter) the children would have an opportunity to meet and play together; and there is no restriction on the girls to play with the boys.

Experiences in Anyuak society and Northern Luo of South Sudan in general, revealed that relationship between wife's children and their mother's brother "Nara" (uncle) is stronger than the relationship between wife's children and father's brother's children "Uwora". It is the mother's brother who usually gives a son of his sister his first spear; the boy also has a right to some of the cattle given as bride-wealth for his sister's daughter, and sometime he could even influence the girl in her choice for a husband.

A young woman does not drink milk under any circumstances, though a woman who is nursing a baby may, as do old women past the reproductive age. Only small boys herd the cattle and milk them, for once a boy has reached maturity there is the danger that he may have had sexual contact, so if he does and milk cows or just walk among the cattle in their pens, it is believed that this might cause cows to become sterile. However, this believe is so strong among the people, in such a way that if a child drinks milk at his neighbour house, when he comes back, he must wash his body before entering into his father's house. Because it is believed that some milk might have remained over his body that could poison the food, and once this food is eaten by unclean person, the person might die. According to Anyuak and Shulluk custom, a man who has sexual intercourse with his wife a night before, he is considered unclear and must not drink milk until the sun sets the following day (Seligman, 1948, P. 73).

In Anyuak communities, like in Shulluk communities, when a woman finishes cooking food for her husband, she is to leave the hut while the husband is eating, but it is not shameful for men and women to eat together; and after a man has eaten the remaining part of a meal is taken and eaten by the woman and her friends (usually her unmarried "Nyiwa". However, in Anyuak society a newly married couple do not eat together for some months or even a year.

Jo-Luo of Bhar El Ghazal Kinship and Family Life

Unlike the other Luo groups of South Sudan, among the *Jo-Luo of Bhar El Ghazal*, there is a clear distinct terms for "father's brother", in the case of the "individual biological father" and the "father's brother" the term is distinguish by adding the word "Wo" before "*Wora*" and that become "*Wo-Wora*" or "*Nera*" (uncle) and this term is applied to every person who is legitimate in the family tree. In Jur Chol kinship the "father's brother's son" is called "*Wad Umi-Wora*". While in Acholi, the "father's brother" is called "*Nera*" (uncle) and the "father's brother's son" is also called "*Nera*" (uncle).

Marriage is forbidden between all persons (males and females) to whom relationship can be traced directly, whether through the father or through the mother. This is correlated with custom governing inheritance and the transfer of bride-wealth, where the cross cousins must stand together, depending on the mutual obligation that exists between the brother and the sister; on the other hand, the children of the sister and the children of the brother are supposed to stand together in case of any family issues, but yet all these depend on the mutual obligation that exist between sister's children and brother's children.

Before we go further with our discussions on kinship and family life, let us see how kinship terms are used among the Jo Luo of Bhar El Ghazal (Jur Chol).

The following are the kinship terms use among the Jo Luo of Bhar El Ghazal (Jur Chol)
   -a person calls biological father as "father" (*Wora*)
   -a person calls biological mother as "mother" (*Mera or Miya*)
   -a person calls father's brother as "father" (*Wora or Wo-wora*)
   -a person calls father's brother's son as (Awora or *Wad Umi Wora*)
   -a wife calls husband's mother as "mother of husband" (*Mi Chwara*)
   -a father calls his boy as "son" (*Woda*)
   -a mother calls her boy as "son" (*Woda*)
   -a father calls his siblings as "children or daughter" (Nyitin (both M. & F) Nyara F."Single)
   -a mother calls her siblings as "children" (*Nyitana (both M. & F. Pl), Nyara (single) nyaa (Pl)*
   -a child calls his/her father's son as "the son of my father" (*Umia (male) Nyimia (female)*
   -a child calls mother's son as "son of my mother" (*Umia or Choka*)
   -a person calls father's daughter as "the daughter of my father "(*Nyimia from* same mother) or (Nyi-Wora from different woman)
   -a person calls wife's brother as "brother-in-law" (*Cumia or Chi-coka or Chi mada*)
   -a person calls mother's daughter as "daughter of my mother" or "sister" (*Ngemia*)
   -a person calls wife's sister as "sister" (Nyimi chiya or Ngemi Chiya or Oraa)
   -a person calls mother's sister as "mother" (*Mara or Moo*)
   -a person calls mother's brother as "uncle" (*Nera*)
   -a person calls mother's paternal grandfather's grandson as (*Kwa wur mia or Kwara*)
   -a person calls mother brother's son as "Uncle" (*Wad-Na or Una*)
   -a person calls mother's brother's daughter as (*Nyinna or Nyinera*)
   -a person calls father's sister as (*Waya*)
   -a wife calls husband's sister as (*Waja or awayi*)
   -a person calls father's sister's son as (*Uwaya* or Awaya)
   -a person calls father's sister's daughter as (*Nyiwaya or Ngwaya*)
   -a person calls father's brother's wife as (*Chi Wora or Cheurra*)
   -a person calls sister's son as (Wod Nyimia or *Wad Nymya*)
   -a person calls sister's daughter as (Ngyar ngmya or *Nya-nyimia*)
   -a person calls father's father as "grandfather" (*Kwa*)
   -a person calls mother's father as (*Kwa*)
   -a person calls father's father's brother as (*Wora*)
   -a wife calls husband's father as "parent-in-law" (*Wora* or *Kwa*)
   -a person calls father's mother as (*Wanga mi-wora or Awong*)
   -a person calls mother's mother as (*Wanga Mi-mia*)

-a wife calls husband's mother as (*Awong*)

-a person calls son's children as (Nyethen *Wada* "Pl" or "*Wod-woda*" (M.) or "*Nya-Woda*" (F.)

-a person calls daughter's children as (Ngethen Ngara "PL" or *Nya-ngara* (F) or *Wod-ngara* (M)

-a husband calls wife's father as "parent-in-law" (*Wur* or *Cwara*)

-a wife calls husband's father's brother as "Abba *or Wora*" or "*Wur-Cwara*")

-a husband calls wife's mother as (Micia)

-a husbands calls wife's mother's sister as (*Mocia*)

-a husband calls wife's brother as "in-law" (*Ora*)

-a wife calls daughter's husband a "in-law" (*Ora*) or (*Chwar-Ngara*)

-a wife calls sister's husband as (*Ora or* Awayi)

-a husband calls wife's mother's mother as "in-law" (*Awong-micia*) .

-a wife calls husband's brother as "brother of my husband" (*Umi-chwara or Anna*)

-a wife calls her co-wife as (*Nyeiga or Thar para*)

-a wife calls husband's brother's wife as (*Nyeiga cumi cwara*)

-a husband calls his brother's wife as (*Cumia or Ora*)

-a wife calls her brother's wife as (*Nyiega or Chomya*)

-a wife calls husband's sister's daughter as (*Nyi-waya or Nya-nyimi cwara or Nya-Nyiega*)

-a person calls father's sister's daughter as (*Nyikwa (female) or Oka (male)*)

The term "*Wora*" is used inclusively to the biological father; but the brothers of the father are called "Wo-Wora"; while the "grandfather's generation" are called "A*kwa or Kwa*" (grandfathers).

"*Mia*" is an individual term used for "biological mother" and in Jo-Luo of Bhar El Ghazal customs, all persons whom "mother" calls "*Cumia*" her children called them "mother's brothers" (*Na or Nera*); and those persons whom the mother calls "fathers" (*wora*), her children call them "mother's fathers" (*Akwa* or *Kwa*); the word "*Akwa or* Kwa" in this sense indicating the generation where the child or children belong and does not indicate clan.

Interestingly, all the children of mother's brothers "*Ngethen*", her children call them "mother's brother's sons" (*Na or Nera*) and if they are girls the children call them "mother's brother's daughters" (*Nyina or Nyinera*). Therefore, all the men and women who belong to the mother's clan are considered as the children of "mother's bother". In Jur Chol, clans are patrilineal and all men under one general are called "Ur-Wora" because their mothers' clansmen are mother's brother's sons "*Na or Nera*". All men from a person mother's generation, the person call them "*Na*" or "*Nera*", while his mother calls them "Ur-*Wora*. In all these lines, only mother's father's sons are important as "*Na*" or "*Nera*" (Uncles). And the person calls mother's or father's son as "*Umia*" (the son of my mother or the son of my father).

Sex taboos and in-laws

When there is intermarriage between village "A" on hand and villages "X", "Y" and "Z" on the other hand, the men in village "X", "Y" and "Z" call men and women in village "A" their 'in laws' (*ora*); vice versa the men and women in village "X", "Y" and "Z" also call men and women in village "A" their in-laws (*ora*). The Sudan Notes and Records (1941) pointed out that the term "in-law" (*ora*) is not applied by the husband to only his wife's kinsmen, but also, strictly speaking, all kinsmen of the wives of all his own kinsmen.....thus, term "in-law" (*ora*) embrace a very large group of people. Like in Shulluk, the word "in-law" (*ora*) in Jur Chol customs has two senses (or meaning): the wider sense and the narrower sense. In the wider sense the term embrace a large group of people but yet limited by the interpretation of the word "kinsmen', this is especially when we come to the sex taboo such as "*buto-ki-dako*". However, in the narrower sense of the word "in-law" (*ora*), it ties the husband to the homestead (*gol*) of his wives only. And throughout the earlier years of the marriage, the men from the village/*pac* ("A") of the husband must pay special respect to the women from villages/*pac* "Y" "X" "Z" of the wives.

Village "A" being reciprocally in-law (*ora*) of villages "X", "Y" and "Z". Thus, if men of village "A" have married women from villages "X", "Y" and "Z", no men from village "X", "Y" or village "Z" may again marry women from village "A" because they call women from village "A" as their in-laws (*Ora*).

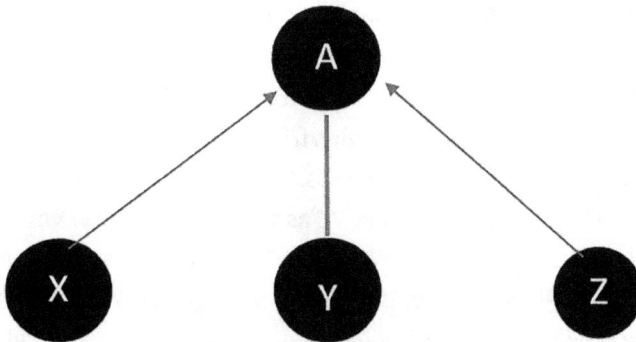

Figure: 96, Simple kinship relationship
Source: Sudn Notes and Records, 1941, P.30

But from a particular group or lineage there could be average of 50-70 married men. In this case, the number of in law (*ora*) will be numerous. The example of the lineage marrying many wives that eventually create more in laws (*ora*) can be seen from the below diagram.

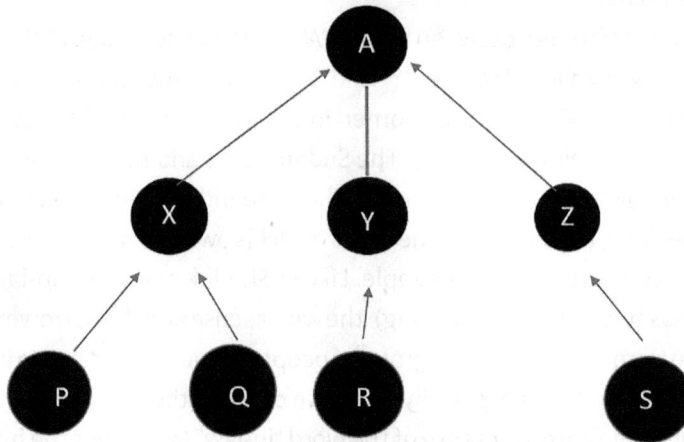

Figure: 97, Complex kinship relationship
Source: Sudan Notes and Records, 1941, P.30

We have seen in the first diagram that there is only one way traffic, that "X", "Y", and Z call "A" their in-laws (*Ora*) because men from "A" having married women from them. In the same way all women and men in village "X", "Y", and "Z" called women and women in village "A" their in-laws (*ora*). Thus, in the second diagram we can see that village/*pac* "X" have married women from village/*pac* "P" and "Q"; men in village/*pac* "Y" have married women from village/*pac* "R"; and men from village/*pac* "Z" have married women from village/*pac* "S". This also implies that other men from village/*pac* "A" are allowed to marry other girls from village/*pac* "X"; as similar with other men from village/*pac* "X" are allowed to marry other girls from village/*pac* "P" because the girls from village/*pac* "P" are said to be from the village/*pac* of the "in-laws"(pa *ora*). In other words, this means that the entire membership of "P", "Q", "R" and "S" are looking at Village/*pac* "A" as the village of the wives of their "in-laws" (*ora*), which called locally as "*myer ora*". In other words the entire members of Village "P", "Q", "R" and "S" become the in-laws (*Ora*) of the people from village "A". Therefore, it is forbidden for any man from village "Village/*pac* "A" to marry or to have sexual relationship with women from the four villages "P", "Q", "R" an "S".

However, if for one reason or the other, another man from village/*pac* "A" seduces another girl from village/*pac* "P", people or men from village/*pac* "X" will be annoyed because they regard the unmarried girls in village/*pac* "P" as in some sense their prerogative. As such this case would be taken seriously, as something that would spoil the good relationships between the in-laws "*ora*" (Sudan Notes and Records, 1941).

For a man, all his wife's clansmen he calls them "*Ora*" (See kinship terms above). Unlike Acholi and Shulluk the Jur Chol the wife does not call her husband's father "ora" but rather call him as "*Wora*" or "*Kwa*" (father or grandfather) and husband does not call wife's father as "ora" but he calls him "Wur-cial". The husband addresses the wife's brothers, like other Luo groups, as *"Ora"* (in-laws). The terms Cousin that is *Nyiwora*, is used to address the sister with different mother but sister with same father and same mother is addressed as "*Nyimia*". The wife calls the husband's brother as "Umi-Cwara" (the brother of my husband) and the husband calls the wife's brother as "ora".

However, in Jur Chol customs, the husband does not call wife's sister as *"Ora"*, but he calls her as *"Nyimicia"* and the husband address the wife's brother's wife also as *"Micia"*. In contrast, in Acholi, the wife's brother's wife is called "mother-in-law" (*Mara*). It must be seen that the term *"Ora"* cannot possibly belong to one clan but the terms mothers, fathers, and mother-in-law, mother, father and brother belongs to clans, they cannot go out of the clans. In other words, they are "copy right" of the clans.

The use of the terms have significant connotation, when a woman speaks of the *"Ora"* this specifically referred to her husband's brother and possibly to the husband's mother's sister. The term *"Ora"* implies respect and ceremonial behaviour when used by the opposite sex.

As we have seen from the terminologies use in Jur Chol kinship, the term *"Mia"* (mother) does not apply to mean "the sister of my mother", as in Shulluk, but a person rather call her/his mother's sister as "Moo". In Jur Chol customs, daughters do not belong to the clan of their mother, they rather belong to the clan of their father, what we called "husband lineage", so a person calls the sons of his/her mother's sister as *"Umia"* and calls the daughters of his/her mother's sister as *"Nyimia"*, subsequently, differentiating them from brothers and sisters, with whom they are usually identified according to the classificatory system.

Anthropologist, Lewis Henry Morgan, defined classificatory kinship system as putting people on the basis of abstract relationship rules, sons of father, daughters of father, daughters of mother and sons of mother are put genealogically although the classes have no overall relation to genetic closeness. For example a stranger may marry from Jur Chol family, when they have children with the wife; all these children will belong to the class or clan of his wife.

In Jur Chol the term *"Ci-wora"* is used for the father's brother's wife. Thus, a person calls his/her father's brother's wife as *"Ci-wora"*; this term *"Ci-wora"* is used throughout the Jur Chol society in the region as well as in Acholi society.

Therefore, from the above illustrations, it becomes clear to us that the Jur Chol systems, whether we considered it as classificatory or not, yet it recognizes clan relationship among the people, it also recognizes family relationship and their differences can easily be seen according to the generation.

While the term *"Yura"* is applied by the wife (woman) to the husband's brothers (*Umi-cwara*). The same term *"Yura"* is also applied by the husband (man) to the wife's sisters. It is important to understand that the term *"Yura"* does not has the same meaning with the word *"Ora"*, although generally it is assumed that the word *"Yura"* might have been derived from the word "*Ora*". However, the term "*Ora*" implies that the man and the woman must show due respect to one another, any breach of the etiquette between them is usually taken as a serious matter.

On the contrarily, the term "*Yura*", among Jur Chol, does not require respect from both sides of the man and the woman; subsequently, many a times one could see a woman target her husband's brothers and her father-in-law's brother's sons (*Uwora*). What Seligman (1948) notes about in Shulluk society, is equally apply to the Jur Chol Society that in case of death, the "*Uwora*" may inherit each other's wives, so that "*Yura*" become legitimate spouses (P.54).

In Jur Chol, like in any other African groups, the biological father is the head of the household; and more importantly, a man belong to his father's clan and all of his father's clans are also his own clansmen. Thus, in a society, once a man is married, he sets up new relationships to his wife's relatives. Which implies that all the relatives of the wife become husband's brother-in-law (*Ora*); these are the people the husband must show great respect, he must talk to them respectfully. In addition the husband is to avoid any misbehaviour that could lead to quarrels or fights before them (*ora*); he will never become familiar with any of them (*ora*). His intimate friends will be chosen from among his (*ora*) clansmen. The relationship to the *yura* is dependent on that between clan-brothers (*Uwora*), who stand a better chance to have access to each other's wives.

Traditionally, a woman who is not yet married can use kinship terms in the same way other people use, but a married woman must use kinship terms according to the manner and attitude expected of her. For example, a married woman may call her husband's sisters as "*Ngimy Chwara*" or "*Awac*" or "*Awayi*" and she calls the husband's father as "*Wora*" or "*Kwa*" and calls husband's mother as "*Anna*" respectively. So, it is not accepted that the wife addresses the father and mother of her husband as relatives by marriage, but should rather address them as her own relatives.

The reason why the wife should now address her father-in-law and mother-in-law as her own relatives is because by virtue of marriage the wife considers herself as a member of husband's family.

However, it is important to notice that although the wife becomes a member of the family, when she is addressing the relatives of the husband she should think of herself as though she belongs to a younger generation than she does in reality as a sign of respect to them.

According to Jur Chol customs, when a father dies, his son or sons takeover his younger wife or wives; and one of the father's brothers can inherit their mother, customarily, no son can inherit his own mother as a wife. In addition, in Jur Chol society, it is not considered disgraceful for a young unmarried son to sleep with his own father's wives.

Traditionally, although a woman is married, she still belongs to her own clan, because of this; it would look strange for the wife to use the same terms her husband employed. In this connection, it should be noted that in all the villages a husband address his wife as "*Nyimia*" and a wife can address her husband as "*Cwara*". Interestingly, the husband calls his wife's sister as "sister" (*Nyimi-cia*) or (*Yura*); and when wife's sister comes to visit them, the husband is not bound by any taboos; he may have sexual intercourse with her, since custom dictates that in case of death a wife's husband could marry one of the wife's sisters. But some people argue that the husband can have intercourse with the wife's sister because he has paid all the bride-wealth of his wife. I do not know how true this is.

Among the Jur Chol, people avoid all persons that they call them "in-laws" (*Ora*), especially the wife's mother (*Micia*). According to customs, the man must avoid entering into the house of the mother-in-law, and above all he should take care of her properly; and when caring for the mother-in-law, the husband is not allowed to look directly into the eyes of the in-law, especially in public places. Like the Shulluk, if a man sees his mother-in-law at a distance, she should take different direction to avoid them meeting faced to face. When he finds a good place, he could then send his friend to greet his mother-in-law on his own behalf.

Yet some informants told me that the husband is not only to avoid meeting his mother-in-law face to face, but he is also not allowed to enter unnecessarily into the village where the mother-in-law lives, until all the cattle for the dowry are paid all.

However, the husband cannot avoid entering into his brother's wife's house; he can enter into the house at any time to eat and drink on the condition that when he wants to enter the house, woman must leave the house. Like Shulluk, when the man enters the house, he is not allowed to sit on the brother's wife's sleeping mate; he is to sit upon bare ground. However, if by mistake the man sits on his brother's wife's sleeping mate, this will be considered as breach of etiquette; and such an act might be taken as he intended to have sexual intercourse with the woman, although the woman is not in the house at that time.

The wife's father's brother's wife (Chi-wora) is also treated as "*Mi-chwara*" or "*ora*" and must be avoided and given due respect. However, if by any chance the father's brother's wife (*Ci-wora*) comes suddenly into the group, so, the person who calls her "*Ora*" or "*Mi-cwara*" must quite the group and site on the other side of the building where the group are, so that he does not see the woman and the woman should not see him. Interestingly, if this woman (in-law) is the man's wife's mother, the man must not remain around the place, he must disappear completely from the surrounding, for otherwise, the mother-in-law might claim that the man has seen her nakedness.

From all the above narratives, we can see clearly that avoidances of in-laws (*Ora*) are of paramount important in Jur Chol society. Therefore, men are checked against avoidances, for example, if the husband of the wife dies and his brother (*Uwora*) wants to inherit the wife, he might meet challenges from the mother-in-law. The mother-in-law may make it difficult for him to inherit his brother's wife, if previously he did not avoid her. However, to enable this man to inherit his brother's wife, he must kill a sheep and ask some of old influential elders to act on his behalf as peacemakers. Because in Jur Chol society, the inheritance of a brother's wife is always interpreted to how much the husband's brother has done to avoid the mother-in-law when his brother was still alive (before his brother actually dies).

It is to be noted that the wife's brother's wife is addressed as "mother-in-law" (*Ora*), so the same avoidance applied to her. She must be looked upon as mother-in-law.

Although the conception of the 'in-law" (*Ora*) relationship is narrow among the Jur Chol society, the omissions from it are as striking as the inclusions. The husband does not call wife's sisters as "in-law" (*Ora*) in any form or shape. As such, the husband always treats them with familiarity; and if he also wants he could take one of them as his second wife. Thus, if he so desires, priority would be given to him but he must give full bride-wealth. It has been reported by some Jur Chol elders that a rich man could marry all of his wife's sisters, which he calls "*Nyimi-cia*".

The man does not call the wives of his father's brother's sons as his "in-laws" (*Ora*); more importantly, in Jo Luo of Bhara El Ghazal customs the wives of the clansmen are divided into two categories. The wives in the first category are the wives with whom sexual intercourse is permissible; and all the wives of the clansmen who are called either "father" (*Wo-wora*) or "father's brother's sons" (*Wod Umi Wora*) fall within this first category. While the second category, are the wives with whom no man is allowed to have sexual intercourse, except their husbands. Thus, having sexual intercourse with the wives of any other clansmen is said to cause death to both the offenders.

It is important to understand that the Jur Chol, like Shulluk people, treat the wives in the second category, we have seen above, with care, as sexual intercourse is not permissible with them. The men have to avoid anything that could cause problem, such as staying in the houses of these women when their husbands are not at home. And if the husbands of these wives are at home, whenever a man goes to visit one of the wives, he should avoid sitting upon the bed or mate of the woman. Failure to do this would be considered as attempt to sexual intercourse.

Staying away from the in-laws is very important especially for the wives or women. Traditionally the wife is forbidden to call her husband's father as "*Ora*"; she is to address him as her own father or grandparents (*Wora* or *Kwa*); she is also to address the husband's mother as mother of my husband "*Mi-cwara*" and treat them with respect. When she is handing something, like water, to her father-in-law, she must kneel in front of him to show respect; and when she is speaking with the father-in-law, she should not see him direct into the face, but must turn her face away as she speaks; however, there is no strict avoidance required.

In Jur Chol villages, like other Luo groups, sometimes mother stays with the her son if her husband dies; in this case she is given small separate hut to sleep in but that hut must be closer to the son's hut in case of any emergency. At breakfast, lunch and dinner time the husband's mother is to eat separately but the wife and the husband could eat together if they so wish. The same things apply if the mother of the wife is staying at the daughter's husband's house.

A man calls his son's wife as grandchild (*kwara*); she falls into the category of wives of a clansman with whom sexual intercourse is inadmissible. However, there are groups of women who are treated as the "in-laws" (*Ora*), these are the wives of clansmen called either "father" (*Wora*) or father's brother's sons (*Ur-wora*). The men of the same class are permitted by local laws to have connection with these wives while their husbands are still alive, the main reason been that these men could inherit the wives after the death of their husbands.

Young people are face to marry the partner of their choice provided that there is no exogamy and no genealogical relationship can be traced on the mother's side. Every Jur Chol, young man wants his sister or his father's sister's daughter to be married properly, with enough money, sheep and goats, because the bride-wealth from these girls are given to him to marry with, and without the money, sheep and goats from his sister's or his father's sister's daughter he cannot marry a wife. The father who uses to marry more wives for himself while his son remains unmarried is often regarded as useless and irresponsible father. In the Sudan Notes and Records (1941) observes that the head of homestead (*gol*) is the sole legal owner of sheep and goats in his homestead (*gol*), he can dispose them in the ways he chooses. He holds them, in fact, however subject to a trust which equity will enforce. The terms of trust are that all domestic animals shall be applied solely towards the marriages of male members of the homestead (*gol*) in order of their seniority, that no domestic animals shall be sold except for the purpose of grain in time of famine, and that the homestead (*gol*) leader shall not himself marry a second time unless there is no unmarried male member of the homestead (*gol*) of marriageable age..........a breach of these terms of trust would be universally censured but would probably not be interfered with by the courts (P.23).

In Jo-Luo of Bhar El Ghazal society, boys build their houses for sleeping, while in a village, where there are many unmarried girls; parents usually give them a house to sleep in. The girls sleep together in this one house and such house is known locally as "*odwaman?*". Thus, when the boys want the girls, they come at night into the house where the girls sleep, and ask the girls if they could allow them into the house. It is not clear, whether customarily; boys are forbidden or allowed to share a house with the girls.

Like any other Northern Luo Groups, courtship (meeting) of boy-friends and girl-friends must take place outside the village and the parents are not supposed to know what is going at that appointed time; although parents are not allowed to meddle in the affairs of the boys and girls. In Jur Chol Society a girl chooses the boy she likes and often the girl breaks several engagements before finally deciding to marry one man, this is not considered disgrace except when the girl has done to excess (too much).

It is reported that the girl's choice of a husband is often influenced by the thought of how wealthy is the boy-friend or the potential husband. Does he have enough money, sheep and goats to provide for her marriage? The girl's thought is often supported by the father, brothers and father's brothers (Na or Nera), they always try to dissuade her from getting married to a poor man.

Nevertheless, we cannot deny that the general opinions of Jur Chol communities are that, they are against forcing a girl to take a repugnant partner for the sake of wealth. Interestingly, the provision of bride-wealth prevents very early marriages; as a result many men appear not to marry when they are still under twenty years of age.

A Jo-Luo man may have sexual intercourse with the wife of any man whom he calls "*Umia*", that is to say he can have sexual intercourse with the wife of his own brother. The elders of Jo-Luo in Melbourne informed me that a brother could sleep with his brother's wife only when that brother has medical problem e.g. he is impotent and does not produce. Such kind of sexual intercourse with brother's wife is usually done in secrete – no boy in the community is supposed to know. He can also have sexual intercourse with the wife of his father's brother's son (*Umia*) and his clansmen of his own generation. The boy may even ask the younger wives of his father to have sexual intercourse with any of them (with exception of his mother). The boy could even ask to have sexual intercourse with the wives of those he calls them "father's brother" (*Ur-wora*). Usually, the father knows that the son is consorting with one of his wives, and as this happen, the father leaves the son to sleep in the hut with the wife, while he goes to sleep in the hut of the other wife; he does nothing about it.

It is said that the Jur Chol men, like the Shulluk men, supported their sons to have intercourse with one of father's wives, because this is better than going to have sex with another woman, where the father would be obliged to pay the fines for adultery (Seligman, 1948).

Thus, in Jo Luo of Bhar El Ghazal cultures, it is acceptable that a grown up boy goes with the wife of his father; or goes with the wife of his father's son (*woda*) or father's brother's son (*Na or Nera*). It is reported that when a man has been out of his house for days or months or years, when he comes back home and learn that his "*Na or Nera*" have had sexual intercourse with his wife, as a human being he might feel very angry, he could even say some bad words to him, but that would be all. In such a case, like in Shulluk land, local chief usually comes and tells the husband of the wife, "This man is your "*Na or Neru*". It is alright for him to sleep with your wife while you are away. On hearing this, the man (husband) would accept the idea, however, if it was a stranger who has sexual intercourse with his wife, the man would try to spear that stranger at all cost. Thus, the young unmarried man, who was exercising his rights towards the wife of his father's brother's son "*Na* or *Nera*", would say to his relative that his father's brother's son "*Na or Nera*" knew before leaving the village that he was interested in the wife; the "*Na or Nera*' made no objection to it, therefore, such a union should not be a surprise to him. This implies that a child born from such a union always the child is considered to belong to the woman's husband and any idea of it belonging to the wife's temporary lover is ridiculous.

Jur Chol traditional custom forbids a man from making an attempt to come closer and closer to the wives of his mother's sister's sons (*Wod-Moo*) wife's brother's sons (*Wod-ora*), mother's brother's son (*Na or Nera*), sister's son (*Wod-nyimia*) as well as the wife of his father's father (Kwa or *Uwora*); if he does, this would be considered more worse than committing adultery, and it is believed that both the wife and the man who are guilty something bad would happen to them (e.g. death, sickness etc. etc.). Not only this, it is also believed that having sexual intercourse with the wife of in-law (Ora) is seen as something evil in the family and society. More interestingly, Alberto Rimo Akot Dimo, one of my informants in Melbourne –Australia, Stated that if your father dies and he has another younger wife, who has a child or two, you as the elder son is not allowed to have sexual intercourse with her, unless on the condition that the dead father has no other brothers who could take care of the woman and help her to produce more children for the family, then, elders from the lineage could advised you as an elder son to take care of the widow (take her for a wife). But having sexual intercourse with the younger wife of your dead father causes problems in the family.

In many cases, married brothers and married sisters happen to live near one another and their children would address themselves as "*Na or Nera*" (uncle's son), "*Nyina Nera*" (uncle's daughter), "*Uwaya*" (father's sister's son) and "*Nyiwaya*" (father's sister's daughter) the children would have an opportunity to meet and play together; and there is no restriction on the girls to play with the boys.

Pari Kinship and family Life

The Pari are divided into some one hundred patrilineal and exogamous clans. Unlike the Collo, the Pari's clans are found around Mount Lafon. The lineages that comprise clans are conceived as village or group. The family homesteads (.....) or given members are grouped together to form hamlets of related kin. This implies that a hamlet of this type may include as many as between eighty and hundred homesteads. Subsequently, these hamlets form larger settlements which are known today as Pari people or Lokoro people.

Like the Shulluk, the Pari people have a system of some relationship terminologies that are commonly known as "descriptive". Studies revealed that the most outstanding feature of the systems in Pari is the absence of a "single" word for a "niece" or a "nephew"; they always described these as "my uncle's daughter" or "my uncle's son". Like the other

Luo groups of South Sudan, among the Pari, there is no distinct terms for "father's brother" they call him also as "*Wuraa*".

At this juncture, it is good to point out that the wide 'classificatory" use of the term for "father" and "father's brother" is correlated with certain Pari customs. Thus, we should not confuse the "individual biological father" with the "father's brother" or "clansman of the father's generation", this term is applied to every person who is legitimate in the family tree.

Among the Pari people, like other Luo Groups, marriage is forbidden between all persons (males and females) to whom relationship can be traced directly, whether through the father or through the mother. This is correlated with custom governing inheritance and the transfer of bride-wealth, where the cross-cousins must stand together, depending on the mutual obligation that exists between the bother and the sister; on the other hand, the children of the sister and the children of the brother are supposed to stand together in case of any family issues, but yet all these depend on the mutual obligation that exist between sister's children and brother's children. Before we go further with our discussions on kinship and family life, let us see how kinship terms are used among the Shulluk.

The following are the kinship terms use among the Pari
-a person calls biological father as "father" (*Wuraa*)
-a person calls biological mother as "mother" (*Miraa*)
-a person calls father's brother as "brother of my father" (*Wuraa*) Umaa is hardly used
-a person calls father's brother's son as (*Uwuraa*)
-a wife calls husband's mother as "mother-in-law" (*Wanga*)
-a father calls his boy as "son" (*Waada*)
-a father calls his daughter as "daughter" (*Nyaraa*)
-a mother calls her "son" and her "daughter" (Waada "M" or Nyaraa "F")
-a father calls his siblings as "children or daughters" (Nyia Pl and *Nyitaa "Single"*) (*nyaraa or Waada*)
-a mother calls her siblings as "children" (*Nyara (single) nyitaa (Pl)*)
-a child calls his/her father's son as "the son of my father" (Umaraa/*Umaroo*). The children of two brothers call themselves "*Uwora*"
-a child calls mother's son as "son of my mother" (Umaraa/*Umaroo*)
-a person calls father's daughter's as "the daughter of my father "or "Sister" (*Nyiwawa*)
-a person calls wife's brother as "brother-in-law" (Oraa)
-a person calls mother's daughter as "daughter of my mother" or "sister" (Nymiraa, Male & Sowaa, Female)
-a person calls wife's sister as "sister" (*Oraa or Orr*)
-a person calls mother's sister as "mother" (*Maraa*)
-a person calls mother's brother as uncle (*Naaro or Naara*)

-a person calls mother's paternal grandfather's grandson as (*kwara/kwaro*)

-a person calls mother brother's son as uncle's son (Wonera male. & Nyinara female.)

-a person calls mother's brother's daughter as uncle's daughter (*Nyinaara*)

-a person calls father's sister as aunty (*Waya*)

-a wife calls husband's sister as aunty (*Waya*)

-a person calls father's sister's son as nephew (Wowaya)

-a person calls father's sister's daughter as aunty (*Nyiwaya*)

-a person calls father's brother's wife as mother (Ciworaa or *Cimada*)

-a person calls sister's son as sister's son (*Nicowa* (single) or *Nyiticowa* Plural)

-a person calls sister's daughter as sister's daughter (*Nyicowa* (single) or *Nyiticowa* (plural)

-a person calls father's father as grandfather (*Kwaraa or Kwaroo*)

-a person calls mother's father as grandmother (*Kwaraa or Kwaroo*)

-a person calls father's father brother as grandfather (*Kwaraa or Kwaroo*)

-a wife calls husband's father as "parent-in-law" (*Kwaraa*)

-a wife calls husband's mother as "mother-in-law" (*Wanga or Wango "general"*)

-a person calls father's mother as grandmother (*Wango or Wanga*)

-a person calls mother's mother a grandmother (*Wango* (general) or *Wanga*)

-a person calls son's children as grandchildren (Nyikwaraa)

-a person calls daughter's children as granddaughter (Nyikwaraa)

-a husband calls wife's father as "father-in-law" (*Orraa*)

-a wife calls husband's father's brother as grandfather (Kwaraa)

-a husband calls wife's mother as mother-in-law (Maara)

-a husbands calls wife's mother's sister as mother-in-law (Maara)

-a husband calls wife's brother as brother-in-law (*Orraa*)

-a wife calls daughter's husband as (Nyinyakaa)

-a wife calls sister's husband as sister- in-law (Wayaa)

-a husband calls wife's mother's mother as mother-in-law (Maraa)

-a wife calls husband's brother as (*Yuaraa*)

-a wife calls her co-wife as Co-wife (*Nyakaa*)

-a wife calls husband's brother's wife as co-wife (*Nyakaa*)

-a husband calls his brother's wife as the wife of my brother (Ci-orraa or Ci-cowa)

-a wife calls her brother's wife as the wife of my brother (Ci Umaraa or Ci-cowa)

-a wife calls husband's sister's daughter as (Nyi-waya)

-a person calls father's sister's daughter as daughter of my aunty (*Nyi Waya*)

-a person call father's sister's son as son of my aunty (*Wowaya*)

The term "*Wuraa*" is used to include the biological father and his brother; and the fathers of all those whom the father calls them "father's sons". Thus, as descent is patrilineal, therefore, all the men of the father's clan and generation are called "*Wuraa*" (fathers); and all the men of the "grandfather's generation" are called "*Kwara or Kwaroo*" (grandfathers).

At this point it is important for us to remember that, in Pari, not all those succeeding generation are called "sons", only a man's own sons are called "*Waada*" (son); the brother's sons are rather called "*Uwuraa*" (sons). The sister's sons are called **"Wo-cowa"**, while the sister's daughters are called "*Nyicowa or Nyiticowa*" they all belong to the clans of their fathers.

"*Miraa*" is an individual term used for "biological mother" and in Pari custom, all persons whom "mother" calls "*Uwuraa*" her children called them "mother's brothers" (*Naaraa*); and those persons whom the mother calls "fathers" (*wuraa*), her children call them "mother's fathers" (*Kwara*); the word "Kwara" in this sense indicating the generation where the child or children belong and does not indicate clan.

Interestingly, all the children of mother's brothers "*Uwuraa*", her children call them "mother's brother's sons" (*Nyinaraa*) and if they are girls the children call them "mother's brother's daughters" (*Naraa*). Therefore, all the men and women who belong to the mother's clan are considered as the children of "mother's bother". More importantly, in Pari, clans are patrilineal and all men under one general are called "*Uwuraa*" because their mothers' clansmen are mother's brother's sons are also called "*Uwuraa*". All men from a person mother's generation, the person call them "*Naaraa*", while his mother calls them "*Uwuraa*. In all these lines, only mother's father's sons are important as "*Naaraa*" or "Uncles".

Sex taboos and in-laws

When there is intermarriage between village "A" on hand and villages "X", "Y" and "Z" on the other hand, the men in village "X", "Y" and "Z" call men and women in village "A" their 'in laws' (*orraa*). The Sudan Notes and Records (1941) pointed out that the term "in-law" (*orraa*) is not applied by the husband to only his wife's kinsmen, but also, strictly speaking, all kinsmen of the wives of all his own kinsmen.....thus, term "in-law" (*orraa*) embrace a very large group of people. But the word "in-law" (*orraa*) in Pari customs' like in Shulluk, has two senses (or meaning): the wider sense and the narrower sense. In the wider sense the term embrace a large group of people but yet limited by the interpretation of the word "kinsmen', this is especially when we come to the sex taboo. However, in the narrower sense of the word "in-law" (*orraa*), it ties the husband to the homestead (*pac*) of his wives only. And throughout the earlier years of the marriage, the men from the village/*pac* ("A") of the husband must pay special respect to the women from villages/*pac* "Y" "X" "Z" of the wives. This is done mainly to avoid two things:

(3) To avoid "tuo" death that kill children after having sexual relationship with married woman from a particular *tung* (lineage or kin). Thus, when a man from *tung* (lineage/kin) "A" has sexual relationship with the woman from *tung* (lineage/*kin*) "X", "Y", or

"Z", who are considered as in-laws (*orraa*), both the woman and the man (if if there is a child) all will be affected by sickness and death. It is believed that if no sacrifice is done to serve them, both man and woman (and child) will die. According to Pari customs, this requires that a man must kill a sheep and put the stomach contents on top of the gourd by the road side, so that people who are passing through the road may carry away the death. Alternatively *atero* (errows) would be tied between the lady knees and ancles. These *atero* would give blessing to the girl and prepare her for marriage. The same death is said to be produced by a man having sexual relation with a woman to whom he is related.

(4) To avoid all possible cause of friction during what may be called "the period of tension". During the tension period the husband will not speak to his in-laws (*orraa*) and if by any chance he meets them on the road, he must avoid them by going off the road to let them pass. After the in-laws (*orraa*) pass, the husband then could return on road and continue his journey. This tension is usually the product of husband not paid all the required bride-wealth. He might have just paid ten head of cattle as a pledge to see the good behaviour of his wife. The full bride-wealth would then be paid as the wife produces children; the husband will be paying five head of cattle for every two baby girls born and two head of cattle for every five baby boys born. This will, bring the total cattle to thirty including the ten head of cattle paid for a pledge. Once the husband has paid this number of cattle, he is said to have completed the marriage and the tension is said to be resolved (P.29).

From the below diagram *tung* "A" (e.g. tung Tong or tung Angulumere) being reciprocally in law (*orraa*) of *tung* "X", "Y" and "Z", because men of *tung* "A" have married women from *tung* "X", "Y" and "Z", as such no men from *tung* "X", "Y" or *tung* "Z" would marry women from *tung* "A", the women from *tung* "A" are considered as mother-in-law (*Maraa* or *Maroo*) to *tung* "A".

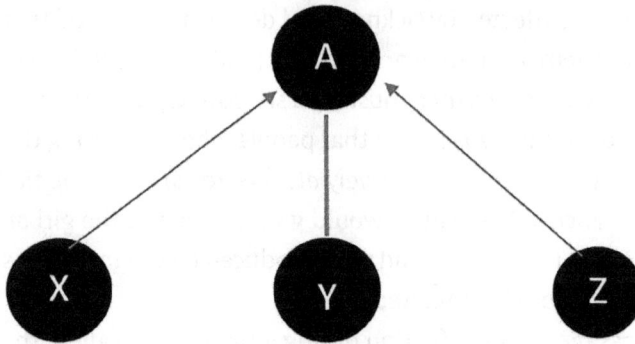

Figure: 98, Simple kinship relationship
Source: Sudan Notes and Records, 1941, P.30

However, from the *tung* "A" there would be average of 50-70 married men, who might have married from *tung* "X" "Y" and "Z". In this case, the number of in-laws (*oraa/Orr*) will be numerous. The good example of the lineage marrying many wives from *tung* and eventually creating more in laws (*orraa*) can be seen from the below diagram.

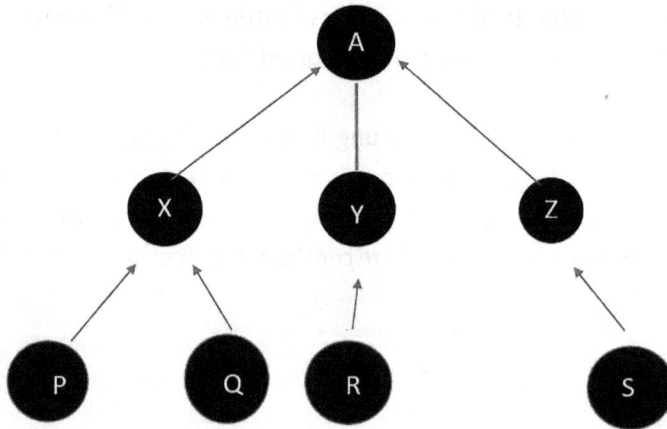

Figure: 99, Complex kinship relationship
Source: *Sudan Notes and Records, 1941, P.30*

From the first diagram above, we have seen that there is only one way traffic, that "X", "Y", and "Z" call "A" their in-laws (*Orraa/Orr*) because then men from *tung* "A" have married women from them.

Thus, in the second diagram we can see that *tung* "X" have married women from *tung* "P" and "Q"; and men in *tung* "Y" have married women from *tung* "R"; similarly, men from *tung* "Z" have married women from *tung* "S". This implies that other men from *tung* "A" are allowed to marry girls from *tung* "P" and village "Q"; similarly men from *tung* "X" are allowed to marry girls from *tung* "P" because the girls from *tung* "P" are said to be from the *tung* of the "in-law" (pa *orraa*). In other words, this means that the entire membership of "P", "Q", "R" and "S" are looking at *tung* "A" as the village of the wives of their "in-laws" (*orraa/Orr*). Customarily, the entire members of *tung* "P", "Q", "R" and "S" become the direct in-laws (*Orraa*) of the people from *tung* "X" "Y" and "Z" as well as the indirect in-laws of people in *tung* "A". Therefore, traditional laws forbid any man from *tung* "A" to marry or to have sexual relationship with women from *tung* "P", "Q", "R" an "S".

For a man, all his wife's clansmen he calls them "*Orraa*" (See kinship terms above); more importantly, the "term" *Orraa*" is a reciprocal term that could be used by either the husband or the wife to address wife's parents (in case of the husband) and to address husband's parents (in case of the wife). The husband can also address the wife's brothers as "*Orraa*" (in-laws). The terms Cousin that is *Nyiwa* or *Nyimia* are all reciprocal terms, which could be used by both husband and wife to address: wife's brother (in case of a husband) or to address sister's husband (in case of a wife).

However, in Pari customs, like other Luo groups, the wife's sister is not called "*Ora*"; it is only the wife's brother's wife who is called "*Orraa*". In contrast, in Pari, like in Acholi, the wife's brother's wife is called "mother-in-law" (*Maraa*). The use of the terms "*Orraa*" have significant connotation, when a woman speaks of the "*Orraa*" this specifically referred to her father's brother or to the father of her husband; and vice versa, when a man speaks of the "*Orraa*", this specifically referred to his mother-in-law's brothers or the father of the wife. The term "*Orraa*" implies respect and ceremonial behaviour when used by the opposite sex.

As we have seen from the terminologies use in Pari kinship, the term "*Miraa*" may also mean "the sister of my mother", if the term "*Miraa*" is used in this sense, it is also classed with the true blood mother, who is referred to as "*Miraa*". In Pari customs, like other Luo groups, daughters do not belong to the clan of their mother, they rather belong to the clan of their father, what we called "husband lineage", so a person calls the sons of his/her mother's sister as "*Umara*" and calls the daughters of his/her mother's sister as "*Nyimara*".

Therefore, from the above illustrations, it becomes clear to us that the Pari systems, whether we considered it as classificatory or not, yet it recognizes clan relationship among the people, it also recognizes family relationship and their differences can easily be seen according to the generation.

While the term "*Yuaraa*" is applied by the wife (woman) to the husband's brothers (Yuaraa). The same term "*Yuaraa*" is also applied by the husband (man) to the wife's sisters. It is important to understand that the term "*Yuaraa*" does not has the same meaning with the word "*Orra*", although generally it is assumed that the word "*Yuaraa*" might have been driven from the word "*Orra*". However, the term "*Orraa*" implies that the man and the woman must show due respect to one another, any breach of the etiquette between them is usually taken as a serious matter.

On the contrarily, the term *"Yuaraa"* does not require respect from both sides of the man and the woman; subsequently, many a times a woman target her husband's brothers and her father-in-law's brother's sons (Uwuraa). Seligman (1948) notes, in Luo groups, that in case of death, the "Uwuraa" may inherit each other's wives, so that *"Yuaraa"* become legitimate spouses (P.54).

In Pari, like any other African groups, the biological father is the head of the household; and more importantly, a man belong to his father's clan and all of his father's clans are also his own clansmen. Thus, in a society, once a man is married, he sets up new relationships to his wife's relatives. Which implies that all the relatives of the wife become husband's brother-in-law (Orraa). These are the people the husband must show great respect, he must talk to them respectfully; in addition the husband is to avoid any misbehaviour that could lead to quarrels or fights before them; he will never become familiar with any of them. His intimate friends will be chosen from among his clansmen.

Traditionally, a woman who is not yet married can use kinship terms in the same way other people use, but a married woman must use kinship terms according to the manner and attitude expected of her. For example, a married woman may call her husband's sisters as *"Waya"* and she may call the husband's father and mother as *"Kwara"* and *"Wanga"* respectively. So, it is not accepted that the wife addresses the father and mother of her husband as relatives by marriage, but should rather address them as her own relatives. The reason why the wife should now address her father-in-law and mother-in-law as her own relatives is that because by virtue of marriage the wife considers herself as a member of husband's family.

However, it is important to notice that although the wife becomes a member of the family, when she is addressing the relatives of the husband she should think of herself as though she belongs to a younger generation than she does in reality.

According to Pari customs, when a father dies, his son or sons takeover his younger wife or wives; and one of the father's brothers can inherit their mother, customarily, no son can inherit his own mother as a wife. But if the father has no sons, after dead his brothers can take over his wives. It is said that in Pari society, it is not disgraceful for a young unmarried son to sleep with his own father's wives.

In Pari society, people avoid persons they call "in-laws" (Orraa and Maraa), especially the wife's mother. According to customs the man must avoid entering into the house of the mother-in-law, and above all he should take care of her properly; and when giving something to the mother-in-law, the husband is not allowed to look directly into the eyes of the in-law, especially in public places.

However, the husband cannot avoid entering into his brother's wife's house; he can enter into the house at any time to eat and drink on the condition that when he wants to enter the house, woman must leave the house. And when the man enters the house, he is not allowed to sit on the brother's wife's sleeping mate; he is to sit upon bare ground. However, if by mistake the man sits on his brother's wife's sleeping mate, this will be considered as breach of etiquette; and such an act might be taken as he intended to have sexual intercourse with the woman, although the woman is not in the house at that time (Seligman, 1948)

The wife's father's brother's wife (*Miraa*) is also treated as "*Miroo*" and must be avoided and given due respect. However, if by any chance the father's brother's wife (*Miraa*) comes suddenly into the group, so, the person who calls her "*Maaraa*" must quite the group and site on the other side of the building where the group are, so that he does not see the woman and the woman should not see him. Interestingly, if this woman (in-law) is the man's wife's mother, the man must not remain around the place, he must disappear completely from the surrounding, for otherwise, the mother-in-law might claim that the man has seen her nakedness.

The man does not call the wives of his father's brother's sons as his "in-laws" (*Orraa*) but rather he calls them as "*Maraa*" or "*Maroo*". It is said in the Pari customs, like in the Shulluk customs the wives of the clansmen are divided into two categories. The first category of wives are the wives with whom sexual intercourse is permissible; and the second of the clansmen who are called either "father" (*Wuraa*) or "father's brother's sons" (*Uwuraa*) fall within this first category. While the second category, are the wives with whom no man is allowed to have sexual intercourse, except their husbands. Thus, having sexual intercourse with the wives of any other clansmen is said to cause death to both the offenders. Most of these wives whose sexual intercourse leads to death are been called by using descriptive terms.

It is important to understand that the Pari people treat the wives in the second category, which we have seen above, with care, since sexual intercourse is not permissible with them. The men have to avoid anything that could cause problem, such as staying in the houses of these women when their husbands are not at home. And if the husbands of these wives are at home, whenever a man goes to visit one of the wives, he should avoid sitting upon the bed or mate of the woman. Failure to do this would be considered as attempt to sexual intercourse.

Staying away from the in-laws is very important especially for the wives or women. Traditionally the wife is forbidden to call her husband's parents as "*Orraa*"; she is to address the husband's father as her own grandfather (*Kwara*); and to address the husband's mother as mother of my husband (*Wanga* or *Wango*); and above all she must treat them with due respect. Thus, when she is handing something, like water, to her father-in-law, she must kneel in front of grandfather (*Kwara*) as well as in front of the husband's mother (*Wanga*) to show respect; and when she is speaking with the father-in-law, she should not see him direct into the face, but must turn her face away as she speaks; however, there is no strict avoidance required.

Sometimes mother stays with the her son if her husband dies; in this case she is given small separate hut to sleep in but that hut must be closer to the son's hut in case of any emergency. At breakfast, lunch and dinner time the husband's mother is to eat separately but the wife and the husband could eat together if they so wish. The same apply if the mother of the wife is staying at the daughter's husband's house.

A man calls his son's wife as the wife of my son (*Chi-waada*); she falls into the category of wives of a clansman with whom sexual intercourse is inadmissible. However, there are groups of women who are treated as the "in-laws" (Ma*raa or Maroo*), these are the wives of clansmen who are called either "father" (*Wuraa*) or father's brother's sons (*Uwuraa*). The men of the same class are permitted by local laws to have connection with these wives while their husbands are still alive, the main reason been that these men could inherit the wives after the death of their husbands.

Young people are face to marry the partner of their choice provided that there is no exogamy and no genealogical relationship can be traced on the mother's side. Every Pari young man wants his sister or his father's sister's daughter to be married properly, with enough cattle, because the bride-wealth from these girls are given to him to marry with, and without the cattle from his sister's or his father's sister's daughter he cannot marry a wife. The father who uses to marry more wives for himself while his son remains unmarried is often regarded as useless and irresponsible father. In Pari tradition the man is the head of homestead and he is the sole legal owner of cattle in his homestead, he can dispose them in the ways he chooses. He holds them in fact however subject to a trust which equity will enforce. The terms of trust are that all cattle shall be applied solely towards the marriages of male members of the homestead in order of their seniority, that no cattle shall be sold except for the purpose of grain in time of famine, and that the homestead leader shall not himself marry a second time unless there is no unmarried male member of the homestead of marriageable age 20 years old. A breach of these terms of trust would be universally censured but would probably not be interfered with by the courts.

In Pari society, like in any cattle keeping society, there are no houses built particularly for boys and girls. The boys normally sleep in the cattle byres (Luak). In a village, where there are many unmarried girls, parents usually give them a house to sleep in. The girls sleep together in this one house and such house is known locally as "*odwaman*". Thus, when the boys want the girls, they come at night into the house where the girls sleep, and ask the girls, if they could allow them into the house. It is not clear, whether customarily; boys are forbidden or allowed to share a house with the girls.

In contrast, Howfmayr states that meeting of boy-friends and girl-friends must take outside the village and the parents are not supposed to know what is going at that appointed time; although parents are not allowed to meddle in the affairs of the boys and girls. He went further by saying that the girl chooses the boy she like and often the girl break several engagements before finally deciding to marry one man, this is not considered disgrace except when the girl has done to excess (too much).

A girl who has more than ten lovers is looked upon as a feast woman (*Nya dei cuo*). It is reported that the girl's choice of a husband is often influenced by the thought of how wealthy is the boy-friend or the potential husband. Does he have enough cattle to provide for her marriage? The girl's thought is often supported by the father, brothers and father's brothers (*Umaa*), they always try to dissuade her from getting married to a poor man.

Nevertheless, we cannot deny that the general opinions of the Pari communities are that, they are against forcing a girl to take a repugnant partner for the sake of cattle. Interestingly, the provision of bride-wealth prevents very early marriages; as a result many men appear not to marry when they are still under twenty years of age.

A Pari man may have sexual intercourse with the wife of any man whom he calls "*Uwuraa*", that is to say he can have sexual intercourse with the wife of his own brother, who is from another mother although they have the same father. He can also have sexual intercourse with the wife of his father's brother's son "*Uwuraa*" and his clansmen of his own generation. The boy may even ask the younger wives of his father to have sexual intercourse with any of them (with exception of his mother). The boy could even ask to have sexual intercourse with the wives of those he calls them "father" (wuraa or *Umaa*). Usually, the father knows that the son is consorting with one of his wives, and as this happen, the father leaves the son to sleep in the hut with the wife, while he goes to sleep in the hut of the other wife; he does nothing about it. It is said that the Pari men supported their sons to have intercourse with one of father's wives, because this is better than going to have sex with another woman, where the father would be obliged to pay the fines for adultery (Seligman, 1948).

Thus, in Pari cultures, it is acceptable that a grown up boy goes with the wife of his father; or goes with the wife of his father's son or father's brother's son (Uwuraa). It is reported that when a man has been out of his house for days or months or years, when he comes back home and learn that his "Uwuraa" have had sexual intercourse with his wife, as a human being he might feel very angry, he could even say some bad words to him, but that would be all. In such a case, local chief usually comes and tells the husband of the wife, "This man is your bother "Uwuraa". It is alright for him to sleep with your wife while you are away instead of other boys to sleep with your wife.

On hearing this, the man (husband) would accept the idea, however, if it was a stranger who has sexual intercourse with his wife, the man would try to spear that stranger at all cost. Thus, the young unmarried man, who was exercising his rights towards the wife of his father's brother's son "Uwuraa", would say to his relative that his father's brother's son "Uwuraa" knew before leaving the village that he was interested in the wife; the "Uwuraa' made no objection to it, therefore, such a union should not be a surprise to him. This implies that a child born from such a union always the child is considered to belong to the woman's husband and any idea of it belonging to the wife's temporary lover is ridiculous.

Pari traditional custom forbids a man from making an attempt to come closer and closer to the wives of his mother's sister's sons (umaraa) wife's brother's sons (Uwuraa), mother's brother's son (*Nyiwora*), sister's son (wo-cowa) as well as the wife of his father's father (Kwara or *kwaroo*); if he does, this would be considered worse than committing adultery, and it is believed that both the wife and the man who are guilty would die. Not only this, it is also believed that having sexual intercourse with the wife of in-law (Orraa) is also looked upon as abomination

In many cases, married brothers and married sisters happen to live near one another and their children would address themselves as "*Nyiwora*" (uncle's son), "*Nyinara*" (uncle's daughter), "Nyiwa" (father's sister's son) and "*Nyiwaya*" (father's sister's daughter) the children would have an opportunity to meet and play together; and there is no restriction on the girls to play with the boys.

Experiences in Pari society and Northern Luo of South Sudan in general, revealed that relationship between wife's children and their mother's brother "*Naaraa*" (uncle) is stronger than the relationship between wife's children and father's brother's children. It is the mother's brother who usually gives a son of his sister his first spear; the boy also has a right to some of the cattle given as bride-wealth for his sister's daughter, and sometime he could even influence the girl in her choice for a husband.

A young woman does not drink milk under any circumstances, though a woman who is nursing a baby may, as do old women past the reproductive age. Only small boys herd the cattle and milk them, for once a boy has reached maturity there is the danger that he may have had sexual contact, so if he does and milk cows or just walk among the cattle in their pens, it is believed that this might cause cows to become sterile. However, this believe is so strong among the people, in such a way that if a child drinks milk at his neighbour house, when he comes back, he must wash his body before entering into his father's house. Because it is believed that some milk might have remained over his body that could poison the food, and once this food is eaten by unclean person, the person might die. According to Pari and Shulluk customs, a man who has sexual intercourse with his wife a night before, he is considered unclear and must not drink milk until the sun sets the following day (Seligman, 1948, P. 73).

In Pari communities, when a woman finishes cooking food for her husband, she is to leave the hut while the husband is eating, but it is not shameful for men and women to eat together; and after a man has eaten the remaining part of a meal is taken and eaten by the woman and her friends (usually her unmarried "Nyiwa"). However, in Pari society a newly married couple do not eat together for some months or even a year.

Shulluk Kinship and family Life

The Shulluk are divided into some one hundred patrilineal and exogamous clans. However, clans are not found in one location, they are scattered all over Shulluk land. Although clans are scattered widely, the lineages that comprises clans are still conceived of as localized groups. As some writers like, Wall (1976) notes that the family homesteads (*gol*) or those of individual lineage members are grouped together to form hamlets of related kin. This implies that a hamlet of this type may include as many as between fifty and sixty homesteads. Subsequently, these scattered hamlets may form a larger settlement with a clearly defined territory. In each hamlet there is an original or "own" lineage, called a *"diel"*.

Unlike the Acholi, the Shulluk have a system of relationship terminology that is commonly known as "descriptive". Seligman (1948) observes that the most outstanding feature of the systems in Shulluk is the absence of a "single" word for a "brother" or a "sister", they always described these as "my father's son" or "my mother's daughter" (P.50). Banydhuro Oyay, one of the Collo intellectuals and resident of Melbourne Australia, argued that in Shulluk there is single word for brother and sister. Thus, in Collo language term *"Omiya"* means a brother and term *"Nyimiya"* means a sister.

Another peculiar feature of the Shulluk system also includes the way in which it holds the balance between the individual family and the clan. Thus, the only "brothers" or "sisters" who are correctly called "father's sons or daughters" (Owa and Nyiwa) are the father's own children, these children could be from same father as well as all paternal cousins in the village.

Like the other Luo groups of South Sudan, among the Shulluk, there is no distinct terms for "father's brother" and "father's brother's son" both are called *"Wa or Wia*. In Acholi, the "father's brother" is called "near" (uncle) and the "father's brother's son" is also called "near" (uncle).

At this juncture, it is good to point out that the wide 'classificatory" use of the term for "father" and "father's son" is correlated with certain Shulluk customs. However, we should not confuse the "individual biological father" with the "father's brother" or "clansman of the father's generation", this term is applied to every person who is legitimate in the family tree. According to Seligman (1948) the various kinds of cousins, found in Shulluk societies are distinguished linguistically.

Marriage is forbidden between all persons (males and females) to whom relationship can be traced directly, whether through the father or through the mother. This is correlated with custom governing inheritance and the transfer of bride-wealth, where the cross-cousins must stand together, depending on the mutual obligation that exists between the bother and the sister; on the other hand, the children of the sister and the children of the brother are supposed to stand together in case of any family issues, but yet all these depend on the mutual obligation that exist between sister's children and brother's children. According to Banydhuro Oyay, marriage in Shulluk is entirely a business of the groom and his father. Any contribution from family members is voluntary. The father is always entitled to bride's wealth. He then decides to help the son with his marriage dowry.

Before we go further with our discussions on kinship and family life, let us see how kinship terms are used among the Shulluk.

The following are the kinship terms use among the Collo
-a person calls biological father as "father" (*Wa, Wi or Wia*)
-a person calls biological mother as "mother" (*Maiya or Mai*)
-a person calls father's brother as "brother of my father" (*Wa or Wia*)
-a person calls father's brother's son as (*Uwa*)
-a wife calls husband's mother as "Grandma" (*Wanga*)
-a father calls his boy as "son" (*wot or pl.. Woda*)
-a mother calls her boy as "son" (*wot or pl. Wode; wad or Wada Pl*)
-a father calls his siblings as "children or daughters" (Nyar "single" and *Nywolla* "Pl")
-a mother calls her siblings as "children" (*Nyara (single) nywola (Pl)*)
-a child calls his/her father's son as "the son of my father" (*Uwa*)
-a child calls mother's son as "son of my mother" (*Umia*)
-a person calls father's daughter's as "the daughter of my father "or "Sister" (*Nyiwa*)
-a person calls wife's brother as "brother-in-law" (*Umia*)
-a person calls mother's daughter as "daughter of my mother" or "sister??" (*Nyimia*)
-a person calls wife's sister as "sister" (*Nyimia*)
-a person calls mother's sister as "mother" (*Nyimia mia*)
-a person calls mother's brother as (*Na*)
-a person calls mother's paternal grandfather's grandson as (*Na*)
-a person calls mother brother's son as (*Unai or Una*)
-a person calls mother's brother's daughter as (*Nyinai or Nyina*)
-a person calls father's sister as (*Waja*)
-a wife calls husband's sister as (*Waja*)
-a person calls father's sister's son as (*Uwaja*)
-a person calls father's sister's daughter as (*Nywaja*)
-a person calls father's brother's wife as (*Mia*)
-a person calls sister's son as (*Okai or Oka*)
-a person calls sister's daughter as (*Nyikai or Nyaka*)
-a person calls father's father as (*Kwa or Kwara*)
-a person calls mother's father as (*Kwa or Kwara*)
-a person calls father's father brother as (*Kwa or Kwara*)
-a wife calls husband's father as "parent-in-law" (*Kwai or Kwa*)
-a person calls father's mother as (*Wang or Wanga*)
-a person calls mother's mother as (*Wang or Wanga*)
-a wife calls husband's mother as (*Wang or Wanga*)

-a person calls son's children as (*Kware*)

-a person calls daughter's children as (*Kware*)

-a husband calls wife's father as "parent-in-law" (*Ora*)

-a wife calls husband's father's brother as (*Ora*)

-a husband calls wife's mother as (*Ora*)

-a husbands calls wife's mother's sister as (*Ora*)

-a husband calls wife's brother as (*Ora*)

-a wife calls daughter's husband as (*Ora*)

-a wife calls sister's husband as (*Ora*)

-a husband calls wife's mother's mother as (*Ora*)

-a wife calls husband's brother as (*Yura or Yure*)

-a wife do you call "uwa" of the husband in English? (*Cimada*)

-a wife her co-wife as (*Nyiaga*)

-a wife calls husband's brother's wife as (*Nyiaga*)

-a husband calls his brother's wife as (*Bong nyimia*)

-a wife calls her brother's wife as (*Nyiaga*)??

- what is words "ba wa" means in English? My father

-what is word "ci wa" means in English? My father's wife

-a wife calls husband's sister's daughter as (*Nyiaga*)

-a person calls father's sister's daughter as (*Nywaja*)

The term "*Wa*" is used to include the biological father and his brother; and the fathers of all those whom the father calls them "father's sons". Thus, as descent is patrilineal, therefore, all the men of the father's clan and generation are called "*Wa*" (fathers); and all the men of the "grandfather's generation" are called "*Kwa*" (grandfathers), but not all men of the same generation of the grandfather are called "*Kwa*", they are rather called "*Uwa*" (father's sons).

At this point it is important for us to remember that, in Shulluk, not all those succeeding generation are called "sons", only a man's own sons are called "*Wode*" (sons); the brother's sons are rather called "*Umadh*" (sons of a friend –"*math*"). The sister's sons are called "*Oka*", while the sister's daughters are called "*Nyaka*"; they all belong to the clans of their fathers.

"*Maiya*" is an individual term used for "biological mother" and in Shulluk custom, all persons whom "mother" calls "*Uwa*" her children called them "mother's brothers" (*Na*); and those persons whom the mother calls "fathers" (*wa*), her children call them "mother's fathers" (*Kwa*); the word "*Kwa*" in this sense indicating the generation where the child or children belong and does not indicate clan.

Interestingly, all the children of mother's brothers "*Uwa*", her children call them "mother's brother's sons" (*Una*) and if they are girls the children call them "mother's brother's daughters" (*Nyina*). Therefore, all the men and women who belong to the mother's clan are considered as the children of "mother's bother". Seligman (1948) notes, In Shulluk, clans are patrilineal and all men under one general are called *"Uwa"* because their mothers' clansmen are mother's brother's sons *"Una"*. All men from a person mother's generation, the person call them *"Na"*, while his mother calls them *"Uwa*. In all these lines, only mother's father's sons are important as *"Na"* or "Uncles" (P.53).

Sex taboos and in-laws

When there is intermarriage between village "A" on hand and villages "X", "Y" and "Z" on the other hand, the men in village "X", "Y" and "Z" call men and women in village "A" their 'in laws' (*ora*). The Sudan Notes and Records (1941) pointed out that the term "in-law" (*ora*) is not applied by the husband to only his wife's kinsmen, but also, strictly speaking, all kinsmen of the wives of all his own kinsmen.....thus, term "in-law" (*ora*) embrace a very large group of people. But the word "in-law" (*ora*) in Shulluk customs has two senses (or meaning): the wider sense and the narrower sense. In the wider sense the term embrace a large group of people but yet limited by the interpretation of the word "kinsmen', this is especially when we come to the sex taboo such as *"dwalo"*. However, in the narrower sense of the word "in-law" (*ora*), it ties the husband to the homestead (*gol*) of his wives only. And throughout the earlier years of the marriage, the men from the village/*pac* ("A") of the husband must pay special respect to the women from villages/*pac* "Y" "X" "Z" of the wives. This is done mainly to avoid two things:

(5) To avoid *"dwalo* sickness", this is a form of sickness that resembles dropsy. Thus, when a man from village/*pac* "A" has sexual relationship with the woman from village/*pac* "X", "Y", or "Z", who are considered as in-laws (*ora*), both the woman and the man (if if there is a child) all will be affected by *"dwalo* sickness". It is believed that if no sacrifice is done to serve them, both man and woman (and child) will die. According to Shulluk customs, this requires that a man must kill a sheep and put the stomach contents on top of the gourd by the road side, so that people who are passing through the road may carry away the *"dwalo* sickness". The same sickness (*dwalo*) is said to be produced by a man having sexual relation with a woman to whom he is related.

(6) To avoid all possible cause of friction during what may be called "the period of tension". During the tension period the husband will not speak to his in-laws (*ora*) and if by any chance he meets them on the road, he must avoid them by going off the road to let them pass. After the in-laws (*ora*) pass, the husband then could return on road and continue his journey. This tension is usually the product of husband not paid all the required bride-wealth. He might have just paid ten head of cattle as a pledge to see the good behaviour of his wife. The full bride-wealth would then be paid as the wife produces children; the husband will be paying five head of cattle for every two baby girls born and two head of cattle for every five baby boys born. This will, bring the total cattle to thirty including the ten head of cattle paid for a pledge. Once the

husband has paid this number of cattle, he is said to have completed the marriage and the tension is said to be resolved (P.29).

Village "A" being reciprocally in law (*ora*) o villages "X", "Y" and "Z". Thus, if men of village "A" have married women from villages "X", "Y" and "Z", no men from village "X", "Y" or village "Z" may marry women from village "A" because they call women from village "A" their in-laws (*Ora*).

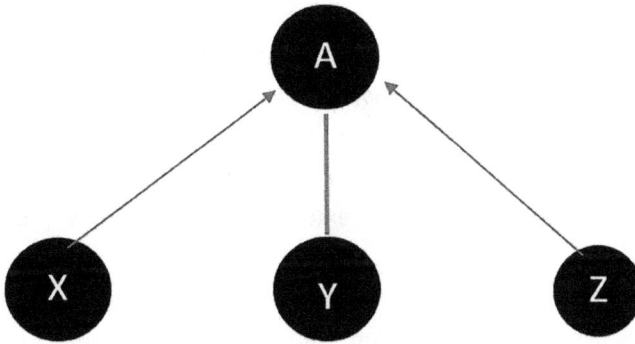

Figure: 100, Simple kinship relationship
Source: Sudan Notes and Records, 1941, P.30

But from a particular group or lineage there could be average of 50-70 married men. In this case, the number of in law (*ora*) will be numerous. The example of the lineage marrying many wives that eventually create more in laws (*ora*) can be seen from the below diagram.

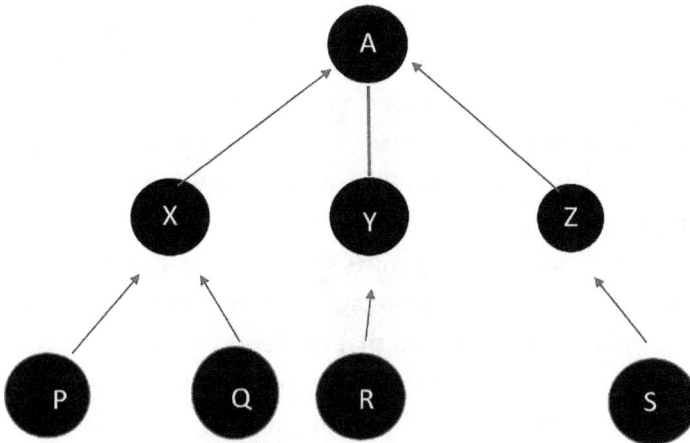

Figure: 101, Complex kinship relationship
Source: Sudan Notes and Records, 1941, P.30

We have seen in the first diagram that there is only one way traffic, that "X", "Y", and Z call "A" their in-laws (*Ora*) because men from "A" having married women from them. Thus, in the second diagram we can see that village/*pac* "X" have married women from village/*pac* "P" and "Q"; men in village/*pac* "Y" have married women from village/*pac* "R"; and men from village/*pac* "Z" have married women from village/*pac* "S". This also implies that other men from village/pac "A" are allowed to marry other girls from village/*pac* "X"; as similar with other men from village/*pac* "X" are allowed to marry other girls from village/*pac* "P" because the girls from village/*pac* "P" are said to be from the village/*pac* of the "in-law"(pa *ora*). In other words, this means that the entire membership of "P", "Q", "R" and "S" are looking at Village/*pac* "A" as the village of the wives of their "in-laws" (*ora*), which called locally as "*myer ora*". In other words the entire members of Village "P", "Q", "R" and "S" become the in-laws (*Ora*) of the people from village "A". Therefore, it is forbidden for any man from village "Village/*pac* "A" to marry or to have sexual relationship with women from the four villages "P", "Q", "R" an "S" above.

However, if for one reason or the other, another man from village/*pac* "A" seduces another girl from village/*pac* "P", people or men from village/*pac* "X" will be annoyed because they regard the unmarried girls in village/*pac* "P" as in some sense their prerogative (men in village/*pac* "A" should not jump over them). As such this case would be taken seriously, as something that would spoil the good relationships between the in-laws "*ora*" (Sudan Notes and Records, 1941).

For a man, all his wife's clansmen he calls them "*Ora*" (See kinship terms above) More importantly, the "term" *Ora*" is a reciprocal term that could be used by either the husband or the wife to address wife's parents (in case of the husband) and to address husband's parents (in case of the wife). The husband can also address the wife's brothers as "*Ora*" (in-laws). The terms used to address female cousins that are *Nyiwa, Nyimia, Nyima* or *Nywaja* and the term Owa, Omia, ona and owaja are terms used to address male cousins.

In Shulluk when we talk of "*Ora*" to mean "mother-in-law", this would include wide range of relationships, in this sense the term "*Ora*" would mean the wives of all men, the married wife calls their fathers her fathers, as well as any woman who stands in that relationship (of mother-in-law).

However, in Shulluk customs, the wife's sister is not called "*Ora*", rather the wife's brother's wife who is called "*Ora*". In contrast, in Acholi, the wife's brother's wife is called "mother-in-law" (*Mara*). It must be seen that the term "*Ora*" cannot possibly belong to one clan but the terms mothers, fathers, and mother-in-law, mother, father and brother belongs to clans, they cannot go out of the clans. In other words, they are "copy right" of the clans.

The use of the terms have significant connotation, when a woman speaks of the "*Omia*" this specifically referred to her husband's brother and term "*mia or wanga*"(my mother or my grandma) is specifically referred to the husband's mother's sister. The term *"Ora"* implies respect and ceremonial behaviour when used by the opposite sex.

As we have seen from the terminologies use in Shulluk kinship, the term *"Ma"* may also mean "the sister of my mother", if the term *"Ma"* is use in this sense, it is not classed with the true blood mother, who is referred to as *"Maiya"*. In Shulluk customs, daughters do not belong to the clan of their mother, they rather belong to the clan of their father, what we called "husband lineage", so a person calls the sons of his/her mother's sister as *"Uma"* and calls the daughters of his/her mother's sister as *"Nyima"*, subsequently, differentiating them from brothers and sisters, with whom they are usually identified according to the classificatory system.

Anthropologist, Lewis Henry Morgan, defined classificatory kinship system as putting people on the basis of abstract relationship rules, sons of father, daughters of father, daughters of mother and sons of mother are put genealogically although the classes have no overall relation to genetic closeness. For example a stranger may marry from Shulluk family, when they have children with the wife; all these children will belong to the class or clan of his wife.

In Shulluk the term *"Mia"* is used for the father's brother's wife. Thus, a person calls his/her father's brother's wife *"Mia"*. I am not sure if the term *"Mia"* is used throughout the Shulluk society, but I understand from Seligman (1948) that the term *"Mia"* is commonly used in Penyikang district. However, Westernmann (1912) give the word *"Mayo-Mai"* and word *"Ma-Mek"* to both mean aunt, what remains a mystery is that we do not understand where he got these words from and where these words were used, whether in Sobat River or at Penyikango districts.

Therefore, from the above illustrations, it becomes clear to us that the Shulluk systems, whether we considered it as classificatory or not, yet it recognizes clan relationship among the people, it also recognizes family relationship and their differences can easily be seen according to the generation.

While the term *"Yura or Ywera"* is applied by the wife (woman) to the husband's brothers (*Uwa*). The same term *"Yura or Ywera"* is also applied by the husband (man) to the wife's sisters. It is important to understand that the term *"Yura"* does not has the same meaning with the word *"Ora"*, although generally it is assumed that the word *"Yura"* might have been derived from the word *"Ora"*. However, the term *"Ora"* implies that the man and the woman must show due respect to one another, any breach of the etiquette between them is usually taken as a serious matter.

On the contrarily, the term *"Yura or Ywera"* does not require respect from both sides of the man and the woman; subsequently, many a times one could see a woman target her husband's brothers and her father-in-law's brother's sons (Uwa). Seligman (1948) notes that in case of death the "Uwa" may inherit each other's wives, so that *"Yura"* become legitimate spouses (P.54).

In Shulluk, like any other African groups, the biological father is the head of the household; and more importantly, a man belong to his father's clan and all of his father's clans are also his own clansmen. Thus, in a society, once a man is married, he sets up new relationships to his wife's relatives. Which implies that all the relatives of the wife become husband's brother-in-law (*Ora*)? These are the people the husband must show great respect, he must talk to them respectfully; in addition the husband is to avoid any misbehaviour that could lead to quarrels or fights before them; he will never become familiar with any of them. His intimate friends will be chosen from among his clansmen. The relationship to the *yura* is dependent on that between clan-brothers (*Uwa*), who stand a better chance to have access to each other's wives.

Traditionally, a woman who is not yet married can use kinship terms in the same way other people use, but a married woman must use kinship terms according to the manner and attitude expected of her. For example, a married woman may call her husband's sisters as "*Waja*" and she may call the husband's father and mother as "*Kwa*" and "*Wanga*" respectively. So, it is not accepted that the wife addresses the father and mother of her husband as relatives by marriage, but should rather address them as her own relatives. The reason why the wife should now address her father-in-law and mother-in-law as her own relatives is because by virtue of marriage the wife considers herself as a member of husband's family.

However, it is important to notice that although the wife becomes a member of the family, when she is addressing the relatives of the husband she should think of herself as though she belongs to a younger generation than she does in reality. Banydhuro Oyay suggested that asserting herself as from younger generation facilitate respect which is always based on age.

According to Collo customs, when a father dies, his son or sons takeover his younger wife or wives; and one of the father's brothers can inherit their mother, customarily, no son can inherit his own mother as a wife. In addition, in Shulluk society, it is not considered disgraceful for a young unmarried son to sleep with his own father's wives (Seligman, 1948).

Traditionally, although a woman is married, she still belongs to her own clan, because of this; it would look strange for the wife to use the same terms her husband employed. In this connection, it should be noted that in some districts a husband can address his wife as "*Nyimia*" and a wife can address her husband as "*Umia*". However, this could be due to the simple fact that the wife is now taken into the family of the the husband as a member. Interestingly, the husband calls his wife's sister as "sister" (*Nyiwia??*); and when wife's sister comes to visit them, the husband is not bound by any taboos; he may have sexual intercourse with her, since custom dictates that in case of death a wife's husband could marry one the wife's sisters. In contrast, the Sudan Notes and Records (1941) argue that the husband can have intercourse with the wife's sister because he has paid all the bride-wealth of his wife (P. 29).

Among the Collo, people avoid all persons that they call them "in-laws" (*Ora*), especially the wife's mother. According to customs the man must avoid entering into the house of the mother-in-law, and above all he should take care of her properly; and when caring for the mother-in-law, the husband is not allowed to look directly into the eyes of the in-law, especially in public places. Thus, if a man sees his mother-in-law at a distance, she should take different direction to avoid them meeting faced to face. When he finds a good place, he could then send his friend to greet his mother-in-law on his own behalf.

Yet some informants told me that the husband is not only to avoid meeting his mother-in-law face to face, but he is also not allowed to enter unnecessarily into the village where the mother-in-law lives, until all the cattle for the dowry are paid all.

However, the husband cannot avoid entering into his brother's wife's house; he can enter into the house at any time to eat and drink on the condition that when he wants to enter the house, woman must leave the house. And when the man enters the house, he is not allowed to sit on the brother's wife's sleeping mate; he is to sit upon bare ground. However, if by mistake the man sits on his brother's wife's sleeping mate, this will be considered as breach of etiquette; and such an act might be taken as he intended to have sexual intercourse with the woman, although the woman is not in the house at that time (Seligman, 1948)

The wife's father's brother's wife (*Mia*) is also treated as an "*Ora*" and must be avoided and given due respect. However, if by any chance the father's brother's wife (*Mia*) comes suddenly into the group, so, the person who calls her "*Ora*" or "*Mia*" must quite the group and site on the other side of the building where the group are, so that he does not see the woman and the woman should not see him. Interestingly, if this woman (in-law) is the man's wife's mother, the man must not remain around the place, he must disappear completely from the surrounding, for otherwise, the mother-in-law might claim that the man has seen her nakedness.

In Shulluk society there are four aspects of the in-law (*Ora*) relationship as follows:-

(1) *The avoidance of the mother-in-law is so common a custom among the Northern Luo of South Sudan.* The aspects of respect to mother-in-law is so complicated that we cannot discuss all of them here but I want to let us know that there are three elements in the custom that we should always remember in relation to the culture. Firstly, it is a MUST that the mother-in-law be given due respect without any condition; Secondly, the in-law (*Ora*) MUST not show any sexual feeling towards her. Thirdly, both the mother-in-law and the son-in-law (Ora) MUST avoid seeing each other nakedness. Among the Shulluk men, who wear garments only knotted over the shoulders, leaving the genitals exposed, the mother-in-law must take a great care to avoid the presence of the in-law, since the eyes travel faster than light. The custom forbids the son-in-law (*Ora*) to have sexual intercourse with his mother-in-law. Thus, if the young man wants sexual satisfactions, he could approach the sister of his wife. However, if the young man in

Shulluk land has not yet collected enough cattle to marry, and he wants to satisfy his sexual feeling, he can exercise his rights and consort with the wife of one of his "Uwa"; he is also expected to treat the mother of the woman with respect as though he is the actual husband, if he does not practice this attitude trouble may arise at the time he wants to inherit his brother's wife after death. But according to Banydhuro Oyay, the above described immoralities do not exist in Shulluk culture. He stressed that the sister of a wife often comes as a visitor and rarely lives in the same village where her sister is married. If she happens to live in the same village for any reason, she is often treated with respect as a sister and likely to attract the attention of the young men in the village to marry her. He concluded by saying that there is a "big no" for Uwa to have for sexuall intercourse with the wife of the brother.

From all the above narratives, we can see clearly that avoidances of in-laws (*Ora*) are of paramount important in Shulluk society. Therefore, men are checked against avoidances, for example, if the husband of the wife dies and his brother (*Uwa*) wants to inherit the wife, he might meet challenges from the mother-in-law. The mother-in-law may make it difficult for him to inherit his brother's wife, if previously he did not avoid her. However, to enable this man to inherit his brother's wife, he must kill a sheep and ask some of old influential elders to act on his behalf as peacemakers. Because in Shulluk society, the inheritance of a brother's wife is always interpreted to how much the husband's brother has done to avoid ten mother-in-law when his brother was still alive (before his brother actually dies).

It is to be noted that the wife's brother's wife is addressed as "mother-in-law" (*Ora*), so the same avoidance applied to her. She must be looked upon as mother-in-law.

(2) *The attitude towards those relatives of the wife to whom she stands in some definite relationship.* Thus, among these relatives we may consider the wife's father and her brother. They are her natural protector, to whom she may complain if her husband is mistreating her. The bride-wealth is always given to the father of the brother of the girl and if the girl dies childless, the father of brother will take this bride-wealth as a compensation for the loss of her children and not as a dowry. The wife's father's brother (Na) and his son receive parts of the bride-wealth, although they do not belong to her clan, however, they are called in-law (*Ora*)

(3) *The attitude towards the husbands of female relatives is the converse of the (1) and (2) above;* the behaviours are reciprocal but not identical. Although the man show respect towards husbands of the female relatives, yet he is more in favourable position. They pay bridge wealth of his sister to him and he can still demand anything he wants from these husbands.

(4) *A woman's attitude towards the relatives of her husband.* The wife looks upon her husband's blood brothers and his mother's sisters' sons as her in-law (*Ora*). By traditional norms she must strictly avoid these people; any show of familiarity would be con-

sidered as an offence. By avoidance her we mean the wife has to avoid having sexual relationships with any of these people; if she avoids them, the sexual intercourse between them can be prevented.

Although the conception of the 'in-law" (*Ora*) relationship is broad among the Shulluk society, the omissions from it are as striking as the inclusions. The husband does not call wife's sisters as "in-law" (*Ora*) in any form or shape. As such, the husband always treats them with familiarity; and if he also wants he could take one of them as his second wife. Thus, if he so desires, priority would be given to him but he must give full bride-wealth. It has been reported by some Shulluk elders that a rich man could marry all of his wife's sisters, which he calls "*Nyimia*".

The man does not call the wives of his father's brother's sons as his "in-laws" (*Ora*); more importantly, in Shulluk customs the wives of the clansmen are divided into two categories. The wives in the first category are the wives with whom sexual intercourse is permissible; and all the wives of the clansmen who are called either "father" (*Wa*) or "father's brother's sons" (*Uwa*) fall within this first category. While the second category, are the wives with whom no man is allowed to have sexual intercourse, except their husbands. Thus, having sexual intercourse with the wives of any other clansmen is said to cause death to both the offenders. Most of these wives are been called by using descriptive terms.

It is important to understand that the Shulluk people treat the wives in the second category, we have seen above, with care, as sexual intercourse is not permissible with them. The men have to avoid anything that could cause problem, such as staying in the houses of these women when their husbands are not at home. And if the husbands of these wives are at home, whenever a man goes to visit one of the wives, he should avoid sitting upon the bed or mate of the woman. Failure to do this would be considered as attempt to sexual intercourse; and that could lead to curse prayer.

Staying away from the in-laws is very important especially for the wives or women. Traditionally the wife is forbidden to call her husband's parents as "*Ora*"; she is to address them as her own grandparents (*Kwa*); and treat them with respect. When she is handing something, like water, to her father-in-law, she must kneel in front of him to show respect; and when she is speaking with the father-in-law, she should not see him direct into the face, but must turn her face away as she speaks; however, there is no strict avoidance required.

Sometimes mother stays with the her son if her husband dies; in this case she is given small separate hut to sleep in but that hut must be closer to the son's hut in case of any emergency. Meals in Collo are only twice per day that is breakfast and dinner; interestingly, breakfast is more private where each household enjoy their breakfast independently of others. Thus, during breakfast, the husband's mother eats separately while the wife and the husband could eat together if they so wish. The same apply if the mother of the wife is staying at the daughter's husband's house; she eats alone while the couple eat together.

In Collo society, dinner is always a group meal when males from the same village gather in one circle and and each household bring their food to the group table. Ladies gather according to the age in a similar way, they gather in one of the houses and share their food.

A man calls his son's wife as grandchild (*kware*); she falls into the category of wives of a clansman with whom sexual intercourse is inadmissible. However, there are groups of women who are treated as the "in-laws" (*Ora*), these are the wives of clansmen called either "father" (*Wa*) or father's brother's sons (*Uwa*). The men of the same class are permitted by local laws to have connection with these wives while their husbands are still alive, the main reason been that these men could inherit the wives after the death of their husbands.

With the above argument in mind, Banydhuro Oyay, one of the Shulluk intellectuals, pointed out that Shulluk customary law forbids all forms of sexual intercourse outside marriage. And should a wife been lured to such act, she will be forced to disclose such mating, and if she fails to disclose it, she would die during delivery. Since this is an embarrassing thing in the community, such occurrences are highly unlikely. However, circumstances may arise when the father feels too old to satisfy a young wife. In such a situation, he may call into his bed his own son to have sexual relationship with his wife. Such practice is aimed at preventing the wife from seeking sexual satisfaction outside the immediate family circle.

Young people are face to marry the partner of their choice provided that there is no exogamy and no genealogical relationship can be traced on the mother's side. Every Shulluk man wants his daughter to be married properly, with enough cattle, because the bride-wealth from these girls are given to him, so that he helps with the marriage of his other sons, without the cattle from daughter, the father cannot assist his son(s) to marry a wife. Groom is entitled to his own sister's dowry and not to his paternal cousin dowry.

The father who uses to marry more wives for himself while his son remains unmarried is often regarded as useless and irresponsible father. In the Sudan Notes and Records (1941) observes that the head of homestead (*gol*) is the sole legal owner of cattle in his homestead (*gol*), he can dispose them in the ways he chooses. He holds them in fact however subject to a trust which equity will enforce. The terms of trust are that all cattle shall be applied solely towards the marriages of male members of the homestead (*gol*) in order of their seniority, that no cattle shall be sold except for the purpose of grain in time of famine, and that the homestead (*gol*) leader shall not himself marry a second time unless there is no unmarried male member of the homestead (*gol*) of marriageable age 18, a breach of these terms of trust would be universally censured but would probably not be interfered with by the courts (P.23).

In Shulluk society, there are usually three huts built in the homestead, one hut is for the couple, the second hut is for the boys/children and the third hut is for kitchen. When the girls reach maturity they are separated and move on to sleep in girls designated group hut within the village, where there are many unmarried girls; parents usually give them a house to sleep in. The girls sleep together in this one house and such house is known locally as *"odwaman"*. Thus, when the boys from another village want the girls, they come at night into the house where the girls sleep, and ask the girls, if they could allow them into the house. It is not clear, whether customarily; boys are forbidden or allowed to share a house with the girls (Seligman, 1948, P. 63). Banydhuro Oyay argues that girls in the same village are not targeted by the boys of the same village; boys of other villages rather targeted girls from this village. The reason had been that boys from the village call girls from the same village as their sisters or cousins, who by Shulluk tradition cannot be sexual mates.

In contrast, Howfmayr states that meeting of boy-friends and girl-friends must take outside the village and the parents are not supposed to know what is going at that appointed time; although parents are not allowed to meddle in the affairs of the boys and girls. He went further by saying that the girl chooses the boy she like and often the girl break several engagements before finally deciding to marry one man, this is not considered disgrace except when the girl has done to excess (too much).

A girl who has more than ten lovers is looked upon as a feast woman (*Nya dei cuo*) and a man who loves many girls or women is known locally as *"wa jal man"* (literarily means a lover of women). It is reported that the girl's choice of a husband is often influenced by parents' thoughts. The parents usually want a wealthy boy-friend or the potential husband. They want to know if groom has enough cattle to provide for the marriage. The parents' thoughts are often supported by the girl's brothers and father's brothers (Na), they always try to dissuade her from getting married to a poor man.

Nevertheless, we cannot deny that the general opinions of Shulluk communities are that, they are against forcing a girl to take a repugnant partner for the sake of cattle. Interestingly, the provision of bride-wealth prevents very early marriages; as a result many men appear not to marry when they are still under twenty years of age.

A Shulluk may have sexual intercourse with the wife of any man whom he calls *"Uwa"*, that is to say he can have sexual intercourse with the wife of his own brother, who is from another mother although they have the same father. He can also have sexual intercourse with the wife of his father's brother's son and his clansmen of his own generation. The boy may even ask the younger wives of his father to have sexual intercourse with any of them (with exception of his mother). The boy could even ask to have sexual intercourse with the wives of those he calls them "father" (*wa*). Usually, the father knows that the son is consorting with one of his wives, and as this happen, the father leaves the son to sleep in the hut with the wife, while he goes to sleep in the hut of the other wife; he does nothing about it. It is said that the Shulluk men supported their sons to have intercourse with one of father's wives, because this is better than going to have sex with another woman, where the father would be obliged to pay the fines for adultery (Seligman, 1948).

Thus, in Shulluk cultures, it is acceptable that a grown up boy goes with the wife of his father; or goes with the wife of his father's son or father's brother's son (Uwa). It is reported that when a man has been out of his house for days or months or years, when he comes back home and learn that his "Uwa" have had sexual intercourse with his wife, as a human being he might feel very angry, he could even say some bad words to him, but that would be all. In such a case, local chief usually comes and tells the husband of the wife, "This man is your "Uwa". It is alright for him to sleep with your wife while you are away. On hearing this, the man (husband) would accept the idea, however, if it was a stranger who has sexual intercourse with his wife, the man would try to spear that stranger at all cost. Thus, the young unmarried man, who was exercising his rights towards the wife of his father's brother's son "Uwa", would say to his relative that his father's brother's son "Uwa" knew before leaving the village that he was interested in the wife; the "Uwa' made no objection to it, therefore, such a union should not be a surprise to him. This implies that a child born from such a union always the child is considered to belong to the woman's husband and any idea of it belonging to the wife's temporary lover is ridiculous.

Shulluk traditional custom forbids a man from making an attempt to come closer and closer to the wives of his mother's sister's sons (*Ma*) wife's brother's sons (*Umia*), mother's brother's son (*Una*), sister's son (*Oka*) as well as the wife of his father's father (*Kwa*); if he does, this would be considered more worse than committing adultery, and it is believed that both the wife and the man who are guilty would die. Not only this, it is also believed that having sexual intercourse with the wife of in-law (*Ora*) is looked upon in the same way.

In many cases, married brothers and married sisters happen to live near one another and their children would address themselves as "Una" (uncle's son), "Nyina" (uncle's daughter), "Uwaja" (father's sister's son) and "Nywaja" (father's sister's daughter) the children would have an opportunity to meet and play together; and there is no restriction on the girls to play with the boys. Nevertheless, segregation of sexes usually imposed later by different lines of duty, where the girl helps mum with household staff and the boy looks after livestock.

As we have seen behind in marriage that when the wife is already six month pregnant she is to leave husband's home and return to her parents' house where the baby would be born. As women often go to live in the "*gol*" of their husbands, the children of the two sisters are unlikely to see each other, in this case what we called "Uma Nyima" relationship become loose, there is no intimacy between them.

Experiences in Shulluk society and Northern Luo of South Sudan in general, revealed that relationship between wife's children and their mother's brother "Na" (uncle) is stronger than the relationship between wife's children and father's brother's children. It is the mother's brother who usually gives a son of his sister his first spear; the boy also has a right to some of the cattle given as bride-wealth for his sister's daughter, and sometime he could even influence the girl in her choice for a husband.

A young woman does not drink milk under any circumstances, though a woman who is nursing a baby may, as do old women past the reproductive age. Only small boys herd the cattle and milk them, for once a boy has reached maturity there is the danger that he may have had sexual contact, so if he does and milk cows or just walk among the cattle in their pens, it is believed that this might cause cows to become sterile. However, this believe is so strong among the people, in such a way that if a child drinks milk at his neighbour house, when he comes back, he must wash his body before entering into his father's house. Because it is believed that some milk might have remained over his body that could poison the food, and once this food is eaten by unclean person, the person might die. According to Shulluk custom, a man who has sexual intercourse with his wife a night before, he is considered unclean and must not drink milk until the sun sets the following day (Seligman, 1948, P. 73).

In Shulluk communities, when a woman finishes cooking food for her husband, she is to leave the hut while the husband is eating, but it is not shameful for men and a woman to eat together with her husband. The tradition of man eating first is derived from the possibility of an emergency (war, etc.).It is perceived men must beat first such that they are ready to defend the family or the village in case of any eventuality. Some people said, after a man has eaten the remaining part of a meal is then taken and eaten by the woman and her friends (usually her unmarried "*Nyiwa*"). But Banydhuro Oyay argues that women are never made to eat the leftover of the men in Shulluk culture. He pointed out that by the time men's foods have being served; women are making theirs and the children's. But since the man's food are often over allocated, there is always left over which may be added to the children's or woman's. Small reed mats called "*paro*" are used to put over pots of food to keep off flies and dust.

Chapter 16:
Arts, Crafts and Music

"Understanding the culture is as importance as understanding the Language"

Arts and Crafts

The Northern Luo Groups, who live in South Sudan, have their distinct artistic traditions as compare to the other ethnic groups in the country, although they share some
common traits among them. Beadwork, pottery, music instruments, and woven goods
are prevalent art forms among the Northern Luo. Women make varieties of necklaces,
bracelets, and corsets from beads. For example, The Acholi, Anyuak, Jur Chol and Collo
make strings of beads that women wear on their foreheads and round the waist.

Pots are made from clay; white baskets are woven from fibres called sisal. Designs
are carved into hollowed calabash gourds with fish hooks, then soot it rubbed into the
depression; these are placed as decorations in houses.

A handcraft has varieties of types in which the objects are completely made by hand
using simple tools. The term "handicraft" usually is applied to traditional techniques of
creating items that are both practical and aesthetic. Handcraft industries are those that
produce things with hands to meet the needs of community members. Many handcraft-
ers use natural and native materials, although some craftsmen in our world of today may
prefer to use modern and imported materials. Whether a person use natural and native
materials or modern imported mated, yet crafts require the development of skill and the
application of patience, and these skills can be learned by virtually anybody.

Art and craft are part of the Northern Luo cultures; crafts have been developed through
the traditions of the people; they come as a result of the feelings of the people responding
to a variety of historical events, and influence the environment in a most spontaneous
manner. Craft is always taken as an occupation especially where the skills or techniques in
the use of the hands are needed. Craftsmanship among the Northern Luo of South Sudan
has been passed down from generation to generation. Craft works among the Northern
Luo Groups varied so much, that one needs to pay attention, if you want to get a taste

of it. In addition one needs to make distinctions of the varieties of crafts and to identify similarities that occur within the crafts of different Northern Luo Groups.

Although there are differences which are prominent, especially as regards the functions and designs of the crafts, yet there are qualities that bind crafts together. In this respect, we should look at various methods people used to produce the crafts. For example, the beads of Anyuak and Acholi are made by using strings that go through them (beads).

In Northern Luo culture, men hardly make pots, because this is considered the work of women; however, men can act as the middlemen and salesmen. When we look at the process of pots building, we find that there are similar methods of building them, but yet we could see that there are still variations in the time taken drying and the method of firing. The local potters in the village build their pots of whatever description and function coming from the base upwards to the rim using the coiling method. In the is way pottery method is the same in Acholi, Anyuak, balanda, Jur Chol, Jumjum, Mabaan, Pari, Shulluk and Thuri/Shaat. It is to be noted that pottery is the work of specialized craftsman.

Wood curving is also another sector that is common among the Luo groups —to certain extend it has been influenced by external forces, especially in terms of design. The commonest domestic woodcraft products include: tools, beds, mortars and pestles, bowls, ladles, trays, wooden canoes and so on.

Traditional craft items among the Luo societies include amulets, necklaces or beads, arm and leg ornaments, bracelets, rings and headdresses. The Luo cultural and domestic can generally be categorised as wood vessels, gourd vessels, smoking pipes potter, basketry, stools, miscellaneous house-hold object, clothing and adornments, animal skins and bark cloth, aprons and tails, hairdressing, girdles and belts, ear and neck ornaments, facial, neck, leg and arm ornaments, dancing accessories, hunting knives and bells, hunting gear, fighting wear such as shields, swords, spears, bows and arrows.

Today, there are classes between cultures, this especial happen when the people move from rural areas to urban areas and more importantly when the people move from South Sudan to the Western World, in places such as America, Canada, Australia and European countries. On one side of the coin it is good the Western World provided protection to these vulnerable refugees but on the other side of the coin it created a wide tendency for men, women, boys and girls to forget and abandon traditional arts and crafts. Eventually, this would lead to the loss of cultures if not rescued.

Music instruments

Musical instruments play a major role among the Northern Luo of South Sudan. Music plays an integral part in rituals of birth and puberty, at marriage and death, in secret-society initiations, and in rituals of livelihood such as hunting, farming and so on. The productions of musical instruments add to the economic, social, psychological, and educational as well as the therapeutic being of the people and communities. The musical instruments form a big part of natives' history and cultural heritage. At a point in life, musical instruments are used in people's daily life activities. It is said that the short story was a well-developed art among the Luo in traditional times. Such stories are often accompanied by music.

In most of land Northern Luo counties, music is used to keep records; musicians compose songs about the present to recall the past. There are songs to make people remember wars, bad weather like rain damages or particular disease that broke out among the communities such cholera, sleeping sickness, measles, small fox to mention a few. Music is also used as forms of mathematics among the society to show wealth counting cattle, sheep and goats; remember trade languages so people can do business with foreigners. People also sing about the number of the cycles of the day, for example, sunrise, sunset, and when the cock crows. Music is an important method to show and teach respect to the children; Acholi cultures require children to be humble and respect elders and this message can be passed through music. Above all music is used to honour and worship the Living and the only God.

Generally, making of musical instruments involve men as well as the use and manipulation of musical instruments such as drums and fiddles is strictly meant for men alone. However, today, due to the external influence from Western World and introduction of schools in both the rural and urban areas of South Sudan, have allowed all sorts of people to learn and participate in the playing of musical instruments.

The traditional instruments of the Northern Luo Groups range from drums (*bul*) bamboo trumpets, horn trumpets, *Lukembe, Adungu* etc. Some of the trumpets can be used to communicate messages from one village to the other in addition to providing music (e.g. drums, horn trumpets, and *Thom* (a guitar like instrument). We shall see this very clearly among the Acholi.

Blacksmiths

In Northern Luo cultures, the making of metallic tools and other utilities by blacksmiths has become a common practice. Blacksmiths among the Northern Luo Groups make cutting tools, spears, arrows, axes, hoes and etc.

To understand how arts and crafts as well as musical instruments are important in the life of the people, let us examine some of the Luo groups and these will represent the whole of the Northern Luo. As I mentioned before that the people of Northern Luo have their distinct artistic traditions and they have much in common.

Acholi

Arts

The Acholi women work on crafts from their homes. The women earn professional wage making beads, pots, wood vessels, baskets, gourd vessels, hair-dressing, ear, neck, leg and arm ornaments and so on and so on.

Figure: 102, Acholi Beads, Women Bag

The Acholi women are professional in making women bag with a circle mouth and upright-walls with slightly pointed centre, and can be pulled into a flatter shape when closed. The bag has flexible body made from soft black wool and coloured shiny raffia-like nylon fibre twined around a series of stiffer vertical elements, made from a twisted blue synthetic cord, it form a blue rim around the top.

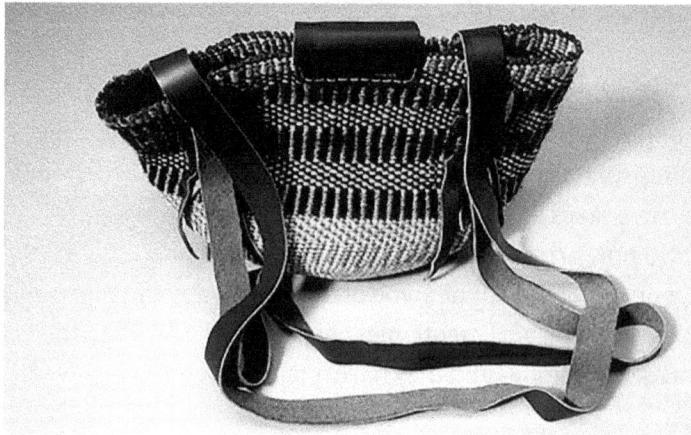

Figure: 103, Women Bag

The weave pattern consists of a series of horizontal bands around the body that alternate between vertical and oblique patterns, with each hand made up of groups of opposing colours. Immediately below the rim is a band of pale green and pink nylon arranged in vertical stripes, follow by three bands of pink nylon and black wool vertical stripes. There is a further band of oblique pink and green nylon stripes below, the same colours in a vertical pattern, and then a final band of green and pale pink oblique stripes; the rest of the base is made up of plain green nylon weave. These patterns are also visible on the interior walls, where there are series of concentric rows of blue nylon cord loops visible. The bag also has two shoulder straps, made from rectangular leather strips, and stitched to the upper body with yet another piece of leather knotted on the inside walls. Further decorative trim has been added below the base of each handle, consisting of three parallel stripes sewn through the body, with the ends hanging as loose tassels. The bag usually has a weight of 434.9 grams and a total length of 660 mm; the handles are 470 mm long, 26 mm wide and 1.5 mm thick (Rachael Sparks, 2005).

Figure: 104, Acholi baskets

Food Cover

Food cover (*latam me umo kwon*) is one of the common arts among the Acholi women. It is sometimes made from European style wool and in other cases it is made from local materials such as stripes of bamboo or sorghum straw. In this book we are not going to discuss on food cover made from local materials, rather, we shall focus on food cover made from European style wool. This type of food cover consists of a square body, crocheted in Fluorescent pink, with a close knight design that divides the area into four sections; leaving the eye like, towards the centre that is stitched with royal blue wool. A border is made around the outside edge of the square, using a much looser type of stitch in the same blue coloured wool. This type of food cover usually weights 57.9 grams, and is 393 mm long and 385 mm wide; the blue border is 30 mm wide and the central upper rosette has a diameter of 32 mm.

These sorts of handicrafts are usually made by women; they could do them individually at home or in group, in self-help development projects, Churches. It is worth mentioning that most of the food covers are sold within the communities only few find their ways in the hands of the tourists.

Musical instruments

The tradition of the Acholi dances is an invaluable treasure which has enriched the history of art culture for generations. Dance has always been the purest manifestation of the Acholi cultural progress and reflects in a unique way the blending of history, the present, future and generally expression of philosophy in the life.

Musical instruments play a major role among the Acholi of South Sudan. As I mentioned before, music add to the economic, social, psychological, and educational as well as the therapeutic being of the people and community as a whole. In the whole of South Sudan, the Acholi are well known for their musical instruments, which serve different purposes. The Acholi, like the other Northern Luo Groups, have musical instruments that differ in sounds depending on particular message they want to communicate to the communities. They have many traditional, musical instruments, all are made from local materials, and these include but not limited to:-

Drums

Drums are the most popular musical instruments among the Acholi of South Sudan as well as among the Acholi in Uganda. Drums are made by a curving tree trunk after it has been shaped into the size and desired shape by the drum-maker. The woods for making drums are always selected from woods that cannot be easily eaten by insects; and the drums must be kept in places offering good conditions to avoid wrapping.

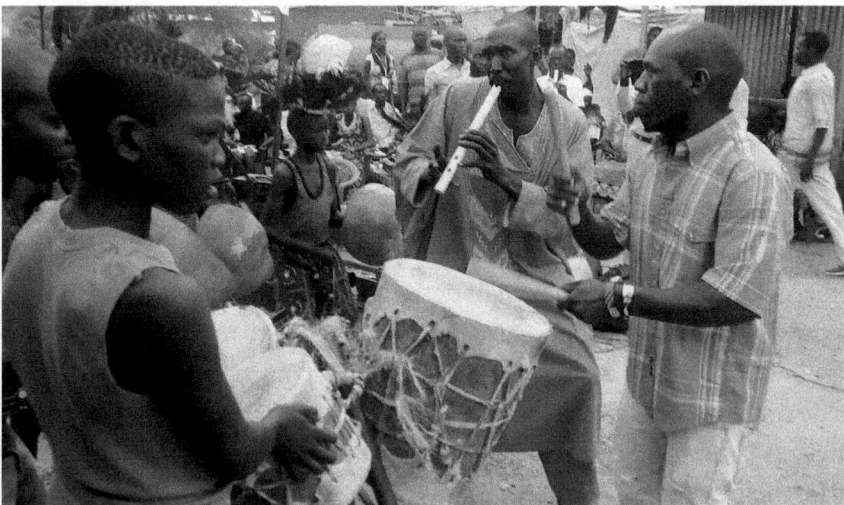

Figure: 105, Acholi Drums

During the manufacturing process, care must be taken in the preservation of wood and skins that would be used. Heavy and relatively thin animals' skins like those from cows, goats and other animals are used in manufacturing small, medium and big drums that are played with aid of wooden sticks and sometimes hands. They are often covered with skins on both sides the top and the bottom. Straps made from animal skin are used in holding the members (top and bottom) to the desired tone and pitch. However, drums are very effective means of communication. In Acholi land, drums are used for various purposes, which include the followings:-

▶ A unique pattern of sounding a drum is used to tell members of community particular events taking place in the areas such as hunting or an elder has just died, or a young person has just died; wedding is in progress; a stranger is in the area and to call upon community members to assemble.

▶ Communal work is to be done such as group cultivation of crops.

▶ A person is missing from the community, and when this music is played, the missing person can follow the sound till she/he reaches home.

▶ Dance is been organised today in a particular village\clan

Unfortunately, many of the Acholi people who are born and grew in towns, cities, refugee camps and Diaspora cannot interpret the messages made by the drums, however, people who are born and grew in the villages are good at interpreting various messages send through the sounds of drums.

Wooden Zither (Nanga)

The Acholi men learned playing zither or Nanga from their fathers or intimate friends, there is no school for harp. Mr. Watmon Mathew, from Gulu-Uganda (2012) made this abundantly clear when he says, "I did not go to school or have any training to play hard; I learned it from my father who was very good at playing the harp" (Acholi Time, 2012, P.1).

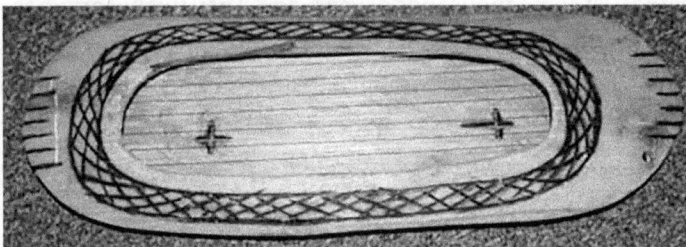

Figure: 106, Acholi Wooden Zither

In the south Sudan, I have come to know Mr. Paterno Lobuk, one of the best payers of harp, who told me that he has learned playing harp from his friends.

The wooden zither consists of eight strings, which run parallel to the resonator that extends along the entire length of the strings. The Acholi zither has a boat-shaped sound box with strings run parallel to the resonator. The instrument is good for telling the past and pre4sent stories of the people. The prayers often compose love songs, prise songs, dirge songs and humorous songs.

Arched Harp (Adungu)

Arched harp is one of the outstanding Acholi handcrafts. It is believed that the arched harp have come from the traditional bow-harp of Egyptians. Michael Levy, a specialist in music of the ancient world, discovered there was a close likeness of the Egyptian arched harp and *Adungu* of the Acholi play today. See Figure---- below

Figure: 107, *Acholi arched harp (Adungu)*

The Egyptian arched harp found today in British Museum is dated between 1534-1296 B.C. However, some scholars rejected the claim made by Michael Levy that African (Acholi) arched harps, have originated from Egyptian, because according to them the Acholi arched harps were created just some thousands of years ago. They further argue that the modern African (Acholi) arched harps (*Adungu*) has tuning-pegs, which the Egyptians did not have. Furthermore, the Acholi arched harp is played with the fingers, while the instrument is placed on the ground, but the Egyptian arched harp is played against the chest.

The modern Acholi *Adungu* consists of an arched neck, a wooden resonator (sound box), covered with goat skin, in which the neck is fixed; the wooden sound box has nine parallel strings of unequal lengths that run over the resonator that extends along the entire length of the neck and the sound box. The strings are fixed along the resonator and run at an oblique angle of the neck, where they are attached and turned with nine pegs.

For a player to produce a desired, ideal sound, he adjust the movable rings which is attached round the neck of the instrument, one to each string, which are pushed into position just close enough below the string to let it vibrate against the string. The Acholi harp is constantly evolving, and while in the past tortoise shells were used to make the harps; but today they are largely made out of wood and goat skins. The Acholi zither comes in a variety of different sizes that is big, medium and small sizes. In many cases they categorised them into mothers arched harp and children arched harps.

At this juncture, it is worth mentioning that in the old days, the harpist was the only musician ever allowed to play in the room of the royal ladies. Today, *Adungu* (arched harp) is played in social gatherings such as dance, in Churches, in Weddings and so on and so on.

It is assumed that the Acholi South Sudan arched harp or string instrument was brought from Northern Uganda. Today, the string instruments are very common among the Acholi villages in Magwi, Panyikwara, Agoro, Omeo, Obbo, Palwar and Pajok.

Figure: 108, Acholi Obbo youths playing Lokeme

This musical instrument is common y known as "Lokeme"; it consists of a series of flexible metal or cane tongues ranging from long, medium to small lengths, which are fixed to a wooden plate sound box. The Lokeme is made from special wood and the tines are made from of high quality spring sheet.

The player holds the instrument in his or her both hands and uses thumbs to pluck the free ends of metal tongues. The instrument is accompanied by songs, sung by boys and girls.

Olere (Flute)

The flute is popular among the Acholi in South Sudan; it has five finger holes, where the player place his fingers on to adjust the required tune. The player blows in the air and he does the adjustment of the tune according to his left and right fingers.

The instrument is used for famous traditional dance called *"dingi-dingi"*. This dance is for welcoming very important persons like Presidents, Ministers, Commissioners and Chiefs. Although I have not seen the other Northern Luo playing traditional wooden flutes, this does not necessarily means that they do not have.

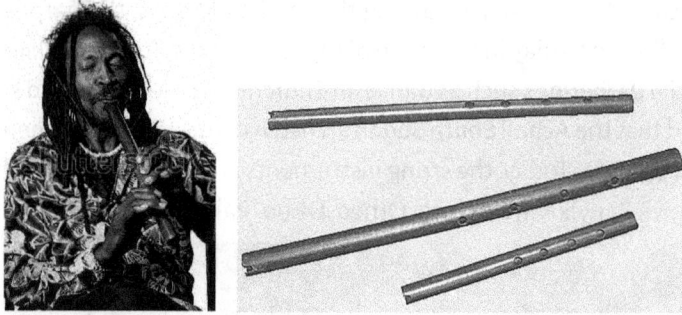

Figure: 109, Acholi man playing traditional flute (olere)

Anyuak Arts, Handicrafts and Music

Arts and Handcarts

The Anyuak main Arts and handcrafts include beads (Tiko), Anyuak marobeats, decorated calabash and traditional dresses for women and girls as shown below:

Anyuak beads consist of beads of different colour with strings running through them and with some nodes approximately 10cm from each other. The women are very good in arranging beads according to the required colour and nodes.

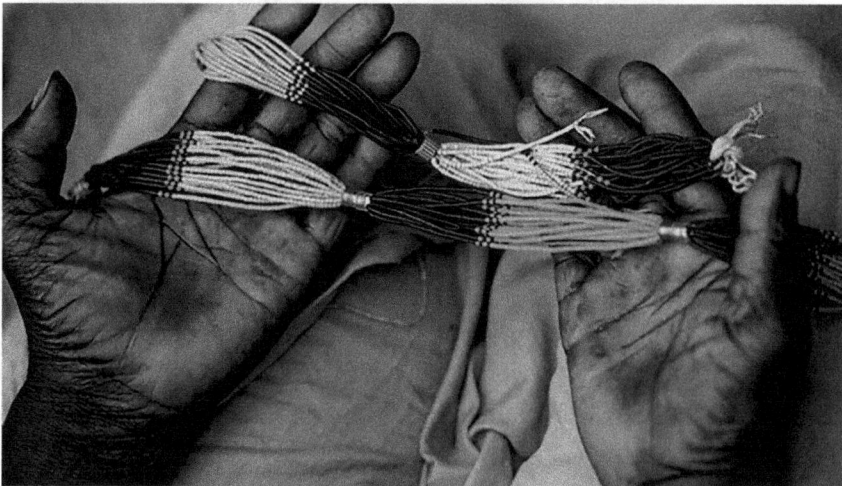

Figure: 110, Anyuak traditional beads

It is believed that the Anyuak who live in the inner land have little knowledge about the money. As a result they hardly accept money for exchange of goods. Conradin Perner (2016) argues that he was informed by Mr. Omot, one of his informants, that beads are vital importance when purchasing flour in the rural areas, especially in the rainy season when flour is very scarce. During the rainy season, exchange rate for beads is always higher as compare to the exchange rate in dry season. (How high and how low the exchange rates are during these periods remain unexplained). The choice of beads its size and colour is of paramount important, because the size and colour of beads are important factor to decide the exchange rates for other things of other value. Many Anyuak girls and women prefer white and black beads, although, we cannot deny that female also used other beads with different colours. However, the Anyuak girls would never hesitate to exchange beads for food, although without necklaces made of beads, girls would look ugly and completely naked. It is not only girls who would never exchange beads for food; boys likewise would do anything in exchange for beads.

Women make some beads that would be dress on the head and neck as we can see from the picture below.

Figure: 111, Anyuak tradition dress beads

The above traditional dresses are normally made by women by using beads of different colour and sizes, and strings. This dress is not use as a common dress of the day but rather it is worn only when there are certain importance occasions such as crowning the king, mourning the dead king and in traditional dance. Interestingly, beads are not only used for the clothing of young Anyuak girls and women, that is for beatification of their physi-

Figure: 112, Anyuak traditional calabash

cal appearance, but beads are also used for spiritual significance, which include taking care of the people from dangers, relief pains and keep away evil eyes or forces.

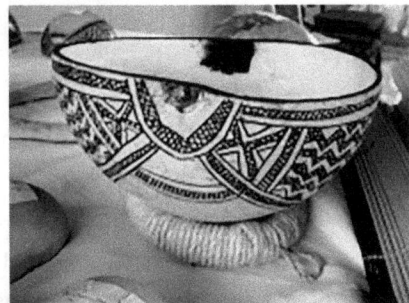

Figure: 113, Pounding and winnowing sorghum

Maize and sorghum are pounded into flour with pestle (*lek*) and mortar (*pany*). For winnowing, a boat-shaped basket (*luur-anywaa*) is used first, then an iron sieve with a rectangular wooden frame (*luur-gaala*).

Anyuak traditional Musical instruments

The musical instrument popular among the Anyuak of South Sudan include Tom or Thom, (there are many variations on this instrument among the Luo), Drums, and animal horn commonly known as "*Tung*".

Tom or Thom

Thom is very popular among the Anyuak people and we could define it as "thumb piano"

The Thom or thumb piano consists of wooden box sound with eight metal strips attached to the wooden box. The player presses his two thumbs on the ends of each metal to produce the required sounds. The longer metal strips, which are in the middle, are use to produce deep sound or

Figure: 114, Picture of Thom

buzzing sound, while the smaller metal strips, which are located on the left and right of the wooden box, are used to produce the required high tones sounds.

Figure: 115, Anyuak Drums

The Anyuak traditional drums, like the Acholi, are the most popular musical instruments among them. Drums are made by a curving tree trunk after it has been shaped into the size and desired shape by the drum-maker. The woods for making drums are always selected from woods that cannot be easily eaten by insects; and the drums must be kept in places offering good conditions to avoid wrapping.

During the manufacturing process, care must be taken in the preservation of wood and skins that would be used. Heavy and relatively thin animals' skins like those from cows, goats and other animals are used in manufacturing small, medium and big drums that are played with aid of wooden sticks and sometimes hands. Lease, see figure 113 above). They are often covered with skins on both sides the top and the bottom. Straps made from animal skin are used in holding the members (top and bottom) to the desired tone and pitch. However, it is believe that drums are very effective means of communication AMONG THE Anyuak like in Acholi land. Drums are used for various purposes, which include the followings:-

- ► A unique pattern of sounding a drum is used to tell members of community particular events taking place in the areas such as hunting or an elder has just died, or a young person has just died; wedding is in progress; a stranger is in the area and to call upon community members to assemble.
- ► Communal work is to be done such as group cultivation of crops.
- ► A person is missing from the community, and when this music is played, the missing person can follow the sound till she/he reaches home.
- ► Dance is been organised today in a particular village\clan

Unfortunately, many of the Anyuak people who are born and grew in towns, cities, refugee camps and Diaspora cannot interpret the messages made by the drums, but people who are born and grew in the villages are good at interpreting various messages send through the sounds of drums.

Figure: 116, Anyuak Tung (Musical instruments

Balanda Arts, Music and Handicrafts

Arts and handicrafts

Balanda, like the Shatt, expressed their culture and social values in songs, poetry, physical arts and dance and folklore which are largely oral. They make tomb totems from hard word.

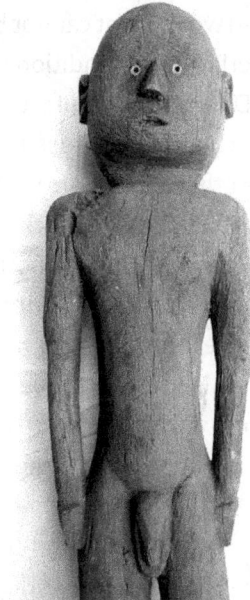

The Statue has been collected on 1958 in Wau, South Sudan by Comboni Father Elvio Gostoli. It belongs to the group of funerary posts sited on the tomb of one of the powerful chiefs. This is a unique Museum piece, not easy to find today and more still not easy to find of this beauty. The Balanda are skilful wood carvers, a craft that endures

Figure: 117, Balanda tomb totem

to the present. They also make basketwork, strainers to filter beer, baskets for carrying things, basket-pots for fishing, basket-work walls of huts and bee-hives. The other crafts the Balanda manufacture are fish-nets, creels and snares, fishing-lines made from fibres of baste.

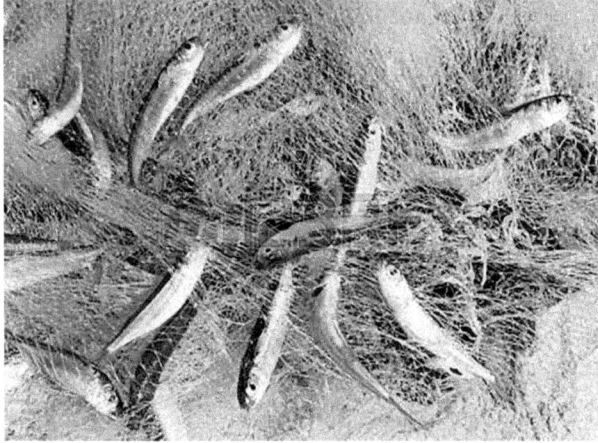

Figure: 118, Fishing Nets

The Balanda were advanced in the smelting of iron derived from the abundant lateritic soil. From the iron the blacksmiths produce axes, spears, hoes for trading with the neighbouring communities (Tim Mckulka, 2011)

The below large sculpture comes from Belanda people of Deim Zubeir, Western Bhar El Ghazal, in South Sudan. This sculpture and others were noted as early as the nineteenth century when travellers reported large sculpted funeral or memorial markers placed above tombs composed of large stones. Among the Balanda, large sculptures recognized and honoured male elites, warriors, and chiefs.

Balanda sculptures associated with funerary practices are carved out of a tree trunk, the base of which is buried sixty to eighty cm in the ground. They are placed either in front of or in the centre of a grave-round and surrounded by stones, the whole measuring as much as three meters in diameter.

Figure: 119, Balanda Sculpture

These tombs are set up near villages, and the figures on them are fully exposed to the weather. Since villages in Western Bhar El Ghazal move when the ground is no longer suitable for cultivation, the older tombs are now mostly lost in the savannah, with no roads leading to them. Given the low density of Balanda population, this makes it difficult for the foreigners to locate many of these tombs, unless the local people are actively involved. The Balanda leave the sculptures in the graveyard, when they move, because they respect their ancestors and their ancestors' tombs, they do not want to disturb the ground by removing the buried part of the sculpture.

Carved from hard woods, the hardness and natural resistance of the wood has lent great durability to these sculptures. The amino damage the valving suffers, then, is caused by moisture, especially during the rainy season. The erosion usually starts in the heartwood, and many of the sculptures become more or less hollow. Bush fire is another thing that causes destructions to sculptures but some of the hard woods are difficult to burn, the sculptures are often only superficially charred.

Balanda Musical Instruments
The Balanda have evolved a culture that honours land and much of this cultural heritage is expressed in songs, music, dance, facial and boy mark. All these are visible during social events such as marriage and funeral ceremonies.

Figure: 120, Drum (Bul)

Jo-Luo of Bhar El Ghazal Arts and Handcrafts
The people of Jo-Luo are essentially expressing their cultures orally. The cultures are usually transmitted in song, music, dance and other bodily expressions. Dance and songs are very important in Wau Luo culture and one distinguishes oneself through them.

The Luo have several dances, and have perfected the art of making whistles and their sounds for different occasions. The Jo-Luo people are famous for iron smelting and they produce hoes, axes, spears and arrows. Their handicrafts include baskets, mats, pottery and chairs.

Figure: 121, Jur Chol Basket

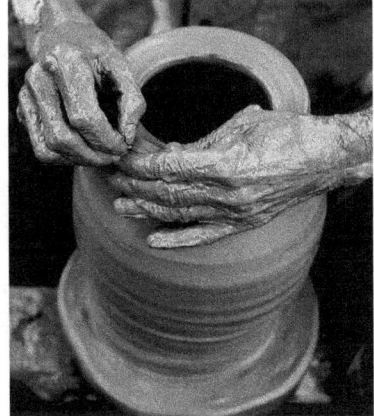

Figure: 122, Jur Chol Potteries

Figure: 123, Jo Luo Wau Dancers

Jo-Luo Height and Dresses

The Jo-Luo is Nilotic tribes; as such they are tall and slender. But, due to their social contact and intermarriage with Bantu tribes like the Zande and the Bongo, they have reduced greatly in height, particularly those who live in the south. Despite this education in height, the Jo-Luo people are very easily identified by the way they dress and behave. They are

513

cautious, conservative and rural oriented people. Men dress their hair like women; they wear a band called "*abonga*". Girls wear "*ajugo*" and women wear and open skirt, known as "*adono*" and "*athor*". It is good to know that in the past women used to wear skins instead of the "*adono*" and "*athor*" of today (Apai Gabriel, 2005).

Shatt/Thuri Arts and Handicrafts

The Shatt people are specialized in metal and wood works, blacksmithing and carpentry. I have been reported that women are also good in arts and craft; they mainly weave baskets, cotton cloths, pots and many others that I cannot mention them all here. Both men and women weave cotton cloths using local technology.

www.shutterstock.com · 381257308

Figure: 124, Cotton cloth weaving using local technology

The men also make beautiful crafts and furniture from erotic wood, which bound in their territory (Gurtong Peace Trust, 2017).

Blacksmith
Blacksmith in Thuri community evolved different kinds of tools and implements people use for agriculture, hunting and fishing activities.

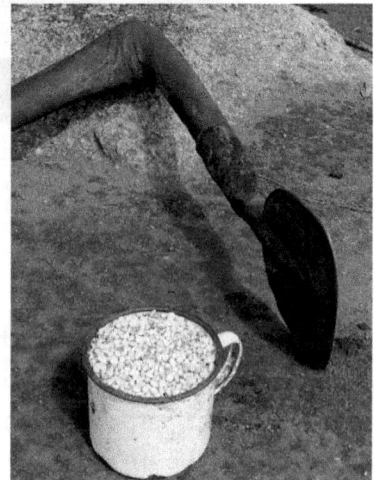

Figure: 125, Local agricultural tool

Musical instruments

The Shatt have also evolved a culture that honours the self and this is always expressed in body and facial marks, speech, songs, music, dance, poetry. The expression of self is more seen especially during social events as marriage and funeral ceremonies. However, we cannot emphasize that Shatt songs carry prises and insults that sometimes could easily invoke conflicts especially among the young people.

Interestingly, the people of Shatt celebrate live from birth to death with especial dedication to music in its totality. Events are celebrated throughout the year.

Shulluk Arts and Music

The Shulluk society has deeply involved in a material and political culture expressions through local musical instruments, which are produced locally in the Kingdom and these, have become part of their daily life activities. They have elaborated system of traditional arts, handicrafts and music that could be traced back to some 500-600 years. According to Diedrick Westernmann (1912) the Shulluk people practice a great number of crafts, and these activities are carried on in families for generation and generation; where the father and mother impart their skills in handicrafts to their children (P.30).

Each village is specialized in particular handicraft some in fishing nets, some in spears, some in making hoes and some in making pots and baskets. Thus, each of these goods are exchanged between the villagers and in this trade bartering or cash are used. However, it is reported that most villagers do their trade by barter, where the use of the word "nyeawo", which means "to sell" as well as "to buy". In all the villages selling and buying are identical actions, one can "buy" anything without at the same time "sell" it (Exchange one thing for the other). Moreover, trading with the Arabs made Shulluk to learn the use of money for exchange of goods and services.

There are varieties of crafts activities in Shulluk society but we cannot discuss them all in this book. However, we think it might be important to focus on some of the following articles:

Blacksmith

The blacksmith is one of the outstanding arts in Shulluk community, they make spears, hoes, axes, harpoons, picks, arm-rings of brass and of iron; they also make bells and chairs. Blacksmiths do their works during the dry season when there is neither cultivation nor harvest. Interesting, blacksmith do not work in one place, they are been hire by different households in different places. The head of the household then pays him in cash or give him one sheep.

Figure: 126, Blacksmith at work

Blacksmiths must always take along his iron for the work. Iron bracelets and other ornaments are made by blacksmiths from imported iron roads. If the rods are too large, they are heated and beaten out and moulded. It is good to notice here that whatever the Shulluk blacksmiths make are always superior to the imported articles, which could be bought in Arab shops.

Figure: 127, Shulluk warriors (1910)

In the old days, the blacksmith from Shulluk used to buy irons from Dinka and Nuer, who bought iron from Jur Chol and Bongo in Bhar El Ghazal region –some iron also came to Shullukland from Darfur through Kordofan (Diedrick Westernmann, 1912, P. 31). But today the modern Shulluk procured their irons from the European countries through Khartoum-Sudan and Juba; and some irons are procured from Ethiopia. As a result, today some of the finished items from the native manufacture are gradually being replaced by European and East Africa imports. Today, one can buy cheap spears, crude axes, mattocks and hoes from the local traders. In Dinka, blacksmiths are given low respect and pay low rate but in Shulluk they are highly respected and they receive good payments for their products.

Pottery

Pots in Shulluk society are made by women; they make them in different kinds and sizes for carrying water, cooking, brewing beer. They also make pitchers, cups for drinking beer and water. Pots are made from clay and they are burned to avoid melting especially when they come to contact with water or liquid.

Brewing of beer

It is reported that large quantity of sorghum cultivated in Shulluk is used in brewing local beer commonly known as "Ombuki" or "Atari". In the preparation of brewing, women put the grain into jars and poured some water into it. The grains is left in the jars till the meal begins to sprout; after this it is spread in the sun to dry; and then pounded or ground into flour. The flour is then put into water and mixed with fresh flour and put back into the jars. The jars are then filled with water and the women stir the flour in the jars until all are mixed well and a little dry flour is sprinkled on the top of it. The jars then covered with small mats and allowed to remain for a day or two until it begin

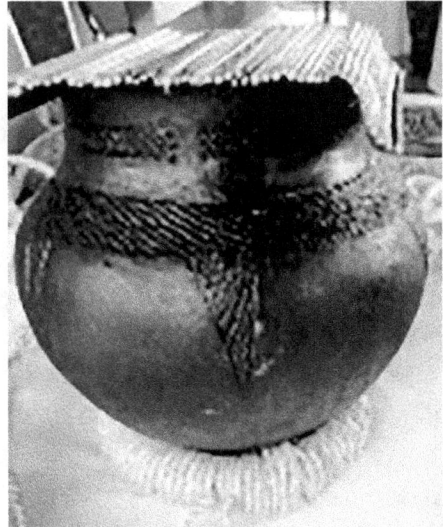

Figure: 128, Local brewing

to ferment, then a little water and additional flour (yeas) are added.

Thus, when the whole mass is well fermented, it is filtered through a grass funnel, and the following day it is ready for consumption. If the beer is not consumed in a day, it can still be kept for a week. Although in a week times the drink would be slightly intoxicating, it does not seem to do harm to human health; it is rather nourishing and the local people regard such drink as a food than a beverage (Diedrick Westernmann, 1912, P.34).

Many of contemporary scholars argue that gourds and calabashes, which are much use in homesteads, are the economic activities of the women.

Smoking-Pipe

The other important craft in Collo society is the production of smoking-pipe, smoking pipe among the Shulluk are made of clay; they are large and rather clumsy and generally decorated with some simple-designs that consist of ring-shaped lines with dots in them. The pipe-stem is long, thick; hollow reed and it measure about one and half to two centimetre in diameter.

The space between the pipe and stick is tightly closed with a leather cover; and on the top end of stem there is small and oblong gourd with a pointed head, it is tighten by a leather cover on the pipe stem. Along the long the stem there run four to five tinny strings which are fastened on the leather cover that is fastened on a small, oblong gourd at the top end of stem. The same four or five strings are also fastened on the leather that joined pipe and stick. The four or five strings are used for carry-

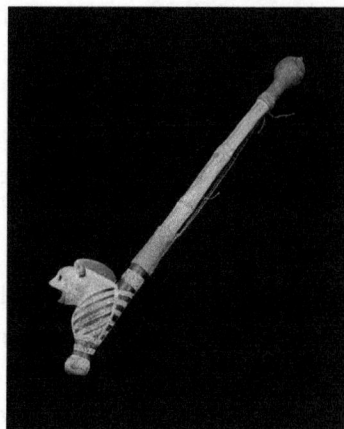

Figure: 129, Shulluk pipe made in the image of one of the totems of the King Figure: Shulluk Smoking-pipe

ing the pipe. There is also a long pointed stick that is tied to the stem; this is used for cleaning the pipe.

There is another pipe found in Shulluk community, it is also made from clay but it is heavy and it uses little tobacco but much charcoal (See figure.....below).

Because the pipe is so heavy, the smokers (men or women) have to sit in circle, each bowing head deep down the pipe while smoking. The rule is that each of the smokers to take one or two draughts (pulls) and pass the pipe over to the next person. Interestingly, tobacco is not only smoked in Shulluk community but some women chew it passionately. The chewing of tobacco is commonly practice by women and girls rather than men and boys.

Neck Supports or Headrest

Neck supports are curved from wood and each neck support is made to resemble the forms of either hippopotamus, giraffe, tiang, camel, ostrich or a particular bird. Making of the neck support is the activity of men; the making of the object does not

Figure: 130, Shulluk heavy smoking pipe

require special skills that are associated with the construction of the neck-supports, any man can make them alongside such activities as house-building and weapon making.

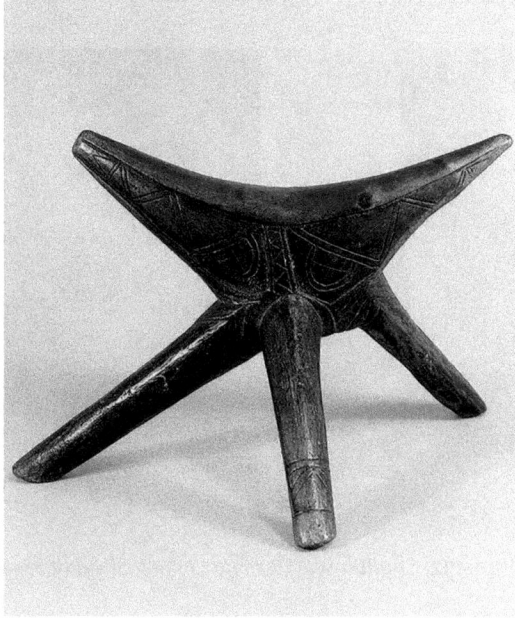

Figure: 131, Neck Support

The symbolic importance of these objects is that the Shulluk believed that neck support was originally invented by Nyikang, their most important ancestor, culture hero and the founder of the Shulluk dynasty. Usually when the old man died, he is buried with his neck-support. But sometimes the old man might decide before his death to pass on the headrest to his heir. The heir in turn is expected to treat it with respect because this wooden piece embodies the spirit of the diseased. It is reported that Anyuak also use neck-support, they also pass it on to their heirs. But according to Anyuak custom, the old man cannot just pass his neck-support just like that as it is done in Shulluk, but the King-elect is to stand on the neck-support, if he is able to balance on a three-legged stool during his investiture ceremony, he would then be acknowledged as the right heir. If he cannot balance on a three-legged neck-support, the neck-support will then be buried with the disease. Neck-supports are used in Shulluk community to help someone to have good sleep. They are also used as a comfort to help protect ceremonial coiffure. In some occasions neck-supports are used as stools. As a personal object, the neck-support has become part of the individual. In the old days, Shulluk men had no chairs to sit on, they used to sit on stools or used neck-supports as chairs; and at that time the young people used to squat or lye on the ground. But today, in the modern Shulluk, one can find chairs in every household.

Shulluk Shield

The shields in Shulluk society are made from wood with two holes for handling and covered up with hide; it measured bout 64 cm long and is oblong in shape.

Figure: 132, Shulluk warrior Figure: 133, Shulluk Shield

The shield has hollowed out cavities inside the handle. It has many functions: it is used to protect warrior from spears during fight, it is used as storage where money, tobacco and other objects could be stores (in the hollowed out handle) and it could be used as neck-support or headrest.

Figure: 134, Shulluk traditional calabash called Lual is used for milk.

Tribal Marks

In the old times and in 1960s tribal marks were very common among the Shulluk; these have been practiced by men, boys, women and girls. The tribal marks for men and for boys consists of two to five rows; and for women and girls it consists of one to three rows of small scares on their foreheads. These are in most or all cases simply caused by wearing bands which are drawn tightly across the forehead. However, tattooing on the parts of the body is not common; nevertheless, some men and women still use crude iron for tattooing dots into their skins. Interestingly, it has been revealed that in the past or present some individuals do not want these traditional marks on their foreheads.

Figure: 135, Shulluk Forehead Female and Male Marks

It is reported that women wear no or only short hair on the head; they love to shave their heads with razor blade or with any straight piece of thin iron, whose edge is sharpened and one side is beaten out to a think sharp edge. Customarily, Shulluk men and women do not like hair on their bodies; as a result they remove any hair on the body by pulling it out with a kind of pincers; men even put out their beard and eyelashes with pincers. Circumcisions are common among men but women and girls hardly circumcised; women and girls who today circumcised are usually from Muslim families.

Painting the body

The Shulluk are very particular about their body cleanliness, culturally, they used to paint their whole body; and to the natives this body painting is seen as beauty but in the eyes of the foreigners, body painting is seen as dirt and primitively. They paint their lower part with ashes and paint their head and breast with red earth or chalk –sometimes, they even rub their bodies with oil or butter. Sometimes the whole body is painted with white or red colour, with lines or figures drawn across the face. Thus, for this reason, Diedrick Westernmann (1912) says, *"What makes the Shulluk look most ugly and most frightful in the eyes of a newcomer is their habit of smearing the whole body"* (P.23).

They wear beads, and other ornaments which include cutting dots on the forehead and tattooing on the body.

Women paint large grain baskets, they also paint small fancy mats for covering food.

Thatching
Other important craft is that of the thatch-maker (see Shulluk building Figure 50, page 159). Thatch-makers make these thatches with neatness which really excites admiration.

Hair-Dressing
Among the young Shulluk men there are two ways of structuring the hair; one way is called hair-dressing and the other is called beaching-hair. For us to understand better let us discuss the two ways of hair structures separately. We are going to start our discussion with hair-dressing among the young Shulluk men. Hair-dressing is more practice by young men rather than adults. The young men dress their hair in different forms or shape; and to do this work from the start to finish it usually take many hours. Hair-dressing takes long process, first the hair is to be loosened with a stick, then it is twisted and brought to the right and desired shape. When the hair is brought to the right shape, the young man then put on it mixture of particular type of mud or cow-dung; and from time to time he pours some oil

Figure: 136, Hair dressing

or better on the hair. This is to ensure that hair remains soft. The hair is constructed into two structures that give the impression that the man has two plates on his head.

More importantly, once hair-dressing is completed, at night when the man is going to sleep he has to support his neck with what they called neck-support or neck-rest.

Once the hair-dressing is completed, it usually does not get damage very quickly; it normally takes weeks or months before it is damaged. Although young men renew their hair-dressing every time there is village-dance, so that they look handsome before girls. Because it is customarily that when there is village-dance, everybody must appear at his/her best. In most case young men were their hair in little knobs that is formed with red mud and fat. They also twist cowries-shells in strings and tie it on the hair. Truly, the young men are very fond of adorning their hair with ostrich feathers or feathers from other birds (Diedrick Westernmann, 1912, P.25).

The other way of structuring the hair is called bleached-hair or hair-bleaching. There are different ways that young men can bleach their hair. Different colours are used to dress the hair; sometimes they use to bleach the hair yellowish, red or grey depending on the needs of individuals. Bleaching is usually done by smearing a thick plaster of ashes, chalk or cow-dung on the hair and leaves it there for about two to three weeks. Yong men can also bleach their hair by rubbing the plaster ash well deep into the hair-root and later on gather the hair from the head and bring it forward to form the shape of a horn. This is to be done with great care for otherwise the shape of the horn will break; once the horn shape breaks, the young man will lose his hair as much of the hair is tied together, to form a shape of a horn.

Figure: 137, Hair dressing

The other way the young man could bleach his hair is by washing the hair continuously with the cow-urine until it become red and stiff. This method of bleaching hair often affects the skin on the head and makes hair to lose the black colour. Often, the young men with bleached hair are tall people with thin body covered with ashes and eyelashes. In many cases, foreigners are afraid and frightened with the appearances of these men.

Shulluk traditional Musical instruments
Shulluk musical instruments consist of large and small drums (bul), a kind of guitar, a stringed instrument called "tom", double horns and lyre -and many others to mark different occasions.

Drums are made from logs of wood; the wood is hollowed out and braced with skin. Drums are usually beaten with the flat hands or with sticks.

While the "tom" is made by splitting in the middle a small section of a long and hollowing out the flat side a little. A piece of raw hide is stretched wet over the "tom", and the flat side becomes the face of the instrument. A round stick some eighteen inches long is fastened at each end. (See Figure...). Some five holes are made across the top end of stick, the holes are large enough in diameter to allow the 18 inches piece of wood to pass through and the stick would still be quite strong. Another five holes are even made on the bottom-end of stick, where strings would be tied and allowed it (string) pass over small wooden bridge covered with shin. Each string is made up of tendons animals or the root park of certain plant and is then wound about across the 18 inches on the piece of wood. The strings are then tightened and the instrument is ready; it can be tuned by wetting these strings and tightening or loosening the strings around the crosspiece. They are tied to this stick, and by winding over themselves, keep them from slipping.

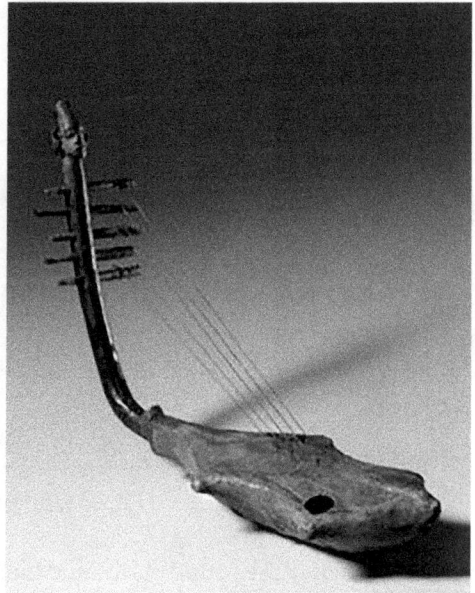

Figure: 138, Shulluk double horns Figure: 139, Shulluk tom

This musical instrument consists of double cow's or antelope's horns. It is also known as wind instrument, it is often used in wars and in war-dances.

Figure: 140, Shulluk man playing Lyre

Clothing and Ornaments

(1) Origin Dress

According to Diedrick Westernmann (1912), in the old days, Shulluk men and unmarried girls, like Dinka and Nuer, they used to walk naked. But the married women used to wear skins of sheep. Goat, calf, gazelle tanned with the hair on and worn with the hair side out; they tied the skin around the waist using one of the forelegs or the hind-legs of the animals. And when a woman walks, the tail and other two legs of skin sheep dangle and flag around her legs as the ornaments. Sometimes instead of wearing the white skin on the waist, women cut them and decorate with beads, brass, and iron-rings and tied them just to cover the front and it will appear as white apron underneath. Women in the old days also wear other skins on the upper part of the body; the skin is supported by tying the fore and hind legs on one side of the hips. The skin then slipped over the head and the other leg of the skin is tied and makes it rest on the right shoulder; while the remaining leg of the skin is passed under the left arm. This is what we call in the old days as "Shulluk woman full dress" (P. 25).

In addition, to make a full dress, women wear as many beads and other ornaments as she could afford. However, these would include strings of beads around the waist, neck, arms as well as brass. A woman could wear as many as ten to twelve brass and iron-rings, which are produced by blacksmith, and that weight several pounds; this ornament usually extend from the hand half way to the elbow. These are not normally loosened, but they are tightened to the flesh. Women and girls are sometimes dressed in cow, calf or antelope skins, which are either wrapped round the body, or hung over the shoulder.

However, the above mentioned ornaments are usually worn by young women, although elder women in addition wear iron-strings on the ankles. The iron-rings worn on the ankles are very heavy and often produce great scares on the flesh.

Similarly, in the old days, little girls used to wear apron only, and when they grow up and become a little bit older she would have to put animal skin on her shoulder; and when she is fully grown up, is then expected to wear skin on her waist. We have just been informed by Diedrick Westernmann (1912) that in the early days, the unmarried women in Shulluk society used to walked naked. The above narratives then demonstrate to us that when a girl reaches the marriage age, she has to abandon wearing skin on her waist and join the unmarried women who walk naked. In other way, as a grown up girl abandons wearing skin, this could be interpreted as the girl is now advertising herself to men. The nakedness and act of joining the unmarried group, is rather to tell the young men, who are looking for girls to marry, that *I am now ready for marriage, if you are looking for a girl to marry, you are well invited.*

It has been reported that dances in Shulluk land offer an opportunity for both boys and girls to advertise themselves. When the time of dance has come, both sexes would be richly dressed. The girls would wear a bundle of heavy iron-strings above the ankles and on the loins they wear large antelope or calf skins with bundle of ostrich egg shell chains suspending downward over the heeds. They also wear large bundle of blue or green beads on their necks, the beads would be hanging down on the breast or chest. While on the wrists, they wear bracelets of beads and around the forehead they wear string of beads. As we have discussed before, when there is dance young men always renew their hair-dress and stick ostrich feathers on it. This is then the right time to renew hair-dress.

During the dance, men wear strip of sheep or goat skin above their ankles; alternatively, they may tie a great skin above their ankles with the hair facing outwardly, and tie knee-bells, sheep or goat skin above their ankles. They also tie on the ankle number of metal bells each consisting in a hollow, oblong piece or iron, in which a small iron ball moves, so as men dance the iron-ball produce a ratting noise.

When people are going dancing, men also wear on their waists bracelets of brass/iron, in addition they wear iron-rings above the elbows; and above this, and they have to wear at least six to eight fold-rings of ambach. And around their necks they wear one or more necklaces of beads; on the head they fix ornament of horse-tail and if there is no horse-tail they could fix on their heads other appropriate animal tails with long hair and with ostrich feathers on their heads. More still men fasten strips of red or white bristles around their foreheads, whereas, each man takes into his hands two lances (which are being adorned with ostrich plumes), two clubs, and one club-shield. Many a time, men also wear skin cloth around their waist and these skin cloths are often adorned with small bells or iron-chains along the edges.

Figure: 141, Iron-balls for dance

Thus, prior to the dance meeting, men smear their faces, arms and upper part of the body with red earth and pour melted butter or oil over their bodies, after this they strip ashes over and make drawing all over the body.

(2) Modern Day Dressing
As we have discussed above, women and girls in the old days dressed in cows, calf or antelope skins, but with the coming of Missionaries in Upper Nile, women and girls started to wear cotton cloth either wrapped around the body or hung over the shoulders. Today if one goes to Malakal, he/she will hardly find a woman or a girl wearing skins. Women and girls of today wear cloths with different colours which measure about two feet long and eighteen inches wide. The cloths are usually made up of two thickness, tied with strings and fastened the two corners around the wait, just below the abdomen falling down to the knee.

Yet today, women and girls wear as many beads and other ornaments as they could afford; these include string of beads around the waist, neck, arms as well as brass. They also wear iron-rings which extended from the hand half way to the elbow. These items are not normally loose ornaments they are rather tightened to flesh. Today these ornaments do not leave marks on the skins because they are wound for a short period of time and removed and kept in a proper place for next occasions.

When going to dance, both boys and girls wear richly. The girls wear a bundle of light iron-rings above their ankles; they even wear large bundle of blue or green or mixed beads on their necks, leaving them hang down on the breasts. Girls even wear strings of beads on their foreheads.

In the twentieth century, men also experienced great change in the way they dress.

Today men wear trousers, shorts with cotton cloth, knotted over their shoulders and hanging down above their ankles. It is reported by some people that cotton cloth that men wear have the same measure with that women wear (i.e. 2 feet long and 18 inches wide.

Today, when going to dance, men still wear t5rousers or shorts with cotton cloth hanging over the shoulder or tied around the waist. Men keep Shulluk tradition by tying animal skins above their ankle, when going to dance. The skin is tied with the hair side facing outward. In addition, men tie bells above their knees as well as sheep or goat skins; they wear iron-rings above the elbow and on their necks they wear necklaces of beads and ostrich feathers on their heads. Therefore, in the modern society, Shulluk men are characterized by wearing adornments of thick and heavy bracelets, armlets of iron, brass, ivory, twisted ambach, cotton cloth and cowries-shells

Chapter 17:
Economy and Livelihoods

The Background to Northern Luo Economy and Livelihoods

The Northern Luo of South Sudan occupies approximately one third of the total areas of the Country. Land is a key factor of production among the people of South Sudan and access rights to these lands inhabited by the Northern Luo are a major flash point for conflict with the neighbouring, National and State Governments as well as with relatives. The people look at land as their only source of survival. Subsequently, limited economic opportunity and the need for survive drive many land disputes. Most land disputes among the Northern Luo with neighbouring affect some 40 precent landholders, and these disputes often resulted into violence.

Before we go ahead with the discussions of lands of the Northern Luo, and for us to have better insight of the land disputes, we would like to categorize the land disputes into two broad categories:-

The first land dispute is always between individuals and families – we could call this as family or individual disputes of the lands. The family or individual land disputes usually involve relatives, neighbours, family members or clans. This kind of land dispute often comes out as a result of lack of clear demarcation of the boundaries. Thus, the common types of these land disputes include disputes between Junior and Senior family members; disputes between widows and members of their late husbands' families since they may claim she has no right to the land entitlement; disputes about land grabbing by neighbouring families, villages, clan or tribe; disputes about land grabbing by government soldiers or National and State authorities; disputes about land could even come about following the selling of the land by individual or dominant tribe without the consent of the land owner (individual or community); disputes about land could also come between landowners and people we called squatters (like what happened at Palotaka in Acholi). Given the low population densities and minimal land pressure, almost anyone who is willing to clear and work on unused land has been welcomed by Religious leaders, lineage heads responsible for such land.

The second category of land disputes is between local people and private sector invest- ments or yet by National and State authorities. This type of land disputes are often related to lack of understanding from the government authorities of how much power they have on traditional owned lands. Such disputes could come about when the government autho- rities use their offices for personal interests. Experiences tell us that government autho- rities often make deal with the private investors where they are given large amount of money as commission after the success of the plan.

Here, I want to point out that the Northern Luo people are not ready to sell their lands to private investors or the outsiders because they know quite well that outsiders have high tendency of land grabbing besides their lack of transparency. Worse still, the .local communities and stakeholders are not always involved in negation over the land that the government authorities may want to give to private sector investors. This sentiment often leads to mistrust among the local population, outsiders and the government of the day.

In order to break this vicious cycle, there is need to develop a two way approach to address land disputes and foster market development through private sector investments. It is believed that if these two way approaches are developed, it will allow aid actors to strengthen markets on one hand and on the other hand it will enable civil society actors to sustainably deliver agricultural support services to the population, instead of depending on external assistance.

Interestingly, it is a common political game in Africa in general and in South Sudan in particular that the governments have tendencies to bar indigenous people from accessing natural resources from their lands. We should not forget that accessing natural resources by indigenous people is as important as the right to a healthy environment. The lives of the Northern Luo people, regardless where they live in South Sudan are closely linked with their environments. Thus, if the environment is not safe and conducive for them, then they cannot afford to continue living on their fathers' lands because they will be exposed to various environmental problems. UNDRIP (2007) pointed out that local community have the right to the conservation and protection of the environment and the produc- tive capacity of their lands or territories and resources. Therefore, this means that the government authorities at all level of the governments should establish and implement assistance programme to local communities for such conservation and protection, without discrimination in any form or shape.

With the above views in mind, every human being should embrace and understand that community's rights to land, territories and natural resources carry far reaching implication as we might think. Since access to these resources also implies access to food, health and a decent natural environment that allows for human development.

The Northern Luo of South Sudan has the rights to lands, territories and resources, which they have traditionally owned and occupied. These people have the right to use, develop and control their lands, territories and resources that they possess by reason of traditional ownership. The South Sudan Government should and must give legal recognition and protection of the lands owned by the Northern Luo in Equatoria, Bhar El Ghazal and Upper Nile Regions. Such recognition should be conducted with due respect to the Kings, Chiefs, customs, traditions and land tenure systems of the local people concerned. Without this recognition and protection of the Northern Luo ancestral lands, it is difficult to maintain the collective identity of the Northern Luo people.

Thus, if the South Sudan government of the day fails to recognise and protect the people and their lands, this will reduce cultivation, grassing land and loss of economic assets in the Luo society.

Agriculture and Sources of Food

Historically the Northern Luo people are either agriculturalists or agro-pastoralists. According to Muchomba (2004) the principal forms of physical capital in Northern Luo areas are cattle and cultivated lands. Households with cattle and farms are considered important persons in the communities because they are perceived as people with better lives, and they have enough cattle and cash to meet their needs, for example they could marry wives to their sons without any difficulty. Importantly, different levels of subsistence and resilience are often associated with different levels of capital accumulation. This means the more you have cattle or farms products the wealthier you are.

Looking back from independent of South Sudan, we could easily see that production is limited to the Northern Luo farmers because agricultural production equipment is largely limited to traditional wooden sticks or metal hand tools for land preparation and weeding. There was no any financial support to the farmers, for the reason known to the government.

The Northern Luo people conduct trades locally and across the borders. The household members travel long distance to access markets in Ethiopia, Uganda, Kenya, The Republic of Congo and Central Africa. According to Muchomba (2004) years of conflict in the country have undermined the social fabric and physical assets of the Northern Luo resident livelihood system, by limiting access to major markets and disrupted agriculture and livestock keeping. With the signing of Comprehensive Peace Agreement (CPA) in 2005, the Northern Luo people hoped that this would provide healthy environment for farmers to re-start farming to eradicate hunger and food insecurity first in the Luo regions and secondly across the South Sudan. The post referendum period as a matter of fact should actually be devoted to the rebuilding of the livelihoods of the people of South Sudan in general and of the livelihoods of the Northern Luo in particular. From the Northern Luo farmers the Shulluk, Acholi, Pari and Anyuak have high potential for the expansion of the commercial and nutritional value gain from fish in the Nile, Sobat, Kinaite, Atebi, Jur and

Ayii Rivers. In the past fish was export commodity in South Sudan and should now be recoverable with sustainable investment and the development of the industry.

As we shall see later in this chapter, some of the Northern Luo people grow both short and long term crops. Quick-growing crops with short cycle include sorghum, maize, okras, peas, beans and green vegetables are planted close to the homestead where they help to ease the effects of the hunger season. These crops contribute as much as 30-70 percent of annual food needs of the people.

Generally, the crops grown by different farmers in different localities of the Northern Luo include sorghum, maize, cassava, sesame, pumpkins, beans, millet, papaya, yams, okras, tobacco, coffee, cotton, cabbages, sweet potato, groundnuts, wheat, bananas, onions, pineapples, sugar cane, tomato and green vegetables. Sorghum is a drought tolerant crop and very important for food security in Africa in general and in South Sudan in particular. PNAS (2014) argues the earliest archaeological evidence revealed that sorghum that the Luo groups eat today started to be grown between the years 1900-8900 B.P. the seeds were excavated (together with the cattle bones) from a site close to the current border between Egypt and Sudan; eventually the livelihoods in the region transformed from hunting and gathering wild foods into agro-pastoralism. More importantly, sorghum cultivation in combination with the cattle herding was successful livelihood adaptation to the dry grassland ecology, and eventually, as the climate changed, the agro-pastoral adaptation spread over large parts of the central African steppes (P. 2). However, I still wonder where the sorghum that was found in the bone of the dead cattle came from? Alternatively, could we conclude that once upon a time there was sorghum o earth but at a certain time it was destroyed by cattle?

The recent development in sorghum diversity was the work of J.R. Harlan and others' work from the 1960s to 1980s. Diversity of sorghum types, varieties and races that we see today has been a result of the movement of the people, disruptive selection, geographic isolation, gene flow from wild to cultivated plants, and recombination of these types in different environments.

Household tends to cultivate two pieces of lands: One is near the homestead usually on upland and the other is farther away from homestead, some 4-20 kilometres from home, and usually on lower ground. This is usually done in order to reduce risks associated with either drought or floods. This means that when there is drought in a year, crops cultivated on lower ground would have low risk associated with the drought. In the same way, if in a year there is flooding, the crops cultivated near homestead on the upland would have low risk associated with the floods.

In contrast, flooding and recession of the waters often enables some households to cultivate two to three times in a year, as well as to benefit from ration cropping.

Drought, floods, pest and birds often affect agricultural productions, and these are the main constraints experience by the farmers among the Northern Luo of South Sudan. Muchomba (2004) argues that drought, floods, pest and birds infestations act as important constraints to agricultural production of the households across South Sudan.

General Sources of Income

The income for households among the Northern Luo varies from society to society and depending on the climate and type of soil they have. However, we can say the main income of the Northern Luo people include labours, livestock sales, fish and oil fish sales, local beer sales, coffee sales, Cassava sales, Sweet potato sales, tobacco sales, cotton sales and handicraft sales which comprise the major income-generation activities. Sorghum is grown by all Northern Luo Groups for foods and sales. Sorghum is a drought tolerant crop with a vital role in the livelihoods of millions of people in marginal areas of the Luo. In accordance to the account made by Product National Academic Science (PNAS) 2014, traditional seed-management practices played an important role in the survival and expansion of agro-pastoral groups in the past and the present. Thus, for the successful of the traditional sorghum, the developers should build on the sorghum seed system developed over the years, rather than seeking to replace the existing traditional seed system with the imported seed systems (P.1).

For the important of income activities among the Northern Luo, we are going to break down later in this chapter, which Luo group grow what and where. Trade in these commodities takes place in local markets, and in towns such as Torit, Juba, Wau, Pachalla, Malakal, Tonga, Renk and sometimes the Shulluk people go with their trading in as far as Kosti.

Readers should understand that livestock sales among Anyuak and Shulluk are a male dominated activity. Bulls are sold and the proceeds used to buy heifers and calves. These transactions are commonly barter-based, although cash-based transactions are becoming increasingly common. But the goat and sheep sales across Northern Luo people of South Sudan are generally in cash because they are cheaper as compare to cattle (Muchomba, (2004). The Northern Luo people are not found in one location or region in the South Sudan. The Acholi and Pari (Lokoro) are found in Equatoria region, the Anyuak, Shulluk, Mabaan and Jumjum are found in Upper Nile region and Balanda, Jur Chol (Luo Wau) and Shatt (Thuri) are found in Bhar El Ghazal. The rains in these areas are not evenly, some areas receive more rains while others receive less rains. As such farmers in the areas where there is less rains cope with unreliable weather. Given these facts, I would like to recommend in this book that the South Sudan Government and other International NGOs operating in the occupied Luo lands should establish a comprehensive training programme to address the climate challenges. Although this requirement may varies according to the farming practice. This implies that when farmers identify delays in rains as risk, they should be supported with the knowledge on alternative water supplies so that they are in position to use seeds suitable for a given condition, and how to planet, when to plant and where to plant. Therefore, there are needs to help Luo farmers develop and adapt better methods of farming through training and input provision.

Acholi Economy and Livelihoods

Writers, like Akena Ponse Otto (2016) argues that before the civil wars, the Acholi land was endowed by beautiful scenery, lush vegetation, rivers, games animals, birds and above all, abundant fertile land which has been communally owned, cultivated and handed down by family members to their descendants from generation to generation. Poverty, famine and drought are hardly heard in Acholi land because of conducive environment for agricultural productions, a traditional ownership of domestic animals and being hard working community.

The Acholi way of life in the old days, people stayed under the rain chiefs; as people stayed under them, each clan (*dog ot or gang*), their force was for the rain makers. The main work for the people at that time was cultivation and protection of the village with its inhabitants. The Acholi people do not drink too much beer as do other tribes in South Sudan. Elders drink beer and young people do not drink because they are said to be too small to drink beer. The Acholi people believe that beer can destroy the good life of the young people.

According to soils and topography of the Acholi land by groups of elders (1974-1986) the soils of Acholi of South Sudan have been classified as :-

- ▶ Soils of high productivity - 45% of the land area
- ▶ Soils of medium productivity - 30% of the land area
- ▶ Soils of fair productivity - 14% of the land area
- ▶ Soils of low productivity - 8% of the land area
- ▶ Soils of negligible productivity - 3% of the land area

Geographically, the Acholi land lies on the western slopes of Imatong Mountains. The farmers divide the seasons into two: dry season and wet season. Dry season is shorter and wet or rainy season is longer. The rainfall in Acholi of South Sudan can be categorised into high and moderate. The high rainfalls are found only in Obbo, Palwar covering up to Upper and Lower Talanga. These areas receive annual rainfall averages 1,200- 1,750 mm, which is 30% of total Acholi land. Moreover, the moderate rainfalls are found in Pajok, Magwi, Panyikwara, Agoro, Omeo and Ofirika, which is 70% of total Acholi land. These areas receive under 1,200 mm of rains per annual. The distributions of rainfalls in the land of the Acholi of South Sudan dictate and restrict range of crops that can be grown as well as qualities of products. In the past and the presence, the Acholi grandfathers predicted rains through cloud formations; direction of winds and when certain trees begin to grow leaves –all these are the signs of rains.

The environment around the land has influenced Acholi lifestyle and economy. The people show themselves both innovative and adaptive in their farming practices over the last seventy to ninety years. Their innovation of various crops varies from place to place especially the land of Obbo and Palwar is good for the cultivation of maize, coffee and bananas, cassavas, pineapples and sorghum; the land of Pajok is good for cultivation of simsim, sorghum, and cotton. The land of Panyikwara is good for cultivation of cassavas, sorghum, cotton and tobacco. The land of Magwi, Agoro, Omeo and Ofirika is good for the cultivation of millets, tobacco, and sorghum. However, it should be noted that cotton, simsim, sorghum and tobacco are grown across Acholi land in both small and large scales. There is potential in minerals like gold and chromite in River Kit.

Traditional land rights are vested in localized patrilineal lineages, which are predominant under the control and guidance of lineage heads and village elders. According to the Acholi customs, an individual male can claim the land of his forefathers and use it for faming for his family members. Such rights pass from generation to generation and no one is expected to interfere with the practices.

The most important economic activity of the people of Acholi of South Sudan is crop cultivation. Almost the entire population of Acholi-South Sudan region produces sorghum, millets and cassava for sales. To the Acholi sorghum, millet and maize are staple foods.

It is been reported that the Acholi have had series of the development of farming systems between 1910 and 1974. Crops such as maize, cotton, tobacco and coffee were introduced into the land between the years 1940-1960. Cotton production increased rapidly in four main villages in Acholi and these include Obbo, Pajok, Magwi and Panyikwara.

Although there are no much quantitative data available for the earliest date, nevertheless, there are descriptions of the crops grown and farming system people used as we shall see later in this article.

Land was normally opened to finger millet, which was and is the dominant crop in Acholi areas. The millet was and is still being interplant with pigeon pea. The Acholi people of one of the luckiest societies in the South Sudan, they have fertile land that they traditionally clear and hoe the new land for each farmer in groups in the exchange of local beer. The ox ploughs was introduced in Acholi land by Norwegian Church Aid (N.C.A.) in 1974 soon after the Addis Ababa Agreement, and following the repatriation of refugees from neighbouring countries.

It is worth mentioning that the farmers were slow to adapt the plough and this probably partly because they do not keep cattle; and secondly, even though many famers might have interest in keeping some imported bulls; the tsetse fly limited the number of cattle people intended to keep in the county.

All-in-all, the people grow millet, sorghum, sesame, cassava, sweet potato, beans, bananas, yams, wheat, groundnuts, cabbages and green vegetables for food. In addition they grow cotton, coffee, maize and tobacco for cash crops. We can say without hesitation that poverty, famine and drought are hardly experienced in Acholi land because of the conducive environment for agricultural production. In other words, we could say the Acholi people are self-sufficient as compare to their neighbouring. The Acholi people are predominantly agriculturalists producing subsistence farming alongside animal husbandry as a major source of livelihood.

Crops Rotation

After the introduction of cottons farmers soon found that it grew very well on newly opened land; the more the farmers keep the cotton clean from weeds the more the crop provide suitable seedbed for other crops such as millet. Eventually, cotton became the first crop in the rotation.

Figure: 142, Acholi Cotton Field

When going through the process of rotation, in the year farmers plant cotton in the newly opened land; when the cotton is harvested, in the second year farmers plant millet. And after the harvest of the millet, farmers often choose to plant cotton again on the land, that would be followed by sorghum, sweet potato; groundnuts interplant with cassava and cotton is planted again. This is followed with cassava and sorghum as the last crop in the same land before the land is given a rest period. Crop rotation has become increasingly important in the recent years. According to Acholi informants, fertilizers are never required in the land in as long as the Acholi farmers continue to practice crop rotation.

However, some farmers instead would preferred to plant groundnuts as the first crop in the newly opened land, instead of planting cotton as we have just discussed; and this will be followed by cotton, sorghum, again groundnut interplant with cassava. When the groundnuts are harvested, it gives way for fresh air for cassava to grow. Although data for crops rotation , in Acholi, land, are not available, nevertheless, there are some indications that between the year 1923-1969 farmers introduced four entirely newer crops which include groundnuts, maize, sweet potato and cassava as food crops.

Land Usage

It is very unfortunate that the Acholi have not fully utilized their fertile land due to recurrent civil wars in the country. Moreover, the land has capacity to grow enough maize, sesame, cassava, sweet potato, tomato, beans, groundnuts, yams, cabbages and green vegetables to feed the population in Capital Juba and Torit town. The farmers lack money for commercial farming. This is a great challenge to South Sudan Government as they failed to consider providing security and rather thinking negatively that farmers might be affected by climate change and would not be able to reimburse the loan (should it be given to them). Above all, almost all the government officials, who have access to country finance, are engaged in personal development that has nothing to do with the general population. The South Sudan Government should review getting loans to farmers to commercialize their farming.

Although agricultural products have been interrupted by civil wars, yet during this period of running up and down, the Acholiland maintains its beautiful environment, high vegetation, rivers, games, animals and birds.

As I mentioned previously, tobacco, coffee, cotton and maize are the major cash crops grown by the inhabitants. However, agricultural production is made difficult for the people to meet their dreams because the areas has no well linked network of roads; the existing roads are poorly built and passable only during the dry season; in the raining season roads in Acholi area are almost impassable. This makes the agricultural products not to meet the market demands in Capital Juba and Torit town. The suppliers are lower and the demands are higher. This is one of the reasons why the Government of South Sudan resorted to imported food that could be grown locally from Uganda. Furthermore, there are no marketing infrastructures such as processing and storing facilities in the county.

Trade

Before the colonization, the Acholi of South Sudan focused mainly on obtaining iron-ore and finished iron products in exchange of products of the farm or herd. At that time nobody think of underground minerals, as such iron-ore deposits were located beyond the western, northern, southern and eastern boundaries of what we call Acholiland today. Iron trade created networks of movement and interaction that helped determine a collective identity within these boundaries. During the late nineteen century, the Acholi became active in the international trade in ivory and slaves; these were used only for exchange for cattle, beads, blankets, cloth and firearms. Soon after the colonisation by the British, the colonial rule brought the penetration of a money economy into Acholiland. The people were able to trade in a money economy with the neighbouring such as Acholi of Uganda, Lotuko in Torit town, Madi in Nimule and Pageri, Bari in Juba and Pari in Lafon Mountain. Today there are number of trading centres in Acholi area and the common place you can find trading centre today are in Obbo, Pajok, Palwar, Magwi and Abara (Panyikwara), where a range of local and imported goods can be found.

Acholi people are great hunters, they hunt either as a group or alone as trappers; the used nets, pits or they hunt the animals into the water and subsequently kill them with their spears.

Anyuak Economy and Livelihoods

The Anyuak are a Nilotic people who live in the Upper Nile Region of South Sudan, they live in the land that lies in the plains below the Ethiopia highlands and in areas bordering South Sudan and Ethiopia. The people live mainly along the tributaries of the Sobat River, which is itself a tributary of the White Nile.

The land has the characteristics of marsh land, rich savannah forest and grassland with rainfall of about 800mm. This has tremendous influence on the economy and lifestyle of the Anyuak people. However, some of the Anyuak people live in lowlands of the Gambella Region in Ethiopia; they are divided by the international border.

Anyuak land lies at altitude between 400m and 500m; four major rivers flow westwards through Anyuakland. They are, from the north, the Baro River (called locally as Upeeno), Giilo River, Akobo River and Oboth River. These four rivers merged and becomes the Sobat River. Many Anyuak people build their villages along these rivers –and many of them located on the very bank of the river (see Figure 34, Page 84)

The eastern part of Anyuak is covered with a dense forest, which marks the western edge of the forest extending from the highlands. The annual rainfall is highest in the eastern part of Anyuakland and going westward the rainfall decreases gradually. The middle part of the land is Savannah woodland and the western part is Savannah grassland subject to inundation (flood) during the rainy season. Therefore, the Anyuak subsistence economy is deeply affected by these ecological differences (Eisei Kurimoto, 1996).

More importantly, the Anyuak classified their natural environment into three major categories according to the vegetation. The first division consists of the environment that runs from west to east, and is known as grassland of seasonal swamps (*bap*); the second division consists of woodland (*wok*) and the last natural environment consists of forest (*lul*). In this forest, trees are sometimes more than 20m high.

The patterns of Anyuak subsistence economy differ considerably in grassland (bap), wooden land (*wok*) and forest (*lul*). The grassland is not very suitable for cultivation, as most of the land is subject to flooding for some months during the rainy season. Cultivated fields are found only in slightly elevated hinterland (*doodo*)) where homesteads are constructed. Hunting, pastoralist and fishing are more important in grassland than in any other areas of Anyuak.

The western part of Anyuakland is flat grassland Savannah flooded during the rainy season. Tress are scare, the soil in grassland is called *"Ukuur"*; it is black clay soil or cotton soil, which becomes very muddy when wet but become hard and cracked during the dry season. However, in the grassland we can find elevated land, which is not flooded and this land is called *"Doodo"*; some people also called it *Thuurr* or *burr"*. The black soil found in this elevated grassland is also call *"Doodo"*.

The weather also varies from land to land or from area to area, for example it is much cooler in the forest than in grassland and woodland. It should be noted that the difference in altitude in Anyuakland is not great. However, many Anyuak people live in grassland. In woodland, farmers cultivate in two different places, some cultivate in the river bank (*bat-nam*) and other cultivate in hinterland (Doodo). The river bank where people built their houses is more suitable for cultivation due to the fertility of the soil and the constant supply of river water, which also renews soil every year by depositing organic matter (Eisei Kurimoto, 1996).

Crops and Cultivation

The Anyuak people are predominantly subsistence agriculturalists, they cultivate sorghum (*bel*), maize (*abac*), sesame (*nyimmo*), tobacco (*thaba*), millet, yam (*bath*), sweet potato (*ajwalle*), cowpeas (*ngoori*), greengram (*ugodi*), groundnut (*apuuli*), pumpkin (*ukonne*), tomato (*atimatimi*), paw-paw (*ulile*), banana (*bale*), mango (*mange*), sugar-cane (*thukeere*), lime (*lemune*), cassava (*ababure*), cotton (*waare*) and gourd (*keene*). It should be noted that not all the above mentioned crops are gron in all parts of Anyuakland; some of the crops are found only in a part of the land. For example, groundnut (athilam) and bambara-groundnut are cultivated only in the forest area (lul), yam is also grown commonly in the forest. While taro (upeela) is only planted on the riverbank in the grassland area (*wok*), taro is not found in the forest (Eisei Kurimoto, 1996).

In Anyuak land, there are many varieties of sorghum, one is called *"gaanga"* with a whitish and compact ear, and this is the most common varity. The other sorghum is called *"abworri"* with a red and compact ear; it is a quick maturing variety. Thus, in the beginning of the rainy season, farmer plant sorghum called *"abworri"* first, they plant *"abworri"* in small garden around the homestead (*atok-oto*). Sorghum called *"abworri"* never planted in the big field. Yet, another sorghum is called *"aburi"*, this is also a quick maturing variety, it is palnted after the *"abworri"*.

We have discovered that there two verities of groundnuts; one variety is called *"apuuli-gaala"* (which means groundnut brought into the land by the white-men); and the other variety of groundnut is called *"apuuli-Anywaa"*, which means the original groundnut of

Anyuak people. The different between the two groundnuts is that the *"apuuli-Anywaa"* is creeping and *"apuuli-gaala"* grows erect.

The Anyuak have three varieties of maize, one is called *"aba-gaala*, which means maize of the white-men; the other is called *"aba-Anywaa*, which means the original maize of Anyuak people; it matures very quickly, taking about three months, and has a short ear with yellowish and round grains. The Anywaa maize and sorghum seem to be resistant to floods because they are not spoilt when the field is covered wioth water for a few weeks. The crops are presumably adapted to the wet ecological condition.

The people of Anyuak prefer "aba-anywaa" to other verities because of its sweet taste; and the last variety is called *"amerika"*, the name suggested that this maize was introduced into the region by American Missionaries. The *"America"* variety of groundnut is the most recently introduced variety around the 1950s; it has a short ear with whitish grains.

Interestingly, Evans-Pritchard (1940a; 1947) argues that the economic life of Anyuak people centres on the cultivation of millet, maize, and other crops. Although their vocabulary gives evidence of a pastoral past, they are a sedentary people today with little interests in cattle or livestock. We understand that although Evans-Pritchard (1940a) says Anyuak have little interest in keeping cattle, nevertheless, today we find a good number of the Anyuak raise cattle, goats, sheep and fowls, which are used for trade and sacrifices to the spirits of the ancestors. Yet, the animals are not of great economic importance. Eisei Kurimoto (1996) pointed out that Cattle, goats and sheep are kept at villages during the rainy season and transferred to the riverine area in the dry season for water and pasture. He (Kurimoto) observed that unlike other western Nilotic speakers such as the Nuer and Dinka, who are economically and culturally depend on animals, the Anyuak people do not have many domestic animals. In fact, in the eastern and middle part of Anyuakland, there is no domestic animal at all because of the tsetse flies.

Most of the work in the subsistence economy is carried out by men. The Anyuak famers who live in the forestland, the hinterland and grassland practice shifting or slash-and-burn cultivation. This field under shifting cultivation is called *"angota"*. The term is driven from the Anyuak verb *"ngot"*, which means "to cut", so, because the farmers are cutting trees and shrubs to prepare the field for cultivation, then people called the slashed-and-burn field as *"angota"*. The shifting cultivation in forest is basically the same as in the woodland and grassland.

In the forest, agricultural work starts in the month of February, this is the time when trees and shrubs in the field are cleared and burnt. The actual planting of sorghum and maize in the gardens around the houses starts in the month of May, and this garden is called *"atok-oto"*, meaning a garden behind the house. The crops planted in the gardens, behind houses, are of quick-maturing types. The field in the forest land usually give one yield of harvest in a year. But the Anyuak generally believed that the red soil (lwala) of forest is so fertile that it produces enough food for a year. The main crops cultivated in the forestland include cowpeas, and green gram, sweet potato, groundnuts and "bambara groundnut". Kurimoto (1996) observed that each of the crops is planted in its own small plot in a field. Weeding (aluudi) is done twice and sometimes accompanied by thinning –this is called *"doi"* (P.43), which the Acholi called *"doo"*. A field in the forest (*lul*) may be continuously cultivated for more than ten years without any manuring or crop rotation. When the land is usually abandoned, the Anyuak say, it is not because the soil is exhausted but rather it is because of weeds and termites. It is right across Africa that the more you cultivated in the same land, the more the weeds increase and the more the field become the world of termites.

The abandoned field in the forest, woodland or grassland become known locally as *"kang"*, or as the Acholi called it *"okang"*. The land thgen becomes covered with tall grass and trees and after a short time, the *"kang"* or *"okang"* become woodland (*wok*).

However, there is more cultivation on the riverbank that has grassland; the farmers may have two harvests in a year if there is adequate rainfall than other part of the land in Anyuak land. The main crops cultivated in forestland include yam, sweet potato, ground-nut and *"bambara* groundnut". The advantage of the field on the riverbank is that it can be cultivated during the dry season when rain is scare, because the soil contains enough water.

In woodland (Wok) agricultural production is mainly obtained by riverbank cultivation and not by slash-and-burnt as we have discussed in the case of farming in the forest. In the flat plains of woodland, the winding rivers become inundated in July through September, turning the riverine area into swamp (*bap*). Seasonal changes in the water level are very drastic, thus, Anyuak riverbank cultivation is an adaptation to this ecological condition. Soils on the riverbank renewed every year by organic matter in river water deposited when the fields are covered with water for some weeks. In the riverbank, in theory, there is no problem of exhausting soil fertility. The filed may be continuously cultivated forever unless the river changes its course (Kurimoto, 1996, P.44).

And the main crops grown in wooden land (wok) include cowpeas and green gram, but most of the crops we have mentioned before are grown in grassland area (riverbanks).

The men use simple agricultural tools, which consist of hoe (*kweri*) and machete (*gajira* or panga) {the cultivation by hoe in Anyuak language is called *"tong"*}; at this point it is should be noted that a hoe and machete are used for many purpose, beside agriculture. Anyuak hoe has a broad iron blade and a wooden shaft measuring approximately 1.5 long. The hoe is used by men to clear the fields before planting or sowing; it is also used for weeding, harvesting root crops and groundnut; even though people use hoe for cultivation it does not dig deep into the soil, it usually crates the surface of the soil. While the machete is used for cutting down the stalks of maize and sorghum after the harvest; it is also used to clear grass and cut smaller trees.

It is also important to notice that the hoe used for weeding is smaller than the hoe used for clearing the field. The hoe used for weeding has a short shaft about 0.75m long. It should also be noted that the angle between a hoe's blade and its shaft is 180 degree. Farmers use it in standing or kneeling position. There are many agricultural tools that farmers use, these include axe (*lei*), which is used for cutting big tree; a digging stick (*apiidhi*), which is used for sowing sorghum; however, when sowing maize, farmers use rather hoe; the other tool is knife (*cakin*), which is used to cut ears of sorghum during harvest; there is also a kind of stick knife (*kulu*), it is used to open and remove the husks of ears of maize.

After the men have completed harvesting, the work of processing the food is left to women. This means women transport crops from the fields to home on their heads; they also do threshing with "*abaiya*" a flat and heavy stick and grinding sorghum and maize beside brewing local beer and cooking for members of families. A widow is expected to do men's work as well as women work, she cultivates the field of her dead husband, harvest the crops, transport the crops home, she trashes and grind sorghum or maize and cook for her children and brother-in-law if applicable. Crops are harvested once a year and this is during the month of August and September.

The population in Pachalla County predominantly rely on maize and sorghum as their staple food, which is obtained from the market or own harvest. Kurimoto (1996) observed that maize has overtaken sorghum as the staple food. He argues that people change to maize because they found that maize require much less labour in scaring away birds. As such the children who used to work in sorghum fields to scare birds now found opportunity to enrol in the schools.

However, food security indices have generally been worrisome. These structural factors are exacerbated by the occurrence of multiple hazards such as unnecessary long dry season that usually last from January to April, pest (mainly birds and domestic animals) among others. These destroy several crops mainly maize and sorghum. As a result food prices become inevitably high; occurrences of food shortages become the order of the day. Consequently, every year there are expectations of reduced cultivation because of hazards, insecurity and seed scarcity.

The water conditions and milk access remain good during the month of January to April, which is the dry season. The land receives above normal rainfall in August and September. People generally experience hunger during the months of May through August. This period is known as hunger season (Mawa Isaac J., 2014),

There are two planting/harvesting seasons that is April to August and September to December. All households cultivate mainly around the settlement and the main crops grown are maize, sorghum, cassava, sweet potato, sesame, papaya, cowpeas, sugarcane and pumpkin. Rainfall continues favourably throughout the growing season. Normally the rain starts in the area in April every year. The Anyuak farmers grow sorghum both as short and long term crop followed by maize. The main harvest is normally done between

Augusts through October but can sometimes stretch into November if rains and planting are delayed. The farmers re-grow sorghum stalks after the first harvest and this often occur with the farmers in the eastern part of Phou and Bieh areas, and the western land areas of Latjor. This second harvest from regrown crops (Maize and sorghum) is called *"upaani"*; this normally happen during the month of January, when rain is in abundant. In case of sorghum, it grows naturally from shoots after the first harvest; but in case of maize, there is need to sow maize seeds for *"Upaani"*. However, the second harvest is usually less in amount as compare to the first harvest.

Cassava is of significance in Pachalla County, which is more agriculturally productive because planting is done twice a year. Other crops grown include sesame, pumpkin, beans, millet and some root crops. Crop productions take place in lowlands, highlands and plateaus in order to mitigate against frequent floods and drought (Muchomba, 2004).

In the past, the people were involved in slave raids against their neighbours and exchange slaves for guns from the people in the highlands in Ethiopia. Persistent Murle cattle raiding remain a great threat to stability and good security in the region. These attacks and insecurity will only cease if the South Sudan Government carries out extensive disarmament of both the Murle and Anyuak pastoralists. Resent updates indicates revealed that tensions persist over increased attacks and cattle raiding from the Murle of Pibor County, they also e4xtended their raids across the border into Ethiopia, where they attacked Anyuak and Nuer tribes in April 2007. The attacks often hampered dry season movements, trades in the eastern part of the Eastern Flood Plains Zone and taking cattle to key grazing areas.

The Anyuak youth pan and extract gold nuggets and dust from the streams that drain the western Ethiopian highlands near Dima and Maji. The gold extracted is used for trade with Ethiopian highlanders or exchange for dimuy (beads), used for settling marriages.

It has been reported by Sudan Peace and Education Development Programme (2014) that ninety percent of households in Pachalla County depend on crop farming, with few on animal husbandry, fishing or forestry for their livelihoods. Productivity across all these sectors is still minimal due to factors we are going to discuss. In Pachalla County, each family depends on his own crop production. There is very little food products in the market, due to recurring civil wars and raiding by Murle. Therefore, family food security and nutrition depends on the resources base that household has and his ability, skills and experience to engage in various food and income activities options.

The Anyuak land has a huge potential in keeping wildlife especially game such as elephants, buffaloes and many others. The animal migration of the white-ear cobs pass through Anyuak land, which become a yearly source of proteins but it has also a huge potential for tourist attraction. In addition, there is potential for exploiting the Shea nuts, gum Arabic and *lalob*, which are in abandon in the forests.

As we have discussed above and under the Northern Luo Groups, the Anyuak people are divided into two groups, some of them are in South Sudan and others are in Ethiopia bordering each other. Whether these groups are in South Sudan or in Ethiopia, nevertheless, the Anyuak have strong traditional and political ties with each other. The two groups have their lands either in South Sudan or in Ethiopia; they have central rights to these lands, and nobody can take away from them *"The right to lands, territories and natural resources is one of the central rights for indigenous people"* UNDRIP ,2007. P. 10).

However, many indigenous people across south Sudan have land problems, where people find it extremely difficult to survive as people and to exercise other fundamental groups' rights. Some of the actors (including many authorities from South Sudan government) fail to understand that indigenous people have the right to maintain and strengthen their distinctive spiritual relationship with their traditionally owned lands, territories, waters and other resources and to uphold their responsibilities to future generations in this regard. According to Chavez (2002) majority of the Anyuak people have strong ties with the lands, territories and natural resources for their survival.

Wild Food Gathering

Besides agricultural products, the Anyuak people also collect wild food from the surrounding bush. According to Anyuak customs collection of wild foods is the responsibility of the women. Many edible plants are collected on the river banks and riverine areas. Today, gathering edible wild plants still plays a significant role in Anyuak subsistence economy. The wild food of the Anyuak can be classified into three categories: (1) herbs and leaves cook in soup, (2) fruits eaten as snacks by children and (3) tubers, fruits and seeds cook as a substitute for staple food. It is commonly spoken that when there is famine the people collect more wild plants to eat.

Household Food Consumption Pattern

Before the civil wars that broke out in 1983, the people of Anyuak have had varieties and enough food to eat in the family, but this was interrupted by all these wars. However, it was hoped that with the independent of South Sudan in 2011 live of the people would go back to normal. The meaningless war that broke out in December 2013 was another blow to the society. The study conducted by Sudan Peace and Education development Programme (SPEDP) in 2014 revealed that there is "poor food consumption patterns among the people of Pachalla County. The study showed that most households rarely consume vegetables and other micronutrient rich food. It has also been observed by SPEDP that households have low vegetable production across the County due to lack of/or inadequate vegetables productions, skills, and scare seeds in almost all the patterns when there is poor harvest.

Majority of the households spend comparable proportions on food. However, the expenditure on cereals is always high when the harvest failed. Meanwhile, there is low expenditure on cereals during the good harvest since it is the staple food in the area.

Fishing Industry

Fishing, as an industry, represents supplementary source of livelihoods next to farming in Pachalla County. In forestland fishing is carried out in small streams but with little catch. In Anyuakland fishing is carried out in both the dry season and rainy season and is an important means for supplying of proteins in people diet. It is said that little fish are caught in the rainy season, but much more fish are caught in dry season.

When people want to go fishing, they first receive invocation (lam) from the owner of the land (*wa-ngommi*). It is believed that land is in heritage together with the fish, because the first ancestor who settled in today Anyuak land initiated and organized fishing. So the owner of the land make invocation so that may fish might be caught; and when fishermen return home they must give him (wa-ngommi) a special share of the catch. According to Kurimoto (1996) this system does no longer exist, it was abolished during the socialist regime during the rule of President Jafar Mohhamed Nimeiri; the former President of Republic of Sudan.

Customarily, fishing is not done during rainy season because at this time there is high volume of water and the water would be running at high speed making fishing difficult. As a result, majority of fishermen preferred fishing in dry season, when the water levels have reduced considerably and this is also the time when fish return from tributaries, flooded plains and pools to the main streams. So fishermen in villages take advantage of this season migration.

Nevertheless, few fishermen go fishing in the rainy season because at this time, fish go with the flows of water into gathered flooded water and some streams or pools. This usually happen during the month of October and November. At this time fish seem to be in reasonable number and the number of fish species is relatively high. In Anyuakland, all types of fish found in the rivers are edible, with exception of swellfish (*tetradon fahaka*) locally known as "*apudo*"; however, it is reported that some people eat it although many people do not eat because of its bad smell (Eisei Kurimoto, 1996, P. 38).

More importantly, there are many methods of fishing among the Anyuak people. I cannot mention them all in this book, if you are interested in more fishing methods of the Anyuak; you need to do more research.

These methods of fishing include weirs (*keek*), which is made of logs and are constructed across the river in order to capture the fish returning to the main streams or fish returning down streams. The other method of fishing is by using fish-baskets or traps (*rwok* or *dipaw*) which are set in a weir. Sometimes the shelf is attached to the downstream side of a weir so that fish, moving near the water surface, may jump into it. However, it is reported that using fishing methods called "Mai" and weirs (*keek*), are often done collectively and never individually.

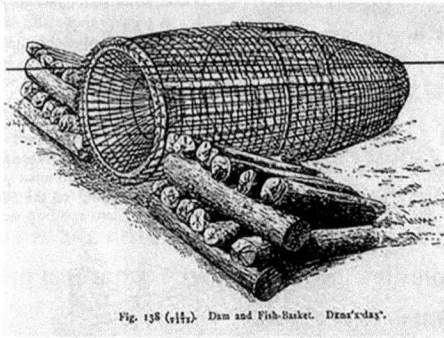

Figure: 143, Fish-baskets

It is reported that fishermen find it difficult to fish in Baro River and Giilo Rivers because they are too large and weir cannot be constructed across them. Nevertheless, at the Baro River experience tells us that a horse-shoe-shape fence called "*diemma*" often constructed across the river, this "*diemma*' is usually constructed with opened mouth upstream. "Many people participate in *diemma* fishing by beating the water surface with sticks to drive fish into the fence. As the fish enter the fence, they are then killed with fishing-spears and sticks.

The other methods of fishing is called "*Mai*" –which means to search or grope for something in dark. As the name suggests, in "*Mai*", men try to catch invisible fish in the water. Furthermore, it has been observed that hundreds of men from the villages also do fish using fishing-spears (*bidhi*), some use harpoons (*aroc*) and other use hand hooks (*goolo*).

Figure: 144, Anyuak boys fishing with fish-net

One of my informants told me that during fishing, men are not left alone, women join them with conical shaped baskets (*thwoj*).There are two types of harpoons one is with a slightly curved shaft and it is called *"aroc"*, the other is with a straight shaft and it is called *"ubeec"*. The *"ubeec"* harpoons are used in individual fishing. According to E. Kurimoto (1996) when fishing with the harpoon, the fisherman takes the head of a harpoon and a hand hook (which are detachable) he tied them to shafts with rope" (P. 37). There is also individual fishing, in this method a man fishes by line and hook (See Figure: 137 above) or by net and basket. A hook (*ubith*) and line is sometimes used without a rod. It is more a form of a play for boys than a means of catching fish.

In Anyuak society, people also used a rectangular shaped net (*ajap*) it is set between two poles erected in the river near its bank. This rectangular-shaped net could be measuring up to about 7m long and 1.5m wide.

There are two types of baskets used by women to catch fish. One is called *"ulitu"*, it is flat basket and sometimes it is elliptical colander whose diameter is about 80cm. It is put into the water and then pulls out with both hands. What the Acholi called *"ogwaa"*. The other type of basket is either a conical or hemi spherical basket with both sides open. This spherical basket is the place into the water with hope that it will catch some fish.

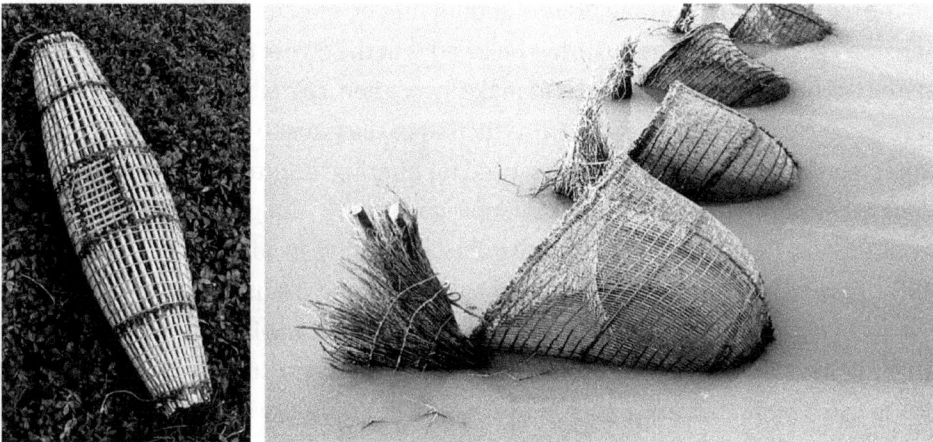

Figure: 145, Fishing Baskets

If fish are caught, they are taken out from the upper smaller mouth. These types of baskets are often used in catching fish from the pools.

More importantly, there are many methods of fishing among the Anyuak people. I cannot mention them all in this book, if you are interested in more fishing methods of the Anyuak; you need to do more research.

There are two ways of smoking fish, one way is smoking it on a wooden shelf (*pem*), and this smoked fish is called *"ugana"*. The other way of smoking is simply by putting the fish close to the fire; and this type of smoked fish is called *"atolla"*. If meat is also smoked on a wooden shelf (*pem*) that dry meat would also be called *"ugana"*.

It is very unfortunate that the recurring civil wars in South Sudan have interrupted with fishing activities in Anyuak land. As a result the fishing activities remain at relatively low scale; the wars have robbed people from fishing equipment and fishing gears. Therefore, this calls for the need to rehabilitate, strengthen and build local capacity of fishermen through training of modern fishing methods and supply of assorted fishing equipment to enhance sustainable livelihoods that substitute people's dietary requirements (Mawa Isaac J., 2014).

Hunting

Hunting in Anyuak language is called "*dwaar*"; the people hunt and eat most of the meats of wild animals. They carry out hunting in the woodland and grassland during the dry season when grass is burnt, and when wild animals migrate to the riverine areas in search of water and pasture. The common animals eaten by Anyuak people are buffalo, giraffe, hartebeest, topi, kudu, bush-buck, reed-buck, white-ear kob, gazelle, duiker, warthog, bush-pig, and cane-rat (Eisei Kurimoto, 1996). There are also wild animals like elephants and pigs in the forest, but the amount of game is small. The Nilotic are good hunters in the open, and not in the forest.

In the old days, each hunting ground should first be blessed by the owner of the land before men go hunting in the bush. It is believed that the owner of the land is the one who initiated the hunt in the area, so he is to make invocation, so that nobody might be caught or harmed by animal. But since the introduction of the guns to the Anyuak at the end of the nineteenth century, they have been used for hunting, eventually, collective hunts are no longer held as they used in the past. People go hunting with guns either individually or in small groups. Two types of traps are also used in hunting, in addition to spears and guns. There are two types of traps: one is called snares (*abiep*) and the other is called gravity traps (*akumma* or *akupa*). Snares are used near the village to catch wild cats, to scare wild cats or to minimise the number of wild cats from eating chickens; snares are also used in the fields to catch birds that eat grains. However, gravity traps are commonly found among the people who live in the forest, they are used to catch genets and mongooses.

Interestingly, Anyuak people hunt elephant mainly for ivory; the guns are used to kill elephants for ivory as well; most people do not eat elephant meat. However, it is said that people who live in forest do eat elephant meat, but the people who live in grassland never eat elephant meat. During the hunt, a large number of men with spears from different villages encircle a hunting ground, close in on the centre, and spear any animal they come across. Today, few big animals such as elephants are found in Anyuakland, it is believed that most of the elephants were destroyed by soldiers of the Sudan People's Liberation Army (SPLA), and this happened when the SPLA set up their headquarters and training camps in Anyuak.

Moreover, animals such as monkeys, foxes, hyenas, lions, and leopards are also found in Anyuakland but people do not eat their meats; although people eat water lizards and tortoises.

On the other hand, beekeeping is rarely practiced in other areas of the County, but it is more common in the forest. Hollowed out logs (*bong-ngo*) are usually set on the tress to make hives; honey is made into mead.

Blacksmith (Remobo)
In Anyuak society the black smoths produce knife (pala), hoe (kweri) and axe (lii) for home use and cultivation.

Balanda Economy and Livelihoods

Crop Cultivation
The land of the Balanda people is low-lying plan dotted with isolated hills perennial streams, which are mainly found in Sue (Jur) and Bo Rivers. The area has tropical climate, the annual rainfall is enough to support a vegetation of thick woodland with tall grass.

The people of Balanda are predominantly agrarian engaging in subsistence agriculture, they also keep sheep, goats and fowl. Their main crops are sorghum, maize, beans, sweet potatoes, cassava and sesame.

Figure: 146, Sorghum Field

The above mentioned products are used mainly as staple foods but if there is any surplus, they find their ways into the local markets. The people of Balanda also produce bee-honey for foods and for sales.

Figure 1: Household food economy of Wau County
(Source: An Introduction to the Food Economies of Southern Sudan, WFP/SCF(UK), WFP FEAU '98)

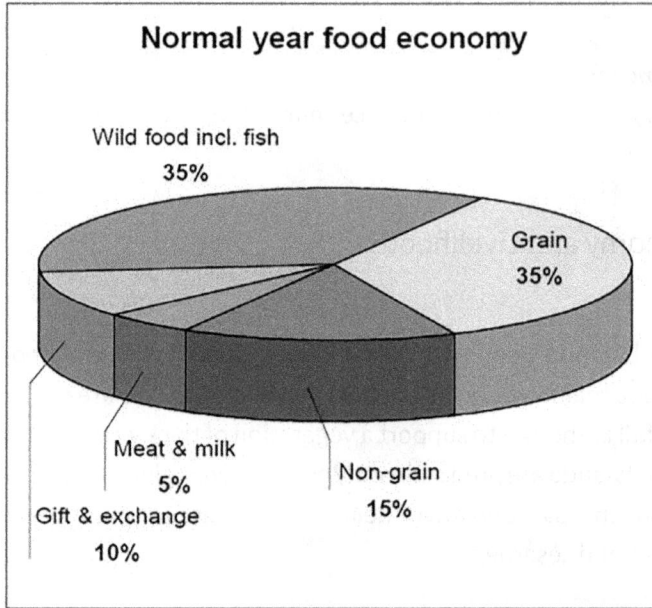

Normal year food economy

Wild food incl. fish
35%

Grain
35%

Meat & milk
5%

Gift & exchange
10%

Non-grain
15%

Figure: 147, Variety of food in Wau County

Hunting

Hunting is one of the economic activities of the people of Balanda; however, the people hunt seasonally, they usually hunt in the dry season. Whenever people go out for hunting, they usually kill animals such as antelopes, gazelles, bush-barks and many more. The meats obtain from hunting are usually consumed with only little amount sold in the local markets.

The Balanda people, like any other Luo Group, collect herbs for curing different kinds of diseases. They also collect roots and tubers and process them in many styles to make different edibles.

Trade

Balanda people are not the exception, they like other Luo groups, was involved in barter trade in the olden days; they also used to exchange item for item. However, very little is known about the trade either within or outside the community. When I was making my research I have never come across that the Balanda people have had border trading with the neighbours, although very little is written about their internal trade.

With the changing world, today the people of Balanda are picking up entrepreneurship and using cash exchange for goods and services. There are few local traders that can be found in balanda land.

Jo-Luo of Bhar El Ghazal Economy and Livelihoods

"Although Jo-Luo people are conservative and like living in rural areas, their lifestyle has been greatly affected by modern development and new technology. Fishing, hunting and collecting wild honey for cash, which was their way of life for decades, has ceased to exist because young men seek cask jobs in city and towns" (Apai Gabriel, 2005, P.62)

The Jo-Luo people of Bhar El Ghazal live in four districts, which include Tonj, Wau, Raga and Aweil. Geographically speaking:- (1) Tonj District is bordered by Gogorial East and Gogorial West in the north; it is bordered by Northern Liech in the north east; it is bordered by Southern Liech in the east; it is bordered by Rumbek and Gok in the south; it is bordered by Wau in the west. (2) Raga District is bordered by Sudan in the north; it is bordered by Aweil North and Aweil Centre in the east; it is bordered by Wau in the south east; and it is bordered by Central Africa in the south; and it is bordered by Southern Darfur in the west. (3) Aweil District is bordered by Raga in the west; it is bordered by Abyei in the north east; it is bordered by Twic and Gogorial in the east and it is bordered by Wau in the south. (4) Wau District is bordered by Awil in the north; it is bordered by Western Equatoria in the south; it is bordered by Warrap in the east; and it is bordered by Central Africa in the west. For the sake of the readers, I would like to inform you that Wau District is more central, the people from different cultural background live in this district.

The land between Luo River (Jur-River) and Kwanga River is typically dominated by the Jo Luo. The Jo Luo land in the region is fertile and provides areas for cultivation. The two rivers (Luo-River and Kwanga River) provide water for fishing.

More importantly, the Luo-River (Jur River) and Kwanga River could be used as key resources for transport, fishing and livestock rearing. The Jo Luo people of Bhar El Ghazal are predominantly subsistence agriculturalist and agro-pastoralist. Agriculture is the pillars of livelihoods of the people of Jo-Luo of Bhar El Ghazal.

The population engaged in various economic activities, which include but not limited to crop cultivation, fishing, black-smith, raising livestock, charcoal making, honey (*kic*) production, Lulu oil production, Kombo production (*pikadi*), and gathering of wild food such as Lulu-fruits (*udeny*), palms (*tuo*), dates (*nywano*) and tamarin (*cwaa*).

Crops and Cultivation

The Jo-Luo of Bhar El Ghazal are predominantly subsistence agriculturalists; they cultivate sorghum (ulel), groundnuts (*abel*), beans (*ngor*), Cassava (*moporo*), okra (*aguoma*), pumkin (*ukono*), sweet potato (*bang-gwe*), paw-paw (*ulile*), maize (*abac*), gourd (*keene*) and tobacco (*thaba*). Apai Gabriel (2005) observes that the Jo-Luo farmers also grow okra and pumpkins as seasonal vegetables. They also grow cassava, a non-grain crop that is acquired from the neighbouring tribes in the west (probably the Azande). Cassava grows very well in Jo-Luo land even in old exhausted land where grain cannot be produced. Its leaves are used as vegetables and roots are dried and made into flour.

Jo-Luo tribe economic activities are geared towards cultivation all the year round. For this reason, the Jo-Luo famers' conversations dominate farming activities. Thus, in April farmers talk about clearing new fields or cleaning old ones to make them ready for cultivation. While in May, when rains begin to fall, their talks are around blessing the seeds; and unleavened seeds are mixed and cooked before part of them are thrown up to heaven as a prayer to God to make the seeds yield well for the next harvest.

Figure: 148, Uluela (Sorghums) and a Jo-Luo woman picking groundnuts

With good rains, cultivations continue throughout May and June. The seeds planted at this time are of a giant grain crop called "*Uluela*" (see figure 138 above), a long-term nutritious grain that takes eight to nine months to ripen. Groundnuts, maize and pumpkins are also planted at this time, so that by late July or early August, they are ready (Apai Gabriel, 2005, P. 67). In July another crop called "black bena" is introduced when "*uluela*" is harvested. The Jo-Luo farmers cultivate garins crops according to the season and timing. The crops ripen at different times, rescuing the people from food shortages.

According to house Economy Analysis (HEA) 2013, there are two main seasons in the Jo-Luo region —that is rainy season and dry season. The first cultivation among the Jo-Luo of Bhar El Ghazal begins in the wet or rainy season, which starts from April through August every year; and the harvest period starts from middle September through January of the New Year.

The Jo-Luo people of Bhar el Ghazal, practice long-term and short-term grain planting; in long-term the framers plant "*uluela*" grain (see figure 138 above). And with short-term grains the farmers plant "*uduro*" grain, and "white bena", which is the last grain crop. The "white bena" grain is usually cultivated in August so that it shoots in December when rains are ending in order to ripen in January with dew.

Meanwhile, cultivation continues in August with a short-term grain, and work goes on to remove weeds from the earlier cultivated fields. Farmers normally weed fields when there are no rains, and they cultivate when there are rains. Grains are harvested in different times but some grains get ready at the same time –they are harvested at same time. Despite changes in seasonal activities in the life of Jo-Luo framers, farming remains the main occupation and is entirely unchanging (Apai Gabriel, 2005).

Figure: 149, White bena grain

The men use simple agricultural tools, which consist of hoe (*kweri*) and machete (*gajira* or *panga*). Among the Jo-Luo farmers, hoe and machete are used for many purposes, beside agriculture. Jo-Luo hoe has a broad iron blade and a wooden shaft measuring approximately 1.5 long. The hoe is used by men to clear the fields before planting or sowing; it is also used for weeding, harvesting root crops and groundnut. Although people use hoe for cultivation, it does not dig deep into the soil, it usually crates the surface of the soil. While the machete is used for cutting down the stalks of maize and sorghum after the harvest; it is also used to clear grass and cut smaller trees. Apai Gabriel (2005) observed that the Jo-Luo farmers have been using the same tools and planting the same seeds their grandfathers have been using in the past.

Figure: 150, Anyuak hoes used for cultivation

It is important to notice that the hoe used for weeding, among the Jo Luo, is smaller than the hoe used for clearing the field. The hoe used for weeding has a short shaft about 0.75m long. It should also be noted that the angle between a hoe's blade and its shaft is 180 degree. Farmers use it in standing or kneeling position. There are many agricultural tools that farmers use, these include axe (*lei*), which is used for cutting big tree; a digging stick (*apiidhi*), which is used for sowing sorghum; however, when sowing maize, farmers use rather hoe; the other tool is knife (*pala*), which is used to cut ears of sorghum during harvest.

About late December, giant "*Uluela*" grain, the main crop, is harvested and heaped on a plate form to dry. The "*Uluela*" is then threshed in the month of March and it is stored; at this time when the Jo-Luo are threshing and storing their "*Uluela*" grains, the Dinka are running short of grains because they have harvested their grains six or seven months earlier. So those Dinka who have cash to buy or bull to exchange come to Jo-Luo land to buy grains. Hence, "*uluela*" has replaced the valuable the Jo-Luo steel hoe of the past (Apai Gabriel, 2005). Most of the work in the subsistence economy is carried out by men in group or on individual basis; but harvest is partly done by men and partly by women.

After the men and women have completed harvesting, the work of processing the food is left to women. This means women transport crops from the fields to home on their heads; they also do threshing with a flat and heavy stick and grinding sorghum and maize beside brewing local beer and cooking for members of families. Crops are harvested once a year, from the middle of September through January as we have just seen above.

A widow, in the Jo-Luo society, is expected to do men's work as well as women work, she cultivates the field of her dead husband, harvest the crops, transport the crops home, she trashes and grind sorghum or maize and cook for her children and brother-in-law if applicable. A widow continues doing men's work til such a time when the elders get for her a right man to inherit her. Customarily, such man is normally selected from among the brothers of the dead husband.

Thanksgiving and base-dances

In Jo-Luo society, the harvest time is usually taken for thanksgiving for a new year and for the good harvest God has given to His people. In addition this time is also normally used by the young people for base-dances. Therefore, on this occasion, farmers group ritually slaughtered rams and chickens for the spirits of the dead to share with the living in the thanksgiving they are offering to God for giving rains and good harvest. The feasting normally last between2-4 days, at this time local beer are brewed from the new grains.

While the men and women feast to celebrate a good harvests, young men and young women would be stagging social dance balls (which commonly known as base-dances) to show their beauty and dancing skill. Such base-dances are usually organized in the house of one honourable person. It is said that the publicity for base-dances is usually done by beating a special drum at dawn so that when young people hear the sound of this drum, they would know that there is base-dance; and the information would be passed from mouth to mouth, especially to those who live in faraway villages who might not hear the drum.

This social base-dance is usually attended by young men and women from far villages. Young men and young women from each village come and circle the courtyard where the dance is taking place. The young men from the host village and the young men from other villages then walk around the courtyard in styles to attract girls. This offer an opportunity for girls to select their partners they admire during the dance. A ball-dance or base-dance last seven days, it starts at noon on the first day and ends at 5.00 p.m. or at 6.00 p.m. in the seven days (Apai Gabriel, 2005). The Jo-Luo people of Bhar el Ghazal have many different types of dances which include pardhala, adhiu, apara, janga, gumo and kanga. From among these dances, apara seems to be fading away with old people; they dance apara at harvest festival dances.

Fishing industries

Fishing, as an industry, represents supplementary source of livelihoods next to farming for the people living in the four Jo-Luo districts. In the Jo-Luo of Bhar El Ghazal land, fishing is carried out in the dry season and rainy season and is an important means for supplying of proteins in people diet.

Customarily, fishing is not done during rainy season because at this time there is high volume of water and the water would be running at high speed making fishing difficult. As a result, majority of fishermen preferred fishing in dry season, when the water levels have reduced considerably and this is also the time when fish return from tributaries, flooded plains and pools to the main streams. So fishermen in villages take advantage of this season migration. Households practice fishing in order to substitute their dietary requirements. The fishing usually happens during the month of October and November. Fish are in abundant in the Jo Luo land and it seems that the number of fish species is high —all types of fish found in the rivers are edible.

The Jo-Luo have different methods for fishing, some people use weirs (*keek*) which is made of logs and are constructed across the river in order to capture the fish returning to the main streams or fish returning down streams, while others use spear (*Badha*) and hooks (*obudhe*). The other method of fishing is by using fish-baskets or traps (*rwok* or *dipaw*) which are set in a weir. Sometimes the shelf is attached to the downstream side of a weir so that fish moving near the water surface may jump into it. The main rivers that fishermen carry out fishing are Kwanga River and Gada River (Jur River). The people who live in the west north that include the following villages: Kanje, Oduci, Get, Kiyongo, Abim and Bar-mayen they all fish in Kwanga River. And people who live in the South of Wau town, which include the following villages: Mopel, Kwajeng, Umbili, Roi-Roi Dong that all fish in Gada River. It is important to notice that the name of Gada River changes to Jur River when the river reaches Wau town; so everyone who lives in and around Wau town, they fish in Jur River.

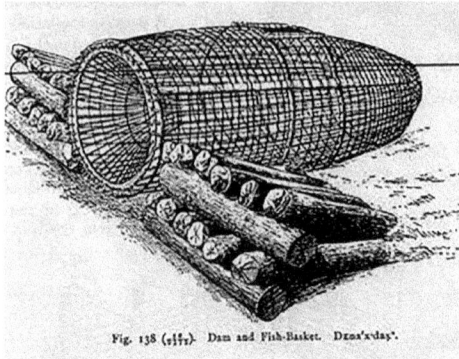

Figure: 151 Fishing shelf

Once the fishermen have done the fishing, they need to smoke them; and it has been reported that there are usually two ways of smoking fish. One way is that the fishermen smoke on a wooden shelf (*pem*), without any fire underneath, and this smoked fish in the Jo Luo language is called "*rei-matal*". The other way of smoking the fish is when the fishermen put the fish on elevated wooden stand, with fire underneath, smoke then goes between fish, and making the fish to shrink and the water in the bodies of the fish evaporate. This type of smoked fish is called by the Jo Luo as "*rei-matal*".

According to Apai Gabriel (2005) the arrival of new imported fishing nets (*boi*), which catch small and big fish alike has reduced the quantity of fish in the rivers, so much so that fishing in some rivers at certain time of the year is useless (p. 63).

My informants in Melbourne Australia pointed out that fishing is important for the Jo-Luo Wau for two main reasons. One reason is that people use fish for food; and the second reason is that they use it for small trade to bring them some income that could be used to purchase other items they do not have.

Trade

In the old days, the Jo-Luo people of Bhar El Ghazal were involved in barter trade within their community and with the neighbours. At that time they exchange item with another item they believed it has the same value with the item in the market. In those days, the black smiths produced hoes, spears, axes and knives, which they sold to the Dinka.

The iron ore is in millions of cubic tones in Jo-Luo land, but a cheap imported iron, which could be easily made into hoes for traditional farming, had monopolised the market. More importantly, the skill of forging hoes, which was a monopoly of the Jo-Luo people of Bhar el Ghazal in the past, shortly many blacksmiths emerged in the neighbouring tribes. In addition mining iron is not a simple job which can be done by one man. It is a hard job and required group work. It takes a long time to accomplish.

The process of iron production starts in with the preparation of accumulation of wood in October; this is the time when the wood is cut and heaped, ready to be burned for charcoal. The burning of the wood into charcoal commences in November to December. In November miners dig the ore by day and by night, some of them crush the ore into pieces, while others continue with burning the charcoal. The charcoal and the crushed ore are then measured into baskets and store in the shelter away from rain.

In December the miners then start the production of iron from crushed ore when rains are light. This is a tedious job that involves building cylindrical furnaces with air-jet struts below furnace. Furthermore, an inlet opening is made beneath the furnace, and fan placed below furnace to fan in the air for the charcoal to burn (Apai Gabriel, 2005).

When this arrangement is completed as miner takes one or two baskets of charcoal and pour into the furnace. After this, the fire is then lit through the base inlet, when fire catches the charcoal; an amount of crushed ore is poured in as the charcoal burns. At the same time, miners line up to fan the inferno non-stop.

When the first flames subside the miners load some charcoal and add more crushed ore. The fanning then continues until the charcoal and crushed ore in the furnace are consumed; this is also when the fanning would stop. However, we should understand that one furnace sometimes take more than 24 hours of fanning and re-loading with charcoal and iron ore.

The experience showed us that sometimes the project fails, this happened especially when the ore quality is not properly tested or if the ore quality that was poured over the charcoal exceeds the heat capacity required to melt it. However, if the project is successful, blocks of crude iron would be pulled out of furnace; they are carried home and given to blacksmith to forge into shapes. The iron is then heated to almost melting point to allow black smith to hammer the iron and forge into shapes the miners want. It is reported by informants that this process is usually done at night with the help of strong men of the village; they would assist in fanning the fire with bellows. When all the pieces of irons have been shaped, blacksmith would then turn them into beautiful steel called Jo-Luo hoe for marriage and some are turned into spears, axes, knives and shaving blades. The rest of the products are sold to Dinka in exchange for cattle and goats but the Jo-Luo marriage hoe remains untouched by the Dinka.

Mining iron ore or forging steel hoes had resulted in a lucrative trade between the Jo-Luo people and the Dinka and Azande; however, this was banned by the British Government. Apai Gabriel (2005) argues that the introduction of money as a dowry has also greatly shifted young men's seasonal occupation to job-seeking in the towns to acquire wealth. When money became token for buying and selling everything, many people resorted to collecting wild honey, since it was the only means of acquiring money without going to the towns. As a result the forest was disturbed and exhausted because bee swarms were disturbed again and again and could not laid eggs to increase the numbers of swarms.

Therefore, honey became scare in the forest (PP. 62-63). Today, the people of Jo-Luo are picking up entrepreneurship and selling their products in cash money; which have been introduced to them by the Arab traders.

Livestock

The people of Jo-Luo keep goats, sheep, chicken (*gen*), dogs (*gwok*) —people keep dos for protection and hunting; they also keep cattle. Farmers do not keep cattle in Jo-Luo land for fear of raiders; they keep them with their friends in Dinka-land. The people keep goats, sheep and cattle mainly for marriages and milk.

Black Smith:

As part of the economy of the Jo-Luo Wau, the black smiths across the region produce hoes (*kweri*) spears (*tong*), axes (*Lei* -single or *lie* –plural), knife (*pala* –single or *pale* –plural). It is reported that most of the items produced by the black smiths are sold to the Dinka —since Dinka do not have black smiths.

Wild food

The Lo-Luo Wau have very strong economy in this that their sources of income do not depend on one branch. Thus, besides selling domestic animals and the products from black smiths, they collect wild fruits such as lulu-fruits (*udeny*), palms (*two*), tamarin (*cwaa*), *nywano* and other *apwoni* (fruits) for consumption and sales. In addition they collect honey (*kic*) lulu-oil and kombo (*pikade*)

Mabaan Economy and Livelihoods

The people of Mabaan live in the plains land between the Nile east of Renk, up to the Ethiopia Highlands. As I am writing this book, I have discovered that what is known about the Mabaan is limited, since I was not able to interview elders from Mabaan County. Secondly, there are few written documents about Mabaan —if any. As a result there is need for Northern Luo Scholars should do more research in the arts, economy and livelihoods of the people.

In Mabaan county, scattered patterns of family settlement are common especially with the married couples; they sometimes build their houses outside family compounds. Villages are situated predominantly in areas which soil is conducive to agriculture. The people are agro-pastoralists, farming food crops and raising livestock in addition to gathering wild foods. However, they produce very little surplus. Most of the Mabaan livelihoods rely on subsistence agriculture with limited livestock mostly in form of pigs, goats and chickens. Being agrarian society most of their social values relate to land rights and agricultural practices.

Crop Cultivation

The most important crops cultivated by the people of Mabaan include sorghum 9which is the primary staple), maize, beans, cowpeas, groundnuts, sesame and okra; but wild plants such as tubers, nuts, seeds, fruits and leaves make up the most important source of food for the members of the families. Before the civil war over 80% of the people have enough food to eat throughout the year. But today less than 20% of the people have enough food to eat throughout the year (DRC, 2014).

The people are also known for their popularity in the production of honey for food and sales, they export the surplus to Khartoum-Sudan, Malakal town and Juba the capital City of South Sudan.

The rainy season lasts from May to October and during this period inhabitants experience severe flooding.

According to Lauren Hutton (2012) the people of Mabaan used to plant, own and inherit plants and trees; although today we cannot determine of how widely is this practice in the community. What we know today is that Mabaan practice communal ownership of plants and trees at family or village levels.

However, following the signing of Nairobi Comprehensive Peace Agreement in 2005, the lifestyles of the people of Mabaan have changed dramatically. As I mentioned above before the civil war over 80% of the people of Mabaan had enough food to eat throughout the year. Unfortunately, the unreasonable conflict that broke out in December 2013 has affected food and land availability for the population in Mabaan County. As a result, the County has less agricultural products; they depend highly on imports airlifted from Ethiopia and Juba. The conflict resulted in the closer of main roads leaving people in the County without food and medicines. Food aid is flown into Mabaan; moreover the prices of flights prevent this practice from being sustainable.

The unreasonable conflict has prevented people from going out for cultivation to make enough food for the families; as a result, this inhibits the population from attaining self-reliance. The survey carried out by Denish Refugee Council (2015) revealed that over 80% of the people of Mabaan do not have enough food to eat, and only 2% of the local people received food aid from refugee camps (P. 7). However, we cannot deny that the influx of refugee from Blue Nile State into Mabaan has created tension between the local people and the refugees, especially in terms of gaps in food and assistance. There have been lack of equal services by the none-government organisations (NGOs), given the fact that due to the conflict the local people cannot produce enough food for themselves, as they have been doing in the past. Subsequently, the local [people perceive themselves to be discriminated against, increasing "negative thoughts and division" and fostering hatred between the groups.

Source of Income

The main source of income for the people consists of farming and livestock; each house-hold, on average, own three cows and six goats. In the 1960s the people used to own fifty to seventy cows and hundred goats and do farming and gardening as their main sources of income, but today, according to DRC (2015) survey, due to conflict only 17.9% of the local people are practicing farming and gardening (P. 13).

Both in the old days and today some people get their income from selling firewood, cutting grass and selling foods; while the next highest source of income in Mabaan community is brewing. However, brewing even has been affected by the conflict; in the past the women have access to materials for brewing and 100% of the women in the county received their income from brewing. But today, due to unwanted conflict only 11.7% of women receive their income from brewing local beer. And yet only 2.3% of the local women identify beverage trade as successful way of making money. The survey also revealed that only 14.8% of young people earn SSP 100-200 per week (DRC, 2015, PP.15, 22, 28).

However, many Mabaan men do not have interest in carpentry; it is assumed that this could be due to lack of education; because of this very few members of community practice carpentry as source of their income.

Small business is another source of income for the members of community, women and men sell tea and do some caterings –although through observation one could conclude that catering has low profits; most people eat in their houses. Before the civil war in 1960s over 50% of men and women used to sell tea but today only 16.3% of the local people are engaged in selling tea DRC, 2015, P. 28). It is believed that the barrier to earning more money is due to lack of education and access to cash by the members of community.

Trade

Cattle have important cultural and financial value; goats, pigs and chicken, livestock can be trade for cash, but are often used in culturally significant ways. Following the Addis Ababa Agreement in 1972, local markets were booming in Mabaan County; but soon after the outbreak of another civil war in 1983, trade in the County dropped dramatically. Nevertheless, in 2005, following the signing of Comprehensive Peace Agreement (CPA) in Nairobi Kenya, market restarted even do they were in small numbers. However, with the coming of UN agencies and non-government organisations (NGOs) and refugees into the County, this has contributed to the growth of local markets. In contrast, the local leaders noted that lack of access to cash money to buy materials was the main barrier to increase market growth as well as income to households. On the other side, the NGOs believed that lit is not lack of cash money but it is rather lack of education that is the major barrier to socio-economic development of the people.

Although trading was picking up in the region, nevertheless, the markets are now vulnerable to road closures due to the conflict of December 2013. Indeed, the conflict has shaken trade routes and disturbed market activities in almost all parts of Mabaan County. Thus, today few formal markets operate in the County. Even where the markets do exist, many goods and services remain unavailable. Despite the fact that there is strong demand for oil, onion, salt, sugar, meat and other non-food items such as clothes, soaps and footwear.

Worst still, the modern Mabaan men have failed to learn small-business skills; as a result most of the current shop owners in the County are people from Ethiopia. Interestingly, some NGOs operating in the areas view Mabaan's market growth as evidence of impro-ved livelihoods for the local population, but as a matter of fact the cash received by the Ethiopian shop owners flows into Ethiopian, instate of circulating for development in the County, consequently leaving the County without cash many (DRC, 2015, P. 13).

However, we cannot rule out that lack of money to buy materials and lack of materials to be bought are main barriers to increase income for the 15.2% of local shop owners. For this reason, NGOs organised training in various location in the County, especially within the refugee camps. The training covered farming, animal healthcare, livestock herding and selling tea. But in actual fact the natives desired to be trained on production of fruits and vegetables and not on selling tea as presupposed by the NGOs. Thus, in the survey conducted by DRC (2015) indicated that despise the training, only 8% of the population in the County benefited from the training.

Environmental Damage

If we compare Mabaan County with other Counties in the Upper Nile, we easily discover that Mabaan County was badly affected by the second civil war 1983, through the unrea-sonable conflict that broke out in December 2013. When I looked through the history of South Sudan, right away from 1991, I have discovered that Mabaan County had become the central to the SPLA/M war; the County was used as operating centre to access support from the then Ethiopian oil installation around Melut and Palioc.

From the year 1991-2000, the areas around the expanding oil infrastructure were subject to air booming, and regular attacks by the Sudan Army Forces (SAF) and SAP aligned mili-tia SPLA-Nasir. Thus, after the split within the SPLA in 1991, Upper Nile became central in the inter-south violence as Lam Akol and Riek Machar's SPLA-Nasir forces were based in Fashoda and Nasir respectively.

According to the people of Mabaan of that time, there was no clear allegiance to either Akol, who is a Shulluk, or Machar, who is a Nuer. As internal politics of the Liberation Movement became increasingly divided on ethnic loyalties, the Mabaan, who are the Luo, neither supported Lam Akol, nor Riek Machar; they decided to side with the SPLA-Torit, which was under command of John Garang De Mabior. John Garang accepted them mainly to preserve place of SPLA-Torit in Upper Nile, since the region is dominated by Shulluk and Nuer people.

From 1997-2002, the oil infrastructure became a military target, so the army of Lam Akol and Riek Machar moved into offensive positions to strike within Melut and Palioc areas. Subsequently, Mabaan County was caught in the middle and the controls of areas have been inter-changed between SPLA, SAF and SAF-aligned SPLA-Nasir.

With the above views in mind, since the year 2006, the Commissioner of Mabaan County has being advocating for the compensation based on abuses that occurred due to the oil expansion and the resultant environment damage. The Commissioner wants the government to pay for the gravel they extracted and the tree that were felled for charcoal, firewood and building.

The foundation of these claims can be traced to Adar, which is a control oil producing area, the Commissioner argues that this place or area belong to the people of Mabaan, who had been forced to move away from the area.

Pari Economy and Livelihoods

Cultivation and Crops

Pari land is wooded savannah and annually receives some 800mm of rainfall. The land is somehow flat, and as a result many places become swampy during the raining season. The people of Pari practice agriculture and animals husbandry. They are farmers and they cultivate sorghum extensively, and the surpluses usually go to the local markets and some-times as far as Torit Town. The age-grade system in Pari plays a central role in the management of sorghum landraces and continues to underpin the resilience of their traditional seed (PNAS, 2014, P.1).

Eisei Kurimoto (1984) characterizes, the livelihood of Pari community as a "multiple subsistence economy", they mainly depend on agriculture with sorghum as the staple crop, supplemented by husbandry, hunting, fishing and collection of wild foods.

Figure: 152, Pari Sorghum (Nyitin)

Other major crops grown include cowpeas, green-grams, pumpkin, Okra, sesame and tobacco. They raise considerable number of cattle, goats, sheep and fowl. Domestic animals are very important to the people of Pari, because the animals are considered to be connecting human beings with our world and the world of God (Juok), besides being commodities. Gathering of wild foods is also an important means of food supply particularly during the hunger period.

Hunting and Fishing

Hunting is one of the activities done in Pari land during the dry season and fishing is always done during the wet season. They hunt with spears, though guns are now available in villages. According to the Pari, hunting (*dwar*) is not only a means to obtain meat. It is an occasion, especially for youngsters who are engaged in specific hunts of four big animals (lions, leopards, buffaloes and elephants) this is the time they can prove their manhood. Killing wild animals is a means to establish one's identity as a man. Unlike other pastoral Nilotic, the Pari have no favourite ox and show much less interest in cattle than in wild animals (Kurimoto, 1984). Pari society is organized in an age-grade system; young men are enrolled in groups based on age, and these age-set Passover different age grades over the course of their lives. The age-grade system remains a fundamentally important institution with political, legal, military, ritual, and economic functions for the people of Pari; and it also plays an important role in the traditional system in Lafon. The ruling age-grade, the mojomiji, decides when sowing commence, and all Pari lineage ritual mix a gourd with their own seeds before sowing. This seed management practice connects sorghum fields and granaries in the villages (PNAS, 2014, PP 4-5).

Fishing and hunting supply protein to the diet of the local people. I could estimate that the people in Torit town receive approximately 30% of their annual protein come from fish brought into the town by the Pari fishermen. It is believed that about 90% of Pari dried fish is sold in Torit market. There is also common saying that Pari land use to be one of the last resorts of wild animals in Africa. A great number of elephants, buffaloes, various antelopes and gazelles were found in Pari land during the old days (possibly before and after the colonial day). The population of these animals radically decreased during the recurrent civil wars (1955-2005). The two rivers, Hoss and Hinaiti, provide a good quantity of fish of various kinds. Dried and smoked fish is an important trade items.

Shatt Economy and Livelihoods

Crop Cultivation
The Thuri are mainly farmers, they live in community with homestead distanced from each other due to the large farms each household own.

Figure: 153, Large farms around the homestead

They grow sorghum, maize, cowpeas, pumpkins, okra, bananas and sesame as their staple foods; the surpluses from the crops are usually for trade. They also grow tobacco for consumption and for trade; in addition they produce bee-honey for foods and sales.

Hunting
Hunting is one of the activities that contribute to the economy of the Shatt. They hunt seasonally; this is the period they also collect bee honey, lulu fruits and nuts to make oil. But when it comes to the professional hunting, the men usually walk a long distance across the country to find animals. Big animals such as elephants, buffaloes, antelopes and water bark does not live around the villages for gear of been killed.

The Shatt people also collect herbs for curing different kinds of diseases. Roots and tubers are collected and processed in many styles to make different edibles.

Trade
In the old days, the Shatt were involved in barter trade within their community as other Luo groups. They usually exchange item with another item they believed it has the same value with the item in the market. But today, the people are picking up entrepreneurship and selling their products in cash money; which have been introduced to them by Arab traders. However, today, bartering is still taking place in villages where they found necessary.

Shulluk Economy and Livelihoods

The Shulluk people live in almost continuous narrow strip along the White Nile and mostly on the western bank with Lake No and Tonga in the south and Thwori Gwang Mountain and Renk in the north. The country composed of five counties and twenty- seven

Payams based on the ten States policy. The five counties include Fashoda, Malakal, Melut, Panyikang and Renk.

Fashoda County borders Sudan to the west, Menyo to the north and Malakal to the south. It composed of four Payams, which include Dethok, Kodok, Kodok Town and Lul. The land has low lying area that consists of savannah grassland, bush and patches of forests. The White Nile is a key resource for transport, fishing and livestock rearing.

Malakal County spans both sides of the White Nile and extends west to the border with Sudan. The land in and around Malakal, like Fashoda, also has low lying area that consists of savannah grassland, bush and patches of forests. Malakal County is composed of six Payams that include Central Malakal, Eastern Malakal, Northern Malakal, Southern Malakal, Leo and Ogot. Like Fashoda County, the White Nile River is a key resource for transportation, fishing and rearing livestock.

Melut is another County that is located in the centre of Upper Nile State. The County is composed of six Payams, which include Bimachuk, Galdora, Melut, Paloich, Panhomdit and Wunamum. The White Nile River flows alongside the western border of Melut with Menyo County. More importantly, Melut borders Renk, Mabaan, Baliet, Akoka and Fashoda Counties. River Adar flows across Melut into the White Nile.

Panyikang County stretches along the western half of Upper Nile State's border with Unity and Jonglei States. Panyikang County is composed of six Payams that include Anakdiar, Dhetiem, Pakang, Panyidwoi, Panyikang and Tongo. However, Panyikang County borders Sudan, Malakal and Baliet Counties. The County is divided by White Nile into East and West before the Nile continues to flow onto Malakal town. In the eastern borders of Panyikang, there are Sobat and Fulus Rivers that flow into the Nile.

Renk is the fifth County; it lays northern most of the Republic of South Sudan, near Kosti. The County borders Sudan to the north, the Blue Nile region to the east, Melut and Mabaan Counties are to the south, and Menyo County is to the west. It is good to know that Renk and Menyo Counties are separated by the White Nile. The County is composed of five Payams, which include Chemmedi, Geger, Jalhak, North Renk and South Renk.

The Collo are predominantly subsistence agriculturalists; historically, pastoralist and agro-pastoralist were and are the pillars of livelihoods of the people in Collo society. The population engaged in various economic activities, which include but not limited to crop cultivation, fishing, raising livestock, charcoal making, hunting and gathering of wild foods.

The livelihoods of the people are located in two main zones, one in Eastern Flood Plains and the other in Nile and Sobat River zone. The Eastern Flood Plains livelihood is located in the north-east corner of South Sudan —it covers Upper Nile and Jonglei areas. But for the purpose of this book we are going to focus only on the livelihoods within Collo who live in Eastern Flood Plains. After having examined the economic back ground of the Shulluk people, let us now discuss in details the economic activities undertaken by the Collo.

Farming

Nearly one half century ago, the Shulluk people were purely a pastoral people. Only within the last decades men began to assume the burden of providing for their families. In those earlier days, the task of tilling the small patch of ground planted annually in dura the responsibilities were taken by the women. The woman hoe was made from the shoulder blade of a giraffe or buffalo, or sometimes from the skin or rib bones of some wild animals (D. Westermann, 1912).

D. Westrmann (1912) in his book entitled "The Sulluk People, Their Language and Folklore" reported that one day, one Shulluk man was asked by a researcher, "What food did you people eat in your earlier boyhood?"

The man answered, "We used to eat grass like cattle".

It was felt that the answer of this man to a researcher carried much truth because in Shulluk society when crop fails, the women go to gather grass seeds from the swamps and plains. They also obtained sugar by bruising and boiling a certain reed, which grows in the swamp.

More interestingly, up to today, the Shulluk people have not yet learned to grow very large varieties of plants –their main crop is dura. Nevertheless, today, the Shulluk also plant few beans, pumpkins, squashes, sesame and occasionally peanuts; they also plant tobacco in small plots on the river bank during the dry season and water by sprinkling the ground from water jar, tobacco also are planted in the beginning of the rainy season. No fruits of any kind are grown in Shulluk land.

They plant dura in same field for years after years until crop fails once or twice. Then a farmer hunts for a piece of high dry ground, preferably in the timber, for the early dura. Farmer then cleans off the timbers and dig up the grass for his new field, which he tills until another failure comes; and when the former field is growing of grass, he will then return to its tillage.

Although there are no new findings by researchers about farming in Collo's land, it seems there are no much changes in farming if any. D. Westermann (1912) observed that Collo farmers' methods of farming are very poor but yet they accomplish good results. The farmers never use ox-plough even though they have cattle, nor do they use spade. They even never use the mattock except to dig up the grass and bushes from new ground. Farmers prepare their ground by raking up old stalks into piles with a palm tree limb and burning them.

As soon as the rainy season opens and sufficient rain has fallen to soften the ground, farmers put their seeds to soak overnight, so it will sprout the quicker, and thus more likely escapes being eaten by the white ants. With a long slender pole that has one end shaped like the bowl of a spoon, farmers open up the ground, and drops in the seeds. As a farmer steps forward to make another hole, he presses down the earth over the seed with his foot. It is worth noticing that hills on which seeds are planted are usually some eighteen inches apart in all directions. Farmer plants a large number of seeds in a hill, and later on thin out and transplants where there are no hills. There are many reasons why farmers have to replant, because there are many enemies of the sprouting grain, and these enemies include rates and white ants. The white ants often destroy the grain in the ground, unless it sprouts quickly. However, under normal conditions the grain comes up very soon after planting. And the warm rains and tropical sun cause the crop to grow up very quickly and nicely.

Needless to say, seeds come on quickly too, as a result a farmer begins hoeing his fields at once. A farmer could have either a thin circular or rectangular piece of iron with a short wooden handle. The farmer, while hoeing sits on the ground can squats on one knee or both, as he chooses, and catching the grass with one hand, cuts it off just under the surface with the hoe. After cutting the grass, the farmer shakes the dirt from the roots of the weeds and throws them into piles, leaving the ground clean and smooth.

It is said that there are about thirty-two varieties of dura in Shulluk society. In my opinion there is possibility that same variety may has different names in different localities within Shulluk society. Subsequently, the actual varieties of dura could be under thirty-two varieties.

In Shulluk society dura are eaten in a variety of ways; one way is that the earliest heads of dura are simply thrown on the fire, roasted and boil till softened and eaten by both the young people and adults. However, a great deal of dura is made into mild beer locally known as "*kwon-chak*" that is used as regular food diet. The regular way of cooking it is to grind it into a fine meal and cook it into a mush and eat it with milk or cook it up with meat. According to Banydhoru Oyay, "kwon-chak" is made with fresh grounded dura in a special way which gives a rather pleasant sweet taste when cooked. It can even be eaten by itself and would take the name of "*kwon*". When it is eaten with milk, it then takes the name of "*kwon-chak*". According to him, the food that is made out of ground pasted dura, cook into a mush and eaten with milk is called, in Shulluk language, "*apodo*" and not "*kwon-chak*". The paste dura used for making "*apodo*" is often mixed with some small bit of fermented substance, to give it a final sour taste. And finally, this is then made into large uneven crumbles as opposed to small even crumble of "*Akelo*". The "*apodo*" is cooked into custard, and that would be its final appearance/presentation. However, "*apodo*" is predominantly children's breakfast meal. As it is easy to make, "*apodo*" sometimes form parts of young men's food, especially when they are being send away with cattle during summer season, to look for pasture around the river banks.

Dura is also sometimes boiled with beans or mix with sesame and eaten; more importantly, dura is often made into dura bread, which is locally known as "*khisira*" and eaten with cooked green-vegetable or meat.

Customarily, in the old day, all Shulluk men were forbidden from riding upon donkeys or carrying food on donkeys, as a result, at that time, Shulluk men never possess donkeys. And cattle were and are never used to carry people or goods –goods are usually carried by women. But today, in big towns such as Malakal, Fashoda and Renk some Shulluk men use donkeys to carry foods or water.

The villages in Shulluk land are full of hunting dogs; like other Northern Luo, and unlike the Nuer and the Dinka people, Shulluk people raise chickens. D. Westermann (1912) observed that the chickens produce eggs, but unfortunately these eggs are eaten only by the women and children but not men. This raised question into my mind, and I asked some of the Shulluk people, "Why do Shulluk men do not eat chicken eggs?" I have received mixed answers to my questions, some members of Shulluk community, like Mrs Terizah Chol Otor, who lives in Melbourne, argues, "Custom that forbid men from eating chicken eggs, as observed by D. Westernmann (1912) is out dated. Today, in modern Shulluk society, men eat chicken eggs. More importantly, some eighty years back, men were eating chicken eggs; because I have seen my grandfather and father were eating chicken eggs". However, people like, Banydhuro Oyah, argues that he is not sure whether, in the old days, Shulluk men were not eating chicken eggs; but if they did, this probably might have been because of the perception of its association with smelly farts.

Crop Cultivation

According to Household Economy Analysis (HEA) 2013, there are three main seasons in Shulluk land, especially in the Eastern Flood Plains livelihood zone. Muchomba (2004) says planting seasons and number of plantings vary with river access and the degree of flooding. The first cultivation may begin in the dry season (January to March), with a second planting after the wet season (April to July). Some households may even plant three crops if they move closer to the rivers as the flood waters recede. Crops can actually be found growing throughout the year when the use of receding waters has been optimised, which has become a significant risk management strategy in some areas. (P. 64).

The area in the eastern flood plains zone is mainly flat, plain covered with savannah grassland and bushes with scattered tress. The soils are a mix of loam, sand and clay-loam. Both areas in Eastern Flood Plains and Nile and Sobat Rivers receive 1,000-1,500mm of rain on average during a single rainy season. Agriculture is rain-fed and the rainy season last from May to October every year. The rainy season in Collo land is followed by a wet-dry season that usually lasts from November to January. The wet-dry season is then followed by dry-season, which lasts from February to April.

The crop cycle begins with land preparation as from April to May, and this soon follows by planting of sorghum and maize at the end of May and into June. By August, maize is usually ready to be eaten fresh or green from the fields. The common crops grown by Collo farmers include sorghum, maize, beans, sesame, pumpkin, gum Arabica, okra, and cowpeas.

The study conducted by HEA (2013) in 2011-2012 revealed that the better-off and middle-income households produced 50-64 per cent of their own food to meet their annual consumptions. They also receive 22-30 per cent of their food from the livestock in term of milk. The study indicated that the better-off and middle-income households do not receive much of their food from the open markets. As a matter of fact they only receive 15-22 per cent of their food from the local markets.

Figure: 154, Sobat River, Source: flickr.com, 2018

With all these views in mind, it is important to notice that sorghum, maize, cowpeas and pumpkin seeds are the principal crops grown for food and cash in the society.

In the two zones (Eastern Flood Plains and Nile and Sobat Rivers zones) the people define the poor and the very poor household on how much the household cultivate and how much land that family own. However, it is reported that the very poor and poor household in Shulluk societies cultivate between 1.5-2 feddans (acres) of land. The very poor and poor households in Nile and Sobat Rivers grow 800 kg of sorghum and maize annually, while the very poor and poor households in Eastern Flood Plains grow 535kg of sorghum and maize. This reflects the different of poor households' productions in the two zones, which means that the poor households in Nile and Sobat Rivers produce more food than the poor households in Eastern Flood Plains. This may be because the poor household farmers in Nile and Sobat Rivers have better and fertile lands. This explains why the very poor and poor households in Nile and Sobat Rivers get better yields than their group in Eastern Flood Plains. In contrast, the better-off and middle-income households, who live in the two zones, have also differences in their agricultural outcomes. It is reported that those who live in Nile and Sobat Rivers grow 2,350kg of sorghum and maize annually, while those who live in Eastern Flood Plains grow only 1,830kg of sorghum and maize annually. The same explanation offered for the poor households above could be apply to the wealth groups living in the two zones —lands along the Nile and Sobat Rivers are rich and fertile, good for agriculture. The same study revealed that the better-off and middle-income households produced 50-64 per cent of their own food to meet their annual food needs. From the above reports, it become clear that the better-off and middle-income households produced sufficient food to meet their annual food needs. In addition, the better-off and middle-income consume 75-80 per cent of the total food productions (mainly sorghum and maize) in the two zones (HEA, 2013).

However, we should also acknowledge that it is difficult to estimate the proportional size of the various wealth groups in each village or zone, because the households keep on moving from rural areas to urban areas and vice-versa and sometime these people move across the international borders.

It is reported that 20-25 per cent of the annual food for poor and very poor households is composed of wild food and fish. Desert dates (*lalop* or *Thou*) are particularly important wild foods. Muchomba (2004) pointed out, "Shea butter nut is particularly grown in the southern parts of the Shulluk land and is important source of income as well as a food source. Desert dates are more common in the northern parts, where a wide variety of grains, fruits, roots, nuts and leaves are both eaten and sold" (P.9). Hamlets are surrounded by gardens of millet, maize and sesame, as well as other species introduced during the twentieth century. The people also cultivate tobacco for personal use and for sales (Burton, J. Et al. 1996). Since the 2013 conflict, both agriculture and trade have suffered from lack of equipment, displacement of the population and insecurity.

1.1 The Rain-Making Ceremony

In Shulluk country there are two versions that cover the ceremony of rain making:

(a) First version of the Ceremony

The first version holds that at the beginning of the cultivation season, elders make sacrifices in all the ten shrines of Nyikang in the country. The King (Reth) gives a cow and a bull to be sacrificed in each of the tomb. The bull is killed, while a cow is added to the herd belonging to the shrine. The bull for such sacrifice is not killed with ordinary spear; it is killed with especial spear of Nyikang called "*alodo*", which is found in every shrine of Nyikang. More importantly, the bull is not killed by ordinary man, it is always killed by the "*bareth* of Nyikang"; it is killed right in front of the shrine.

In Fashoda the King (Reth) takes part in the ceremony, as this been is capital and residential place. When the bull is brought in front of shrine, the King comes and stands near it, as the *bareth* members kill it. He stands there while praying, loudly to God (*Juok*), through Nyikang for the rains. As he is praying for the rains, he holds spear (*alodo*) with his right hands, and the spear is pointing upwards.

However, in other shrines where the King is not there, one of the *bareth* member represent the King, he also stands by the bull during the sacrifice; he stands with a spear (*alodo*) in his right hand, while pray loudly to God (*Juok*), through Nyikang for rains. He would be saying pretty much the same words that the King at Fashoda is saying in the ceremony.

After the prayers, the meat of the bull is cut, cooked and eaten by all the *bareth* of Nyikang. When they finish eating, the bones and as much blood as can be collected in gourd are to be thrown into the river; and the skins of the animals are left as a mat for Nyikang to sit on.

We have seen before that at the harvest time, farmers donate millets to Nyikang and they are kept in the hut Nyikayo in the shrine. Thus, in this annual ceremony much of the millets preserved in the shrine is made to beer (*ombugi*) for the occasion (C. G. Seligman, 1976). Still today, the King sends animals to various shrines, to be given to Nyikang.

(b) *The Second Version of the ceremony*

According to C.G. Seligman, 1976 and Dietrich Westernmann, 1912, the second version of the ceremony for rain starts by old women cleaning the ground within and outside the fence of the shrine enclosure. They pull up every weed and blade of grass; plastering mud, painting the walls with ashes and generally tidying the shrine.

When they finish the decorations of the walls, drums are brought out and placed in the open space near the centre of the village. Then all the Shulluk living in the area come out to dance. Both men and women dance vigorously, men holding spears and other weapons, at the same time they raise their hands into the air and singing praise to Nyikang.

Shortly, the sacrificial bull is brought and tied under the tree next t the dancing ground. The people continue with dance, and when night falls, they continue to dance and pray to Nyikang or any other ancient King, to whom the temple is dedicated.

In the middle f the night, one person gets up to pray and praise Nyikang saying, "I beg you for something (food), put it into my mouth. The earth has been spoiled by the people; Lenydaro (the army of Daro) is travelling on the earth. I raise up my heart to our grandfather, the chief of the daughter of Nyidwai, the children of Nyikang" (Diedrick Westernmann, 1912),

After dance and prayers, the chief of the district pours water, recently brought from the river, into the hollow of his hand, spits into it, and sprinkles the bull with the water. When the chief finishes sprinkling g of the sacrificial bull, women go to grind dura flour, soon the bull is brought to the shrine and speared high up in the flank, so that the wound is not immediately fatal, and is allowed to walk freely. In theory, the bull should go to the river and come to die in front of the shrine. Traditionally, the sacrificial bull is usually killed using special spear of Nyikang called "*alodo*". People do not use ordinary spear to kill sacrificial bull, this will be considered as abomination to God and Nyikang.

The animal is then skinned and cut up and the meat is boiled in pots and eaten by chiefs and elders present at the ceremony. The head, one forelimb and the bowls are cooked separately and eaten by the attendants (*bareth*) of the shrines as well as other members of the community. But the pregnant women and their husbands, and men and women who know they have had sexual intercourse last night, they are forbidden to eat the meat of sacrificed bull because they are considered unclean. Interestingly, when eating the meat nobody is allowed to break the bones. Therefore, care must be taken and after the meal, all bones are collected and thrown into the river. The skin of the animal is prepared and used as a mat for Nyikang to sit on.

The Harvest Festival

The preparations for harvest festival usually start in November and the actual ceremony is conducted in December of every year. At the harvest festivals farmers bring ears of ripening millets and entrust them to the hut *Nikayo* dedicated to the mother of Nyikang. This means that when a millet is cut from the fields, everyone brings a portion to the "*bareth* Nyikang" (the servants of Nyikang).

According to C. G Seligman (1976), the harvest ceremony is usually to be performed at the royal grave shrine as well as at shrines of Nyikang, though it is recognized that this is not absolutely necessary (P. 90). The millet entrusted to hut *Nikayo* is then ground and made into porridge with fresh water brought from the river; some of the porridge is poured out at the threshold of the "hut *Kwayo*" (the hut dedicated to Nyikang); and some of the porridge is poured on the ground inside the "hut *Kwayo*", where Nyikang is believed to reside. Until this is done, nobody may eat of the new crops.

This is then followed by the really harvest festival celebration, elders and chief of the district are asking Nyikang to save millets and other crops from being attack by birds and other wild animals.

Before come to dance around the shrine, some 30-40 women clean the area by pulling out every weed and blade of grass. As women clean the area, they also sing; and when they finish cleaning, they then dance slowly round the outside fence of the shrine –singing and clapping their hands.

The cleaning of shrine may take between two to four days, depending on amount of works to be done. During this time women will be doing the work of cleaning, plastering mud, painting walls with pictures and generally tidying the shrine.

When all works are done, the King then provide a bull for sacrifice, a large and small drums are beaten and the people assembled and dance around the shrine. The dancers surging in spirals round and away from the drum; while number of old men and women would be sitting near the huts on the edge of the dancing ground; hundreds of spears are tilted (??) against the thatches, although all the men come to the ceremony with their spears, no one is allowed to dance with his spear, except the King (Reth). When dancing, the Reth will be carrying on his left hand a narrow pointed spear of Nyikang called "*alodo*", in his right hand carries a whip with a double lash and wearing around his head a fringe of black goat-hair. In a certain period, the King dances alone inside the enclosure of the shrine, and occasionally comes out and dances round outside the shrine. During the dance (in December), many young men are also allowed to dance with the old men and the King. During the dance no young man is allowed to dance with his spear, with exception of the Reth as I have just mentioned.

The young men usually wear leopard and other skins suspending from their waists and hanging behind, they also wear decorated anklets and armlets of different kinds, belts of disc beads are worn round the waists as well, and the plumes (???) suspending from the forearms. The head dresses of the young men vary a great deal but the common one is called "Tam O" (Shanter) –this is a dominant fashion for young warriors. The more popular style being a flat, circular mass of felted hair on either side of the head; many of the young men have their heads shaved and wear a wig or crown of sheep wool. They also carry millet or flowery grass stalks, some also hold clubs and parrying shields and the ceremonial bow called *"dang"*. The children can also take part in the dance, imitating the elders.

Dance is not only for men and young people, women also dance but in a separate place not far away from shrine, all carrying hippo hide whips. As soon as the drums stop, everybody stop dancing; at this moment only sound of a *kudu* horn trumpet would be heard. Shortly, drums sound and people go back to dance.

After the dance, two to three men enter and take spears (*alodo*) from hut *Kwayo*; the sacrificial bull is then brought and led across dancing ground and release to wander around the shrine enclosure. As the bull is wandering about unmolested, the men again start beating drums, and after this, the people go back to dance. At this time the Reth (King) takes his spear and join the dance, he dances within the enclosure of the shrine and from time to time comes out to dance around the drums, while selected men run to the river to wash the spears. When the men, return from the river, the bull is then caught and hold outside the entrance of the shrine.

As soon as the people see bull being dragged to the shrine, all leave the dancing ground. The only people who will remain behind in the vicinity of the shrine are the Reth (King), group of chiefs and elders from the Golobogu and Pameiti clans. The men sit in a row, while the King stands beside the bull, at this time the bull is held on each side by ropes and one man grasping its tail. Before the bull is actually killed, the men, who are sitting in a row, get up one by one, each makes a brief speech.

Interestingly, all these short speeches are appeals to Nyikang, asking him not to allow natural disasters to befall on the people; and more importantly, not to let their millets suffer from the birds and other wild animals. When these people finish prayers, one of the *bareth* Nyikang takes spear, strikes the bull's flank, after stabbing he first washes the blood on spear in the holy water that is kept in the hut *Kwayo,* and later the runs to the river, to wash the remaining blood (C. G. Seligman, 1976).

A bull is then guided into the enclosure of the shrine, there it is stabbed again with another spear; and it is led to die in the enclosure of hut *Kwayo* (Nyikang's hut), while the King and the chiefs remain just outside the shrine. Shortly, the King and chiefs follow the bull in the hut *Kwayo*. The chiefs kill the bull in Nyikango (Kwayo). It is the customs of the Shulluk that such sacrificial animal must be stabbed outside the enclosure of the shrines and led to die in the hut dedicated to Nyikang (*Kwayo*).

The bull is then skinned and the right forelimb is cut off and shown around the village and later to be taken to Ayuriel; while the rest of the meat is divided and eaten by the King and men from Golobogu and Pameiti. After the meal, a band of warriors, in their full dress bearing their spears return to the dancing ground. They dance only once around the sacred enclosure of the shrine, get down on their knees, clapped their hands and leave the place quietly (C. G. Seligman, 1976).

It is believed that during the sacrifices at the shrine, Nyikang often appears in different forms, sometimes he appears like snake locally called *red* or like a little white bird or like a bull. However, it is very rare that Nyikang appears to his people in form of a bull.

Livestock, Fishing and Hunting
(1) Livestock
Shulluk, unlike the other Luo groups, their wealth and social position are estimated in how many cattle a person has. The majority of the Shulluk are pastoralists who combine the herding of cattle, sheep and goats with growing crops. The common food crops grown include: maize, dura, beans, sim-sim, millet, gourds, pumpkins and tobacco is grown both for use and to sell.

The main incomes of the people come from their animals and what they grow. Traditionally, they barter the items among themselves; however, in this point and time, it is good to note that selling for profit is not in the minds of the Shulluk. Women and children also gather kinds of wild fruits and leaves of a certain tress.

Herding assumes prime important part of life among the Shulluk; therefore, their life revolves around their cattle. Thus, cattle besides being source of wealth, they are also source of elements to be used for rituals. They are commonly slaughtered for sacrificial purposes.

The Shulluk hardly use cattle as source of meat and animal proteins. However, the important nutritional contribution from cattle, in Shulluk society, is milk, which is consumed by all people in the community—primarily children. The people live beside the river so there is no need to migrate to cattle camps for summer grazing. In the summer, the cattle are taken across islands and the young people accompanying them build temporary cattle camps. Traditionally, women and girls are not allowed to milk the cows, the cows are milked only by boys and old men (Anders Breidlid, 2014).

The issue of women and girls not allowed to milk cows, became of interest to me, I then asked question to some Shulluk men and women: Why women and girls are not allowed to milk the cows in Shulluk society? Mrs Tarazah Chol Otor, said the question that why women and girls are not allowed to milk the cows is not a question to be asked by researchers, because entire Shulluk people believe that Nyikang has told women and girls not to milk cows and even not to drink milk. And this is what Collo say, "*Kwong Nyikang*", which means "Forbidden by Nyikang". However, she admitted that with the modern development and civilisation in Shulluk land ,women and girls, who live in the cities, now drink milk, although yet they do not milk cows. Mr Banydhuru Oyay, an informant, took the matter further by saying that maturing girls are not allowed to consume milk because milk is perceived in Shulluk society as something sexually stimulating; so to keep the girl subdued sexually , they are prevented from consuming milk.

The Collo keep small herds of cattle, in addition to lager flocks of sheep, goats and poultry. The wealth and social position of Collo men are estimated in number of cattle and land a person own. Muchomba (2004) reported that the principal forms of physical capital in the livelihood zones of Collo consisted of cattle and cultivated land. Households with these assets are able to participate in all-important social occasions. According to HEA (2013) livestock production complements farming, fishing and gathering wild foods in both zones. Cattle, goats and sheep are kept by most households; the animals are kept mainly for milk and for sales when cash needs arise in the family (P.8). In addition cattle are for food in the context of ritual and ceremony occasions. Moreover, livestock sales increase during the rainy season that is from July to August, this is the time households needs cash to buy food.

Milk production is also highest during the rainy season, when grass is plentiful. Shulluk people live along the river, so there is no need to migrate cattle to cattle camps for summer grazing. In summer the cattle are just taken across into the islands, there the young people build temporary cattle camps.

Although cattle are the measure for wealth and positions, grazing areas is increasingly becoming common problem for the cattle owners. In the dry season, grazing becomes impossible around the settlements, so cattle owners have to make sure that their cattle are moved to places where cattle can find grass. During this particular time, village elders assign young men and girls to travel with the herds to swampy areas (*toic*). The study revealed that livestock migration occurs during the dry season within the livelihood area; they migrate from elevated areas (*gok*) to swampy areas (*toic*). The Shulluk along the White Nile cross over to the eastern bank; likewise the cattle of the Sobat Shulluk descend to the lagoons south of the Sobat. So in both cases they have to cross the rivers, which is, on account of the many crocodiles living in them; the crossing is done with much care as we have mentioned somewhere in this book that weighty ceremonies would be done before the cattle cross the rivers, this done in order to keep the crocodiles away.

Livestock return to their settlement (*yom*) during the first rains (HEA, 2013). When cattle move away from the homestead, children may have less access to milk and this has often been noted as a chronic seasonal problem in nutritional surveys (Muchomba, 2004).

Traditionally, men herd and milk the animals—boys and old men milk cows. According to Shulluk traditional customs, women and girls are not allowed to milk cows or goats. Contemporary scholars stated that livestock play important role in the lives of the villagers, especially in term of social and economic functions as we shall discuss late about marriage in Chapter 18.

(2) Fishing Industries

As we have mentioned somewhere, the most significant natural assets in Shulluk land is the Nile and Sobat Rivers. The rivers are key resources for transportation and fishing as well as livestock production. According to Muchomba (2004) fishermen in Shulluk society are people who do not own cattle, these are the people Dinka called them the "Mony-thany". Interestingly, most river fishing takes place in dry season by men, women and young people. The wealthy groups are not involved in direct fishing they rather practice indirect fishing as we shall discuss later. We say the fishermen prefer carrying out fishing in dry-season because this is the time when yields are at their highest peak. Fish catches are consumed while fresh and surpluses are usually dried for later consumption or sales (Muchomba, 2004).

Fishing from the rivers usually starts from November to April. During the rain, when the rivers overflow their banks, the families' fishing activities switch to swamps and ponds instead of direct fishing from the river. Muchomba (2004) confirmed this when he says, "When excessive flooding occurs, shallow water fishing and water-lily collection become common economic activities of the day. When water levels are high, hooks, nets and spears are used; as the water recedes (gets lower) baskets are used to trap the fish. While in very wet years, fishing takes place in streams and ponds. And after the dry season, lung fish, which have been buried since the previous year, emerge from the ground and are also caught by fishermen. Burton, J. (1996) argues that Shulluk aggressively and successfully exploited the rich resources of the White Nile and its numerous tributaries and distributaries. They regularly catch many species of fish with fishing nets and spears.

However, when we compare South Sudanese standard with other developing countries' standards, we could say that the enormous fish resources in the White Nile and Sobat Rivers are not fully exploited. Truly speaking, not all households have sufficient tools for fishing. The work of fishing is mainly carried out by very poor and poor households. This then implies that those households without fishing equipment have to rent tools from the wealthy groups through village networks. The poor households do not have direct link with the wealthy groups, in this case they have to use their friends or neighbour, who have good link with the better-off to rent fishing equipment. The better off and middle-income in turn will receive portions of the catches from poor households for using their fishing equipment -the owners are paid in-kind (HEA, 2013).

Fish is an important commodity in Shulluk society, it is sold locally and sometimes they are even exported. Fresh fish is available throughout the year in almost all the local markets. Dry fish is only sold during the dry season; and the main market outlets of the dry fish from Malakal town to the neighbouring, such as Sudan, is only through Nasir markets. It is reported that fish traders come from across Sudan into Nasir town to buy dry fish. Once they have purchased a good quantity of dry fish, they they pack and transport to Sudan by river. While the traders from Juba purchase dry fish in Malakal town, they pack and transport them to Juba capital either by river of by road (land), since there are good communication networks between Juba and Malakal. In and around Malakal, the Nine and its tributaries provide good market access in the absence of threats to physical safety. Traders use canoes to travel along the major rivers and sell non-food items such as clothes, soap and cooking utensils in exchange for fish and cereal from communities living along the riverbanks.

Needless to say, fishing is a particular important source of food for poor households all year round; and an expandable option for all socio-economic groups during times of stress. Interestingly, the combination of fish, wild foods and game meat regularly contribute approximately half of poor households' annual requirements in Shulluk society.

(3) Hunting

The Shulluk go hunting by surrounding a large area with animals inside it; as the hunters surround the area, the animals then spread all over the area. As animal tries to escape it is speared and killed. But if a man wounded an animal, and it runs away, the people follow until the animal is tired and falls down. Interestingly, game obtained from hunting form only a small part of people's diet.

Any leopard killed during the hunting, the skin is to be taken to the King or in some cases kept by the man and worn only in especial occasion with the permission of the King. While for any lion killed in the hunt, the skin is tanned but not worn by anybody because it is considered evil animal.

In the old time people used to hunt hippopotamus by harpooning the animal and dispatching him with spears, when he comes to the surface of the water to breathe. The natives are famed for their skill in hunting the hippopotamus. They approach the animal in a perfect silence, in their light canoes, and plunge their sharp spears, with thorns atta-ched, into the back of one of the huge hippopotamus. The animal then dashes in the water, towing the canoe at a rapid rate. The animal continue dashing while hunters hold on the rope, till the strength of the hippopotamus is exhausted, when, they call for help from other hunters if possible. The hunters tow the animal into shallow water, and carrying the rope on shore, tie it on any shrubs available, and continue attacking the animal until it is dead. But today, South Sudan Government forbids people from hunting this animal again. But yet people hunt hippopotamus illegally.

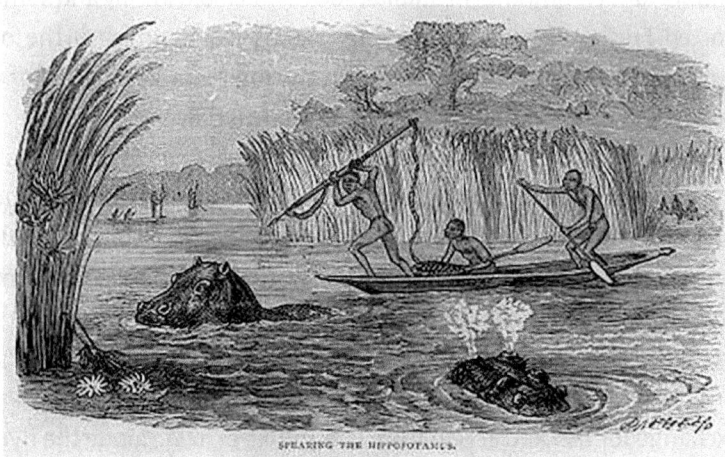

Figure: 155, Shulluk men hunting hippopotamus

As we shall see later, any man who let the animal escapes from him will be fined. For this reason, most of the animals surrounded are usually killed.

The men use spears, clubs, and traps in killing or catching animals. The animals that are commonly caught in hunting include: antelopes, gazelles, leopard, waterbucks and lions.

Moreover, in Shulluk society, there are two kinds of hunting: one being that people hunt for their food, where man kill animal such as gazelles and waterbucks for meat in the families. The second kind of hunting being is usually organised occasionally for the King, where men hunt to procure animal with skins that form part of the revenue of the King. In this second kind of hunt people hunt specifically for antelopes (gyiek) and leopards. The skins for killed leopards and antelopes belong to the King, no one can temper with them. If a man let antelope escapes, he would surely be fined. Traditionally, before men go out to hunt for the King, the district chief addresses them saying, "Oh! You people, hear a commandment of the King about antelopes (gyiek), which belong to him, if any man let the game escape, he shall surely be fined". After the chief has given the instructions, he then prays to Nyikang saying, "Oh! Nyikang, this matter is under your control! Do not let your people suffer fines from escapes of animals. Your grandfather always supports us to kill antelopes for the King, so that we are not fined! Oh! Nyikang be with us so that our spears fall on the animals!" (Diedrich Westernmann, 1912, P. 126).

After the prayer, a goat and a cock are brought, the goat is speared, skinned and cooked; for the cock, the head is cut off. When both the goat and cock are slaughtered, hunters start to move in different directions. When antelopes are killed, the people bring the meat and skins to the chief, who distributes the meat to various chiefs and their people. The skins

578

of antelopes are tanned and later on brought before the chiefs to examine them, to make sure that the works have been done properly and in accordance with the Shulluk customs.

If the people killed over sixty antelopes, only sixty skins of antelopes will be taken to the King at Fashoda. Traditionally, the skins are taken with a cow by the chiefs. When the chiefs reach Fashoda, they would be welcomed by the King, who orders his servants to kill big steer for the chiefs. Usually the meats of steer are eaten together with the meat of the cow that chiefs took along. After meals chiefs are given some local beer before they return to their villages. When the chiefs return to their respective villages, each chief would gather his people and tell them, "Oh! You people! We have returned from Fashoda, the King has received your offer of sixty skins of antelopes; may the country now live in peace" (P.126).

Wild Foods Gathering

The wild foods consumption is relatively high in the very poor and poor households. It is said the annual needs of poor and very poor households composed of 12-16 per cent of wild foods. There are many different types of wild foods gathered from the bush; but the most prominent include desert dates, dried fruits from the ziziphus tree (*lang* or *olango*) and tamarind seed (*Cuei/Koat* or *cwaa*). Customarily, wild food gathering is done by women and girls and sometimes by boys. It is true that both the poor and better off households eat fish and wild foods but it is important to understand that the proportional used of wild foods varies. Wild foods and fish provide between 21-25 per cent of the very poor and poor households annual food needs; and other hand the wild foods and fish provide between 5-10 per cent of better-off and middle-income households annual food needs. The very poor and poor households produce between 39-47 per cent of their own food through crop cultivation to meet their annual needs. They also receive in-kind between 9-11 per cent of the fish for using nets and hooks of the wealth groups.

Sources of income

The study conducted in 2011-2012 revealed that milk and livestock sales are the primary sources of income for the wealthy groups in the society. It is found that in a year, both the better-off and the middle-income households sell 25-33 per cent of their milk produce from 4-7 cows and 6-10 goats. Then whenever the better-off want to raise additional family cash for their basic foods they sell 1-2 bulls, one sheep and one goat in a year.

The study also found that very poor and poor households earn from the sales of fish approximately SSP 125-130 per person per year; and the better-off households earn SSP 700 from the sales of crops and fish per year. Since the wealthy groups have many people living with them in the house, when this amount is divided among the average household member, each person will be said to earn SSP 105 per year. The earning of an individual from wealth groups is low, as compare to the very poor and poor households, because there many individuals in the house and some may not even be working. This is the common practice of some Southern Sudanese, where relatives tend to leave their homes and come to live with a better-off household whether in the towns or in the villages. One informant told me that in a better-off household there could even be between 15-30 persons living in the household; while in the poor and very poor household there could only 3-6 persons.

The Collo households also get their income from sales of sale of firewood and grass, although this economic activities is ran mainly by poor households. The activity usually conducted in dry season. The other sources of Collo households come from sales of wild foods such as desert dates, ziziphus fruits and tamarind seeds. It is reported that the sales of wild foods starts in November to April every year. Sales of wheat flour and other non-staple food also form another source of income for the wealth groups.

Wealth Breakdown

As we mentioned somewhere before, Collo define wealth primarily with how much live-stock and land does an individual has. Thus, people with many animals and have access to land are said to be wealthy. Although livestock ownership is correlated with wealth, actual herd sizes fluctuate from year to year depending on wealth outcomes, disease, incidence and the incidence of raiding. As I mentioned before, the better off may own between 50-100 head of cattle. It is reported that in the bad period when animal reproduction is low, when there is outbreak of disease, and worst still when the area has been raided, the households, head sizes may drop from 50-100 to typically in the range of 10-30 head of cattle (HEA, 2013). Other scholars like Diedrick (1912) also argue that the favourite occupa-tion of the Shulluk is cattle-breeding; cattle mean wealth and social position, while the cultivation of the crops is merely the means of procuring daily food. The cattle are of the Zebu race, with hump behind the neck; they are tall with rather long legs, a slender body, and large horns. The horns, while young, are dressed into most manifold strange forms; this is usually done by a particular craftsman, called the "dresser of horns". Many a time, in a large herd, one hardly sees any cattle with the horns in their natural shape (P.28-29).

Poverty is associated with few and consequently with lower production and income as well as with cattle raid and animal pests. This would affect communities at all levels even though the effects may be received at various degrees. Muchomba (2004) puts it clearly that, on one hand better-off groups may be more directly affected by a cattle raid or crops distribution, by wild animals such as elephants, locusts, and birds, and on the other hand, the poor households will also be affected indirectly as they will lose access to whatever surplus their richer relatives may have had to share.

In terms of assets such as fishing nets and other gear, the poor households own very little. Subsequently, the poor collect fish mainly from swamps where they fish using basic spears. It is unfortunate that the very poor households do not own any cattle, but it is said that the poor households own small herds of cattle that range from 3-5 head of cattle, 2-10 goats and 1-5 sheep.

In both Eastern Flood Plains and Nile Sobat Rivers, communities define the poor and very poor households as people who have 2-4 cattle and cultivate 1.5-2 feddans (acres), this implies that they have lower agricultural productions. According to the communities, the poor households also own very little fishing nets and gear. Subsequently, we could conclude that this is one of the reasons why the poor households collect fish mainly from swamps and ponds, where they fish using basic spears. The very poor and poor household produce 40-50 per cent of their annual food needs; they do not own any cattle at all, but may have 1-5 sheep and 2-10 goats and poultry; although some people argue that the poor households normally have between 3-5 head of cattle. Still in Eastern Flood Plains, better off and middle-income households are defined as people who produce sufficient crops to meet about 50-65 per cent of their annual food needs. Of this, 75-80 per cent of own crop production are sorghum and maize. They have some 50-100 head of cattle, 60-150 sheep and 60-100 goats and poultry.

National cattle movements have been significantly disrupted by the violence, with large scale and long distance displacement of livestock from the conflict affected areas –to Melut; and abnormal livestock movements into Melut area have stressed natural resources and contributed to the outbreak of livestock disease in the area (FAO, Dec. 2014).

Cattle are regarded as personal beings, and as a matter of fact in Shulluk society, they have approximately 40 names for cattle; these names could be give according to their colours, the configuration or the size of the horns of that particular bull or cow. In other words we could say that the Shulluk are more interested in the colour and shape of the cow's horns. Nevertheless, here below are some of the names Shulluk call their cattle:

Abach is a cow with horns directed straight sideways

Agwognom is a cow with horns directed straight upwards, like a goat's horns

Ayokak is cattle with black body with white tail

Bany is a cow with one horn directed upward, and the second horn directed downward

Nabek is cattle with one leg white, and the rest of the body is yellowish

Najak is a cattle with yellow head, brown spots on the back; the rest of the body is white

Najok is a cattle with black head, black spots on the back; the rest of the body is white

Naker is cattle with flanks black belly and back white

Nading is cattle with brown back, and small spots throughout the body.

Niyom is a cow with white head; the rest of the body is either black or yellowish

Nyabong is with white colour throughout the body

Nyenyari is cattle with striped white and red colour throughout the body

Nyat is a cow with the horns cut off

Nyakwach is cattle with black and white sport throughout the body

Onugo is a cow with two horns directed straight backwards

Obyiech is cattle with ordinary, non-dressed horns

Ochodo is a cow with no horns

Odielo is a cow with horns all turned down

Odiulo is a cow with horns pointing forwards

Ogak is a cow with white belly and neck, and with black head and the rest of the body is black.

Ogwiel is an ox with horns turned towards the eyes

Oleng is a cow with large spotted brown and white

Olek is a cattle with grey-white spotted through the rest of the body

Olut is a cattle with brown-white small spotted throughout the rest of the body.

Tabur is cattle with ash-colour throughout the body

Teduk is cattle with grey colour throughout the body

Takyiech is cattle with flanks white, white and the nest of the body is black

Tedigo is cattle with red-brown colour throughout the body

Tetany is cattle with black colour throughout the body

Tyiel Riek is an ox with white feet and the body can be any colour

Wariegot is an ox with one horn directed forward and the other horn is directed backward.

Warnamtai is an ox with two horns directed straight backwards

Source: (Deidrick, 1912, P. 108)

Expenditures Patterns

The Collo, like the other Luo groups, spend their incomes in many different ways, which include the purchases of: Staple foods, non-staple foods, drugs for animals, fishing equipment, agricultural tools, seeds, labour and transportation. The study conducted by HEA (2013) revealed that the very poor and poor households spend 33% of their annual income on staple foods. Most of them cannot afford access to basic health services. They spend 5% of their income on non-staple foods such as oil, sugar, soap, salt, beans and wheat flour; whereas the better-off and middle-income households spend only 14% of their annual income on staple foods. The better-off and middle-income households also spend 10% of their annual income on non-stable foods. The report clearly reflects that there are significant differences of expenditures between the poor and wealth group households. The study indicated that the cost of living for the better off households, who are living in Eastern Flood Plains zone, spend SSP 275-320 per capital per year on livelihood inputs; and they spend SSP 765-1,125 on their staple foods, and SSP 145-690 on non-staple foods annually. Muchomba (2004) pointed out that the better-off have almost three times the income of the poor and so their expenditures patterns vary greatly. The poor spend a high proportion of their income on sorghum and whatever remains covers essential non-food items. The better-off have more choice, spending a lower proportion on staple foods, a little more on basic essentials, and whatever remains is spend on veterinary supplies. Yet, if there is anything left over, it is used to buy more cattle (P.64).

The same study compared the households in Eastern Flood Pains and Nile and Sobat Rivers and discovered that the very poor and poor households in Nile and Sobat Rivers spend SSP 492-569 on their stable foods; and spend SSP 90-90 on their non-staple foods; they spend SSP 10-120 on agricultural equipment and SSP 55-75 on seeds. In addition the poor households spend SSP 0-0 on labour, the reason been that they do not hire labour. Yet they spend SSP 0-75 of their income for the purchase of livestock and spend SSP 30-65 on animal drugs, meanwhile they spend SSP 0-0 on transportation. The poor households spend between SSP 70-85 (in-kind) in return for using fishing equipment. The poor spend SSP 0-0 on labour since they do not hire labour for land preparation, weeding and harvest. They do the work by themselves. The even spend just SSP 0-75 of their income on purchase of livestock and spend SSP 30-65 on animal drugs– as we have seen behind many poor and very poor households do not keep cattle.

The study also revealed that the better off and middle-income households, who live in Nile and Sobat Rivers zone spend SSP 230-260 of their income on fishing nets, hooks and spears. They spend SSP 1,300-1,380 of their income on staple foods and spend SSP 1,436-1,664 of their income on non-staple foods. They spend SSP 365-630 of their income on agricultural tools, SSP 100-140 on seeds. They spend SSP 230-675 on labour (i.e. for land preparation, weeding and harvest). They spend between SSP 1,115-1,400 of their income for the purchase of livestock. The better-off and middle-income household spend some 10-20% of their annual income on social services. These low expenditures on social services may be due to limited and medical services available in the livelihood zone. However, both poor and wealth group living in Nile and Sobat Rivers spend literally SSP 0-0 on fertilizer, since the land is fertile and does not need imported fertilizers (HEA, 2013, PP. 13-16)

The above narratives inspired me to conclude that here are great differences of income and expenditures between the households depending on where one lives. It is also interested to learn that both the better off and poor households in both zones (Eastern Flood Plains and Nile and Sobat zones) do not incur transport costs. This is amazing to learn that areas where Collo live in the two zones are relatively isolated from good road networks; as a result people travel on foot and sometimes by boats. It is also interesting to learn that wealth groups in Nile and Sobat Rivers zone spend just SSP 80-140 of their annual income on transportation, the reason might be that because they live closer to Malakal town, the capital of Upper Nile State. As such, they have better access to the transport network associated with the Capital Hub leading to less transport expenses.

Trade

This takes place using either cash or barter, however, the Shulluk trade has improved considerably as from the old days to the present. In the past the Shulluk trade trips of hippo-hide for making whips for the Arabs and they trade spears, pots, hides and other vegetables to the neighbouring by just bartering for cattle, iron, grain and cloth. It is reported that in the old days, Shulluk did not understand the important of salt because at that time they substituted salt with cow's urine. But iron was important element as the iron was widely used for manufacturing spears (South Sudan, Ministry of Education, Science and Technology, 2014).

However, today trade is well developed in Shulluk land, there is no more bartering but customers purchase cattle, salt, iron, pots, grain and cloth with cash. –which is a clear indication of trade development. HEA (2013) revealed that the population who live in Shulluk towns purchase commodities in cash from the local markets; although food purchases are affected by market access and food availability. It is reported that access to markets is typically poor in Upper Nile because there are few all-season roads in the region. During the rainy season roads are often washed out and villagers find it difficult to access the places by roads. The roads are inaccessible in the region for at least seven months in a year and road is only accessible in the region for five months in a year. Muchomba (2004) the poor households currently access most of their income through petty trade activities, the sales of fish products and tobacco. The lack of markets, seasonal access problems, lack of infrastructure, cash and the mixed currencies in the border of Ethiopia continue to pose greatest constraints to local trade.

However, before civil wars that broke out in 1955 -1972 foods were plentiful in all local markets in Shulluk land. Soon after the Addis Ababa Agreement in 1972 there were again good supplies of foods in the local markets. This did not last long in 1983 the supplies of foods were interrupted until 2005, following the signing of Comprehensive Peace Agreement (CPA). From the year 1983-2005 availability of foods in Upper Nile were at their lowest level. This affected the flow of the market as well as rising of the costs of goods.

The sorghum is the main staple food in the area; it is usually supplied by local farmers in and around Malakal. In the good year sorghum find their rout to the neighbouring (Ethiopia) from Renk to Melut and from Melut to Malakal and from Malakal to Nasir; and from Nasir to Ethiopia. The rout follows the White Nile and Sobat Rivers. Most of the Northern traders purchased sorghum direct from Renk and transport them to Sudan. It is reported that customers purchase sorghum from Renk and Nasir in Sudanese pounds or Ethiopia Birr (ETB). However, in the bad year when there is poor supply of sorghum, local traders imported sorghum from Sudan. In good rears maize and cowpeas are supplied locally, although these items are not unusually available for consumption from March to August. Prices for staple foods fluctuated greatly during the year due to poor condition of road infrastructure combined with limited access to cross-border trading with Sudan and Ethiopia. The recent study revealed that some households paid up to SSP 7-12 per kg of sorghum between the month of June and August in 2012.

Hazards

The livelihood in the two zones, Eastern Flood Plains and Nile and Sobat, are subject to a number of hazards. Some hazards are chronic and undermine food security every year. Crops pests, livestock disease and flooding are the principal chronic hazards in the zones. Livestock diseases are a serious hazard to livestock production every year affecting all live-stock owners regardless of their wealth status. One of the most serious livestock diseases in the zone is pneumonia. Interestingly, pneumonia mainly affects cattle. It reduces milk production and may lead to death.

With respect to crop pests, birds (quelea quelea birds) and rodents are typically a problem at harvest time, particularly for sorghum. Flooding and heavy rains are another chronic hazard that affects crop production. Excessive rain causes leaf rust disease. Extremely heavy rainfalls during the main harvest period also reduce crop outcomes and affect all wealth groups as well as the very poor and poor groups. Flooding itself is often caused by heavy rainfall in the highlands of Ethiopia, which then causes the Sobat River to over-flow (HEA, 2013, P.24). The frequencies of the flooding in the zones often result in heavy rains making crops loose the productivities.

Cattle raiding and clan tensions over water and grazing occasionally occur during dry season and this could be counted as one of the hazards. Chapter 18: Marriage and Ceremony

"So God created man in his own image, in the image of God he created them (Genesis 1:27). And God blessed them, and said to them, "Be fruitful and multiply and fill the earth and subdue it, and have control over every living thing in the seas and on earth (Genesis 1:28-29). "Therefore a man shall leave his father and his mother and join with his wife, and they become one flesh" (Genesis 2:24, Ephesians 5:31). Wives have to submit to their husbands, for the husband is the head of the family. In the same way, the husbands should love their wives as their own bodies (Ephesians 4: 22-23, 28). The husband should give to his wife her conjugal rights, and likewise the wife should do the same to her husband. For the wife does not have authority over her own body, but the husband does; likewise the husband does not have authority over his own body; but the wife does" (1Corinthians 7:2-4).

Chapter 18:
Marriages

What is marriage?

"Gender relationships in South Sudan are shaped by the social and economic realities of being one of the world's Least Developed Countries and by decades of conflict. There are more men than women in south Sudan: 52% male to 48% female compared to the global average of 51% female to 49% male. In south Sudan it is a marriage that shapes a woman's experiences, her status and her responsibilities" (Care, 2013, South Sudan: Gender, P. 1). Therefore, the acquisition of several wives is seen as important for socio-economic advancement, as many wives can bear many daughters, who in turn can bring in many sheep, goats, cattle, money from bride-price when they eventually marry. Polygamy is the rule; the number of wives is only limited by the ability of the husband to support them.

"A marriage without sex isn't really a marriage".
From the beginning of time, as recorded in the Book of Genesis, God planned for man and woman to unite in love and harmony for continuity of his creation, the human race. This is the primary purpose why God created a man and a woman. But with the current Trent of education and philosophy, do all human beings understand the meaning and purpose of marriage? I leave this question to be answered by readers.

Given the fact that we are not in the twenty-first century, the above question will have varieties of answers. This is a complex issue that need biblical reflections. Thus, some people may believe that it is not important to marry, while others may say marriage does not necessarily means union between a man and a woman, it is an individual choice whether to marry the opposite sex or to marry the same sex or whether to remain without marriage altogether. The question is if we marry same sex or remain without marriage completely, how we would feel the commandment of God that says, "Be fruitful and multiply and fill the earth". When we look around the world, we can see that some Churches today have approved same sex marriage and before long they will also approve abstinence from marriage. This philosophy is not in accordance with what the Living God tells us through the Holy Bible.

Any rational human being should understand that marriage is the intimate union and equal partnership of a man and a woman. It is not something created by us, but it comes to us from the hand of the Living God, who created male and female in his own image, so that they might be fertile and multiply. However, by the word "equal partnership of a man and a woman", used here, doesn't necessarily implies that both man and woman are of equal status in everything whether it be physically, emotionally, socially or morally –that they identical to one another. Although man and woman are equal as God's children, they are still created with important differences –biologically they are not the same- that allow them to give themselves and receive each other gifts.

According to Friederike Bubenzer and Orly Stern (2011) agues, "South Sudanese society is strong traditional and deeply patriarchal, and this is clearly seen in the institution of marriage. The roles and positions of both men and women within a marriage are clearly defined and strictly enforced: men are the heads of households, holding families; women are subservient to their husbands, with their roles focused on the home and the rearing of children. While not equal, this division of roles and responsibilities was intended to ensure a clear allocation of tasks, and to guarantee that all were taken care of, protected and supported" (P.2). Northern Luo culture is deeply patriarchal; men are heads of their households, and hold positions of authority and power within their families. The husbands have final decision-making power in the family on all matters of importance. Women are expected to be subservient to their husbands, and to be obedient to their husband's male relatives, even if these are young children (Seligman & Seligman, 1932). The responsibilities and duties of Luo wives are enormous. Wives tend to work an 18 hours a day that revolves mainly around food production, domestic work, fetching water and fuel for the households, and caring for their families, including being sexually available to their husbands. Husbands are expected to be the providers: a husband is supposed to take care of his wife, build her a house, cultivate for her –although some women do also cultivate the fields, provide her with money, meat and care for livestock (City Gachago et al, 2003.)

There is strong pressure from the community for husbands and wives to fulfil their duties within marriage. However, the sanctions for not fulfilling one's obligations differ for men and women. A woman who fails in her wifely duties faces a variety of punishments that range from being reprimanded to being beaten by the husband. Meanwhile a man can easily get away with neglecting his responsibilities and is seldom censured. I often like mixing with women and often hear some say that their husbands do not adequately fulfil their role, as a result they leave women to do bulk of the work. I quote, one woman was saying, "Husbands generally do not do much, traditionally they supposed to cultivate farms and care for animals as well as family members. But today men have lost their roles as men, they sit under big trees or bars to talk and drink local beer. So, women do the hard work". However, I have observed that there is rapid changing circumstances that resulting shifts in gender dynamics affecting marriage among the South Sudanese people, some couples now share responsibilities in the house –specially women and men in Diaspora who are being exposed to many new and Western influences –the roles and responsibilities in marriage are being shifting and being adapted.

Many aspects of the traditional Northern Luo way of life began to shift during the civil wars (1983-2005). Among these shifts are changes in the roles that men and women play within marriage; a number of men and their wives left South Sudan and took refuge in the Western World, such as Australia, America, Canada and Europe. Women who lost their husbands in wars now acting as heads of their households, holding positions of independence and responsibility that are new to them. The same with the women in Northern Luo regions, in South Sudan who has lost their husbands in wars are also acting as heads of their households and holding positions of independence and responsibilities they did not have. Women in South Sudan and in Diaspora are today working in Government or Organisation in order to fill the gaps left by men in essential services and to ensure that money and food are available for their families.

These changes, caused by necessity, gradually began to affect the rigid division of responsibilities between men and women, and husbands and wives. The war created avenues for women to assure greater levels of responsibility. Many women contributed to the war effect, largely in support roles, with certain women rising to leadership positions in the independent movement. The opportunities for increased female independence and power began to affect the rigid power dynamics within marital relationships. Interestingly, the war also created some alternatives to life as women previously never had such opportunities. Some men are even happy that women now occupying the positions they could not hold before. This contributed to the development of a crisis of masculinity, which could be argued as playing a role in the rising rates of domestic and sexual violence (Friederike Bubenzer & Orly Stern, 2011, P.7).

Marriage is both a natural institution and a sacred union because it is rooted in the divine plan of creation. For this reason, in every marriage the spouses make agreement with each other; however, this agreement could be done before religious Minister or before parents and relatives of the couples –all these carried same weight before God. In my opinion, it is the free consent of the spouses that makes a perfect marriage before Almighty God, regardless of where the marriage was solemnly conducted in the Church or it was agreement reached before parents and relatives. All-in-all the form of the ceremony consists of either a woman finger pointing the husband to her parents and relatives or by mutual exchange of vows by a couple in the Church. In the Church there is mutual exchange of rings between a man and a woman. Therefore, the Minister himself is the one marrying the couple by pronouncing the couple a wife and a husband after the exchange of vows and rings.

However, in the Middle Ages when there was persecution of Christian in Europe and other British colonies, when public religious ceremonies were forbidden, a traditional or clandestine form of marriages were considered valid. Jesus Christ stressed the importance of the marriage bond in his Ministry, from the beginning of creation, "God made them male and female. For this reason a man shall leave his father and mother and be joined to his wife, and the two shall become one flesh. So, they are no longer two but one flesh. Therefore, what God has joined together, let no man put asunder" (Mark 10:6-9). Today, clandestine marriages are still considered valid, especially in developing countries; although the Churches do not consider they are no longer valid. This remains to be seen when we get to the Kingdom of God, because the clandestine marriages shall continue until the second coming of our Lord Jesus Christ. The bond of marriage between a man and a woman lasts for all days of their lives on earth.

Marriage recognizes the interpersonal relationship of man and woman, in which the well-being and self-realization of each partner become a priority for the other. Marriage provide framework for the mutual love and self-giving of a man and a woman to each other in human sexuality, and in doing, provides for continuity of the human family. Without marriage, society will fail; -in other words the existence of human society would come to an end. In our human language, we can say, "*The man gives to the woman and the woman receives the man; in this way the love and friendship between a man and a woman grow into a perfect and sustainable marriage. Therefore, both grow into a union of heart, body and soul to provide stability for them and their children as children are the fruits and bond of a marriage*".

Among the Northern Luo in South Sudan, young women are typically marry around the age of 18 years, while young men usually marry between the ages of 20 years and 25 years. Most young men and women live with their parents until they marry. Many married couple stay with their families until they have one or two children of their own. This will become clear as we discussion marriage in individual Luo groups.

John P. and Wayne G (2016) argue, "*God's gift of complementary manhood and womanhood was exhilarating from the beginning (Gen. 2:23). It is preciously beyond estimation. But today it is esteemed lightly and is vanishing from much of modern society. We believe that what is at stake in human sexuality is the very fabric of life as God wills it to be for the holiness of his people and for their saving mission to the world Bible teaches that God intends the relationship between husband and wife to portray the relationship between Christ and his Church. The Husband is to model the loving, sacrificial leadership of Christ, and the wife is model the glad submission offered freely by the Church. By submission we do not mean that it is absolute surrender of a wife's will. It is rather her disposition to yield to her husband's guidance and her inclination to follow his leadership. The absolute authority is Christ and not her husband. The wife should never follow her husband to wrong doing or to sins"* (P.20-22).When Paul tell us of the headship of the husband, here he has in his mind the picture of a body with a head and this headship signifies some kind of leadership. In some societies, people think that the man is the head, but the woman is the neck and she can turn the head in which ways she pleases. But according to the Northern Luo Cultures, woman is not allowed to direct the man wherever she pleases.

God's design is for "a man to be united to his wife, and they will become one flesh" (Genesis 2:24). God created sex in marriage to be shared, not withheld. Our God-created physiology, he creates a physical need for regular sexual release. If this need isn't satisfied, we are less emotionally engaged with our wife, more emotionally tuned out to her needs and the needs of our family, and just overall – quite irritable. And when romance, tenderness, and sex are not shared, a sense of loneliness sets in that can ultimately result in emotional and sexual temptation.

I quite agree with Ms Chima Ezeh (2017) when she said sex is food, it is not all about children. A woman must be different to her husband every time; the woman must seduce the husband and not expect that husband to ask her for sex every time the husband wants sex. There should be no time timetable for sex; the wife must be creative and gives him what he wants to avoid losing influence over the husband. And if a woman loses influence over her husband, this means that she has also lost womanhood. A wife should be part of the husband, and whenever the husband is sexually satisfied, he becomes emotional stable. A wife should not tell her husband, "Why do you like too much sex, is it a food?" The wife should not be too outspoken, she must know when to talk, when to listen and when to be quiet. Some people are not happy that you are happy in your marriage, proof them wrong that you love you yourselves.

Having said these let me say something briefly about the husband. Some men think that they might never, in time turn, down the sexual advances of their wives. This is often not true, if so it is very rare. However, some of the working men sometime feel too tire after work and would like their wives to leave them alone (no sex). This is not a good thing for the man to do because it often brings frustration to wife. However, most men do not go to such drastic measures to avoid intimacy in marriage. According to Robert Byrne (2017) Sex is very, very important to men. Research consistently shows that between 80 and 90 percent of men view sex as the most important aspect of their marriage. When asked what one thing they would like to change in their marriages, they wish that their wives would be more interested in sex and more willing to initiate physical intimacy.

One of the biggest differences between a wife and her husband is the fact that he experiences sex as a legitimate physical need. Wife should understand that just as her body tells her when she is hungry, thirsty, or tired; the same way the husband's body tells him when he needs a sexual release. Your husband's sexual desire is impacted by what's around him but is determined by biological factors, specifically the presence of testosterone in his body. Immediately after sexual release, men are physically satisfied. But as their sexual clock ticks on, sexual thoughts become more prevalent, and they are more easily aroused. The physical need for sexual release intensifies as sperm builds in the testicles. The body continues to produce and store sperm, although sperm production fluctuates based on levels of testosterone and the frequency of sexual release.

More importantly, a woman's sexual desire is far more connected to emotions than her husband's sex drive is. While a man can experience sexual arousal apart from any emotional attachment; he can also look at a naked woman and feel intense physical desire for her, but at the same time he may be completely devoted to and in love with his wife. For most women, this just doesn't work. A fundamental difference in the wiring of male and female sexuality is that men can separate sex from a relationship while for a woman, the two are usually intertwined.

Here below are some living examples, from the women who experienced difficulties in sexual relationships with their husbands, but due to confidentiality, I will not mention their names:

Example 1: *"I try to get him aroused and interested in sex, but he is never really in the mood nor is he affectionate to me. He expects me to let him know when I want to be intimate, and I need to do the seducing. This is really hurting our marriage and I am resented of his lack of interest. I try to be as attractive and sexy as I can, but nothing seems to work".*

Example 2: *"My husband and I have been married for five months. I am 40 and he is 46 —both first time marriages. However, intimacy in our relationship is almost non-existent. He seems pretty much disinterested and 99% of the time rejects me when I try to initiate lovemaking. I have tried to talk to him about it but he says there is no problem. The rejection I am experiencing has become almost too much to bear".*

Example 3: *"My husband has no desire to make love to me. I have to initiate all of the encounters, most of the time unsuccessfully. I felt rejected on a nightly basis, so I took a night shift job so I would not continue this rejection".*

The top ten secrets one need to know about marriage

It is not a surprise that very few people actually know what they are getting into when they are getting marriage. It is important that we should ask such people, "What do they know marriage is all about?" Each couple have different perception of hopes and expectations of marriage. Our perceptions of marriage are often influenced by movies we watch, by information we receive from our friends, by watching TV shows; many young people do not understand what is meant by "Marriage" until they get into it with little or no knowledge at all.

When I was doing my research on "what are the top ten secrets of marriage we need to know!" I read different opinions but I did not agreed with any of them, but I only became convinced and agreed with the opinion expressed by Debra Fileta (2017), a Professional Counsellor and author of the book "True-Love-Dates". I am made to believe that in marriage there are things one could do and know and there are also other things one cannot do and would not know because they are deep within us.

Marriage alone cannot heal us, it cannot even makes us complete in ourselves, but God can use marriage to transform us into a new person. Let us now examine the ten top secretes in marriage.

(1) *Marriage is more intimate than sex:-*When people think about marriage the first thing that comes into mind is "sex" In many societies, sex is viewed as cement that closes potential gap in the relationship. Even though we think sex closes the potential gap between relationships, nevertheless, we value closeness within the sexual relationship as good marriage makes for good sex and not good sex makes good marriage. Before marriage many a times we do not realise the intimacy that comes with committing to allow another person have a look inside your life. It is amazing opportunity to allow somebody you love to look inside your life, your mind, your heart and your very soul during your marriage period. This is what we called "true intimacy".

(2) *Marriage reveals selfishness, but can also cultivate selflessness*: - Many people know the ability of being selfish, but they do not actually know what is selfish. These types of people will not know the really meaning of selfish, they only come to know the meaning once they are into it. A wife may ask the husband to give her a knife or she may ask him to move away from where he is currently sitting because she wants to clean the place. The wife may also ask the husband to apologize for the wrong done to her; in all these, a man may see as though the wife is trying to dominate him, here is when the selfishness is revealed. From this point of view we can see that selfishness is something that has to be lived out with. I have learned from the Scripture that God has given out his only beloved Son to die for my sin and to make Salvation absolute. Here God selflessly gave up his only beloved Son for ransom to many. Therefore, we should learn from God to give selflessly to our wives the things we value so much in our lives. This is not easy it requires determination and commitment.

(3) *Oneness in a true sense of the word (which means one):-* We often talk about the deep spiritual and physical benefits of oneness, but have we ever consider: one house, one bed, one bank account, one bathroom, one toilet and one budget for our home? The reality is that we hardly think about these things but in marriage, we need to learn how to "share" and these come without pre-education. We need to learn about "giving away" the basic words such as "mine" and "yours" and receive word such as "ours" because in marriage everything is truly ours; there should be no word "mine" or "yours" in marriage life. The term "mine" and "yours" pre-supposed individual ownership, while the term "ours" pre-supposed collective/family ownership. I know giving

away word "mine" or "yours" is something hard but still there is something beautiful about it, because as love deepens and grows stronger and stronger, at the end of the day what is "mine" become "yours" and eventually become "ours"; how wonderful is this? However, some informants expressed concern that it is difficult to have one bank account and one budget with a partner, they think that it would be better to have separate account and have one mutual account. Some men do not feel they need to ask their wives when they want to buy another pair of shoes; they feel they can manage their own budget. Nevertheless, people like Joellen appreciated the idea of "sharing". Joellen said she has been married for 37 years, and found it difficult to share things with her husband during the first 20 years of their marriage but now she shares everything, although it was not always easy to share, but once they got the correct path, this made a difference in their lives –sharing everything is the secrete to a happy marriage, I can attest this –said Joellen.

(4) *Marriage has no smooth path, at some points you will be disappointed:* - I know this is a hard reality to comprehend and believe. But let us ask ourselves, "Why do we find it difficult to believe that one day or another we shall displease one another in marriage life?" Both husband and wife understand their human weaknesses but they cannot know "this simple truth" until they experience it in the house. In the home a husband and a wife could love themselves dearly but yet they could also hurt each other very deeply. When you allow somebody you love to bury his or her heart in yours, without doubt one day, you will the given pressures, and these could be in form of unkind word, a thoughtless action. This then tells us that marriage is not a Paradise it will hurt. Marriage is not just and lived happily ever after fairy tales or just a bed of rose, but it comes with its challenges that help to improve the lives of the individuals involved. But if you lift up your heart and entrust marriage life to God, in this way the Grace of God will pave way for forgiveness and restoration of true love.

(5) *Marriage offers opportunity to learn the meaning of forgiveness:*-The fact that couples know they hurt each other, despite their deep love, this reality should cultivate in the hearts the need of learning "forgiveness". But the biggest lesson to learn is that true forgiveness comes not because the person standing before you is deserving forgiveness, rather, it comes out from the heart that understands how much God has forgiven our sins, through the blood of Jesus, though we, too, were undeserving.

(6) *Marriage is costly:* - In this context I am not talking about the cost of wedding. The cost of wedding is nothing in comparison to the emotional costs that come with becoming one. The bottom line is that you lose a part of yourself within the glory of marriage. You exchange a little bit of who you are for a little bit of who your husband is. You have to learn to give and take, and you must learn to give up some of the things that really don't matter in your previous life. At the end of the day, you will realize

that what you have given is far less than what you have ultimately received. This is the fruit of love.

(7) *Love is not a feeling; it is rather a series of decisions:* - During courtship and before marriage one can never comprehend the feeling going around his/her heart and what is going around in the mind. You can realize that your feeling of love toward the person is getting higher and higher and you want to be together any time of the day or of the week. Love cannot be trusted because sometimes you feel you should break up or sometime you may not like each other person. Feeling come and feeling also go, this is normal in life; in other word we could say feelings are a compass, and sometimes a guide, but they are never to be followed. Love is easy when you feel like it, but when you do not like it, that should be taken as a test of real love. If you have a real love, you will preserve the feelings, but when you do not have true love, you will break up. You have chosen to love, to give, and to serve because of the commitment you have made. You choose other person to be part of you; therefore, you are not choosing yourself to be yourself. This is the very definition of "Love" in its truest form.

(8) *Marriage requires good communication:* - Whether you like to talk or you don't like to talk, it does not matter, but in as soon as you are married you must learn skills of good communication with your partner. Even though you are a poor communicator, marriage will force you to bring your insides out. It will requires you to polish your words before you utter them out; it will also requires that you examine and reflect on your opinions, beliefs, ideas and feelings before you share them with your partner, neighbours or in-laws. Marriage will cause you to answer hard questions, and speak the difficult truths, because communication is the lifeline between two people. There is no beating around the bush when communicating each partner must call spade, spade. Communication will cause you to take responsibility for not just what you say to others but how you say it, which means your body language, tone and sarcasm will all be taken into consideration to interpret your inner feelings and ideas. Not only does sex set you and your spouse apart from simply being roommates, it also requires a deeper level of communication that you don't normally do with just anyone. Sex requires you to talk to each other about intimate, emotional things. For example, to have a truly intimate experience with your spouse, you need to tell your spouse where you like to be touched, and make requests for certain things. This requires that you both feel a comfort level with each other that you've never felt with anyone else before. It requires you to both become very vulnerable by asking, receiving and giving sexually. And it requires you to reach a deeper level of trust that your spouse will respond to your requests without judgment. Poor communication in marriage is number one complaint from ladies around the world. As a result many women ask, "How can I help my husband to communicate with me?" According to

Dr. Wayde Goodall (2013) men are natural communicators, they are more doers; but women are natural communicators. Therefore, this implies that men are challenged to learn how to communicate and to understand the needs of their wives like the Bible tells us in Ephesians 4:15, which reads, "Rather, speaking the truth in love, we are to grow up in every way into him who is the head, into Christ. Ephesians 4:29 "Let no corrupting talk come out of your mouths, but only such as is good for building up, as fits the occasion, that it may give grace to those who hear". Colossians 4:6 "Let your speech always be gracious, seasoned with salt, so that you may know how you ought to answer each person.

Thus Communication is a two ways traffic, where a husband communicate words to his wife, the wife receives the message and responses to the husbands in words. What matters much in this form of kind of communication is that each must communicate understandable words to other. It is interested to note that poor communication in marriage usually starts with uttering harsh words and talking rudely. This poor communication creates room for misunderstanding and eventually makes the other party to feel negative. This often follows with accusations on different ground against the other, and if no conflict resolution is offered, either from within or from outside, it leads to separation and finally divorce.

(9) *Marriage requires continuous Conflict Management:* As we shall see later under article "healthy marriage", it is important to note and understand that marriage is composed of *"commitment to each other"*. Thus, for a man and a woman to remain in marriage all their lives, they must communicate in a family; and where there is communication, there often arise conflict. Conflict issues in marriage have been there since the beginning of human race. The most important thing for couples to do in this situation is to learn how to manage conflict in their home as it arises. However, my advice to couples is that if you do not want to learn conflict management, you should stop yelling at each other, stop calling bad names to each other, and don't raise your hands to hit the other –because all these acts are abusive behaviour. God created man and woman to live peacefully; he (God) does not want bad behaviours in family life. He wants couples to remain intimate and supportive to each other; but we are fighting with our human nature that often goes contrarily to God's plans or wishes. Some people intend to avoid each other, so that they do not create conflict. Thus, avoiding each other because people are afraid of conflict is not the answer to conflict resolution. Rather each partner should honour and respect the other because they have decided freely to marry.

(10) *Marriage is not an end in itself (the end of your destination):-* Before one is involved in marriage it is easy to see marriage as the grand final. It is the thing that one dream of and intends to live for. The dream of marriage is the force behind destination of

life; as such it is propelling us forward into this destination we call life. Finally we get married in life, and then what next? The next thing is that once you are married, there is a strange movement within you, making you to realise that this relationship has not come by chance, it is God who has blessed you with a fraction of the grand scheme God has for your life. Subsequently, your purpose and passions will extend far beyond the reach of your relationship with your spouse. More importantly, you will see God is at work within you because of the strong relationship he has given you and your spouse, and this love is reflected to the world around you. Therefore, we can say that marriage is not an end; it is the begging life that leads to eternal life if one keeps the commands God has given them.

Marriage gives you the opportunity to learn much more in life:-Some of us never learn about God when they are rubbing up against someone day in and day out. There are reasons why God uses the analogy of marriage to describe his love for his Church —because no relationship will ever compare to the intimacy that is exchanged within this earthly connection. Not only is God's love for us magnified through the lens of a healthy marriage, but God also uses this marriage to shape us, refine us, and put us through the fire, making us more and more like him along the way of our marriage life. It is only Jesus who will keep our marriage alive. It is said there are many routes to holiness, therefore, we should realise that marriage is one of those routes. We should be thankful for the blessing of marriage, and continue to look forward to what God has for us in the store after marriage. We should thank God for teaching us along the way in our marriage life.

Understanding Healthy Marriage: - The perception of healthy marriage varies from society to society and from region to region, as a result there is a lot of confusion about "healthy marriage". The world is changing, so today, the final decision to marry or not to marry relies with an individual and according to human laws, and we are forced to respect the decision of an individual. So what does a "healthy marriage" look like, and why there are confusions about marriage life? In marriage we enter into commitment (covenant), and this commitment is for the rest of our lives, we must continue living together whether in pains or in happiness. Both man and woman get married because it is something they have chosen after a long studying of each other characters. However, I would like to say that "*the true love in marriage*" is not based just on emotional feelings, because the emotional feelings come and go but the true love remains. Sincerely speaking, I would like to conclude that emotional feelings are more associated with husband and wife sleeping on bed at night or with what couples eat in a day. In healthy marriage, a wife will be investing her life into the husband's life and vice versa. I personally believe that for couples who believe in the Lord and allow God to involve in their daily activities, they will make better family life.

Purposes of Marriage

Many men and women get marriage without understanding the purpose of their marriage. This is like going on the journey where you do not know where you are going and how to get there. In marriage, understanding and faithfulness are essential because they foster and protect the purposes of marriage. Therefore, the purposes of marriage are of two folds the first is *"growth in mutual love"* between the couples and the second is *"education of the children"*. This implies that the mutual love of married couple should always be open to the new way of life. The couple should be ready to express their sexual union without fear or shame. Most of us know and understand that the ultimate end of sexual intercourse is to create or conceive a child. In other words, we could say mutual love is a mean to cultivate fertility, however, this does not necessarily means that couples who are not able to produce children should abandon sexual intercourse, they should rather continue to express their openness to live. If they have no children, they can share their love with other children or family members and the wider community.

What constitute true marriage?

In our world today, where marriages and family lie are being challenged or threatened, it is important for the couples to understand and know what marriage all about and what is its essence. Above all, people who intend to marry should ask themselves this important question: "Why is it crucial to understand that a true marriage is only between a man and a woman?" Without understanding of this basic question, all other types of marriage become possible, as it is today in our times. While for those who believe that marriage is a covenant between a man and a woman, this type of people should teach others about the true nature of marriage. The term covenant suggests an agreement that has been reached between the two, must be respected by other people whether relatives, non-relatives or friends. The agreement reached forms the essence of a true marriage. But some people think that this is just a mere contract that can be broken anytime by either of the people involved. Thus, if this is what marriage is, then this type of marriage is not a genuine and true marriage. We understand that the term "covenant" and "contract" are interchangeable, but to the believers in Scriptures (words of God), with covenant, what is exchanged is "essence of self" or "essence of human being; while with the "contract", what is exchanged is a mere material.

D.J. McCarthy (2013) defines *covenant* as a means by which the ancient world took to extend relationships beyond the natural unity of blood. Indeed, covenant is a type of familial bond based on an oath. The Hebrew word for covenant, *"berit"* means to "bind" –in its etymology the word means to bind together by blood. We have seen in the Scriptures God made a series of Covenants with his people. Thus, through the covenants God "binds"

Himself to the human family. This type of relation is sacred because it is based on God's "binding" love for his family.

Having seen the over view of marriage, its purpose and secretes, let us now examine traditional marriage among the Northern Luo of South Sudan. I preferred to deal with the Luo groups one by one, instead of generalising. As usual, let us start with Acholi group.

Polygamy

The practice of polygamy is prevalent, legal and widely accepted among the clans of Northern Luo of South Sudan, which include Acholi, Anyuak, Balanda, Colo, Jur Chol, Jumjum, Mabaan, Pari, and Shatt (Thuri). A Luo man can marry as many women as he can afford to pay bride prices for, so the number of wives often depends on a man wealth. Where a man has several wives, each of them often has their own house, kitchen and fields, making each wife effectively a self-sufficient economic sub-unit.

The husband is supposed to support each of these homes, providing financially and ensuring that there is shelter and sufficient money, cattle and land for food production. However, some men take on several wives even when they are not in a position to adequately support them, leaving some or all of their wives poor and struggling to provide for their children.

Something seldom documented is how difficult life can be for a woman in a polygamous union. Apart from problem around support and maintenance, polygamous marriages can be extremely difficult and unsatisfying for women in more personal ways. Men with several wives often do not regularly see and satisfy their wives. Generally a woman who is one of three wives would sleep with her husband no more than ten nights a month, while a woman is one of the six wives might get no more than five nights a month with her husband, although this is subject to variation (Friederike B & Orly Stern, 2011).

Writing in the 1970s, E.E. Evans-Pritchard speculated that these acts of absence of the husbands from their wives led wives to committing adultery (Evans-Pritchard, 1970).The negative effect of polygamy is the spread of sexually transmitted infections (STI), especially if other wives are having sexual intercourse with other men who might have STI. Thus, if one of a polygamous family network contracts sexual transmitted infections or HIV, the rest of the family network is put at risk.

With modernisation, some people mainly in upper classes have moved away from polygamy. More interestingly the South Sudanese who are in Diaspora have moved some tens of thousands of kilometres away from polygamy. In particular the educated Northern Luo in South Sudan have become less likely to accept polygamy, refusing to allow their husbands to take on additional wives. Today, some educated men and women believed that modernisation would probably have ended the war of polygamy, should it not been for the recurring wars in South Sudan. The wars halted the modernisation process in some ways because, when people are attacked it poses a direct threat to their way of life; women were obliged to be aligned to men for protection, this is also the time when many men were killed. Therefore, polygamous marriage became an appealing practical option for some women. However, there is a perception among the South Sudanese and the Northern Luo in particular that there are more men in the country than women because many men are killed in civil wars. Therefore, men should take several wives to ensure that all living women have husbands (Friederike B & Orly Stern, 2011).

More importantly, polygamy, in African context, is a culture and exists all over Africa continent. It is strongly believed that these types of marriages have been throughout the old and modern histories of Africa. One of the reason why the African have resorted to polygamy is because societies see children as a form of wealth, subsequently family with more children is considered to be more powerful.

However, some people argue that polygamy has not originated from Africa; it was imported from European and Asian Colonists, because at that time the colonists have many issues in property ownership that required more men-power. Thus, polygamy first became popular in the west part of Africa, and when the Muslim started to enter the region, the prevalence of polygamy started to continuously reduce since restriction was put more on number of wives a man should has. According to the Koran (Holy Book for the Islam) a man should has maximum of four wives and number of wives beyond that is an unacceptable.

Polygamy is very widespread across South Sudan; as a result some men have as many as five wives or even more. Even the current president of South Sudan: Salva Kiir Mayardit, although he has not openly sustained that polygamy or multiple marriages are one of the option to increase South Sudanese population, he is currently marriage to 3 wives (one of them lives in Adelaide, South Australia, the other lives in Nairobi Kenya and the third wife lives with him in Juba. The wives are placed in different countries due to insecurity in the New Nation).

Al-in-all, the polygamy in South Sudan and Africa is very common practice; it is believed that this kind of practice is very common among the animist and Muslim Communities. For example in Senegal almost 47% of the marriages are polygamists, although no statistic is done, today, ii is estimated that 35%mof the marriages in Luo groups are polygamists.

What should a wife stop doing if she wants to improve her marriages?

The above question was asked by Mary May Larmoyeux (2015), a writer and editor of FamilyLife, to a number of girls and women; I intend to agree with the answers given to her question. The girls and the women responses to the questions were as follows:

1. *Stop thinking your way is "the only right way".* So if a man does something different

from what the woman thinks, this does not necessarily mean that he is wrong. Whenever a wife insists on her own way, this implies that the wife wants to tell the husband, "I am in control".

2. *Don't put others before your husband.* God designed companionship in marriage so that a husband and wife can meet one another's need for a close, intimate, human relationship. God even says in Genesis 2:18, "It is not good that the man should be alone. I will make him a helper fit for him". Whenever a wife put another person before her husband, she is actually taking a step of isolation in marriage.

3. *Don't expect your husband to be your girlfriend.* Most men and women not only look different physically, but also have unique ways of processing things in life. One example is that a partner may talk something which does not interest the other such as introducing herself, how she first met the husband. If the husband was to be a girlfriend, all of these details would definitely matter!

4. *Don't dishonour your husband.* Respect your husband for what he is; do not correct her in front of others. Do not be rude to your husband by saying, "I don't really care about what you always say".

5. *Stop expecting your husband to fail you as your Dad might have failed your mother.* Don't think that what your father has done to your mother, your husband will also do to you. People are different in all aspects of family life. Don't generalise men's actions your husband is unique.

6. *Don't put your husband on defensive.* A man may be searching for a solution to family issue and obviously he has not yet found a solution; does it really help for a wife at this time to tell the husband that "he is not creative and useless"? A wise woman keeps quiet in such a situation; however, if she wants to make comments and suggestions, she weights her words by asking herself these questions before uttering out a sing word: "Are my words needed? Would they be helpful and encouraging". Proverbs 10:19 says, "When words are many, transgression is not lacking, but whoever restraints his lips is prudent".

7. *Never use sex to bargain with your husband.* Some women intentionally or unintentionally say to their husbands, "When I get what I want, you get sex but if I don't get what I want, there is no sex". Truly speaking, 1Corinthians 7:4 says, "For the wife does not have authority over her own body, but the husband does. Likewise the husband does not have authority over his own body, but the wife does". Thus, the Holy Bible reminds husbands and wives that their bodies are not their own —therefore, do not deprive one another.

8. *Stop reminding your husband about things over and over.* Do not make your husband feel guilty every now and then. One model wife said that when wives

constantly remind their husbands about diet, weight, medication, picking up the dirt from the floor, swishing off lights etc. they are actually acting more like their mothers rather like their wives.

9. *Don't make your husband earn your respect.* Some women today in the towns and cities think that they will respect the husbands only when they earn it. We are going contrarily from the teaching of the Holy Bible, what does Ephesians 5:33 tells us? It says very clearly and with loud voice, "Let the wife see that she respect her husband". Somebody emphasized, "A woman could learn to understand that respect is a man's native tongue, that it absolutely heals his heart and ministers to him like nothing else, it would make the biggest difference in the world".

10. *Stop giving your husband your long term to-do list.* All men under the sun do not want to be overwhelming with too much information. This may cause a man to feel like he is going to fail the test; the man will be thinking that wife's long list of what to do in a family, implies that the wife is discounted with everything in the family. As a result, out of fear, a man may think that his wife want him to do something immediately, but since the list is too long which one should he start with and which one should he ends with?-causing panic.

11. *Don't act like your spouse is a mind reader.* Some women think that their husbands should know and understand their needs even though they don't say them out. Instead in your marriage, be specific about your requests. One busy wife said that she used to feel overwhelmed with household chores (works), wishing her spouse would help her. The same wife also said one day she lost her watch and expected the husband to buy her one for the replacement. She later realized that the only way the husband knew her needs was the time she told him. One day she told the husband, "Honey, will you tuck the kids in tonight while I get the kitchen cleaned up?" The man gladly helped. This is how the wife discovered that a few words are all it takes "to change a resentment-filled, stressed-out night into a team-effort bonding time".

12. *Stop putting housework ahead of hubby.* One night a young wife told her husband that she didn't want to make love because she has just changed the sheets and she wanted them to stay clean. What do you think was running in the mind of the husband when he heard such commanding words? Another lady, who put her husband ahead of the housework, answered to this question as she said, "Do not leave the unfolded laundry on your marriage bed but do know that great marriages don't just happen; be intentional with yours".

13. *Put an end to taking the lead because you think he won't take it.* Many wives are frustrated that despite the fact they have lived with their husbands for many years, the husbands do not take charge of things to be done at home. Some of them

are even wondering of how long would it take for them to live under the same situation? However, the wise wives begin to learn that men do not take leadership because women are quick to jump in and take care of it all". On the other side another woman realized that men do not take leadership because their wives often threaten them of being sent out of the house through Police Force. As a result such men wonder why they should take leadership while tomorrow they will leave all behind the good things he has done in the house. This especially applied for couple who took refuge in the Western World, where the law puts child first, woman second, dog third and man last in family structure. In the country side of South Sudan Husband always take leadership.

14. *Do not expect your husband to be Prince Charming*: After all. The perfect husband only exists in fairy tales and your marriage exists in real life. Do not keep on focusing on your husband shortcomings; there are many good things about him. So, it is better to keep on learning, on day to day basis, the good things about your husband. By so doing you will be encourage even to do more for your husband of your dreams.

15. *Never look first to a self-help book, a plan, or a person to fix a problem in your marriage*: Every day reflect on God's words and believe and act on the things that he tells you in your heart. 2 Peter 1: tells us that God has already given us everything we need in our life and godliness, but we have to live according to the promises and expect him to show up for us.

Source: www.familylife.com/marriage (23/3/2017)

What should a husbands stop doing if he wants to improve his marriage?

Having seen what a wife should do if she wants to improve her marriage, let us then see what the husband should do as to improve their marriage life? Thus, here below are some practical advices given by Dave Boehi on what the husband should not do as to improve their marriages. As human beings, we men often need encouragement from our wives, friends and relatives; we also need continue support from God. To encourage us sometimes, we need information and sometimes we need training; sometimes even we need a mentor —someone who will show us how to be godly men, how to love our wives as Christ loves the Church. Above all we should know what to be done and what not to be done in family's life.

I am sure that when some women read the above fifteen things wives should stop doing, they will definitely ask fundamental questions, "What are the things the husbands should also stop doing?" To balance our ideas, here below are the lists of things given by Dave Boehi that the husbands should not do, if they want better marriage life. However, I want readers to understand that not all the items listed below apply equally to every man, although some of the items may hit male readers. Secondly I want to make it abundantly clear to readers that the items listed below are global, they could be applied in Europe, United States of America, Canada, Asia, Australia and Africa.

Therefore, according to Dave Boehi the things the husbands should stop doing include the followings:-

1. Stop acting like the battle is won in pursuing and getting to know your wife. Have fun together, just like you used to do before you walked down the aisle.
2. If your wife is a stay-at-home mom, stop treating her like her work during the day is somehow less strenuous or less important than yours.
3. Stop coming home from work and plopping in front of the television for the night, leaving your wife to bear the responsibility for everything else going on in the home.
4. Stop working so much. Find a healthy balance between work and family. Your wife would rather have you than a big house, nice car, etc.
5. Stop acting like you're listening when you're really watching TV.
6. Stop allowing the spiritual leadership of the family to default to your wife.
7. Stop being passive when it comes to disciplining and training your kids.
8. Stop saying you know and understand what your wife is saying or feeling when you haven't even listened to what she has to say.
9. Stop being a closed book. Open up to your wife. Don't be afraid to show emotion.
10. Stop allowing your role as leader in the home to be an excuse for selfish behaviour. Don't forget that a true leader also serves.
11. Stop dishonouring your wife by criticizing her in front of your children or in public.
12. When your wife irritates you, don't answer right away. Instead count to 10 and remember that she is a gift from God.
13. Stop using your size and strength and anger to intimidate your wife and children.
14. Stop using the word "divorce" in your vocabulary.
15. Don't shy away from difficult conversations with your wife.
16. Stop saying you'll do something and then procrastinating.
17. Don't purchase any major item without first discussing it with your wife.
18. Don't allow your eyes to linger on beautiful women who pass by. You can't help the first look; it's that second, longer look that you need to avoid. (And if your wife is with you, don't lie to her and say you didn't see that woman. Just admit you looked.)
19. Stop thinking, *I know more than my wife.* You and your wife will each have more knowledge than the other in certain areas.
20. Don't assume you know what your wife is thinking. Ask her how she is feeling and why.
21. When your wife tells you about a problem she's having, don't immediately try to solve it. She may just need you to listen to her.

22. Stop the sarcasm. You may be trying to sound funny, but you're only cutting down your wife.
23. Stop treating your wife like a child. Remember that God has given her a wealth of experience and information that you need.
24. Stop acting like God and trying to control your wife.
25. Stop pointing out her mistakes and asking for explanations. Doing these things can make her feel like a failure.
26. Never casually or disrespectfully talk to other guys about sex with your wife.
27. Stop telling your wife that she is supposed to "submit" to you. If she is not following you that means you're not leading her as Christ loves the church.
28. Stop feeding your sexual desires from any source other than your wife.
29. Don't be alone with any woman who is not your wife or related to you.
30. Stop discussing deep-level issues with a woman who is not your wife or related to you.
31. Stop deceiving your wife about your finances.
32. Don't look up old girlfriends on Face book.
33. Stop putting a number on how often you should enjoy sexual intimacy.
34. Stop acting as if you have a GPS programmed into your brain. Before you go somewhere with your wife, get the right address and find out how to get there. If you are lost, don't hesitate to get directions—from your Smartphone map, even from a person.
35. Don't make fun of your wife to other guys.
36. Don't allow guy-only activities (like playing golf, basketball, etc.) to rob you of leisure time with your wife and kids.
37. Stop expecting your wife to do all the housework.
38. Stop saying, "Honey ... can you get the kids to be quiet?" when the kids are being monsters. Get up and go quiet them down yourself!
39. Stop putting all your stuff in the laundry basket and then acting as if you "did the laundry."
40. Stop acting like picking up a gallon of milk is equal to the martyrdom of St. Stephen.
    Sources: www.familylife.com/marriage (23/3/2017)

What are the things a man should not do after getting married?

The below suggestions are collected from the writing of Kelly Rencher, who wrote on 22 march 2016 an article on marriage and family life. Here below are things what husbands should not do after getting marriage:-

1. Don't ignore or dismiss her. Women can be easily made to feel invisible. Don't ever let your wife feel you don't notice her.

2. Don't belittle her. I don't just mean by talking down to her in an obnoxiously patronizing way. If you make snide comments about the TV shows she watches, the books and magazines she reads, the causes she supports, the friends she keeps, etc., it will hurt her. It makes a woman feel you think she's silly, unintelligent, superficial or boring.

3. Don't criticize her family and friends. They may not be perfect, but they are hers.

4. Don't be irresponsible or stingy with money. Don't make a big purchase without at least warning her first and don't prevent her from occasionally buying something she really wants that you think is unnecessary. NEVER refer to the household income as "my money".

5. Don't assume all of the household chores are hers alone to tackle. Whether or not she works outside of the home, try to help out around the house, maybe cooking, washing dishes or folding laundry together. It makes her feel valued and gives you a great opportunity to talk.

6. Don't stare at your Smartphone, computer or television while your wife is talking to you. Look at her. Make eye contact. You'll miss subtle expressions if you aren't looking at her and she will feel disrespected and devalued.

7. Don't volunteer her for things without her permission. Maybe she's not comfortable hosting a holiday dinner for your entire family at your house. Maybe she doesn't want to be a chaperone on your kid's 2nd grade field trip to the zoo. Maybe she doesn't want to cook her lasagne for your office party potluck dinner. Ask her first.

8. Don't assume she should have sex with you whenever you want just because you're married. Be affectionate with her even when you aren't trying to seduce her and she'll be much more likely to want to be intimate with you sexually. Make her feel attractive and loved on a regular basis in non-sexual ways, like hugs, kisses, cuddles, shoulder rubs, compliments, flirtatious behaviour, etc.

9. Don't swoon over other women. We don't care if they are huge celebrities you have no chance of ever actually meeting much less getting it on with, we just hear, "Ooo, look how hot *she* is. I wished *you* looked like that." We wish we looked like that too, so doesn't rub it in our faces.

10. Don't start sentences by saying, «You always...» or, «You never...» in the midst of an argument. Such absolutes are rarely true and will only put her on the defensive.

11. Don't be controlling, be interested. There's a huge difference in wanting to know what your wife is doing and telling her what she is allowed to do. Suggestions are fine. Demands are not.

12. Don't lie to her. We usually know when you're lying and your denials just make us feel like you think we're stupid or crazy. Don't even like when she asks if you think a certain outfit looks good on her, just say, «It's alright, but I think the other one is so beautiful on you.»

13. Don't tell her you will do it later and then keep putting it off. Wives nag mainly when our husbands procrastinate.
14. Don't avoid talking about important issues like careers, children, housing, insurance, health issues, etc. Delaying these conversations will only lead to stress, confusion, chaos and resentment.
15. Don't forget to tell her you love her and know how lucky you are to have her. Sure, you could be with someone else, but so could she. Don't make her wish she were. Make her feel like the most important person in your world.

    Source: www.quora.com (23/03/2017)

What should I do when my wife doesn't want to have sex with me?

Many marriages are in trouble when one of the partners does not want to have sex. In such cases, I would like to suggest that the couple have to seek for some professional help before their marriage actually comes to an end. Many men are usually frustrated and even angry when their wives repeatedly rejected them to have sexual intercourse. As a matter of fact many marriages end in divorce because of this issue.

With these views in mind, I want to give the men the inside of how to get more of what they want and need from their wives. However, I understand that the wife is totally responsible for her behaviours whether they are good or bad.

According to Dr Carol (2017) instead of feeling sorry or looking outside to other women for sexual satisfactions, men should answer and reflect on some of the following questions that comes out with practical things men could do to improve sexual connection between them and their wives:-

(i) *Does she feel loved by you?* The number one thing a woman wants in marriage is that there should be unconditional love, just like it is for men. This is an important question to consider very carefully, and one should not take it to himself as blame. Many women will only be able to engage sexually if the emotional temperature between them and the men is warm enough. Similarly, a woman will have hard time if a man starts sexual intercourse and ejaculates early before she reaches her climax. Husbands make sure your wives know you love them unconditionally. And if there is conflict in the family be the first to take initiative in working through them together.

(ii) *Are there physical problems affecting your wife sexually?* Women go through series of hormonal changes during their life stages, some of which significantly affect their sexuality. Other medical problems or when constantly taking medicines the side effects can also affect them in this way. A woman's sexual response is more complicated than a man's, and it is advisable to get medical doctor to evaluate the underlying physical problems that may be affecting her. You as a husband should encourage her to see a physician and if possible offer to go with her—sometimes she might not even want you to accompany her.

(iii) *Is she too distracted, worried, depressed, or tire?* Therefore, when work, children, worry, finances is wearing your wife out, she may find it mentally and physically difficult to connect with you sexually in bed, even she wants to. It may be hard for her to put down those worries. Nevertheless, make your wife know that you miss her intimacy very much. If possible take over some of her duties.

(iv) *Is she having sexual intercourse with other men elsewhere?* However, this is not a possibility any husband would like to think about, but it is a reality far too often. Not all women go out with other men rather than their wives but the reality is that some of them do cheat—they go out with other men. If you know your wife is looking for sexual intercourse elsewhere, seek help or make a conscious choice about what to do about it.

(v) *Are you romancing your wife?* I want to tell husbands that foreplay don't start when you crawl into bed at night, it starts with what you do with your wife in the kitchen and what you say to your wife as you are sitting in the living room. A woman is always conscious, she wants to know if you desired her and care for her. A woman does not want to feel that a husband just want her body and does not care for her. God speaks to husband in Ephesians 5:25, 28 that says, "Husband, love your wives, as Christ loved the Church and gave himself up for her.....In the same way husbands should love their wives as their own bodies. He who loves his wife loves himself". If you are not romantic to your wife, stretch out yourself and find a way to romance your wife. Think back the first day you met with your wife. I bet you will be surprised at her response to you at that time.

Reasons why a wife doesn't want to have sex

When a wife does not want to have sex with the husband, it would be good for the husband to think back to the first night, as well as, early months of their relationship. The husband should realise how close they were and they belong in their own world. How comes now that she is depriving her man of sex when she said yes to his marriage proposal. Let us understand that this can happen in marriage because of the breakdown of some sort that is causing her to turn away from being intimate with the husband. It is worth mentioning that this break down could be due to number of facts:

This could be due to issue direct related to the man: The husband might be rude, thus the woman may decide not to sleep with a rude husband. Secondly, the husband may be treated her like a child and not like a wife. Thirdly the wife might have lost interest in the husband for a number of reasons known best to her. Fourthly, the husband may not being satisfying her sexually and lastly she may feel emotionally disconnected from him. Fifthly, when the husband's body does not smell good. Women are supper sensitive to small –so a man should brush his teeth and shower before bed and in the morning. Sixthly, the man may be boring to her. If the wife has made some changes in body or hair, please, acknowledge and tell her that I really like that dress or hair colour. Never make fun of your wife in public also do not make fun of her in private.

This could also be due to family life: The wife may be stressed, depressed or worried of something. Most of the women are often worried about finance. If a woman is suffering from depression, this can affect her mood, energy and sex drive. Secondly the wife could be mentally exhausted. Thirdly she does not wants the children to hear what is going on between her and the husband, this usually happen when children are sharing bed with the parents. Fourthly, the wife might be pregnant and does not desire sex. Fifth, the woman may be nursing and she does not want to get pregnant while the child is still too small. She then needs her space so that she can breathe.

This could be something ion her own inner demons: She may feel guilty of having sexual experiences in the past, and this may appear to her as sexual abuse. Secondly, she may have small vagina and she fears damage to her body.

This could be due to physical issues: this means that she may be experiencing pains while performing sex. Secondly, she may be physical exhausted, as such she loses interest in sex. Thirdly the wife may be chronically sick and giving into sex for somebody who is chronically sick is a problem in itself. Fourthly, the wife may have a low sex drive. Fifthly, she sleeps less –a recent study published in The Journal of Sexual Medicine revealed that for each additional hour of sleep a woman has, it increased the likelihood of her having sex by 14%. They also found women with longer average sleep duration reported better genital arousal than women with shorter average sleep length.

Narcissistic Partner

Figure: 156, Narcissistic husband

Before we enter deep into our discussion about what qualify a person to be a "narcissistic partner", I think it would be better for us to know first what does the term "Narcissism" really mean. And once we know the meaning of "narcissism", we shall then attempt to explore whether "narcissistic partners" really exist in some of the Luo marriages.

According to Oxford the Australia Dictionary, the term "narcissism" means *self-love* or *self-admiration*. And the adjective "narcissistic" is drawn from the name of a young Greek (called Narcissus) who catches his own reflection in a pool of water and falls in love with it. Subsequently, the term "narcissistic" is a word usually used by mental health experts to describe extreme (pathological) self-admiration.

Who are narcissists: Can they be found in Luo marriages?

From the definition above we can now understand that a narcissist is a person who thinks that he or she is the only person, in marriage, who knows all things and the world revolves around him or her. Such a person often lashes out at the people who try to give him or her criticisms. Therefore, this type of person does not like criticisms, such a person rather likes imposing his or her ideas on others.

StopTheNarcissistsNow/facebook

Narcissists are notorious for placing blame on other people and not on themselves. Even when they clearly and definitely did something wrong, they cannot- and will not- accept responsibility. They almost always deflect the blame elsewhere. Narcissists either ignore their contribution to the situation or insist that the other person (spouse, child, co-worker or etc.) made them do it. Narcissists know right from wrong, they just cannot allow something bad to be their fault. It is another manifestation of their supreme self-centeredness as well as a protection for their fragile ego. It is also a primitive method of avoiding external repercussions.

~thenarcissisticlife.com   StopTheNarcissistsNow/facebook

Source: Google, 13/11/2017

It was unfortunate that, during my research, I did not have much discussions with the Luo elders of South Sudan about "narcissistic partners" or "Narcissism". Nevertheless, as I previously moved among the Luo families in South Sudan, and now associating myself with Luo families in Australia, I have come to believe that narcissists exist among in Luo's marriages. But unfortunately, they do not know that they are living with narcissistic partners. My conclusion on narcissistic partnerships among the Luo people is based on what I have heard and is being hearing between husbands and wives.

One may wonder why I was encouraged to come out with such a "preposition". The reason is simple, in my research I am made to understand that narcissistic relationships are found across the globe, therefore, the Northern Luo families would not be the exception. In addition, what I have observed and is observing among the Luo couple is a clear manifestation of narcissistic relationship. Although some husbands may take these as forms of "disciplines".

However, having made that preposition, the difficult part is that how academically we could substantiate the argument, given the fact that there are no documents in Universities of South Sudan, and other public libraries, to support these arguments. Despite all these difficulties, still from our experiences and observations, we could deduced that "narcissistic partnerships" are being practiced in Northern Luo Societies. South Sudan government lacks mental health experts as well as facilities to diagnose "narcissism. This explain why many families, in Luo groups, live with narcissistic partners without knowing if they really married narcissists.

Am I married to a Narcissistic partner?

According to Mayo Clinic, not all signs of narcissism are obvious — some people have narcissism but they are not aware until it is diagnosed. Interestingly, some people may show traits of narcissism but in actual fact they are not narcissists, unless such a person is diagnosed

Clinical Definition

Must have 5 or more qualities for diagnosis:

- Reacting to criticism with anger, shame, or humiliation.
- Taking advantage of others to reach own goals.
- Exaggerating their own importance, achievements, and talents.
- Imagining unrealistic fantasies of success, beauty, power, intelligence, or romance.

with narcissistic personality disorder. More importantly, such a diagnoses must be done by qualified mental health worker. After the diagnoses, and to qualify that the client has narcissism, that client must has at least five of the below nine symptoms:

(1) Has the feeling of self-importance
(2) Is preoccupied with unlimited success, power, beauty or ideal love
(3) Believes that he or she is special and unique from other people and such person feels he or she could associate only with people of high-status.
(4) Wants people to admire him or her better than others
(5) Wants to boss for what is not compliance with his or her expectations
(6) Is exploitative of others, for example takes advantage of others to achieve his or her own ends.
(7) Lacks empathy (unwilling to recognize or identify with the feeling of others).
(8) Is often envious of others or believes that others are envious of him or her
(9) Regularly shows arrogant, haughty behaviours or attitudes.

What should I keep on watching for from my partner?

As we have discussed above that not all signs of narcissism are obvious, this then implies that a partner must keep on watching on the other party to identify signs of narcissism. We can assert that identifying signs of narcissism in a groom or a bride, during courtship and interrogation, is very difficult and impossible. As such, lovers usually go ahead with their marriage arrangement. And once you married a narcissist, your life will be sooner than later difficult.

However, in this book I am trying to raise an awareness of narcissism and at the same time attempt to help my readers, especially the people from Northern Luo of South Sudan, to revisit their memories after reading this book, to reflect whether they are living with narcissistic partner or not. As all of us may know, no one ever wants to marry a narcissist, if symptoms of narcissism were detected in the earlier period of marriage process. But the reality is that narcissists exist and they are very good at trickery and manipulation; and once you are locked into and get married, it might takes months, if not years, to unlocked yourself from hell and ascend into heaven.

Therefore, if you want to know whether you are married to a narcissist or not, you need to keep on watching on your partner to identify at least five or more of the following 18 top signs of narcissism:-

(1) A person is extremely likable when you first met. Was charming and personable at first glance and making you believe in his or her wonderfulness.

(2) A person takes his or her times, but eventually shows you his or her true colour.

(3) A person who likes to talk about himself or herself –preoccupied with self. Thus, whenever you are in a conversation such kind of person often turn the conversation to himself or herself.

(4) A narcissist always exaggerates self-importance –wants always to talk big and show off.

(5) A person who thinks his or her outward image is more importance than his or her inner reality. Subsequently, such a person like nice things and will only buy expensive items such as cloth to show off.

(6) A person who spends a lot of time on his or her appearance –wants to look perfect since outward appearances matter a great deal.

(7) A narcissist loves social media and claims as many friends or followers as possible.

(8) A person who thinks his or her needs must be met first. Such a person expects a partner to plan his or her life around his or her needs.

(9) A narcissist loves to give in the beginning of love affairs but as time passes by wants to take more than give. Such a person thinks he or she deserves all the attentions under the sun and could care less if the other party is without.

(10) A narcissist wants a partner always to agree with all that he or she says; and when-

ever a partner disagree, that would create problem, he or she will put the other party on the wrong side of the equation.

(11) A narcissist is completely insensitive towards the feelings of others –even about so-called love-one. Narcissist today can be on the side of a partner but tomorrow would care less – he or she would not care whether action done could hurt a partner or not.

(12) A narcissist believes he or she is above everybody (superior to everyone) and has a high level of self-importance. Such a person usually believes that many people do not understand his or her intellectual capacity, and resent people who do not treat him/her as superior.

(13) A person who thinks he or she is never wrong and if you try to point out his or her mistakes would turn it on you –would say you are the one wrong. Such kind of a partner will talk about how you are to blame until you agree with him or her –or you eventually give up and stop defending yourself. Such a person never says, "I'm sorry".

(14) A person-sensitive to criticism. Such a person is hyper-sensitive to criticism and get angry quickly when criticized. The "ego" of such a person is so fragile; not only this, even his or her skin is so thick that it takes every negative comment in a form of abuse.

(15) A partner who is a narcissist usually does not know that he or she has a personality disorder. Although the person reads all what we have discussed above, especially the signs or symptoms of narcissism, yet he or she would think other people (partner) is narcissist and he or she is not narcissistic person.

NEVER ARGUE WITH SOMEONE WHO BELIEVES THEIR OWN LIES.

(16) Often tells lies and does not trust anyone

(17) Lacks empathy (unwilling to recognize or identify with the feeling of others).

(18) Is often envious of others or believes that others are envious of him or her

Narcissistic Relationship Wheel

Figure: 157, Bad family life

(5) Once you are convince beyond doubt that your husband or wife has at least five or

more out of the above symptoms, then you need to take action by visiting mental health experts or a psychologist for support. Above all I would leave the final decision to you readers. But remember, if you are married to a narcissist, you will likely feel more like an object to be used and manipulated to meet the narcissistic partner's goals or needs. At the end of the day you will realize that your partner does not see the real you. This could be said as heart-breaking, when you discover that you have been trapped or hooked by someone who formerly you trusted and loved.

Case Study: Narcissistic Partner

Mrs Teresa Akello met Mr Christopher Oyoo through arranged marriage from South Sudan on 20th November 2014. She came to Melbourne-Australia in December 2014. The first few months of marriage were brilliant. The husband of Teresa seemed to extremely care for her, which made her felt loved.

In February 2015, Mrs Teresa asked the husband if he could allow her visit her parents in Juba-South Sudan. The husband did not refuse; what he did surprise his wife with very beautiful and expensive gifts to be given to mother-in-law.

The following week, Teresa left for Juba, the capital of South Sudan. When Teresa was still in Juba, Mr Christopher brought another person to stay with him in the house. In June 2015, Teresa returned from South Sudan, she found a person in the house and asked the husband, "Why did you bring this person to stay in our house?"

The husband answered, "This is my house, it is not your business". When the husband finished speaking, he threw the glass of beer, which he was holding, on the face of the wife. Shortly, Mrs Teresa started to uncover some hard rules which are laid out for her to follow. These rules include but not limited to: (1) Don't question or confront me. (2) Don't socialise with other women and don't make friendship. (3) I am private person and you must follow my food steps.

In July 2015, the husband was adamant that he and Teresa should both detox. Interestingly, the husband did not detox but forced his wife to fast all day and after that to use castor oil as a laxative. Teresa did, and when she was at her weakest, exhausted and dehydrated, the husband raped her. He forced Teresa to accept anal sex; he then put his hand on the mouth of the wife and forced his penis into her anus, pushing Teresa's anus open, while asking her to keep calm.

The next day, Teresa could not sit nor walk properly; she then asked the husband if he could take him to family doctor. Christopher told Teresa, "You are not eligible for Medicare because you have not yet received your permanent resident (PR)". When Teresa understood that the husband was reluctant to accompany her to family doctor, she cries throughout the night. The husband was staying in the house at that time as nothing is happening.

In the morning, the husband decided to take her to an after hour clinic. There (in the clinic) the husband did not allow his wife to talk to the doctor; the husband instead spoke on her behalf and not telling the doctor exactly what has happened to Teresa, (the wife was supposed to tell the doctor, what took her to the clinic! But she was not allowed to do so. Teresa, due to cultural bond, could not tell the doctor anything.

After seeing the doctor at the after hour clinic, the couple returned home. The husband took excellent care of Teresa; he gave her hot water bottle and made her some delicious soup. In two days, Teresa received kicks on her stomach, and her face was smashed against the floor, strangled, pinned down, bruised and stepped on her breast with the shoes.

Christopher repeatedly told Teresa not to get job because he feels that this would undermine his status as a senior contractor. All these days, the husband kept the wife's passport in his drawer at the workplace for fear that she might return to Juba.

On 20th November 2015, Teresa invited her elder brother and sister to come to Australia to witness their first year wedding anniversary. The two siblings received entrance visa to Australia; they came to Australia and on the first day they have found that their sister has been hurt, where she has bruises on both forearms and face. The siblings were very angry and upset; they then advised their sister to report the act of violence to the Victoria Police.

Teresa was not contented with the piece of advice; subsequently, she did not accept the advice of her brother and sister on the ground that when she reports to the police, this would make the matter worse —it would be more a problem than a solution.

The behaviour of Christopher left Teresa in an altered state of mind and body —she could not understand anything. She even does not know how to tell anyone in Melbourne about the violence she is going through. One day, Teresa took courage and suggested to the husband if they could visit a counsellor, the husband agreed to the suggestion, and they went to the counsellors and had two sessions. Christopher did not like the sessions at all and things got worse.

With all these happenings, Teresa thought there was something wrong with her that must be fixed. She did not know what to do, but one evening she took courage and asked the next door neighbour about counselling for herself without the knowledge of the husband. The neighbour advised her to attend either Victoria Foundation House or City Life Community care. Following the advice, Teresa made appointment with City Life Community Care for 28th April 2016 at 10.00 a.m. When the time came Teresa went and had the guidance and support from the counsellor. It took a few counselling sessions before the counsellor shows her that there was nothing wrong with her and that Christopher Oyoo is narcissistic. He thrives on control and is very abusive towards her physically, emotionally and financially. In addition, Counsellor explained to her the cycle of violence and abuse.

The wife told counsellor that she could not break her marriage and go back to South Sudan or even to marry another husband here in Australia. She told counsellor that is she breaks her marriage, culturally, she would become the reason for a failed marriage and this would bring shame to herself and to the entire family members. Teresa said she would like to work harder to keep her marriage, which implies coming up with strategies to help the husband.

After all these counselling sessions, Teresa went back to her husband's house and tried to work out what to say and how to behave in front of her husband. Despites all these efforts, there was no change in husband's behaviour. The husband continues to kick, strangle, bit the wife in the face and throwing her around in the room and often held her neck forcefully against the walls.

On January 15th, 2017, the husband took Teresa to a bar, there he got extremely drunk. The wife then tried to care for him; but the husband did not want, he started beating her outside the Bar. The husband called taxi and both couple were taken home that cost them $200. The following morning Christopher accused Teresa of making him spend too much money. He blames her for every wrong thing that he does.

One day Teresa thought of getting out of being helpless. She read books about narcissism, abuse and domestic violence and felt more confident to start taking control of herself.

One evening both Teresa and Christopher phoned their parents in South Sudan, and told them they are no longer going to live together as husband and wife. The parents of Christopher were quick to agree on the proposal but the parents of Teresa did not agree on their proposals —they said Teresa and the husband should work out things. When Christopher heard this he changed his mind and agreed to go with Teresa's parents' suggestions. Christopher said he would do everything to make their marriage continue. A week later, Christopher told Teresa that he would return her to South Sudan because he does not want to waste his time and money on her.

Teresa became worried that if her husband returns her to South Sudan she would feel shame in the eyes of community and more importantly, she would not be safe and would be abandoned by community. She then decided to separate with the husband; and now Teresa lives in a two bedroom rental property in Northern Melbourne.

Acholi Traditional Marriage and Family

Traditionally in Acholi land, when a boy wants to marry a girl it involves many steps that begin from courtship *"cuna"* (a boy seeing a girl and starting to court her) to the time the girl is introduced to the parents, uncles and relatives. During the courtship the girl is expected to be firm and hard and not to give in easily to what the boy says and wants, in order to protect her morally upright reputation. After long discussions and arguments the boy eventually wins her over and this lead to the boy giving her bracelet and in turn she gives to the boy beads and handkerchiefs and sometimes bed sheets to show her love. This is a final sign that they have agreed to marry each other. This pursuit is known by the Acholi as "*yee*" or "*luk*" (getting to love each other). However, some boys and girls express their love through "blood packing", this is when they cut their bodies (usually left arm) and each suck the blood of the other. This is the Acholi version of connecting a relationship. Customarily once the girl and the boy tasted the blood of each other and at the end the marriage fail, this will affect mainly the girl she will not give birth with another man but the boy can marry another girl and they can have children. The reason behind why this act affect girl along remain unexplained that needs more research. And in the event of a death, the surviving partner demands that a sacrifice be offered in form of goat to the spirits of ancestors and the corpse is taken for burial through the back door but if the house as only one door, the surviving partner leaves the room after sacrifice so that the corpse is taken for burial. This partner never sees the corpse again; she is not allowed to go to the cemetery or burial place. Thus, the sacrifice offered released the surviving partner from the bond of tasted blood.

Family background is a key issue in selection of the right girl; this is usually made possible because Acholi people live together in villages or towns and they know themselves. In the old days, marriage involved the boy's aunt to interrogate the girl and recommend accordingly whether she is the right girl to marry or not. Today, interrogation of the girl is done by boy's mother, sisters and friends. Before both boy and girl declare marriage, their mothers go around inquiring about them and their clans to verify that they are not related by blood. Friederike B. and Orly Stern (2011) cited Benesova (2004) as saying that marriage in South Sudan (and particular among the Northern Luo) is not understood as an arrangement between two individuals (boy and girl) and the culmination of a love affair, it is rather a social institution, involving whole families. That ties together separate kinship groups. And since marriages are families' agreements, families have a say in who their daughter or son should marry (P.3).

Similarly in the past a young man who wants to marry depended upon his father and uncles to give him permission to marry, he also received the material required for marriage from either his father or his uncles. But today, young man who wants to marry depends on his own incomes; he also chooses the girl without prior approval from lineage head or uncle; he pays bride-prices by himself -with little support from the father and uncles if any. However, the requirements of bride-prices differ from non-educated girl to educated girl; and from girl with low education and girl who might have graduated from the university (Saturnino Onyala, 2014, *History and Expressive Culture of the Acholi of South Sudan*, PP-488-496).

After the girl has been interrogated by the boy's mother, sisters and friends, and they agree for him to marry the girl, this is seen as a green light for the boy to marry the girl. He then informs his father and the father in turn informs the wife (mother of then boy). Soon, the father spreads the message to his brothers and invites them to come for a meeting on a fixed day to discuss the matter and what should be done. The items to be delivered as bride price are discussed and a specific date set for the delivery. The members would then ask the father of the boy to write a letter to girl's father. The boy's father writes a letter and in the letter he tells the girl's father that especial visitors are coming on set date e.g. 5th March 2017 at 5.30 p.m. And in the letter he must enclose some money that usually varies from $100.00 to $500.00. This money is said to be for reading the letter.

As the girl's father receives the letter, he immediately invite relatives both men and women. In preparation for the meeting, the wife prepares food for the visitors. When the invited guests arrive, the envelope is then opened in their presence; the money is taken and shown to the members. One man would be selected to read the letter. After the letter is read, the members, in collaboration with the girl's father, draw up the list of b ride-prices. When the bride-prices are completed food is brought and some women are given either whole or part of the above mentioned money, they go and buy local beer from the neighbours. Is there is balance of money; this is usually divided among people who do not drink local beer.

Sometimes, the girl's father does not send the list of bride-prices to the boy's father; he waits for the appointed time and presents the bride-prices to the delegations. But in many cases, like today, girl's father sends bride-prices lists to the boy's father prior to their coming for marriage. If it happen that the girl's father sends the bride-prices lists to boy's father, preparations are made, meeting are held after the other, goats, sheep and money are collected and shown to the relatives, so that they know their material and financial strength. In addition, other items such as father-in-law's suit and shoes and mother-in-law's clothes etc. are bought.

While from the side of the groom, the mother will be doing preparation to receive visitors; the girl's mother and other women smear the walls of the houses in homestead with mud mixed with cow dung or goat dung.

Figure: 158, A woman smearing house for the preparation of special visitors

The women make sure that any damage on the wall is fixed and any broken fence is raised prior to the arrival of the visitors.

When the day comes the mother of the boy invites the entire family members and together with the boy, his father, his uncles, his sister and brothers and his friends they go to groom's house. The boy team must make sure they keep the appointed time (5.30 p.m.); and if they think they might be late they will have to send at least 2-3 members ahead to avoid fines of not less than $200.00 for late arrival. In Acholi customs, people going to marry must arrive the homestead of the groom before darkness, for otherwise they will be fined $200.00.

On arrival at groom's gate, boy's delegations are not allowed to enter home. At the gate there would be three to four sisters of the groom blocking the entrance. They will demand some $400.00 to $1,000.00 for opening the gate, (this is known as cost for opening the entrance). If the demand is not met, nobody from boy team would be allowed to enter homestead. However, if the demand is met, delegations are then allowed to enter the homestead.

Once the money is paid, the gate is then opened and before long every members from boy's side remove shoes from their feet and crawl on their knees as they enter the mother-in-law's house; this is done simply to show their humility.

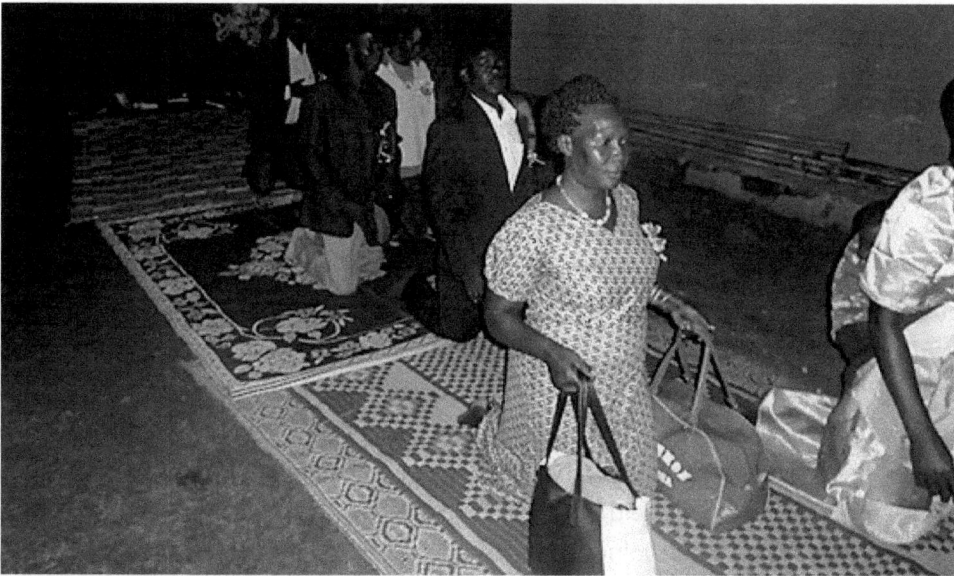

Figure 159,

Groom's team crawling into house designated for negotiation According to Acholi tradition, people from boy's side who come to marry, do not even enter mother-in-law's house with their shoes; shoes are to be removed and put at the entrance of the house. When everybody has entered the house, two or three girls from groom side come with a big bag (usually empty sack) and put all the shoes of delegations into this beg. A bag with the shoes could be left at the entrance but often bag filled with shoes are taken and hidden in a nearby place.

Figure 160, Shoes are left in front of the house

Similarly, people from boy's side who come to marry do not just enter the house while walking, they have to keep crawling each time they enter another room that might be allocated to them during the occasion. Throughout the negotiation period, everybody is expected to remain humble and calm until the negotiation is over. Inside the house, the people from boy's side are seated on one side of the room, they are seated on large mat; and the people from the girl's side are seated on the other side of the room, they are seated on chairs.

When the delegations are all seated both sides show their speaker and 3-4 representatives (who assist speaker during negotiation). Interestingly, in Acholi culture women are never and would not be selected to be speaker nor could they be selected to be representatives in such event, but two or four women from both sides could be allowed to attend the negotiation, yet they have no voice. I assume that the reason could be that *because it is a man who is marrying the woman and not the woman marrying the man.*

Soon the speaker from bride side stands up and asks the delegations of the boy, "Who are you? Why have you come to our home today? –and sits down.

The speaker from boy's side then stands and answers the questions. "We are the relatives of ......(name him while pointing at the boy). "We have come to seek the hand of one of your daughters". –he then sits down.

The speaker from bride side stands and asks, "Who is the girl you are seeking?" –he sits down.

Broom's speaker gets up and says, "We are seeking one of your daughter by name...... (naming and pointing at the girl)". He then sits down.

Once the name of the girl has been mentioned, a speaker from bride side turns to the girl and asks her, "Do you know these people seated before you? Can you point to us the boy that wants to marry you?" –the speaker sits down.

In some families the girl keeps quiet until the groom's relatives put an envelope of money in front of her. As soon as the delegations of groom puts an envelope of money in front of the bride, the girl tells names of the groom, the names of his parents of this boy are. Not only that, the girl also tells her relatives the clan or lineage where the boy comes from. But in other families, money is not presented to the bride; she answers straight away the above answers.

The speaker from the bride side again stands up and asks the girl, "Where did you find yourselves with this boy (which means where did your courtship took place, did you meet on the road or in public places)"?

The girl answers with solemn and respectful voice, "The boy found me in my mother's house" (Please read Saturnino Onyala, 2014, History and Expressive Cultures of the Acholi of South Sudan, P. 491".

After this, the negotiation begins that is always based on the bride-price lists drawn up by girl's father and uncles. But if the bride-price lists were not sent to the boy's father, prior to their coming, the speaker from girl's side then introduce the bride-prices. They negotiate one item at ago; and before long, one girl from bride side, come and switches or blows off the lights to create total darkness. The lady then jokingly asks the visitors, "Do you pay the electricity bills of this house?" Or "Do you buy the kerosene for the mother of this girl?"

As I mentioned above, groom's delegations usually come with their own lamp, matchbox and kerosene and once the mother-in-law's lamp has been blown off, they light their own lamp. A groom's uncle then takes the lamp they have brought, lights it and puts it in between the two groups. The uncle of the groom also takes a 20 litters of kerosene, gives it to the bride's brother or any relative in the house. However, if by any mistake the boy's delegations have not brought their lamp, matchbox and kerosene, they girl from bride side would request delegations of boy's side to pay some $150.00 in order for them to relight the house (lamp). The negations now go ahead; for some hours the two sides haggle. At one point, the groom's delegation would be encouraged to go outside to hold a caucus after falling to make headway. They come in and pay some more money and once the bride prices has been accepted, aunt makes an ululation, which is a signal to the people sitting outside that the bride prices has at last been accepted.

Figure: 161, The bride's and groom's team are negotiation for bride-wealth

The bride-prices are usually given in different forms some pay in form of sheep, goats and money; but other people (especially Acholi in Uganda) pay in form of cows, sheep, goats and money. During the negotiation of the bride-prices both the groom and bride sit there to see and listen, but they have no voice. The items being asked in marriage include the followings:

(1) Fines for having sex before marriage $2,000.00
(2) Mother's milk $5,000.00
(3) Fines for pregnancy if the girl is pregnant $1,500.00
(4) Opening the in-law's door $ 500.00
(5) Delegation Allowances $ 500.00
(6) Fine for interruption of education $1,500.00
(7) Mother-in-law's clothes $ 1,000.00
(8) Lamp $100.00
(9) Kerosene $ 200.00
(10) Soap two pieces $ 30.00
(11) Cigarettes $100.00
(12) Matchboxes two pieces $ 20.00
(13) Arrows (Bows)   In kind
(14) Goat for blessing for a woman with a child from outside, in kind
(15) Sheep in kinds
(16) Goats in kind
(17) He goat and she goat for blessing the bride

(18) One goat for aunty in kind
(19) One or two goats for uncles in kind
(20) Dowry amount varies $ 25,000.00
Grant Total   **$ 37,450.00**
　Source: Saturnino Onyala, 2014, PP.497-498

　　Once the bride price is paid, the couple is officially married and the bride becomes a member of the groom's clan. The bride prices are then often been used by the bride's clan to enable one of her brothers to marry another girl using the same bride price.

　　Originally, people give about 50 sheep, 40 goats and between $10.000.00 and $20.000.00 but today groom is expected to pay 10-20 sheep, 15-30 goats and between $20,000.00 and $50,000.00. The negation usually takes about 3-5 hours; if no agreement is reached, the speaker from boy's side and his team would be asked to go out and to reorganise themselves. This means people of the girl are discontented with the amount of money put on the mat, by so doing, they are asking delegation from groom side to top up the money so that the negotiation proceed. Friederike B. & Orly Stern (2011) pointed out that significant sums are paid in b ride price, which can be important sources of income for families, and the need to afford bride-price payments gives men an important motivation to accumulate wealth. One of my informants describes the central relationship between marriage and wealth as, "Lives are structural around money, marriage and children: animals and money give you marriage, marriage give you children –therefore, there is a circle".

　　Seligman (1942) observed that a rain maker gives 80 sheep, 50 goats and about $10,000.00 and $30,000.00 to marry ordinary wife. But to marry "Dak-ker" (royal-wife) a rain maker gives 100 sheep and 70 goats without payment of any cash money.

　　In addition to the above animals, the husband is expected to give 5 sheep and 5 goats to wife's uncles. However, if the mother of bride was not fully married, part or all of the bride-wealth receives by the father of the bride will be given to the uncles or grandfather of the bride.

　　It is interested to note that some of the uncles (*nero*) of the bride may never receive the 5 sheep and 5 goats. The common reason could be that the father of the girl might have not received 20 sheep and 30 goats as bride-wealth for his daughter, and because he has not married the mother of the bride, he could then give all the animals he received to his father-in-law (the father of his wife).

Figure: 162, Goats brought for dowry

Sometimes after 3-5 hours of negotiation, the two parties might have reached agreement where all the required items are paid; if this happen the people then go for entertainment. In other cases, despite the long negotiation, no agreement might have been reached between the two parties (which mean that not all the items in bride price lists have been paid). When this happen, the speaker from the groom side and his team would be given food to eat and beer to drink that would follow by traditional marriage dance. Men, women, boys and girls from both sides dance till morning. Customarily, no man or a woman from groom side is allowed to sleep that night. But should any of them fall asleep, the girls from bride side would cover the head and the body with flour.

It is not usual that all the items listed in bride price are paid the same day. Things like animals and part of the money for the dowry could be put in writing; which means that groom and his father and uncles agreed to complete the payment in future.

In the morning around 8.00 a.m. or 9.00 a.m. negotiation resume; this second negotiation should not take long, but sometimes it might 6 hour to 10 hours, if both sides have strong negotiators. Once the items have all been paid, the bride aunt, who is also in the house, yells with loud voice, this is to inform the people waiting outside the negotiation room that all the money, animals and other items that were brought for marriage have been accepted —in other words, dowry has entered mother-in-law's house. This is what mark and seal the Acholi traditional marriage. When the items are accepted by the bride parents and uncles, the girl's status changes at that time immediately from been *"Anyaka"* (girl) to *"Dako ot* (wife). Often the girl's dowry is not consumed or spent but saved to marry another girl for her brother.

To the foreigners payment of dowry is seen as purchasing or buying the girl, this is not true for the Luo, because the Luo believed that purchasing of human beings has ended with slave trades. According to Siwan Anderson (2014) in South Sudan, bride-price is interpreted explicitly as the recognition and valuing of women's productivity and it often serves to limit women's control over their bodies. However, the reality is that primarily the dowry is paid to the girl's parents as a token for the upbringing of the girl, and from the secondary point of view the dowry is paid to the parents so that they release and bless their daughter to be productive. Truly speaking, the payment of bride-price is not unique to the Northern Luo Groups; the custom of bride-price dates back as far as 3000BC, and existed during the history of many societies in African countries such as Egypt, Zaire, Uganda and today it is pervasive in Syria, Iran and rural areas of China.

Soon after this, the people continue with dance celebrating the new marriage; and in our time today, this is followed by cutting of marriage cake. Whereas, in the past, there was no cake, so people dance and disperse to their houses. Needless to say, at this point the girl and the boy are traditionally married; the girl could even go with the boy to his house soon after marriage celebration. Seligman et al (1932) observes that when the dowry for the girl is completed the girl is allowed to come to her husband's house. She sleeps with the husband in the husband's sleeping hut called *"Juan"* or *"Otogo"*, but during the day she spends all of her time in the house of her mother-in-law (the mother of the husband) P. 120.

After a few months the husband builds a house for his wife next to family home, the mother of the husband then invites the women (relatives only) of the entire village and they make feast before the wife moves to her house. During the feat, the mother of the husband gives some furniture and utensils to the wife of her son. When this is done, the wife moves to her house there she prepares food and bring it out at fire-place (to men) and other food she takes it to women group that gather in front of particular house to eat.

In some cases women are been taken without paying bride-wealth, if this happen the groom is expected to work hard to produce the bride-prices; this may takes ten years or more in which sometimes the bride might be married with the bride-wealth of her daughter and for more information, please, see Saturnino Onyala (2014), *The History and Expressive Culture of The Acholi of South Sudan*. In the old days, if by bad misfortune the husband takes the wife without paying bride-prices and the woman dies in her first childbirth or without children, the husband would have to pay a fine of fifteen sheep to his father-in-law. If the fifteen sheep are not paid, the uncles and the father-in-law have the right to stop the burial of the woman. This eventually forces the relatives of the husband to look for the required fifteen sheep. If the whole number of the sheep is not found and two third of the sheep is paid, the relatives of the husband would be allowed to bury their dead wife (Seligman & Seligman, 1932, P.120).

Similarly today, if the husband takes the wife without paying bride-prices and the woman dies in her first childbirth or without children, the husband would have to pay the dowry as in new marriage to his father-in-law. If the dowry is not paid, the uncles and the father-in-law have the right to stop the burial of the woman. This eventually forces the relatives of the husband to look for some money to pay the dowry. If the relatives of the husband get some money which is acceptable by the uncles and the father of the bride, the relatives of the husband would then be allowed to bury their dead wife. The process of getting dowry of a dead wife often results to fighting between the in-laws and groom's relatives, and if not handled well would result to another death.

Divorce

In Acholi customs divorce is not accepted; it is accepted only when the husband has reached the stage of cutting the wife with axe or beating her to death. Alternatively divorce is accepted by the Acholi people only when the woman become s prostitute and reckless in the house. Whether it is in the past or present, if there is divorce, and the woman and the man decided to leave each other, the father of the girl will give back the bride-wealth to the father of the boy and the father will hand the bride-wealth back to the son to marry with a new wife. Sometimes, stubborn in-laws refuse to pay back the dowry that often makes the husband to conceal some of her personal belongings. Seligman (1932) observes that if the husband conceals her personal belongings, especially the "*cip*" or "apron", the wife will not bear a child after divorce or else she may even dies.

Where there are children the father of the woman would not hand over the entire dowry, parts of the dowry will be left behind because of the children –this is considered as the dowry for the children.

Anyuak Traditional Marriage and Family

According to Anyuak tradition marriage has many steps; it starts with courtship and ends with marriage ceremony. Customarily courtship is not allowed to take place in the house of the girl, therefore, courtship starts in dancing ground and ends in dancing ground, where many boys and girls gather to fight for their opportunities to look for partners. During the dances the boy always dress very smart, walks and dances with romantic style because girls want to be partner of boys who look smart, handsome and talk politely. One of my informant from the Anyuak elders, who live in Melbourne Australia, says, "The first place young people meet for courtship is at traditional dances, which are organised by the youth; one of the dances is commonly known as drum dance (*buul* dance) and the other is known as stick and hands clapping dance where girls clap their hands and boys beat sticks and everybody sing with one loud voices. These dances offered and opportunity for boys to talk freely to the girls they are interested in them. As the boy meets the girl for the first time, he asks her, "What is your name?" (*Nyingi nga?*) He also asks the girl, "From which clan are you?" (*Ina nyijoka*, which literarily means: "Where are you living?"). The boy is obliged to ask these questions for the purpose of doing preliminary assessment to ensure that they are not related". However, such questions are usually asked to the girls who come from another area, because the boy knows the name and clan of the girls he live with them in the same area.

In the village, *agwage* games usually start from 7.00 p.m. when it is somehow dark, and ends at 9.00 p.m. The Anyuak young people usually dace when there is bright moon light. When the dance stops at 9.00 p.m., no one is allowed to continue with the courtship. The Anual people believe that young people should go to sleep by 9.00 p.m. and every young person must respect this policy. But the (*buul*) drum dance starts between 7.00 or 8.00 p.m. and can go through to late night.

Mr Henry Opio, informant, stated that when a boy sees a girl that he is interested in, traditionally, he is not allowed to approach the girl directly; instead he tells his friends and asks them to pass the message to the girl. So, two or three of his friend will then pull the girl by the dancing ground and tells her, "Girl, we have call you here because we have very important message from our friend (they name him); he told us that he loves you and would like you to be his wife. But he would like to know your name and clan".

The girl then answered to the groom's friends with soft and respected voice, "I have heard what you have said, my name is Miss X and I m come from the clan of......"; -more importantly the girl tells the friends of the potential groom, "I want you to give me some times so that I can think about the proposal and after that I will let you know my opinion".

Since girls are expected to be strong during courtship, so in many cases the girl would not answer softly as above, but she would rather answer, "I have heard what you have said, but I want you to tell the boy that I do not have interest in him. Let him forget about it". This answer does not necessarily means that the girl has no really interest in the boy, she has interest in him but she wants to hide her feelings at this stage. The bride is supposed not to show her interest in the groom for the few months of courtship.

The Anyuak girl, like the Collo girl, after receiving this message from potential partner, she goes and tells her friends the story; and from that time onward groom's friends are not allowed to discuss love matters directly with her, they would then discuss any love matter with bride's friends. Such discussions may take up to three months or even more. It is to be noted that the meeting between the bride's and groom's friends often offer an opportunity for the boys and the girls to get advantage of falling in love with each other.

Source: Kwekudee-tripdownmemorylane.blogspot.com
Figure: 163, Friends of bride and groom are for discussions

Friederike B & Orly Stern (2011) observes that in rural areas courting takes place during traditional dances. A boy would spot a girl and would approach her through the friends in the dance (P.9).

After the dance or in the same day of the dance, the girl will inform her friends about the intention of the boy. As from that time onward the girl and the boy will not be allowed to talk face to face; instead the bride's friends and groom's friends will be discussing among themselves about this love affairs.

According to Anyuak tradition, the friends from both sides discuss the love matters between two to four months until the individual contact is developed. Once the bride's friends and groom's friends reach agreement, then the door will be opened for the bride and groom to continue discussing their love affairs face to face. Henry Opio, an Anyuak elder who lives in Melbourne Australia says, "After the friends have reached agreement, the door is then opened for the bride and groom to meet face-to-face and discuss their love affairs".

The bride and groom meet as many times as they could to finalise their love affairs. Mr William Dimo says that the frequencies of meetings depend on the boy and girl; they can decide "*when*" to meet and "*where*" to meet and each keeps this appointed times in mind. Besides meeting in the dancing ground, the bride and groom often prefer meeting under big tree and quiet place that is a distance from the village. The boy may decide to meet the girlfriend alone or he could decide to meet her with one or two of his friends.

It is reported that in Anyuakland direct courtship between the bride and groom takes between two to six months before the bride actually give in to the groom. This is also the time when friends and mothers from both sides will be doing interrogation: the bride's friends and mother will be interrogating the groom and groom's friends and mother will also be interrogating the bride. The main reason for interrogation of the two proposed lovers is to find out whether there is any relationship exists between the two, whether the boy would make good husband and whether the girl would make good wife (in other words right person to marry). Family back ground is a key in selection of the right bride and groom; this is usually made possible because the Anyuak people live in villages and they know each other. Even those Anyuak who live in towns and cities they know themselves. Therefore, it is not difficult for friends and mothers to carry out the interrogation.

Once the two sides are satisfied with the background of the potential couple, they then give green light for them to go ahead with their plan for marriage –should they wish to do so.

Friederike B & Orly Stern (2011) cited SMLS (2008) as saying that among the North Luo marriage cannot take place unless a woman is willing to marry the man, although in some cases, women often placed under considerable social ad parental pressure to opt for their preferred candidate (P. 9). Seligman & Seligman (1932) noted that the marriage process in selecting a partner and negotiating a marriage traditionally in South Sudan allow the young people their choice of spouse. Although in theory, a bride needs her family's approval, young women are free to marry any man who can produce suitable bride-prices.

Once the bride and groom has reached an agreement (fallen in love with one another) and developed confidence in each other, this is the time the girl will open her heart to the boy and says in low voice, "I have no problem now". Customarily, girls cannot say openly to the boy that *I love you* but when she tells the boy "I have no problem now" this means "*I love you with all my heart*". Unlike the Acholi, the Anyuak do not change gifts between the bride and the groom instead when the girl has accepted the love this will follow with sexual intercourse because the girl is now ready for the boy. In contrast, William Dimo argues that the girl usually does not accept sexual intercourse before marriage processes are completed. But generally, it has been reported that the girls do have sexual intercourses with boys even though the process of marriage is not completed, provided that the girl has given her love to him. This simple fact is also another explanation of why there is pregnancy before marriage.

More importantly, once the girl has given her love to the boy, they are not allowed to have another friend. If by any mistake, the bride or the groom develops another friendship, this will generate big fight between the groom and the new boy-friend or between the bride and the new girl-friend. Many a times this may lead to excommunication of the couple by both side of the family should the bride and groom insist on going ahead with their marriage.

Soon after the bride and the groom have reached an agreement, the groom informs his father and the father in turn informs the groom's mother. In the words of William Dudimo, "After the boy has reached an agreement with the girl, he (boy) approaches the father and tells him that he has got a lady and he is willing to marry her; it is only after this that the boy can now go back and ask the girl's hand in official marriage". On the other side the bride also informs her mother and the mother in turn informs the husband (the father of the groom). As the information spread among the relatives of the bride and relatives of the groom, the father of the bride invites uncles and relatives to talk about clan and culture, he also forms delegation to welcome the important guests; while on the other end, the bride's father also forms his delegation. Friederike B. and Orly Stern (2011) says once both parents received information that their son and daughter want to get married, the girl's parents as well as the boy's parents invite the friends and relatives to speak about the clan, tribe and culture of the groom and the bride. In this family meeting, they want to know if there is blood relationship, laziness, witchcraft and riches are there in the groom's or bride's homestead (P. 12).

In Anyuak society marriage consists of two main stages that giving the hoes and paying bride-prices. Before the relatives of the bride enter into stage one of the marriage, the family of the bride sits and draw up the bride-prices (although this is usually fixed). Once the bride prices are drawn, the date is set and the delegation from bride's family invites the relatives of groom's family to come home to make the payment; this is known as the first visit or payment. This first meeting consists of few selected people (mainly father, mother, aunts, uncles and some young ladies and men) from both sides, who are to deliver and receive a "hoe"-we should note that "hoe" very important in Anyuak traditional marriage, it is a symbol of opening door for cultivation, since the Anyuak use hoes for cultivation. Henry Opio says, "A hoe brings food at home". William Dimo, an informant, said that besides a hoe, groom's delegation has to bring with them some goats and cows as gifts, this is the most important traditional requirements to please different gods in the bride's family; it is a condition to give the girls and boy health and happiness in the marriage life". When the hoe, goats and cows have been received, the delegation of the groom then go back home to prepare themselves for the payment of dowry, which usually takes place after six months following the delivering of the "hoe", "goats" and "cows".

Generally, the groom is expected to work hard to produce parts of bride-wealth; although in most cases if the groom has sister who has been married that bride-wealth would be given to him. In other cases father and uncles do or do not contribute to bride-wealth. After six months bride-prices are then delivered into bride's home, the groom's father and other closed relatives sit to fix the date and time for the delivering of the bride prices. After this the boy's father send message to the girl's father informing him of their coming. On receiving the message, the girl's father also reply to the message to confirm the date for groom and his family members to come for marriage discussions.

However, it is important to note that dowry is not a work of one person it is fixed by traditional laws; nobody can change the amount of items required for bride-wealth. The bride prices usually consists of four "Dumoi" (beads) if all the four "Dumoi"i are paid this means the dowry is completed and if the "Dumoi are not paid in full, payment in instalments are accepted. It is said that one "Dumoi" is equivalent to four cows or 30 combine sheep and goats. Seligman, C.G. & Brenda Z. Seligman (1932) observe that an average transfer of bride-wealth might include: One bead necklace (*shauweir*) [*shauweir bead is a largish read head*], five waits-rings of *tet* beads [*tet bead is of a dark greenish colour*], two waist-rings of "*dumoi*" (beads) [*the dumoi beads have blue colour and they vary in shade some look watery blue and oter look dark blue. They are very scarce, difficult to find in open market, they are in high demand for marriage. One dumoi is worth one cow or 30 okwen spears*], four cows, one bull, 70 spears of the *okwen* type, two spears of the *dem* type, one spear of the *jo* type [both the *dem* and *jo* types of spears are very scarce, ten brass wrist bangles. The payment of this bride-wealth may be, and usually spread over a number of years, and defalcation is common and gives rise to many disputes. Should a woman die after marriage and before she has given birth to a child then the bride-wealth or part that is paid by the husband is returned to him (P.111).

When date for delivering bride prices come, the young men are always the first to arrive in the house of bride's father, they usually come in the evening, as the sun is setting. The young men enter the homestead with songs and dance, they are receive at the entrance, by the girls; they would dance at the entrance for some minutes, after this, the girls then take them into the big house; there they dance and chat the whole night – no sleeping for both young men and girls. Early the following morning, the elders, groom's father, mother and other people follow the young men (they usually arrive at around 7.00 a.m.). They are also welcome at the entrance of homestead by the girls, who show them place where to sit while waiting for the negotiation to commence.

When the groom team come they are put in one big house next to bride's mother's house and the bride's team site outside the house (see figure132); before the negotiation begins both sides is expected to select their speakers and introduce them. Unlike the Acholi, the Anyuak do not fine groom's team for late arrival at bride's homestead. William Dudimo said that where there is no big house that can accommodate groom's team, negations may be conducted under a big tree where the bride's group site on one side of the tree and the groom's team site on the other side of the tree. Shortly, the speaker from bride team stands up and asked the speaker from groom team to show them what they have brought for dowry. Speaker from groom gets up in turn and shows the bride team what they have brought for dowry. Interestingly, as each speaker speaks, their speeches are always being translated by selected interpreter that come from the bride family.

Customarily, woman can't be selected to the position of the speaker. William Dimo stated, "Women are not allowed to be a speaker at all in our culture, during marriage ceremony; this is something we have inherited from our ancestors and we have been practicing for generations and will continue to practice for generation and generation". Negotiation during bride-wealth may take 12 hours or more.

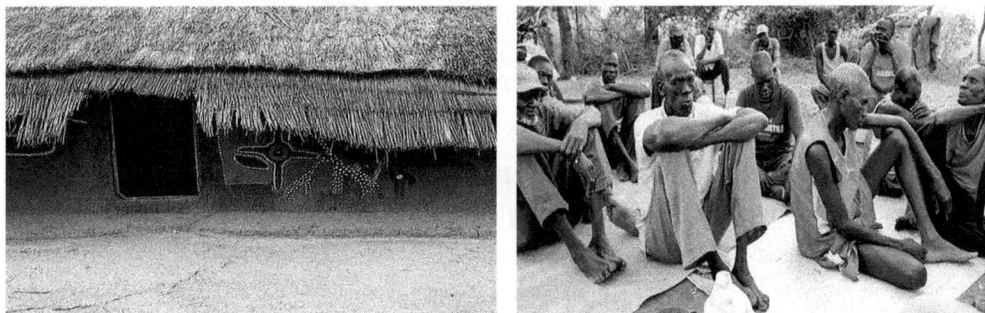

Figure: 164, Bride team waiting for the negotiation that is taking place in this house

The bride's team will then come for negotiation where the especial visitors are housed. As negotiation continue, the speaker from bride's side will keep coming out to inform the father of the bride and other relatives, who are waiting outside, on what they have reached upon; and when the two teams cannot reach agreement, consultation is made with the bride's father.

Friederike B. and Orly Stern (2011) cited Benesova (2004) as saying that marriage in South Sudan (and particular among the Northern Luo) is not understood as an arrangement between two individuals (boy and girl) and the culmination of a love affair, it is rather a social institution, involving whole families. That ties together the kinship groups that have separated. And since marriages are families' agreements, families have a say in who their daughter or son should marry (P.3).

In Anyuak society, payment of dowry in instalments is accepted and these payments may take years or decades. Once the beads are paid, from that date onward, the groom has total claim over the bride and her fidelity is guaranteed, although the groom may not take the bride the same day, parents drives away any boy who might come to attempt the bride. The bride wealth is usually divided into two: part of the dowry is consumed by the father, uncles and aunts and part is saved for future marriage of the brother (or brothers). Gurtong (2017) noted that the Anyuak have no ceremonies attached either to birth, graduation into adulthood and marriage between *Nyako* (girl) and boy (*Wadmara*). The bride stays in her parents' home until the dowry or half of it has been paid, after which she moves to her husband's home. Sometimes a poor groom may raise up to two children with his wife while she is still staying with her parents. Mr William Dimo pointed out that it is not true that the Anyuak have no marriage ceremonies, they do. He emphasizes that when the half dowry or full dowry have been accepted this usually follows with two traditional dances. Firstly, the bridegroom is asked to dance three times in front of the delegates; after that the people dance *Adhuu*, in which the bride is then invited. The *Adhuu* dance always follow with food and drinking local beer; when the people eat and drink they continue to dance until such a time they feel it is enough, then people disperse.

Customarily the bride is not allowed to go straight with her husband after the dowry has been paid. She is to remain behind and prepare her things that she wants to go with to husband's home. When the bride completes her preparation, she will be accompanied to her husband's home by her aunt and some of her friends, the role of these ladies is to encourage the bride to have confidence.

On arrival at groom's home the bride will not stay immediately in her own house, she will stay with the groom's mother for some months or even years before she moves into her own house to start a new life.

Although it is unlikely that the bride may dies, after full or half dowry is paid, while she is still in her parents' home, however, should this happen the bride's parents' could return the paid dowry to the groom or alternatively the groom's family may ask one of the bride's sisters to be given to the husband to replace the dead sister. When the groom's family ask for dead bride's sister, they are always given according to Anyuak customarily law.

Thus, should the husband dies after the bride has been brought home and she has one or more children, after fourty days the family of the deceased hold meeting and in this meeting they ask the widow, "What did your husband tell you before his death?" It is expected that the husband tells his wife or wives before death, which of his brother should take care of the children after his death; which means that the very brother should inherit the wife. And if it happened that the husband did not tell his wife or wives whom he wants to look after his children, then there will be candidates from which the widow or widows could choose from –these candidates could be young men or married from the family of the deceased husband. Once the wife or wives have chosen her/their new partner(s), according to Anyuak tradition they are not allowed to have sexual intercourse until after one year following the death of the husband. Both the woman and the man are avoiding sexual intercourse as a sign of respect for the diseased; and above all, to give times for the widow to forget the pains she has experienced due to the lost of her husband.

Divorce is not acceptable in Luo group of South Sudan and in particular among the Anyuak people. Should divorce happen in a marriage, this could cause several (may be up to ten or more marriages) could be broken by breaking just one marriage in the line. It is to be noted that "*dumoi*" have become rare; as a result they are circulated and hence could even come back to the original owner in the course of several marriages.

The marriages of King's (Nyie) daughters differ from ordinary arrangement of marriage we have discussed above. According to Anyuak custom King (*Nyie*) gives his daughters to wealthy grooms without prior courtship. Indeed, to behave lovingly and to amuse yourself with Anyuak King's daughter could invoke his wrath resulting in confiscation of one or abduction of three girls from the village that the groom lives.

Seligman, C.G. & Brenda Z. Seligman (1932) observed that the Anyuak are polygamous, but probably the great majority have only one wife. William Dimo, an informant, confirms

Anyuak people practice polygamy but the number of wives a man should has depends on his financial situations. She states that most of the men marry up to three wives, but others marry up to five wives and yet others marry more than five wives.

Balanda Traditional Marriage and Family

Balanda perceive marriage as a permanent relationship between a woman and a man. It also ties the two families together that enable them share social events. Like the other Luo groups such as Acholi, Anyuak, Jo-Luo, Jum-jum, Maban, Pari, Shatt and Shulluk marriage in Balanda society involves many steps that starts from courtship, introduction of both girl and boy to their respective parents and relatives and into marriage ceremony. To under-stand the process among the Balanda people, it is good for us to examine the step one by one.

(1) *Courtship*

It is generally believed that in Balanda land courtship takes places in dancing ground as we have seen among other Northern Luo Groups. Dances are usually organised by young men; like the Maban, at this time animals are slaughtered, food and local beer served.

During dancing ceremony, girls and boys are expected to wear their best dresses to attract the opposite sex. According to Balanda custom, this is the time the when the boy can identify a girl of his choice.

From this point, it remains unclear as to whether the courtship is done on individual basis or in group as we have seen among other Northern Luo Groups. However when the bride and groom reach agreement (that have fallen in love), this then leads to the second step that is introduction of both bride and groom to their respective parents. It is impor-tant to note reaching agreement or falling in love between the bride and groom does not necessarily mean acceptance of marriage. The really acceptance of marriage would be seen only during the payment of dowry as we shall discuss later.

As we have just mentioned above, in Maban-land courtships are carried out during the harvest feast (*Gatti*), which are normally performed in December of each year. This is also the time when all food products have been harvested and stored in granaries. Generally, in South Sudan the month of December is when the moon is bright and young Luo people could organise dances. During harvest feast celebration animals are slaughtered, food and local beer are served and dances performed. The dances usually being organised by young people and where boys and girls appear in their best dresses and they must have smart look; wearing necklaces (Linyan) to attract the attention of the opposite sex.

However, what remains a mystery for me, as I am writing this book, is whether the groom approaches the bride directly during courtship, like in Acholi, or he approaches her in the first stage through his friends, like in Shulluk and Anyuak societies. This requires further research.

(2) *Introduction of partners to respective parents*

Once the bride and groom have reached agreement, the boy will then inform his father and the girl in turn will also inform her mother. The groom's father will in turn inform the groom's mother and the bride's mother will inform bride's father. On receiving these messages, both parents will make their interrogations: the parents of the girl will interrogate the boy, while the parents of the boy will interrogate the girl. When doing these interrogations the parents of the boy will be taking into the account whether the girl is hard working, disciplined and coming from family with good background. The parents of the girl, when doing interrogation about the groom will be taking into account on whether the boy is responsible, respectful, hardworking, brave and ready to defend his wife, he is coming from a reach or poor family and whether the groom is coming from family with good background (where there is no poisoning, no witchdoctor, no evil eyes etc.). It is only after these interrogations that the two parents would either approve or disapprove the marriage based on their findings. Approval of the marriage by the parents usually paves way to the third step, which is final, this is delivery dowry.

It is assumed that in Maban-land direct courtship between the bride and groom takes three months as we shall see below that marriages are normally performed in March. Once the girl and the boy reached agreement, the boy then introduces the girl to his parents; the girl also introduces the boy to her parents. Once the parents have been informed of the proposed marriage, the parents of the groom will be interrogating the bride and the parents of the bride will be interrogating the groom. The main reason for interrogation of the two proposed lovers is to find out whether relationship exists between the two, and whether the boy would make good husband or the girl would make good wife (right person to marry). Family back ground is a key in selection of the right bride and groom; this is usually made possible because the Maban people live in villages and they know each other. Even those Maban who live in towns and cities they know themselves. Therefore, it is not difficult for fathers and mothers to conduct interrogation. During the interrogation should one of the party is found unfit for the marriage, the boy and the girl have to break their relationship; the boy to look for another girlfriend and the girl to look for another boyfriend. But should they persist to continue with their relationship this may lead to an excommunication of the couple by either of the family.

Bur once the two sides are satisfied with the background of the potential couple, they then give green light for them to go ahead with their plan for marriage. Since the Maban society is matrilineal clan the matter of the bride-price is then referred to the groom's maternal uncle. The bride parents draw the bride-wealth and the groom's uncle pays the dowry. The groom's uncle then set day and time for delivery of dowry, where the father, mother, relatives and friend of the groom will attend.

(3) *Marriage*

As soon as the approval of marriage has been received, groom's father sets date and day for delivery of bride-prices. He sends written message to the bride's family. It is believed that bride's father does not draw any bride-prices because all the bride-wealth has been agreed by community on fixed condition.

I have noted that in the past, in Balanda society dowry had consisted of spears, bows and arrows but today dowry has undergone dramatic change, where dowry now consists of two clean hoes and some money if the groom's family can afford. The most important thing is that the hoes for dowry must be of high quality metals to avoid risk of rejection.

When the day comes, the groom's family members, relatives and friends go to bride's homestead. On arrival the visitors will be put in one of the big house of bride's mother, no greeting is allowed at this stage.

Shortly a speaker from bride's team come into the house and asks, "Why have you come in this home today?"

Speaker from groom's side answers, "We have come to this home today because our boy wants to hold the hands of one of your daughters".

Speaker from bride's side again asks, "Have you married one of our girls before?"

Speaker from groom's side answers, "We have not married one of your girls before. This is why we feel honour to be in this house". Customarily a man is not allowed to marry two wives from the same family or lineage. This is one of the reasons behind interrogations.

Speaker from bride's team again asks, "Has there been any fight or misunderstanding between this family or clan and your family or clan?"

Speaker from groom's team answers, "There has never been any fight or any misunderstanding between bride's family or bride's lineage and groom's family or groom's lineage".

After bombarding especial visitors with many questions, the speaker from bride's team leaves the house and joins the other people waiting outside the negation house.

Groom team then select two married men and two married women age 50+, who have been in marriage life without any trouble. These men and women must be close relative of the groom. The men and the women then take the "two clean hoes wrapped in white sheet" on a tray together with some cash money –if any- and place them outside the door of bride's parents (this is the same house where the groom team are seated).

The parents of the girl would ask their daughter to pick the hoes and take them into the house if she consents to the marriage. If the girl declines to take the hoes into the house, the marriage would be considered cancelled but if she picks the hoes and take them into her mother's house, this is indication she has accepted the marriage.

Once the bride has picks and takes the hoes into her mother's house, this also paves way for her relatives to enter the house. As the relatives enter the house, they are welcomed by groom's family. This is follows by eating, drinking local beer and dancing to mark marriage ceremony.

Contrarily to other Northern Luo Groups, the Balanda do not demand more goats, cows, pigs and money for dowry; they only need two, clean and high quality hoes to marry a wife. These two hoes are not to be used for digging they are rather saved for future marriage; although today many girls in balanda do not appreciate using hoes paid for her dowry to be used for marrying another wife, because the wife may cause many problems that may lead to divorce and withdrawal of hoes.

Traditionally, after the handover of the hoes, the groom has to remain behind to labour for the in-laws for at least four years. During this period the groom cultivates various crops and builds or rebuilds houses for the in-laws. However, it has been reported that today these practices have been minimized in community.

Even though the reasons behind labour have not been explained, it is absolutely clear for the Luo people that the unpaid labour, for four years, contribute additional bride-wealth in the bride's family.

Thus, in modern time, more and more money is being paid as dowry in additional to the two hoes, which previously made marriage almost impossible for the young people of Balanda.

It is reported that after four years of services to the in-laws, the husband would has the right to take his wife to his father's home. It is also hoped that during the four years, the wife may has one or two children to accompany her to the new home. Balanda customs have not mentioned whether the husband will receive fine for having one or two children in in-laws homestead. Therefore, the husband will take his wife without a goat to b less the in-laws homestead.

Maban society is more loosely structured than some of the Northern Luo Groups. According to Maban tradition, the primary aim of the marriage is for the production of children and expansion of the family.

Marriage among the Maban, like in Acholi, Anyuak, Balanda, Jur-Chol, Jumjum, Pari Shulluk and Shatt, involves many steps that starts from the time of courtship, introduction of the girl by the boy to his parents and relatives, through to payment of the dowry. It is important to note that mature boys and girls prepare their marriages in the month of December, this is the month when the people celebrate harvest feast (*gatti*) and become known as month of courtships.

Marriage among the Maban people is done according to the age-group; and if by any misfortune a boy misses the marriage of his age-group, he could miss it forever. The courtship as we have discussed above, begins in December during the harvest feast. It is important to note that if marriage begins with elopement or pregnancy, it is considered against the customs, illegal and shameful to the parents of the bride. Thus, if this happen, the situation must be handled carefully by the two families, especially the boy's side for otherwise this could result into conflict and violence.

According to the Luo customs, elopement or pregnancy of the girl is commonly known as *"coming into the bride's family through the window"*. However, when the groom pays dowry prior to elopement and pregnancy, this is considered by the Luo as *"coming through the door into the house of the bride's mother"*.

In Maban society, the main marriage ceremony takes place in March; during which the maternal uncle of the groom pays 5 pigs, 10 goats, hoes and axes (Gurtong, 2017). Once the dowry has been paid, the girl is then officially ready to be accompanied to the husband's home. Interestingly, unlike other Luo groups, the Maban people do not build house for the bride in her husband's father's home, the house for the bride is built in the groom's maternal uncle's home –this is the man who pays the dowry of the bride.

The bride is then expected to leave for husband's home in as soon as the dowry paid in full or half. But if for one reason or the other she rains in her parents' home, customarily the husband is allowed to visit her and have sexual intercourse during the visits, and if she became pregnant and delivers a child at her parents' home, the groom would be fined –he will pay one pig to the bride's parents before taking his wife to his home.

But in many cases, after the dowry is paid, the bride usually leaves for the husband's home. When the bride arrives at husband's home, customarily, they have to spend four months in "wedding room" (*Chanyo*). During this period, they only get out for toilet and bathing, no one is allowed to enter the room and interfere with the new life of the couple; food is cooked and given to them at appropriate time. This is the period when the parents and relatives of the groom expect the bride to get pregnant –this period is commonly known among the Maban as *big wedding ceremony*. After four months, the groom dresses very smartly, wears a special necklace and starts walking solemnly within his village while the bride remains in the wedding room. After this, the groom then goes back to his wife in the weeding room with big smile in his face. Shortly, the bride and the groom would then be allowed to get out of wedding room and join public life.

Maban have no public ceremonies attached to marriages; the only custom linked to marriage is the payment of 5 pigs, ten goats, hoes and axes.

Jo-Luo of Bhar El Ghazal Traditional Marriage and Family

In Jo-Luo society, there are different forms of marriages, which include arranged marriage, force marriage or marriage through abduction and normal marriage. However, the focus of this book is on normal marriage, as such, I will discuss briefly about arranged marriage, force and marriage through abduction, before we discuss the normal marriage in Jo-Luo society.

Arranged marriage

In the past, the Jo-Luo of Bhar el Ghazal, in South Sudan, practiced arranged marriage, but in many occasions some girls refused the marriages but others were persuaded by their relatives to obey their parents' will —so they got married. Arrange marriage is the work of the parents of the bride and groom, with no involvement from the girl and the boy.

It is said that when the father of the young boy sees that the wife of his neighbour gives birth to a beautiful girl, he (father) would arrange with the parents of the new born girl, so that when the girl grows she would marry his son. In most cases marriage arrangements do not meet any resistance from bother parents; the problem may come only when the girl grows up, as we shall see later in this chapter.

Unfortunately, no Jo-Luo elders told me how the inauguration of arranged marriage is conducted. But according to Jo-Luo oral history, after the inauguration, the young men from groom's side make visits to the proposed bride from time to time. Thus, such visits continue until the girl is big enough to be taken to husband's house. The society made strict rules, which states, "If the young girl is still growing at her father's home, the groom must not attempt to have sexual relationship with her". Importantly, the visits of the young men from the side of the groom provides an opportunity for them to find girlfriends from the village, as girls are always called from the nearby houses to help with services to the guests (Apai Gabriel, 2005).

However, the time the girl is still at her father's house, the groom could make visits to her and during these times, he would do for the mother-in-law many things which are not included in the bill. He may be called to build a house for the mother-in-law or to help in cultivation or cleaning. As tradition goes the groom does not do work like building and cultivating alone, he takes with him 4-5 friends (men) to assist him. When the groom and his friends are asked to do something at mother-in-law's house, they perform the work decently with enthusiasm as a way of winning over anyone in the family who may not be happy with the arranged marriage. When the girl grows up, since both parents have agreed for the marriage, engagement ceremony would be held and rings placed on the girl's wriest; after this, she would be taken to her husband's house.

In arranged marriage, when a woman died without children and the father of the late woman has no way to repay the dowry, or when the father of the late woman wants to maintain their relationships, the father would offer his other daughter to the man to replace her dead sister. Sometimes, the girl that is offered may be too young for marriage, if this happen they girl would be left to stay with her parents until she reaches maturity the she can join he proposed husband.

However, sometimes the girl offered is too young for the man; in this case groom's father would look for his younger son, and assigned the offered girl to him. When both the girl and the boy grow up, engagement ceremony would be held and rings placed on the girl's wriest; after this, they would then live together as husband and wife. In this way, the relationship between the two families is kept alive (Apai Gabriel, 2005).

Force or Marriage through Abduction

Among the Jo-Luo communities, there are force marriages, when a wealthy man request to marry the daughter in a family, and the girl rejects the man. The girl's father, desiring the wealth of the man, may force his daughter to marry him. In other cases, forced marriage happens when a girl conceives out of wedlock. The Jo-Luo people consider a girl getting pregnant before formal marriage, she is bringing shame to the family; they also consider act of the boy that lead to pregnancy is an abuse to the family of the girl and entire community. Therefore, parents in a mood of anger would simply girl her to the boy without asking her opinion.

The Jo-Luo people practice certain marriage procedure known as "abduction". In this situation, the man usually engage secretly with the girl that he wants to be his future wife; but a man knows quite well that he and his family have no bride-price to pay. Yet, the man would approach the parents of the girl, as he calculates what help his relatives may give him if he abducts the girl. Before long, the man would abduct the girl and takes her into the house of his elder (in most cases it is the uncle). This situation often put an elder in dilemma where on one side he would think on one hand that, if he tells the boy to take the girl back to her father's house, the parents of the girl would think that it is he who does not want the marriage. On the other hand, if he keeps quiet, the family of the girl may learn later on that the girl is in his house.

As the elder is in dilemma, he would call the family with all the associated relatives, in the first night, to discuss and decide on what could be done with the girl. In such cases, relatives often are sympathetic towards the young men, especially when many of his agegroups are married. Subsequently, they relatives would decide to raise fund to support the young man. Once the fundraising is successful, the elder, with whom the girl stays, informs girl's parents of her whereabouts (Apai Gabriel, 2005).

However, abduction sometimes has negative impact on the groom's family. This means that if the groom has a young sister, who is still in the house, and nobody applied to marry her, the relative would ask her if she has anybody who is interested in her. If she says she has nobody, and the relatives have failed to raise fund, they will enforce her to get marry to any man available at the time —who has dowry. In doing this, the parents would offer the girl three options, by naming three young men whom they think are coming from rich families. The girl would be asked to choose one of them. The girl would then be forced to choose one of the young men. At this point and time, the abducted girl would be kept in a secrete place to wait for the approval from the parents of the chosen boy. Once the parents of the boys accept the offer, the parents of abducted girl would be informed. The bride price is then brought to the house of groom's sister's mother's house.

In some cases the boy may not have sister, but yet he may want a girl for a wife and the parents of the girl have accepted their marriage. In this situation, the bound, cooperation and unity of the Jo-Luo speaks, especially when the boy visit the girl openly and frequently. The associates of girl's father will advise him to allow the young, poor man to marry his daughter. The hope is that she may bring forth children who will be married and, with the bride-price, the unfinished dowry for their mother would be completed according to the Jo-Luo tradition and culture.

After having discussed arranged, force and abducted marriages, let us now focus on our main topic—normal marriage.

Normal Marriages

"Among the Jo-Luo of Bhar el Ghazal, marriage is a goal of life. Everybody must marry as all the people learn how to read and write in developed world" (Apai Gabriel, 2005, P. 28)

The Jo-Luo of Bhar el Ghazal, like any other Northern Luo Groups, rear average of four to six children in one family. However, some women give birth to ten or fourteen children.

Due to poor medical services, in the entire areas of the Jo-Luo, fewer children survive and grow to maturity. These children die from preventable or curable diseases such as malaria malnutrition and diarrhoea. The other cause of low birth rate among the Jo-Luo of Bhar el Ghazal is due to the fact that a man is not allowed to have sexual intercourse with his wife, while the baby is breast-feeding, that is, until the baby reaches two or three years old.

The community consider it a great shame for a woman to become pregnant before the two or three years elapsed. Unfortunately, if the woman become pregnant, in any way, before the two or three years elapsed, the relatives of the husband would say that the wife has been committing adultery, since the husband has been abstaining from sex.

In Jo-Luo society, marriage is arranged according to the seniority of birth. This means the eldest son marries first, and followed by the second elder son —and so on, and so on.

The Jo-Luo of Bhar el Ghazal are not exceptional among the Northern Luo Groups, their marriages also takes series of steps, which starts with courtship, payment of *"gikwer"* (part of dowry), *"Atoja"* (the oath for weeding) and finally accompanying the bride to groom's house. To understand these processes, let us examine them one by one as follows:-

(1) Courtship

Traditionally, in the Jo-Luo society, courtship could be initiated by the individual or by age group. According to Apai Gabriel (2005), "There are bold contacts between men and girls who are not yet engaged to anyone. In such case, a man may come to visit a girl lat in evening. He would be given a seat at the courtyard and the girl would. They would converse and he may eventually put his case to her as a future partner. This marks the beginning of courting which may actually lead to engagement" (P.42).

It is said that, even the courtship is initiated by an individual, eventually will end up with involvement of the age group. Thus, when a boy is interested in a particular girl, he approaches the girl individually and tells her, "Dear girl, my heart loves you so much and I want you to be my future wife. What do you think?"

The girl would always answer, "No".

Since the boy has shown interest in a girl, he then presents himself in a handsomely appearance and demonstrates to the girl that he is professional dancer. In addition, the boy may formulate poems and signs them out in attractive methodology as do many Jo-Luo young men. These songs are songs of praise aimed at attracting girl and convincing her family of the feelings of the man who like to engage her as his future wife. These songs are usually composed in a tactical way, so that when sung and heard by the girl and her relatives, they would understand how serious the singer would like to be part of their family in marriage.

Figure: 165, Jo-Luo girl

When the boy has done self-introduction, he then informs his age-group or friends about the discussion and love he has with the girl.

The group courtship is done during the dancing ceremonies, where boys and girls meet in the *"Tiar bul"* (dancing ground). Thus, when the day for dance comes, the age-group or friends would approach the girl in a group of 5 to 10. Unlike the Shulluk, the Jo-Luo age-group approach the girl in the presence of the boy, who is interested in the girl.

Customarily, the age-group calls the girl from the *"Tiar bul"* (dancing ground) and meets her just outside the dancing ground. There, the age-group tells the girl, "We have called you here because Mr. X – (mentioning the name) is interested in you; and we want you to marry him. What do you think about it?"

But if this girl was formerly approached by the boy who wants her, the group would then say, "We are here to follow up your last discussion with Mr. X – (mentioning the name). We hope that you have been reflecting on the matter in the last days. What can you tell us today?"

The girl answers, "I heard what you said but I don't want".

Such answer usually invites many questions such as "Why don't you want to marry him/ me? Do you already have a boyfriend? What have you seen wrong in him/me?"

The most interesting thing with the introduction in courtship, in Luo culture, is that whether the courtship is initiated on individual basis or on group base, the answer of the girl would always be, "No. I don't want. I don't have interest in you/him".

The age-group doesn't give ways to the girl; they continuing asking her many questions and eventually the girl will say, "Please, I have heard enough. Do not give me more headaches; give me time to think about the matter".

Customarily, the girl is supposed not to be too soft to the boy or age-group and show her love and admiration for the boy.

After the introduction, the girl usually goes back to the "Tiar bul" (dancing ground) and informs her friends about the proposal. At this stage friends do not say anything about the matter nor would they give advice of where to go from there.

According to Apai Gabriel (2005) the Jo-Luo young men often

Dances often end at midnight, but sometimes the young people dance till morning and they return to their respective houses.

When the second day for the dance comes, the boys and girls again meet at "Tiar bul" (dancing ground). According to the Jo-Luo of Bhar el Ghazal culture, like other Northern Luo cultures, dancing is considered conventional courtship time, because such meeting at the "Tiar bul" (dancing ground) are still unofficial. For this reason, such courtship is to be done away from the sight of the family members. The parents are not even informed of what is taking place at the "Tiar bul" (dancing ground), though they would not meddle in the affairs of the young people (boys and girls). In the second meeting, the age-group call the girl aside as usual and talk to her saying, "We hope you have reflected on our last discussions, regarding the love affairs between you and Mr. X - (name the boy here). Today, we have come to hear your response and to review the matter. Sincerely speaking, we want to let you know that since the day of introduction, when you rejected the proposal of Mr X (Name here), he did not sleep most of the nights. In addition, we want to reassure you that the boy is ready to marry you, because he loves you from the bottom of his heart. On top of this, the family of the boy believe that you are the right lady for their son. So, what do you think?"

In Jo-Luo society, courtship could last between three to six months, before the girl could give her love to the boy. The three and six months periods, are the periods when the girl carefully observes the boy on whether the boy is sincere in his words or not.

The boys and the girls usually meet in the intervals of dance ceremonies (see *Thanksgiving and base-dances* under Jo-Luo economy), which are organised regularly in every two months during the summer season (dry season). The dance runs on daily basis for two or three consecutive weeks. I want to assure the readers that the courtship can last as long as possible until the groom and the bride reach agreement.

Once the bride and groom have established their relationships, the bride gives necklace beads (*Tii*) to the groom to demonstrate her sincere love for him. In other words, this is a manifestation that the bride has now accepted the love of the boy. In turn the groom would also give the *"magala"* (ear-rings) to the bride; this is a demonstration that love has been exchanged.

When the boy and girl finished exchanging their gifts, the bride takes the *"magala"* (the ear-rings) to her mother; and shows her as she says," My mother, this *"magala"* was given to me by Mr. X – (name here). He wants to marry me and I also want to marry him". The mother would not say anything to her but she would be very proud and happy that her daughter at last has found somebody whom she loves.

On the other hand, the groom would also take the *"tii"* (necklace-beads) to his father, and shows him as he says, "Father, this *"tii"* (necklace) was given to me by Miss X – (name here). She wants to marry me and I also want her to be my future wife". The father also would not say anything to him but would be very proud and happy that his son at last has found a lady he could loves.

When both the girl and the boy have shown their gifts to their respective parents, the parents would start interrogations.

Interrogation

Shortly, after the parents have seen and received the gifts, the mother of the girl and her aunties would commence the interrogation of the groom. On the other end, the father and uncles of the groom would commence interrogation of the bride. The interrogations usually cover the following areas:

▶ Is there blood relationship between the two proposed couple (bride & groom)?
▶ Does the bride's family or the groom's family suffer from leprosy?
▶ Do the family members of the bride or groom practice evil eyes?

Apai Gabriel (2005) observes that during this period, a lot of consultations would be conducted among relatives of the girls and the boy for any possible blood relation between them. Based on the interrogations, if one side is found to have being having one or two or all of the above, the marriage would be cancelled. When this happen, the mother of the bride always asks her daughter to return the *"magala"* (ear-rings) to the groom. The father of the boy also would ask the groom to return the *"tii"* (necklace) to the bride; in this way, the marriage arrangement is considered cancelled. Marrying blood related relatives is strongly avoided among the Jo-Luo because they believe that sexual contact with relatives produces a disease called *"adual"*, which affect normal body growth and causes overweight or extreme skeleton-like thinness, particularly to offspring.

Having said this, I would like to mention that in the Jo-Luo of Bhar el Ghazal Cultures, if any family member practices evil eyes or suffer from leprosy, it is often difficult for that girl or boy to find a partner. In this situation, the girl, whose family are practicing evil eyes or have leprosy, will only get married to the boy whose family have evil eyes or has leprosy.

However, if the interrogation is very successful in which no blood relationship, no leprosy and no evil eyes is found in the families of the bride and groom, the marriage arrangement would then proceed. It is reported that the groom usually confirms marriage arrangement by asking the bride, "What have your parents discovered with my family?"

The girl would answer, "My parents said they have found nothing wrong with your family, they said, you have good family". Which, literarily means, your family members do not practice evil eyes, do not having leprosy and there is no blood relationship.

The bride in turn would ask the groom, "What have your parents discovered with my family?" The boy would answer, "My parents said they have found nothing wrong with your family, they said, you have good family". Which, literarily means, your family members do not practice evil eyes, do not having leprosy and there is no blood relationship.

When the girl and the boy finish the consultation with each other, the bride then invites the groom to meet her parents. The groom receives the invitation with all his heart, he then goes to his father and tells him, "Father, my girlfriend has invited me to visit her parent in regards to our love affairs".

Acceptance of Marriage

According to Apai Gabriel (2005) once the parents have formally agreed, the groom goes officially to the bride's mother's house. The groom goes to the house in the evening and sleep there. In the morning, the girl would call her young women relatives and other girls in the village to come and hear what the groom has come for. Even though it is known that the groom has come for engagement, the procedure is that the one woman from the side of the bride would be asking the groom and the bride. The woman will start asking the groom and later on asks the bride as follows:

Woman: Who are you?

Groom: My name is —tells his name.

Woman: Why have you come here?

Groom: I have come here because I am interested in one of your girls called...... Both of us have agreed and promised to marry each other.

Woman: On hearing this one of the women would go out and coveys the news to responsible men, who are just waiting outside for the purpose. She would tell the men that the man (groom) has come here to marry our sister........ The men then tells the woman, "Oh! If that be the case, we have no objection. We accept his idea.

Woman: The woman comes back into the house and tells the other women and girls that the elders have accepted the marriage. The women/ girls and the groom would then negotiate a date for his return, it may be a week or two weeks, to put a wrist ring onto the wrist of his girl. In the past, groom used to put closed copper ring on the wrist of the girl. Today, it is replaced with a black and white bead necklace, which is very popular among the Jo-Luo.

After this, the groom would go back and report to his male elder that the engagement had been accepted, and he is told to go back within one or two weeks to put the ring on the wrist of the girl.

Soon after this good news, after a week or so, the father of the groom selects between eight and ten boys (who are very close to the boy) to deliver $5,000.00 to the father of the bride and a ring-necklace for the bride. Thus, before going to bride's father's home, the selected boys meet to decide who would be their team leader. Team leader is usually selected on merit of their being good in leadership. When the delegations have finished selection of their team leader, and as evening comes, the delegations then proceed to the bride's father's home, and spend night there.

According to the Jo-Luo of Bhar El Ghazal customs, the $5,000.00 is never delivered to bride's father's home by day, it is usually taken in the evening, before darkness. When the groom's delegations arrive at the house of the father-in-law, they are welcome by number of girls from the village. The girls give them food, water and local beer to drink; this may follow with some games and they are given a place to sleep. Traditionally, the groom's delegations are not expected to give the $5,000.00 to the father-in-law on that very night, they usually give it in the morning. Very early morning, as the sun raises in the east, the father of the girl sends two of his members to the groom's delegations. It is the custom of the Jo Luo that the bride has to be there when the father's delegate is asking groom's delegations. The father's two men delegations then ask groom's delegations, "Why are you here?"

The team leader from groom's side answers, "We are here because our brother Mr. X (gives name) wants to marry one of your daughters called Miss X (give name here)".

Before the $5,000.00 and ring-necklace is given to the bride, the groom and bride make serious vow; the vow is usually performed by intelligent old woman with experiences in marriage life. For us to understand the Jo-Luo vow ceremonies well, let us give name to the groom and bride. In this case we called groom as Uliny and we called bride as Achan. The woman leading the vow ceremony would start asking questions (**Q**) to Uliny and he is to answer (**A**) that would be followed by questions and answers to Achan.

Q. You are Uliny; am I right?

A. Yes, I am.

Q. You have come to marry Achan. Is that right?

A. Yes, it is right

Q. Who told you to come?

A. I decided it myself.

Q. Achan does many activities. If she loses her eyes, would you not reject her and say that you do not want a one-eyed wife?

A. No, I would not do that.

Q. Pound ding grain is a hard work to do. If the wooden crusher falls onto her leg and breaks it, would you not reject her when she is lame?

A. No, I would not reject her.

Q. Achan has much hair. If she falls sick and loses her hair, would you not reject her?

A. No, I will not reject her.

Q. Suppose you discover in your house that she does not know how to cook well, would you not send her back to the parents?

A. No, I will never do that nasty thing.

The woman the n turns to Achan and asks the following questions:-

Q. Achan, have you agree to marry Uliny?

A. Yes, I have agreed.

Q. Who told you to accept him?

A. Me alone.

Q. men have many accidents in life. If he becomes blind later, would you not abandon him?

A. No, I wouldn't.

Q. Uliny is a handsome man. If he sees that you have become a mother to children and he marries another wife, would you not abandon him, together with the children?

A. No, I wouldn't.

When the groom and bride finished completed the vow, the groom takes the $5,000.00 and hands it over to the bride, in the presence of the two delegates (sent by the father). Interestingly, the bride does not hand the money to the two men delegations, she takes the money and gives it to her father; he then receives the money from the hands of his daughter. It is generally believed that such act of receiving the $5,000.00 from bride's hands, by the father, is an indication of acceptance of marriage. The bride's father then sends message to groom's team leader saying, "The money has been received, and marriage is accepted".

According to Apai Gabriel (2005), after the vow and the $5,000.00 has been received by the bride's father, the broom then put the ring or places bead necklace on the neck of the bride; this is follows by and outburst of ululation from other women. The marriage is now considered officially inaugurated. Soon after this, the bride and the groom would be asked to open the ground for traditional dance. The traditional dance is done right in front of the bride's mother's house. This is then followed by second round of dance where the girls rush in and join their sister (bride) and the young men on the other hand also rush and join their brother (groom). During the dance the young men and girls match up. It becomes vigorous and attractive, making onlookers shake their heads to the rhythm (.P.38).

The people dance in front of the bride's mother's house because eth it is meant for the mother-in-law to see her son-in-law's dance at the close range, while she sits inside the house watching. The dance is ceremonial and usually does not last long, because sometimes such dance is performed in the midday when the sun is hot. Food and drinks would be served after the dance, when young men, young girls and women and taken their seats.

Customarily, after this traditional dance, the mother-in-law and the son-in-law keep a distance from one another as a sign of respect.

Jo-Luo oral history tells us that dancing on inauguration day was introduced in the society in the late 1950s. Jo-Luo elders are thinking that the dance started in either Amac section or in Logo section and by 1960s the traditional dance in inauguration engulfed the whole of Jo-Luo region.

After receiving the message from the bride's father, and after the traditional dance, the team of groom do not go back immediately; they wait to hear from bride's father on the number of animals required for the *"gikwer"*. As the groom's delegations are waiting for important message from the bride's father, they would be entertained by some girls from the village. The requirements of the "gikwer" is fixed by traditional law, and base on this, the father of the bride would then sends list of animals required to be paid as the "gikwer" to the team leader of groom and his group. At the same time the bride would be asking groom's team leader to bring the *"gikwer"* (first dowry) in three weeks.

The groom's team on receiving this message, discuss the proposed period for delivery with the delegation from bride's father. The team of the boy may find that three week is too short for the groom's father and uncles to prepare for requirements of the *"gikwer"* (first dowry). The two sides will take some minutes to discuss the issue related to period given, until they reach common ground, which could accommodate both parties. From the proposed three weeks, they may agree that the *"gikwer"* will have to be brought in two months' time. Apai Gabriel (2005) noted that when young men are going back after inauguration, the list of bride price is given to them to carry to their elders (especially groom's father). Some of these bride-prices may be urgently needed, while others are only to take note of (P. 39).

In as soon as the agreement is reached, the bride's team leader and his members withdraw from the house; but the girls would continue with the entertainment to the groom's delegation, until time of their departure (from bride's father's home). When the boy's team leader and his members arrive home, they go straight to the father of the groom, and inform him of what they have reached upon with bride's father's team. According to Apai Gabriel (2005) in the old days, after inauguration, a male elder from groom's side would go for a visit, to confer with the father of the bride about the marriage. He would then be given the bill containing all that is wanted. He would take the list and when the groom's elders are called for the official settlement, they would bring all the things along (p. 39).

Payment of Gikwer (first dowry)

Shortly, the bride's father convinces morning meeting with his brothers and uncles of the girl, to talk about part the bride wealth that would be brought, since marriage is already accepted. As I mentioned above, the requirements of the *"gikwer"* are always fixed by traditional laws. The payment of the *"gikwer"* (first dowry) usually consists of fifteen goats (*diek*), one bul (*mwor abutha* —literally the term *abutha* means long stick that old women use to support them in walking). However, in some cases bride's father may even require two *"rom"* (male sheep) and three *"mal"* (female sheep).

Before two months elapsed, the boy's father take the fifteen goats and one bull and give them to the team leader of the former delegation, to deliver to the house of the bride's father. The father of the girl usually receives the *"gikwer"* without any complain. Traditionally, there is usually no written documentation required when receiving the *"gikwer"*; but the delegations from both parties are considered the eye witnesses. If in future there is problem with the marriage, they will be consulted.

Moreover, it should be noted that the payment of *"gikwer"* (part dowry) does not necessarily mean that the marriage would go on smoothly without any problem. This is not true, because experiences have shown us that although the bride and groom have taken an oath, nevertheless, there is often possibility of marriage break down. There are many reasons that lead to marriage breakdown as we shall discuss below.

Break down of Marriage before Atoja (oath)

It is sometimes happen that despite the vows and cordial inaugurations, occasionally marriage breakdown before the final oath (*Atoja*). Apai Gabriel (2005) observed that, "There are many factors that lead to the breakdown of marriages. Some of these factors could be "disagreement" between the relatives of the bride about groom been accidently engaged to kin (*kith*) which often produces gossip. Many at time, gossips damage relationship between the proposed bride and proposed groom. This often become worst if a wealthy young man is interested in the girl, he can pressure the opposing group to approach the girl on his behalf.

Secondly, like in arranged marriage, the girl may be too young and when she grows up and when the relatives begin telling her the parents have made wrong choice for her, she would begin to show signs of disrespect to the proposed groom.

Thirdly, the engaged groom has no wealth to pay the bride-price and other competitors may use this situation to influence the relatives to speak to the girl and her family about breaking the engagement so that they can step in. Such development in a marriage is often followed by the abduction of the girl by secrete organized, unknown man. When such man is traced and brought before the court the girl would stick to him, thus, forcing the court to ask the father of the proposed bride and the abductor to pay back all what the previous groom has paid as part of the dowry" (P. 40).

Needless to say, this kind of marriage breakdown is part of the Jo-Luo life and culture; and this explain why in the inauguration ceremony, a woman was asking the girl many questions; that was to find out if she freely, without being induced by anyone, agrees to marry the proposed groom. Question: "Who told you to accept him?" and answer: "Me alone" are the core of future family life.

The "Atoja" (the oath or weeding)

After receiving the *"gikwer"* from groom's delegations, the father of the bride convinces meeting to discuss what would be brought for the *"Atoja"*. The Jo-Luo of Bhar El Ghazal pay dowry according to the capacity and ability of the groom. In the old days, the Jo-Luo paid dowry using beads, hoes, spears, axes and other iron products. But today, the Jo-Luo pay dowry with some 16 cows, between 15-30 goats (*diek*), 12 spears (*tong*) and between $7,000.00 and $15,000.00 cash money. According to Apai Gabriel (2005) in the past, the official Jo-Luo dowry was three cows, two bulls, two she-goats and three he-goats.

"These living things were required; in addition to the main item, which consists of 400 Jo-Luo hoes, as time went on, other items like beads, guns and even slaves were introduced into the bride-price. However, the demand for beads, guns and slaves were shortly dropped because such items were not available locally, but hoes and cattle remained the key items in Jo-Luo marriage. Hoes were also dropped later when the British administration in South Sudan forbade iron mining between the year 1929 and 1930. The British administration also forbade hoes for marriage; this was

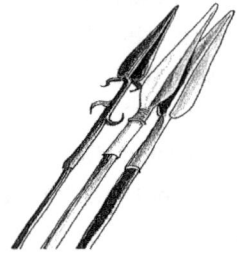

Figure: 166, Spears

done to discourage the Jo-Luo from mining iron core. As a result, today, the Jo-Luo bride-price is processed on money merit, cattle and goats" (Apai Gabriel, 2005, P.29).

When the day of *"Atoja"* comes, the young men from groom's side always go a day before the actual day for *"Atoja"*. When they arrive there, they are received by selected girls from the village; bride's girlfriends bring their friends to play and chat with the young men the whole night. According to William Dimo, the young men are always the first to arrive in the house of bride's father; they usually come in the evening, as the sun is setting. The young men enter the homestead with songs and dance, they are receive at the entrance, by the girls; they would dance at the entrance for some minutes, after this, the girls then take them into the big house; there they dance and chat the whole night—no sleeping for both young men and girls. Early the following morning, the elders, groom's father, mother and other people follow the young men (they usually arrive at around 7.00 a.m.). They are also welcome at the entrance of homestead by the girls, who show them place where to sit while waiting for the negotiation to commence.

As stated by William Dimo above, in the morning, huge number of people (men and women), from groom's side, follow the young men to bride's father's home to attend the "*atoja*" (oath). When these huge numbers of people arrive, and have taken a short breath, the team leader of the groom then handover the above mentioned items (*for Atoja*) to the bride's father in the presence of everybody. When the father of the bride receives the animals, spear and money, the groom and bride enter into oath (otoya) vide which they pledge to remain together when in good or bad times.

During the marriage ceremony, an elder man cuts the ear of the goat, which is brought by the parents of the bridegroom; pieces of cut ear and bead would be tied around bride's and groom's necks. The team of the groom remain in the home of bride's father for three days and three nights, eating, drinking local beer and dancing. It is reported that during these three nights, some of the young men from groom's side, take an opportunity to play sex with the girls who are there to serve them. The act of sexual intercourse is usually done without the knowledge of bride's elder men. One of my informants, Mr. Alberto Primo Akot Dimo says, "Yea! In the nights any possible thing can happen". Mr. William Gol confirmed the statement made by Alberto, and says, "In *Atoja* day, relationships are developed between girls and the young men, therefore, there is no surprise that young men, from the groom side, would take an opportunity of the three nights of weeding to have sexual intercourse with their beloved girlfriends".

In the last day of the "*Atoja*" (weeding), if the dowry payments are not completed, elders from both parties agree on the year for the completion of the dowry. However, if what is left from the dowry is small number, the groom's team may ask for the bride to go to the house of the husband. And the balance of the dowry will be completed while she is in her husband's house. But if the remaining dowry is slightly great, the groom's parents must pay more before the bride could be released.

More importantly, it should be noticed that in Jo-Luo of Bhar El Ghazal society, although the dowry is paid in full, the bride is not allowed to go immediately with the husband. She will have to stay behind for at least two to three weeks, to provide an opportunity for the mother and clan women to prepare the required items for her to settle in her new house.

It is reported that as the bride is not allowed to go with the husband, the Jo-Luo of Bhar El Ghazal customs, forbid the husband from having sexual intercourse with her, in the period she is waiting in her father's house. This is completely contrarily to Acholi and Pari cultures. As we have seen among the Acholi and among the Pari marriage and family life, husband and wife is allowed to have sexual intercourse when the bride is detained in her father's house for one reason or other.

Accompany bride to husband's house

"When the marriage is settled and the b ride-price is paid, the man would arrange to take his fiancée (girlfriend) home as a wife" (Apai Gabriel, 2005, P.42).

When the parents of the bride and other relatives are satisfied with the payment of the final dowry, they fix the day for the aunties, uncles, sisters and brothers of the girl (bride) to meet and prepare for the day of accompanying the bride to her husband's house. The mother of the bride, the aunties, and the clan women assist in the preparation of beads, local beer, new cooking pots, pots for storage and any other requirements.

When the preparation is completed, the groom is expected to pay between $500 and $1,000 to the aunties before they finally released the bride.

However, the Jo-Luo focus group advised me that the amount to be paid to aunties, for the release of the bride, varies from family to family, depending on the wealth of groom's father. As such the poor and the middle classes can pay between $500 and $800; the rich family can pay between $800 and $1,000.

When the bride is released, she is accompanied by two or three hard working girls. Unlike the Shulluk, aunties do not accompany the bride to her husband's house. According to Apai Gabriel (2005), the wife is usually taken secretly to the home of the groom, this is done to avoid evil eyes person seeing her at departure. A woman would be chosen by the girl among her relatives to make arrangements with the groom or his agent about when to leave and when to arrive. The parents of the bride would not know about this arrangement, especially if part of the dowry is not completed. Thus, to avoid discovery of preparations, a woman usually facilitate the elopement and the girls who are going to accompany the bride may be from different villages. Yet this woman and the girls would coordinate their preparations, so that they all meet in a given place and day to depart together. The departure is always at night and the girl is usually taken to the house of the male elder or groom's uncle. She would stay in this house for about seven days; and during these seven days, evening dances would be held to publicise the arrival of a new wife and for people to see her face for the first time –if she has been brought from a far village. During that time the male elder would slaughter a he-goat as a blessing for the new couple. Thus, on the first seven nights that the bride is accompanied to husband's home, she is not allowed to share bed with her husband; she sleeps with the girls who accompanied her. The eight nights, the bride is finally allowed to sleep with her husband; she is then taken to the husband's hut that is located within groom's uncle's compound (P. 43).

As the sun rises in the east, the girls get out from their hut and help in domestic work, such as cleaning the compound, cooking and fetching water from the river. In the meantime, the groom goes to the nearest market to buy some clothes and beads for them. And as the sun sets in the west, in the old day, aunty of groom or old woman takes pieces of "kali" (strong rounded and hard beans) into the hut of the groom and places them under

"*Abel-tek*" (the animal sleeping skin where the wife and the husband are to sleep). Please, see the pictures of the hard beans which are placed under "*Abel-tek*" in Figure: 152 below.

Figure: 167, Kali (Kola) nuts

In the old day, people used to sleep on the "*Abel-tek*"; the aunty or old woman places the "*kali* seeds" under the "*Abel-tek*". But today, this practice has faded way because boys in villages sleep on mattresses; secondly, today, the idea of virginity (of the girl before marriages) is no longer important in Jo-Luo society. Alberto Rimo Akot Dimo, an informant, argues, "Today, a Jo-Luo young man would blame his wife if he finds her a virgin. The young man may tell his wife that she was not admirable and beautiful to other boys, nobody was interested in her, and as a result, she remains a virgin.

In old days, when the aunty has placed the "*kali*" under animal sleeping skin (*abel-tek*), and soon after darkness, the two or the three girls accompany the new wife to her husband's hut and they leave her there. When the time for sleeping comes, the husband closes the door and shortly, they are on business.

In the morning, as the groom and bride get out of the sleeping hut, the same lady who put the "*kali*" (nuts) under the "*Abel-tek*" enters the hut to inspect if the "*kali* nut seeds" have been grounded or broken into pieces. When the *kali*-nut seeds are broken to pieces, this is an indication to the groom's relatives that the groom has done good job at night but if the *kali* nut seeds are not broken, people become worry and doubt about the boy or suspect the girl. However, in most case, "*kali* nut seed" are broken, which is good news for all.

According to Jo-Luo of Bhar El Ghazal customs, when a wife is brought to husband's house she does not go straight to her mother-in-law's house but rather she stays with groom's uncles or male elder in the family for seven days and seven nights before going to work in the house of the mother-in-law. After seven days, the wife then moves to work in the house of her mother-in-law, where she is expected to live and work there during the days and at night she goes to sleep with the husband in their own hut; this practice continues until the wife delivers her first child, the wife and the husband would then move to live in their separate house near uncle's and husband's mother compounds.

Apai Gabriel (2005) observed that in the new life of the bride, she would become an apprentice under her mother-in-law but sleeps with her husband in their hut. When she is with the mother-in-law, she manages the house under directives from the mother-in-law. For example, if she cooks the food, she calls the mother-in-law to watch how the food is divided. While the bride is learning how things are done in the family, the mother-in-law would also being observing her how the bride is focusing on speed, accuracy, behaviour and patience, which are considered very important for household management (P.44). The bride may spend three to four months with the mother-in-law, after this she may she may ask her husband to allow her to visit her mother for a short period of time. There (at mother's house) she arranges with her mother what utensils would be brought to her later. She returns quickly, according to the time given her by the husband.

After the bride has returned to her husband's house, the mother may take the utensils to her daughter by herself or she may delegate a younger woman with some girls to take the utensils. The delegated woman and the girls then set everything, they brought, in place and the woman and the girls return to their respective houses. At this stage the bride would be cooking independently with her own utensils. When the bride becomes independent in her household management, the husband builds a house somehow away from his parents' house —but close to his father. Although the groom moves away from the parents, he keeps on coming to see them from time to time —he may even assist his parents in routine work. The wife may cook food in her house and take part of it to her father-in-law and mother-in-law. The husband often shares food with his younger brothers.

However, if the groom is the last son in the family, the bride may continue to live with her mother-in-law for life. She is expected to support the mother-in-law and care for her as she (the mother-in-law) is growing older.

It is generally reported that the two or three girls, who accompanied the bride to the groom's house, may stay with the bride between three and four months. It is also reported that these two or three girls often find their boyfriends from the village during their three or four months of staying with the bride. When the three or four months elapsed, the groom buys more cloths and beads for the girls before they depart to their respective houses.

As the wife lives happily with her husband in the new home, the mother-in-law and other elderly women in the family would observing her very closely for the signs of fertilization. They every moment take note of her face, breast changes and her taste for special food and when these signs are seen in her, they take these signs for pregnancy, but if these signs disappear as the cycle fades away without conception, then the woman like this is said to have "*thobba*" (abortive sensitivity to pregnancy. The wife is then taken to an old woman who no longer has menses to "return her inside with *thobba*. This means the old woman performs the ritual of *thobba* on the wife —as we shall discuss later.

Unlike the Acholi and the Shulluk, the Jo-Luo of Bhar El Ghazal tradition does not allow the brother of the husband to have sexual intercourse with brother's wife. In Jo-Luo society, a brother is only allowed to have sexual relationship with his brother's wife when the brother dies or when the brother proved impotent.

More interestingly, the elder brother is not allowed to have sexual relationship with the wife of his younger brother, rather the younger brother is allowed to have sexual relationship with the wife of his elder brother in case of impotency. In the same way, the elder brother is not allowed to inherit the wife of his younger dead brother, but the younger brother is allowed to inherit the wife of his elder dead brother.

Once the marriage of the elder son is settled, the father and the younger son start to find money and cattle for him. If this man is lucky to have girls and boys among his children, it would be a matter of matching boys with dowry brought after girls as bride-price to marry with.

However, there are occasions when matching dowry with boys is not applicable. For example, if an elder son is born, followed by two or three girls and younger boys arte born last, it would not be possible to match, because the boys would still be small and too young to marry when their sisters are married. In such situations, the elder brother uses the dowry to marry the second wife, bearing in mind that when his brother attains marriage age later, he would be responsible for the dowry. If, by good luck, his first wife happens to bring forth a baby girl in future would be given to his younger brother in turn (Apai Gabriel, 2005).

Sometimes, it is the father who may have so son or younger brother to use the dowry brought after his daughter. In this case, the man can marry the second wife or third wife, as a method of Jo-Luo family planning. The man then takes this wife and places her in the hands of the woman whose daughter's dowry was paid as bride-price for this second or third wife.

According to Jo-Luo tradition, this is usually marks the beginning of family division. The woman attached would see herself always as part of the original wife. And the children, too, would consider the young wife of their father as part of their mother, because the bride-price paid after their sister was used as dowry. Therefore, here, a degree of family relationship right with affiliation unveils itself in all the family asset-sharing from then onward (Apai Gabriel, 2005, P.46)

The Ritual of Thobba

As we have seen above a wife has what they called "*thobba*" but in other cases the woman may not have a "*thobba*" but yet she may not conceive. In both cases, the wives would be taken back to her father's home where general prayer (*kock*) would be performed for her. This prayer is conducted because the Jo-Luo believe that when the girl was still growing in the family, she might have come into conflict with an elderly woman who may angrily cursed her to be barren for what she said or did to her (old woman). In the cause of time, both the elderly woman and the young grown up girl may forget about it, but the anger and curse stand before God (Juok). Therefore, in order for the young woman to conceive, the father would arrange to get a bull or ram and call a spiritual leader to come and lead the prayers of sacrifice, repentance and forgiveness (Apai Gabriel, 2005).

When a ram is being offered to God, the spiritual leader puts in a bowl of water pieces of white sell, and this bowl of water is passed from person to person. As the bowl passes from person to person, each person is expected to spit into it a little saliva; while the prayers continue until the ram urinates, and this would be taken as a sign that God has accepted the offer. After this, the wife, for whom the prayer is offered, jump over or sit on the ram so that the blood of the ram blot out her iniquities or any other grievances she might have done to anybody in her life that has prevented her from conceiving.

Thus, when the ram is sacrificed, spiritual leader takes the blood of the ram and drops it into the consecrated water in the bowl. And the spiritual leader ties the Jude on the neck of the wife while saying: "Go and conceive".

The people who are in the ritual of *thobba* then go one by one and pick a Jude, tying it on the neck of the wife as each says: "If it is me, you will conceive". In the end, the pieces of Jude become like a nest around the head of the bride.

After the general prayers have been conducted, the bride goes back to her husband's house. The people would wait for a month or two, and when she conceives following this general prayer, that child would be named "Lau", which means saliva; this name can be given to any child no matter of sex.

Persistent death in marriage
When a man marries and his wife dies soon after marriage; and he marries another wife, and that wife also dies soon after, the third time he marries, and before they enjoy sexual relationship in their married bed, both will spend night in a mock-hut, which is erected somewhere near the house. When time of sleeping come and everybody has gone to sleep, the couple would also go to sleep in their mock-hut. In the mock-hut they would be naked and pretending to sleep. Elder would designate somebody to act in this occasion; so the designated person comes and sets fire on the mock-hut. The woman and the man get up and run out naked, leaving everything in the mock-hut to burn. The Jo-Luo people believe that by so doing, the spirit of misfortune, which lurks over the man will be scared or frightened and would run away from the body of the man, and it will never follow the man again. In this case, the third wife will not die and the man would enjoy the fruits of marriage.

Divorce
Traditionally, divorce is generally not accepted by all the Jo Luo of Bhar El Ghazal. Nevertheless, this does not mean that divorce is not practice, it is being practiced but in

severe cases such as family violence that is life threatening. In the Jo Luo society, divorce has been experienced here and there, but it is very rare among the community. All the Jo Luo women are expected to submit to their husbands and never think of divorce even though she is often been beaten by the husbands. Apai Gabriel (2005) observed that divorce is the most difficult and rare to happen in Jo-Luo marriage, especially when children are involved. In Jo-Luo society, people speak of husband bringing a wife at home as he reserves the right for reproduction with her for children but as matter of fact if anything happens to her, ever member in a family/community would be concern.

It is generally believed that in the Jo Luo of Bhar El Ghazal society divorce is amicably arranged by the wife, since a man only seeks divorce when his wife has left him for another man. Mr Heasty states that it is very rare a husband to divorce his wife, it is usually the wife or her family who often initiates the separation.

Traditionally, the breaking of a marriage means the return of cattle, sheep, goats and money to the bridegroom. Experiences tell us that this is the most difficult matter in family life, since the bride-wealth is paid to bride's father and the bride-prices might have married ten other wives. Once a family breaks down occur with the first woman married in the family, this will affect the whole chains of marriages; in other words it will play part in other marriages. Seligman & Seligman (1932) notes that when the wife decides to divorce the husband, she is to hand over bride-wealth to her father, who would in turn return the original bride-wealth to the first husband. It is said that if there is a grown up daughter, and the wife's father cannot return the bride-prices, the husband has right to take the daughter instead of the bride-wealth. For this reason, every man wants the marriage of her sister to be stable.

The bridegroom is usually anxious to please his bride's relatives so that they in turn do not influence her to break off the engagement. He does this not only because he wants to preserve marriage but also for the fear of the woman losing prestige after marriage. Seligman & Seligman (1932) informs us that there are many reasons why the bridegroom should seek to keep on good terms with his wife's relatives, and by his correct behaviour, because these prevent the possibility of rupture.

It is to be noted that if a wife leaves the husband, the husband will demand the sheep, hoes, fish spears, and cows to be returned.

Mabaan Traditional Marriage and Family

Being one of the Northern Luo group, we assume that the people of Mabaan have the five steps we have discussed in other Northern Luo group of South Sudan. I cannot discuss it here step by step because I did not meet any of Mabaan elder that would have narrate to me various steps taken by the people in marriages.

However, we are made to understand that marriage among the Mabaan is for reproduction of children and for the expansion of the family. The people in Mabaan conduct marriage according to the age-group and if one person miss to marry at his age group, he will miss it for ever.

The courtship begins at the wake of the harvest feast and is usually accompanied by two ceremonies. The Mabaan dowry consists of 5 pigs, 10 goats, hoes and axes; the dowry is usualy paid by maternal uncle of the groom.

Once the dowry is paid, the first ceremony begins in which both the bride and groom spend the first four months of their lives in the weeding room locally known as "chanyo". During this period small ceremony is conducted in which the groom wears a special neck-lace and moves a short distance within the village; he moves alone, leaving the bride in the weeding room. After 4 months, the big marriage ceremony is then conducted. This usually occurs in the month of March in which the bride is given new harvest and she is taken to her husband's house. Traditionally, the housband's home is usually built near the maternal uncle's home.

Pari Traditional Marriage and Family

According to Pari local tradition, marriage usually consists of seven steps, which starts with courtships; delivering of hoe to bride's father's home; Interrogations; Groom telling the bride's parents and uncles number and types of cows he is going to deliver for his wife; Negotiation on actual number of cows to be delivered; accompanying bride to her husband's home and finally the actual delivering of agreed number of cows to bride's parents.

(1) *Courtship*

In Pari society, courtships take place in the compound of one of the old wise woman, who is a widow. Every day at around 7.00 p.m., when every family member have eaten supper, and girls have washed all cooking pots and plates and put them in their respective places, girls of the same age in a village, gather in the compound of an old woman. They gather in different places within the village. The house of the old woman does not only provide an opportunity for girls to gather and chat but it also provides a sleeping room for the very girls. The girls sit on special seat made of wood locally known as *Padho* and elevated about 50cm high above the ground. They sit and chat, while waiting for the opportunity to find boys who could have interests in them. Customarily, courtships are not allowed to be conducted outside the compound of old woman in the village. The reason has been that elders are suspicious that when boys and girls meet outside the old woman's house, there are possibilities of sexual relationship, which is for dibbed by the customs. And if this happen, the brothers of the bride would organised themselves and beat the groom.

As girls leave for their usual gathering places, the young men of the village also gather in a certain point within the village. The boys also gather according to their age groups that numbered between 40 and 50 young men. Shortly, the group would divide themselves into small groups of ten; and each small group moves under the leadership of a young man from a royal family.

As I mentioned above that girls gather in many different houses in the village; this also provide an opportunity for the young men to move around these houses in search of a lover. The reason been that a boy may not find a girl of his interest in a house of one old woman but there is high chances that he can find one from other groups.

It is worth mentioning that in these small groups of tens, some of them might have already girlfriends. Such young men travel with the group not to look for another girl-friend but to assist their friends in courtships as some girls proved to be difficult to approach. In addition this group movement also may provide him an opportunity to meet their dated girl-friends.

Thus, when the young men arrive at the compound of an old woman, each boy would look around for the girl of his interest. If the boy has spotted one girl that attracts him, he then calls her aside. David Akwai, one of the elders of Pari, who live in Melbourne Australia, said, "When a girl is called by the boy in this occasion, she should not get up immediately and attend the call, but she must stay for about five minutes or more and attend the call". And in Pari custom, whether the girl has already a boy-friend or not, she must attend the call of the boy.

When the girl comes to the boy, he then tells her, "My dear, I have called you here because I want to let you know that I have much interest in you; and would like you to be my future wife. What do you think?" The answer to this question depends on whether the girl has a boy-friend or not.

If the girl has a boy-friend, she answers, "Thank you for your interest in me. I want to inform you that I already have a boyfriend, so I do not have interest in you". On hearing this, the boy must stop courtship and never force the girl to change her mind. The girl goes back to the group immediately.

However, if the girl has no boy-friend, she answers, "Thank you; I have heard what you said but I have no least interest in you".

The boy continues, "Why don't you have interest in me? Please, I want you to understand that I love you so much and I feel your love in the marrow of my bones. When you go back, after this talk, I want you to think very deeply about my sincere love for you and when we meet in two weeks, I would like to know how far you have gone". Courtship per night takes usually three hours, and after that the bride and groom can go back to the group.

Telling me an informant who preferred anonymity says, "If the girl has no boy-friend, although she tells groom that she has no interest in him, this does not necessarily means true. The groom has to continue meeting her until she gives in".

Traditionally, girls are expected to be strong during courtships, she is supposed not to answer the groom softly but rather with tough voice like No. I don't. I can't etc. For example,

"I have heard what you said, but I want to let you know that *I am a girl of a man* (meaning don't think I am an easy going girl). I do not have interest in you. You better look for girls you can play with". Is this true?

As reported by one of the informants, such answers do not necessarily means that the bride has no really interest in the groom; but she just wants to hide her feelings from the groom at this stage. The bride is supposed not to show her interest in the groom during the first few months of courtship.

Customarily, after first meeting, the best friends of bride would ask her and she tells her friends the intention of the groom and what they have discussed. Friends usually are expected to give her suggestions on how to deal with the case (love affairs).

Subsequently, after two weeks the groom comes back with his age group to the previous group. He then calls the girl; and this time the groom may be accompanied by one of his best friends because it is believed that bride often prefers opening her heart to groom's friend.

Thus, when the bride comes, the groom would say, "How are you? I hope you have been reflecting on our last discussions. Today, I want to hear what your thoughts are about this matter".

Bride answers, "Yes, I have reflected on the matter but there is no change in my heart. I then came into conclusion that I have no love for you".

Groom asks, "Did you find a boy-friend".

Bride answers, "No. I did not find a boy-friend".

Groom asks< "If you did not find a boy-friend why do you keep on saying, *you do not love me?*"

Bride answers, "I don't know".

The groom and his friend continue to meet the bride for another one month or twelve months.

One day, after discussion on love affairs with the bride, in the presence of his friend, he (groom) moves away leaving the girl alone with his friend.

Groom's friend says, "Dear girl, I have witnessed your discussions in this matter for a long time. I trust and respect you; I hope you trust me and if so I would like you to be opened to me. I want to reassure you that my friend loves you so much and he is ready to marry you. I know also that you have been thinking deeply about the matter. Please, tell me, what is your final opinion on this matter?

Bride answers, "Thank you. I also want to let you know that I trust you. I know your friend has taken his times to discuss this matter. Today, I want to tell you that I am in deep love with your friend.

I have made this decision after long consultation with my friends. But there is one important thing I want to tell you: *Let your friend not give me shame. He must show himself to my parents*".

David from Melbourne Australia, said, "It is the custom of Pari that when a girl has loved the groom, she would ask him to show himself to her parents. If the groom does not show himself soon to bride's parents the brothers of the bride would gather and beat him. Because it is expected that once the girl has given her love to the boy, this often follow with sexual relationship. Above all, the final decision for marriage lies with the bride's parents following intensive interrogation". We shall discuss about interrogation later on.

After some minutes groom comes back and says to his friend, "I hope you have reached to some conclusions!"

Groom's friend answers, "My friend, congratulation. This girl has accepted your love with two hands. But she does not want you to give her shame" Groom hears such word with smile and excitement on his face.

According to Pari tradition, once the bride has given her love to the groom, on one hand the bride is not allowed to have another boy-friend, and on the other hand the groom is also not allowed to have another girl-friend. If by any mistake, the bride or the groom develops another friendship, this will generate big fight between the groom and the new boy-friend or between the bride and the new bride. Many a times this may lead to excommunication of the couple by both side of the family should the bride and groom insist on going ahead with their marriage.

Like the Anyuak and unlike the Acholi, the Pari don't change gifts between bride and groom. Instead, once the bride accepts groom's love this is usually followed by sexual relationship because the bride is now ready for the groom.

After this good news, the three people go back to the group and from there the groom and his team mates return to their sleeping place. Seligman & Seligman (1932) notes that the marriage process in selecting a partner and negotiating a marriage traditionally in South Sudan allow the young people their choice of spouse. Although in theory, a bride needs her family approval, young women are free to marry man who can provide suitable le bride-wealth. Since the bride has already shown her love to the groom, according to Pari custom the next step is for the groom to inform his father before showing himself to bride's parents.

So, one evening groom goes to his father in the fire place or in the house when he is alone and tells him, "Father, I want to inform you that I have found a lover. She is the daughter of Mr and Mrs (name them)". The father in turn will inform groom's mother.

From the other side a bride also goes and tell her mother, "Mother, I want to inform you that I have given my love to Mr X (named) son of Mr. & Mrs from such a clan. The mother of the bride in turn will also inform bride's father.

Unlike the Anyuak, the Pari parents do not conduct interrogations of the bride and groom prior to the delivering of the "hoes". They only conduct interrogations after the hoe has been delivered.

(2) *Delivering of hoe (kweri) to bride's mother's house*

Once the groom has informed his father, the father then selects one of his friends to deliver the hoes to bride's mother's house.

At around 7.00 p.m. the delegation arrives at bride's father's home. The Pari people believed that this is a good time to deliver the hoes, because by then everybody would have eaten and taken their seats. This time is also good because it comes after darkness, in Pari custom delegations from groom should not arrive at bride's father's homestead when there is light (when the sun is still up in the sky). As he arrives at the entrance, he knocks on the wooden fence and the father of bride sends messenger to see who is knocking on the fence at this time of the night.

Messenger meet delegation at the gate and without saying a word the messenger goes back to bride's father and says, "There is a visitor at the gate". A messenger normally does not talk to the delegation/visitor.

When bride's father hears this, he immediately understand that such visitor is coming because of his daughter. He then tells messenger, "Go and call the visitor and give him cow skin to sit on". Messenger calls the visitor and gives him cow skin to sit on in front of bride's mother's house.

As the visitor is seated, bride's father sends one man to find out why he (visitor) has come to this house. Such man must be related to brides, he then goes and asks the visitor, "Why have you come to this house after darkness?"

Visitor/delegation answers, "I have been asked by groom's father to deliver this hoe to this house, because one of his sons wants to hold the hand of your daughter (name here)".

The man goes back to bride's father and tells him, "The visitor said he has brought a hoe for your daughter (name here)".

Bride's father then says to the man, "Go and take the hoes and let the visitor go back to his home".

The man comes and picks the hoe from the delegation and wish him good bye. A delegation goes and tells the father of groom that the hoe has been received by bride's father. The father of groom then thanks him and says, "Let us wait for their response after their interrogation".

(3) *Interrogation*

Soon after the delivering of the hoes, bride's parents commence interrogation on the groom. The interrogations cover but not limited to: (a) how many cows does the groom has? (b) Is he a hardworking person? -Because parents of the girl do not want their daughter to suffer hunger after marriage. (c) Is there blood relationship between groom and bride or not? (d) How much land does the groom has for cultivation to avoid future hunger.

Friederike B. and Orly Stern (2011) says once both parents received information that their son and daughter want to get married, the girl's parents as well as the boy's parents invite the friends and relatives to speak about the clan, tribe and culture of the groom and the bride. In this family meeting, they want to know if there is blood relationship, laziness, witchcraft and riches are there in the groom's or bride's homestead (P. 12).

David Akwai, one of Pari elders and intellectual, says, "Some families have no cows but if they have big farm and groom is said to be hardworking, in this case, bride's parents may allow their daughter for marriage. This is because the parents believed, beyond doubt, that groom would cultivate enough crops, has good harvest and sell parts of the harvests to purchase cows.

Once the parents have found that groom has good family back ground, they accept the marriage and wait for groom to tell them of how many cows he would actually deliver. However, if the parents found out during interrogation that groom has bad family back ground or he is related to the bride, they would send for the man who brought hoes to come and return the hoe to groom's father. This means that marriage arrangement is cancelled. Subsequently, the groom and the bride would be forced to search for another partner. In Luo culture, family background is a key to marriage.

(4) *Groom discloses his wealth to bride's parents*

The third steps of Marriage in Pari society consists of groom disclosing his wealth to bride's parents. In this step groom takes time one day and goes to bride's father's home. There he meets bride's parents alone and discloses to him how many cows he has, he goes as far as demonstrating the colour of each cow e.g. white, black, black and white spotted etc.

Marriage is very important in the life of the Pari young men, as such, he is expected to work hard to get his own cows. He can do this through cultivation, where he works hard and produces good crops for sales and the money receive from sale would be used to purchase cows. In the Pari community, uncles and other relatives hardly assist groom with cows for dowry. However, in the case of groom who may has his elder sister, and that sister gets married prior to groom's marriage, part of the cows obtained may be given to the young man to add to his cows for marriage. It is generally reported that groom is usually assisted on the basis of his capability to obtain his own cows and good behaviour with the uncles. David Akwai, an informant, said, "The uncle of groom may help with some cows if he (groom) has been working hard in the fields of uncles or has been looking after the cows of the uncle, but uncle is not compel to give him cows".

When the groom finished providing required information to bride's parents, he then returns to his home and tells his father what he has told bride's parents. The father in turn sets day, date and time to negotiate on bride-prices to be delivered although the final decision on the bride wealth lies with the bride's father and relatives. The groom's father then sends groom's mediator to inform bride's fathers of the day and time of negotiation.

(5) *Negotiation on Number of Cows for Marriage*

Once the day comes, bride's family invites the family of groom for negotiation on numbers of cows needed for marriage. On this day the bride's parents will know how many cows would really be delivered. This is also the time bride's parents would approve whether groom was telling them the truth about his wealth.

Moreover, it is important to note that on this very day of negotiation, groom's father is not expected to bring along the cows but he rather confirms how many cows would be delivered.

Customarily, cows for marriage in Pari society are not delivered on the same day of negotiation; rather, cows are delivered to bride's father's home after one year following the negotiation.

Interestingly, dowry is not a work of one person, it is fixed by traditional laws; nobody can change the number of animals required for marriage.

Customarily, the bride's wealth usually consists of sixteen cows and twenty-five mixed sheep and goats. In many cases, money is not required in marriages but sometimes if groom cannot get all the required number of cows, sheep and goats, he could be asked to pay the remaining number of animals in money value. Currently the value of one cow is at SSP 3,000 and one sheep at SSP 1,000 and one goat is also at SSP 1,000. It is worth noticing that the above prices for cow, sheep and goats are only prices given in marriages, however, when one wants to buy these animals, he may finds the costs of these animals are lower or higher depending on individual trader and the environment.

After reaching agreement during negotiation, from that date bride is considered the legal wife of the husband. Nevertheless, although bride is considered legal wife of the groom, she is not allowed to go with the groom on this day. She remains in her father's house for a period of at least three years.

In Pari society, girls are usually married at the age of 15 years; and boys get married at the age of 17 years. Thus, any girl who remains unmarried after the age of 20 years, she would be considered by community as somebody out of the "marriage-ring". Girls between the ages of 15 to 19 years are considered to be within "marriage-ring". This means that a girl of 20 years would not be married by young men, she would be married to men who have been rejected by girls; alternatively such girls would be married to married-men as second or third wife. It is generally reported that in Pari-land, by age of 18/19 years, every grown up girls are married.

During the three years bride is staying in her father's home, sexual relationship between her and the husband is permitted. However, they are not allowed to practice such sexual relationship within bride's father's compound, but, if groom wants sexual relationship with his wife, he informs members of his hut-mates that he has visitor that night. When the hut mates hear this, each person looks for a place to sleep that night. While, from the other end, the bride also informs her closes friends, whom they share room, that tonight I am going to visit my husband.

Soon after darkness, groom goes to the house of an old woman, where bride sleeps, he picks her and bring her to his sleeping room. The old woman would know that the girl has been picked by a boy-friend, but she says nothing, because it is normal. The girl sleeps with the boy-friend in his hut till 5.00 a.m. when the boy would escort her back. When the girl arrives at the door, she makes sign, such as coughing, so that her friends would know she has just returns and one of them would open the door for her.

At this juncture, I want to point out to readers that such sexual relationship could go on in as long as the bride is still in her father's home.

One informant who preferred anonymity tells me, "Sometimes the girl may get pregnant while at parents' home; and sometimes bride may go to her husband's home when she has either one or two children. This is normal and acceptable by the society".

More importantly, during these three years, when bride is still at her father's home, groom is expected to cultivate and harvest crops for her parents. During cultivation, groom takes his friends to the field and during this time girls, from bride's side, are expected to cook food, brew local beer and take the food and local beer to cultivators. David... said, "Girls bring food to the groom's friends when cultivating or harvesting crops. The girls sometimes also bring water and wash the bodies of these young men after cultivation and harvest".

Furthermore, traditionally, during these three years, when people go hunting and bride's father gets meat, groom is supposed to carry the meat for him. After the people finished hunting, groom takes the meat of his father-in-law till home. Not only this, groom is also expected to build or repair houses at his parents'-in-law's home. On the other hand, the three years bride is at her parents' home, it offers an opportunity for him to expand his farm and build new house where they will later on live with his wife.

Customarily during the three years, bride also attend to her mother-in-law's house from time to time where she does some grinding of sorghum, collecting firewood, cleaning houses and compound as well as cooking. Not only have these, during weeding, bride also helped her mother-in-law in farm's work.

What I find fascinating is that during the three years when bride is still at her parents' home, groom is not allowed to see the faces of his parents'-in-law. When walking within the village, and should it happen that groom meets his mother-in-law in one of village winding foot-paths, he must make a U turn, for otherwise mother-in-law has the right to beat him with her walking stick, should groom persists to pass by.

(6) *Accompany Bride to Her House*

Weeks before day of accompanying bride to her husband's home comes, mother of the bride prepared beads, beer, new big pots to be used for storage, small pots for cooking and all that the new wife required to move with to her new home. It is to be noted that bride's mother is usually assisted by clan's women in the preparation of these items.

After three years, the girls would be turning 18 years of age or 19 years -depending on the age when she was married. Bride would then be accompanied by her friends to husband's home. As we have discussed before, by this time groom would have built new house, put enough food into granaries and has acquired large farm.

When wife arrives with her friends, they would be welcomed at the entrance by groom friend. The young men dance with the young women throughout the night and this is a sign of welcome. While boys and girls are dancing, parts of beer would be distributed to groom's parents and relatives who might be there at the time.

(7) *Delivery of Cows to bride's father's home*

As we have discussed above, during negotiation no physical cows would be needed on the day. They are usually delivered to the bride's father after one year following the negotiation.

When the day for delivery of cows comes, the young men go first into the bride's mother's house. The young men usually go to the house in the evening when the sub n is setting. The young men enter the homestead with songs and dance; they are received at the entrance by girls. They dance with these girls for some time at the entrance and then enter homestead. The girls take and seat them in one of the big house within the homestead. There, the girls provide them with local beer called *"kongo or konyitar"*, the young people continue with their dance throughout the night; and in the morning girls bring them water to wash and then bring some more beer.

Just early morning, while young people are still dancing, groom parents and other close relatives follow the young people. They usually arrive at about 7.00 a.m. On the arrivals the parents and relatives of groom are welcomed at the entrance of homestead by the girls of the village. They are given a place to sit while waiting for the formal transfer of cows from kraal to kraal. In Pari customs cows are not brought to the home of bride's father but they are transferred from groom's family's kraal to bride's family's kraal. The exact agreed number of cows must always be transferred.

Friederike B. and Orly Stern (2011) cited Benesova (2004) as saying g that marriage in South Sudan as an arrangement between two individuals (boy and girl) and the culmination of a love affair, it is rather a social institution, involving whole families. That ties together separate kinship groups. And since marriages are families' agreements, families have a say in who their daughter or son should marry (P.3).

Thus, when time comes for the transfer of the cows from one kraal to another kraal, witnesses are chosen from both sides, because they play key roles in the transfer of cows. This is to say, in future, when something happen in the marriage, the witnesses would be summoned to testify. The speaker from groom side does not speak direct to bride's father, but he speaks to bride's father's witness who communicates the matter to bride family.

Witnesses are highly respected in Pari society; after the transfer is complete, when people go back to their respective places, groom is expected to call the witness to his house any time his wife has local beer. The witness would be offered a special place to site and drink.

It is important to note that in Pari society, payments of dowry in instalments are accepted; and these payments may take years or decades. Once the agreed cows have been transferred, from that day onward the groom has total claim over the bride and her fidelity is guaranteed, although the groom may not take the bride on the same day. As bride is remaining behind, parents drive away any boy who might come to attempt the bride.

What happen when the wife dies?
Although it is unlikely that bride may dies, after full or half dowry is paid, while she is still at her parents' home, if it happens that she dies, the bride's parents will have to return the paid dowry or alternatively, the groom family may ask one of the bride's sisters to be given to the husband to replace the dead sister. I cannot overemphasize that such kind of request is usually respected and implemented in Pari community.

However, should the husband instead dies, after the bride has been accompanied home, and she has a child or not, after fourty days following the burial, the family of the deceased hold meeting and in this meeting they start with asking question to a widow, "What did your dead husband tell you before his death?" If the husband told her what to do after his death, then the widow would narrate to family members.

In Pari society, like in Anyuak society, it is expected that husband tells his wife or wives which of his brothers he would likes to take care of the children when he dies. In other word, whom he wants to inherit his wife after his death? This is the main reason why family of deceased husband always ask the widow.

However, if it happened that the deceased husband didn't leave message behind that is whom he wants to look after his children, then, there will be candidate from the brothers of the deceased or from the clan.

Once the widow has chosen her inheritor, according to Pari traditional laws, like the Anyuak, he is not allowed to have sexual intercourse with her until after one year. While waiting for the time to mature, he cultivates crops for the widow.

In some cases a young widow would refuse to be inherited by the brother of the deceased husband or by any clan man.

When this happen, and if the widow remarried outside the Pari culture, all the children begotten by the new husband will belong to the deceased husband, which means that children remain properties of the deceased. But if the new husband has returned all the cows which were paid, then the children begotten by the new husband will belong to him, only children that widow went with in her care, would belong to the deceased husband's family.

Polygamy and Divorce

In Pari society polygamy is practiced, however, a man is permitted by traditional law not to take more than three wives. Divorce is generally not accepted by all the Northern Luo, and Pari is not an exception. Nevertheless, this does not mean that divorce is not practice, it is practice but in severe cases such as family violence that is life threatening. In Pari society divorce has been experienced here and there but it is very rare among the community. Many families do not want divorce because it could cause many family breakdowns as we have seen in other Luo groups.

Shatt/Thuri Traditional Marriage and Family

In Shatt land, like in any other northern Luo groups, marriages are conducted through many steps, which usually start with courting to introduction of the partners and finally payment of the dowry to bride's father. As I mentioned somewhere in this book that steps taken in marriage process seem complex, but when we discuss them one by one it eventually become clearer and clearer. However, the most important I want readers to understand is that the Northern Luo people of South Sudan value marriage.

(1) *Courtship*
Marriage in Shatt society, like other Northern Luo Societies, starts with courtship, where the boy makes his choice of the girl in dances or in any other appropriate social gathering. In Shatt there are many types of dances, which include Bull, Gwenh-dhela, Rong, and Guma and so on. The dances are usually being organised by young people and where boys and girls appear in their best dresses and they must have smart look, to attract the attention of the opposite sex. Participation in these dances offer and opportunity for the boy to get the right bride he wants to marry. Once the boy has sported the girl he wants to be his future wife, the courtship then starts. Like in other Northern Luo groups, courtships are done either individually or in group.

(2) *Introduction of the partners to the respective parents*
The courtship may take three to six months or even more; after the bride and groom have reached agreement each of them is to inform their respective parents. On one hand the boy informs his father about the agreement they have reached with the girl. The father in turn informs the groom mother about the love affairs. And on the other hand, the girl informs her mother about the agreement they have reached with the boy. The mother of bride in turn informs bride's father about the love affairs the daughter has with the boy.

Once the parents have received the information, the parents of the bride starts interrogation about the groom; and the parents of the boy also starts interrogation about the girl. As we have seen with other Northern Luo groups of South Sudan, family background is a key to selection of good wife and good husband.

When the parents are doing their interrogation they always take into account whether the boy is responsible, respectful, hard working, brave and ready to defend his wife, he is coming from a reach or poor family and whether the groom is coming from family with good background (a family there is no poisoning, no witchdoctor, no evil eyes etc.).

Once both parents have completed their interrogations, they can either approve or disapprove the marriage pending on their findings. If the marriage is disapproved, like in other Luo groups, marriage will be considered cancelled and both the boy and the girl have to go back in the dancing ground for further selection and competition.

(3) *Marriage*

But once the marriage have been approved by both parents, the father of the groom takes this a privilege; he calls for relatives and uncles of the groom to come, discuss and contribute to the dowry payment. In the meeting they decide who is to contribute *what, by how much* and *when* according to what one can afford. The relatives and uncles of groom then contribute their share in the payment of dowry as agreed in the meeting. Gurtong (2017) notes that, in Shatt society dowry contribution is shared out among the relatives.

It is important to know that Shatt have customs held in common with the Shulluk and the Anyuak. When agreement have been reached by the couple, interrogation done successfully and contributions have been done by the relatives of the boy, the father of the groom would be charged with responsibility of setting date for delivery of dowry and sending message to the bride's father. When bride's father receives the letter, he then confirms the date by send another letter to the groom's father.

When the date comes, the groom, his father, mother, relatives and uncles leave for bride's father home. On arrival the visitors will be put in one of the big house of bride's mother, no greeting is allowed at this stage.

Shortly a speaker from bride's team come into the house and asks, "Why have you come in this home today?"

Speaker from groom's side answers, "We have come to this home today because our boy wants to hold the hands of one of your daughters".

Speaker from bride's side again asks, "Have you married one of our girls before?"

Speaker from groom's side answers, "We have not married one of your girls before. This is why we feel honour to be in this house". Customarily a man is not allowed to marry two wives from the same family or lineage. This is one of the reasons behind interrogations.

Speaker from bride's team again asks, "Has there been any fight or misunderstanding between this family or clan and your family or clan?"

Speaker from groom's team answers, "There has never been any fight or any misunderstanding between bride's family or bride's lineage and groom's family or groom's lineage".

After bombarding especial visitors with many questions, the speaker from bride's team leaves the house and joins the other people waiting outside the negation house.

Groom team then select two married men and two married women age 50+, who have been in marriage life without any trouble. The men and women then give two hoes to the bride's father, among the Shatt the hoes given as dowry are not to be used for digging but to be kept for future marriage occasions. The elders and chiefs of the village say that all the twenty seven Shatt tribes, with only exception of three villages, accept only two hoes be paid for dowry.

After receiving the two hoes for the dowry, the marriage celebration is then marked with special dedication to music in its totality. Before I go off this topic, I would like to inform my readers that the Shatt value marriage. Soon after the marriage celebration the bride is surrendered to the groom. What is not clear to me is whether the groom takes his wife along as he goes back to his home or he comes back to collect her on certain agreed date. This is something that may need further research in future. Even though some people argue that the father of the bride surrendered his daughter to the groom on the same day.

Shulluk Traditional Marriage and Family

The average number of children reared in one family varies between four and six. The number of children born by one woman is not low; there are some women with ten to fourteen children. However, as a general rule, the poor health care systems affecting communities in developing countries, often results in fewer survival rate no more than four to six children grow up to maturity. As a result, a family blessed with ten live births, could be reduced to 6 children. The rest of the children may die from preventable/curable diseases such as malnutrition and malaria. If in course of time, the south Sudan government build a number of hospitals in Shulluk-land and supply the required medications to these hospitals, the population of the natives would no doubt increase.

The cause of the low birth rate in many families is the fact that a man is not supposed to have sexual intercourse with his wife, while a baby is nursing, that is, until the baby reaches one year or two and half years old. The community consider it a great shame for a woman to become pregnant before this time has elapsed. If such a case happens, they generally will say that the woman has committed adultery because her husband is obtaining from sex.

Like other Northern Luo groups, marriage in Shulluk society takes series of steps that starts from courtship, fish festival, through to payment of dowry and accompanying the bride to the groom's home.

(1) *Courtship*

Traditionally, in Shulluk society, there are two types or forms of courtships and these are known as group courtship that takes place in the dancing ceremonies and house court that take place in in the house known locally as *Goltong*. For us to understand the different between the two forms of courtships, it would be good to discuss them separately. Let us start by examining group courtship in dancing ceremonies.

(a) *Group Courtship in Dancing Ceremonies*

The group courtship usually is done in group by age groups or friends during the dancing ceremonies, where boys and girls meet. This is what Banydhuro Oyay (one of the informant, who lives in Melbourne-Australia) describes it as, "Group Business".

When a boy is interested in a particular girl, he usually present himself in a handsomely appearance and to demonstrate to the girl that he is professional dancer. Banydhuro Oyay says in addition to all these, the boy may formulates poems and signs them out in attractive methodology as do many Collo young men. He went further to say that all decorated appearance, song and professional dance are done by the boy to attract attention of the girl, and above all, this is indirect message to the girl that the boy has really interest in her.

After this, the boy then inform his age-group or friends that he loves the girl and would like her to be his future wife. On receiving this message, the friends or age-group then approach the girl in group. They usually approach the girl without the presence of the boy who is in love. Customarily, the friends or age-group call the girl from dancing ground and meet her just outside the dancing ground. There the friends tell the girl, "Mr X –give name here- has advised us that he has special interest in you and he would like to hear your response on this". In response, the girl tells the boy's friends or age-group, "I have heard what you have said but I want you to give me time so that I think about it".

After this introduction, the girl goes back to the dancing ground, and like what we have seen among the Anyuak, the girl then informs her friends about the proposal. Customarily, as from that day the girl is not allowed to meet with these group of friends/age-group again face-to-face; instead the girl's friends will meet with boy's friends to discuss the proposed love affairs between the bride and bridegroom to be. The boy's friends and girl's friends will continue to meet until they reach common agreement, which may take three to four months or even more before the two sides would reach agreement. Once agreement is reached by both sides, it is said individual lines are now developed and green lights are given for the girl and the boy to meet face to face during dancing ceremonies; the bride and groom continue meeting till they reach agreement.

However, it is important to remember that when the girl and the boy are discussing their love affairs, at the same time the boy's mother and girl's mother, plus other relatives from both sides, will be conducting interrogation about the boy and the girl. The bride's friends, mother, aunty and family in general, will be interrogating the groom; while the groom's friends, mother and aunty, as well as the groom's family in general, will be interrogating the bride. The main reason for interrogation, of the two proposed lovers, is to find out from one hand whether the boy would make good husband and from the other hand whether the girl would make good wife (in other words right person to marry). Family back ground is a key in selection of the right bride and groom among the Collo; these interrogations are usually made possible because the Shulluk people live in villages and they know each other. Even those Collo who live in small and big towns and cities know themselves. Therefore, it is not difficult for friends, aunties and mothers to carry out the interrogation. The interrogations usually include looking at good characters of the boy and the girl, whether their families practice watch doctors or whether their families have evil eyes, and more importantly, from side of the groom whether he has enough animals that would be required and so on, and so on.

However, in their first meeting, the bride is supposed not to show her love or interest to the boy; I want to say this in the words of Banydhuro Oyay, "In the first meeting the girl exhibits a rather shy attitude; no return of admiration is usually given at this stage".

The courtship could last for weeks or months before the girl give in to the boy. The girl and the boy usually meet in the intervals of dancing ceremonies, which are done regularly in every two months during the summer season; and the dances run on daily basis for two or three consecutive weeks. The courtship can last as long as possible until the boy and girl reach agreement.

According to Collo culture night is considered the conventional courtship time, because such meetings are still un-official. As such the courtship is to be done away from the sight of family members. Holmayr tells us that during the courtship the girl assumes a tone of considerable hauteur; the meetings must be taken outside the village and the parents must not be told, though they would not meddle in the affairs of the young people. In many cases such courtships do not lead to genuine marriage. Nevertheless, the girl chooses whom she likes although she often breaks several engagements before finally accepting a suitor. In Shulluk culture this behaviour is considered normal unless the girl has done it to access (too many times). Thus, a girl who has more than ten lovers is looked upon by the society as "*nya dey cuan*" or "*nya dei cuo*", which literality means "a woman for men"; and if a boy who breaks engagement many times, and has more than ten girlfriends, is looked upon by society as "*wacal-man*", which literarily means "a man for woman".

A Shulluk lady from Melbourne-Australia, who does not wants her name to be mentioned, agreed with the writer Holmayr and she emphasized that approximately 90% of the courtship carried in the dancing ceremonies never lead to genuine marriage because the courtships are faked and deceitful. She went on to say that about 98% of courtships carried in the house (*goltong*) –as we shall see below- lead to genuine marriages. Seligman & Seligman

(1948) observes that boys who come by night in *"odwaman"* (girls' house) and take the girl to *"goltong"* for discussions about love affairs, such courtships usually lead to marriages.

(b) *Group Courtship In The House (Goltong)*

Goltong" is an official hosting and meeting house, nominated by the bride's family after acceptance of a marriage proposal. It then becomes the designate hosting/meeting place for the groom's family/delegates. Adult girls usually serve male delegates/guests from the groom's side at "Goltong". As such, opportunities may eventuate for a birth of new love/s and eventually, a possible marriage/s out of courtships which may take place between respective age group of the girls from the bride's family and boys from the groom's family.

Many Shulluk people believe that courtships which are carried out in houses (*goltong*) always lead to genuine marriage. However, this is something that remains unspoken of in the society, experiences tell us that the Collo rarely discuss publicly about it. Even this be the case, we shall still ask ourselves, "How is courtship organised in the *goltong* (house)"?

It is reported that sometime boys find their lovers in the same village but in other time they find lovers in another village. Interestingly, whether boys find their lovers in the same village or in another village, the bottom line is that boys are not allowed to meet girls in their mother's houses and discuss love affairs. When boys want to meet girls they are interested in, they usually go to the house of the aunties or any good neighbours.

So when a boy wants to introduce himself to the girl, he and one or two of his friends go to the house of aunty, sister-in-law or any good neighbour. This house where the boy meets the girl to discuss their love affairs is called *"goltong"* across the nation or region. When the boys are given seat, the boy in love tells the woman of the house, "We have come to your house to tell you that I –name here- is interested in a girl called (name here), who lives in that house (pointing at the house). Therefore, we would appreciate if you could talk to the girl's family and to the girl herself".

In this circumstances, the woman would first approach the mother of the girl and tells her, "Neighbour, Mr X –name here- told me that he is interested in your daughter and would like to arrange meetings with her if you agree. I personally know this boy very well and is a good boy".

If the mother of the girl agrees, the woman then proceed to the girl and tells her, "Mr X (name here) came to my house some days ago and he told me that he is interested in your and would like to me you. I have already told your mother about the case and she has no objection in you meeting with the boy. What I want to hear from you now, what is the suitable day for you so that I can arrange for your meeting in my house".

In Collo customs, when the intention of the boy is discussed with the mother, and the mother gives green light, the girl rarely reject the idea of meeting with the boy. This is also a sign of respect to her mother.

The girl would then answers the woman, "I think if possible for the boy, we could meet with him on (she mentions the day)"

The woman would then go back to her house and sends message to the boy, telling him the agreed day for meeting. Thus, when the day comes, the boy and one or two of his friends go to the woman's house; the woman goes and brings the girl. It should be remembered that at this time there is no need approaching the mother again because she knows about the matter. The girl comes and enters the house (*goltong*); there the boy introduces himself as well as his friends. During introduction the boy is expected to tell the girl his name, his lineage and clan; he also tells the name of his friends. This very important because lineage and clan are the keys to decide about marriage; although we understand that love and bride-wealth are the corner stone for marriage. When the boy finishes introduction, chance is given to the girl to introduce herself, where she is expected to tell her name, lineage and clan.

Sometimes, soon after the introduction the girl would ask the boy, "Why did you call me here?" But in other case, soon after the introductions, the boy tells the girl, "Dear, I have called you here because the first time I saw you (he names the place), since then I have developed interest in you and was looking for when I could meet you. Today, I want you to know that I love you with all my heart and would like you to be my wife".

In response the girl says, "I have heard what you have said, but I want you to give me time to think about it". According to Banydhuro Oyay, in the first meeting the girl is supposed to hide her love and she should rather show shy attitude without admiration in any form or shape. This is in line with what Hofmayr has just said above that a girl assumes a tone of considerable hauteur. This eventually brings the end of the first meeting.

The girl then goes back to her mother's house; after two to seven days, the boys arranges for the second visit. In the second visit, the boy goes into the same house (*goltong*) and the woman calls the girl to the house. During the second and third and so and so, the girl will be called without information to the mother because the mother is already aware of what is going on.

It is very common, but not usually, that a boy may say to the girl, "In our first meeting you have asked us to give you time so that you think about my proposal. So, how far have you gone with your homework?"

The girl answers, "I have been thinking about it but unfortunately your proposal could not enter my mind". As we have said earlier in this chapter, the girl is expected to be soft and give in very easily even though she has interest in the boy but she must be hard and difficult. Therefore, the above answer, from the girl, indicates how difficult she is. For the bride and groom to reach agreement, it might take between six and twelve months or sometime even longer.

However, today, some of the educated Collo who have moved to live in big towns or in City do not carried courtships in house (*goltong*), they preferred to exchange letters between the boys and the girls. The first day the boy could meet the girl face to face in a secret place and they established lines of communications. The boy will tell the boy where she will get her letters and at same time the girl is expected to reply the letters by putting in the same place. This is known as *"courtship letter box"*, which could be located along the road or under particular tree that is to be checked on regular basis. The boy writes love letter and gives it to his sister, who delivers into a *"courtship letter box"* the girl comes and takes the latter. In return the girl writes the letter and delivers it in the same place; the boy then comes and collects it. Traditionally, such correspondences should not be known by the bride's parents or brothers, if it is discovered that the girl doe corresponds with the boy, she could be beaten. Interestingly, the girl never gives her letter to a friend or to a sister to deliver to the *"courtship letter box"* she does it herself; and the boy never ask his sister to check the "courtship letter box" he check the *"courtship letter box"* by himself.

It is reported that beside the exchange of letters between the boy and the girl, they occasionally meet face to face to discuss their love affairs.

In Collo society, like other Northern Luo societies, marriage cannot take place unless a girl/woman is willing to marry the man, although in some cases, women often placed under considerable social and personal pressure to opt for their preferred candidate (SMLS, 2008). Once the bride and the groom have established their relationships, the bride informs her mother and the mother in turn informs the bride's father. On the other side the groom also informs his father who in turn informs groom's mother.

When both sides of the parents have been informed, then from that moment onwards the girl's mother, friends and aunties will interrogate the groom; while the groom's mother, sisters and friends will also be interrogating the bride. Seligman & Seligman (1932) notes the in-law relationships among the Luo actually begins at the time of betrothal negotiations. Betrothals are usually arranged when the girl and the boy when they are still young; but experiences tell us that majority of the girls, when they grow up, do not marry the boys they were arranged to marry.

A Shulluk girl called Rose, who lives in Juba City; one day says she takes a man home if it is serious enough that they are considering marriage. If she is serious about someone, she will first inform her mother about this, and her mother will begin to find out about him to ascertain whether he is from a good family and who his kin are. When they are ready to negotiate marriage with the boy, she will inform her aunty that the man is interested in marrying her, and that she loves him in return. He aunty will then arrange a meeting with her mother at the aunt's house during which her aunt will inform her mother about the proposed marriage. Her mother then informs her father and uncles, who plan a meeting with the man's family to start negotiation around the bride-prices (Friederike B. & Orly Stern, 2011, P. 9)

Like any other northern Luo groups, the interrogations of the bride and the groom are based on: good characters, watcher or evil eyes, wealth etc. When both sides of the parties are satisfied with their findings, the team from bride side give green light to her to go ahead with discussions, and the team from groom side also give green light to the groom to continue with their discussions (about their love affairs). The final decision of whether to marry or not to marry is the responsibility of the bride and the groom. In theory this is true, but in reality the girl's choice of a husband may at times be influenced by the material status of her father, particularly if he is indebted to his own in-laws, for uncompleted dowry. In these exceptional circumstances the wealth of the family of the young man could be considered seriously. Needless to say the girl's father, uncles and other relatives would try to dissuade her from a poor marriage (Seligman & Seligman, 1932).

Sometimes after the interrogation one or both parties may not recommend the bride and groom for marriage due to their respective finding. Charles Oyo Nyawelo (1992) says that if the boy and the girl insist on their choice of one another, they may run away to unknown place. Once the man has steal the lady in such un-agreeable manner, he is liable to pay one bull known as "*waath tyel* (journey bull) and a sheep to the girl's father as a compensation for the troubles entailed by searching up the lady, even if such a search has not actually taken place. If the lady becomes pregnant in the meantime, then he is to pay additional three cows, three sheep, one spear, one fishing spear and a hoe. Should the lady die during delivering or immediately after delivery, the man's family would be liable t pay additional five cows to the deceased's family. But should the death occurred after the lady has recove-red, that is, after fourty days from delivery date, then the wrongdoer/ husband is free from such liability. On the other hand should the woman continue in sickness after the delivery and dies after fourty or within fourty days, the man is liable to pay some cows because it is considered that sickness that has caused her death was due to pregnancy conditions. And should the man decide to marry the lady he has stolen and the marriage is accepted by the parents, then the one bull that was paid before for running away with the girl, would not be included in the bride-price but rather the bull is taken as a fine for wrong doing (PP.15-16).

In Shulluk customs, once the girl and the boy has reached agreement, the boy's father takes 14 goats to the bride's parents, this is a sign that the groom is ready for marriage. This is soon followed fish festival. The number of goats or sheep is set by the bride's family and is pertinent to the family tree and spirits of the family if any.

(2) *Fish Festival*

Banydhuro Oyay, a Collo intellectual and elder living in Melbourne Australia said, the groom family is always informed of a preliminary number of sheep for the spirit of the bride's family tree. And these must be paid before the dowry negotiation meeting. Thus, soon after the groom has taken the 14 sheep to the bride's parents, the next thing he does is to take fish to the girl's family and fish festival is soon organised by the mother of the girl. Customarily, fish festivals are usually organised in late evening.

Figure: 168, Women are dancing in Fish Festival

When the mother of the girl receives the fish, she then invites relatives and friends to celebrate fish festival and to give advice to the daughter before she gets married (See figure 134 below). During this celebration the main food eaten is fish, although certain food such as *"Akelo luum"*, *"goat meat"* and *"kola-nut"* are also eaten during the celebration.

Figure: 169, Elder women are giving advice to the bride

It is to be remembered that fish festival is one of the most colourful festivals in the Northern Luo Culture. Indeed, this celebration also offers an opportunity for women from different villages to meet and blend with traditions.

During the fish festival bride wears green neckpiece that symbolizes her clan.

Figure: 170, Bride wears green neckpiece in fish festival

Seligman & Seligman (1932) observes that when a young man shows interest in the girl, the prospective bridegroom starts providing fish for the future parent-in-laws early enough; the boy goes to the river every day and when he catches fish he takes them to the house of girl's parents; and whenever, he catches the big fish, he could keep part of the fish for himself while give the other part to the girl's parents. If he gets on well with his parents-in-laws and they encourage their daughters to marry him, he may continue to do so for a long time even after the marriage; however, the young man is not bound by traditional laws of Collo to do so (P.64).

According to Collo culture, after the girl gets married she will no longer wear the green neckpiece but rather wears yellow neckpiece, which symbolizes that she is now joining the boy's clan (See Figure 135 above). The change of colour of necklace, signals the change of costumes from that of her village to that of her groom's village, which is yellow.

When the women and girls have eaten, and are fully satisfied, the fish festival is then followed by late evening and night dances, but before women involved in evening or night dances, the mother of the girl gets up and thanks all the women who turned up for the celebration, and to wish her daughter the best; as she will soon be moving into husband's home after the wedding is done.

Figure: 171, Woman dancing in late evening and into the night

Payment of bride-wealth

During the payment of bride-wealth, bride's family invites three groom's delegation and prepare a small banquet. The boy is expected to contribute some head of cattle for his marriage, while the father of the boy and his uncles are also expected to contribute some head of cattle. The groom may ask politely or demand privileges from his sister's husband and this often not refused. The groom may even decide to show much zeal in doing services for his wife's relatives as part of his contributions towards the marriage, but this is not absolutely necessary. According to Shulluk tradition, the father's sister (*waja*) contributes two head of cattle to the bride-wealth of her brother's son, as women usually do not own cattle, the "*waja*" demands these cattle from her husband, who usually does not refuse. But should the husband refuse to give the cattle; the woman may threaten to leave him. Such strong decision always forces the husband to give the cattle to his wife to pass to her brother's son. Contribution of paternal aunt through her husband is either in kind or a demand for completing a reminder of her dowry. No such thing as demand to leave her husband. Seligman B. & Seligman (1932) notes that the provision of the bride-prices often prevent very early marriage among the young Collo, as the young man is expected to contribute to the bride-prices, this explains why many young Shulluk men do not marry under the age of twenty years (P.64).

In Northern Luo Customs, generally it is the bride's father that fixes or draws bride-wealth. He does this with the assistance of the girl's paternal uncles (his own brothers). When the bride-wealth has been arranged, the bride's father appointed his delegation and sets a date for the coming together of the two families. Banydhuro Oyay said, "When the date comes, the team of the groom then go to the house nominated by the bride's family in her village or neighbours in case of modern Shulluk societies and they hold negotiation and meetings in that dedicated house called *"Gol tong"*.

According to Seligman & Seligman (1932) when the groom team arrives at girl's homestead and before they enter, the girl's father brings a goat and gives it to the father of the boy, who cuts goat's ear at the entrance and binds it to the girl's ankle. After this, both the father of the bride and groom pray to Nyikang and Juok to intercede in the marriage. However, some of the Collo elder disputed Seligman (1932) argument, on the ground that in such occasion neither the father of groom nor the father of the bride is never part of the delegations.

The bride-wealth usually consists of cows, sheep and spears as shown below:

Dhoggi- nywom (dowry in way of cattle) usually consists of ten to fifteen cattle which are given for marriage. The Cattle are always taken by the bride's father; however, the father cannot keep all these cattle for himself, he gives out some of the cattle to girl's uncles (Na), some to his in-laws, if he has not finished dowry of his wife before, and he may reserve some of the cattle for his son; indeed, in some cases the father might not even keep one cow. At this point, it is important to remember that most of the head of cattle given in *dok num* usually go to the girl's mother family. These cattle are distributed among the bride's family members, because it seals and bonds the two extended families and furthermore, making the marriage a communal arrangement. As indicated before, the dowry wealth is a property of the father in the first place. Any bit that might go to the bride's uncles is definitely a completion of an overdue dowry. The father may distribute some to his brothers or cousins in kind. But to the eyes of the foreigners, distribution of these cattle to extended family members is problematic, by so doing; this will encourage women to ignore marital problems that making the continuation of the marriage more important than ensuring the wellbeing and protection of the wife.

But if the girls are twins (Nywol-jwok), according to Seligman (1932), the distributions of the bride-wealth would be rather different. For example, let us assume the bride-wealth received were 20 head of cattle, the distribution would be as follows: 4 cows will go directly to the girl's elder brother (especially when the father is dead), 5 bulls will go to bride's mother's brother (uncle or Na), 3 bulls and 2 cows will go to the cousin brother of the girl (Una). The number of cows given for marriage in Shullukland is not fixed; it varies from family to family and from status to status. In Shulluk society, distributions of the bride-wealth to the respective people are done by the father but if the father died it is done by the mother of the girl. From this simple example we can clearly see that the term "Na" can be used in a classificatory sense, but with regard to the bride-wealth this aspect of the relationship is not considered. Subsequently, the cattle from the bride-wealth which are supposed to be given to the "*Na*" are given to the mother's brothers or her blood sons, and no cattle are given to the "*Uwa*" of the mother (boys who share same father with girl's mother but from different mothers).

Dyeg-nywom –this usually consist of seven-15 sheep, or even more, given for marriage. These sheep are usually given to the bride's brother (*Uwa*). However, if the bride has no blood brothers (from same mother and same father), the sheep are therefore divided among her father's brother's sons (Uwa) and if there is remaining of the sheep they are then given to her father's son (*Uwa*).

Jam nywom –this usually consists of four hoes, two large fish spears and two small fish spears, a basketful of tobacco and a bundle of firewood. The four hoes, two large fish spears, two small fish spears, a basket of tobacco and a bundle of firewood are usually kept by the parents of the girl; according to Collo traditions, these items are not to be shared with other relatives. Friederike B. & Orly Stern (2011) argue that traditionally, the Northern Lou families did not need hard currency; however, money is increasingly being used among the Northern Luo (P.5).

When cows and sheep have been paid the negation team may move to request materials such fishing net, a canoe, agricultural tools and if possible some additional cows. It is very uncommon that the groom's side object to these requests, however, the additional request does not form essential part of the dowry. If groom's side decide to give and they have given, this would be taken as a sign of good intention and establishment of good relationship with the in-laws.

In Shulluk traditional marriage, the *Reth* (King) observes the same procedures I have described above, the King is also prohibited to marry women who are blood related to him, and King is usually allowed to marry the commoners. However, the *Reth* (King) will seldom marry a wife from the "Ororo" unless that woman is an unusually attractive women, the reason been that when a King marry an Ororo woman her son will never become the next King in Shulluk Kingdom. Every Royal son is recognized as a potential Reth and is treated with the utmost indulgence and respect; the royal daughters also have privileges to become Queen, but unfortunately they are not so highly regarded. Queen Abudok was the only female ruler in Shilluk Kingdom as we have seen before. However, Royal daughters are treated with great respect, they may exercise considerable influence in District affairs, and when they grow up they may help and advise the local chiefs, but they never become guardians of shrines as do the King's wives and widows.

In the old days the daughters of the King were not allowed to marry at all (because it was unfit for a woman from royal birth to marry a commoner and she could not even marry a "*Nyireth* or descendants of one as this would be incestuous). Nevertheless, the Royal daughters may have lovers, they could have normal sexual intercourse but she must not bear children. The Royal daughters may select their lovers, which could be from "Kwareth" or Commoners, they can even select close relatives including half-brothers, the half-brothers sharing same father with the Royal girl are not permitted (Seligman $ Seligman, 1932)

But today, the practices are changing as the Collo are being influenced by the foreign ideological ways of marriage. It is reported that when the King (Reth) agrees for his daughter to be married by the commoner, he (Reth) fixes the bride-prices by himself and the wealth would include all what we have just described above. However, if the daughter dies after bride-prices have been paid, there will be no negotiation for the husband to recover cattle already paid. According to Seligman & Seligman (1932) The King (Reth) has no respect to the in-laws and the relatives of the husband by marriage, but they are expected to respect the *Reth* (P.48).

Hasty informs us that one woman is always appointed by the Reth to be head in Royal life; she is treated with respect by all including the King himself. The woman appointed in this position must be an elderly widow from the previous King; she will hold her position until the King displaces her for a particular reason.

Interestingly, the Shulluk people sometimes defer dowry, in time when cattle are scare, they either defer the whole or part of the bride-wealth. By doing this, it does not necessarily mean that the debt is cancelled, no it is not. This is done to give an opportunity for the husband to wait until he receives bride-wealth from his daughter's marriage; this is then when he will give the original number of cattle for his wife to her father or to the father's next of kin. Payment of bride-wealth in instalments is accepted by the Collo. The transfer of the bride-wealth might take between two to three years or even more; the first instalment usually consists of two cows (Seligman & Seligman, 1932, PP. 64-67). In contrary, if a married woman dies without leaving a child, the husband demands from the bride's father all the cattle he paid as bride-wealth.

Wedding Feast

The wedding feast usually takes place at the house of the bride's father, after the whole or part of bride-prices have been paid to the girl's father. In the wedding feast the bride's mother usually prepare much food and local beer, and where the girl's father can afford, he could bring a bull to be killed for the celebration. In the good days, people may eat, drink and dance for three days. It is to be remembered that the marriage feast is to mark the fact that the bride is now ready to leave her father's home and go to join husband's family.

In Collo culture and customs bride is not taken immediately after the marriage feast. Some people say that after the wedding feast the groom retunes to his village, leaving the bride in her father's home. It is said that in Shulluk tradition bride cannot be accompanied officially to her husband's home; for this reason the husband arranges with some of his friends, following the wedding feast, to fetch his wife away from her father's house. There is no formality in fetching the bride. It occurs at random. However, in modern day marriage, this tradition has disappeared as there is usually a church wedding following the traditional one and the bride moves to her husband's home peacefully!

But according to writers like Seligman & Seligman (1932) they say that after the bride-wealth is paid in full, and after the wedding marriage, the bride will have to stay in her father's home for at least two years. Most of the Collo elders do not agree with this statement, they say that following the wedding, it becomes legitimate for the groom to fetch the bride in the old tradition.

Interestingly, Seligman & Seligman (1932) insist that it is only after these years that the bridegroom takes decorated ox to bride's father. The bridegroom does not take the ox by himself; he is usually accompanied by his age mates from both sexes. When they arrive at the skirt of bride's homestead, and before they enter the village, there a mock battle takes place and the bride's friends eventually seize the ox, while the bridegroom and his companions storm the bride's homestead and throw spears into the courtyard; in the end the bride gives herself up and starts running to the dancing ground; while the groom and his companions retreat from bride's homestead. Shortly, bride's mother takes a shield and spear and throws the spear at the retreating bridegroom; he dodges the spear and retaliates by throwing especially good spear at his mother-in-law. The mother-in-law then takes this good spear and keeps it as a memento. Finally, drums are beaten and this is a demonstration that people are now ready to dance; on hearing the drums, the bridegroom dresses in a leopard skin, he then moves to the dancing ground and dance with his wife when her people surround them. After dancing for an hour food is brought; the bride's father's sister (aunty or *Waja*) comes and wash the hands of bridegroom and bride. But in many cases bride and bridegroom wash each other hands ceremonially (PP 64-65). When the ceremony is finished, the bridegroom can now take his wife solemnly.

Friederike B. & Orly Stern (2011) observes that even after the above ceremonies have been completed, marriage is considered not finalised. Marriage is considered finalised only when the bride has given birth to at least two children. Upon the birth of a third child, the marriage is then considered "tied" and at this point, the wife and the children are accepted as full members of her husband's clan (P.13). Banydhuro Oyay, one of the Collo intellectuals in Melbourne, disagreed with Friederike B. & Orly Stern (2011) and argues that such claim, if practiced, could depict the hostile families who may render marriage hostage of birth of children as the norm! He went on to say that many families accept their fate as a divine sacred. Some may choose to marry a second wife without divorcing the current one in attempt to bear children.

Wife in Husband's House

After the wedding feast the bride lives her father's home and goes to live with the husband, where the husband has already built a house with the aids of his friends or age mates. When the bride first comes to live in her husband's homestead, she is usually accompanied by her father's sister "*waja*" (Aunty). The aunty then initiates the bride in her roles as a house wife.

In the first night the couple enter their house and the "*waja*" (father's sister) sleeps in the house nearby. In the morning, the father's sister (*waja*) comes very early in the house where the couple sleep and wash the body of both the husband and the wife, after washing; she anoints their bodies with oil. This action is repeated for several mornings, when the bride's father's sister (*waja*) is still in the husband's homestead (Seligman & Seligman, 1932, P. 65).

A young lady is supposed to be virgin at the time she makes the first sexual intercourse with her prospective husband and no man shall interfere with her sex after that contact. Marriage consummation ceremony is made in the early hours of the first morning of her stay in the husband's home. A sheep is brought, family elders are called, and a prayer is made to God and the ancestors of the husband. The father of the groom begins the prayer, and after he finishes his prayer, he spits into a bowl of water prepared for the occasion. After this, turn is given to the each elder to pray and spit in the bowl; this goes on and on until all the elders have pray and spitted into the bowl. Each person who prays and spit into the bowl is assumed to have added blessing into their container. After this a small part of the animal ear is cut off, divided into two smaller parts. Each part of the divided ear is pierced in the middle and is added to two lines of small white beads prepared for the occasion. After this a round of such beads is tied at the right leg ankle of each of the couple to symbolise authority and respect they have acquired by virtue of marriage. The blood shaded by the sheep as a result of ear cutting consecrates the union so created by the matrimonial rite. After the tying of the beads the couple is ordered to enter their matrimonial bed-room and sit on the bed. Traditionally none of them should look behind as they leave the prayer group and entering the matrimonial bed-room. As the couple moves toward their bedroom an elder sprinkle them with some water from the bowl. When they are in, the room is then closed behind them to signify the fact that they have been firmly united before God and men, and that, no any man apart from husband "*uwa*" shall be allowed to have sex with the bride in the time they are living together as husband and wife. Should it happen that the wife has gone with another man rather than "*uwa*" she would be spiritually defiled and to purify her, a man must kill a goat to purify her and to allow her reunite with her husband. The wrong doer (man) is liable to compensate the husband for loss of prestige. The seep whose ear has been cut is sent to the mother of the girl at some later date, usually after three months when the wife is pregnant. In the night following the consummation ceremony, the wife is bond to disclose the names of other men with whom she has had sexual intercourse. The wife usually confesses to her husband only after they have two subsequence sex act on matrimonial bed-room in that night.

According to Collo customary law, once the wife has mentioned the names of the men who have had sex with her before marriage, such men are reliable to pay one bull and three sheep (each) to the husband who has been deprived from the virginity of his wife. The three sheep one is known as banquet ram (*Kwan*) the other sheep is known as child honouring sheep "*gedhi*" and the third sheep is known as mother purification sheep "*Kaagi*" (Charles Oyo Nyawelo, 1992, P.12).

During the first week with her husband's family, a bride is known as the "guest wife" and during this time she is served by her new family members and the wife is not allowed to serve others at this stage. As days pass by, she gradually takes on more of the housework, until she is fulfilling her full duties as a wife (Benesova, 2004).

Shulluk, like other Northern Luo groups, are polygamists; after the first marriage, there is no objection for a man to marry another wife or wives. In Shulluk customs, a man is allowed to marry two or three sisters from the same family, however, this often happen when the first sister has proved barren, so the husband can take second or third sisters in place of the first sister; if the husband takes the sisters for wives, he will be asked to pay another bride-wealth for each (Seligman & Seligman, 1932, P.68).

Interestingly, a Shulluk man may have sexual intercourse with the wife of his own brother, or brother they share father with but have different mothers (*uwa*); the man can also have sexual intercourse with the wife his father's brother's son or the wife of his clansmen, of his own generation. A man can also claim the same privilege from the wives of his own father but not his mother; he can even have sexual intercourse with the wives of those men he addresses as father (*Wa*). As I have mention somewhere in this book that a father will know that his son is consorting with his wife, and will go to another hut and say nothing about it. The Collo believed that it is better for the son to have sex with his father's wife rather than going to another woman, that would make the father oblige to pay the fine for adultery.

Although brother (*uwa*) realised later that his unmarried brother had sexual intercourse with his wife, when he was away on journey, he could be angry but he could do anything about it. They may exchange some bitter words but that would be all. According to Charles Oyo Nyawelo (1992) such a brother "Uwa" is not liable to pay compensation to the husband of the woman. The reason lies in the collective nature of Shulluk family (P.18). But if it was a stranger who had sex with his wife, the man would try to fight or even kill him. Some elders may even tell the unmarried man that he has done the right things because that is the wife of his brother "*Uwa*". A child from such union always belongs to the bride's husband, any ideas for the child belonging to the wife's lover is ridiculous.

However, any attempt towards intimacy with the wives of "*Uma*", "*Umia*", *Una*", "*Oka*" or "*Kwa*" is considered worse than adultery, and it is believed that the person who has sex with the wives of these people may die. In this juncture, it is important to note that having intercourse with the wife of brother-in-law (*Ora*) would be looked upon in the same way (Seligman & Seligman, 1932, P.69).

Pre-Marital Sexual Intercourse

Similarly, if a potential husband has sexual intercourse with his lover, before marriage is conducted, and this has resulted into conception, he has to pay three cows and three sheep to the family of the girl. After this the head of lady's family sends a three-man delegation to convey the incident of conception to the man's family. This information must be sent to the boy in the second month of pregnancy. However, if the girl's family do not inform the boy in two months of pregnancy and rather inform the boy in the six months of pregnancy, this often throw doubts into the mind of the groom and the boy could even use this as a good defence in court, should he denies the incident. It is believed that boys usually deny incident where no female witness was there when the boy was calling the girl for sex. For this reason many Shulluk girls refuse direct call by a man if the issue to discuss involve sex. In response to such call a Shulluk girl usually comes with her friend just to bear witness should it ends up in sexual intercourse.

If the wrongdoer accepts to marry the lady ha has conceived, his request must be honoured, unless there exists a blood relationship as to prevent the marriage between them. So, once the marriage is accepted, the three cows and three sheep paid before to the lady's family constitute part of the dowry so that he would be required to pay the remaining seven cows and agreed number of a combination of sheep and goats. In case the man involved is a non-Shulluk whose tribe pays more cows as bride prices than the Shulluk, for example the Dinka and Nuer, he would be required to complete the bride-prices on the basis of his tribal custom.

When the delegations are sent convey the incident of conception to the man's family, they return back after meeting their journey costs, and usually they are asked to return on specified date. In the meantime the involved man is asked to response to the message. He may admit or he may even deny the incident.

However, if he accepted the message, the family of the boy make arrangement for the payment for full or part of the dowry. This admission is usually conveyed to the girl's family after the delegations return and immediately the boy's father will pay three sheep and one sheep will be killed as order for a banquet. In addition he is expected to pay child honouring sheep that will be killed to cool the anger of ancestors and to give good health to the baby.

If the child is born and it dies before the payment of child honouring sheep (*gedhi*), the guilty man is liable to pay four cows and four sheep in case of *Kwanyireth* (royal clan) child. From these animals two cows and two sheep go to the King the head of the royal families, while the remainder goes to the father of compensation for his child. But in case when the lady is not from royal family and the child dies before the payment of child honouring sheep (*gedhi*), the wrong doer is liable to pay two cows and two sheep to the father of the dead child. The cause of the death of the child does not matter provided that it took place before the payment of child honouring sheep (*gedhi*).

The wife is also bonds to give the list of those men who she has had sexual intercourse with them to the traditional birth attendant (TBA) soon after her first delivery; this is a condition for putting the baby to the breast. Unless the mother discloses the names of all such men to the TBA, the child will not suck milk from the breast and if she continues holding the information away from the TBA eventually the child will die of hunger. This is the reason why a Shulluk lady has to give birth of her first born in her parent's home, for this eliminates shyness entailed by confession (Charles Oyo Nyawelo, 1992, P.12). From the above statement it follows that if a man performs sex with a young unmarried woman, and at the end did not marry her, this is considered a dishonour to the potential husband, not only this, it also puts the life of girl's first born at risk.

Should the same or another man cause a second pre-marital conception of the same lady, he would be liable to pay four cows and four sheep to the lady's family for increased damage to the family's prestige. Should a third conception occur from any person including the previous men, additional four cows and four sheep would be payable. Interestingly, if the lady again conceived for the third time, whether it is from the previous men or from somebody different no further compensation can be recovered. However, should the first three conceptions occur from the same man, and then the woman becomes his legitimate wife by operation of the law. In the eye of the traditional law, he stands in the same position as the man who married her, paid full price, produced children, and then abandoned the woman without announcing divorce. He is free to take the woman with her children. Sometimes if the man wants he may also take the children without their mother, and if this happen he is considered to have divorced his wife.

Moreover, if the unmarried young lady, who is conceived, is a princess, the man will have to pay ten cows to the King on the ground that the living King takes over all the living wives of previous Kings his wives in levirate. In theory the King is the father of all princes and princesses. It is to be remembered that King marries with free cows of the royal treasury (Bwori maj), therefore making the children of Bwori maj. Every prince marries on the account of the King, from the ten cows those were allotted to him from the time of birth; and if prince is in trouble of getting more cows, the King is obliged to meet his financial difficulties from Bwori maj.

In the same way no payment accrues for other conception of a princess, whether from the same man or another man, since the Shulluk bride price has been covered the first time. In spite of completion of bride price, it is up to the princess on whether she wants to stay with the same man that pays the bride wealth as husband and wife or she wants to live with another man; however this is often depends on the treatment the princess receives from the man. The requirement to pay tens cows at a go, coupled with the princesses' freedom to desert the man at any time, for this reason many Shulluk men refrain from associating with them. This is one of the reasons why the princesses intend to live with the wealthy Shulluk men.

Adultery

It is generally believed that although a woman is married to the husband, sometimes she is allowed to have sexual intercourse with the husband's brother "*Uwa*"; indeed, sometimes married women do go around having sexual intercourse with other men. If this happen and she becomes pregnant by the husband, at the time of birth she must disclose to the mid-wife all the names of the men with whom she has had sex. The confession of adultery by a married woman is an evidence of facts, and the men whose names are mentioned can't deny, dispute or even disproving the fact in court. But if the man thinks that he is innocent, he may be ask to pay more bull for the administration of oath by both parties on the Holy Spear. The oath by definition is a presentation of facts to God and the demand upon Him to punish the untruthful person. But in case where the adulterers are caught while performing sex the witness's testimony alone is enough to prove the case against the wrong doer. However, it should be noted that oath is usually administered in the absence of the witness.

Once a disclosure has been made, it is communicated to the husband as soon as possible. The husband then arranges with the head of his family to send delegation to the family of the man who committed adultery with his wife. They will be directed to collect two cows, three sheep, one spear, one fishing spear and a hoe. Of the three sheep one of them must be a ram that is locally known as "*Onywog dwalo*" or spiritual contamination ram. The other sheep is called locally as "*gedhi*" or child honouring sheep; and the last sheep is locally called "*Kagi*" or adulterers separation sheep. The ram is killed to purify that ceremony.

Where ever adultery leads to conception, the man who pregnant the woman pays four cows, four sheep and *Jami kwer* (which include spear, fish spear and a hoe) to the husband as compensation. One may ask, "What is the theory underlying the compensatory payments to the husband? The answer is as clear as day light, the compensatory payments are given to the husband because marriage gives him exclusive right to the sex of his wife and whatever child she produced before she divorced the first husband. This implies that even if the conception occurs from another man rather than the husband.

More importantly, after collection of the compensation from the wrong doer, the husband may decide to divorce his wife. Should he do so, then the wife's family would be liable to pay back the number of cattle they have received as bride-wealth. Customarily, the off-springs of the cattle "*nywoli dhog*" usually remain behind on the ground that they correspond to the off-spring of the woman. It is to be noted that where the husband calls back the off-spring of his cattle, this then means that he is to surrender all the children to the family of the wife, where they will be adopted by the maternal uncles. Having said this, I would like to remind the readers that this is only theory and nothing happen in practice, because no family refuses children, especially where the number of the children is more than one.

Divorce

Traditionally, Luo people generally do not want divorce. But there are certain circumstances that may lead to divorce. All Shulluk women are expected to submit to their husbands and never think of divorce even though she is often been beaten by the husbands. More importantly, when a King wants to beat his wife for wrong doing, she must kneel and bow her head to receive a whipping from the King.

It is generally believed that in Shulluk society divorce is amicably arranged by the wife, since a man only seeks divorce when his wife has left him for another man. Mr Heasty states that it is very rare a husband to divorce his wife, it is usually the wife or her family who often initiates the separation.

Traditionally, the breaking of a marriage means the return of cattle to the bridegroom. Experiences tell us that this is the most difficult matter in family life, since the bride-wealth is paid to bride's father and the bride-prices might have married ten other wives. Once a family breaks down occur with the first woman married in the family, this will affect the whole chains of marriages; in other words it will play part in other marriages. Seligman & Seligman (1932) notes that when the wife decides to divorce the husband, she is to hand over bride-wealth to her father, who would in turn return the original bride-wealth to the first husband. It is said that if there is a grown up daughter, and the wife's father cannot return the bride-prices, the husband has right to take the daughter instead of the bride-wealth. For this reason, every man wants the marriage of her sister to be stable.

The bridegroom is usually anxious to please his bride's relatives so that they in turn do not influence her to break off the engagement. He does this not only because he wants to preserve marriage but also for the fear of the woman losing prestige after marriage. Seligman & Seligman (1932) informs us that there are many reasons why the bridegroom should seek to keep on good terms with his wife's relatives, and by his correct behaviour, because these prevent the possibility of rupture.

It is to be noted that if a wife leaves the husband, the husband will demand the sheep, hoes, fish spears, and cows to be returned; but he cannot ask for tobacco, fish and firewood to be return because these items have already been consumed. Nevertheless, in some cases when the husband is very and very angry about the separation, he may even ask bride's father to return the consumed items. When the husband asks for the return of the already consumed items, the community would not be happy about such demands but they can do nothing to prevent it.

Yangjok, a former chief of Tonga, stated, in one of his courts, that if a woman leaves her husband, one bullock must be retained, but if the wife is pregnant at the time of divorce, one cow and one bull should be retained by the father of the bride (Seligman & Seligman, 1932, P. 67).

Chapter 19:
Reproductions
· ·

Introduction to Reproduction Process

Reproduction can be defined as the biological process that produces new organisms. It is a basic function of every organism on Earth and passes on the building blocks of life from one generation to another. Every culture in South Sudan has traditional rules, and ceremonies which preside around reproduction. This implies that the Northern Luo groups in South Sudan are not the only exception; they also have ceremonies which preside around reproduction as we shall see later in this chapter. Therefore, this book will particularly focus on reproduction related to the Northern Luo Groups.

The Northern Luo cultures have norms governing sex and reproduction premarital sex. In their cultures women are expected to have as many children as they could produce, and communities prefer to have sexual methods of reproduction as oppose to asexual methods of reproduction.

In asexual reproduction man's semen is planted into the woman and she bears a child without physically having contact with the man. In the history of the Northern Luo, no person has been known to be able to reproduce asexually. Sexual reproduction is the only method of reproduction. Sexual intercourse is the productive act in which semen from a man can expelled into the female vagina through male ejaculation. During sexual intercourse, the man inserts an erect penis into the vagina of a woman, and their bodies keep on moving together until a man reaches ejaculation.

Once semen is expelled into the vagina, sperm swim through the service and uterus until they reach the Fallopian tubes. At this time, if the woman ovulated, the mature egg is waiting within one of the Fallopian tubes.

A single sperm then penetrates the egg and fusses together. This process is called fertilization. The fertilized egg move into uterus; and once the fertilized egg is in uterus, it begins to implant into the wall of the uterus. Subsequently, if the egg successfully implants, the woman is said to be pregnant and the egg is referred as an embryo.

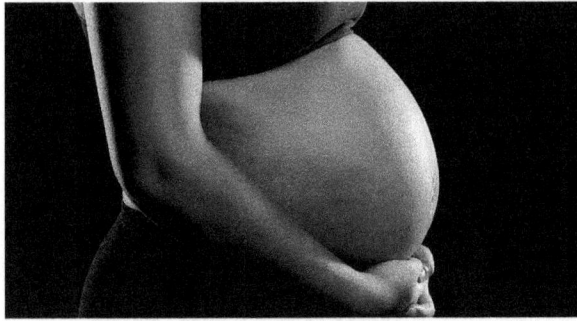

Figure: 172, Woman expecting a baby

Scientifically, the normal pregnancy lasts between 38 to 42 weeks. During this period, the embryo develops its organs that will let the foetus live after it leaves the womb. In any woman, the embryo is always attached to the body through what we called umbilical cord; and it is through this umbilical cord that the foetus receives all its nutrients needed while still in the womb.

However, once the pregnancy reaches nine weeks, the embryo becomes known as a foetus until birth; and at birth the foetus is called a baby. Birth is when a woman expelled the baby through the vagina and sometimes by surgery suzerainty. The hormones of the woman usually develop a process generally called 'labour', this process slowly pushes the baby out of the vagina. Labour is often recognised by strong and painful muscle spasms that come frequently and maintain a regular pattern. Once the birth occurs, the foetus become known as a baby or infant, and this is considered the final step of reproduction.

According to the Encyclopaedia of Anthropology, human reproduction refers to "the process by which new social members are produced" –specially, the physiological process of pregnancy, birth and child raising. Although since the 1990's the anthropological studies of reproduction have focused on the new reproduction technologies such as intrauterine devices and birth control pills, these have never been practice by the Northern Luo women. The women rather practice what we called "Natural birth control" were both husband and wife abstain from sexual contact for at least two years as the wife breast feed the baby. This is also where polygamy plays an important role in the life of the people. However, we cannot deny that some women in the Luo society have no control over their fertility, and that is why some of them get pregnant within one year (soon after birth). For us to avoid complications and misunderstanding of reproduction among the Luo women, we are going to discuss the reproduction systems, births and child naming ceremony, in accordance to the practice of each group. During our discussions we shall clearly see the similarities and the differences of reproduction among the Northern Luo. Let us first see child birth, naming ceremony among the Acholi tribe.

Acholi Reproduction Process

The Acholi people recognized two types of birth customs namely the normal birth and the twin or abnormal birth called *Jok Anywala*

(a) *Normal Birth*

Child birth varies for women in all different cultures of South Sudan; for the women of Northern Luo Societies, who live in rural areas, traditional birth attendant (TBA) handle birth but for the Luo women who live in big cities like Juba, Wau and Malakal, when a woman goes into child birth, she usually rushes to the hospital, where she would be assisted by either a doctor or a nurse.

It is good to remember that in all environments, the new born infant is not completely safe from diseases. In the village when the women go into labour, they simply go to a quiet place like a house, squat and proceed to deliver their children with the help of TBA, locally known as *Lajol*.

Source: Figure 173, Maternity in Africa by **Paolo Patruno**

A child is usually born head first and facing upward; the ATB then cuts off the umbilical cords by using new sharp knife, bamboos, slices of reed cane and ties the cord with fibre. The placenta would be buried outside the home under the woman's granary or sometime the placenta is buried in the bush. The ATB washes the body of the new born with normal fresh water, the ATB also clean the mother's body and the breasts squeezed, so that the first drops of milk fall on the ground. The child is then put to the breast. After all these, both the mother and the baby are confined to the mother's hut which is surrounded by a fence' poles or a symbolic rope. When the mother is in the house, she is to abstain from certain things as such as cooking, cleaning, grinding as well as sex. The mother leaves the house only when she wants to use toilet or to wash and no one is allowed to enter the house —even the father of the baby. Normally a young experience woman would be assigned to the mother and this woman is responsible to address the mother needs, which means she cooks food for her and carry out some important message to the people outside mother's world —although customarily, she is encouraged to minimise her direct contacts with others.

Majority of the people in Acholi, especially the elders believe that the seclusion is meant to protect the newborn baby from potentially harmful influences. The Acholi clans believe that if a person with evil eyes enter the house and touches the baby during the time of the seclusion the baby will definitely dies; and if the uncle also enter the house although he does not touch the baby, the baby will has bended neck. Others also believe that touching the genitals will cause the baby to be infertile when it grows.

In this situation complications are very common; there is often risk of infection, risk of excessive bleeding, and overall lack of medical attention. However, when the complications arise it is advisable for the mother and the newborn be taken immediately to the nearest hospital. But if there is no complication, the mother and the newborn are kept in the house; the seclusion period varies according to the sex of the baby —for the boy the mother has to stay in the house for three days, and for the girl, the mother has to stay in the house for four days. In case either the mother or the baby is sick, this period can be extended for two or three weeks (Glenday, 1980). It is only after the third or the fourth days that the elder woman calls mother's friends and other close relatives to come for a meal. Before the people eat, women and men give names to the newborn. As we shall see later, names are given with the meaning; they are often driven from some events connected with the birth. Soon after this people eat, drink local beer and dance —this is also the time the baby is free to go out of the house. Shortly, the father hangs on its neck number of various charms which are thought to be against diseases.

As I mentioned early in this book that the Luo people (Northern, Central and Southern) have migrated from the land of Canaan -therefore, there are similarities between Acholi naming and Jews naming. The writing of Glenday, who was a Jew teacher in Northern Uganda, he confirmed my research when he writes,

"*Several years ago when I was teaching religion in a boys' secondary college in northern Uganda among the Acholi people, I was struck by the way in which the students I was teaching were able to describe the beliefs and customs of the traditional Acholi religion. They were especially familiar with what their families and clans did, think and said when a child was born. There appeared to me similarities between some traditional Acholi birth rites or ceremonies, the same Acholi birth ceremonies have been found to be practiced among Jews in the Middle East. Not only these Luo names like "Acan" or Achan are in the Bible*" (Glenday, 1980). It is true; you can find the name Achan in Joshua 7: 18-20, 24.

When the child has grown sufficiently to be able to take water or porridge to its father, it is regarded as old enough to be weaned and the mother to prepare for another conception.

(b) **Twin Birth**

Like in normal birth, when labour starts, the women send for ATB and if there is no ATB at the time, two women would be called to assist the woman in labour. If the expectant woman happens to be in a hut or a house, and she proves to be fearful of delivery, she will be advice to hold the centre post in the house known as *wir* or *acwer* or *awinyo* for support. A woman would support her from behind. The ABT kneels in front of the woman and received the baby in outstretched arms However, if the woman is not fearful, she would be asked to lie down on her back with opened legs and the ATB or one of the women kneels in front of the opened legs and receives the baby in her outstretched arms. The TBA then cut the umbilical cords. The umbilical cords are then cut and buried in the space round the house, usually the cords are buried under tree of God (*Jok*) called "*Okengo*". However, if the umbilical cords are not buried round the house, they normally buried them under granary (*dero*). So every year and in the same month the burial place become the scene of the sacrifice of a white rooster or a sheep. Sacrifices of white rooster or sheep always follow by eating, drinking local beer and dancing what they called "*myiel rudi*" or "twin-dance)". If the twins are of opposite sex, the mother will have to stay in the house for four days. After four days, the father builds a special shrine, in appearance resembling a "*Kac*" but it is rather called "*Jok –Rud*" or "*Kac pa rud*" (Seligman & Seligman 1932).

In the fourth day, a goat or sheep and a chicken are killed, and the older people of close relatives eat the flesh before the shrine together with vegetable food. After the ceremony, the shrine is used for some years; when the father wants to go hunting, he often resorted to it and if he kills animals during the hunt, he will not eat the meat of the animal until some of the meat has been roasted before the "*Kac pa rud*" It is only after this that he moves aside to eat the rest of the roasted meat.

Similarly, at the time of harvest, then parents roasted the first millet and eaten it in front of the "*Rud*"; however, if the twins are grown up, the father can then eat the roasted millet together with the twins. This is always done to avoid sickness that might come across.

It should be noted that if the twins or twin dies young the bodies are put in a pot and buried with two small openings in it but with the mouth completely closed. The pot then is taken and buried in the bed of some small streams. Some of the Acholi families practice erecting tree of God (*Okengo*) over the grave, they also put some ashes over the grave so that no one can overlook its existence by the stream.

More importantly, if the parents move to another site to live, the pot would be dug and removed, taken and buried again at the bed of the stream near to the new home. This practice seems to be interrupted by the current civil wars taking place in South Sudan. Many parents have fled to the neighbouring countries, some are displaced within their own country and yet some have migrated to Western World as refugees leaving the dead and buried twins behind. I, therefore, wonder if peace comes to South Sudan, whether these parents would go and dig the pots from where they were buried and move them to another stream where they will settle. However, this remains to be seen and more research may be required.

Moreover, the birth of twins, whether of the same or opposite sex, is recognized as dangerous because the babies are regarded as a menace to the lives not only to the lives of the parents but also to the elder brothers and sisters; they will certainly bring bad luck in hunting season. At the same time Fr. Crazzolara (1951) informs us that twins' births are not disliked, they are rather being regarded as a special manifestation of *Jok* (God). It should be noted that if only one twin survives there is no great danger to the parents, elder brothers and sisters but if both babies live, then the parents, elder brothers and sisters may be expected to die any time.

Child Naming and Ceremony

The *"nying-kwon"* (name given during birth ceremony when eating food) is the real name and is given after three days for male and four days for female. Acholi names are meant for identification of the person. The prefixes "Wod" or "Nya" mean "son of..." and "daughter of..." respectively. For example, if there is a man called "Wod-otim" this means "son of Otim" and if there is a girl called "Nyaomal" this means "daughter of Omal".

Most Acholi names are intended to ensure memory of the past. Interestingly, Acholi remember the past in a combination of ways that includes songs, oral traditional history and children names. This may be something that might have happened to one of the parent in courtship, marriage process or in a family life. For example, if during the court- ship parents and other relatives tried to discourage the lady from accepting the love from the man, but yet she ignore them and proceeded to marry the disliked man, her first born would be name "Amito" if the baby is a girl and if the baby is a boy it will be called "Omita".

The parents may also want to remind themselves about what they went through, for example if they delayed in getting a child, the first born will be named "Olur" if the baby is a boy; and if the baby is a girl, she will be named *"Alur"*. The word *"Lur"* in Acholi language means impotence or sterility. Therefore, this name is given to remember the nonsense talk people were talking about the parents that they were unable to produce a child.

Like in every society, bad things are remembered more than the good ones. In the same way, the Acholi remember bad conduct or misfortune more than the good things. Similarly, the Acholi remember war more than peace.

In the old days, marriages among young people were very common, a girl could even conceive before seeing her first menstruation. Furthermore the girl would not even know she is pregnant until at Advance stage. Such a child would be named *"Okumu"* if the baby is a boy; or will be named *"Akumu"* if the baby if a girl. The mother when she sees that she has taken longer months than the expected nine months of pregnancy, after birth the child will be named *"Oruni"* if the child is a boy and if the child is a girl; she will be named *"Laruni"* (Dr. Charles Amone, 2014).

For us to understand the naming of a child in Acholi, let us clarified the names in Jok Names, names given to show place of birth, names showing time of birth, names related to death during delivering and names that show hardships. Let us start with the Jok names.

(1) **Jok Names**

The jok names are given to children born under particular circumstances, which the Acholi people consider to be abnormal circumstances. This could be due to the circumstances of birth or due to physical marks on the baby. All these abnormalities are attributed by the Jok (God).

Male	Female	Meaning of the name
Oruni	Laruni	The birth was delayed by Jok
Okumu	Akumu	The woman conceived before menstruation by power of Jok
Opiyo	Apiyo	The name shows the first born in twins
Ocen	Acen	The name shows the last born in twins
Okello	Akello	First follower of twins
Odong	Adong	Second follower of twins or the mother dies when the child was still an infant
Ojara	Lajara	Born with more than five fingers or toes
Odoc	Adoc	The baby came out from the womb with the leg first
Ouma	Auma	Born facing down
Oyite	Ayite	Born with unusual mark on the ear/ears
Ojok	Ajok	Born with some unusual mark or feature on some parts of the body
Owino	Lawino	The umbilical cord was covered around the neck during delivery, and the child was almost strangled.
Okot	Akot/ Lakot	Some unusual liquid was found in the umbilical cord; or the child was born when it was raining.
Olaa	Lalaa	Born after blessing from elder when the mother has stayed long without pregnancy. It also means a child born after spiting saliva on the mother that allowed the baby come out

(2) *Names showing place of delivery*

Male	Female	Meaning of the names
Oyoo	Ayoo	The child born on the Road
Ongwec	Angwec	The child was born when the parents were running due to war
Otim	Atim	The child was born away from home e.g. in towns
Odwar	Adwar/Lodwar	The child was born during hunting or the father is a hunter
Lowila	Lawil	A child was born in the market when the mother was shopping
Okullo	Akullu/Lakullo	The child wa born in the water point such as stream or river
Olum	Alum/Lalum	"*Lum*" literarily means "grass". This means the baby was born outside the house on the grass
Ogang	Langang/Latwa	The child was born in bride's home before formal marriage was conducted
Odera	Adero	"Dero" literarily means ""granary", thus this child was born near the granary. Alternatively this child was born after many tolerances
Wokorac/Woko	Lawoko	The child born in foreign soil e.g. in exile
Olal	Alal	Born after the parents live for long in towns, another nation without contact with family members
Ogot	Agot	Born on the mountain

(3) *Names showing time or period of delivery*

Male	Female	Meaning of the name
Okec	Akec	Child born during famine
Oceng	Aceng	Born during clear sunshine such as around Midday

698

Male	Female	Meaning of the name
Owor	Awor	Born at night or in total darkness such as at Midnight
Otyieno	Atyieno	Born in the evening
Owot	Awot	The mother is found of travelling. Alternatively born when the mother is on journey
Onyango	Anyango	Born when the sun was rising between 7.00 a.m. and 8.00 a.m.
Olaroker	Alaroker	Born during succession when there was dispute over the throne
Obaloker/Obalo	Abaloker/ Abalo	The mother wrestled the favour of the man from co-wife
Oryiem	Aryiemo	Born when the mother had been chased away from home.
Mwaka	Lamwaka	Born on the eve of Christmas or soon after Christmas
Obonyo/ Obwonyo	Abonyo/ Labwonyo	Born when there was locust swam
Ogwen	Langwen	Born at the time when the people were catching white ants
Okongo	Akongo	Born when the mother was selling local beer

(4) *Names signifying death of relative during the time of delivery*

The names below are related to death; the baby might have been born during the time of mourning (Ocola/Acola) or when there was life threatening situation (Aloyotoo).

Male	Female	Meaning of the name
Otoo	Atoo	Born when relative has just died
Oburu	Laburu/Nyaburu	The child born during the funeral rights
Ocola	Acola	Born during mourning
Ogen-too	Agen-too	Pray that death takes me away otherwise I will do it
Lamtoo	Latoo	Pray so that death does not strike
Lam/Kilama	Lalam	A child that survived from the curse
Ongom	Angom	This child is likely to be buried again

Male	Female	Meaning of the name
Oyik	Ayiko	I have buried too many children before this child
Obol	Labol	This child will be thrown to God /Jok again
Otika/Too-otika	NA	Death follows me all the time
Okema/ Too-okema	NA	Death is directed to me only
Toolit	NA	Death is painful
Oloya	Aloyo/Aloyotoo	I survived death narrowly
Obwot	Abwot	I survived death

(5) *Names related to hardships*

Male	Female	Meaning of the name
Ocan	Acan/Achan	Born during hardship or poverty
Ocira	Aciro	The parents were enduring some form of hardship
Okanya	Akanyo	This child shows that he/she has tolerated me
Onen-can	Anenocan	I have face problems all through my marriage life
Ocora	Acoro	I have pushed myself to marry a person of undesirable quality
Onyala	Anyalo-koma	Born when the marriage was disliked, gut yet begged for marriage
Onen	Aneno	Born after long dispute about pregnancy, the father denying that the child was not his but people would see with the picture after birth
Omita	Amito	Borb after hard family life

(6) *Common names depicting gender*

Male	Female	Meaning of the name
Omara	Lamaro	I am highly loved
Ogena	Ageno	I trusted him and she trusted me
Obwoya	Abwoyo	I have defeated him/her

Male	Female	Meaning of the name
Odoki	Adok	She intends to divorce and go back home
Oyenga	Ayengo	I looked earnestly for the right partner
Okwera	Akwero	He/she rejected me
Kibwota	Abwoto	I was abandoned
Obalo	Abalo	I am blamed falsely for spoiling the goodness of this home
Olweny	Lalweny	Short tempered father or mother who often picks a fight
Orac	Arac	One of the parents is hold he/she is a bad person
Akera	Laker	The Royal child

Anyuak Reproduction Process

Birth:

Anyuak like the Shulluk, in the ordinary birth, when a woman is six months pregnant, she leaves the house of her husband – for intercourse should now cease – and goes to her parents' home, where the baby will be born.

I have not been informed that at delivery the woman (girl) would be asked to confess (*kwano*) all the sexual relationships she has had as a girl. The reason for confession being that the child could die if the husband doesn't know his wife's ex-boyfriends.

Source: Figure 174 , Maternity in Africa by **Paolo Patruno**

When the time of deliver comes, an experienced woman, known as traditional birth attendant (TBA) assists the mother at birth; when the baby comes out the TBA receives the baby, cuts its umbilical cord with a silver of millet stalk and buries the afterbirth outside the house on the right of the doorway. The place where afterbirth is buried is not marked in any way; this is to avoid a person with evil eyes from digging afterbirth and burning it. It is believed that if a person with evil eyes digs the afterbirth and burns it, the baby will die or otherwise it will prevent the mother from further pregnancy.

In the old days, after birth the woman will remain in the house with the new born baby for one month (both for female and male) and the house where the mother is would be fenced; a man with two wives or a pregnant woman is not allowed to come into the house or even pass by the house. The people of Anyuak believe that if a man with two wives or a pregnant woman passes by the house where the new born is, that baby will die. It is only after one month, when the baby is brought out ceremonially, and then the fence would be removed. It is said that on the day the new born would be brought out, the baby would immediately handed to a pregnant woman, who holds the baby and turn it around her stomach (womb) so that the baby inside her gets blessing and comes out in peace (as this baby was born in peace).

However, in the modern Anyuak society, this old custom/tradition have been abolished; as such the first child can now be born in husband's home without a woman confessing (*kwano*) all the sexual relationships she has had before marriage. Secondly, the mother and the new born baby may stay in the house for a period of seven days or less, and there she would be fed by other women. When seven day is finished the child is given a name. In Anyuak society the first child is always name Omot (Ochan) if a boy, and if a girl she is either named: Amot, Aryiet or Achan (the name Ochan or Achan is given only to the first born twins).

In the old days, after the birth of the child, the wife would remain in her parents' home for another six months or event two years. It is said that the wife does this simply to avoid unnecessary pregnancy. In Anyuak customs, like in other Luo customs, it is shameful for a woman to conceive during lactation. It is generally believed that if a woman becomes pregnant during the lactation period, this is an indication that she has been having sexual relationship with other man rather than with the husband. Because, the entire community knows that the husband is allowed to visit his wife at this time but he is not allowed to have sexual intercourse with her. Seligman and Seligman (1932) observed that at the time of birth the husband can make enquiries about his wife, but he must avoid the house of his parents' in-law (P.70).

According to Anyuak tradition, only the first born can be born at wife's mother's house but other children can be born at husband's house. When the wife gives birth to other children, after the first born, she is expected to stay in the house with the new born baby for seven days; but in other cases, the wife could even stay in the house with the baby for two weeks, after that the child would be brought out for naming. The new born is usually name after circumstances surrounding him/her or the family as we shall see below.

Seligman and Seligman (1932) also tells us that in Anyuak land there are certain tress, grown both in the bush and in village, that are credited with supernatural powers, and these trees are connected with child birth and naming of children. As a matter of fact, in the old days, the people regularly offer sacrifices and food plus tobacco to these trees.

Moreover, today the young Anyuak tend to overlook the credibility of these trees, although some still put some big beads under the roots of the trees, but yet the old men observes the customs. It is generally believed that anyone damaging the trees he/she would die, not only this, if the person has children, the children would suffered.

Child Naming and Ceremony

(1) *Juok names*

Male	Female	Meaning of the names
Ujuok	Ajuok	A child born with physical marks on toes or fingers (Atum-tum)
Upiew	Apiew	The first born twins
Ochan	Achan	Also first born twins
Okello	Akello	The child that follows twins
Achuil	Achuil	The third born twins
N/A	N/A	The fourth born twins
N/A	N/A	The fifth born twins
N/A	N/A	The six born twins
N/A	N/A	Child that comes from the womb with legs first

(2) *Names showing place of deliver*

Male	Female	Meaning of the names
Uyoo	Ayoo	Child born on the road
Okuc	Nyakuc	Child born when the father is away from home
Odwar	Adwar	Child born during hunting
N/A	N/A	Child born in the market when the mother is shopping
Upii	Apii	Child born in stream when the mother when to fetch water
Onamo	Nyinamo	Child born in River Nile when the mother went to fetch water
Olum	Nyilum	Child born in the grass. "Lum" (Path) is grass.
Udik	Nyadik	Child born near or under the granary
Okidi	Nyakidi	Child born on no hills or mountains
Oyamo or Okout	Nyiyamo or Akout	Child born on windy day
Opoudh	Apuodha	Child born in the field

(3) *Names showing period of delivery*

Male	Female	Meaning of names
Ukech	Akech	Child born during famine

Uceng	Aceng	Child born in clear sunshine
Uwar	Awar	Child born in mid-night
Okouth	Akouth	Child born when it is raining heavily
Chang Juok	Chang Juok	Child born during Christmas. "Kudmi" means Christmas
Ubaa?	Nyabaa?	Child born in evening
Chang Juok	Chang Juok	Child born in Easter days
Chang-Juok	Chang-Juok	Child born on Sunday
Opwoya	Apwoya	Child born at 6.00 p.m.

(4) *Names related to hardship*

Males	Females	Meaning of the names
Uchan	Achan	Child born during hardship or poverty
Uling	Nyaling	Child born during the war/battle
Uling	Nyaling	Child born when the mother and the father like fighting
Okier	Akeir	Child born when relative do not want the marriage but yet couple decide to marry
Mayong?	Nyayong?	Child born when the people of the home do not want the wife/husband
Ubech	Nyabeng	Child in small bag of water
Ukuc or Agwa	Akuc or Agwa	Child born when the father died before the baby was born
Chol (usually the give the name of the diseased relative)	Achol (usually they give the name of the diseased relative)	Child born after somebody has died
Upudoka or Muc	Aweely or Nyamuch	Child born after mother has been awaiting for pregnancy

Jo Luo of Bhar El Ghazal Production Process

Like the Acholi, Anywak and the Pari, the Jo Luo of Bhar El Ghazal recognised two ways of birth customs, and these mainly include ordinary birth and abnormal birth (twins) called "kwoi". The Jo Luo people of Bhar El Ghazal do not have Kings as such they do not have birth of King's children. As such their traditional cultures never talk about the birth of King's children. Let us now examine the two types of birth in the Jo-Luo of Bhar El Ghazal society.

Normal Birth:

In ordinary birth, when a woman is six months pregnant, she leaves husband's house and goes to her mother's house, where the baby will be born. Mr. Lau Uling Bouk and Mr Andrea Lual, members of Bhar El Ghazal who live in Melbourne-Australia, informed me that in Jo-Luo society only the first born have to be delivered at bride's mother's house, but the rest of births will have to be delivered in husband's house. Like other Northern Luo groups, when the time of deliver comes, an experienced woman, known as *"Ngat-gam"* (traditional birth attendant TBA) assists the mother at birth.

However, if the delivery at mother's house is prolonging, traditional birth attendant would call the husband and asks him to swear, so that the baby comes out. The husband then says, "Child, if you are my really child come out now, now, now". It is generally believed that if the child is of the husband, soon after the husband swears, the child would come out into the hands of TBA. But if the child is conceived by another man (–with whom she has sexual relationship), the child would not come out. The *ngat-gam (traditional birth attendant) then tells the wife, "Please, as you have seen your husband has sworn to the God of his father, yet the child did not come out. This means that you have been sleeping with boyfriends before your marriage. For this reason, and according to Jo-Luo of Bhar El Ghazal customs, you are now to confess (kwano) all the sexual relationships you have had before marriage". Traditionally, it is believed that if the wife does not confess the names of the boyfriends she have had sexual intercourse before marriage, child could die, if the father does not know the previous boy-friends of his wife.*

The woman lists all the names of men she have had sexual relationship. When the young mother finished confession, the TBA discloses the names of the previous lovers of the bride to the bridegroom. Such confession angers the husband and this often lead to fight between the husband and the men their names are listed. Mr. Alberto Primo Akot Dimo argued that confession of the wife is good because it makes the girls to avoid sexual relationship before marriage". When all these histories have been told to the husband, he cannot reject the wife after marriage because the Jo Luo traditional laws do not permit this.

However, in Jo-Luo society, some women are made pregnant by unknown men and when ask to tell the name the men they refuse to name them. One may bluntly says, "I got it from the grass or in the field where she used to go to chase away small animals and birds from destroying crops. In this situation, the mother usually delivers the baby, even though after a long labour.

According to Apai Gabriel (2005) the Jo-Luo people do not consider virginity a necessary virtue in a girl, but yet they detest a married woman who practices adultery in the community. Therefore, when a woman is in labour and the child does not come out quickly, people would suspect her of hiding a serious crime, like adultery. When the child fails to come out quickly, the midwife (TBA) asks the woman who are with her to leave the place —leaving her alone with the labouring mother. The TBA then turns to the woman and tells her, "As you can see the child is not coming out after long labour. I am asking you to tell me, if you have had sexual intercourse with other men before marriage. You must tell me their names, otherwise the child will not come out and you will die". So, because of pains and fear of death, the woman would confess to TBA the names of men she has had sexual relationships (P. 49).

Shortly, the baby comes out, as the baby comes out, the TBA receives the baby, cuts its umbilical cord with a silver of millet stalk or razor blade. If the child is found to be a monorchid or a deformed, the baby is immediately killed by old women assisting the *ngat-gam* (TBA) they take the body of the baby and buried it in smallest ant-hill commonly known as "*tuk*". In this case the father of the baby and other men would be informed that the child had been shocked with blood. But if the child is born normal, the *ngat-gam* (TBA) then washes the baby before burying the afterbirth; and gives the baby to the young mother to lactate her first child. After this, the TBA burries the afterbirth outside the house; if the baby is a boy, she burries its afterbirth on the right side of the doorway, but if the baby is a girl the *ngat-gam (TBA) burries its afterbirth on the left side of the doorway*.

Customarily, the place where the afterbirth is buried is not usually marked —this is to avoid a person with evil eyes from removing the afterbirth and burns it. It is generally believed that if a person with evil eyes removes the afterbirth of a baby and burns it, the baby will die or otherwise it will prevent the mother from further pregnancy.

The baby is kept with the mother in the house for three days, if the baby is a boy, and if the baby is a girl the mother will keep her in the house for four days. So, after three days or four days, birth ceremony would be conducted. Customarily, in all birth ceremonies, women give the name of the baby; they also eat, drink and dance (where appropriate).

Source: Figure 175, Maternity in Africa by **Paolo Patruno**

According to Apai Gabriel (2005) among the Jo-Luo a baby is kept in for 3 or days, because during these days, the baby would be preparing important tools required in life. In this respect, a baby-boy spends three days inside, making a spear on the first day, a hoe in the second day, and an axe on the third day, and he is brought out on that third day. In respect to the baby-girl, she spends four days in the house, she would be making a pot for her cooking on the first day, decorating her drinking cup on the second day, making a broom on the third day, and mending or platting her hair on the fourth day, and on the very fourth day, she is brought outside (PP. 17-18).

During ritual ceremony of both baby-boy and baby-girl, women and young teenagers perform tribal dances as part of the ritual. They usually dance in the evening; when the ritual is for the baby-boy the boys would be performing a mock fight between the Jo-Luo and the Bongo. In the end a Bongo man is killed and a bow and arrows are taken away from the dead man. The bow and the arrows are then hung up at the door to show any visitor that the newborn in this house is a boy. And if the ritual ceremony is for a baby-girl, a woman would be pretending to sweep compound, and another woman would come and take away the broom from her hand. The woman then takes the broom and places it on top at the entrance to the house. This is to show visitors that a new baby born in this house is a girl.

Interestingly, the Jo-Luo of Bhar El Ghazal decline to ask and tell what sex the newborn baby is. So, to show to the people the sex of the child without telling them, the mother would carry a fishing spear with a tip pierced into a piece of wood for a baby boy; and she carries a spear likewise for a girl.

Apai Gabriel (2005) pointed out that when the child is born nobody is allowed to mention the sex of the child; it is only the TBA who declares the sex of the baby, after tharal inspection. Customarily, whether the child is born in the bright day light or at night, the fire must be lit to have a better view of the child's body parts. More importantly, the mother is to taste the blood when the afterbirth is being separated from the navel as a vow for her to avoid bitter anger and curse against the child in the future.

The ritual ceremony goes as follows:

At sunset: a mother takes the baby into her lap and sits at the door inside the house while the door is closed. An old woman, who has been given authority to give the name of the baby, knocks on the door three or four times, depending on whether the baby is a boy or a girl. Te woman would then be calling the mother with honourable name and the mother expected to answer.

Old woman: The mother of Arach!

Mother: Yes, here I am

Old woman: You, the new baby, come out!

Mother: Brings the baby outside

Old woman: The old woman then leads a procession carrying an axe (if a baby boy) or a broom (if a baby girl). She would be followed by the mother with the child and other women behind them.

If the baby is a boy an old woman then cuts the ground with the axe that is in her hand; but if the baby is a girl an old woman would then sweep the compound with a broom. It is only for the baby boy that the old woman utters set words, as she cuts the ground and the mother plus other women respond accordingly as follows:-

Old woman: Building a house for your wife!

Mother and others: Man's duty

Old woman: Hunting animals to eat!

Mother and others: Man's duty

Old woman: Denying a crime you committed!

Mother and others: Man's habit

Old woman: Walking at night in the dark!

Mother and others: Man's habit

Old woman: Preparing your spears and shield!

Mother and others: Man's power

Sources: Apai Gabriel, 2005, PP. 19-20)

This procession goes from road to road which lead into the house and within the courtyard, the old woman continuous to tell the baby-boy what he will be doing when he grows up on earth.

After the ritual naming or birth ceremony of the child, the midwife calls the husband aside and reveals to him the confessions of his wife. However, when the confessions are of serious consequences, like, a man being implicated in adultery with the wife of his brother, an elder man is usually called in. Such confessions are then revealed in the presence of an elder and the husband; this is normally done to avoid any violence which may arise.

Thus, if it is learned that a man has had sexual intercourse with this woman, the man would be ordered to bring a she-goat by himself, and take it to the house of woman's father, to perform a ritual of *"adual"* as a remedy for the disease. During the ritual ceremony, the she-goat is laid on its back and the partners sit on opposite sides, each holding one side of the she-goat. After the operator has slit the she-goat from head to tail, the man and the woman run in opposite directions, each running with half the she-goat, to symbolise that they would never again be joined together in sexual relationship. Meanwhile, the parents of the partners would be watching their children (daughter and son) with shame, as if they have committed cannibalism –they have broken the taboo. After the operation, the woman and the man are served a nasty drink from juice pressed from roots of the "allwy" tree, which cause them terrible vomiting, and they would be asked to lead the place (Apai Gabriel, 2005).

Interestingly, the birth confessions always happen to women who are giving birth for the first time, because most confessions cover the period of girlhood. However, we understand that in Jo-Luo society, women who are married to elderly men in polygamy tend to have secret sexual relationships with other men outside legal wedlock. Subsequently, the confession for the Jo-Luo women does not end at birth. Thus, if it happens that in the two years, a woman is suckling, a child falls ill, the mother would be asked to confess, because the Jo-Luo believe that having sexual relationship during this period would cause illness. They also know very well that during this period the husband is abstaining himself from sexual relationship with his wife. Thus, if the wife does not confess the child would die of "*ting*" (a disease of hidden illegal sexual intercourse). Because of the love the woman has for the child, even the most strong-minded woman would confess to any relationship with a man, although it is one that she had so far managed to hide.

After having said these, the community also know that a child's sudden illness is not always attributed to the mother's secret sexual relationships with other men, because the Jo-Luo people also believe in evil eyes. Therefore, new born children are always kept way from public places; and when a child suddenly falls sick, parents secretly enquire who has passed through their house that day. If it is confirmed that a suspected evil-eyed woman has passed by, then the cause of the illness is attributed to her and the parents go to a witchdoctor for the treatment and advice.

Following the birth of the first born, the wife remains in her parents' home for one month. After one month, the young mother is accompanied to her husband's house by two or four girls and possibly one young girl who would be left behind as baby seater.

Mr. Deng Apiny Alany and Mr. William Gol, who both live in Melbourne-Australia, categorically said, "There are four main reasons why the first child is always born at wife's mother's house" These reasons include the followings:-

The first reason been that it provides an opportunity for the wife's mother to teach her daughter how to wash and care for the "*nipilal*" *(new born baby) as the bride has no experiences in caring for the new born baby.*

The second reason has been that traditionally it is not allowed for the husband's mother to see nakedness of her son's wife in the first delivery.

The third reason has been that the period of absence of the wife, trains the husband to control his sexual motives and to reframe from sexual intercourse.

The fourth reason has been that customarily the name of the first born must belong to the bride's family. If the baby is a girl, she will be given grandmother's name; and yet if the baby is a boy, he will be given man's name from the grandmother's side. Although, according to Jo Luo of Bhar El Ghazal society the name of the first born child must bear the name from wife's family, but some Jo Luo elders do not accept this practice. They argue that the name of the first born could bear the name of groom's mother or grandfather.

Twins Birth:

In Jo Luo of Bhar El Ghazal society, like in Acholi and Shulluk, twins are welcomed and they are spoken of as "*Kwoi*", which literarily means "children of God/Juok). It is unfortunate that no elder nor a woman from Jo Luo of Bhar El Ghazal told me, during my research, where the afterbirth of the twins are buried; but Deng Apiny Alany reported that if the twins are male child and a female child, the mother will remain in the house for four days, which means the three days for the male baby is included. After four days, the mother would then be brought outside. The ritual ceremony would be the same as we have described in normal birth ceremony above. The different between the ritual ceremony for normal birth and ritual ceremony for twins (*kwoi*) is that, in the fourth day, when the two babies would be brought outside, two goats would be slaughtered, that is, a she-goat and he-goat. The she-goat stands for the female child and a he-goat stands for male child. The he-goat would be slaughtered on the left side of the door, and the she-goat would be slaughtered on the right side of the door. Alberto Rimo Akot Dimo stated that during this time, women who are coming to see the twins must put some coins in each of the calabash provided (one basket place on the left side of the door and the other basket placed on the right side of the door); and any woman, who wants to raise any of these babies, must put additional coins in each of the calabash, for otherwise she would not raise the baby.

However, what we know is that after the twins and the mother have been discharged from the house, there would be no segregation of the sexes, brother and sister grows up together and they play together when they are still young children. But once they have grown up to big children, the boy may play with the boys and the girl may play with the girls. At this stage, the boy may beat his sister if she does something wrong in the family, customarily this is acceptable and cannot be viewed as domestic violence.

Jo Luo of Bhar El Ghazal Child Naming

According to Apai Gabriel (2005) naming a child for the Jo-Luo of Bhar El Ghazal is a ritual ceremony, a reception of a newborn into the community. This is also considered as a tribal baptism (P.17). In Jo Luo society, the new born (*nipilal*) is given milk name; if the baby is a boy he is given the name after three days abut if the "*nipilal*" (new born) is a girl, the name is given after four days following birth. Naming of the new born is performed with a ceremony three days for boy and four days for girl. The meanings of the names often refer to the experience or circumstances of the parents, close relatives, some incidents that occurred in the family or what the elders expect from the new born baby.

More importantly, in the ceremony of a baby boy, the elders feast and shout traits saying, "We wish you courage in your future life. Let God (Juok) gives you strength so that you work hard in future to help your family. Whenever you go hunting let your spear, bow and arrow falls on animals-so that the family have plenty of meat. Let the hoe you will be using for cultivation produce more crops".

While in the ceremony for a baby girl, the elders feast and shout traits saying, "We wish you a good housekeeping when you grow up. We also wish you good caring for your potential children; and above all, we wish you good luck and take care of your future husband together with his relatives".

Today, an influx of Dinka has affected the Jo-Luo naming system. For example, today, we find children with name Majak, Mayen and Macar, which are names of bulls the Dinka give to their children as a remembrance of favourite bulls that they paid for women (wives).

In Jo Luo of Bhar El Ghazal, the first born is usually named after the grandfather if a boy or named after the grandmother if a girl as we have seen before. Apai Gabriel (2005) says, "Although names are often given according to the parental grandparents, first born children are traditionally given women to name after their fathers or mothers; except in rare cases such as twins or when the father has died leaving the woman pregnant" (P. 20) The names of other children usually describe the situation of the parents or the environment of the birth as we shall see below:

(1) *Juok names*

Male	Female	Meaning of the names
NA	NA	A child born with physical marks on toes or fingers (Atum-tum)
Ochan	Apio	The first born twins
Upio	Ocen	When both twins are boys
Ochan	Achan	The second born twins
Ochan	Achan	The third born twins
Ochan	Achan	The fourth born twins
Ochan	Achan	The fifth born twins
Ochan	Achan	The six born twins
NA	NA	Child that comes from the womb with legs first

(2) *Names showing place of deliver*

Male	Female	Meaning of the names
Uyoo	Ayoo	Child born on the road
NA	NA	Child born when the father is away from home
NA	NA	Child born during hunting
NA	NA	Child born in the market when the mother is shopping

Okul	Akul	Child born in stream when the mother when to fetch water
Onam	Anam	Child born in River Nile when the mother went to fetch water
Olum	Akuek	Child born in the grass. "Lum" is grass.
NA	NA	Child born near or under the granary
NA	NA	Child born on plan land as there are no mountains and hills in Collo land
Oyom	Nyayom	Child born on windy day
Bola	Abola	This child will be thrown to God/Juok
Obwot	Abwot	I survived death
Ukello or Bol	Akello or Nyibol	Child born following twins

(3) **Names showing period of delivery**

Male	Female	Meaning of names
Okech	Akech	Child born during famine
Oceng	Aceng	Child born in clear sunshine
NA	NA	Child born in mid-night
Ukot	Nyakot	Child born when it is raining heavily
NA	NA	Child born during Christmas. "Kudmi" means Christmas
Ongom	Angom	No hope for survival
Odhuot	Nyadhout	Child born in Easter days
Chang-Juok	Nyachang-Juok	Child born on Sunday
NA	NA	Child born on the day an important person arrives home

(4) **Names related to hardship**

Males	Females	Meaning of the names
Orach	Arach	Child born during hardship or poverty
Uling or Ongwec	Nyaling or Angwec or Teng	Child born during the war/battle
Uling	Nyaling	Child born when the mother and the father like fighting

Males	Females	Meaning of the names
Ubeng	Nyabeng	Child born when relative do not want the marriage but yet couple decide to marry
NA	NA	Child born when the people of the home do not want the wife/husband
NA	NA	Child in small bag of water
Ukic	Nyakic	Child born when the father died before the baby was born
Uchola	Achola	Child born after somebody has died
NA	NA	Child born after mother has been awaiting for pregnancy
NA	NA	Child born during the visit of important person in the society
Ocwur	Acwur	One of the parents has blindness
Orac	Arac	One of the parents is considered bad
Otoo	Atoo	Child born following the death of another child
Obur	Abur	Child born during the funeral rite
Apiny	Apiny	Child born ....
Alany	Alany	Child born in abuse environment
Gol	-	Child born........
Uyat	Ayat	Child born when the mother conceived aws she went chasing small animals from destroying crops
Unguec	Anguec	Child born after the mother has run away with another man because the former husband could not pay the dowry
Unguec	Anguec	Child born after the mother has been taking refuge to other places due to family violence
Abanyo	Ubanyo or Aduek	A child born during the period of locusts
Kang	Akang	Child born after the death of another child
Wino	Awun	Hild born after the death of another child
Luo	N/A	Revolution

The Jo-Luo of Bgar El Ghazal regards calling a non-relative by name as disrespectful behaviour and would, therefore, avoid using the official given name (family name) when addressing a person. For example, a woman is not to address her husband by name, but the husband reserves the right to address the wife by calling her with her family name.

Furthermore, other members of the family are not allowed to address the wife and the husband wirth their names. The members of the family and neighbours, if they want to address the wife or the husband, they have to use honourable name. For example if the husband has a daughter called Abur, they will call him father of Abur (Uchalla Wabur); and the mother can be called by the name of the son e.g. mother of Olum.

This brings us to a question: *"How does one come to know an honourable name of a person who has no daughter or son?" Apai Gabriel (2005) argues that to know an answer to this important question, we need to look back to the traditional system of giving the children the names of grandfathers and grandmothers. The best example illustration is for us to examine the name Uchalla —this name is understood in the community as "Uchalla Wabur" (which means father of Abur), the honorary daughter. When it comes to name Achalla, which is the opposite sex of Uchalla. Thus, the name Achalla is understood in community as "Achalla Minyikuac (means mother of Nyikuac). However, this does not mean that any man could address ant woman unrelated to him by her honourable name. It is only the people from the clan of a man to whom she is married and other connected relatives of her husband can call her by her honourable name (P.21).*

In fact, Jo-Luo people can easily detect an uncultured urban Jo-Luo man when he calls an unrelated woman by the honourable name. This means that, that man is either only a Luo-speaking person or one who has been brought up outside the community, because calling a woman by her honourable name gives a connotation of being one of the relatives eligible to become a competitive genetic husband in case of misfortune (deatho. Therefore, this is the true hidden core of Jo-Luo culture.

Apai Gabriel (2005) argues that in the Jo-Luo society, there are two ways in which people can identify their genetic origins. The first way to identify one genetic origin is by boasting. This boating happens often when there is dispute between two Jo-Luo nations. Thus, if a true Jo-Luo wants to discriminate the adopted Jo-Luo, and to show him or her that he/she is not a true Jo-Luo, that person utters the names of his/her paternal ancestors and ending the list with the name of the last known ancestor. This basic Jo-Luo cultural practice is difficult for any non-Jo-Luo to comprehend, because there are two or more forms of utterance. Usually, the first utterance involves only the paternal ancestors. The second utterance, which is well ornamented, involves grandmothers, as the grandmothers might have come from different clans and sections by marriage. This sophisticated cultural background has indeed enabled the Jo-Luo to know their family (kith) and kin in blood relationship to other sections. Among the people of Jo-Luo of the present day, the original people of Alur (known as Pingor), Piyua, Bongo of Gir retained their genetic to Gir.

The second way for one to know his/her genetic origins is by using clan or sectional emblem. This emblem may be a special tree, which people of the section revere and avoid cutting or burning; it can even be an animal, which the people from a particular genetic origin do not kill or eat because early legendary historical ties with it.

Having said this, let us now see the examples of genetic boasting. To make our illustration possible and simple, let us take three men: Anei from Abat, Deng from Athiro and Lau from Alur sections. In this illustration, each person wants to show to other that he is the true original Jo-Luo by birth.

(1)   Anei from Abat stands up and boasts saying:
- Ater Uta (my father) ujo (of) Tedom Tango
- Lual kuo (my grandfather) ujo (of) Tedom Tango
- Dut kuo (my garnd-garndfather) ujo (of) Tedom Tango
- Akot kuo (my grand-garnd-garndfather) ujo (of) Tedom Tango
- Nibango kuo (my grand-ngarnd-grand-garndfather) ujo (of) Tedom Tango

(2)   Deng from Athiro gets up and boasts saying:
- Agany Uta (my father) ujo (of) Pinhadho Tango
- Mahuei kuo (my garndfather) ujo (of) Pinhadho Tango
- U cu juo (my grand-grandfather) ujo (of) Pinhadho Tango
- Piem kuo (my garnd-grand-grandfather) ujo (of) Pinhadho Tango
- Majok kuo (my garnd-garnd-garnd-garndfather) ujo (of) Pinhadho Tango

Thus, from the above monotonous counting, we cannot be sure whether Anei's and Deng's mothers were from different sections or clans. If this be the case, then Anei and Deng would not be called true and original Jo-Luo. Let us not see the third person (Lau) boasting, as he adds in the grandmothers and their origin.

(3)   Lau from Alur gets up and boasts saying:
- Ball Uta (my father) ujo (of) Gir, Aliny wonga (my drandmother) wonga nyijo (coming from) Gol Tango
- Ateny kuo (my grandfather) ujo (of) Gir; Akuac wonga (my grandmother) wonga nyijo (coming from) Pinhadho Tango
- Tong kuo (my grand-garndfather) ujo (of) Gir; Angomo wonga (my grand-grand-mother) wonga nyijo (coming from) Tedom Tango
- Bidho kuo (my grand-grand-garndfather) ujo (of) Gir; Aluel wonga (my grand-grand-grandmother) wonga nyijo (coming from) Tedom Tango
- Ujango kuo (my grand-grand-grand-grandfather) ujo (of) Gir; Apieu (my grand-grand-grand-grandmother) wonga nyijo (coming from) Demo
- Ukumo kuo (my grand-grand-grand-grand-grandfather) ujo (of) Gir; Atuol (my grand-grand-grand-grand-grandmother) wonga nyijo (coming from)Tedom Tango

Source: Apai Gabriel, 2005, P.9

As we can see from the above illustration, Lau is challenging Anei and Deng that their grandfathers might be from slaves or woman picked up during tribal war, not married officially, or of unknown origin. Therefore, from such genealogic practice, the genetic knowledge is passed on from on e generation to another generation. It counts so much when grandmothers' sides are involved; it becomes a genetic network.

Mabaan Child Naming

The new married wife gives birth at her husband's home. If it so happens that she delivers in her parent's home, then husband is fined one pig before taking her home. The children are name after family names but sometimes animal, birds, tress' names may also be given to the child. Twins are treated differently and are named *Keta-Buto* (boys) and *Jote-Butta* (girls). In Mabaan society naming is performed after 10 days from birth in a ceremony in which a goat is slaughtered and people eat meat, drink beer.

Shatt Reproduction Process

We have mentioned before that Shatt and Shulluk have many customs held in common. According to Shatt customs, the first born baby must be delivered in the bride's parents' home. Thus, when the wife reaches her fourth or fifth months of her pregnancy, she leaves husband's home for parents' home. And when the time of delivery comes, the wife is assisted by traditional birth attendant (TBA); few children are born in the hospital, as hospitals are usually located in the big towns and city. Birth attendant receives the baby, cut its umbilical cord and clean any dirt on the body.

We have seen with the Collo society that after the TBA cut umbilical cord of the baby, before she hands over the baby to the young mother, she usually asks the mother to confess of her previous love affairs —for otherwise the child may die. Given the fact that Shatt/Thuri and Shulluk have many customs in common, this therefore, implies that the TBA should do the same with a young Shatt mother. However, there is no confirmation on whether the young mother confesses her previous love affairs to the TBA or not. This requires further research by the young Northern Luo writers.

Source: Figure 176, Maternity in Africa by **Paolo Patruno**

Child Naming and Ceremony

In Shatt society naming ceremony is performed after three days if the baby is a boy; but if the baby is a girl, naming ceremony is performed after four days. During the naming process, if the child is a boy, it is laid on skin mate or reed mate and is made to hold three tiny pieces of grass in its hand. The person giving name then picks these three pieces of grass from the hand of the child, dips them into warm water and sprinkles the water on the baby while giving the name. And if the child is a girl, in the same way, it is laid on skin mate or reed mate, she is made to hold four tiny pieces of grass in its hand. The person giving the name comes and picks the four pieces of grass from the hand of the baby, dips them into warm water and sprinkles the water on the baby while giving the name. However, it should be noted that each of the tiny pieces of grass represent particular names. Then person giving the name will not pick all the three or four tiny pieces of grass from the baby, so the one tiny piece of garass that remains in the baby's hand, the mname that tiny piece carries will be given to the baby. I was not able to find out what name each of the tiny pieces of grass represent—this area needs more research.

The naming ceremony is also the time when relatives and friends come together to eat, drink and dance. I would like to advise my readers that I cannot give list of different names according to Shatt way of naming the children here, because I have no Shatt men or woman who could tell me these. I know the common Luo names, but I cannot create one or two for the Shatt, although they are Luo groups.

Shulluk Reproduction Process

Like the Acholi, the Shulluk recognised two ways of birth customs, and these mainly include ordinary birth and abnormal birth (twins) called (*Nuole*). The Shulluk people differ a bit from the Acholi and other Northern Luo Groups of South Sudan in this that they also recognized the birth of King's children. Viewing birth customs from these three ways of birth customs seem to be complex, but as we discuss them individually they become more clear and clear.

(1) *Ordinary Birth*

In the ordinary birth, when a woman is six months pregnant, she leaves the house of her husband and – for intercourse should now cease – and goes to her parents' home, where the baby will be born. Traditionally, when a woman has a baby of one or two months, her husband or somebody who had just buried a relative are not allowed to come closer to her because they are believed to be ceremonially polluted. They are not even allowed to be ten metres away from the new born baby (Chris Morton, 2004).

When the time of deliver comes, an experienced woman, known as traditional birth attendant (TBA) assists the mother at birth; when the baby comes out the TBA receives the baby, cuts its umbilical cord with a silver of millet stalk and buries the afterbirth outside the house on the right of the doorway if a baby boy, but if the baby is a girl, the umbilical cord would be buried on the left side of doorway. The place where afterbirth is buried is not marked in any way; this is to avoid a person with evil eyes from digging afterbirth and burning it. It is believed that if a person with evil eyes digs the afterbirth and burns it, the baby will die or otherwise it will prevent the mother from further pregnancy. One informant told me that according to the Shulluk custom if a person takes menstrual blood and burn it, the woman from whom the blood have flowed would become sterile.

Source: Figure 177, Maternity in Africa by **Paolo Patruno**

When the TBA receives the baby, she cuts the umbilical cord of the baby and washes the baby, before she give the baby to the young mother to lactate her first child, she makes sure that the wife confesses all her previous lovers by telling their names, for otherwise the baby may die. After confession of the wife, the TBA discloses the names of the previous lovers of the bride to the bridegroom. When the husband hears this list of names, he can do nothing about it; because this is allowed by the customs. He cannot complain nor reject the wife after marriage because Shulluk traditional laws do not permit this, since these confessions are referred to the time before marriage was accepted.

Soon after the confession, the TBA takes the baby and shows to the father and other men, who are waiting to hear that the child is not a monorchid or deformed. If the child is found to be a monorchid or a deformed, the baby is immediately thrown into the river.

But if the baby is normal, it will be kept together with the mother in the house for at least one or two month –regardless whether the child is a boy or a girl. During this period no one is allowed to enter the house where the baby and the mother are kept. During this period the mother would eats special foods to assist in a quick and easy recuperation. Whenever the mother wants to get out of the house, she must always carries a stock of sorghum and wears a cross marked of ash on her forehead. However, at the birth of the child a goat is killed and its skin rubbed smooth to form a mat to be used for carrying the baby later.

Following the birth of the child the wife would remain in her parents' home for another six months or event two years. It is said that the wife does this simply to avoid unnecessary pregnancy. In Shulluk customs, like in other Luo customs, it is shameful for a woman to conceive during lactation. It is generally believed that if a woman becomes pregnant during the lactation period, this is an indication that she has been having sexual relationship with other man rather than with the husband. Because, the entire community knows that the husband is allowed to visit his wife at this time but he is not allowed to have sexual intercourse with her. Seligman and Seligman (1932) pointed out that at the time of birth the husband can make enquiries about his wife, but he must avoid the house of his parents' in-law (P.70).

It is reported that only in the first pregnancy the woman delivers at her mother's house but other pregnancies that follow are deliver at husband's home. However, there is no doubt that the husband practices abstinence from sexual relationship with his wife during the later month of pregnancy and during lactation —should the wife be in husband's home.

(2) *Twins Birth*
In Collo society, twins are welcomed and they are spoken of as *"Nuole Juok"*, which literarily means "children of God/Juok). It is unfortunate that no elder Shulluk woman told me during my research where the afterbirth of the twins are buried; and how many weeks or months could the twins and their mother remain in the house following birth day. However, what we know is that after the twins and the mother have been discharged from the house, there would be no segregation of the sexes, brother and sister grows up together and they may play together when they are still young children. But once they grow up to big children, the boy may play with the boys and the girl may play with the girls. At this stage, the boy may beat his sister if she does something wrong in the family, customarily this is acceptable and cannot be viewed as violence.

(3) *Birth of Royal Child*
The King or Reth of Shulluk lives in Fashoda with many of his wives (Please, see figure 60). When the wife of the King is four or five months of her pregnancy, the expectant mother is sent with attendants and cattle to distant village, and there she remains under the protection of the village chief until the child is weaned, customarily, Reth's (king's) children are never born in Fashoda; after this the wife of the Reth may return to Fashoda.

During birth, king's wife, like any other Shulluk woman, is assisted by well experienced woman at birth. The woman receives the baby; she cuts the umbilical cord with either spear, silver of millet stalk or razor blade and buries the afterbirth outside the house on the right of the doorway. I am not sure whether the king's wife also confesses of her previous love affairs to the TBA. Soon after birth, the mother is allowed to carry her new born baby. One of the things that remain absurd to me to me is what happen to the baby if he/she is monorchild or deformed. It is also not clear of how long will the Reth's wife remains in the house —one month or two months as we have seen in ordinary birth.

It has been reported that during each subsequent pregnancy the King's wife has to go to a different village with attendants and cattle —where she gives birth to another child. The child is born and brought up in this native village and the cattle brought by the mother remain as wealth of the child when he/she grows up. King's child born in a particular village remains in that village where he/she also gets married and dies. Every royal son is recognized as a potential king and is treated with utmost indulgence and respect.

Child Naming and Ceremony

In Collo society, new born is given milk name and this usually given soon after birth and the meanings of the names often refer to the experience or circumstances of the parents, close relatives or to some incidents that occurred in the family. The birth name is usually retained throughout life; but D. Westernmann (1912) states that a man often acquires a second name later, usually that of a cow or an ox. According to writers like Hofmayr (1925) names play an important role in the life of a Shulluk, just as they do among the other Northern Luo Groups. A girl keeps her name for life, but a boy is continually adding names; they play the part of decorations, he may take one name at his first dance and later on take the second name in another occasion that he may wish to celebrate.

The prefix *"nya"* denotes usually a female but is sometimes shared by male. Some children are also named following the visit of Reth (King) in that community. "Okach" or "Nyakach" refers to famine; the name "Oyoo"(male) or "Nyayoo"(female) is given to a child been born on the road; the name "Acwanyo" (both sexes) is given to the child that is born on the day that coincides with the arrival of an important person or relative; and the name "Ronyo" (male) or "Aban" (Female) is given to the child that is born on the day that coincides with death of somebody either in a village or in towns. The first born of twins are called "Anga" (Female) "Ngor" (male), the first born twins are sometimes also called "Chan" (male) and "Achan"; and the second born of twins are called "Bol" (male) "Nyabol" (female). The third born of twins are called "Achuil" (male) "Achuil" (female). The fourth born twins are called "Otuu" (male) "Nyatu" (female). The fifth born twins are called "Olawi" (male) and "Malawi" (female). The sixth born twins are called "Ulit" (male) and "Nyalit" (female). The Shulluk by tradition do not name a child after a living person.

(1) **Juok names**

Male	Female	Meaning of the names
Nyikang	Nyikang	A child born with physical marks on toes or fingers (Atum-tum)
Ngor	Anga	The first born twins
Chan	Achan	Also first born twins
Bol	Nyabol	The second born twins
Achuil	Achuil	The third born twins
Otuu	Nyatu	The fourth born twins
Olawi	Malawi	The fifth born twins
Ulit	Nyalit	The six born twins
	Alwuong	Child that comes from the womb with legs first

(2) *Names showing place of deliver*

Male	Female	Meaning of the names
Oyoo	Nyayoo	Child born on the road
Okuc	Nyakuc	Child born when the father is away from home
Ulai	Nyalai	Child born during hunting
Ocuk	Nyacuk	Child born in the market when the mother is shopping
Olui	Nyalui	Child born in stream when the mother when to fetch water
Kiir	Nyakiir	Child born in River Nile when the mother went to fetch water
Olum	Nyalum	Child born in the grass. "Lum" is grass.
Odhong	Nyadhong	Child born near or under the granary
Okwoc	Nyakwoc	Child born on plan land as there are no mountains and hills in Collo land
Oyom	Nyayom	Child born on windy day

(3) *Names showing period of delivery*

Male	Female	Meaning of names
Okach	Nyakach	Child born during famine
Oceng	Nyaceng	Child born in clear sunshine
Umut	Nyamut	Child born in mid-night

Okot	Nyakot	Child born when it is raining heavily
Okudmi	Nyaudmi	Child born during Christmas. "Kudmi" means Christmas
Agud	Nyikai	
Odhuot	Nyadhout	Child born in Easter days
Chang-Juok	Nyachang-Juok	Child born on Sunday
Acwanyo	Acwanyo	Child born on the day an important person arrives home

(4) **Names related to hardship**

Males	Females	Meaning of the names
Orach	Nyarach	Child born during hardship or poverty
Uling	Nyaling	Child born during the war/battle
Uling	Nyaling	Child born when the mother and the father like fighting
Ubeng	Nyabeng	Child born when relative do not want the marriage but yet couple decide to marry
Mayong	Nyayong	Child born when the people of the home do not want the wife/husband
Ubech	Nyabeng	Child in small bag of water
Akic	Nyakic	Child born when the father died before the baby was born
Chol or Ronyo	Achol or Aban	Child born after somebody has died
Muc	Nyamuch	Child born after mother has been awaiting for pregnancy
Acwanyo	Acwanyo	Child born during the visit of important person in the society

Pari Production process

Like the Acholi, the Pari recognised two ways of birth customs, and these mainly include ordinary birth and abnormal birth (twins) called (*Nuole*). The Pari like the Acholi, they do not practice birth of King's children since they have no King. Viewing birth customs from these two ways of birth customs seem to be complex, but as we discuss them individually they will become more clear and clear.

(1) *Ordinary Birth*

In the ordinary birth, when a woman is six months pregnant, she leaves the house of her husband – for intercourse should now cease – and goes to her parents' home, where the baby will be born.

When the time of delivery comes, an experienced woman, known as traditional birth attendant (TBA) assists the mother at birth; when the baby comes out the TBA receives the baby, cuts its umbilical cord with a silver of millet stalk and buries the afterbirth outside the house on the right of the doorway. The place where afterbirth is buried is not marked in any way; this is to avoid a person with evil eyes from digging afterbirth and burning it. It is believed that if a person with evil eyes digs the afterbirth and burns it, the baby will die or otherwise it will prevent the mother from further pregnancy. One informant told me that according to the Pari custom if a person takes menstrual blood and burn it, the woman from whom the blood have flowed would become sterile.

When the TBA received the baby, she cuts the umbilical cord of the baby and washes the baby, before she gives the baby to the young mother to lactate her first child. She makes sure that the wife confesses all her previous lovers by telling their names, for otherwise the baby may die. After confession of the wife, the TBA discloses the names of the previous lovers of the bride to the bridegroom. When the husband hears this list of names, he can do nothing about it; because this is allowed by the customs. He cannot complain nor reject the wife after marriage because Pari traditional laws do not permit this, since these confessions are referred to the time before marriage was accepted.

Soon after the confession, the TBA takes the baby and shows to the father and other men, who are waiting to hear that the child is not a monorchid or deformed. If the child is found to be a monorchid or a deformed, the baby is immediately thrown into the river; because there is no hospital to treat this deformed baby.

Source: Figure 178, Maternity in Africa by **Paolo Patruno**

But if the baby is normal, it will be kept together with the mother in the house for at least one or two month –regardless whether the child is a boy or a girl. During this period no one is allowed to enter the house where the baby and the mother are kept. However, at the birth of the child a goat is killed and its skin rubbed smooth to form a mat to be used for carrying the baby later.

Following the birth of the child the wife would remain in her parents' home for another six months or event two years. It is said that the wife does this simply to avoid unnecessary pregnancy. In Pari customs, like in other Luo customs, it is shameful for a woman to conceive during lactation. It is generally believed that if a woman becomes pregnant during the lactation period, this is an indication that she has been having sexual relationship with other man rather than with the husband. Because, the entire community knows that the husband is allowed to visit his wife at this time but he is not allowed to have sexual intercourse with her. Seligman and Seligman (1932) pointed out that at the time of birth the husband can make enquiries about his wife, but he must avoid the house of his parents' in-law (P.70).

What remains unclear to me and possibly to other scholars is whether the wife leaves her husband's home during every pregnancy or she leaves him only in the first pregnancy. However, there is no doubt that the husband practices abstinence from sexual relationship with his wife during the later month of pregnancy and during lactation –should the wife be in husband's home.

(1) *Twins Birth*

In Pari society, twins are welcomed and they are spoken of as "*Nuole Juok*", which literarily means "children of God/Juok). It is unfortunate that no elder Pari woman told me during my research where the afterbirth of the twins are buried; and how many weeks or months could the twins and their mother remain in the house following birth day. However, what we know is that after the twins and the mother have been discharged from the house, there would be no segregation of the sexes, brother and sister grows up together and they may play together when they are still young children. But once they grow up to big children, the boy may play with the boys and the girl may play with the girls. At this stage, the boy may beat his sister if she does something wrong in the family, customarily this is acceptable and cannot be viewed as violence.

Child Naming and Ceremony

In Pari society, names are given to the child soon after birth and the meanings of the names often refer to some incidents that occurred in the family. The birth name is usually retained throughout life; but Wassermann (1912) states that a man often acquires a second name later, usually that of a cow or an ox. According to writers like Hofmayr names play an important role in the life of a Shulluk, just as they do among the other Northern Luo Groups. A girl keeps her name for life, but a boy is continually adding names; they play the part of decorations, he may take one name at his first dance and later on take the second name in another occasion that he may wish to celebrate.

The first born of twins are called "Apio" (Female) "Upio" (male), the Pari do not have names for the second born of twins, the third born of twins not the fourth born twins as we shall see among the Collo Community. Interestingly, in Pari society the first born male child is called "Bongo" and if the child is a female is called "Bong". The second child who usually follows the first born is called "Julu" if male and if the child is a female she is also called "Julu" Moreover, the last born in the family is usually called "Agudo" if a boy or called "Angudo" if a girl.

(1) Juok Names

Male	Female	Meaning of the names
Ujuok	Ajuok	Children with physical marks on the toes/fingers
Ukuc	Akuc	The child conceived before menstruation
Upio	Apio	Twins "Ukuo" names
Udoc	Adoc	Child born with legs first
NA	NA	Child born after twin
NA	NA	Child with face downward
NA	NA	Child born in bag of water

(1) Names showing places of delivery

Male	Female	Meaning of the names
Uyoo	Ayoo	Child born on the road
Ugala	Agala	Child born in towns/city
Udwar	Adwar	Child born during hunting
Ulum	Alum	Child born in the grass
Udero	Adero	Child born near or under granary
Ukidi	Akita	Child born on the mountain or Hill

(2) Names showing period of delivery

Male	Female	Meaning of the names
Ukec	Akeca	Child born during famine
Uceng	Aceng	Child born in clear daylight
Uwar	Awar	Child born in mid-night
Mwaka	Mwaka	Child born during Christmas
Ubaa	Nyabaa	Child born in evening
Unyango	Anyango	Child when sun is rising
Ukot	Akot/Akota	Child born when it is raining heavily
Ukongo	Akongo	Child born when mother was selling local beer

(3) Names showing death of relatives

Male	Female	Meaning of the names
Ubur	Abur	Child born just after the death of a relative
Ubur	Abur	Child born when people are mourning

(4) Names related to hardships

Male	Female	Meaning of the names
Ukak	Akak	Child born during war
Ukech	Akeca	Child born in poverty/hardship
Ukak	Akak	Child born when parents are found of fighting
Ukwer	Akwer	Child born when relatives do not want the wife
Udong	Adong	Child born when father died before it was born
Utoo	Atoo	Child born after then death of its brother or sister

Chapter 20:
Female Genital Cutting (F.G.C.)

Definition

Many scholars may have different ways of defining female genital cutting (FGC). This depends on where each scholar is coming from. Generally, in Africa and Middle East communities, people who practice TGC defines it as *"Traditional and customary removal of part or total of woman's or girl's clitoris, Labia Minora and Labia Majora"*. The World Health Organisation (2014) defines the cutting of woman's organs as, "Female genital mutilation (FGM) also referred to as "female circumcision" or "female cutting" comprises all procedures involving partially or total removal of the external female genitalia or other injury to the female genital organs for non-medical reasons". The FMC is recognised internationally as a violation of the human rights for girls and women. FMC involves an elder woman slicing off all or part of a woman's clitoris, Labia Minora and Labia Majora as part of the ceremony that is often conducted around the time that the woman reaches puberty. The female genital cutting (FGC) is also known as female genital mutilation (FGM) or female circumcision –all have the same meaning.

Interestingly, it is reported that FGM is not a new thing, it has originated from Egypt before Jesus Christ was even born. Herodotus wrote about FGM being practiced in Egypt as early as 500 BC, while the Greek geographer Strabo who visited Egypt in about 25 BC, reported that one of the Egyptian customs was "to circumcise the males and excise the females". According to the U.S. Department of Health and Human Services, FGM is actually practiced by Muslim, Christian and Jewish groups. There are countries, such as Nigeria, Tanzania and Niger, where the prevalence of FGM is even greater among Christian groups. In Egypt, FGM is also practiced on Coptic girls, while in Ethiopia, the Beta Israel or Falashas, a Jewish minority, subject their girls to genital mutilation. While, according to the Hebrew bible, circumcision is required for all male Jewish children in observance of God's commandment to Abraham (Genesis 12-17), female circumcision was never allowed in Judaism (African Journal of Urology, 2013, PP. 127-129).

UNICEF (2013) estimates that at least 200 million woman and girls are victims of female genital cutting. It is said, globally, a woman or a girl is violated by FGM every ten seconds. This then means that approximately 8,500 women and girls are harmed by FGC daily. WHO (2014) reported that about 3 million girls in Africa are at risk of undergoing female genital cutting each year.

Tools used for cutting genital tissues include knives, scissors, scalpels, pieces of glass and razor blades (See Figure 142)

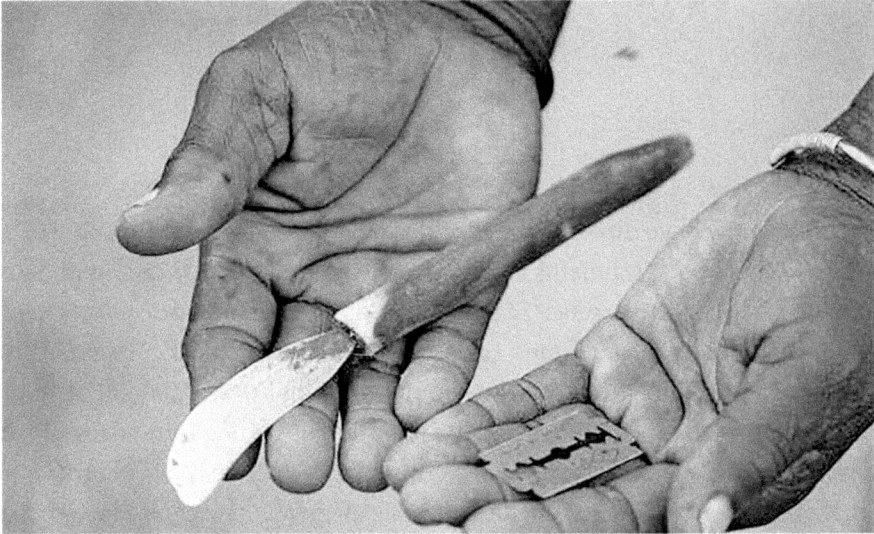

Figure: 179, Tools used for cutting genitals tissues

The female genital cutting is the practice of intentionally cutting or altering the female organs. In most societies, where female genital cutting happens, it is seen as a cultural tradition and is deeply rooted in equality between the sexes.

However, the people who practice FGC believe that it has connection with health of the child and above all it prepares girls for marriages. On the 11th August 2017, during Female Genital Cutting Workshop in Lower Templestowe (Melbourne), I was amazed to hear one presenter, from Family and Reproductive Rights Education Program, says, "People who practice FGC believed that if the clitoris touches the head of the baby during birth that baby would dies"; subsequently, this clearly explained the connection between FGC and child's health.

Female genital cutting is a universal problem, primarily concentrated in 29 countries in Africa and Middle East, but is issue across the globe. In Egypt, Ethiopia and Somali, nearly half of the female population are survivors of FGC. Although there is no statistic taken, it is assumed that more than half a million women and girls in developed countries are at risk of undergoing or have undergone female genital cutting –given the fact that FGC is illegal in UK, USA, Australia, to mention a few.

As a result, today many programs have been set up in developed countries such as Australia to work with women, youth, men and key community members about female genital cutting.

The Short and Long Term Health issues Associated with FGC

According to many who practice female genital cutting, God has made a gross mistake by creating unwanted or extra skins (clitoris, Labia Minora and Labia Majora) which are not needed and these can only be cleaned by cutting them off. They have very little or no knowledge about the use of clitoris, Labia Minora and Labia Majora, as well as the effects of FGC. The immediate effects of cutting women's genitals are many we cannot mention them all in this book; however, some of the short and long term effects are:-

- ▶ Severe and Constance pains
- ▶ Difficulties in passing urine
- ▶ Bladder and urinary tract infections
- ▶ Difficulty in vaginal examination during pregnancy and child birth, which could be life threaten to the mother and the baby
- ▶ A woman who has her genital cut finds sexual intercourse painful. This could also result in reduction of sexual feeling or desire and often a lack of pleasure sensation
- ▶ Repeated infections which can lead to infertility
- ▶ Bleeding, cysts and abscesses
- ▶ Female genital cutting can cause emotional difficulties for a woman throughout her life (these would include depression, anxiety, flashback to the time of cutting, night mares, self-harm and other sleep problems)
- ▶ Painful periods
- ▶ Painful urination
- ▶ Neurogenic shock
- ▶ Some women who have their genitals cut may find it difficult to become pregnant, and those who do conceive can have problem in child birth.
- ▶ Some girls die from blood loss or infection as a direct result of the procedure
- ▶ Injury to nearby genital tissue and sometimes leads to death. Death usually occur when there is severe bleeding leading to haemorrhagic shock
- ▶ Scarring

Manfred Nowak, a UN Special Reporter on Torture, says "The pains inflicted by female genital cutting does not stop with initial procedure, but often continues as ongoing torture throughout woman's life".

Almost all women and girls who have undergone female genital cutting experience pains and bleeding (as we have just discussed above), as a consequence of the procedure. The event itself is traumatic as girls are held down during the procedure. Girls are often told by their parents that the procedure of female genital cutting is a secret and she is not allowed to talk about it to anybody apart from family members. After female genital mutilation, sometimes the girl's legs are tied together so she can take only small steps to prevent tearing any stiches or opening her wounds (See Figure 143)

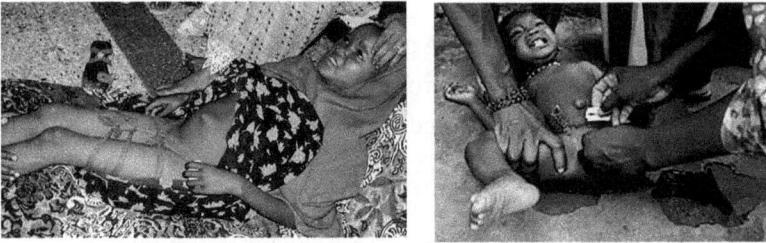

Figure: 180, cutting genital organs. Source Chapman (2017)

In some countries, female genital cutting is a rite of passage to womanhood and ways to control girl's involvement in sexual intercourse. On the other hand female genital cutting is also considered as a preparation for marriage. This is often motivated by the beliefs of sexual behaviour and virginity as well as chastity. It is believed that once a woman has undergone genital cutting, this would create fears in her not to have sexual intercourse with a man because the covered vagina will open with much pains, thus, this often discourage a woman from sexual intercourse.

There are large numbers of women and men who no longer favour this practice of cutting female genitals, but yet they have their daughters cut regardless. Somebody may ask why this is continuing. The answer is that in many cases FGC is not an individual decision, it involves number of family members such as mother, mother-in-law, husband, friends, father and entire community -in other words we can say FGC carries with it social pressures. While about 80 % of the younger women in Northern Luo are going uncut in Africa, such countries as Egypt and Chad more than 80% of the teenagers still undergo the procedure.

In the old days, the girls between 15-19 years of age and the women between 45-49 years of age were circumcised. Today, female genital cutting is practiced in girls between ages 6-10.

Types of Female Genital Cutting

Generally speaking there are only three types of female genital cutting that women practice (although some people speak of type four female genital cutting).

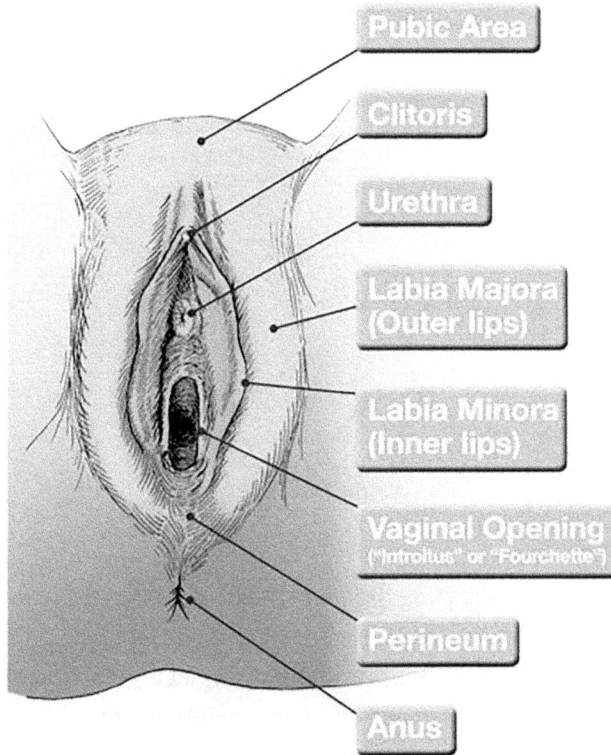

Figure: 181, Normal Vaginal

Type 1: The female genital cutting type one is when a part or all of the woman's clitoris is removed, without cutting any part of Labia Minora or Labia Majora. Female genital cutting type one is also known as clitoridectomy. The term clitoris means a small, sensitive and erectile part of the woman's genitals. Interestingly, some people referred to clitoris as woman's penis. (See figure 145)

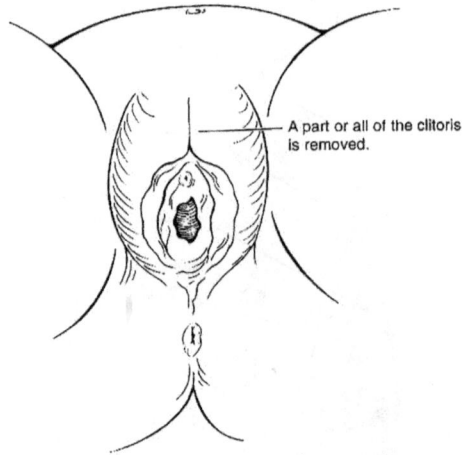

Figure: 182, FGC Type One

Type 11: The female genital cutting type two is when most of the woman's clitoris and some or all of the Labia Minora are removed, with or without cutting part of the Labia Majora. By definition then Labia Minora and Labia Majora are "*the lips*" that surround the vagina (or skin and tissues around then vagina). The female genital cutting type two is also known as "excision" (See Figure 146).

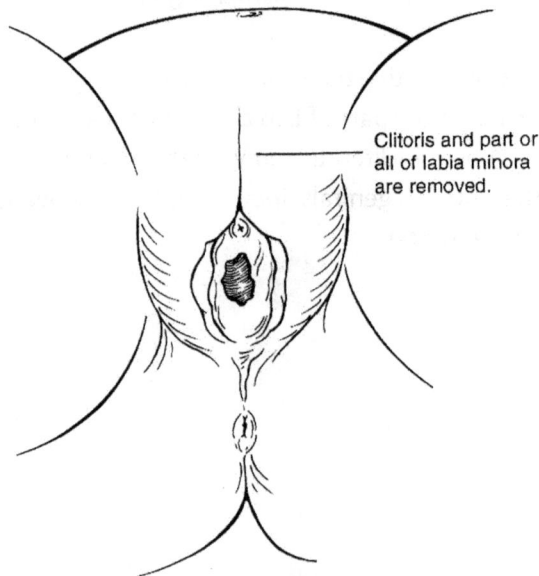

Figure: 183, FGC Type Two

Type 111: The female genital cutting type three is when most or part of the clitoris, Labia Minora and Labia Majora are removed. After cutting clitoris, Labia Minora and Labia Majora, an old woman then stitches together Labia Minora and Labia Majora are by bringing them together to cover the urethra, which makes the vagina opening very small to allow for the flow of urine and the menstrual blood. The female genital cutting type three is also known as "infibulations". Mathews (2011) argues, "The fundamental aim of this excision and infibulations procedure is to transform the girl into what is perceived to be more feminine form, and ultimately render her sexually-inactive until marriage" P.139 (See Figure 147).

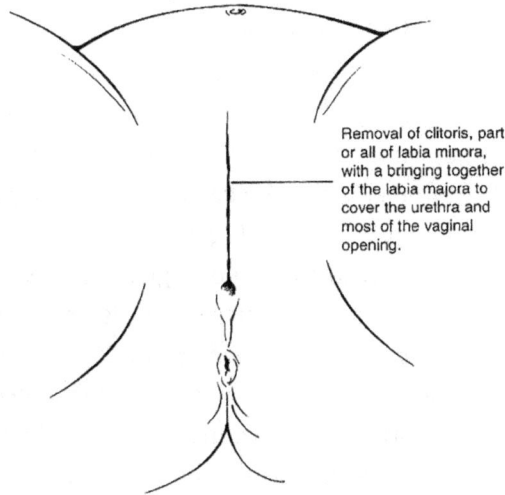

Removal of clitoris, part or all of labia minora, with a bringing together of the labia majora to cover the urethra and most of the vaginal opening.

Figure: 184, FGC Type Three

Practices of Female Genital Cutting among the Northern Luo

As we have discussed that female genital cuttings are very common among the African societies, but the history tells us that the Northern Luo women never practice female genital cuttings in the Stone Age. The idea of female genital cutting is something new to the Northern Luo. It is reported that the practice of female genital cutting penetrated the Northern Luo societies of South Sudan through marriages and Islamic religion. Although female genital cutting penetrated the Northern Luo groups through marriages and Islam, the bottom line is that not all the Northern Luo women and girls practice female genital cutting. The Northern Luo who did not intermarried with the Arabs or Nuba and those who are not converted into Islam never practice female genital cutting.

It is reported that female genital cutting practices are carried out among the Balanda,

Shatt (Thuri), Mabaan and Jumjum. However, some people say a hand full of women and girls in Shulluk society practice female genital cutting. But generally, the Collo do not accept nor do they appreciate female genital cutting. The only Northern Luo groups who do not practice female genital cutting are the Acholi, Anyuak, Jur-Chol and Pari (Lokoro).

Practically speaking, in the Northern Luo communities where female genital cutting are practiced, young women and girls do not have right to control what happens to their bodies and the right to refuse female genital cutting.

Like in other African communities, female genital cutting is carried out among the Northern Luo women and girls for various religious and social reasons within the families. The people who practice FGC believe that female genital cutting brings some benefits to girls; and these benefits include preparation for marriage as well as preserving virginity.

However, it is important to note that in Northern Luo women groups female genital cutting usually happens to girls whose mothers, grandmothers or extended female family members have had female genital cutting themselves. Fundamentally, FGC happens to girls whose their mothers come from a community where female genital cutting is being practiced.

From our discussions about the Northern Luo religions, we have discovered that some of the Luo groups are been converted into Islam. There is a common saying among the Muslim that *"a good wife is a woman who has undergone FGC and remains virgin throughout her life till the time for marriage. A woman can keep her virginity only when her genitals are cut"*. Although nothing is clearly written in the Quran about the holiness of FGC, the Muslim assumed that the above preposition justify the Quran. Furthermore, some Sunni Muslims legitimate FGM by quoting a controversial *hadith* (a saying attributed to the Prophet Mohammed) in which the Prophet allegedly did not object to FGM provided cutting was not too severe. And that the least invasive type of FGM (partial or total removal of the clitoris and/or the prepuce) is also called "Sunna Circumcision". Consequently, FGM widely became known as something associated with Islam.

According to Chapman (2017) some members of the community who have spoken out against the practice, they said the purpose of this cutting is to suppress female sexuality in attempt to reduce sexual pleasure and promiscuity. While Momoh (2004) pointed out that in many parts of the world the ritual of genital cutting remains critical to re-affirming her adult gender/role/status with her community and in strengthening the intrinsic moral structure of that community,

Strangely enough, some Islamic scholars disagree on FGM they said there is no obligatory rules exist but some argues that there is mention of female circumcision in the Hadith. According to Sami A. Aldeeb Abu Sahlieh, the most often mentioned narration reports a debate between Muhammed and Um Habibah (or Um 'Atiyyah). This woman, known as an exciser of female slaves, was one of a group of women who had immigrated with Muhammed. Having seen her, Muhammad asked her if she kept practicing her profession.

She answered affirmatively, adding: "unless it is forbidden, and you order me to stop doing it." When Muhammed heard the reply of the woman, he said female genital cutting is allowed in Islam. Abu Sahlieh further cited Muhammad as saying, "Circumcision is a *sunna* (tradition) for the men and *makruma* (honorable deed) for the women."

The Law and the FGC

It is worth mentioning that the practice of female genital cutting has become a human and health issue in western countries where the practice is continued by immigrants from countries where FGM is commonly performed. For example, the German organization called "Terre des Femmes" estimates that about 30,000 girls and women living in Germany have undergone or are at risk of being subjected to female genital cutting.

Although it is legal to practice FGC in some parts of the African countries,(e.g. South Sudan and Kenya) it is illegal to practice it in most of the developed countries such as UK, USA and Australia. The authorities say it is an offence (a) to perform female genital cutting and this including taking a child a broad for the circumcision; (b) to help a girl to perform cutting her genital tissues or to help anyone perform female genital cutting; (c) it is an offence for a care taker of a girl to fail to protect her from FGC. The practice has no place in modern society and those who perform FGC on minors (girls) will be held accountable under federal law. According to Caflish and Hohlfield (2009) although the Western Laws forbid female genital cutting there is evidence that the practice of FGC is continuing among many of the African, Asian, Middle Eastern refugee and migrant groups who now reside in Western Nations —a key issue for public health practice.

Thus, in Australia anyone who performs female genital cutting is liable to undergo 15 years imprisonment; while in Europe and USA anyone who performs FGC can be put in prison for 14 years. And anyone found guilty of failing to protect a girl from female genital cutting cane face up to 7 years in prison.

Despite the fact that people understand the consequence of practicing FGC, yet people with Islamic back ground find it difficult to follow and stop the practice. They are ready to go to prison rather than not helping girls with female genital cutting. For example, Chapman (2017) reported in his paper titled *"Detroit Doctor Charged With Female Genital Mutilation, Alleged Victims 7 Years Old"* that in United State of America, Female Genital Cutting is a serious federal felony but yet professional people like Dr. Jumana Nagarwala, who is employed as a doctor by the Henry Ford Health System in Detroit was arrested in April 2017 for performing FGM on girls between the ages of 6 and 8 years old. Although the Acting Assistant Attorney General Blanco claimed that the Department of Justice is committed to stopping female genital cutting in using the full power of the law to ensure that no girls suffer physical and emotional abuse yet practices are going on.

Chapter 21:
Inheriting the Wife of the Dead Brother

"Widowhood practices are closely ties to cultural and traditional beliefs about death, ghosts, inheritance, feminine roles, family structure and family relationships. The overpowering belief in the ability of the ghost of a dead person to come back to dispute and haunt all kind of things and relatives has reinforced and perpetuated the age-old practices of widowhood in Africa....The belief that death brings corruption and the dead still have contact with the living, especially their closest partners in life, is one of the reasons used for subjecting widows to inhuman and humiliating customary practices, rituals and practices is therefore believed to help restore the balance and security, which the death had sought to overthrow. The people, due to superstitious beliefs, rationalise these practices with the argument that they perform important functions, such as giving the widows protection from their powerful deceased husbands" (Sossou, 2002). The levirate union arises out of a covenant, a solemn pledge made between the bride and groom at the time of marriage by the lineage of the groom. It is part of the mutual agreement, based on love and respect, to do all to continue the family of the groom if he should be removed by death (Kirwen 1979: 205).

Inheritance of Widow of Deceased Brother

In African societies and in the Northern Luo in particular, it is natural for one to grieve for the death of a loved one. This is due to the fact that death creates many social problems between families, this include caring for the widows and the children. The survivor's helplessness is made worse if the loss of capacity to negotiate is extended to the activities those are meant to restore normal family life of the widow after the death of her husband.

However, in Luo community, this matter is taken seriously and the people want to make sure that a widow gets an opportunity for care, support and empowerment. Thus, when a man dies, his wife or wives become the wife or wives of his son or bother with some ceremony. It should be note that the caring for a widow, according to the Luo people it is based on two fundamental things that is immortality of the husband (or wife) through the children and on natural responsibility in society.

Let us explain what we exactly mean by *"immortality of the husband/wife through children"*. According to Northern Luo cultures, the only person who is said to be *dead* is the person who dies childless. This is when the Acholi people say, *"Otoo labong wiye"*, which literarily means a person who dies childless. But a person who dies leaving a child or children behind is said *not a death person*. This is when the Acholi say, *"Otoo oweko cogo woko"*. Word *"cogo"* in Acholi means *bones*; this literarily means a person who died leaving a child or children behind –these children are taken as physical representation of the deceased person. Therefore, in Luo community, a person who dies leaving a child or children behind is said to be living, therefore, he/she is not dead. Whenever the children perform responsibilities in the societies, they are said to be doing these things because of the physical power that a deceased father or mother has left them.

At this point, it is important to note that in Northern Luo culture, being alive and acquiring social status imply the fulfilment of social responsibilities. This means that a person or a child who does not perform his/her responsibilities is considered by the society to be socially worthless –a human being without any social value. Such person usually referred to by the Acholi as *"Dano ma kony mere peke i wi lobo kany"* (which means a useless person in this world). The Southern Luo of Kenya referred to this kind of a person as *"Ng'ane to ok dhano"*, which means –"so and so is not a person". Oriare Nyarwath (2012) observes, "To live a meaningful life is to have children and to care for them. This is considered a sacred duty of everybody, but having children calls for responsibilities that go beyond physical death (P. 93).

And when we talk of mutual responsibility in a society, this in many cases related to immortality of the father or mother through the children. As a matter of fact, many a time we do not realise the continuity of the society that is made possible through reproduction and mutual care for members of the society. I do not want to over-emphasize that the Luo people value love so much that they expressed this love in different ways of conduct in life. For example, one general rule of conduct found across Luo community is, *"Don't despite any person because that person is a stranger or from different ethnical or racial groups. Love your kin as you love yourself; and care for his property and life, as you would care for your own"*.

Significant of Death

Death is something feared by people globally regardless of our race and religious background. Archaeological evidence shows that death has influenced cultures, behaviour, beliefs and attitudes of the people since pre-historic times. Given the fact that many people on earth love the world, and they do not know what would be after death, they do not want to die (Samson O. Gunga, 2009). Indeed, in our contemporary world we have medical doctor specialists yet they cannot stop death nor prolong life of a person, which would be against the will of the Living God if they could. Ecclesiastes 3:1-2, which says "For everything there is a season, and a time for every matter under heaven: a time to be born and a time to die.....".

Furthermore, when we look back in our pre-life, which is before birth, we can see with the eye of our mind that there was no fear of death, but in as soon as we are born in this world life creates a special bond –that is fear of death and responsibility thereto. In Luo society, brothers are only considered usual if they are able to provide care for the widows. Death does more harm to than good to us, it takes away our beloved ones at the time we need them most. It takes them to unknown destination by many.

The feelings of loss by the people left behind are usually express in grief. They find it difficult to control the impact caused by death on their lives. Although we do not want to die, it is irrational to believe that someone may live forever on this earth. We are visitors to this world. The reality is that every intimacy with, or dependency on, another person will be terminated by death of one of the partners; many people hardly prepare for death. Instead they have hope of that nothing of the sort should come soon.

Derriba (2005) argues that the basis of all friendship is mourning. The fundamental law of friendship and intimacy is that one must die before the other –that there is no friendship without this inevitability. Consequently, marriage is a temporal institution –it ends at the death of one of the spouse; and this is a certainty that must be endured.

However, by the end of the day, death serves as the true expression of love, because it acts as an intensified and ennobler of love lost, hence, when we come to the point of no longer loving lives, some people start to wish for death so that they go to rest in eternal peace.

As we have mentioned above that many people are not usually prepared for death of their spouses, parents, relatives, and children and so on. This statement brings us to another important question, "What does the preparation for the inevitability of a partner's departure consists in?" Different people may give different answers to this question, depending on where they are coming from, but in my opinion the preparation for a departure of our beloved ones should not has big impact in our lives if we know about *life after death*. Secondly it gives us an opportunity to reflect on the good works the deceased has done when he/she was alive and more importantly, the good relationship that existed between the deceased and survivors. With these in mind, we should not be very much concern about death, because this will affect our coping mechanisms. All-in-all, at the bottom of this come the necessity for inheritance of a widow. In case of the death of the husband the brother-in-law of a widow must take responsibility and to shape the life of a widow and her children.

As soon as a spouse (husband) dies, people launch investigation to check if the couple were properly married. If they found that they were not properly married, the parents of the widow would then demand for the remaining bride-prices to be paid before or soon after inheritance. If the inheritor has money or animals he could pay the remaining bride-wealth, but if he has no money or animals at that time he could still continue with the inheritance and ask the parents of the widow to wait till when the elder girl in a family is married and the money or animals will be paid to them. This request, in Luo custom, is often accepted. In Northern Luo culture bride-wealth serves various purposes. Samson O. Gunga (2009) informs us that the various purpose of bride-wealth among the Luo includes: (1) It gives husband power to congratulate his wife for her good behaviour during courtship (2) It enables the husband to appreciate his wife's dignity and worth and (3) It gives the wife the assurance of her husband's continuous recognition and respect (PP. 172-173).

No marriage is recognized by both families if the man never paid the bride-wealth. In the event of that a husband dies before he pays bride-prices of his wife, his brother or uncles must pay the unpaid bride-wealth, form otherwise the woman might be withdrawn from the deceased husband's home. And once the unpaid bride-wealth is paid the widow is now has a position to find a close relative, usually a brother of the deceased, who will perform various duties for the widow in her new life. However, if a wife dies before the husband, the bridegroom would be asked to the remaining bride-wealth before the husband is allowed to bury his dead wife as his wife.

But once it is approved that marriage was properly consummated, the widow then requires permanent acceptance in the home; and if it is the husband who survives her, he is then now allowed to undertake the widowhood rights. The Luo social structure shows a gender division of roles and responsibilities. Certain duties such as building a house and cultivation of crops must be undertaken by a man, while duties such as cooking, cleaning, weeding and washing clothes for the family must be undertaken by a woman. Among the Northern Luo of South Sudan, gendered roles do not change in widowhood.

However, in case of a woman, if the home of inheritor is far away from the home of the deceased husband, he (inheritor) has the right to relocate the residence of the widow —she may move with her children in the village where the surrogate husband lives but a widow would live in her own house. The man can build a house for the widow in the same village he lives. This is usually done to enable an inheritor to carry out his duties and responsibilities effectively and on time. Lopata (1979) observes that in order for widows to be integrated into the support system of their communities, they need a friendly culture which specifies dignified ways in which the community expects them to behave and how they should be treated by their kin and those of their deceased husbands.

Restoration of widow's new life and her children begins with the moving into the deceased house soon after the widow has removed the black dress from her body.

Mourning in the Northern Luo Culture

As in many African communities, death in Northern Luo is seen as a great and unavoidable tragedy; death does not care whether a person is young, middle age or old. Saturnino

Onyala (2014) says the people do not see death as something natural; as a result they always try to find the cause of death (PP 533-535).

However, among the Northern Luo of South Sudan, people feel the pains of death more when a young person dies than when an old person dies.

Widowhood is a natural occurrence associated with grief; interestingly, in Northern Luo societies, widowhood is characterized by rituals, loneliness, loss of status, fear of the future and inheritance.

Removing the black clothes from the survivor's body, which usually takes place in a year following the burial of the deceased, marked the end of mourning and beginning of a new life with a surrogate spouse. This is known as the last funeral rite. It is been said that the ceremony of inheritance (*Lako/Laku*) takes place in Luo societies only once in a lifetime. After a widow has been inherited, she may has subsequent surrogate husbands, when this happened, traditionally she cannot go through ritual of wearing black clothes as we have discussed above (Oriare, Nyarwath, 2012).

After mourning ceremony, there are no more activities to do with the mourning since the brother-in-law of a widow would move into the house as inheritor (*Kalako/Lalaku*) and assumes full responsibility for the widow and her children.

Widow Inheritance and Levirate

The practice of inheritance of widow, around the globe in general, and among the Northen Luo of South Sudan in particular, is not something new. The practice of inheritance of widows have been practiced before the birth of Jesus Christ and after his death, resurrection and ascension into heaven. Gospel Translation Bible, English Standard Version, Deuteronomy 25:5 tells us, "If brothers dwell together, and one of them dies and has no son, the wife of the dead man shall not be married outside the family to a stranger. Her husband's brother shall go into her and take her as his wife and perform the duty of a husband's brother to her.

The patrilineal institution of care is community known by Luo people as "Lako" or "Laku" and is often referred to as inheritance of a widow. The brother who takes up the roles of the deceased is referred to as *"Jalako or Lalako"*. The person who become *"Jalako or Lalako"* must be a brother-in-law (*yuero*) of the widow. Northern Luo Cultures accept that a *"lalako or Jalako"* could be either the brother of the deceased or a paternal first cousin or a clan cousin.

Some writers like Nzomo (1994a) criticized widow inheritance, he says human being should not and cannot be inherited; only property and other objects can be inherited. He stressed that widow inheritance is wrong because a human being cannot be inherited, therefore, if inheritance is allowed, this would paves way for men to dominate women and exploit them. A woman has individual autonomy, she should be accorded condussive environment and resources to exercise and enjoy that autonomy. We are now in the twenty first century where there is great recognition and respect for autonomy and individual rights. If a widow decides to remarry or not to remarry it should be gracefully permitted. But if she decides to marry it would be appropriate to get a guardian from the deceased husband to avoid the dislocation of children. Nevertheless, JSTOR (2011) argues that the inheritage of the wife is the means by which Luo people attempt to deal with the socio-economic issues caused by death –filling the gaps created by the death of the husband. The practice of *"Lako or Leyo dako"* is so central to the Luo culture that is because a taboo and difficult to violate it. Once it is violated it can bring into the family serious misfortunes (P. 95).

Kirwen (1987) puts it across through his Luo mouthpiece: *"It is through our marriage rites that the adult world is entered, controlled, shaped and lived. It is the central institution that sets the tone; for our whole society without marriage, we would be a people without roots, without stability, and without responsibility. We would be little better than animals that breed promiscuously. And this is why marriage is the concern of more than just the bride and the groom –it is the concern of the lineal families of both partners............You see, when people marry, there is the unspoken agreement that this relationship will survive his untimely death. That is, the man's family pledges to carry on the marriage by caring for the wife and children in the event of the husband's death. This is very sacred and essential part of our marriage. This is why families of both the bride and the groom are very concerned that the partners are properly selected and carefully scrutinized prior to any agreement for marriage. Indeed, it is through the marriage that one becomes an adult that one takes responsibility before the community for the passing on the gift of life, the greatest gift that one has received from the great Creator (Juok). Life is not as personal possession that one can manipulate for one's own purposes. No, life is a shared reality received from the ancestors to be passed on to next generation"* (P. 59-60).

From the above quotation it should be clear that, according to Luo Culture, marriage is intended to be an everlasting contract whose purpose and function extend beyond the physical death of one or both of the spouses. As we have discussed in Chapter 22 of this book, for the Luo people, a person physical dead is still considered alive, present and cable of influencing the living. Therefore, if a man refuses to care for a brother's wife, it shows that he doesn't love his brother. And such a breach of trust could cause the deceased to put the evil eye on him and bewitch him because the brother has treated his wife (the widow) and family unjustly.

Interestingly, the act of inheriting (*Lako/Laku*) among the Luo does not necessarily entail sexual intercourse. For example, in case a man is to inherit an old-widow who is childless or with children and has no longer has the desire or capacity for sexual intercourse but she still requires the husband's support in her home; in this case, she would get a inheritor (*Jalako/Lalaku*) to take care of her and perform the required roles except sexual intercourse. However, should the old widow has interest in sexual intercourse the inheritor (*Lalako/ Jalako*) could spends nights in her house to satisfy her needs.

According to writers like Cunnison (1959) and Kirwen (1979) the word Levirate is a term derived from Latin word "Levir", which means a "husband's brother" or "brother-in-law of a widow". Thus, in this context by caring it means the brother of the deceased husband enters into union with the widow to substitute the caring roles and responsibilities the deceased husband used to performed (Cunnison, 1959, P. 96-97; Kirwen, 1979, P. 3).

As we examine very closely the meaning of the word "Levirate", we find that this understanding fit with the concept of "Lako/Laku" in Luo groups.

When a man dies with or without children, his brother (or nephew) may take over his widow and raise up children to the family. But some Luo groups said when a man dies, the brother can take his widow and raise children to the deceased husband, who are recognized as his legal children and who come from him. However, in Luo Culture, the widow directly remains the wife of the deceased husband, while indirectly she remains the wife of the inheritor. Luke (2001) states that the inheritor serves widow's sole legitimate sexual partners; while Samson O. Gung (2009) emphasized that the inheritor functions as a husband in all aspect including acting as the father figure during the marriage of the widow's children (P. 171).

According to the Anyuak culture, when a man is seriously sick and he knows that he is going to die, before his death, he calls his brother and the wife into the hut to give his "will". And if the sick man has a son with this woman, the son is also expected to attend the giving of the "will". The sick man then tells the brother and his wife how many cows he has, to whom he owes debt and how much, when he dies, his brother should take care of the wife and the son (or children), which means he (bother) is to inherit his wife. In according to traditional law, if the woman has a son, the woman and the son will take over the cattle and pay any outstanding debt. The man would inherit the woman to just sexually satisfy her needs; however, if the sick man has only a daughter, the inheritor and the widow to be will take over the cattle and pay any outstanding debts.

Among the Luo, only those culturally defined as *brother* of the deceased man can be surrogate husband to his wife (Ojwang, 2005). Generally, Luo widows are not permitted to formally remarry or to take other sexual partners in addition to the surrogate husband (Potash, 1986a). Inheritance of a wife is limited to widows of own father, own brother or step brother—said Banydhuro Oyay.

The word "*Lako*" is derived from the proverb "*Lago*", which means to care or to provide for. As an Institution, *Lako* or *Laku* means assumption of the roles of a deceased husband, hence caring for the family of the deceased: wife/wives, children and property. Whoever assumes the roles is referred to as "*Jalako* or *Lalako*"

Moving into the house of a widow by inheritor and sharing bed with widow symbolized the sacred mission of passing on the gift of life. If the couple are fertile, after few months they would have a child. The peace, harmony and prosperity of new family life depend on the relationships between a woman and a man. Thus, in case of an old widow, the guardian would still be required to create a symbolic minimum condition for partaking in creative act of life and to perform the roles that her deceased husband would otherwise have performed (Oriare Kirwen, 2012).

Evans-Pritchard (1951O observed that inheritance of a widow usually change the social identity of the widow and her children; because the children in the new marriage belongs to the new husband and not to the deceased (P.112). Therefore, it is a misfortune and a mistake not to marry or remarry.

The main reason why a woman must never remarry after death of her husband is that the death of a spouse does not dissolve marriage according to Luo traditional culture. In a patrilineal society, an already married woman cannot be remarried unless the former marriage is nullified. Thus, marriage, for the Luo, is only dissolved through divorce, and divorce is not a matter solely for the spouses, but for their lineages as well (Wilson, 1961, P. 122). Therefore, despite the death of her husband, the woman remains a wife of the grave (*Ci liel* or *Chi liel*) and to the clan into which she is married. Since she remains a wife, she cannot be married again but can be inherited by her brother-in-law.

The principle of love requires that the family of a deceased man must be properly taken care of. If a man dies and leaves behind a wife and children, then the brothers of the deceased must ensure that they are well taken care of. The most important thing to the institution of inheritance (*Lako/Laku*) is not for the sexual intercourse with the widow (*Chi liel*) but to take good care of the widow and children and property (wealth) so that the family of the deceased prospers and the children do not suffer (Oriare Kirwen, 2012, P. 98)

Challenges of Inheritance of a Widow

In the institution of Inheritance (*lako/Laku*) there has aroused Sharpe debates among African anthropologists about African cultures, and in particular among the African scholars who are currently in Developed countries. These debates come around in the period when the HIV/Aids have become the daily story of sexual life of the Africans. Many scholars alleged that the institution of inheritance violates a wife's rights of freedom of choice, that is the choice of whether to marry or not. When the husband dies, without understanding that according to Luo or African cultures, choosing to remain unmarried is considered against the law (order) of the Living God. Who says in Genesis 1: 28, God blessed them and said to them, "Be fruitful and increase in number; fill the earth and subdue it....."

From the prospective of the above scholars, they think that inheritance of a widow would spread HIV/Aids among the communities. Truly speaking, not every Luo or African living under the sun has HIV/Aids; this is a wrong hypothesis. On the other side, to say that inheritance of a widow denies her from her rights to choice to remarry or not; at this point we should note that in the Luo cultural universe, the encouragement phrase "to choose to marry, remarry or not" is in itself, does not arise in the African societies. Otherwise a woman choosing not or to marry/remarry is seen as self-annihilation and abdication of the sacred duty of partaking in reproduction of human being.

"Man is not born free. He cannot be free. He is incapable of being free. For only by being in chains can he be and remain human. What constitutes these claims? Man has a bundle of duties which are expected from him by society, as well as a bundle of rights and privileges that the society owes him. In African belief, even death does not free him. If he had been an important member of society while he lived, his ghost continues to be reserved and fed; and he, in turn, is expected to guide and protect the living. No human being today and in the past has ever had absolute liberty. Not even autocratic Kings and Presidents ever have absolute freedom; Absolute freedom is only in the Living God" (p'Bitek, 1986, P. 19)

Widowhood Empowerment programme

Empowerment is the means through which people reduce their helplessness and alienation in order to gain greater control over all aspects of their lives and social environment (Simond, 1990). It involves freedom in tackling of something unacceptable in human lives that come along personal pathway. These could be psychic, physical, cultural, sexual, legal, political, economic and even spiritually.

Subsequently, solutions to these inhibitions require a multi-dimensional approach that must improve psycho-social, cultural, educational, emotional and spiritual development. Widow needs assistance to recognize and understand the potential disadvantage they may be facing and what steps to take. Social education is very important in this situation because it enable both men and women to collectively and eventually take full control over their lives.

Although the family should be encouraged to play a more central role in assisting widows in particular, alternative support systems should be developed to help the widows in case the family is not cooperative. It is personal and collective empowerment and transformative approaches that can help widows to take control of their own lives and destinies. Transformative approaches involving people of influence, who understand the situation, village elders, individual women and groups are necessary in the struggle for the elimination of negative widowhood practices in the Northern Luo cultural systems.

However, it is important to note that empowerment process is complex; it requires a holistic approach to dealing with personality and the social environment. It entails providing conducive environment for the bereaved to gain her role in the restoration of normality after losing her husband.

Interestingly, it is becoming more clear and clear that the Northern Luo widows with secondary school education and University degree tend to ignore the cultural requirement of inheritors. On the other hand, widows in rural areas with no education or with primary school education, who have no economic empowerment, more readily submit to such cultural demands. Therefore, we can say that intervention through education has helped the Northern Luo widows and widowers to refrain from life threatening widowhood rituals.

Another empowerment approach is legal literacy, which aims at creating awareness through human rights education and among different stakeholders in the society. Apart from education and economic empowerment, there is need to improve legal structures so that widows can inherit the property of the deceased directly, instead of owing it through their sons, as in the case in the Northern Luo community today. This will prevent widow from risking their health for the sake of property. Legal literacy may be used to enhance the rights of widows in the society by engaging the participation of stakeholders who are received to perpetuate such rights. These would include heads of families, village elders (men and women), widows and widowers among others. To avoid misinterpretation of legal literacy, those who popularize the change to embrace new ideas must be people who are respected by the community members. Such persons will provide assurance that the need for change is timely, and that it is for the benefit of the community itself (Samson O. Gunga, 2009).

Chapter 22:
Deaths and Burial Ceremonies among the Luo of South Sudan

Humans' particular experiences of sickness and death depend largely upon their respective cultural and religious beliefs. They also depend greatly upon the social practices of their community, which define expectations for how members of society are to respond when people are sick, dying, and dead. This grounded theory ethnography drew data from sixteen home-based interviews and examined the experiences of sickness and death as recounted by members of the Luo tribe from a rural community on the Nyakach Plateau in western Kenya. The study identified and described Luo responses to sickness and death and demonstrated that members of this community tend to face sickness and death not as individuals, but as a community. The Luo communal response to suffering is rooted in the Luo practice of gathering to tell stories. The content of shared stories, which are often spiritual at the core, are important in providing a common basis from which tribes people communally identify and respond to each other's health needs. Furthermore, the physical act of coming together for storytelling prepares the community to likewise come together in material, spiritual, and emotional solidarity in order to confront sickness, death, and other hardships as a united body of people.

Source: - Department University Scholars -http://hdl.handle.net/2104/9777

Background

The Northern Luo people of South Sudan perform a series of rituals and many feasts for the dead because they strongly respect the dead. Hey believed that life is sacred and that every person has a right to proper treatment and care in life and death. Therefore, all the Luo people believe in treating the ill, the injured, the dying and the dead with care. As you continue reading through this book, you will discover that Luo people do not conduct their rituals in the same ways other ethnic groups of South Sudan conduct theirs.

The Northern Luo generally believed that human beings are expected to live and enjoy a normal life until death in old age; in addition they believed that early death is not a natural occurrence. They believe that early death is caused by sorcery or evil spirits.

I cannot overemphasized that the Luo in South Sudan and those in Diaspora (regardless wherever they are) are known as a people who are seriously concerned with their burial place as oppose to other ethnic groups of South Sudan. Today, decisions are taken in cities, towns (and developed countries) when a Luo Adult dies he or she must be buried in his or her homeland. This action makes the non-Luo to recognise how deeply Luo people are preoccupied with their burial place.

Having said this, we can conclude that the Luo attitude towards their burial place evidently shows how they fear and respect the deceased, regardless of ongoing modernization and urbanization in the South Sudan.

Being a Luo man, I have witnessed a number of death cases and have attended whole series of these rituals. I hope this will help me share the insight of my knowledge of how the Luo people conduct rituals for the dead.

Traditionally, Luo people perform a total of 5-6 rituals for one deceased as we shall discussed later on. Rituals are performed when adult or a baby/child died; however, with the changing world certain numbers of rituals are omitted depending upon ages, sex and marital status of the deceased. Interestingly, rituals include death announcement, vigil (*guro tipu pa latoo*) grave digging, burial, shaving heads, funeral party, departure of relatives and others to their home and last funeral ceremony. We shall discuss these topics one by one.

The Luo's Concept of Death

Although death is seen as a dreaded event, yet the Northern Luo perceived it as the beginning of a person's deeper relationship with all of creation; it is the completing of life and the beginning of the continuous communication between the visible and the invisible worlds. To them (Luo people), the goal of the life is to become an ancestor after death —which is a prestige. This explains why every person who dies must be given a correct and respectful funeral, supported by religious ceremonies if the deceased is a Christian or parents are Christians.

It is believed that if this is not done, the dead person would be angry and may become a wandering ghost in the village, unable to live good life after death, and therefore, a danger to those who remain alive in this world. For this reason, some people say that proper death rites are more a guarantee of protection for the living than to ensure a safe passage for the dying. It is believed that the dead have power over the living.

Antonius C.G. Robben (2004) observes that in African cultures and beliefs death does not occur at one moment in time but is a drawn out process. When a person dies he/she is considered part of society, as he/she continues to live among people, this obliges them to provide food, engage in conversation, and show respect as if though the deceased is still alive. In other words the deceased's spirit or soul does not depart for the land of the dead but wanders in the vicinity of the corpse and frequents the places where the deceased used to dwell. As the bereaved relatives fear the soul's wrath for past wrongdoing, they appease the soul through sacrifices, taboos and mourning. These circumstances make the mourners stand apart from society. They cannot participate in its daily routines, and wear distinctive clothing and ornaments. The second and the final period of death begins when the body has disintegrated sufficiently, the soul has detached from the deceased, and the mourners have properly expressed their grief and carried out their social obligation (P.9).

As we discuss more and more about the death rituals, we shall discover that the burial rituals of the Northern Luo of South Sudan differ significantly from the other 63 tribes of South Sudan. Their main aim is to ensure the safe passage of the spirit into the afterlife, and to prevent the spirit from returning and causing mischief.

According to Robben (2004) the Luo people tend to talk about a "good death", a "bad death" and a "tragic death". Accordi9ng to the Luo people the "good death" is fast, without suffering, and is accepted most readily by the relatives. The "bad death" involves a long, painful dying process exacting a heavy toll on both the sufferer and the family members. The "tragic death" takes place suddenly and strikes healthy human beings, often in the mid of their lives. These types of deaths are hard on the surviving relatives. Suicide is a tragic death that stands at the extreme, opposite of natural death because it involves a violation of cultural I n his own demise (P. 5).

However, most anthropological studies on death are concerned with natural death, although scholars like Durkheim (1966) developed interest in studying suicide and James Frazer (1976) developed interest in studying murder and sorcery but no study has ever been conducted on death caused by animals or water. In general term suicide, murder or sorcery are known as uncommon death.

People have always died at certain points in history and the living have always mourned the death of loved ones with some type of ceremony. It is reported that whether it be in the past or present the Luo families, relatives and friends care for the body of the dead person; as such burial usually follow a traditional ways.

All Luo people practice the custom of removing a dead body through the main door and not through the back door if there is any; many houses of the Luo people consist of one door. However, with the modern development some of the Luo people have a house with more than one door. The reason for removing the dead body through the main door is to make it easy for the dead person to find his/her way back to the family whenever he/she wants to return to the house for a visit. Many people want to make sure that the dead are easily able to return to their homes, this is why dead person is buried next to their homes.

All the Northern Luo believed that death is a separation of a soul/spirit from the body.

Now a day, as many Luo people become Christians, they came to know and understand that the separation of body and soul is not much different from a Western dualism that separates physical from spiritual. However, the Luo believed that when a person died, there is not only some part of that person that lives on –rather, it is the whole person who continues to live in the spirit world, receiving a new body identical to the earthly body, but with enhanced powers to move about as an ancestor. Robben (2004) argues that in African communities burial is a collective affair through which the deceased joins the ancestors and the community bids him/her farewell.

According to the Luo of the ancient time, the soul of the dead person goes to live either under the earth, in groves, in the homes of the earthly families or on the side of a river. However, the modern Luo believe that the soul of a dead person goes either straight to heaven, where God lives and it stays in some unknown place waiting for the resurrection when Jesus Christ returns. This group of Luo people believe this because they are been influenced by bible teaching. Robben (2004) observes that in many cultures, life does not end with biological death because of the belief in an eternal life "spirit" (P.11). In contrast, an anthropologist and writer like Mbiti (1969) argue that the African in general believed in the continuation of life after death does not constitute a hope for a future and better life. He went on to say that the African believed that to live in this world is the most impor-tant concern of their religious activities and beliefs.......Even life after death is conceived in materialistic and physical terms. As such, there is neither paradise to be hoped for nor hell to be feared in hereafter (PP. 4-5).

The Luo sons and daughters, who now live in the Western World such as Australia, United States of America, Canada, Europe and Asia often make a choice to take the dead person i n his/her country of origin. I believe beyond doubt that many readers of this book have witnessed these events, regardless where they live. It is very costly to transport the dead body to Africa, as well as refrigerated the body, yet the people are committed to contributing financially from the little they have in their accounts. However, some of the Luo people who live in the Western World do also choice to bury their beloved ones in the host country as they believed that soil is the same whether is in the developed or undeveloped countries.

More importantly, Luo funerals are community affairs in which the whole community feels the grief of the bereaved and shares in it. The purpose of the activities preceding the funeral is to comfort, encourage, support and heal those who are hurting. This transition during the mourning period is sometimes accompanied by cleansing rituals by Church leaders in which the bereaved are assured of their acceptance and protection by God. Robben (2004) cited Emile Durkheim (1995) as saying the individual grief experienced at death of another human being is expressed collected in culturally prescribed ways of mour-ning; crying relates in the same way to grief as weeping and wailing relate to mourning.

Mourning is not a spontaneous emotion but a collective obligation manifested in appeasement rites (PP.7-8). In weeping or crying for the deceased, no individual Luo would be demanded to cry or weep, it is a personal expressions of sorrow. These cultural practices are in the blood of any Luo man or woman; they automatically express their sorrow by weeping or crying.

In Northern Luo of South Sudan, the dead are never burned but buried in the earth. Cremation, as an attractive alternative in the Western World is not acceptable by the Luo people who live in developed countries. Cremation is practiced in Australia, Canada, United State of America and Europe. Thus, cremated remains could be buried in a cemetery, placed in a columbarium niche, scattered at sea or on land, kept by the family (Jay D. Schvaneveldt, 1989).

Death in Acholi

Death is a matter of importance in Acholi, which explains why the Acholi highly respect the dead. The death of persons between one year and 60 years is more painful, than the death of persons who are 61 to 100 years. One day when I was still in Juba, I asked three Acholi elders who were seated under a mango tree, "If people respect death, then why do some people quarrel or fight during funeral rites?" The elders responded that, there are many reasons that explain the cause of quarrels and fights and they trying to give reasons. But from all their talks I realized that there are only two main reasons why people always quarrel or fight among themselves during funeral rites. First, people do not understand that Satan is the cause of death on earth, much as the Acholi always say during burials, "It is God who gives and takes away." Secondly, the Acholi think death befalls a person because of actions of a fellow man. Hence, even if one dies during hunting, the Acholi still say he was killed intentionally. According to the history of Acholi, death takes two forms. The first is referred to as *skin death* (natural death). The second one is known as *hand death or wilderness death*. To better understand the two forms of death, I want us to look at each of them separately. We shall start by looking at *hand death or death in the wilderness*, and later handle natural *death* (Saturnino Onyala, 2014, PP. 622-623)

Acholi Rituals for the Dead

The rituals that mark a traditional to a new phase in life, is commonly known as rites of passage, and are important in Acholi culture. Although these traditional have been passed down from general to generation, they are part of the history of the community. Some of the Acholi people in the modern times tend to down play or give up the practice. This is due to assimilation policies of the Western cultures; as such they think some rites of passage are no longer important and safe in today's cultures. It should be understood that proper mourning guarantees the spirit a peaceful resting place. If anyone thinks that this is all folly, I would ask him/her to pay a visit in the Northern Luo Communities, when they are making burial ceremonies and bereaving. It is said that the Luo people who now live in the Western World such as Australia, America, Canada and Europe, tend to or have forgotten to do burial in their respective places according to living traditions. Although the mourners accompany the dead body as it leaves from the Church or from the mortuary heading to the cemetery.

(1) *Death Announcement:* - Acholi conceive of death as an inevitable, personal defeat and tragedy, against which there is no ultimate defence. The personal and group loss resulting from death is acknowledged as real and permanent.

(2) In many cases in Acholi, the members of community come to know of death when they hear the women's long, quivering wail; sometimes follow by the sound of drums. In Acholi land, the death is always announced at dawn or late evening as many people die at this time of the day. One can rarely hear about death announcement in broad day-light. In Acholi society children are not allowed to get near the dead body is, in addition they are not even allowed to attend the funeral services.

(3) The wife of the deceased is meant to sit next to her dead husband and sleep by him until the dead body is buried.

(4) As opposed to the Luo of Kenya, the Acholi, like the other Northern Luo groups, when adult or baby dies their death is announced immediately at the time the man or baby passes away. In contrast, the death announcement among the Luo of Kenya varies according to the age, sex and occupation. Thus, if a baby dies in the morning, its death is announced immediately and its baby is buried the following morning. However, if a man or a woman dies his/her death must be announced after the sunset. As women are tools of communication, they will have to wait for the right time to start wailing (Shiino W. 1997).

(5) *Grave Digging (Kwinyo-bur):*- Tradition ally, a grave is dug as soon as a person has died, following with a small and brief ceremony prior to burial. All procedures are conducted with care, to attempt to ensure that the spirit of the departed does not become angry. In Acholi land, grave is prepared collectively, young and middle-age male, relatives, in-laws and several neighbours join forces in digging.

(6) The history of Acholi tells us that prior to the digging of the grave, an elder first measure the length of the dead, so that the body fits in the grave. Before the boys start digging the grave, the bereaved family makes incantations using spears, and the stick used to measure the grave and spear or knife. A member of the bereaved family who made the first incantation picked the stick and measures the length of the grave and made a mark on the ground three times (if the dead was a man) or four times (if the dead was a woman), while saying, "There! If the death which befell this person came

from our family, it should meet a hole." He or she throws down the sick. And another maternal family member takes the stick and marks the ground three times (man) or four times (woman) while praying, "There! If this death came from our family, it should meet a root." He throws the stick down Saturnino Onyala, 2014, P.637).

Shortly, the men would start digging the grave that take between 1-2 hours, depending on the type of soil. Digging usually begins around 6.30 a.m. and complete about 8.30 a.m. And if the burial is in the evening, the digging usually starts around 3.00 p.m. and complete at about 5.00 p.m. of the burial day. However, digging of the grave for a child takes less hours, but it also depends on the type of soil. Where the soil is rocky it may takes more hours and where the soil is not rocky it may takes lass hours. The Acholi do not bury the dead in a shallow grave. The grave is sunk about one and a half to two meters as the dead is buried by the doorway, the grave is dug deep so that the stench does not come out.

If the cause of death originated from the paternal relatives, they will find a hole in the grave as they asked in their prayer. But if the cause of death came from the maternal relatives, then they find a tree root as they had prayed for. Sometimes, people find both roots and a hole in the grave. This means the cause of dead is from both the maternal and paternal relatives. Most times, people find roots or a hole only when digging a grave. But sometimes they do not find either roots or a hole. If this happens, it proves to the clan elders that the death was caused by outsiders. Although the root and hole are not found, the clan members of the deceased still investigate the cause of death by setting a trap or scattering sorghum for the chicken.

Clan elders are watching a cock or hen that would be the first to eat the sorghum (Figure: 185)

If a hen is the first to start eating the sorghum, it proves that the death was caused by a woman. But if a red rooster starts eating the sorghum first, it means a man caused the death.

Pouring sorghum or sesame to chickens: On the second performance of *teka*, the maternal relatives get sorghum or sesame, and pray, "The people of my ancestors! If this death was really caused by us, then a hen must be the first to eat the sesame or sorghum poured out here. But if it was caused by paternal relatives, we pray that a cock should be the first to eat the sorghum and sesame poured down here." A paternal relative would then pour the sesame or sorghum on the compound, before resuming their seat. The elders take their seats in a shelter, or by a granary, watching whether a cock or hen would be the first to eat the sorghum or sesame.

Acholi history states that if the death was caused by maternal relatives, a hen would come first to eat the sorghum poured on the compound. Conversely, if the death was caused by paternal relatives, a cock would come first to eat the sesame or sorghum. If a hen ate the sesame or sorghum first, the clan members would conclude right away, "The maternal relatives were the ones who brought death upon their own person." Yet, if a cock ate the sesame or sorghum first, the clan members would also say, "This death came from the family of the husband." Shortly, the family members would be blamed for the death. As we can see from the above Figure 175, it is a cock that starts eating the sesame or sorghum; this indicates that the death has come from the side of the husband's family (Saturnino Onyala, 2014, P. 645).

(7) *Burial Rites*: - In Acholi land, the dead are buried at the doorway. If the deceased was a man, he would be buried on the right, but if she was a female, the body would be laid to rest on the left side of the doorway. However, those who lived in town are buried in the cemetery.

In Acholi society, the human body is prepared in some fashion before it is finally laid to rest. After the boys and men have finished digging the grave, they inform the elderly men and women seated near the dead body. According to Acholi culture, if the deceased was male, aged between 30 and 100, his body is cleaned by aged men. But if the deceased was a boy, aged between one day and 29 years, elderly women are in charge of cleaning the body. If, however, the deceased was a girl or woman, elderly women would are the ones to clean the body. Men are not supposed to clean the body of a dead woman or girl. Unfortunately, the Acholi elders did not tell me why Acholi culture prohibits men from cleaning the body of a dead woman or girl. After the dead body has been cleaned, the men or women smear the body with Shea nut oil (Saturnino Onyala, 2014).

Prior to burying the dead, the girls ask for their money which is usually paid for taking the dead body out of the house. If the request of the girls is not met (if any), so girls have the rights to stop the burial, while they remain sitting by the entrance.

When the burial fee is paid to the girls, they release the dead body and an elder man instructs three men to enter the grave to receive the dead body. He also chooses three or four other men to lower the body to the grave. Tradition has it that no one is expected to cry when the dead is being placed in the grave. Culturally crying when a dead person is being lowered in the grave is an abomination. It means the person crying expects another person to die. In other words, crying at this time bring bad spirits on the living. After all the rituals have been performed, and the dead has been buried, those who took part in the burial moved aside after washing their hands and tools on the grave; there the complete the washing of their hands. Therefore, this is the time, women, men and young people can now cry (Saturnino Onyala, 2014).

During the burial itself the immediate family members of the deceased stay together on one side of the grave at a designated place. They are forbidden from speaking or talking any vocal part in the burial.

Figure: 186, Acholi men carrying the coffin for burial

The burial for deceased adult or a child both burial rituals normally start at 8.30 a.m. or 5.00 p.m. One elder from the relatives of the deceased person preside over the proceeding of the ritual. As we have seen above the program of the ritual include receiving the dead body, lowering the deceased into the grave, throwing soil on the dead person and burying the body instantly.

According to Acholi culture, the dead person is buried within a day or two from the day the person passed away. The other one day usually given for the relatives, who live far away from where the deceased live, to come and attend the burial. The Acholi people never keep dead bodies for lengthy days.

Clan members pouring soil on the dead (Figure: 187)

Soon after the burial, if a husband is the one who died, a widow immediately wears black dress. And is a wife who dies, there is no customary law for the husband to wear black cloth. Wearing black cloths for men is optional. The black cloth is worn by a woman as a sign of mourning, and removing black dress from the body after a year is a cleansing rite. After 4 days or a week the grave would be tidied up and smoothing.

When people finish burial all those who participated in the burial have to wash their hands. The Acholi believed that the people who have physical contact with a corpse are regarded as unclean; this is the main reason why they have to wash their hands to cleanse them.

It is also believed that if the dead person is buried out of the deceased compound, there at the grave the spirit of a dead person would be hovering on the earth restlessly until it the spirit is brought home: such a situation is extremely dangerous for the family. To do this all people who are in the burial yard have to come home after burial, which is known as "kelo tipu wii latoo paco" (bring the spirit of the dead person home). The spirit is brought home to look after the family as an ancestor. And if the dead person is a grown up man, one or two men would be blowing animal horns, while holding a spear in his/their left hand(s). The woman would be carrying an oiled leaf with an egg and she keeps on calling the name of the dead until they enter into the house where the spirit is meant to rest. The following morning the people who dug the grave would gather to wash their hands again; they also wash the tools that have been used for digging. The water is taken and spear on the grave. The family member believe that the dead person is always going to be with them, where he will be listening to every conversation with answer; and sometimes people even give him his previous favourite food in the place people believe the spirit is living.

(8) *Vigil (Kumo too)*:-someone has died in a house, the women removed all pictures, and mirrors are removed from walls and the bed in the deceased's room. All bereaved person sit on the floor usually on skin of sleeping mats/metres. After the burial, everyone sits under the granary or tree in the compound. Those who performed the burial put the tools used in digging the grave on the veranda (or near the grave, if it was near the door way).

However, if the burial takes place in town, people go to the home of the deceased, to take the spirit of the dead home. And the tools used for digging the grave remain on the veranda until the day of performing the funeral rite.

Acholi custom says, if the deceased is male, people mourned for three days, but if the dead is a female, people mourned for four days. The days are counted starting on the day the dead was buried. Tradition also says on the first day of burial, the clan members do not hold any meeting until the evening of the second day, that is when the maternal and paternal relatives of the deceased would seated around the fire place, to discuss issues relating to the death.

Figure: 188, Acholi women sit during vigil

The Acholi take issue of death quite seriously and in an orderly manner. Before the maternal relatives ask what could have caused the death of nephew or niece, they first investigate the form of death that befell their nephew or niece. As people gathered around a fire place an elder from the side of the deceased, would tell the maternal relatives the cause of death of their sister's man/child. He narrates how the illness started, the number of days or months the sickness lasted, until the death. After the people have listened to the elder and understood what he narrated, the kin of the mother of the deceased then asks for payment for the death. For more information about the death fees, please (read Saturnino Onyala, 20140).

The matters of death are not only discussed by the maternal and paternal relatives alone, but mediators also attended. The history of Acholi states that mediators did not take part in the discussions, but only listen to what has been agreed upon or rejected by either party. They however intervene and give their opinions, when the meeting turns rowdy. But the maternal relatives preside over the entire meeting. If tensions begin to flare among the maternal and paternal relatives, then the people shake a leaf – which means the meeting would be adjourned and resumed the following evening. In some cases, such meetings take two days, especially if the maternal and paternal relatives are in disagreement. If the maternal relatives are dissatisfied with the discussions, they table the matter before all, on the final day of the funeral rite. Before everyone has eaten or started drinking alcohol, they first discuss the matter that is unresolved between the maternal and paternal relatives. If the issues are resolved well, the mourners would then start indulging.

Today, the night vigil sometimes becomes a time for pastoral care, where the Pastor or Priest (or Christians' Choirs) come to comfort and encourage the bereaved family members.

(9) *Shaving hairs and Funeral Ceremony*: - After three or four days, on the day of funeral, the clan members gather to perform a funeral rite. The first thing they do in the morning is to shave the heads of family members of the deceased. Relatives shave their hair because life style is concentrated in the hair, so shaving the hair symbolizes death, and its growing again indicates the strengthening of life. If the deceased is a man, the heads of the widow, children and close relative such as brother and sisters would be shaved. Hair shaving is usually performed by an old woman or middle-age woman.

On this day, the father of the deceased slaughters a billy goat on the grave, to cleanse the hands of those who buried the dead. The goat is slaughtered and skinned by a nephew, and no one else. After finishing his work, the nephew gives the meat to women to cook. The nephew takes the head and neck of the goat and cooks it by himself and eats it all alone, if there are no other nephews around. One of the elders takes the dung of the slaughtered goat and smears it on the hands of those who buried the dead, and sprinkle some on the tools used for digging the grave. He puts some of the dung on the chests and legs of members of the bereaved family.

In Acholi land, there are two types of hair shaving, the first shaving is normally done on the last funeral day; and the second shaving is usually done after a year following the burial of the dead person. If the dead person is a man, shaving and other funeral services are done on the third days following the burial, but if the dead person is a woman, shaving and other funeral services are done in the fourth days. The family members usually shave their hairs using razor blade. It is said that the first shaving ends the morning after burial and marks the beginning of the new life. While the second type of hair shaving is said to be performed after one year following the burial of the dead person, and this ends the bereaving period –which usually last for one year. During the bereaved (in 3 or 4 days) the close relatives stay at home and do not socialize or have sexual intercourse.

However, it is important to note that in the modern Acholi, many people, especially the young ones, do not heed the two types of hair shaving; they rather preferred the hair from their foreheads to be shaved in a symbolic manner.

The shaving rite is the time for big feast and to be attended by great number of relatives and other interested people. This is also known as the funeral dancing day. The size and nature of this occasion depend on the age and status of the deceased.

Figure: 189, Acholi funeral dance

During the time of mourning and funeral ceremony, customarily, the things that belong to the deceased such as eating utensils, cloths, shoes, blankets and so and so are not used. After the funeral all the belongings of the deceased are divided among older women or older men depending of the sex of the deceased; some of the belongings are divided among family members and in extreme cases the belongings of the deceased are destroyed by burning.

(10) *Departure for Homes*: - In the evening when food is already prepared, people gather to eat and drink. If the dead were an old person, a married man, or a married woman, after eating and drinking, people would start dancing funeral dance. The Acholi love funeral dance very much and usually enjoyed it till morning. Soon after this, surviving family members, relatives, in-laws and friends return to their respective home.

There is no any order for departure set about by the customary laws; any one can leave the compound of the deceased any time after this ceremony.

In the morning, while other people leave the deceased place, the maternal and paternal relatives remain behind. They wait till evening, and then the maternal relatives ask for a funeral goat –a hen and cock to wave on the grave. The bereaved family hands over all these items to the maternal relatives, so that the spirit of dead is appeased wherever it would be. The maternal relatives slaughter the funeral billy goat and cook it that very night. The next morning, after the third cock crow, the maternal relatives get up to eat. They eat the whole goat, but leave the head. They hide the head of the billy goat, and after eating, one of the elders waves the cock on everybody on the compound, while silently saying, "God, we want to go back home. We pray that you pour your blessings upon the members of this family. Give the dead a cool heart in his grave."

After this, the maternal relatives do not go back to sit, but pick the head of the billy goat they had kept together with the bones left after eating, and a billy goat, and head home. Acholi culture states that whenever the maternal relatives are going back from a burial, they are not supposed to look back. Secondly, Acholi culture prohibited a man holding the sacrificial cock from stopping to urinate on the way, even if the urge is strong. If by bad luck the man urinates on the way, the chicken he is holding dies. Acholi culture also states that if the sacrificial cock dies, it indicates that the cultural norm has not been observed, and may lead to bad omen on women, children and men from the maternal side.

After the maternal relatives have moved a distance from the burial home, and reached the first stream, they sit by the side of the stream and eat the head of the billy goat here. After eating the head of the billy goat, they leave the skull and bones in the stream. Reports show that the maternal relatives also leave the utensils they used for eating in the stream. This is a rule that applies whether the deceased was an elder or a child (Saturnino Onyala, 2014, PP. 647-648)

In Acholi, the period between the time of death and departure for home is called "kumo", which usually refers to the mourning period, and literally means men and women at the funeral place are sitting without doing anything as some women cook and feed them.

Traditionally, a widow had to remain in mourning for a year after her husband's death and if it is a child that dies, the mother will have to remain in mourning up to two months.

Acholi funerals are community affairs in which the whole community feels the grief of the bereaved and shares in it. The purpose of the activities preceding the funeral is to comfort, encourage, and heal those who are hurting. There after the family members bereaved make the transition back to normal life as a smoothly and as quickly as possible. This transition during the mourning period is sometimes accompanied by cleaning rituals by which the bereaved are assured of their acceptance by community and protection by the spirit of ancestor.

Anyuak Rituals for the Dead

The Anyuak, like the other Northern Luo groups of South Sudan, perform a total of six rituals for one deceased. All rituals are performed only when adults die; certain numbers of rituals are omitted depending upon age, sex, and marital status of the deceased.

(1) *Death announcement*: - The Anyuak, like the Acholi, conceives of death as an inevitable, personal defeat and tragedy, against which there is no ultimate defence. When a person dies, members of the community come to know when they hear women wailing and crying, if the chief who dies, the people know about his death with the sound of drums.

In Anyuak society, when a person dies death announcement is made immediately. The death is always announced at dawn or in the evening as many people die at this time of the day. In Anyuak society members of community rarely hear about death (women wailing) in broad daylight.

(2) *Grave Digging (kuny)*:- In Anyuak society grave is dug collectively by grown up men who are age between 30-55 years of age. Customarily, young men ages 20-29 years are not allowed to dig the grave. The place of grave is defined by culture. What I am not sure of is whether men with pregnant wives are allowed to participate in grave digging or not. More interestingly, when a father or a relative of a young person (ages 20-29) dies, the young person does not attend the funeral. It is reported that friends of the young person sometimes come to chat with him/her in the house of one of his/her friends next to the home where death has taken place.

Digging of grave usually commence in the morning or in the evening depending on the decision of the elders. The grave is sunk about one and half to two meters, the grave is dug deep so that the stench does not come out. Seligman (1948) cited Colonel Bacon as saying, "Men dig grave about 3-4 feet deep and usually within two paces of the door of the deceased's hut" (P.112).

(3) *Burial Rites(Yik)*:- In Anyuak society, like the Acholi society, the dead person is buried by the doorway; burial usually takes place within a few hours of death, as soon as the grave is ready.

The burial of deceased adult or a child both burial rituals normally start at 8.30 a.m. or 5.00 p.m. According to the customary laws, it is stated that one of the elders of the deceased must presides over the proceeding of the ritual. Seligman (1948) observes that sometimes a person can express a wish of where he/she is to be buried when a person dies.

Figure: 190, prepared dead person for burial and men digging grave

As the dead person is brought by the grave (in village), a skin sleeping-mat is laid on the bottom of the grave and the corpse is laid on this at full length, but lying on its side with one hand under the head and the other stretched out along the body. A second skin mat is then put over the body to keep off contact with the earth.

The programme of the ritual, as we have just discussed above, starts with the lowering the deceased into the grave, and this is followed by throwing soil on then dead person and burying the body instantly. The soil is then pounded down. The water is sprinkled on the grave to harden the surface which is finally sanded. Antonius C.G. Robben (2004) argues, "The funeral ceremony has three objectives: to give burial to the remains of the deceased, to ensure the soul peace and access to the land of the dead, and finally to free the living from the obligations of mourning" (P.10).

It is reported that if the deceased is a wealthy person a bull is killed and the fresh skin is put at the bottom of the grave, the meat being eaten by those who have dug the grave and otherwise assisted at the funeral. In rare cases when a man has lost all his relations or children, his body would be covered with branches of thorn bush, and may be exposed on a raised wooden platform far away in the forest (Seligman, 1948, P. 113).

After the burial, the surviving family members and other relatives sit down to eat and drink local beer. The relatives will sleep in the dead person house between 3-7 days, which is followed by removal of ashes known as "keto buru or buru".

More importantly, if the dead person is buried by the doorway, after a month the grave mound is beaten down again to reduce it to ground level and once more the surface is treated with sand. The grave is now inconspicuous in the clean sanded floor that is maintained in all Anyuak homesteads (Seligman, 1948).

Unfortunately, the Anyuak of today, especially those who live in Western World and big cities and towns in South Sudan do not follow the traditional way of burying the dead person. They rather bury in coffin, without skin sleeping mat, and the dead person lies on his/her back with the two hands stretching along the stomach. The body is covered with modern Western burial cloths and with some flowers on the coffin.

(4) *Vigil (budho):*-Soon after the death and before the burial of the dead person family and friends gather at the dead person's house. There, people cry on one side and on the other side, they comfort and support the dead family members.

The closer relatives of the dead person such as spouse, uncles, aunties, parents, step-mother, sisters and brothers stay in the home of the deceased from day one till the departure time as we shall discuss later. In the first or second days, they stay awake throughout nights singing or praying. These one or two days passed before burial is put aside to enable relatives who live in a faraway villages and towns to attend the burial. Men and women continue crying and singing their lamentations and often war song if the dead person is a man or an old man.

As people approach the deceased home, women raise a strange voice before entering the compound to announce their arrival. Men entering the compound play their whistles and singing their own elegies. These visitors or mourners go straight into the house where the dead body is laid without greeting other people; there they cry or pray in their own ways. As the visitors enter the house, those who were in the house leave the room to accommodate the visitors. After a while visitors come out of the house, greet other people and join them in conversations.

During the days of vigil, no food would be cooked in deceased person's home; however, relatives and friends usually bring food from their houses to the people who are mourning. This is when they say to the neighbours, "Acakarthou", which literarily means, "I am taking food to the funeral"). During vigil period people can eat food which are brought from outside but the mourners are not allowed to drink local beer and dance- dancing is not permitted at this stage. Today, Christian songs are also sung during vigil; however, most of the vigil visitors follow traditional ways of expressing condolence.

Outside the house, some people sit on stools, logs and other sleep on the mat or sleeping skin. As the sun set, men make two fire places this is known as "wang mac" or "magenga". The fire places are usually made near the house, one fire for women and the others for men to warm themselves, because at night it becomes very cold. The vigil usually continues up to some days after burial.

From the day following the death, the surviving family members and relatives of the deceased busy themselves to prepare for the burial services, building the shade, and preparing for the coffin and cloth. During this period people leave their usual duties and come to attend the funeral preceding rituals.

(5) *Shaving Hair (liedo) and Funeral Ceremony*: - After 4 days for male and 3 days for female, following the burial, close relatives of the deceased such as spouse, parents, sons and daughter (if any) shave their hair using razor blade.

The Anyuak society, like the Acholi society, has two types of shaving; the first one is done soon after death to end mourning and to mark the beginning of their new life. The second shaving is normally done after 30 or 40 days, following the funeral. This is done to commemorate the passing of loved one, and sometimes accompanied with Christian prayer or Mass –this is usually known as the first anniversary of the dead person, which is called in Anyuak language *"gonyo cholla"*. If the funeral is in the village, after the prayer people eat drink and dance. Robben (2004) observes that among the Luo groups the bereaved relatives join for a meal, sing songs and celebrate the final passage of the deceased. This eventually b rings mourning to an end, where the social order is restored and people pick up their everyday life (P. 10).

However, as we have seen with the Acholi of South Sudan, modern Anyuak also do not want their hair shaved. As there is no reason given for the refusal to shave, I then call it *"change in culture that will lead to the loss of Anyuak traditional cultures"*. It is important to note that the first anniversary for a dead person, if a man, is done to free widow from mourning taboos, and allows her to choose a man who would inherit her after *"gonyo-cholla"*.

(6) *Departure for Home:* - In Anyuak society, soon after the last day of mourning, surviving family relatives, friends and in-laws may return to their respective homes. However, the closer relatives such as uncles remain behind and continue mourning for another 7 days or even more and depart to their respective homes.

According to Henry Upio, Anyuak elder who lives in Melbourne-Australia, stated that in the last day of the funeral, tomb is smeared, two bulls are killed one bull would be skinned and the meats are distributed to the people who are not related to the deceased. And the other bull would be killed, skinned, the meat cooked and eaten by the people in the funeral. Women usually contribute local beer; after food, people eat, drink local beer and dance from evening till morning. In the morning, people who have come from far away depart for their homes.

In Anyuak community, like in any other Northern Luo groups, there is no any order for departure set about the customary laws regarding funerals; any one can leave the compound of the deceased whenever he/she wishes. The period between the death and the departure for home, in Acholi is called *"kumo"* and in Anyuak they call this period "Ksar-kim" or Kar-too", which usually refers to the mourning period, and literarily means men sitting without doing anything, while women cook to feed them.

Balanda Rituals for the Dead

The balanda people are known for their artistic testimonial for funeral styles, which are usually erected on the grave after one year after the death of an important tribal member.

. The fixing of artistic testimonials are usually done with solemn ceremonies and large stones settings are piled up over them. However, there are two types of graves or funeral styles among the Balanda society. The first one is made up of simple wood posts, in which disc-form, round subdivisions are incised (See figure: 155). It should be noted that these discs symbolise the possession of the cattle and the profit the deceased has organised during his life on earth. Such types of graves are generally crowned with a stylised head only at the top.

The second types of graves are those which are rarely found among the Balanda. This is also called naturalistic-figural type. It is curved out of hard woof of a single trunk: the head is represented as distinct with small slanted eyes, a flat nose, rectangular mouth and round ears, set close to the head of both sides.

Figure: 191: The Balanda naturalistic-figural type

On the head on the top a hollowed cavity (for sacrificial offering) is preserved, and the legs are straight and stand firmly (without feet) on a round plinth, which is anchored into the stone layer on the grave. An interesting, old rare object, with heavy traces of weathering and cracks, especially on its right side. This is because the funerary figures of the balanda stand outdoors for many years, always exposed to the wind and weather. Height is about 90 cm, and width is about 14 cm.

(1) *Dear Announcement*: - The Balanda, like the other Luo groups of South Sudan, conceive of death as an inevitable, personal defeat and tragedy, against which there is no ultimate defence. The personal and group loss resulting from death is acknowledged as real and permanent.

When a person dies, the first thing the people of Balanda do is that the death is announced and this is usually accompanied by the wails and cries of women. In Balanda society, announcement is done very early in the morning or in the evening. Customarily, it is believed that this is the time people die in communities. Thus, inside home of the deceased relative lit lamp to enable people see and move. Death announcement is rarely announced during the broad day light or mid-day.

Interestingly, in as soon as the death announcement is made, close relatives rush to the deceased's home and they remain there until the dead person is buried.

(2) *Grave Digging (Kunyo bur)*:- Grave is usually dug collectively, it involves young men, male relatives, in-laws and several neighbours join forces in digging it. I am not sure whether there are some other people who might not be allowed to participate in grave digging; this is something that needs further research.

Digging of grave usually commence in the morning or in the evening depending on the decision of the elders. The grave is sunk about one and half to two meters deep, it is dug deep so that the stench does not come out. When the grave is ready, the people involved then informed a person that presides over the burial ritual rites.

(3) *Burial Rites (iko):* In Balanda society Burial takes place before 12.00 (mid-day) or late evening but never after the sun set.

The Balanda, like the Acholi and the Anyuak, bury the deceased adult or a child soon after death and burial rituals normally start at 8.30 a.m. or 5.00 p.m. According to the customary laws, it is agreed that one of the elders from the deceased's family must presides over the proceeding of the rituals.

In Balanda society, during the burial, the corpse is bound in a crouching position with the knees drawn up to the chin; men are placed in the grave with the face to the north, and women with the face to the south. The form of the grave is peculiar, consisting of a niche in a vertical shaft, recalling the mastaba graves of the ancient Egyptians. The tombs are frequently ornamented with rough wooden figures intended to represent the deceased—see figure 152.

The programme of the ritual starts with the lowering the deceased into the grave, as we have seen above and this is followed by throw-

Figure: 192, Statue to be placed in front or centre of the grave

ing soil on then dead person and burying the body instantly. The soil is then pounded down. The water is sprinkled on the grave to harden the surface which is finally sanded.

We are also not sure whether the burial of a wealthy person, in Balanda society, is the same like other ordinary burials or do they have especial treatments.

The above statue is carved from a type of wood called Khaya grandifoliola, which is the same like African mahogany. The sculpture in figure----- measures 102 cm in height, 23 cm, in width, 21.5 cm in depth and it weighs 10.5 kilos. The sculpture was collected by Comboni Father Elvio Gostoli from Wau, Bhar el Ghazal, in 1958.

In Balanda society, a sculpture associated with funerary practices is usually carved out from a tree trunk and the base is normally buried 60-80 cm into the ground. The statue is fully exposed to the mercy of weather and white ants and other insects.

Traditionally, this statue is always placed either in front of or in the centre of a grave-mound and surrounded by stones. The statue is usually erected with a ceremony during which dishes of food are left at the grace of the figure, and post are also said to protect against sorcerers spells.

Although Balanda are from the Northern Luo of South Sudan, tradition tells us that they do not bury their dead ones by the doorway as we have seen among the Acholi and the Anyuak. Tombs are usually set up near villages and along feeder roads.

What remains a mystery to me is whether the Balanda people buried their dead ones on a skin sleeping-mat or on cloth.

As people often move in search of a new cultivation land and settlement, the tombs are usually left behind. It is reported that many a time, some family members find it difficult to find the tombs of their loved ones as environment might change from time to time. However, despite the fact that most of the statues places on the tombs are been disturbed by fire, the reality is that the hard wood is difficult to burn.

It is said that such sculpture, as we can see above, is only erected to recognised and honoured male elites such as warriors and chiefs. The tombs of ordinary men and women are covered with only large stones and sometimes they place just a curved head to indicate that somebody has died and was buried here.

(4) *Vigil (budho):-* Vigil is the period that follows immediately after death announcement has been made to the community. During this time, people from different walks of life from the communities are expected to pay a visit to the house of the dead person. At this time also the close relatives of the deceased such as wife or husband, parents, stepmother, uncles, aunty and in-laws gather and stay in the compound of the deceased throughout the night until the deceased is buried. However, during this time, whenever the new visitors come into the house to pay their condolence, some people who are in the house must step outside to make room for visitors—as the hut is too small to accommodate many people.

As mourners come to the home of the deceased, women usually raise a strange voice before they enter the compound of the dead person; this is to announce their arrival. While men entering the compound of the dead person blow their horns (if the deceased is adult, married and over 40 years of age). These visitors go straight into the house where the dead body is laid without greeting other people who are already there. In the house the visitors continue crying or praying in their own ways. After a while they come out of the house and start to greet people who are sitting in the compound of the deceased, and join them in conversation.

However, during the vigil period, the mourners continue to cry, sing their lamentation as well as war songs. War songs are sung only for dead adult; and the war songs are sung to indicate that death has defeated the warriors (them) and they have no ultimate defence. While the women cry and call out the name of the dead person as if they are talking to the dead body.

At this stage, relatives usually prepare the funeral services by building a shade, preparing the coffin and the clothing for the deceased. As people continue to come to the compound of the dead person to pay their condolence, the women, from the side to the dead person, continue preparing tea and meals to feed them. Outside the house men make fire place called "*wang-mac*" or "*magenga*" near the house to warm the visitors as it usually get very cold at night. Vigil usually continues between 3-4 days and sometimes longer following the burial.

(5) *Shaving Hair and Funeral Ceremony*: - Like the Anyuak the Balanda shave their hair after 4 days for male and 3 days for female, following the burial. Not everybody can shave their hair, but only close relatives of the deceased such as spouse, parents, sons and daughter (if any) they shave their hair using razor blade.

The Balanda are not the exception in Northern Luo of South Sudan, they also have two types of shaving; the first one is done soon after death, to end mourning and to mark the beginning of their new life. The second shaving is normally done after 30 or 40 days, following the funeral. This is done to commemorate the passing of loved one, and sometimes accompanied with Christian prayer or Mass —this is usually known as the first anniversary of the dead person, which is called in Balanda language as "*gonyo cholla*".

In Balanda society, the shaving of hair means day of big feast which is to be attended by good number of relatives and other interested people. This is also known as the funeral dancing day which is accompanied with food and drink. However, the size and nature of this occasion depend on the age and status of the deceased.

Figure: 193, Balanda Funeral drums

Needless to say, as we have seen with the Acholi and Anyuak of South Sudan, modern Balanda as well do not want their hair shaved. There is no concrete reason given for the refusal to shaving hair, we could still say that they young Balanda now want to forget their culture and burial rituals. As I mentioned above, I n Balanda society the first anniversary for a dead person, if a man, is also done to free widow from mourning taboos, and allows her to choose a man who would inherit her after *"gonyo-cholla"*.

(6) *Departure of Home*: - In Balanda society, like in any other Northern Luo groups, soon after the last day of mourning, surviving family relatives, friend and in-laws may return to their respective homes. However, the closer relatives such as uncles remain behind and continue mourning for another 7 days or even more and depart to their respective homes

In Balanda community, like in any other Northern Luo groups, there is no any order for departure set about the customary laws regarding funerals; any one can leave the compound of the deceased whenever he/she wishes. The period between the death and the departure for home, in Acholi is called *"kumo"* and in Balanda is called "--------?", which usually refers to the mourning period, and literarily means men sitting without doing anything, while women cook to feed them.

Jo-Luo of Bhar El Ghazal Rituals for the Dead

Death to the Jo-Luo people is a tragic thing and is mourned accordingly. When a dynamic strong man dies, it is a shock, not only to his family and relatives, but also to the whole community. Traditionally, when someone is ill, his/her closest relative comes to nurse him/her and it is this person who will convey the will to the relatives on the last day of lunar month of mourning. Such a person must tell the truth and only truth without adding or leaving some parts of the will, otherwise, the spirit of the dead person would summon him/her to eternity for telling lies in his/her name (Apai Gabriel, 2005).

In Jo-Luo land, like the Acholi, relatives mourned the death of a boy or a man for 3 days, but the relatives mourn the death of a girl or a woman for 4 days. During the three or four days, worldly acquired things are to be returned to where they were found before the spirit proceeds back to God in heaven. Therefore, before the end of the three or the four day rituals, people believe that the spirit of the deceased is dwelling among them, and returning the acquired secular tools one by one. Apai Gabriel (2005) observes that when a child dies, the mother usually wears ropes on her neck instead of wearing beads, as a sign of mourning. More interestingly, if the woman conceives during the period of mourning, the ropes are removed and when that child is born, he/she would be called Wino for a boy or Awun for a girl (P. 15). For us to understand this complex case, let us examine more in depth the death or burial rituals among the Jo-Luo of Bhar el Ghazal.

(1) *Death announcement:* The Jo-Luo people, like any other Northern Luo groups of South Sudan, conceive of death as an inevitable, personal defeat and tragedy, against

which there is no ultimate defence. The personal and group loss resulting from death is accepted by the society as real and permanent.

However, when a Jo-Luo person dies, the first thing the relatives do is to announce the death through wails and cries of women. In Jo-Luo society, death is usually announced at dawn or late evening as many people die at this time of the day. Death announcement is rarely made during the broad day light, although sometimes we hear announcement of death at mid-night.

Once death announcement has been made, close relatives, friend, in-laws and neighbours rush to the deceased's house; some of them remain there until the deceased is buried. If a dead person has killed wild animals such as lions or elephants, drums are beaten soon after the announcement and this would continue throughout the three or four days of mourning. Deng Apiny Alany, one of my informants, stressed that no drum is beaten in the funeral of a person who has never killed wild animal. Apai Gabriel (2005) argues, "Before the burial, when the corpse if lying inside, there would be a lot of wailing and crying from the relatives and friends, mainly women and girls. At this very time, men would be arranging to dig a grave" (P.56).

(2) *Grave Digging (kuiny-bur):-* In Jo-Luo society, when a person dies and before men start digging the grave, one elder measure the height of the dead person with dried sorghum straw (*teng*); and if in the town, the old man measures the height of the dead person with robe (*tol*). The old man then come and lays the dried sorghum straw (*teng*) on the ground, where the grave is to be dug. This measurement is impotent, in this that, if no measurement is taken, the people may dig grave that is shorter than the dead person, which would make burial difficult.

The digging of the grave is symbolically started by the son or a close relative of the dead person. Once the son or a close relative starts digging the grave, shortly he leaves the work to be completed by others. At this juncture, it is important to note that grave is prepared collectively, where young and middle age male relatives, in-laws and neighbours join forces in digging the grave. However, where there are large numbers of men in a village, young men between the ages of 18-25 years are not allowed to dig the grave.

More importantly, in Jo-Luo society, some of the dead persons are buried by the doorway while others are buried in the highland next to the house. It is the custom of the Jo-Luo that, when a man wants to build a house, he must always build it near the high land with ant-hill in it. When there is no death yet, in the family, the highland would be used for cultivation, but in as soon as a person dies in the family; this highland turns to a cemetery.

The digging of grave is usually done according to the occurrence of death. For example if a person dies late evening or at night, the grave would be dug in the morning; when the weather is cold, at around 6.30 a.m. and buried in as soon as the digging is completed. But if a person dies at around midday, the digging of grave would commence immediately, so that it is finished and the dead person is buried before dark; digging of grave usually takes between 1 to 1.5 hours. However, digging of grave for a dead child may take less than an hour, yet this depends on the type of soil where the grave would be dug. The Jo-Luo people of Bhar el Ghazal, like other Northern Luo groups of South Sudan, do not bury the dead in a shallow grave. The grave is dug about one and half to two meters deep. Since some of the dead bodies are buried in the highland near the house, and others are buried by the doorway, the graves are sunk deep so that the smell from the decaying body should not come out. Apai Gabriel (2005) observed that grave is dug on the left side of the house in case of a man, and it is dug on the right side in case of a woman, and it should be dug one and half metres deep.

In my research, one elder man called Alberto Rimo Akot Dimo, from the Jo-Luo community, who lives in Victoria-Melbourne, explained that in Jo-Luo society, no other ceremonies are performed during the digging of the grave, with exception of the measurement of the grave, which we have discussed above.

(3) *Burial Rites: -* As we have just discussed, in culture of the Jo-Luo, some of the dead persons are buried by the doorway, while others are buried in the highland next to ant-hill; the dead body is usually washed before burial and at this time, there is usually a mock warming of water with a little grass to symbolise the normal way of preparing water for bathing. After washing the dead body, it is wrapped in a white cloth, leaving part of the cloth that covers the head untied and loose; the body is then lower into the grave. The Jo-Luo people believe that if the face and the mouth of the deceased were wrapped up tightly, the dead person would not be able to open his/her mouth to appeal to God about who caused his/her death (Apai Gabriel, 2005).

Inside the grave, the body is laid on the right side facing the east, so that the spirit of the deceased always rises with the sun and dwells among the people during the day until sunset. When all internal covering has been done, close relatives like wives, and children come and squat along the grave throwing the earth backward into the grave, as a last farewell to the beloved deceased.

However, if a person dies in the cities or town like Wau and Tonj, the body would be buried in the cemetery, unless the relatives decide to transport the corpse in the village for burial. Children who die in the towns are usually not buried in the cemetery; they are rather buried in the compound where family live or along the fence if the house has a fence. Unlike urban people, rural people bury their dead at night to avoid youngsters seeing the dead body being buried, which may traumatise them. For a Jo-Luo family, who live in Government house, when a child dies, the government do not allowed them to bury their dead child in the compound. For this reason, the relatives of the dead child would be forced to bury the child in the cemetery.

Whether the burial is in the village or in the town, the head of the dead person must always be buried with the head facing the direction of home. After the burial is completed, one elder takes two wooden sticks and places one on the head and the other stick on the feet. Saturnino Onyala (2014) observes that the burial of the deceased of adult or a child, among the Northern Luo of South Sudan, and in particular, among the Acholi, starts at 8.30 a.m. or at 5.00 p.m. And during the burial rites, Luo tradition says one of the relatives elder of the deceased must preside over the proceeding of the ritual. Interestingly, the programme of the burial includes lowering the deceased into the grave, throwing soil on the dead person and burying the dead person instantly.

Tradition has it that no one is allowed to cry when the dead person is being placed in the grave. It is reported that culturally, when a person cries while a dead body is being lowered in the grave, is an abomination. This means that the crying person wants another person to die.

After all the rituals have been performed, and the dead has been buried, those who participated in the digging of grave and burial, move a distance to wash their hands. Washing of hands after burial means a lot for the Jo-Luo, they believe that the dead person died because of bad luck, for this reason, when the burial is finished they wash the bad luck from their hands. This is also the time women, men and young people are allowed to cry. It is reported that some women cry for ten days and ten nights starting from the day of death announcement; they cry in the interval of 3-4 hours.

Thus, after the burial, the mourners would gather in the house of the dead person for burial ritual mourning and counselling. If the dead person is a male, they would stay in the house for three days and then perform a smock-cleaning ritual of hands and feet, implying that they would not shake hands or visit the dead man again. If the dead person is a female, after four days they would do likewise.

A day after the burial, the relatives of the deceased slaughtered a goat and the old women tidy-up the grave; while the people beat the drum of war.

(4) Vigil (*budho*):- In Jo-Luo society, soon after the death announcement and burial, vigil nights are spent in the house of the deceased by close relatives. The vigil usually takes between 3-10 days depending on the relatives of the deceased. Jo-Luo tradition states that the period from burial to the last funeral day (which is 12 months), is considered the period for vigil for a partner who has lost a partner. During this period a widow is not allowed to apply perfumes on her body, she is not allowed to wear beads (tii), and above all she cannot eat together with other people whether be a member of family or not. And when she wants to visit a friend she takes along plate for her to eat on and a cup for drinking water. During the first three or four days of vigil, many people would come from far away to the house of the dead person to pay their condolences to the relatives of the deceased. Those who are unable to come at

774

the time of death would do so on the 40 days celebration.

After three or four days, this would be followed by funeral ceremony and departure of the mourners to their respective homes (as we shall discuss below). It is said that after the departure of mourners vigil nights may continue for another two or three days.

(5) *Shaving hair and funeral ceremony:* - Soon after the burial, if the deceased is a husband, the wife and her youngest son or daughter would have their hair shaved; however, if the dead husband has two or three wives only the head of the first wife and her youngest son or daughter would be shaved. The other wives would shave their front-head or protruding hairs. According to Apai Gabriel (2005) the hair of the widow would be shaved as well as the heads of first and last born children would be shaved as a sign of mourning. The other grown up children or married children their hair would be cut short and backwardly to disguise their beauty (P. 58). Thus, if the deceased is a wife (whether be the first, the second or the third wife), the man alone would shave his hair and not the children. It is usually the old woman who shaves the head of the widow and her daughter or son; and it is a middle age man who shaves the head of the husband. The family members shave their heads using razor blade (*angudho*). When the hairs are shaved, the used "*angudho*" (razor blades) are thrown away in the grass. Interestingly, soon after shaving hair the wife or the man is not allowed to have sexual relationship for a period of twelve months or even more. Apai Gabriel (2005) observed that after the hair or heads have been shaved the widow would wear ropes around her neck and waist instead of beads. She would also avoid attractive ointment like perfume-she has to avoid beauty. While the closest female relatives of the deceased would wear simple dresses and they must not attend dances for one lunar month.

While waiting for the last funeral rites; it is reported that a person who has his/her hairs shaved must keep it shaved for 12 months, which means every time the hair grows the person must cut it off.

(6) *Departure of mourners to their respective homes:-* In Jo-Luo society, like in any other Northern Luo groups of South Sudan, after the last day of mourning, surviving family relatives, friends and in-laws may return to their respective homes. However, the closer relatives such as uncles remain behind and continue mourning for another 2 to 3 days or even more and eventually depart to their respective homes. In Jo-Luo community, there is no any order for departure set by the customary laws regarding funerals; therefore, anyone can leave the compound of the deceased at any time he/she wishes. The relatives of the deceased go back homes and come back to the house f the dead person after 40 for what they called "Challa" ceremony.

(7) *The Gonyo Challa or Forty Day:* In 40 days the family members of the deceased and other relatives gathered to perform the ceremony of "Challa" or "Gonyo Challa. The

performance of "*Challa*" must be done in lunar month. In Jo-Luo society, lunar month always begins when the new moon appears in the east. Thus, is a person dies when the moon is full, people will not start to count lunar month from that time, instead they begin counting lunar month when the new moon appears.

Interesting, lunar month for mourning, in Jo-Luo society, apply only to men and women who are married and have left behind children, wives or husbands.

As mentioned above, the mourning celebration must coincide with the appearance of the new moon – this occasion is called "Challa". This is the time when the close female relatives of the deceased take off their mourning cloths; they plait their hair and put on official and attractive cloths. At this time, relatives of the dead person slaughter a he-goat or a bull, brew local beer, and food cooked for the ceremony; people dance and drink, while the old women come to tidy-up the grave. Men beat war drum and dance if the deceased is an adult but if the deceased is a child or a young person, people normally do not beat war drum and do not dance. The relatives dance in the funeral of an elder man or woman because they want to praise them for the long live they have lived on earth. The people dance for 3 days for a death of a man and they dance for 4 days for a death of adult woman – they would be praising (mwoch) the departed and his/her ancestors. Robben (2004) pointed out that during the last funeral services the bereaved relatives join for a meal, sing songs and celebrate the final passage of the deceased. Thus, the mourning has become to end, the social order is restored, and the follow up of everyday life is picked up again (P.10). Subsequently, this ends the mourning period after burial and marks the beginning of the new life; the following morning everybody departs to his or her respective house; they will return in twelve month, for the final funeral ceremony.

However, the main mourners: widow or widower and the mother of the deceased would continue to mourn for another three months after this ceremony. Although the widow shall remove the black clothes in three months, she is not still allowed to neither put perfumes on her body nor wear beads (tii) for some months. During all mourning days, the widow is not allowed to eat with other women; even after three months when the clothes have been removed, she would also continue not to eat with people, she eats alone in her plate; she also has her own cup for water, nobody is allowed to tamper with the widow's plates and cups.

According to Apai Gabriel (2005) after three months, the elders would suggest to the head of family to allow the widow to remove the ropes from her necks. Subsequently, the headman would call the elders of the family, to come to the house of the dead husband to discuss his family affairs. In the gathering, the widow would be informed that she has mourned her husband enough and that it is time to forget about him and remove the ropes from neck and waist (P. 60).

However, in rare cases, when a man dies during his honeymoon, the protocol of mourning is usually changed because a young bride has no mourning experience and elder women would not allow her to practise. Because the Jo-Luo believe that for a young woman who is newly married and started regular sexual relationship with the husband, if suddenly she is asked to stop sexual relationship, she may be affected with what is locally known as "Kang" (fertility sensitivity in which she becomes numb in her conception because she has not been sexually active for a long period).

The Jo-Luo of Bhar el Ghazal believe that "Kang" may continue to affect the young widow later when she is inherited by a man, after a long time gap, as a result she might not conceive quickly. Therefore, to avoid "Kang" affecting the young widow, she must be given to another genetic husband and this is arranged on the day of *"Challa"*. The other important reason behind giving the young widow soon to a man is that she may feels nothing is binding her to the dead husband's family except the dowry paid after her. Thus, to avoid any inconvenience, the young widow must be matched up with somebody to keep her bound to the family; otherwise people who want to swim in dirty water may come around trying to take her and risk the family reactions. So when she is given

Figure: 194, Young Woman in honeymoon

to somebody and she conceives, the relatives of the dead husband would say, "She is chained to the family". Subsequently, everybody is relieved and happy ((Apai Gabriel, 2005, PP. 59-60).

(1) *Last Funeral Rites*: In 12 months relatives and members of family from both sides gather to perform the last funeral rites. At this time a bull is slaughtered as a sacrifice. People eat drink and dance; this is also the time a widow shall be given chance to choose from the close relatives of the deceased husband (usually brothers) she wants to inherit her (*lak*) and take care of the children. Generally elders sit to discuss who should inherit the widow. The man who shall be given to the widow should be of the same age with her dead husband or must be younger. If the man is older than the husband, customarily is not acceptable. However, the widow is not obliged to accept the man she knows would not take care of her with the children. In this case she may choose her own man from the family of the dead husband or she looks for another man outside the dead husband's family. The *lak* or inheritance ends the bereaving period.

In a normal death, there is always a will left behind by the dead person, detailing who should be responsible for his children, wives and wealth. Apai Gabriel (2005) pointed out that, the part concerning wife or wives is very important because it specifies the man who would rehabilitate the wives for generic fertilization for children in his name. The "will" would also confirm the affiliation relationship the deceased had already established in wealth-sharing among his children. The "will" is normally oral, since many Jo-Luo people in rural areas do not know how to read and write. The will is usually told to the members of the family the next day when the sympathisers have gone. Members of family usually dispute a will which leaves a widow under the care of a cousin who is not genetically related to the family blood. For example, a son of a sister of the dead husband is not welcomed by his brothers, while the third paternal cousin is not discriminated against.

According to the Jo-Luo customs and norms, when it comes to the inheritance, the widow is always referred to the will of the dead husband; if she has no objection a small occasion would be arranged for a male to come and slaughter a he-goat as a purification and blessing for the dead husband's bed, on which the new genetic husband would come to sleep. If a widow in this situation conceives and gives birth later, the child would be called "Uchalla", if the child is a boy; but if the child is a girl' she will be called "Achalla", these names are given in memory of the "will" disclosed at the time when the widow is removing the ropes from the neck and waist (Challa). However, if the widow refuses the "will" of her dead husband, the elders would give her options to choose from the brothers of the dead husband or from the cousins of the dead husband. Moreover, if the widow is in her thirties, she would be strongly advised not to flirt with young brothers and cousins of her dead husband, who are not yet married because that would result in a terrible blunder, a "buol" (a disease caused by sexual relationship between a widow and a man older than the dead husband). In Jo-Luo society, sexual intercourse, between a widow and man older than her dead husband is strictly forbidden (Apai Gabriel, 2005). Genetic blood trace among the Jo-Luo of Bhar el Ghazal plays an important part in inheritance of a widow, this is to avoid discrimination to children born later, and who carry the name of the dead husband.

It is said that the Jo-Luo talk of ghost fathers. According to Jo-Luo tradition, ghost father is a man who inherited a widow, but that man is not from immediately family member, he could be from next of kin. Alberto Rimo Akot Dimo and Deng Apiny Alany, who are my informants in Melbourne Australia, the practice of ghost father is borrowed from the Dinka tribe. The ghost father looks after the wife of the deceased so that his (dead husband) name remains on earth —what they called "to keep the family alive". Many anthropologists also believed that the idea of ghost father has come to Jo-Luo of Bhar El Ghazal society from the Dinka, which are their neighbours.

Mabaan Ritual for the Dead

The Mabaan is the only exceptional Northern Luo that has only 5 rituals for one deceased. However, all rituals are performed for both adult and a child, although certain number of rituals are omitted depending upon age, sex and marital status of the deceased.

(1) *Death announcement*: - The Mabaan, like other Northern Luo of South Sudan, conceives death as an inevitable, personal defeat, and tragedy, against which there is no ultimate defence.

When a person dies in Mabaan society, members of the community come to know about it when they hear women wail and crying. It remains unclear on whether the Mabaan people also use sound of drum to spread sad news. However, death is usually announced at dawn or in the evening. This is the time for the day when people are expected to die. It is said that the Mabaan people rarely announce about death in the broad day light or mid-day.

As soon as the relatives, friends and neighbours of the deceased hear of this news, they hurry to the home of the dead person to share the sadness and to console family members.

(2) *Grave Digging and Hair Shaving*: - In the old day, when a person died, the grave was dug in the house and the body buried there instantly. Today, the people of Mabaan have abolished this old tradition. So, when a person dies, the members of family immediately shave their heads and the men dig grave outside the deceased's compound. The grave would be measuring 2 meters deep and one meter wide.

Interestingly, in Mabaan society, when men are digging the grave the women prepare the corpse by washing and covering with cloth. We have seen in Acholi tribe that when people are digging the grave, no one is allowed to cry; but contrarily, in Mabaan society, when men are digging the grave relatives continue to cry and sing songs of sorrow until the digging of the grave is finished. The burial ceremony in Mabaan community is called *"pumko"*; this is when women prepare the corpse and men dig the grave during which people sing songs of sorrow as we have just discussed above.

(3) *Burial Rites*: - In Mabaan society, the traditional of burying the dead in the house has been abolished. Today, the dead body is buried in a grave almost two meters deep after shaving the head. The dead are not buried by the doorway, as we have seen among the Acholi and the Anyuak; the dead are rather buried outside the compound not far away from the deceased's house.

Figure: 195, Dead body waiting for burial

As the dead person is brought by the grave (in a village), a skin sleeping-mat is laid on the bottom of the grave and the corpse is laid on this at full length, but lying on its side with one hand under the head and the other stretched out along the body. The white cloth that was used to cover the body is then put over the body to keep off contact with the earth. The burial usually includes receiving the dead body, lowering the deceased into the grave, throwing soil on the dead person and burying the corpse instantly.

(1) *Vigil*: - As we have seen with other Luo groups, and Mabaan is not an exception, they have vigil period that follows immediately after death announcement. When people hear about the death announcement, they gathered in the house of the deceased person and other people are expected to follow them sooner or later. During vigil period close relatives of the deceased such as wife or husband, parents, step-mother, uncles, aunty and in-laws gather and stay in the compound of the deceased throughout the night until the deceased is buried. However, during this time, whenever the new visitors come into the house to pay their condolence, some people who are in the house must step outside to make room for visitors – as the hut is too small to accommodate many people.

As mourners come to the home of the deceased, women usually raise a strange voice before they enter the compound of the dead person; this is to announce their arrival. While men entering the compound of the dead person blow their horns (if the deceased is adult, married and over 40 years of age). These visitors go straight into the house where the dead body is laid without greeting other people who are already there. In the house the visitors continue crying or praying in their own ways. After a while they come out of the house and start to greet people who are sitting in the compound of the deceased, and join them in conversation.

At this stage, relatives usually prepare the funeral services by building a shade, preparing the coffin and the clothing for the deceased. As people continue to come to the compound of the dead person to pay their condolence, the women, from the side to the dead person, continue preparing tea and meals to feed them. Outside the house men make fire place called "*wang-mac*" or "*magenga*" near the house to warm the visitors as it usually get very cold at night. When a dead person is buried relatives and uncles remain in the house of the deceased for 4 days if the deceased is a male, and 3 days if the deceased is a female. By the end of 4 or 3 days relatives prepared food and local beer; after eating and drinking, people dance a funeral dance.

However, a widow or widower usually mourns for one month during which she/he wears a certain plan material locally called "*tanyan*" on the arms and the neck. Thus, after 30 days, a ceremony called "*muka duran*" is performed by a woman. This ceremony is generally called "*gonyo-cholla*" in Luo, but because the Mabaan have lost much of the Luo language, they then called the ceremony as "*muka duran*" which is equivalent to "*gonyo-cholla*". During this ceremony pigs and goats are slaughtered and people feast to cleanse the home.

(1) *Departure for Home:* - In Mabaan society, like in other Northern Luo groups of South Sudan, after the last day of mourning, surviving family relatives, friends and in-laws may return to their respective homes. However, the closer relatives such as uncles remain behind and continue mourning for another 7 days or even more and depart to their respective homes. In Mabaan community there is no any order for departure set about the customary laws regarding funerals; any one can leave the compound of the deceased whenever he/she wishes.

Pari Rituals for the Dead

The Pari, like the other Northern Luo groups of South Sudan, perform a total of six rituals for one deceased. All rituals are performed only when adults die; certain numbers of rituals are omitted depending upon age, sex, and marital status of the deceased.

(1) *Death announcement:* - The Pari, like the Acholi, conceives of death as an inevitable, personal defeat and tragedy, against which there is no ultimate defence. When a person dies, members of the community come to know when they hear women wailing and crying, if the chief who dies, the people know about his death with the sound of drums.

In Pari society, when a person dies death announcement is made immediately. The death is always announced at dawn or in the evening as many people die at this time of the day. In Pari society, members of community rarely hear about death (or women wailing) in broad daylight,

(2) *Grave Digging (kunyo):-* In Pari society, grave is dug collectively by grown up men and some strong elders; young men rarely join forces in digging of the grave. When a person dies his/her grave is dug in the compound of the homestead. I am not been informed whether men with pregnant wives are allowed or not allowed to participate in grave digging. Digging of grave usually commence in the morning or in the evening depending on the decision of the elders. The grave is sunk about one and half to two meters, the grave is dug deep so that the stench does not come out. Seligman (1948) cited Colonel Bacon as saying, "Men dig grave about 3-4 feet deep and usually within two paces of the door of the deceased's hut" (P.112).

(3) *Burial Rites (Yik or iko):-* In Pari society, unlike the Acholi society, the dead person is not buried by the doorway; the dead is rather buried by the side of the hut. Burial usually takes place within a few hours of death, as soon as the grave is ready. When the burial is completed the grave is closed with a wooden fence and an ox or a bull if the dead is an adult, or a goat if the dead is a child, is slaughtered and the contents of

its stomach (*weny* or *wee*) are sprinkled on the grave and some of stomach contents are smeared on the body of attendants (Kurimoto E. 1992).

If an animal is not available, cucumbers are crushed and their juice is smeared on the body of attendants. When the juice of cucumber is smeared, its seeds would remain on the skin, sticking to the body for some time. This, as well as stomach contents of animal that is smeared on the body, serves as a mark which indicates that the person has attended the burial ceremony. In some cases the stomach contents and cucumbers may be used at the same time. In this case, the body parts and other part which both substances are smeared include step or entrance into the hut, knees, middle of the chest, forehead and top of the head. Kurimoto E. (1992) in his *"Am Ethnography of Bitterness: Cucumber and Sacrifice Reconsidered"* reports that on various occasions the Pari sacrifice wild cucumbers which are called *"akalajo"* –it fruit is covered with blunt tubercles. Kurimoto observed that many a time "akalajo" are sacrificed in the burial ceremony. According to Evans-Pritchard (1956) *Akalajo* could be identified as *Cucumis figarei* (P. 203). Akalajo is a wild creeping plant that is very common in Luo land, especially around homesteads and in cultivated areas. The plant tastes very bitter and it is not edible.

The burial of deceased adult or a child both burial rituals normally start at 8.30 a.m. or 5.00 p.m. According to the customary laws, it is stated that one of the elders of the deceased must presides over the proceeding of the ritual. Seligman (1948) observes that sometimes a person can express a wish of where he/she is to be buried when a person die.

As the dead person is brought by the grave (in village), a skin sleeping-mat is laid on the bottom of the grave and the corpse is laid on this at full length, but lying on its side with one hand under the head and the other stretched out along the body. A second skin mat is then put over the body to keep off contact with the earth.

The programme of the ritual, as we have just discussed above, starts with the lowering the deceased into the grave, and this is followed by throwing soil on then dead person and burying the body instantly. The soil is then pounded down. The water is sprinkled on the grave to harden the surface which is finally sanded.

After the burial, the surviving family members and other relatives sit down to eat and drink local beer. The relatives will sleep in the dead person house between 3-7 days, which is followed by removal of ashes known as "keto buru or buru".

Unfortunately, the Pari of today, especially those who live in Western World and big cities and towns, in South Sudan, do not follow the traditional way of burying the dead person. They rather bury the dead person in coffin, without skin sleeping mat, and the dead person lies on his/her back with the two hands stretching along the stomach. The body is covered with modern Western burial cloths and with some flowers on the coffin.

(4) *Vigil (budho):*-Soon after the death and before the burial of the dead person family

and friends gather at the dead person's house. There, people cry on one side and on the other side, they comfort and support the dead family members.

The closer relatives of the dead person such as spouse, uncles, aunties, parents, step-mother, sisters and brothers stay in the home of the deceased from day one till the departure time as we shall discuss later. In the first one or two days, they stay awake throughout nights singing or praying. These one or two days passed before burial is put aside to enable relatives who live in a faraway villages and towns to attend the burial. Men and women continue crying and singing their lamentations and often war song if the dead person is a man or an old man.

As people approach the entrance of deceased homestead, women raise a strange voice before entering the compound to announce their arrival. Men entering the compound play their whistles and singing their own elegies. These visitors or mourners go straight into the house where the dead body is laid without greeting other people; there they cry or pray in their own ways. As the visitors enter the house, those who were in the house leave the room to accommodate the visitors. After a while visitors come out of the house, greet other people and join them in conversations.

During the days of vigil, visitors take tea, eat food and drink local beer but dancing is not permitted at this stage. As the family of deceased is considered to be in mourning, the women from the dead person's home are not allowed to cook; other women who came from outside do the cooking and serving the food to the visitors. Today, Christian songs are also sung; however, most of the vigil visitors follow traditional ways of expressing condolence.

Outside the house, some people sit on stools, logs and other sleep on the mat or sleeping skin. As the sun set, men make two fire places this is known as "wang mac" or "magenga" The fire places are usually made near the house, one fire for women and the others for men to warm themselves, because at night it becomes very cold. The vigil usually continues up to some days after burial.

From the day following the death, the surviving family members and relatives of the deceased busy themselves to prepare for the burial services, building the shade, and preparing for the coffin and cloth. During this period people leave their usual duties and come to attend the funeral preceding rituals.

(5) *Shaving Hair (liedo) and Funeral Ceremony*: - After 4 days for male and 3 days for female, following the burial, close relatives of the deceased such as spouse, parents, sons and daughter (if any) shave their hair using razor blade.

The Pari society, like the Acholi society, has two types of shaving; the first one is done soon after death to end mourning and to mark the beginning of their new life. The second shaving is normally done after 30 or 40 days, following the funeral. This is done to commemorate the passing of the loved one, and sometimes accompanied with Christian prayer or Mass –this is usually known as the first anniversary of the dead person, which is called in Pari language *"gonyo cholla"*. If the funeral is in the village, after the prayer people eat, drink and dance. Robben (2004) observes that among the Luo groups the bereaved relatives join for a meal, sing songs and celebrate the final passage of the deceased. This eventually b rings mourning to an end, where the social order is restored and people pick up their everyday life (P. 10).

However, as we have seen with the Acholi and the Anyuak of South Sudan, modern Pari also do not want their hair shaved. As there is no reason given for the refusal to shave, I then call it *"change in culture that will lead to the loss of Pari traditional cultures"*. It is important to note that the first anniversary for a dead person, if a man, is done to free widow from mourning taboos, and allows her to choose a man who would inherit her after *"gonyo-cholla"*.

(6) *Departure for Home:* - In Pari society, soon after the last day of mourning, surviving family relatives, friend and in-laws may return to their respective homes. However, the closer relatives such as uncles remain behind and continue mourning for another 7 days or even more and depart to their respective homes.

In Pari community, as we seen with other Northern Luo groups, there is no any order for departure set about the customary laws regarding funerals; any one can leave the compound of the deceased whenever he/she wishes. The period between the death and the departure for home, in Acholi is called *"kumo"* and in Pari is called **"--------"**, which usually refers to the mourning period, and literally means men sitting without doing anything, while women cook to feed them.

Shulluk Rituals for the Dead

As the ultimate end of every living moral thing, the Shulluk accept and respect death. It is celebrated when it is the aged or important person passing.

(1) *Death announcement:* The Shulluk society, like the Acholi and other Northern Luo Societies, conceive a death as inevitable, personal defeat and tragedy, against which there is no ultimate defence. The personal and group loss resulting from death is acknowledged as real and permanent.

Moreover, in Shulluk society, community members come to know of death of an ordinary person when they hear the women's long quivering wail, sometimes followed by drums. But in regard to the death of the King (Reth), the community members come to know of his disappearance (death) after two months following the day of his actual disappearance. Diedrich Westernmann (1912) observes that the community members only come to know about the death of the King (Reth) after two years following his disappearance; and this is when some of the funeral ceremonies are already done and the announcement is made publicly, "The King (Reth) has disappeared "died" (1912, P136).

It is reported that death announcement of an ordinary person is usually made at dawn or late evening as many people die at this time of the day. Among the Collo people, death is rarely announced during the broad day light (mid-day), although sometimes death announcement may be done in the middle of the night.

Once the people hear about death announcement (of ordinary person) relatives rush to the house of the deceased; they come to the house and begin vigil nights. But in case of Reth's death, when people hear the death announcement after two months, each person cries or weeps in his/her house, only chiefs that would gather in Fashoda to mourn the King as we shall discuss later.

(2) *Grave Digging (Kwinyo-bur)*:- In as soon as community members hear about death announcement, they immediately prepare for grave digging. The people begin to collect digging-tools and the relatives select place where the grave would be dug.

Grave digging is usually done collectively, and once digging-tools are collected and place to dig the grave has been marked, young and middle age males, relatives and several neighbours join forces in digging. It is customarily that when people have gathered to dig the grave, a cow must be killed. As a cow is killed, some men dig the hole while other men skin the cow. After skinning, the hide is brought and cut into strips and a bier is tied together with these strips of skin, and the bier is put on the ground.

Digging of grave usually starts at around 6.30 a.m. and complete at around 8.30 a.m. or starts from 3.00 p.m. and complete by about 5.00 p.m. of the burial day. The Collo, like other Northern Luo groups of South Sudan, do not bury the dead in a shallow grave. The grave is sunk about one and half to two meters; this is done to avoid the stench from coming out of the grave.

(3) *Burial Rites and Vigil*: - According to Collo tradition a man whose wife is in early pregnancy (1-2 months) he is not allowed to participate in burial rite but if under any circumstances he participates in burial he would be considered polluted; and he is also not allowed to stand within ten metres of the compound where the newborn is. Burial rite is something complex among the Shulluk, thus, before we enter deeply into our discussions about burial rites, I would like to inform the readers that in Shulluk society not everybody is buried so ceremoniously as we shall see below. It is reported that only the old, respected or rich people, chiefs and Reth (King) could have ceremoniously burial. Although it is said that only the old, respected or rich people, chiefs and King could be ceremoniously buried, the true of the matter is that the ceremony for the old, respected or rich people, chiefs and King differ significantly.

To avoid confusion about the complexity of these burial rituals, I am going to discuss the burial ceremony of the old, respected or rich, chiefs separately from that of the King

(Reth).To do this well, I shall first discuss the burial rite for the old, respected or rich and chiefs; secondly I shall discuss about the burial rite for the King. Thus, in almost every village one could see the horns of an ox buried projecting from the ground; this is the grave or shrine of such important people. A child or uninitiated male is buried without much ceremony.

Truly speaking, Collo mourning could be divided into 5 parts: the first part is inter-mediary period when the dead body is temporarily kept in the house, when the soul of the deceased remains near the corpse and the bereaved relatives have suspended their daily routines as they enter into mourning. The second part is when burial is done and people are allowed to cry and weep. The third part is when people burnt eating tools and belongings of the deceased and rub the ashes on their foreheads and depart for their respective houses for a period of 4 days. The fourth part is when people gather again after four days to besmear the grave with wet mud and smoothing the surface of the grave. And the fifth part is when people meet after one month for final festival.

(a) *Burial of old, respected or rich people and chiefs:* In Shullukland, when a grown up man dies, he is buried in or just before his hut. The chief is buried in a hut. But if a woman or a child dies, they are buried in the bush not far away from the village. The first day of the funeral of the burial of a man, an ox is killed as a funeral feast (Diedrich Westernmann, 1912). The head of dead person is shaved and the body is dressed according to status in society. An adult male is dressed in war regalia and the burial is accompanied by the war dance and wailing by women, usually with mock war against the "jwok/juok" that is assumed to have killed him.

However, today, the Collo people have abolished burying dead body in the hut; they bury the dead by the doorway. Only chiefs are being buried in the huts. The modern Collo also do not bury women and children in the bush, they are rather buried by the door way or in a place within the compound. But for the modern Collo who live and work in the city or towns they bury their dead persons in the cemetery.

(1) **The first part of mourning:**

In Shulluk society when a person dies, the people of the village would be asked to go and hunt for fowls. They catch the fowls and bring them home alive. When they arrive home, one fowl is caught and thrown onto then ground and it dies. One elder takes the dead fowl and throws it into the corner of the hut where the dead person lies. The second fowl is taken and thrown onto the ground it dies. An elder person takes this second dead fowl and put it on the head of the head of the deceased person.

However, as the dead body is kept in the house and before burial takes place, if the dead person if a man, one of the middle age males, from the deceased family washes the corpse and wraps it with cloth. But if the dead person is a woman or a child, one of the old men from the deceased family washes the dead body and wraps it with cloth.

It is not clear to me whether the Collo also smear the body of the dead person with oil after the body has been washed.

Figure: 196, People wait outside the hut as the body of the dead is being washed

(2) ***Second part of mourning:***

The dead person is then carried into the yard; there, adze, spear, hoe, beads and a skin cloth are brought. Some people would be asked to go and cut thorns. Before these men are asked to go and cut thorns beads would be tied round their feet; and when they cut the thorns, they have to bring them and give to the grave-makers. The grave makers after receiving the thorns they ask for adze and when adze is brought they ask someone to measure the length of the dead person. The measurement of the dead person is then taken and place on the ground to measure the size of the grave.

After measurement of the size of the grave, some people then dig the grave, while others kill and skin the cow. The hide of the skin is brought and cut into stripes and these strips are used for tying the bier. When the preliminary works are finished, and before the corpse is put into the grave, the wife of the deceased is called, she comes and holds the feet of her dead husband, and the dead body is then laid on bier. As the corpse has been laid on bier, the wife (widow) or any other woman from the deceased family sweeps the place where the feet of the dead person lies.

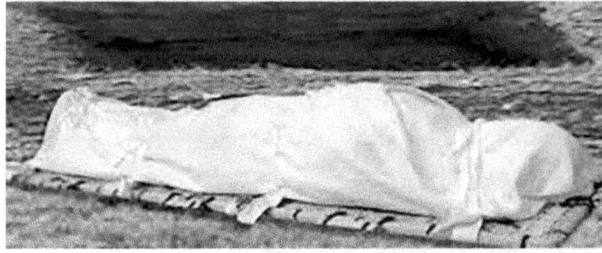

Figure: 197, Dead body waiting for burial

Shortly, a female relative of the deceased is also called, she would be asked to throw away any hearth-stones that may be lying around there. Interesting, at this time, no person is allowed to cry or weep. The dead body is then lowered into the grave, the soil is thrown on and the dead person is buried instantly.

After the burial, drums is brought, a goat is brought and killed with a club; the goat is then left behind in the yard. Customarily, a goat for such event is not normally slaughtered but beaten with a club until it dies. When the goat is killed, the men then beat the drums and this is an indication that men and women, who are around the burial place, are now allowed to cry and weep, which is usually followed by mourning dance. After this the people go home and assemble at the deceased home, where they go around in a procession dancing (See Figure 188 below). There an ox is brought and killed and skinned and cooked. More importantly, this is also the time when the Collo would say the burial of the man is finished (Diedrich Westernmann, 1912, P. 111).

Figure: 198, Shulluk performing procession dancing at funeral

Soon after burial, a fence is built around the grave, when the men finished building the fence they go to wash their bodies in the river and come back to the village. It is observed that when people come from the river, as they arrive in the house of the deceased, a plant called "oboyo" and a fowl are brought; each person is then touched with the "oboyo", which is a sign of blessing, so that God protects whatever a person cultivate with his hands that he touched with the dead body.

(3)   *Third part of mourning:*

The people then sit down; at this time, when everybody is seated in the compound of the deceased, after coming from the river, one man takes the eating-tools of the deceased out of the hut and he burnt them instantly in front of everybody. When the eating tools are completely burnt, people take the ashes of the burnt tools and rub it on their foreheads. After this third burial ritual ceremony people scattered, they go to their respective homes for four days. However, not everybody will leave the dead man home, some of the close relatives would be left behind (Diedrich Wassermann, 1912). I was wondering why Shulluk rub ashes from burnt eating tools of the dead person on their foreheads in the third part of mourning. Mrs Tarizah Chol Otor told me that this is an old practice, the modern Shulluk people no longer practice due to modern civilisation. As such, the modern Shulluk do not know why Shulluk in the old days rub ashes on their foreheads.

(4)   *The fourth parts of mourning:*

After four days, people gather again, and old woman takes mud, makes it wet with water and plaster it on the grave. When she completing plastering the grave, an old woman again takes a gourd of beer and pour it on the grave.

This is done in accordance to traditional norm that says, "After four days, following the burial of the dead person, the grave must be smeared with wet mud and the surface of the grave must be smoothened". It is to be noted that the work of smearing and smoothing the surface of the grave is usually performed by old women. On this same day, an ox is killed and cook, more beer brewed; however, at this time some clans kill two oxen instead of one; but this is something not compulsory for everybody in the country. After eating and drinking, drums beaten, women bearing spears go round and round the grave, dancing, while other women beat drums and shook bells. Every man takes arm and makes war-plays by going around in procession, they dance with spear and shield. According to Seligman (1932) dancing women bearing spears, they go round and round the grave, while others play bells and drums. This scene shows the attendance of men in war regalia, spears and shields, along with a tribal flag. After the performance of the fourth parts of the mourning, people again depart for their respective homes.

(5)   *The fifth and final mourning:*

When one month has passed, the people start to preparation for the final mourning festival. In this preparation, people collect dura from the house of each family member of the deceased. Dura is then pounded, fermented and women brew beer and relatives are invited for mourning festival. Thus, in the afternoon drum is beaten and an ox is killed, in addition, number of goats are also killed for the feast. The invited relatives then come eat and drink beer, the relatives may even dance, and when night comes, they go to sleep.

The next morning, the actual mourning then begins, where 4 cows are killed in the yard and another 4 cows are killed in the middle of the village. Everybody is then invited to eat, drink and dance. In the afternoon of the mourning day, another 4 cows are killed in the bush not far away from the village. Traditionally, the meats of the 4 cows killed in the bush are not brought and cooked at home, but rather the meat is cooked in the bush in the place where the cows were killed. On this same day, the cooking pots and other household materials of the deceased are carried out from a hut of the deceased. A hole is dug in the compound near the place where the head of the deceased is facing. When the hoe is ready, people take two pots, one gourd and small pot used for beer, a mat for covering food and two dishes formerly used by the deceased, they break these items into pieces and throw then into the dug-hole (Diedrich Westernmann, 1912, P. 113).

When this ceremony is finished, people take two horns of a cow and other horns from the deceased's cattle; they plant them on the grave —so that any passer-by may see that this is a place where a person was buried. Soon after this, the relatives of the deceased hold meeting commonly known as *mourning-meeting*. In this meeting the participants would distribute the butchered cattle as follows: they give a shoulder of a cow to one of the family members. They give fore-legs, and ribs to other member of the family. They give bowels, head and the feet and neck of the ox to people who participated in the digging of the grave. These are the five stages of burial ritual rites in Shulluk land. Having seen how ordinary people are buried, let us now examine on how the burial rituals of Reth (King) is performed.

(b) Burial of the King

When the King dies (disappears) his body is taken and laid in a hut. Shortly, a cow, under three years of age, is brought, speared; its skin removed and cut into long thin strings. While some men are cutting the skin of a cow into pieces, other men go into the bush to chop three with axe. When the men finished chopping the tree, the branches are brought to the site and driven into the earth. After this, the skin of the cow and logs of tree are tied with strings to form a stand (bier) on which the body of the King will be placed before burial.

When the bier is ready, the body of the King (Reth) is removed from the hut, without people knowing it; the body is then laid on the bier and put into the hut. There, the body of Reth is dressed (adorned) with a leopard skin; when the people finish dressing the King, two girls are then brought into the hut, where the dead body of the King is. One girl holds his head and the other girl holds his feet. Interestingly, each of the girls is given a tobacco and a pipe.

Now the entrance of hut is then walled, and all openings are closed with mud, so that there is no way for the air to enter. The two girls are left in the hut to die there, due to lack of oxygen.

When the burial is completed, according to Shulluk traditions, the people must wait for two months, before opening the walled hut. This is done simply to give enough time for the warms to eat up the flesh of the Reth and the two girls. Thus, after two months, the walled hut is then opened, as men are opening the hut, worms come crawling out through the roof. This is indication that the flesh is already consumed by the worms and that only bones are remaining. After this, the hut is left opened for a night, to allow most of the worms to get out of the hut. The following morning, the high chief summons all chiefs of Shulluk country to assemble at Fashoda—all must come; no chief is allowed to remain behind.

After two months, all the chiefs from Tango are summoned to move to Mwomo; and chiefs from Mwomo are asked to gather in a given place at Fashoda, where they will meet with the chiefs coming from Tango Districts. Each chief of Tango and Mwomo districts brings with him a cow. When they reach near Fashoda, all the cows are gather together and speared. As they spear the cows, at the same time they say publicly, and on top of their voices,

"The King has disappeared!
The King has disappeared!
The King has disappeared!
The King has disappeared!"

In Collo traditions, the word "die" is never used for the King. They preferred using the word "disappears" rather than using the word "dies". When people hear this public announcement, and the sad news, they weep and start to mourn Reth.

From the number of cows killed, a chief skinned one cow; the skin is soaked into water and put up to dry in the sun to form leather which is later on made into a bag. When the bag is ready, 2 to 4 chiefs enter the hut and collect the bones of the king, put them into this skin bag. Many elders told me that this secrete place is normally in the middle of the River Nile.

However, after the burial, the mourning continues, everybody mourns for the disappearance of the King. After a few days, chiefs send out messengers to collect many spears from the villages—these spears are tied together and put into the boat; other items such as cattle-bells, beads pots, dishes and gourds are also loaded on the boat.

Before the boat take off, one man and one woman are brought and put into the boat. The two persons are made to lie in the boat with their hands and feet tied. One person is laid in the back seat of the boat and the other is laid in front seat of the boat (with the corpse of the King in the middle). Suddenly, one man rows the boat with the King and the two people; and the other man rows an empty boat that will be used by the two men to return home. The two men row the boats into the middle of the water, this is a place commonly known as a secrete place for the burial of the King. When the boat carrying King's bones reaches the middle of the river, the man pierces the boat, so that water enters into it. As the boat starting to be filled up with water, the man jumps into the other boat and two of them start to row the boat back home. Therefore, the boat which was pierced sinks down with the bones of the King, together with the bodies of a man and a woman, with all other items in the boat perish in the river.

In contrast writers like Evans Pritchard (2011) argued that the history of Kings being walled up in a hut is a confusion arising from the usual walling up of the remains of a dead King; where the bones later buried after decomposition of the flesh, he therefore, concluded that the ceremonial putting to death of Kings is probably a fiction. According to him this story probably arises from the dual personality of the King, who is both himself and Nyikang, both an individual and an institution, which accounts also for the linguistic convention that a King does not die but disappears, in the same way Nyikang is said not to have died but to have disappeared in a storm (p. 415).

However, writers like Crazzolara (1951) observed that today there is much change in Collo tradition when burying the King. He argues that when a King dies, he is placed on a kind of bed-frame in an ordinary hut and cover with a pile of blankets or cloths. As the body of the King is laid on bed-frame, elders order for a girl of Kwa-Abaaka clan, from Alaal (Pachodo), to be brought. When the girl if brought she is wrapped up so that she cannot move and put to lie alive by the corpse of the King. After this, elders walled the entrance of the hut and the girl is left to die near the corpse.

From the two narratives, we have learned that in the old time two persons a woman and a man were left to drawn as to accompany the bones of the King to his resting place. Meanwhile Crazzolara (1951) tells us that in course of time the number of people ordered to die by the King has dropped from four to only one girl. It is disturbing hearing human sacrifices are being practice in Shulluk Kingdom. I then asked some Collo elders from Melbourne (Australia) and I wanted to know whether these human sacrifices are still today being done. However, some competent men claimed that such practice has been dropped, and now human sacrifices are been replaced by heifer (a cow under three years of age), it is laid beside the King and left to die there. After two months as usual, bones of the dead King are removed and taken into the middle of the River Nile. They alleged that people of Collo reached this decision because they felt that such practices finish human beings.

Conclusion

In conclusion, we can say that the Northern Luo of South Sudan have originated from one grandfather called "Okwa", who is the father of Nyikang, Dumo or Duwaat, Anyuak, Gilo, and Achol. In the earliest times, the Northern Luo of South Sudan was one of the tribes who settled in ancient Israel. The Bible tells us that the people who migrated into ancient Canaan were named in the bible as "six people" each having its own identity and territory. Having said this, we could also assert that the grand-grandfather of "Okwa" was born in ancient Canaan, and might have died at the eve of Luo migration from Canaan to Lower Egypt.

Moreover, the Luo people in ancient times (when they were still in ancient Canaan) were known as "Phoenicians" or Semitic people", who lived in ancient Israel. They were speaking their own language known as "*Dho-Luo*" or "Luo Language".

According to Gen. 9:18 Canaan is considered the fourth son of Ham, who is the father of the black races of Africa. Gen. 10:6 tells us that Ham has had four sons namely: (1) Cush, (the father of the Phoenicians/Luo), the Phoenicians (Luo people) moved from Canaan Land (unknown year) and settled in Lower Egypt, and became known as "Cush" or "Kush". They occupied three towns called Kerma, Napata and Meroe. Here they formed what was called "Kush Empire" and they made Meroe their capital city. (2) Egypt, the father of the Egyptians, who moved from Canaan and settled in the current Egypt land; (3) Put, the forefather of the Libyians, who moved from Canaan and settled in the current Libyain North Africa; and (4) Canaan, the forefather of Canaanites, who remain in the original land (Canaan) of their father Ham.

The oral history said before long, the Phoenicians/Luo fought the Egyptians and conquered them. When the Luo conquered the Egyptians the man called Menya became the first King of Egyptian Dynasty 1. Since the Europeans speak different language, they found it difficult to pronounce Luo word with the consonant "ny", for this reason the Europeans omitted completely the letter "y" and the word "Menya" came to be read as "Mena" or "Menes". Houston Dunjee (1985) confirmed that the first King of Egypt was not called "Mena" or "Menes" as is written in many European texts, but he is called "Menya" (P. 69).

Luo oral history tells us that Rwot Menya (Chief Menya) was the first King to build the first temple in ancient Egypt. He built many temples and one of the temples was called

"*Ptah*", which was named in the honoured of the "Sun god". Luo ruled Egyptians for many years (probably between 5 and 10 years), and the last Luo King to rule Egyptian Dynasty was called Dimo or Dhimmo. Some of the Luo elders think that this King Dimo was also the ruler of Pothethura (Wau) at that time, whose daughter was married to Nyikang and they have a son called Dak.

It is reported that Kush Empire had tremendous natural wealth such as gold mines, ivory, incense and iron ore. Thus, due to this richness, some of the kingdoms neighbouring Kush, like Egypt, became very jealous of the wealth and often tried to conquer Kush land so that they take over the wealth. Unfortunately, they found the Luo people were great fighters and have many warriors, as such the Egyptians could not conquer them.

Interestingly, when the Phoenicians (Luo) settled in Lower Egypt, they abandoned their true God and started worshipping gods of the Egyptians. These Egyptian gods include god Apolo, god Dianysas and god Osiris.

However, it should be noted that the Phoenicians/Luo shared the worshipping of gods with Egyptians, so that the Egyptians would feel that they are part of them. But in actual fact the Egyptians did not recognised them (the Luo) as Egyptians.

By the 4th century, Kush Empire was attacked by the kingdom of Axum of Ethiopia and eventually they were defeated. Following this attack, the Luo people moved away from Lower Egypt, they went southwards –and this was the end of Kush Empire. The Luo moved, and came and settled in Wau (Bhar El Ghazal in Sudan). The Luo stayed here for some years, after that they divided themselves into three main groups, and started to move southward. They came to Uganda, and some of them settled in Northern Uganda, and some moved and settled in Congo (Alur of Congo); this group today are known as the Central Luo. The other group of the Luo continued to move southward and entered Kenya and Tanzania -this group of the Luo today are known as Southern Luo. While the third group of the Luo moved westward and settled near Lake Albert (in Uganda, the place was commonly known by the name Karoo), they stayed at Karoo for some years and later on returned to Sudan, this group is known today as the Northern Luo.

The Genealogy of the Northern Luo is as follows: - Omara begotten Okwa, Okwa begotten Dumo or Duwaat, Nyikang, Anyuak, Gilo and Acholi; Nyikang became ancestor of the Shulluk and the Shaat; Dimo became the ancestor of the Jo-Luo of Wau, Balanda Bor and Balanda Viri; Anyuak became the ancestor of Anyuak; Gilo became the ancestor of Pari; Acholi and Olum became the ancestors of the Acholi.

The Northern Luo stayed near Lake Albert for some years; as years passed by, the relationship between Nyikang and Dumo became unhealthy. Subsequently, quarrels broke out between Nyikang and his brother Dumo, over the kingdom, following the death of their father. However, some writers like Crazzolara (1952) noted that the group (Nyikang and Dumo) broke away because of spear and dead issue –see Chapter 4.

Nyikang decided to leave Karoo (Lake Albert) around 1550 A.D. he came and settled in Potherthura, near Wau in Bhar El Ghazal. Chief Dimo was the ruler of Pothethura at that time. There is a conflict report about Rwot Dimo, some of the elders stated that Rwot Dimo did not moved with the three groups of Luo we have discussed above but he and his people remain in Wau. Other group of elders state that Dimo moved together with the Luo groups and settled with the third group of the Luo near Lake Albert. He did not stayed long in Lake Albert (Karoo), shortly he decided to return to Sudan with his people. There is no evident to show us in which year did Dimo returned to Sudan and what his motive of return to Sudan was. These are all due to lack of documentations. However, the bottom line is that Nyikang came from Lake Albert in Uganda and found Rwot Dimo with his people in Pothethura (Wau). It was reported that Rwot Dimo had some Divine Power.

Collo tradition revealed that Nyikang on arrival at Pothethura (Wau), he married the daughter of Chief Dimo, who gave birth to a son called Dak. It has been reported that the relationship between Nyikang and his father-in-law (Dimo) was getting fragile for considerable number of reasons. Due to this unhealthy relationship, Nyikang decided to move away with his people from Wau (Bhar El Ghazal). He (Nyikang) came with his people and reached Blue River Nile, where he found the River was covered with plants and grass and it was impossible to pass through. It is said that the Blue River Nile was covered with plants and grass just at the meeting point of Sobat River and Blue Nile River. Nyikang and his people have cows, goats and sheep, and they could not cross the Blue River Nile with their animals because of these floating plants and grass. Nyikang did not know what to do, as to take his people across the River. Eventually, they were blocked, and from here, the people could not go further.

One of his men called Ubogo then asked him (Nyikang), if it would be possible for one of the men to be slaughtered to appease the River, so that the way opens for the groups to cross with their animals. When Nyikang heard this suggestion, he did not know who could be slaughtered; so he asked Dak if he could give one of his servants to be slaughtered, but Dak refused the requests of his father.

When Obogo heard that Dak has refused to give one of his servants to be slaughtered, he (Ubogo) offered himself to be slaughtered on the condition that Nyikang has to take care of his wife and children. Nyikang did not reject the request of Ubogo; he accepted the idea and proposal. So Ubogo was then slaughtered by the River bank, his blood poured into the water. As the blood flows over the water, the plants and grass, that were covering the river opened and the people with their animals crossed the Blue River Nile. Thus, when Nyikang crossed the Blue River Nile with his people, they went straight to El Duem, which is on Khartoum/Kosti Road –in North Sudan.

When Nyikang and his people arrived in El Duem, they found the land was inhabited by Funj who are the aboriginal of the place. Oral history tells us that before long Nyikang fought the Funj at El Duem, conquered them and wanted to bring them under his kingdom; but the Funj people declined; they (Funj) moved northward and settled in their present day home in North Sudan. It is said that the Funj people are Luo group who might have separated from the main Luo groups during mass migration as people moved southwards.

From El Duem, Nyikang moved with his people to Malakal, through the today Renk. Thus, when Nyikang reached Malakal with his people, there he also found a large number of Funj people, who are said to be the aboriginal of the land. Traditional history revealed that Nyikang arrived in Malakal with few people and few warriors. Although the warriors were few, they fought Funj people and driven them to Sennar, north of the present Malakal. From Sennar, the Funj expanded rapidly at the expense of neighbour States (see Chapter 2).

As regards to cultures, the study revealed that all the Northern Luo of South Sudan practice three components of religion, which are clearly distinguishable, with exception of Shulluk who practice the fourth component of region. The study found that the first component of all the Northern Luo of South Sudan is the religious practices to "Juok" or "Jok"; they believed in supreme beings called Juok or Jok, who resides in the sky together with the spirits of the ancestors. The second component is the religious practices used through the witch-doctors or sorcerers. The witch-doctors are believed to have some power to cure the sick people and to protect the evil eyes from causing damage to the members of the family. The third component is religious practices through today Christianity. They believe that there is only one true God, who created heaven and earth; who loves the world so much that He sent His only believed Son to redeem the people from their sins, through his death on the Cross. They also believe that only people who accept Jesus Christ as their Saviour and Redeemer can enter the Kingdom of biblical God. The fourth component of religious practices is being practice only by the Collo, which is a religious practice to Nyikang; the Collo believe that Nyikang is a divine King or semi-divine being. However, it is important to note that these three or four religious practices among the people of Northern Luo do not exist separately from each other, but they have many relations among one another.

The study also revealed that the Northern Luo of South Sudan are very good in Traditional Justice and Reconciliation (TJAR). They settle conflicts in the community through well-developed mechanism for the prompt resolution of conflicts in as soon as they arose. Luo traditional conflict resolutions are usually done through well experienced and influential elders in the communities. Interestingly, in Northern Luo society women are not allowed to be chairperson in any family mediation. Majority of the Luo think that the reason for choosing man to be a chairperson is that men are usually considered, by communities, to have better knowledge and skills in conflict resolutions.

In this book, we have discovered that in the Luo societies, kinship is something considered to be of importance because family structures bind the Luo people together. It is the kinship that ties and determines the rights, responsibilities and behaviour of a person.

Northern Luo culture is deeply patriarchal, where men are considered to be heads of the households, they hold position of authorities and power within their families. The husbands have final decision making power in the family and all matters of importance. Women are expected to be subservient to their husbands and to be obedient to their husband's male relatives, even if they are young boys. Thus, the Luo males who are in the Western world when they hear of "Gender Equality" they think that this is more about assimilation or indoctrination than gender equality. They are forced to shallow the bitter pills because they just want to survive and wait to see what God has for them in the store.

Traditionally marriage among the Northern Luo of South Sudan involves many steps that usually begin with courtship through introduction of the bride to the parents, interrogations, negations, paying bride wealth, wedding feast and accompanying bride to her husband's home. The ultimate aim for marriage is reproduction, the Luo considered marriage as intimate union of a man and a woman. The union between a man and a woman is not something invented by the Luo people, but marriage is something created by God. In Gen. 1:27, 28 the bible tells us that God created male and female in his own image. He blessed them and said to them, "Be fruitful and increase in number; fill the earth and subdue it". We read in Gen. 2: 18, that God said, "It is not good for the man to be alone. I will make a helper suitable to him". This is a creation of marriage, where in Gen.2: 24 God says, "For this reason a man will leave his father and mother and be united to his wife, and they will become one flesh".

Marriage has two purposes: the first one is "growth in mutual love" between the couple. This mutual love should open ways to new life. The second purpose is "education of the children", which means man and a woman have share responsibilities for upbringing of their children both at home and in school. This research attempted to find out whether the Luo male and females do marry with the understanding of "the purpose of marriage or not". I could not make this discovery because elders who offered themselves to assist with the book they have had no enough time, due to workload in Australia, as such this topic required more research from Luo young scholars in future.

According to the tradition of the Northern Luo of South Sudan, young women are typically married around the age of 18 years, while young men usually get married between the age of 20-25 –this is the age a young man when he is considered to be mature. In Luo, most partners live in their marriages for live; however, it is important to know that polygamy is accepted among the Luo across South Sudan.

The practices of female genital cutting (FGC) or female genital mutilation (FGM) are not common among the Northern Luo of South Sudan. The study revealed that a handful of females who practice FGC/FGM are from Islamic background –for more information, please read Chapter 20.

Bibliography

Anne Storch, 2003, *Dynamics of Interacting populations*: Language Contact in the Lawoo languages of Bhar El Gahzal, studies in African Linguistics, Volume 32, Number, Johann Wolfagang Goethe-University

Anne S. Straus, 1978, *The Meaning of Death in Northern Cheyenner Cultures*: Journal of the Plains Conference, Volume 23, No. 79, Nairobi Kenya

Acaye Genesis, 2016, *Acholi traditional marriage*, BMS World Mission, Uganda

African Journal of Urology, 2013, *The Jewish and Christian view on Female Genital Mutilation*, Pan African Urological Surgeons Association

Anders Breidlid & Avelino Androga Said, 2014, *A Concise History of South Sudan*, New and Revised Edition, Fountain Publisher, Kampala-Uganda

Aguilar, Mario I, 2011, *From Age-Sets to Friendship Network in Comparative Sociology*: The continuity of Soda among the Boorana of East Africa, University of St Andrew, Scotland.

Akwoch Dok Kwanyiyek, 2008, *The Installation of the Shilluk Reth (King)*, Chollo Community News, South Sudan

Akena Ponse Otto, 2016, Research Chapter 111: *Background of Acholi cultures of South Sudan*

Apai Gabriel, 2005, *Our Inheritance for New Horizons: Sudanese Luo Traditional Culture*, TAFE Queenland Government, Department of Education, Training and the Arts, Australia

Arop Madut-Arop, *The Genesis of Political Consciousness in South Sudan*, Khartoum Sudan

Arop, M. Arop (2012): *The Genesis of Political Consciousness in South Sudan*, Charleston, SC, USA

Bathwell A. Ogot, 1967, *History of the Southern Luo*, Vol. 1. Immigration and Settlement 1500-1900. University College of Nairobi, Kenya.

Bethwell Ogot, J.A. Kieran, 1968, *Zamani A Survey of East Africa History*

Breasted, James Henry, 1937, *A History of Egypt*: From the Earliest Times to the Persian Conquest, New York, Charles Scribner and Sons.

Burton, J. (1996) *Encyclopaedia of World Cultures*, Oxford Dictionary, The Gale Group, Inc.

Charles Oyo Nyawelo, 1992, *Customary Law in The Sudan*: The Shilluk Kingdom from a Legal Perspective, Fashoda Cultural Association Khartoum Sudan

Charles Lerche, 2014, *Peace Building Through Reconciliation*,

Collins, Robert O. , 1956, *History of the Anuak to 1956*, Anyuak Justice Council, in website

Cressida Marcus, 1991, *Insperial Nostalgia, Christian Restoration and Civic Decay in Gondar*

David Graeber (1996) *The Divine kingship of the Shulluk on Violence, Utopia*, and the human condition, or elements for an archaeology of sovereignty, University of London, U.K.

Diedrich Westernmann, 1912, *The Shilluk People: Their Language and Folklore*, Philadelphia, P.A. The Board of Foreign Mission of N.A., Berlin

Douglas H. Johnson, 1990, *Fixed Shrines and Spiritual Centres in the Upper Nile*, , Rift Valley Institute Juba, 73 Publication, South Sudan

Dr. Wayde Goodall, 2013, *Healthy Marriages*, Family-Relationship.com

Dr. Charles Amone, 2014, *Identity, Memory and Gender in Child naming Among the Acholi people of Northern Uganda*, Gulu University, Uganda

DRC (2014) *Food Security and Livelihoods Assessment in Mabaan County*, Upper Nile, UNHCR, Juba-South Sudan

Frunkfort H. 1948, *Kingship and the Gods*, The University of Chicago Press, London

International Rescue Committee, 2017, *No Safe Place: A lifetime of violence for conflict-affected women and girls in South Sudan*.

Jwothab Wanh Othow, 2009, *Historical Perspective of Nyikango Early Political Movement*, Life Magazine.

P'Bitek Okot, 1985, *Acholi Proverbs*, Nairobi, Heinemann

P'Bitek Okot, 1989, *White Teeth*, Nairobi, Heinemann

Danish Demining Group, 2013, *Displacement, Disharmony and Disillusion*: Understanding Host Refugee Tensions in Mabaan County, South Sudan.

Derriba, Jacques, 2005, *The Politics Friendship*. New York: Verso

Diop C.A. , 1974, *The African Origin of Civilisation*, Chicago, Lawrence Hills Books

Dr. David Thabo, 2010, *Reth Anei Kur*, Pachodo.org, Community News, South Sudan, Juba

Diop C.A. 1981, *Civilisation or Barbarism*, Brooklyn, New York, Lawrence Hills Books

Crazzolara, 1950, *The Lwo Part 1*, Lwo Migration, Missoni Africana, Verona, Italy

Crazzolara, 1951, *The Lwo Part 11*, Lwo Traditions, Editrice Nigrazia, Verona, Italy

Crazzolara, 1954, *The Lwo Part 111*, Clans, Editrice Nigrazia, Verona, Italy

Cunnison, I. 1959, Tje Luapula People's of Northern Rhodesia, Manchester University Press-London

Evans Pritchard, E.E. 1934, *Ther Nuer Tribe and clan*" Sudan Notes ands Records 17, 1-57

Francais, 2015, *History of Sudanic and Nilotics in Sudan and Uganda and their political impacts*, African Time.Com, Namibia.

Godfrey Lienhardt, 1999, *African Worlds*: Studies in the cosmological ideas, Shilluk of the Upper Nile, Transaction Publishers, Rutgers University, North America.

Hegel, Georg Wilhelm Friedrick, 1956, *The Philosophy of History*, New York, Dover

Holt P.M., 1963, *Funj Origins*: A Critiques and New Evidence, Journal of African histories, Cambridge University Press, London

Holmayer Withelm , 1923, *Die Shilluk, Geschicihte, Religion und Leben Ein Niloten-Satammers*

Houston Drussilla Dunjee, 1985, *Wonderful Ethiopian Of The Ancient Cushitic Empire*, Black Classic Press, Baltimore

J.O Awolalu (1976) *What is African Traditional Religion*, World Wisdom, Inc.

Jay D. Schvaneveldt, 1989, *Remembering at Death*: Funeral and Related Rituals, Utah State, University.

Journal of Social Relevance and Concern, Volume 2, Issue 5, May 2014

Jommo, R.B. 1994, *Wife inheritance a Male creation*, Sunday nation, Nairobi-Kenya, March 20

Lars Christoffer Skongrand (2015), *Reconciliation and Identity in South Sudan*: The Church's roles in reconciliation processes and the challenges reconciliation efforts face in South Sudan, Norwegian School of Theology, Oslo-Norway

Lederach, J. P. (2013). *Building Peace: Sustainable Reconciliation in Divided Societies* (10th Edition.). (Washington, D.C: United States Institute of Peace Press (USIP Press)

Linda James Myers and David H. Shinn, *Appreciating Traditional Forms of Healing Conflict in Africa and the World*, Black Diaspora Review

Llewellyn, J.J.,& Philpott, D. (Eds.). (2014). *Restorative Justice, Reconciliation And Peacebuilding*. Oxford: Oxford University Press, USA.

Katabaro Miti, 1980, *South Africa and conflict Resolution in Africa*: From Mandela to Zuma, Southern African Peace and Security studies, Volume 1, University of Pretoria, South Africa

Kirwen, M.C. 1987, *The Missionary and the Diviner*. New York. Orbis Books

Hirwen, M.C. 1979, *African Widows*. New York, Orbis Books

Kuel Maluil Jok, 2010, *Animism of the Nilotics and Discourses of Islam Fundamentalism in Sudan*, Sidestone Press, Leiden

Kurimoto E. 1992, *An Ethnography of Bitterness*: Cucumber And Sacrifice Reconsidered, Journal of Religion in Africa XX11, Tokyo University of Foreign Studies

Kurimoto, 1996, *The Year Hunting Ritual of the Pari*: Elements of Hunting Culture among the Nilotes, Japan Association for Nilo-Ethiopian Studies

Kurimoto, 2012, *Copying with enemies*: Graded Age system among the Pari of South Sudan

McCosker Laura Kate 2009, *Body-politic in the context of Female Genital Mutilation* –an analysis of structure and agency, Student in Nairobi University Kenya

Machar Riek Speech, 2004, *Upper Nile Youth Association for Development*, Panyagor

Madut, 2008, *Luo in Sudan*, Khartoum Sudan

Mathews, B.P. 2011, *Female genital mutilation*, Australian Law, policy and practical challenges for doctors, Medical Journal of Australia, 194 (3) 139-141

Mason Simon A. 2007. *Learning from the Swiss Mediation and Facilitation Experiences in Sudan*, Centre for Security Study, Switzerland

Mawa Isaac J., 2014, *Report on Food Security and Livelihoods Assessment in Pachalla Country*, Published by Sudan Peace and Education Development Programme (SPEDP)

Missionary Atlas Project Africa-Sudan: Snapshot section (2016)

Momoh, C. 2004, *Gender and daily life in Ethiopia*, Contemporary Review, 285 (1663) , PP 97-101

Mtullah, W.V. et. Al. 1995, *Women and Inheritance Laws and Practice in Kisumu District*, Nairobi-Kenya

Muchomba, 2004, *South Sudan Livelihoods..............*

Nyawelo, Charles Oyo, 1992, *Customary Law in the Sudan*; The Shilluk Kingdom from a Legal Perspective, Khartoum-Sudan

Nzomo, M. 1994a, *Interview in Sunday nation*, Nairobi-Kenya, January 16th

Nzomo, M. 1994b, *Wife Inheritance Pert of Our culture*, daily Nation, Nairobi-Kenya, February 23rd

Ochieng Daudi, 2013, *History is a Mirror of Society*, www.acholitimes.com/cultureacholi/index on 12/02/2914

Ochieng David 2013, *Luo Origin of Civilisation*: Towards a positive identification of the Ancient Itiyo-pi-Anu Peoples, Word Press.com – online 3rd August 2015

Ochieng Philip, 2011, *South Sudan may as well be called Luoland*, The pilgrimage –online 3/08/2015

Ocholla_Ayayo, 1976, *Traditional Ideology and Ethnics among the Southern Luo*

O. Deng, Albino, 2016, *The Sudan Vision*, Oxford Dictionary of Rhymes, UK, London

Ojulu Odola, 2013, *History of Gambella*, Gambella Voice, Tribune and Fearless Monitor, Ethiopia.

Ojwang, H.J. 2005, *Towardfs a Social Philosophy of the African*, Leviratic Custom: How the Luo Marriage Survive Death. In Gutema, Bekele and Daniel Smith eds.

Okello Paito, Terence, 2011, Luo *Origin of Civilisation*, Acholi Times –online 3/08/2015

Okello Paito, 2011, quoted in A.C.A Wright (1940) who reviewed J.P. Crazzolara about "*A study of the Acholi language*, Uganda Journal, 1940, Vol V11 No. 4

Okot P'Bitek, 2013, *Acholi People*: The Famous East African Spiritually and Martially Power People with Culturally Unique *Mato Oput* Justice and Reconciliation Rituals Ceremony, Uganda

P'Bitek, O. 1986, Aritist: *The Ruler-Nairobi*: Heinemann Kenya Ltd

Okoth, Dennis, 2002, *Luo Animistic Beliefs and Religious Practitioners and How to Reconcile them to Christ,*

Omolo Leo Odera, 2013, East Africa: *Brief Correction about Luo family tree and lineage,* and immigration from Sudan to Kenya, www.blog.jaluo.com on 12/02/2014

Onyango-ku-Odongo J.B. Webster, 1976, *The Central Lwo During the Aconya*, East Africa Literature, Bureau, Nairobi Kenya

Potash, Betty, 1986a, *Widows in African Societies*, Stanford: Stanford University Press, pp. 1-4

Rachael Sparks, *The Acholi Bag from South Sudan*, Equatorial Civic Fund Women's Group, Kiryandongo Refugee Settlement Camps, Uganda

Robben Antonius C.G. M., 2000, *Death Squad*: The Anthropology of State Terror, University of Pennsylvania Press, USA

Robben Antonius C. G. M., 2004, *Death, Mouring, and Burial, A Cross Cultural Reader*, Blackwell Publishing, Malden, USA

Saturnino Onyala, 2014, *Tekwaro Pe Acholi Me South Sudan*, Trija Press Melbourne, Australia.

Seligman, C.G. & Brenda Z. Seligman, 1932, *Pagan Tribes of Nilotic Sudan*, George Routledge & Sons, Ltd, Broadway House, London.

Shiino W. 1997, *Death and Rituals among the Luo in South Nyamza*, Tokyo Metropolitan University

Smith Adam, 1776, *The wealth of Nations*, London, Penguins

Steve Cornell, 2012, *How to Move from Forgiveness to Reconciliation*, Millersville Bible Church, Millersville, Penn

Sossou, M.A. 2002, *Widowhood Practices in Western African*: The Silent Victims, International Journal of Social Welfare, 11, pp. 201-209

South Sudan, Ministry of Education, Science and Technology, 2014, Foundation Publishers, Oxford OX1 9En, UK

Sudan Notes and Records Vol 28, 1941, McCorquodale and Co, (Sudan) Ltd

Wallis Budge E.A. 1994, *The Egyptian Book Of The Dead*, Brooklyn, New York, A & B Books Publishers

Tanaka, Jiro, 1996, *The World of Animals* : View by the San Hunter Gatherers in Kalahara, Kyoto University, Textversion, China

Government of South Sudan (2009): *Laws of Southern Sudan, the Local Government Act, 2009*, Juba, Southern Sudan: Ministry of Justice

Tim Mclulka, 2012, *A Shared Struggle*: The People and Cultures of South Sudan, UNMISS, Durham university, London

Toby Collins, 2013, *South Sudan's Kings and Chiefs visit the courts of Ghana*, http://centrum. humanitasafrika.cz

Torben Andersen, 1988, *Downstep in Pari*: The tone system of a Western Nilotic Lnaguage, Volume 19, University of a Aalborg

Debra Fileta, 2017, *True-Love-Dates*, www.truelovedates.com, extracted 20/02/2017

Wasonga, 2012, *The Luo Rout from the North*........

WASSARA S. (PhD) (March 2008): *Traditional Mechanism of Conflict Resolution in South Sudan*, Bergdorf Foundation for Peace Support, www.berghofpeacesupport.org

Wamari Elly , 2006, *Anyuak People of Sudan and Canada*, Canada

Wilson, G. 1961, *Chik Gi Tim Luo*, Nairobi, The Government Printer

World mark Encyclopaedia of cultures and Daily Life, 2009, Shilluk, Khartoum Sudan.

Appendix

· · · · · · · · · · · ·

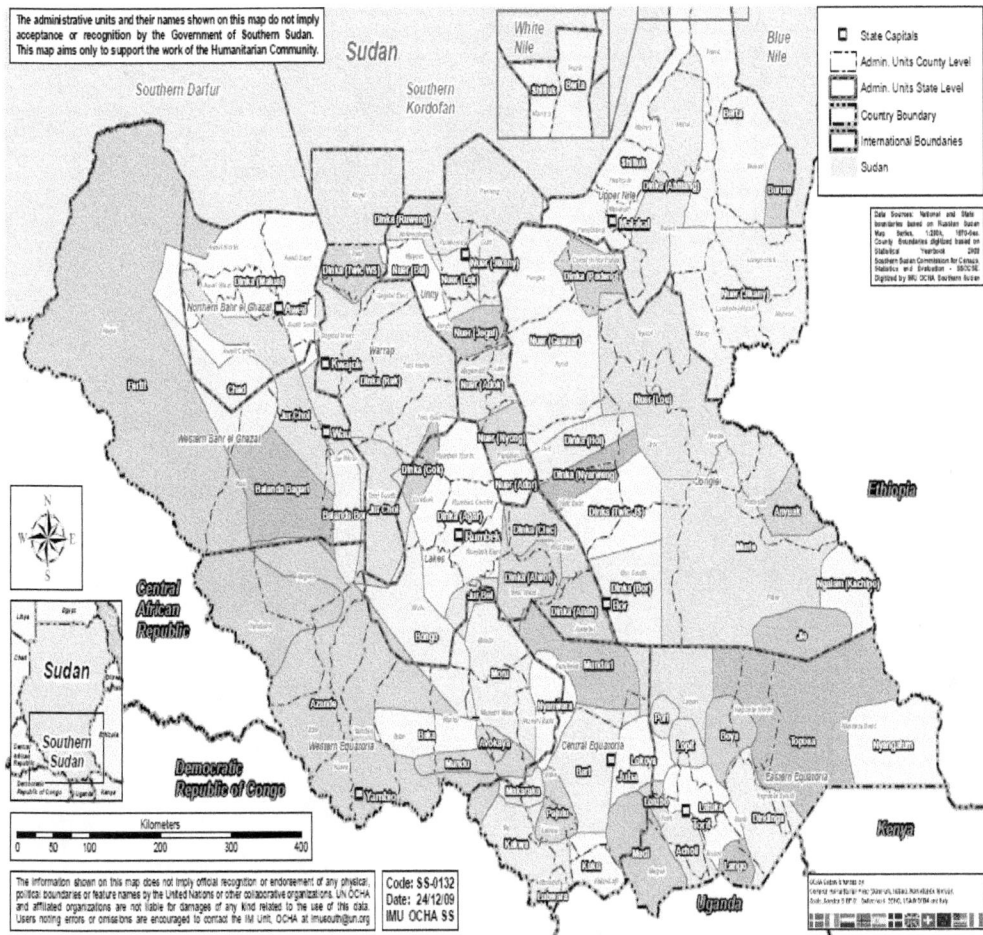

Ethnics' locations in South Sudan

www.ingramcontent.com/pod-product-compliance
Lightning Source LLC
Chambersburg PA
CBHW052127020426

42334CB00023B/2634